T0189704

# Advances in Intelligent Systems and Computing

Volume 553

**Series editor**

Janusz Kacprzyk, Polish Academy of Sciences, Warsaw, Poland
e-mail: kacprzyk@ibspan.waw.pl

*About this Series*

The series "Advances in Intelligent Systems and Computing" contains publications on theory, applications, and design methods of Intelligent Systems and Intelligent Computing. Virtually all disciplines such as engineering, natural sciences, computer and information science, ICT, economics, business, e-commerce, environment, healthcare, life science are covered. The list of topics spans all the areas of modern intelligent systems and computing.

The publications within "Advances in Intelligent Systems and Computing" are primarily textbooks and proceedings of important conferences, symposia and congresses. They cover significant recent developments in the field, both of a foundational and applicable character. An important characteristic feature of the series is the short publication time and world-wide distribution. This permits a rapid and broad dissemination of research results.



More information about this series at http://www.springer.com/series/11156

Sanjiv K. Bhatia · Krishn K. Mishra
Shailesh Tiwari · Vivek Kumar Singh
Editors

# Advances in Computer and Computational Sciences

## Proceedings of ICCCCS 2016, Volume 1

*Editors*
Sanjiv K. Bhatia
Department of Computer Science
University of Missouri
Columbia, MO
USA

Krishn K. Mishra
Department of Computer Science and Engineering
Motilal Nehru National Institute of Technology
Allahabad, Uttar Pradesh
India

Shailesh Tiwari
Department of Computer Science and Engineering
ABES Engineering College
Ghaziabad, Uttar Pradesh
India

Vivek Kumar Singh
Banaras Hindu University
Varanasi, Uttar Pradesh
India

ISSN 2194-5357 ISSN 2194-5365 (electronic)
Advances in Intelligent Systems and Computing
ISBN 978-981-10-3769-6 ISBN 978-981-10-3770-2 (eBook)
DOI 10.1007/978-981-10-3770-2

Library of Congress Control Number: 2017931526



Printed on acid-free paper

This Springer imprint is published by Springer Nature
The registered company is Springer Nature Singapore Pte Ltd.
The registered company address is: 152 Beach Road, #21-01/04 Gateway East, Singapore 189721, Singapore

# Preface

The ICCCCS is a major multidisciplinary conference organized with the objective of bringing together researchers, developers and practitioners from academia and industry working in all areas of computer and computational sciences. It is organized specifically to help computer industry to derive the advances of next generation computer and communication technology. The invited researchers present the latest developments and technical solutions.

Technological developments all over the world are dependent upon globalization of various research activities. Exchange of information and innovative ideas are necessary to accelerate the development of technology. Keeping this ideology in preference, Aryabhatta College of Engineering & Research Center, Ajmer, India has come up with an event—International Conference on Computer, Communication and Computational Sciences (ICCCCS-2016) during August 12–13, 2016.

Ajmer is situated in the heart of India; just over 130 km southwest of Jaipur, a burgeoning town on the shore of the Ana Sagar Lake, flanked by barren hills. Ajmer has historical strategic importance and was ransacked by Mohammed Gauri on one of his periodic forays from Afghanistan. Later, it became a favorite residence of the mighty Mughals. The city was handed over to the British in 1818, becoming one of the few places in Rajasthan controlled directly by the British rather than being part of a princely state. The British chose Ajmer as the site for Mayo College, a prestigious school opened in 1875 exclusively for the Indian Princes, but today open to all those who can afford the fees. Ajmer is a perfect place that can be symbolized for demonstration of Indian culture, ethics and display of perfect blend of wide plethora of diverse religion, community, culture, linguistics, etc., all coexisting and flourishing in peace and harmony. This city is known for the famous Dargah Sharif, Pushkar Lake, Brahma Temple and many more evidences of history.

This is the First time Aryabhatta College of Engineering & Research Center, Ajmer, India is organizing International Conference on Computer, Communication and Computational Sciences (ICCCCS 2016), with a foreseen objective of enhancing the research activities at a large scale. Technical Program Committee and

Advisory Board of ICCCCS include eminent academicians, researchers and practitioners from abroad as well as from all over the nation.

In this volume, selected manuscripts have been subdivided into various tracks named 'Intelligent Hardware and Software Design', 'Advanced Communications', 'Power and Energy Optimization', 'Intelligent Image Processing', Advanced Software Engineering', 'IoT', 'ADBMS & Security', 'Evolutionary and Soft Computing'. A sincere effort has been made to make it an immense source of knowledge for all and includes 147 manuscripts. The selected manuscripts have gone through a rigorous review process and are revised by authors after incorporating the suggestions of the reviewers.

ICCCCS 2016 received 429 submissions from around 729 authors of 12 different countries such as USA, Iceland, China, Saudi Arabia, South Africa, Taiwan, Malaysia and many more. Each submission has been checked with anti-plagiarism software. On the basis of plagiarism report, each submission was rigorously reviewed by at least two reviewers with an average of 2.45 reviewers per review. Even some submissions have more than two reviews. On the basis of these reviews, 140 high-quality papers were selected for publication in this proceedings volume, with an acceptance rate of 32.6%.

We are thankful to the speakers, delegates and the authors for their participation and their interest in ICCCCS as a platform to share their ideas and innovation. We are also thankful to the Prof. Dr. Janusz Kacprzyk, Series Editor, AISC, Springer and Mr. Aninda Bose, Senior Editor, Hard Sciences, Springer for providing continuous guidance and support. Also, we extend our heartfelt gratitude to the reviewers and Technical Program Committee Members for showing their concern and efforts in the review process. We are indeed thankful to everyone directly or indirectly associated with the conference organizing team for leading it towards the success.

Although utmost care has been taken in compilation and editing, however, a few errors may still occur. We request the participants to bear with such errors and lapses (if any). We wish you all the best.

Organizing Committee
ICCCCS 2016

# Organizing Committee

**General Chair**
Dr. Amit Shastri, Chairman, Aryabhatta Academic Society, Ajmer, India

**Program Chairs**
Dr. Krishn K. Mishra, Motilal Nehru National Institute of Technology Allahabad, India
Dr. Munesh C. Trivedi, ABES Engineering College, Ghaziabad, India

**Conference Chair**
Dr. Shailesh Tiwari, ABES Engineering College, Ghaziabad, India

**Conference Co-Chair**
Mr. Ashish Guwalani, ACERC, Ajmer, India

**TPC Chairs**
Prof. Nitin Singh, Motilal Nehru National Institute of Technology Allahabad, India
Dr. Vishal Bhatnagar, AIACTR, Delhi, India

**TPC Co-Chair**
Dr. Sanjay Mathur, ACERC, Ajmer, India

**Publication Chairs**
Dr. Deepak Kumar Singh, Sachdeva Institute of Technology, Mathura, India
Dr. Pragya Dwivedi, MNNIT Allahabad, India

**Publication Co-Chair**
Mr. Gaurav Phulwari, ACERC, Ajmer, India

**Publicity Chairs**
Dr. Anil Dubey, Government Engineering College, Ajmer, India
Dr. Deepak Kumar, Amity University, Noida, India
Dr. Nitin Rakesh, Amity University, Noida, India
Dr. Ravi Prasad Valluru, Narayana Engineering College Nellore, AP, India

Dr. Sushant Upadyaya, MNIT, Jaipur, India
Dr. Akshay Girdhar, GNDEC, Ludhiana, India

**Publicity Co-Chair**
Mr. Surendra Singh, ACERC, Ajmer, India

**Tutorial Chairs**
Prof. Lokesh Garg, Delhi College of Technology & Management, Haryana, India

**Tutorial Co-Chair**
Mr. Ankit Mutha, ACERC, Ajmer, India

# Technical Program Committee

Prof. Ajay Gupta, Western Michigan University, USA
Prof. Babita Gupta, California State University, USA
Prof. Amit K.R. Chowdhury, University of California, USA
Prof. David M. Harvey, G.E.R.I., UK
Prof. Madjid Merabti, Liverpool John Moores University, UK
Dr. Nesimi Ertugrual, University of Adelaide, Australia
Prof. Ian L. Freeston, University of Sheffield, UK
Prof. Witold Kinsner, University of Manitova, Canada
Prof. Anup Kumar, M.I.N.D.S., University of Louisville, USA
Prof. Prabhat Kumar Mahanti, University of New Brunswick, Canada
Prof. Ashok De, Director, NIT Patna, India
Prof. Kuldip Singh, IIT Roorkee, India
Prof. A.K. Tiwari, IIT, BHU, Varanasi, India
Mr. Suryabhan, ACERC, Ajmer, India
Dr. Vivek Singh, BHU, India
Prof. Abdul Quaiyum Ansari, Jamia Millia Islamia, New Delhi, India
Prof. Aditya Trivedi, ABV-IIITM Gwalior, India
Prof. Ajay Kakkar, Thapar University, Patiala, India
Prof. Bharat Bhaskar, IIM Lucknow, India
Prof. Edward David Moreno, Federal University of Sergipe, Brazil
Prof. Evangelos Kranakis, Carleton University
Prof. Filipe Miguel Lopes Meneses, University of Minho, Portugal
Prof. Giovanni Manassero Junior, Universidade de São Paulo, Brazil
Prof. Gregorio Martinez, University of Murcia, Spain
Prof. Pabitra Mitra, Indian Institute of Technology Kharagpur, India
Prof. Joberto Martins, Salvador University-UNIFACS, Brazil
Prof. K. Mustafa, Jamia Millia Islamia, New Delhi, India
Prof. M.M. Sufyan Beg, Jamia Millia Islamia, New Delhi, India
Prof. Jitendra Agrawal, Rajiv Gandhi Proudyogiki Vishwavidyalaya, Bhopal, M.P., India
Prof. Rajesh Baliram Ingle, PICT, University of Pune, India

Prof. Romulo Alexander Ellery de Alencar, University of Fortaliza, Brazil
Prof. Youssef Fakhri, Université Ibn Tofail, Faculté des Sciences, Brazil
Dr. Abanish Singh, Bioinformatics Scientist, USA
Dr. Abbas Cheddad, (UCMM), Umeå Universitet, Umeå, Sweden
Dr. Abraham T. Mathew, NIT, Calicut, Kerala, India
Dr. Adam Scmidit, Poznan University of Technology, Poland
Dr. Agostinho L.S. Castro, Federal University of Para, Brazil
Prof. Goo-Rak Kwon, Chosun University, Republic of Korea
Dr. Alberto Yúfera, Instituto de Microelectrónica de Sevilla (IMSE), (CNM), Spain
Dr. Adam Scmidit, Poznan University of Technology, Poland
Prof. Nishant Doshi, S.V. National Institute of Technology, Surat, India
Prof. Gautam Sanyal, NIT Durgapur, India
Dr. Agostinho L.S. Castro, Federal University of Para, Brazil
Dr. Alok Chakrabarty, IIIT Bhubaneswar, India
Dr. Anastasios Tefas, Aristotle University of Thessaloniki
Dr. Anirban Sarkar, NIT Durgapur, India
Dr. Anjali Sardana, IIIT Roorkee, Uttarakhand, India
Dr. Ariffin Abdul Mutalib, Universiti Utara Malaysia
Dr. Ashok Kumar Das, IIIT Hyderabad
Dr. Ashutosh Saxena, Infosys Technologies Ltd., India
Dr. Balasubramanian Raman, IIT Roorkee, India
Dr. Benahmed Khelifa, Liverpool John Moores University, UK
Dr. Björn Schuller, Technical University of Munich, Germany
Dr. Carole Bassil, Lebanese University, Lebanon
Dr. Chao MA, Hong Kong Polytechnic University
Dr. Chi-Un Lei, University of Hong Kong
Dr. Ching-Hao Lai, Institute for Information Industry
Dr. Ching-Hao Mao, Institute for Information Industry, Taiwan
Dr. Chung-Hua Chu, National Taichung Institute of Technology, Taiwan
Dr. Chunye Gong, National University of Defense Technology
Dr. Cristina Olaverri Monreal, Instituto de Telecomunicacoes, Portugal
Dr. Chittaranjan Hota, BITS Hyderabad, India
Dr. D. Juan Carlos González Moreno, University of Vigo
Dr. Danda B. Rawat, Old Dominion University
Dr. Davide Ariu, University of Cagliari, Italy
Dr. Dimiter G. Velev, University of National and World Economy, Europe
Dr. D.S. Yadav, South Asian University, New Delhi
Dr. Darius M. Dziuda, Central Connecticut State University
Dr. Dimitrios Koukopoulos, University of Western Greece, Greece
Dr. Durga Prasad Mohapatra, NIT-Rourkela, India
Dr. Eric Renault, Institut Telecom, France
Dr. Felipe RudgeBarbosa, University of Campinas, Brasil
Dr. Fermín Galán Márquez, Telefónica I+D, Spain
Dr. Fernando Zacarias Flores, Autonomous University of Puebla
Dr. Fuu-Cheng Jiang, Tunghai University, Taiwan

Prof. Aniello Castiglione, University of Salerno, Italy
Dr. Geng Yang, NUPT, Nanjing, People's Republic of China
Dr. Gadadhar Sahoo, BIT-Mesra, India
Prof. Ashokk Das, International Institute of Information Technology, Hyderabad, India
Dr. Gang Wang, Hefei University of Technology
Dr. Gerard Damm, Alcatel-Lucent
Prof. Liang Gu, Yale University, New Haven, CT, USA
Prof. K.K. Pattanaik, ABV-Indian Institute of Information Technology and Management, Gwalior, India
Dr. Germano Lambert-Torres, Itajuba Federal University
Dr. Guang Jin, Intelligent Automation, Inc
Dr. Hardi Hungar, Carl von Ossietzky University Oldenburg, Germany
Dr. Hongbo Zhou, Southern Illinois University Carbondale
Dr. Huei-Ru Tseng, Industrial Technology Research Institute, Taiwan
Dr. Hussein Attia, University of Waterloo, Canada
Prof. Hong-Jie Dai, Taipei Medical University, Taiwan
Prof. Edward David, UFS—Federal University of Sergipe, Brazil
Dr. Ivan Saraiva Silva, Federal University of Piauí, Brazil
Dr. Luigi Cerulo, University of Sannio, Italy
Dr. J. Emerson Raja, Engineering and Technology of Multimedia University, Malaysia
Dr. J. Satheesh Kumar, Bharathiar University, Coimbatore
Dr. Jacobijn Sandberg, University of Amsterdam
Dr. Jagannath V. Aghav, College of Engineering Pune, India
Dr. Jaume Mathieu, LIP6 UPMC, France
Dr. Jen-Jee Chen, National University of Tainan
Dr. Jitender Kumar Chhabra, NIT-Kurukshetra, India
Dr. John Karamitsos, Tokk Communications, Canada
Dr. Jose M. Alcaraz Calero, University of the West of Scotland, UK
Dr. K.K. Shukla, IT-BHU, India
Dr. K.R. Pardusani, Maulana Azad NIT, Bhopal, India
Dr. Kapil Kumar Gupta, Accenture
Dr. Kuan-Wei Lee, I-Shou University, Taiwan
Dr. Lalit Awasthi, NIT Hamirpur, India
Dr. Maninder Singh, Thapar University, Patiala, India
Dr. Mehul S. Raval, DA-IICT, Gujarat, India
Dr. Michael McGuire, University of Victoria, Canada
Dr. Mohamed Naouai, University Tunis El Manar and University of Strasbourg, Tunisia
Dr. Nasimuddin, Institute for Infocomm Research
Dr. Olga C. Santos, aDeNu Research Group, UNED, Spain
Dr. Pramod Kumar Singh, ABV-IIITM Gwalior, India
Dr. Prasanta K. Jana, IIT, Dhanbad, India
Dr. Preetam Ghosh, Virginia Commonwealth University, USA

Dr. Rabeb Mizouni, (KUSTAR), Abu Dhabi, UAE
Dr. Rahul Khanna, Intel Corporation, USA
Dr. Rajeev Srivastava, CSE, ITBHU, India
Dr. Rajesh Kumar, MNIT, Jaipur, India
Dr. Rajesh Bodade, Military College of Telecommunication, Mhow, India
Dr. Rajesh Kumar, MNIT, Jaipur, India
Dr. Ranjit Roy, SVNIT, Surat, Gujarat, India
Dr. Robert Koch, Bundeswehr University München, Germany
Dr. Ricardo J. Rodriguez, Nova Southeastern University, USA
Dr. Ruggero Donida Labati, Università degli Studi di Milano, Italy
Dr. Rustem Popa, University "Dunarea de Jos" in Galati, Romania
Dr. Shailesh Ramchandra Sathe, VNIT Nagpur, India
Dr. Sanjiv K. Bhatia, University of Missouri—St. Louis, USA
Dr. Sanjeev Gupta, DA-IICT, Gujarat, India
Dr. S. Selvakumar, National Institute of Technology, Tamil Nadu, India
Dr. Saurabh Chaudhury, NIT Silchar, Assam, India
Dr. Shijo. M. Joseph, Kannur University, Kerala
Dr. Sim Hiew Moi, University Technology of Malaysia
Dr. Syed Mohammed Shamsul Islam, The University of Western Australia, Australia
Dr. Trapti Jain, IIT Mandi, India
Dr. Tilak Thakur, PED, Chandigarh, India
Dr. Vikram Goyal, IIIT Delhi, India
Dr. Vinaya Mahesh Sawant, D.J. Sanghvi College of Engineering, India
Dr. Vanitha Rani Rentapalli, VITS Andhra Pradesh, India
Dr. Victor Govindaswamy, Texas A&M University-Texarkana, USA
Dr. Victor Hinostroza, Universidad Autónoma de Ciudad Juárez
Dr. Vidyasagar Potdar, Curtin University of Technology, Australia
Dr. Vijaykumar Chakka, DAIICT, Gandhinagar, India
Dr. Yong Wang, School of IS & E, Central South University, China
Dr. Yu Yuan, Samsung Information Systems America—San Jose, CA
Eng. Angelos Lazaris, University of Southern California, USA
Mr. Hrvoje Belani, University of Zagreb, Croatia
Mr. Huan Song, SuperMicro Computer, Inc., San Jose, USA
Mr. K.K Patnaik, IIITM, Gwalior, India
Dr. S.S. Sarangdevot, Vice Chancellor, JRN Rajasthan Vidyapeeth University, Udaipur
Dr. N.N. Jani, KSV University Gandhi Nagar, India
Dr. Ashok K. Patel, North Gujarat University, Patan, Gujarat, India
Dr. Awadhesh Gupta, IMS, Ghaziabad, India
Dr. Dilip Sharma, GLA University, Mathura, India
Dr. Li Jiyun, Donghua Univesity, Shanghai, China
Dr. Lingfeng Wang, University of Toledo, USA
Dr. Valentina E. Balas, Aurel Vlaicu University of Arad, Romania
Dr. Vinay Rishiwal, MJP Rohilkhand University, Bareilly, India

Dr. Vishal Bhatnagar, Ambedkar Institute of Technology, New Delhi, India
Dr. Tarun Shrimali, Sunrise Group of Institutions, Udaipur, India
Dr. Atul Patel, CU Shah University, Vadhwan, Gujrat, India
Dr. P.V. Virparia, Sardar Patel University, VV Nagar, India
Dr. D.B. Choksi, Sardar Patel University, VV Nagar, India
Dr. Ashish N. Jani, KSV University Gandhi Nagar, India
Dr. Sanjay M. Shah, KSV University Gandhi Nagar, India
Dr. Vijay M. Chavda, KSV University Gandhi Nagar, India
Dr. B.S. Agarwal, KIT Kalol, India
Dr. Apurv Desai, South Gujrat University, Surat, India
Dr. Chitra Dhawale, Nagpur, India
Dr. Bikas Kumar, Pune, India
Dr. Nidhi Divecha, Gandhi Nagar, India
Dr. Jay Kumar Patel, Gandhi Nagar, India
Dr. Jatin Shah, Gandhi Nagar, India
Dr. Kamaljit I. Lakhtaria, Auro University, Surat, India
Dr. B.S. Deovra, B.N. College, Udaipur, India
Dr. Ashok Jain, Maharaja College of Engineering, Udaipur, India
Dr. Bharat Singh, JRN Rajasthan Vidyapeeth University, Udaipur, India
Dr. S.K. Sharma, Pacific University Udaipur, India
Dr. Akheela Khanum, Integral University Lucknow, India
Dr. R.S. Bajpai, Ram Swaroop Memorial University, Lucknow, India
Dr. Manish Shrimali, JRN Rajasthan Vidyapeeth University, Udaipur, India
Dr. Ravi Gulati, South Gujarat University, Surat, India
Dr. Atul Gosai, Saurashtra University, Rajkot, India
Dr. Digvijai sinh Rathore, BBA Open University Ahmadabad, India
Dr. Vishal Goar, Government Engineering College, Bikaner, India
Dr. Neeraj Bhargava, MDS University Ajmer, India
Dr. Ritu Bhargava, Government Womens Engineering College, Ajmer, India
Dr. Rajender Singh Chhillar, MDU Rohtak, India
Dr. Dhaval R. Kathiriya, Saurashtra University, Rajkot, India
Dr. Vineet Sharma, KIET Ghaziabad, India
Dr. A.P. Shukla, KIET Ghaziabad, India
Dr. R.K. Manocha, Ghaziabad, India
Dr. Nandita Mishra, IMS Ghaziabad, India
Dr. Manisha Agarwal, IMS Ghaziabad
Dr. Deepika Garg, IGNOU New Delhi, India
Dr. Goutam Chakraborty, Iwate Prefectural University, Iwate Ken, Takizawa, Japan
Dr. Amit Manocha Maharaja Agrasen University, HP, India
Prof. Enrique Chirivella-Perez, University of the West of Scotland, UK
Prof. Pablo Salva Garcia, University of the West of Scotland, UK
Prof. Ricardo Marco Alaez, University of the West of Scotland, UK
Prof. Nitin Rakesh, Amity University, Noida, India
Prof. Mamta Mittal, G. B. Pant Engineering College, Delhi, India
Dr. Shashank Srivastava, MNNIT Allahabad, India

Prof. Lalit Goyal, JMI, Delhi, India
Dr. Sanjay Maurya, GLA University, Mathura, India
Prof. Alexandros Iosifidis, Tampere University of Technology, Finland
Prof. Shanthi Makka, JRE Engineering College, Greater Noida, India
Dr. Deepak Gupta, Amity University, Noida, India
Dr. Manu Vardhan, NIT Raipur, India
Dr. Sarsij Tripathi, NIT Raipur, India
Prof. Wg Edison, HeFei University of Technology, China
Dr. Atul Bansal, GLA University, Mathura, India
Dr. Alimul Haque, V.K.S. University, Bihar, India
Prof. Simhiew Moi, Universiti Teknologi Malaysia
Prof. Vinod Kumar, IIT Roorkee, India
Prof. Christos Bouras, University of Patras and RACTI, Greece
Prof. Devesh Jinwala, SVNIT, Surat, India
Prof. Germano Lambert Torres, PS Solutions, Brazil
Prof. Byoungho Kim, Broadcom Corporation, USA
Prof. Aditya Khamparia, LPU, Punjab, India

# Contents

# About the Editors

**Dr. Sanjiv K. Bhatia** received his Ph.D. in Computer Science from the University of Nebraska, Lincoln in 1991. He presently works as Professor and Graduate Director (Computer Science) in the University of Missouri, St. Louis. His primary areas of research include image databases, digital image processing, and computer vision. He has published over 40 articles in those areas. He has also consulted extensively with industry for commercial and military applications of computer vision. He is an expert in system programming and has worked on real time and embedded applications. He serves on the organizing committee of a number of conferences and on the editorial board of international journals. He has taught a broad range of courses in computer science and was the recipient of Chancellor's Award for Excellence in Teaching in 2015. He is a senior member of ACM.

**Dr. Krishn K. Mishra** currently works as a Visiting Faculty, Department of Mathematics & Computer Science, University of Missouri, St. Louis, USA. He is an alumnus of Motilal Nehru National Institute of Technology Allahabad, India, which is also his base working institute. His primary area of research includes evolutionary algorithms, optimization techniques, and design and analysis of algorithms. He has published more than 50 publications in international journals and proceedings of international conferences of repute. He has served as a program committee member of several conferences and also edited Scopus and SCI-indexed journals. He has 15 years of teaching and research experience during which he made all efforts to bridge the gaps between teaching and research.

**Dr. Shailesh Tiwari** currently works as Professor in Computer Science and Engineering Department, ABES Engineering College, Ghaziabad, India. He is also administratively heading the department. He is an alumnus of Motilal Nehru National Institute of Technology Allahabad, India. He has more than 15 years of experience in teaching, research, and academic administration. His primary areas of research are software testing, implementation of optimization algorithms, and machine learning techniques in software engineering. He has also published more than 40 publications in international journals and proceedings of international conferences of repute. He has served as a program committee member of several conferences and edited Scopus and E-SCI-indexed journals. He has also organized several international conferences under the banner of IEEE and Springer. He is a Senior Member of IEEE, member of IEEE Computer Society and Executive Committee member of IEEE Uttar Pradesh section. He is a member of reviewer and editorial board of several international journals and conferences.

**Dr. Vivek Kumar Singh** is Assistant Professor at Department of Computer Science, Banaras Hindu University, India. His major research interest lies at the area of text analytics. Currently, he is working on scientometrics, sentiment analysis, social network analysis, altmetrics, which are the broader research area of text analytics. He has developed and coordinated a text analytics laboratory, which works in various text analytics tasks. He is an alumnus of Allahabad University, Allahabad, India. He has published more than 30 publications in international journals and in proceedings of international conferences of repute. He has also served in South Asian University, Delhi, India as Assistant Professor for more than 4 years. He has also associated with several research projects such as Indo-Mexican Joint Research Project funded jointly by the Department of Science and Technology, Government of India, along with the National Council for Science and Technology (CONACYT) of the United Mexican States.

# Part I
# Intelligent Hardware and Software Design

# Biogeography-Based Optimization for Cluster Analysis

**Xueyan Wu, Hainan Wang, Zhimin Chen, Zhihai Lu, Preetha Phillips, Shuihua Wang and Yudong Zhang**

**Abstract** With the aim of resolving the issue of cluster analysis more precisely and validly, a new approach was proposed based on biogeography-based optimization (abbreviated as BBO) algorithm. (Method) First, we reformulated the problem with an optimization model based on the variance ratio criterion (VARAC). Then, BBO was presented to search the optimal solution of the VARAC. There are 400 data of four groups in the experimental dataset, which have the degrees of overlapping of three distinct scales. The first one is nonoverlapping, the second one is partial overlapping, and the last is severely overlapping. BBO algorithm was compared with three different state-of-the-art approaches. We ran every algorithm 20 times. In this experiment, our results demonstrate the maximum VARAC values that can be found by BBO. The conclusion is that BBO is predominant which is extremely quick for the issue of clustering analysis.

**Keywords** Biogeography-based optimization · Genetic algorithm · Cluster analysis

X. Wu · H. Wang · Z. Lu · S. Wang (✉) · Y. Zhang (✉)
School of Computer Science and Technology, Nanjing Normal University, Nanjing, Jiangsu 210023, China
e-mail: wangshuihua@njnu.edu.cn

Y. Zhang
e-mail: zhangyudong@njnu.edu.cn

X. Wu
Key Laboratory of Statistical Information Technology & Data Mining, State Statistics Bureau, Chengdu, Sichuan 610225, China

X. Wu
School of Computer and Information Engineering, Henan Normal University, Xinxiang, Henan 453000, China

Z. Chen
School of Electronic Information, Shanghai Dianji University, Shanghai 200240, China

Z. Chen
Key Laboratory of Symbolic Computation & Knowledge Engineering of Ministry of Education, Jilin University, Changchun, Jilin 130012, China



S.K. Bhatia et al. (eds.), *Advances in Computer and Computational Sciences*, Advances in Intelligent Systems and Computing 553,
DOI 10.1007/978-981-10-3770-2_1

## 1 Introduction

In the context of a group, cluster analysis is defined as a case of substance with the mode that a lot of targets in a large cluster which are closer to another side than others in the additional clusters [1]. Cluster analysis is a way of unsupervised studying, and in lots of areas, it is also used for statistical data analysis in a common technique, consisting of breeding value [2], food quality monitoring [3], gene engineering [4], pediatric immunization distress [5], chronic rhinosinusitis [6], community analysis [7], etc.

Currently, various algorithms were proposed for cluster analysis. They can be basically classified into the following four categories: centroid-based clustering, distribution-based clustering, density-based clustering, and connectivity-based clustering.

In the research, the most attractive to us is centroid-based methods. In this type, there are two representative algorithms, one is fuzzy *c*-means clustering (FCM) [8], the other is *k*-means clustering [9]. These are iterative methods and affected by a lot of factors, for example, if the initial partition is not determined properly, they may stop at local best instead of global optimal solution.

To solve above the problems, Cotta and Moscato [10] proposed that the best way to detect the global optimum clustering is branch and bound algorithm. But it requires too much calculating time. In the past few years, to solve the problem of clusters, evolutionary algorithm has been proposed, because it is insensitive to initial values. Jarboui et al. [11] put forward an original clustering approach on a foundation of the combinatorial particle swarm optimization (CPSO) algorithm. Gelbard et al. [12] proposed cross-cultural research to cluster analysis with the methodology named Multi-Algorithm Voting (MAV). Niknam and Amiri [13]

Z. Lu
Guangxi Key Laboratory of Manufacturing System & Advanced Manufacturing Technology, Guilin, Guangxi 541004, China

P. Phillips
School of Natural Sciences and Mathematics, Shepherd University, Shepherdstown, WV 25443, USA

P. Phillips
West Virginia School of Osteopathic Medicine, 400 N Lee St, Lewisburg, WV 24901, USA

S. Wang
Department of Electrical Engineering, The City College of New York, CUNY, New York, NY 10031, USA

Y. Zhang
State Key Lab of CAD & CG, Zhejiang University, Hangzhou, Zhejiang 310027, China

Y. Zhang
School of Computing, Mathematics and Digital Technology (SCMDT), Manchester Metropolitan University, Manchester M156BH, UK

thought the $k$-means algorithm is extremely dependent on its initial state and integration of the local best solution. Hence, they put forward to a fresh hybrid evolutionary algorithm. Huo [14] raised chaotic artificial bee colony (CABC) as one method for the issue of cluster division. Abul Hasan and Ramakrishnan [15] underwent a survey on cluster analysis about hybrid evolutionary algorithms. They believed that optimization algorithms are required for cluster analysis to obtain better results. Zhang and Wu [16] proposed bacterial foraging optimization (BFO) for cluster analysis. Kuo et al. [17] put forward an original approach named dynamic clustering on the basis of PSO and GA (DCPG) algorithm. Firefly algorithm (FFA) was used by Zhang and Li [18]. They took 400 data and divided them into four groups for testing. Yang et al. [19] proposed and explored the idea of exemplar-based clustering analysis optimized by genetic algorithms (GA). Wan [20] made a full use of the assembling analysis and researched $k$-means with particle swarm optimization (KPSO). Palaparthi et al. [21] researched a multi-objective optimization (MOO) channel in relevance with cluster analysis to research the vocal folds' form-function connection. On the basis of the algorithms of improved genetic, Cao and Mu [22] put forward to a weighted $k$-means clustering algorithm. Ozturk et al. [23] proposed a modified binary artificial bee colony algorithm. Zhou and Zhang [24] proposed a very advanced algorithm called quantum-behaved particle swarm optimization (QPSO).

Nevertheless, those above approaches are vulnerable to several disadvantages: (1) They may converge to local optimal solutions, corresponding to suboptimal results. (2) They may cost too lengthy time and cost large memory. Hence, a new algorithm, viz., biogeography-based optimization (BBO) [25] was suggested in settling the issue with this research.

The remaining portion of this paper is as follows: In Sect. 2, we define the mathematical model of partitional clustering, as well as propose the clustering criterion and encoding strategy. In Sect. 3, we describe the novel BBO method. In Sect. 4, we describe the experiment, this section includes three types of artificial data which have different overlapping degrees. In Sect. 5, we discuss about the experiment results. Finally, in Sect. 6, we conclude this paper.

## 2 Model Definition

**Hypothesis**: there are $n$ samples $\boldsymbol{Q} = (q_1, q_2, \ldots, q_n)$ in a $d$-dimensional metric space. Every sample belongs to one of the $m$ sets. Every $q_i \in R^d$ indicates a particular feature vector that is composed of $d$ dimension. The clusters are denoted as $\boldsymbol{F} = \{f_1, f_2, \ldots, f_k\}$, where $k$ represents the number of clusters. They should submit three declarations as follows:

$$\begin{aligned} &f_i \neq \phi \quad \text{for} i = 1, 2, \ldots, m \\ &f_i \cap f_j = \phi \quad (i \neq j) \\ &\cup f_i = \{q_1, q_2, q_3, \ldots, q_n\} \end{aligned} \tag{1}$$

The aim of this study is to look for the most appropriate partition $\boldsymbol{F}^*$ that corresponds to the best sufficiency among all possible solutions [26].

The search space is set to be $n$-dimension, which is consistent with $n$-objects. Every dimension not only expresses a sample but also expresses the $i$-th individual $Y_i = \{y_{i1}, y_{i2}, \ldots, y_{in}\}$ that conforms with the affection of $n$ samples, such as $y_{ij} \in \{1, 2, \ldots, m\}$, in which $j$ indicates the $j$-th sample [27].

Suppose $n = 6$ and $k = 2$. Suppose the first cluster A contains the first, second, and third object, and the second cluster B contains the fourth, fifth, and sixth object. Figure 1 illustrates the encoded representation of the cluster solution.

Some criteria have been put forward to estimate the sufficient where a predefined dataset could be assembled. Among the partitional clustering strategies, "VAriance RAtio Criterion (VARAC) [28]" is one of the most popular used, which is defined as

$$\text{VARAC} = \frac{E}{R} \times \frac{n-d}{d-1}, \tag{2}$$

where $\boldsymbol{R}$ and $\boldsymbol{E}$, respectively, represents the transformation between intra-cluster and inter-cluster. Their definitions are as follows:

$$R = \sum_{j=1}^{d} \sum_{i=1}^{n_j} \left(q_j^i - \bar{q}_j\right)^T \left(q_j^i - \bar{q}_j\right) \tag{3}$$

$$E = \sum_{j=1}^{d} n_j (\bar{q}_j - \bar{q})^T (\bar{q}_j - \bar{q}), \tag{4}$$

where $(.)^{\mathrm{T}}$ represents the transpose operation, $n_j$ stands for the cardinal of the cluster $f_j$, $q_j^i$ represents the $i$-th target allotted to $f_j$, which is one of the clusters, $\bar{q}_j$ represents the center of the $j$-th cluster, and $\bar{q}$ represents the center of the whole data, $(d - 1)$ denotes the between-cluster variation freedom degree, while $(n - d)$ represents the within-cluster variation freedom degree.

| Cluster | Object |
|---|---|
| A | 1, 2, 3 |
| B | 4, 5, 6 |

→ $Y_i$

| $y_{i1}$ | $y_{i2}$ | $y_{i3}$ | $y_{i4}$ | $y_{i5}$ | $y_{i6}$ |
|---|---|---|---|---|---|
| 1 | 1 | 1 | 2 | 2 | 2 |

**Fig. 1** Encoded representation

Generally speaking, it is expected that tight and divided clusters have important value of small $R$ and large $E$. Therefore, if data partition result is much better, the value of VARAC is much larger. In order to avoid the rate to increase monotonically as the quantity of the clusters, the term $(n - d)/(d - 1)$ is normalized, which make VARAC by means of an optimization of the clusters, it can achieve maximize criterion.

## 3 Biogeography-Based Optimization

We gave an analogy in Fig. 2 with the well-known GA as a counterpart for making it easy to understand. As is shown in Fig. 2, the population of candidate solutions is represented as habitats in BBO [29–31]. The solution vector components are considered to be suitability index variables (SIVs). Those good solutions are considered habitat with high habitat suitability index (HSI), and vice versa [32–34]. BBO has been proven to give better performance than genetic algorithm (shorted as GA) [35], artificial bee colony (shorted as ABC) [36], firefly algorithm [37], bacterial chemotaxis optimization [38], particle swarm optimization (shorted as PSO) [39], and ant colony optimization [40]. Table 1 demonstrates the pseudocodes of BBO.

## 4 Experiments

The experiments were performed on a computer with i7 processor and 16 GB memory, under Windows 7 operating system.

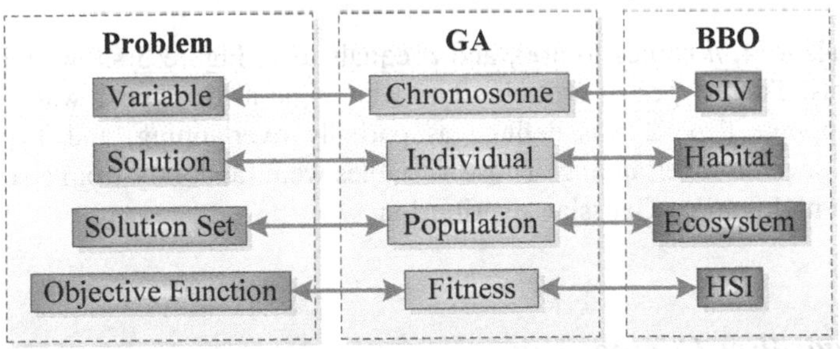


**Fig. 2** The connection of optimization problem, GA, and BBO

**Table 1** Pseudocodes of the proposed BBO

| | |
|---|---|
| Step A | Parameters Initialization. It covers a problem-oriented approach that maps solutions to habitats and SIVs. The maximum number of species $S_{max}$, the modification rate $P_{mod}$, the maximum mutation probability $l_{max}$, elite count $p$, and the maximum migration probabilities |
| Step B | Initialize a set of habitats randomly |
| Step C | Calculate each habitat's HSI, and calculate $S$, $\lambda$, and $\mu$ of every habitat |
| Step D | Update the whole ecosystem using migration according to $P_{mod}$, $\lambda$ and $\mu$ |
| Step E | Implement the mutation procedure on the ecosystem according to mutation rates |
| Step F | Do elitism implementation |
| Step G | Once satisfying the termination criterion, offer the optimal habitat obtained. Otherwise, return to Step C |

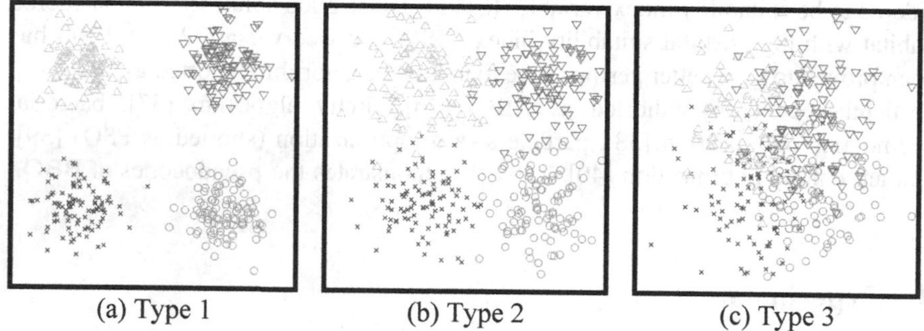
(a) Type 1 (b) Type 2 (c) Type 3

**Fig. 3** Three different types of artificial data [18]

## 4.1 *Artificial Data*

Let $k$ equals to 4, $n$ equals to 400, and $d$ equals to 2. Figure 3 shows their distribution plots. Three types of artificial data were generated. Type 1 was defined as nonoverlapping, Type 2 was defined as partially overlapping, and Type 3 was defined as severely overlapping. Those variables were randomly from a source that adheres to multivariate Gaussian distribution.

## 4.2 *Algorithm Comparison*

Our BBO was tested, and we made a comparison with four state-of-the-art algorithms: chaotic particle swarm optimization (CPSO) [11], genetic algorithm (GA) [19], and firefly algorithm (FFA) [18]. Every approach ran 20 times for fair comparison. Table 2 shows the experiment results.

**Table 2** The results found by VARAC among 20 runs

| Type | VARAC | CPSO [11] | FFA [18] | GA [19] | BBO (our) |
|---|---|---|---|---|---|
| 1 | B | **1683.2** | **1683.2** | **1683.2** | **1683.2** |
| | M | 1534.6 | **1683.2** | 1321.3 | **1683.2** |
| | W | 1023.9 | **1683.2** | 451.0 | **1683.2** |
| 2 | B | **620.5** | **620.5** | **620.5** | **620.5** |
| | M | 607.9 | 618.2 | 594.4 | **618.7** |
| | W | 574.1 | 573.3 | 512.8 | **592.8** |
| 3 | B | **275.6** | **275.6** | **275.6** | **275.6** |
| | M | 203.8 | 221.5 | 184.1 | **240.1** |
| | W | 143.5 | 133.9 | 129.0 | **159.3** |

B = Best, M = Mean, W = Worst
The bold values means the best among all four algorithms

## 5 Discussion

In Table 2, the consequents show that all four algorithms could reach the maximal VARAC of 1683.2 at least once in the case of Type 1 instances. Additional to it, both FFA and BBO succeed for all 20 runs. However, there are still Type 2 instances exists. In these cases, all of the four algorithms: CPSO, FFA, GA, and BBO could reach the 620.5 VARAC within 20 runs, which is the best. The average VARACs of these approaches are 607.9, 618.2, 594.4, and 618.7, respectively, while the lowest VARACs are 574.1, 573.3, 512.8, and 592.8, respectively. This indicates that the BBO outperforms the rest three algorithms. Finally, for Type 3 instances, though the four algorithms can reach the same maximal VARAC of 275.6, the average and worst VARACs of BBO does better than CPSO, FFA, and GA. This further validates the superiority of BBO.

## 6 Conclusions

In this paper, our group put forward a novel clustering method with the combination of VARAC and BBO. The experiments have proved that the proposed BBO outperforms the currently most advanced approaches.

Future research includes the following aspects. First, we shall retest the proposed algorithm when larger datasets are available for us. Second, the performance on severely overlapping data is less stable compared to other results, which may be further improved. Third, the proposed algorithm will be further optimized to suit various application scenarios. Fourth, more advanced optimization methods, such as Jaya algorithm [41], chaotic immune PSO [42], and real-coded BBO [43], shall be tested.

**Acknowledgements** This study was supported by Open Fund of Key Laboratory of Statistical information technology and data mining, State Statistics Bureau (SDL201608), Natural Science Foundation of Jiangsu Province (BK20150982, BK20150983), Open Project Program of the State Key Lab of CAD&CG, Zhejiang University (A1616), NSFC (61602250), Open Fund of Key laboratory of symbolic computation and knowledge engineering of ministry of education, Jilin University (93K172016K17).

## References

1. Pasupuleti, D., et al., *Classification of biodiesel and fuel blends using gas chromatography - differential mobility spectrometry with cluster analysis and isolation of C18:3 me by dual ion filtering.* Talanta, 2016. **155**: pp. 278–88.
2. Cruz, D.A.C.d., et al., *Cluster analysis of breeding values for milk yield and lactation persistency in Guzerá cattle.* Ciência Rural, 2016. **46**(7): pp. 1281–1288.
3. Arana, I., et al., *Monitoring the sensory quality of canned white asparagus through cluster analysis.* Journal of the Science of Food and Agriculture, 2016. **96**(7): pp. 2391–2399.
4. Bodei, L., et al., *Measurement of circulating transcripts and gene cluster analysis predicts and defines therapeutic efficacy of peptide receptor radionuclide therapy (PRRT) in neuroendocrine tumors.* European Journal of Nuclear Medicine and Molecular Imaging, 2016. **43**(5): pp. 839–851.
5. Pedro, H., et al., *Pediatric Immunization Distress: A Cluster Analyses of Children's, Parents', and Nurses' Behaviors During the Anticipatory Phase.* The Clinical journal of pain, 2016. **32** (5): pp. 394–403.
6. Soler, Z.M., et al., *Cluster analysis and prediction of treatment outcomes for chronic rhinosinusitis.* Journal of Allergy and Clinical Immunology, 2016. **137**(4): pp. 1054–1062.
7. Liu, D., et al., *Estimating the optimal number of communities by cluster analysis.* International Journal of Modern Physics B, 2016. **30**(8), Article ID: 1650037.
8. Verma, P. and R.D.S. Yadava, *Polymer selection for SAW sensor array based electronic noses by fuzzy c-means clustering of partition coefficients: Model studies on detection of freshness and spoilage of milk and fish.* Sensors and Actuators B-Chemical, 2015. **209**: pp. 751–769.
9. Garcia, M.L.L., et al., *K-means algorithms for functional data.* Neurocomputing, 2015. **151**: pp. 231–245.
10. Cotta, C. and P. Moscato, *A memetic-aided approach to hierarchical clustering from distance matrices: application to gene expression clustering and phylogeny.* Biosystems, 2003. **72**(1–2): pp. 75–97.
11. Jarboui, B., et al., *Combinatorial particle swarm optimization (CPSO) for partitional clustering problem.* Applied Mathematics and Computation, 2007. **192**(2): pp. 337–345.
12. Gelbard, R., et al., *Cluster analysis using multi-algorithm voting in cross-cultural studies.* Expert Systems with Applications, 2009. **36**(7): pp. 10438–10446.
13. Niknam, T. and B. Amiri, *An efficient hybrid approach based on PSO, ACO and k-means for cluster analysis.* Applied Soft Computing, 2010. **10**(1): pp. 183–197.
14. Huo, Y.K., *Chaotic Artificial Bee Colony Used for Cluster Analysis*, in *Intelligent Computing and Information Science, Pt I*, ed. R. Chen, 2011, Springer-Verlag Berlin: Berlin. pp. 205–211.
15. Abul Hasan, M.J. and S. Ramakrishnan, *A survey: hybrid evolutionary algorithms for cluster analysis.* Artificial Intelligence Review, 2011. **36**(3): pp. 179–204.
16. Zhang, Y. and L. Wu, *Bacterial Foraging Optimization Used in Cluster Analysis.* International Journal of Digital Content Technology and its Applications, 2012. **6**(22): pp. 345–354.

17. Kuo, R.J., et al., *Integration of particle swarm optimization and genetic algorithm for dynamic clustering*. Information Sciences, 2012. **195**: pp. 124–140.
18. Zhang, Y. and D. Li, *Cluster Analysis by Variance Ratio Criterion and Firefly Algorithm.* JDCTA: International Journal of Digital Content Technology and its Applications, 2013. **7**(3): pp. 689–697.
19. Yang, Z., et al., *Exemplar-Based Clustering Analysis Optimized by Genetic Algorithm.* Chinese Journal of Electronics, 2013. **22**(4): pp. 735–740.
20. Wan, S.A., *Entropy-based particle swarm optimization with clustering analysis on landslide susceptibility mapping*. Environmental Earth Sciences, 2013. **68**(5): pp. 1349–1366.
21. Palaparthi, A., et al., *Combining Multiobjective Optimization and Cluster Analysis to Study Vocal Fold Functional Morphology.* Ieee Transactions on Biomedical Engineering, 2014. **61** (7): pp. 2199–2208.
22. Cao, Y. and X.W. Mu, *Weighted K-means Clustering Analysis Based on Improved Genetic Algorithm.* Sensors, Mechatronics and Automation, 2014. **511–512**: pp. 904–908.
23. Ozturk, C., et al., *Dynamic clustering with improved binary artificial bee colony algorithm.* Applied Soft Computing, 2015. **28**: pp. 69–80.
24. Zhou, X. and G. Zhang, *Cluster Analysis by Variance Ratio Criterion and Quantum-Behaved PSO*, in *Cloud Computing and Security*, 2015, Springer. pp. 285–293.
25. Guvenc, U., et al., *Performance analysis of biogeography-based optimization for automatic voltage regulator system.* Turkish Journal of Electrical Engineering and Computer Sciences, 2016. **24**(3): pp. 1150–1162.
26. Papathomas, M. and S. Richardson, *Exploring dependence between categorical variables: Benefits and limitations of using variable selection within Bayesian clustering in relation to log-linear modelling with interaction terms.* Journal of Statistical Planning and Inference, 2016. **173**: pp. 47–63.
27. Goncalves, K.C.M. and F.A.S. Moura, *A Mixture Model for Rare and Clustered Populations Under Adaptive Cluster Sampling*. Bayesian Analysis, 2016. **11**(2): pp. 519–544.
28. Zhang, Y.D., et al., *Chaotic Artificial Bee Colony Used for Cluster Analysis*, in *Intelligent Computing and Information Science, Pt I*, ed. R. Chen, 2011, Springer-Verlag Berlin: Berlin. pp. 205–211.
29. Garg, V. and K. Deep, *Performance of Laplacian Biogeography-Based Optimization Algorithm on CEC 2014 continuous optimization benchmarks and camera calibration problem.* Swarm and Evolutionary Computation, 2016. **27**: pp. 132–144.
30. Zhang, Y., et al., *Pathological brain detection in magnetic resonance imaging scanning by wavelet entropy and hybridization of biogeography-based optimization and particle swarm optimization.* Progress In Electromagnetics Research, 2015. **152**: pp. 41–58.
31. Ji, G., et al., *Fruit classification by wavelet-entropy and feedforward neural network trained by fitness-scaled chaotic ABC and biogeography-based optimization.* Entropy, 2015. **17**(8): pp. 5711–5728.
32. Krishnasamy, U. and D. Nanjundappan, *Hybrid weighted probabilistic neural network and biogeography based optimization for dynamic economic dispatch of integrated multiple-fuel and wind power plants.* International Journal of Electrical Power & Energy Systems, 2016. **77**: pp. 385–394.
33. Sangeetha, S. and T.A.A. Victoire, *Radio Access Technology Selection in Heterogeneous Wireless Networks Using a Hybrid Fuzzy-Biogeography Based Optimization Technique.* Wireless Personal Communications, 2016. **87**(2): pp. 399–417.
34. Rashid, A., et al., *A dynamic oppositional biogeography-based optimization approach for time-varying electrical impedance tomography.* Physiological measurement, 2016. **37**(6): pp. 820–42.
35. Lu, S., *A note on the weight of inverse complexity in improved hybrid genetic algorithm.* Journal of Medical Systems, 2016. **40**(6), Article ID: 150.
36. Wu, L., *Optimal multi-level Thresholding based on Maximum Tsallis Entropy via an Artificial Bee Colony Approach.* Entropy, 2011. **13**(4): pp. 841–859.

37. Wu, L., *Solving Two-Dimensional HP model by Firefly Algorithm and Simplified Energy Function.* Mathematical Problems in Engineering, 2013.
38. Zhang, Y., *Stock market prediction of S&P 500 via combination of improved BCO approach and BP neural network.* Expert systems with applications, 2009. **36**(5): pp. 8849–8854.
39. Ji, G., *A comprehensive survey on particle swarm optimization algorithm and its applications.* Mathematical Problems in Engineering, 2015, Article ID: 931256.
40. Ji, G.L., *A Rule-Based Model for Bankruptcy Prediction Based on an Improved Genetic Ant Colony Algorithm.* Mathematical Problems in Engineering, 2013, Article ID: 753251.
41. Cattani, C. and R. Rao, *Tea Category Identification Using a Novel Fractional Fourier Entropy and Jaya Algorithm.* Entropy, 2016. **18**(3), Article ID: 77.
42. Jun, Y. and G. Wei, *Find multi-objective paths in stochastic networks via chaotic immune PSO.* Expert Systems with Applications, 2010. **37**(3): pp. 1911–1919.
43. Wu, X., *Smart detection on abnormal breasts in digital mammography based on contrast-limited adaptive histogram equalization and chaotic adaptive real-coded biogeography-based optimization.* SIMULATION, 2016. **92**(9): pp. 873–885.

# BTpower: An Application for Remote Controlling PowerPoint Presentation Through Smartphone

**Md. Asraful Haque, Abu Raihan and Mohd. Danish Khalidi**

**Abstract** This paper presents an interesting android-based application "BTpower" which turns our phone into a remote. The app lets us control PowerPoint presentation from across the room, so we can walk around freely during presentations. The ppt/pptx file will be stored on the mobile. Bluetooth is used for connectivity purpose. The application provides a user-friendly interactive interface by which we can interact with Microsoft Office PowerPoint on our PC. With BTpower, we can start our PowerPoint presentation, jump to the next or previous slides, resume, or exit the slide show with a touch of our finger—all from our phone.

**Keywords** Android app · PowerPoint tools · Bluetooth · Remote desktop · Mobile communication

## 1 Introduction

The term "BTpower" has been derived by combining two words Bluetooth and PowerPoint. The use of smartphones, tablets, and other touch screen devices is gaining popularity with an unbelievable pace over the years. The smartphone applications can transfer commands to PC using the device communication mechanisms such as Bluetooth and Wi-Fi [1]. Controlling electronic devices and computers wirelessly is an important aspect of the technology [2]. Bluetooth has a tremendous potential in moving and synchronizing information in a localized setting. One can interact with electronic devices using Bluetooth of a smartphone. Many applications have come up in the market in Google play store, IOS play store,

Md.A. Haque (✉) · A. Raihan · Mohd.D. Khalidi
Computer Engineering Department, Aligarh Muslim University, Aligarh 202002, India
e-mail: md_asraf@zhcet.ac.in

A. Raihan
e-mail: aburaihan.mnb@gmail.com

Mohd.D. Khalidi
e-mail: abradan12@gmail.com



S.K. Bhatia et al. (eds.), *Advances in Computer and Computational Sciences*,
Advances in Intelligent Systems and Computing 553,
DOI 10.1007/978-981-10-3770-2_2

and for windows phone for serving this purpose. These applications provide options like streaming audio, video files of your computer controlling mouse and keyboard. One of the most widely used mobile OS these days is Android. Android is a powerful operating system supporting a large number of applications in smartphones. BTpower is an android-based application used to control PPT presentations remotely inside a house, an office, or a conference room. Many times while giving presentations either one has to be dependent on other person to change slides or he has to change it manually [1]. With BTpower, we can start our presentation, jump to the next or previous slides, resume, or exit the slide show with a touch of our finger —all from our phone. The connectivity is made between mobile and computer using Bluetooth.

The rest of the paper is organized as follows. The next section, Sect. 2, briefly describes some related work. Section 3 explains the implementation process of BTpower. Section 4 provides a user manual with some snapshots. Section 5 mentions the required technologies to execute the application. Section 6 concludes the paper with some future remarks.

## 2 Related Work

Many mobile applications are available in the market of Google play store, IOS play store, or Microsoft Windows store for controlling a PC from a smartphone. Some of them are Office remote, MyPoint etc. Office Remote is an application of Windows phone for controlling Microsoft Office, providing convenient touch-based control of Word, Excel, and PowerPoint documents projected from our PC. MyPoint is a PowerPoint remote application compatible with iOS. BTpower is an android-based application. The functionality of MyPoint is similar to our work. Yenel et al. developed a useful mobile software, called PocketDrive to access PC applications in 2007 [2]. Additionally, PocketDrive supports zooming and presentation mode with user-friendly GUI for fast forward and backward jumping on a presentation. The last few years have seen a growing interest in developing different mobile applications for controlling PC and thus making smartphone smarter. Chintalapati and Rao [3], Yang and Li [4], Mishra et al. [5], and many more researchers also suggested similar type of applications, which enable a cell phone to act as a remote controller device for desktop PCs and their applications. Our application is slightly different in this context. The application's purpose is to display the PowerPoint presentation on a computer or a projector and also to provide a user-friendly interactive interface in the mobile. The ppt/pptx file will have to be stored on the mobile. There is no need to copy the file into the computer. So we can present the data on anyone's computer without sharing it. The application will be accessible to the MS Office of the target PC over a Bluetooth connection.

## 3 Implementation

The purpose of the proposed system is to use a mobile phone as a remote for PowerPoint presentation. The ppt/pptx file stored in the mobile will be projected on a PC monitor or projector. The connectivity is made using Bluetooth. For establishing connectivity a server code in java is required to be run on PC. One way to send and receive data in Java Bluetooth is to use the RFCOMM protocol [6, 9]. Using RFCOMM, the application establishes a serial communication between smartphone and PC. After making the connections through Bluetooth, commands are transferred from android application to the computer and the computer responds according to the commands send through the mobile. So the system consists of two parts: an android application for our mobile and a server code to run on computer. Android application is implemented by using Bluetooth adapters, socket, and threads. The server code in PC is implemented using java bluecove directory. This bluecove package is used for connecting the PC Bluetooth to other Bluetooth devices. A waitthread and a process connection thread are used in server code for interacting with mobile's Bluetooth device by accepting requests and data. Java robot class is used for PC control (Fig. 1).

We used Eclipse for development purpose since it includes a base workspace and an extensible plug-in system for customizing the environment. With the help of Android SDK emulator we prototyped, developed, and tested our application. Android applications are composed of one or more application components, i.e., activities, services, content providers, and broadcast receivers [7, 10]. Each component performs a different role in the overall application behavior. There are mainly four modules in our application as follows.

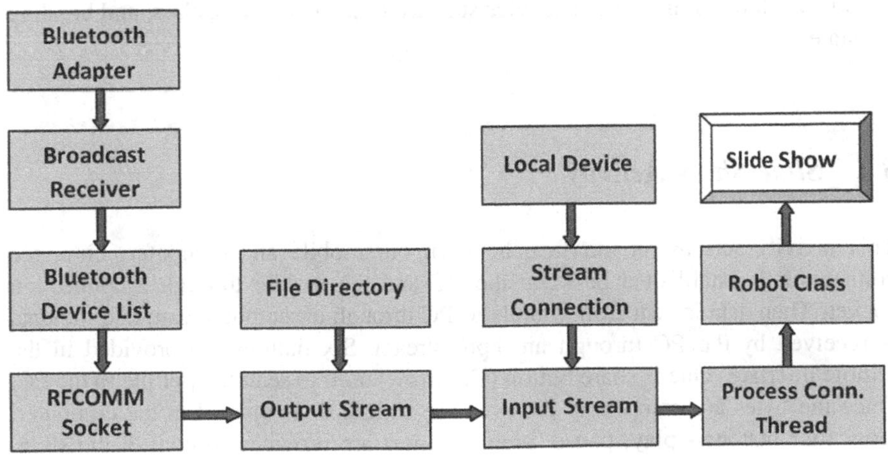


**Fig. 1** Block diagram of the application

### 3.1 File Selector

It allows a user to select the required file among the .ppt/.pptx files stored in the android mobile. The files of .ppt/.pptx extension are added to the file list view. When user selects the file, its name is passed on along with its address in the directory to the next activity.

### 3.2 Bluetooth Device List Activity

It provides an interface to the user to select the target device among all Bluetooth devices present in his surroundings. First, the Bluetooth adapter is turned on. The list of paired devices is displayed. Through the scan button the user can search for nearby unpaired Bluetooth devices. To search new devices, a broadcast receiver is used. If a new device is found it is added to the available device list view. Now the user has to select a device from the list of available devices. Once a device is selected, the next button is pressed. Now two arguments, the address of the selected Bluetooth device and the filename along with its address are passed on to the next intent. The components used in the module are:

1. BluetoothAdapter: It represents the local Bluetooth adapter, i.e., Bluetooth radio which is the entry point for all Bluetooth interaction. Using this, we can discover other Bluetooth devices, query a list of paired devices, instantiate a BluetoothDevice using a known MAC address and create a BluetoothServerSocket to listen for communications from other devices.
2. BluetoothDevice: It represents a remote Bluetooth device. It is used either to request a connection with a remote device through a BluetoothSocket or to query information about the device such as its name, address, class, and bonding state.

### 3.3 Slide Show Activity

This activity acts as an interface between our mobile and computer. First, the connection is established between the PC and the mobile through a connection socket. Then data is sent from mobile to PC through an output stream and the data is received by the PC through an input stream. Six buttons are provided in the remote interface. One is share button (Up-arrow) used to send the ppt file to the PC. Once the bytes are completely received the slideshow is opened in the computer. Now four buttons—play, pause, previous, next are provided to control the slideshow on the pc. These buttons send signals to the computer in integer values and the computer responds to it accordingly. Once the slideshow is completed the file can be deleted by using exit button. Components used are:

1. BluetoothSocket: It is similar to a TCP Socket. It allows an application to exchange data with another Bluetooth device via InputStream and OutputStream.
2. UUID: A Universally Unique Identifier (UUID) is a 128-bit number used to uniquely identify information. In this case, it is used to identify our application's Bluetooth service. To generate a UUID for our application, we can use one of many available random UUID generators on the web, then initialize it with from String (String).
3. FileInputStream: An input stream that reads bytes from a file.
4. ByteArrayOutputStream: It implements a specialized OutputStream for class in which the data is written into a byte array. The buffer automatically grows as data is written to it. When the writing is considered to be finished, a copy of the byte array can be requested from the class.

### *3.4 Server Code*

Program listings or program commands in the text are normally set in typewriter font, e.g., CMTT10 or Courier. Once the connection is established, remote Bluetooth server class starts the Waitthread class which then initiates the process connection thread. The server code is run through the bluecove-2.1.0 java package. This package provides tools to manage the PC connection through other remote Bluetooth devices by assigning a UUID to the PC. Components used are:

1. Waitthread: This java file starts a thread that waits for a connection that has the same UUID as the computer. It sets the local Bluetooth device discoverable. If the device is found and paired, the process connection thread is started. If a new device tries to connect a pairing request is made before establishing connection. Once wait is over the object of the process connection thread is passed to a newly created processor thread.
2. Process Connection Thread: It contains commands and functions that respond to the data send by the mobile application. First, all the bytes of the ppt/pptx file are read and then the file is opened in slideshow mode. The commands are read as integers. This thread uses Robot class to press the computer right and left arrow keys to move to the previous and next slides. It uses Runtime class to open the PowerPoint file in slideshow mode.

## 4 User Manual

The application is very easy to use. Following are few simple steps to make the application operational:

Step 1: Execute the jar file on PC.

Step 2: Open the app on the smartphone. First choose the file stored in the mobile phone for presentation. User has to select .ppt/.pptx file only.

Step 3: Now turn on Bluetooth. It will show the available nearby Bluetooth devices which are turned on. User has to select the target PC from the list (Fig. 2).

Step 4: Now application will provide a remote interface to the user. It has total six buttons for controlling presentation. Share button is used to temporarily transfer the data to the remote device. Play button is used to start the presentation. Next and Previous buttons are for slide movements. Pause button is used to stop the slide show and back to the editable window. One power button is also there to exit from the application and delete the ppt/pptx file from the remote device.

Step 5: Smartly present your information through a projector/monitor (Fig. 3).

## 5 Requirement Specification

The proposed application was successfully tested on a personal computer with the use of a smartphone. During the testing stage, first devices were connected via Bluetooth after executing the server code on PC. Application successfully enables user to remotely access their presentation in a convenient way. We identified the following as the requirements of a system that turns a smartphone into a remote.

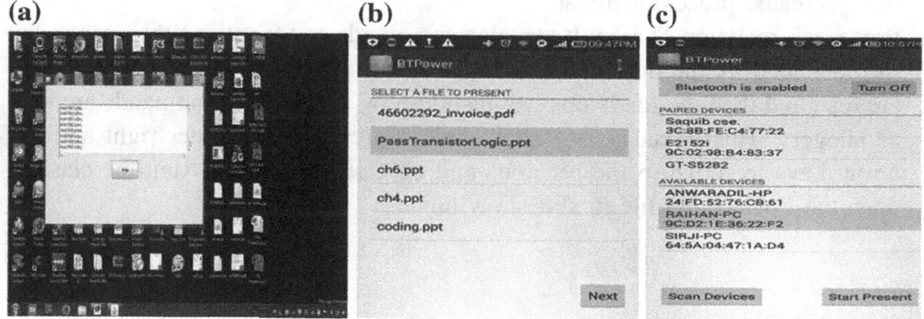


**Fig. 2** **a** Step 1: jar file execution, **b** Step 2: file selection, **c** Step 3: device selection

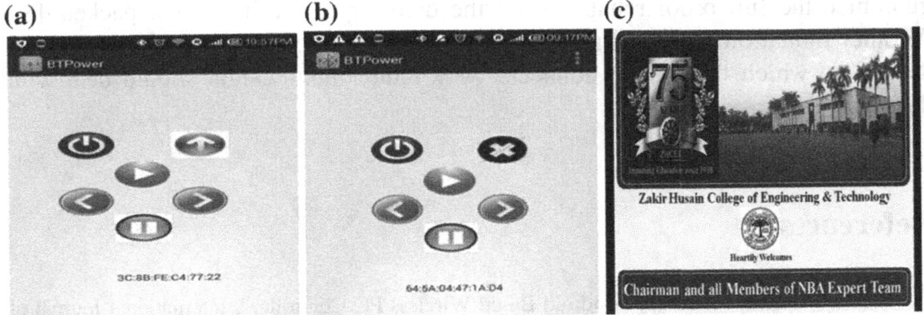


**Fig. 3** **a** Step 4: remote interface for data sharing, **b** Step 4: remote interface for presentation, **c** Step 5: slideshow

**Mobile**:

1. Bluetooth version 3.0 or above.
2. Android version 4.0 or above.
3. 2 MB space for installation.

**Computer**:

1. Windows 7 or above operating system.
2. Bluetooth version 3.0 or above.
3. Microsoft office 2007 or above to open ppt file.
4. Java jdk 7u45 or above for running the server code jar file.

## 6 Conclusion

With this application user can display the presentation stored on his/her mobile through a projector/monitor by connecting his/her mobile to a computer wirelessly via Bluetooth. The interface with next, previous, pause, and play buttons provide user's full control on the slide show. We can roam around the room carrying our mobile and giving our presentation (Obviously room size should not exceed the Bluetooth range, i.e., 10 m). This application can be used in colleges, offices, business meetings, and community interaction programs. There is no need of any additional hardware devices. But it needs to be mentioned that the application is in its initial version of development. So there is a huge scope for further enhancement. Functionalities of BTpower can be extended by including other application interfaces such as whole MS Office package, audio–video players, or virtual keyboard. One major limitation of BTpower is that the application establishes a serial communication using RFCOMM sublayer of Bluetooth for message exchange between smartphone and PC. The ppt/pptx file is actually temporarily transferred at the time of presentation. The RFCOMM sublayer of Bluetooth and the layers below it do not guarantee the full proof reliability for the delivery of the bytes and packets [8]. Another limitation is the running of server code which does not ensure security of the PC to which the mobile connects. Any future modification should deal with these issues.

## References


1. Mahesh Deshmukh et. al., "Android Based Wireless PC Controller", International Journal of Computer Science and Information Technologies, Vol. 6 (1), 2015.
2. Yenel Y. and Ibrahim K., "PocketDrive: A System for Mobile Control of Desktop PC and its Applications Using PDAs", International Symposium on Computer and Information Sciences-ISCIS, 2007.
3. J. B. Chintalapati, S. Rao, "Remote computer access through Android mobiles", IJCSI International Journal of Computer Science Issues, Vol. 9, Issue 5, No 3, September 2012.
4. Y. Yang and L. Li, "Turn Smartphones into Computer Remote Controllers", International Journal of Computer Theory and Engineering, Vol. 4, No. 4, August 2012.
5. P. S. Mishra et. al., "Controlling PC Application through Mobile Phone", International Journal of Advanced Research in Computer Science and Software Engineering, Volume 3, Issue 1, January 2013.
6. A.G. Villan and J.J. Estev, "Remote Control of Mobile Devices in Android Platform" IEEE transactions on Mobile Computing, 2011.
7. H. Hyun et. al., "PC Application Remote Control via Mobile Phone", International Conference on Control Automation and System, 2010.
8. Brent A. Miller, Chatschik Bisdikian, "Bluetooth Revealed: The Insider's Guide to an Open Specification for Global Wireless Communications", 2nd Edition, Prentice Hall, 2001.
9. Bluetooth special interest group. https://www.bluetooth.org/.
10. Android. http://www.android.com.

# Crowd Monitoring and Classification: A Survey

Sonu Lamba and Neeta Nain

**Abstract** Crowd monitoring on public places is very demanding endeavor to accomplish. Huge population and assortment of human actions enforces the crowded scenes to be more continual. Enormous challenges occur into crowd management including proper crowd analysis, identification, monitoring and anomalous activity detection. Due to severe clutter and occlusions, conventional methods for dealing with crowd are not very effective. This paper highlights the various issues involved in analyzing crowd behavior and its dynamics along with classification of crowd analysis techniques. This review summarizes the shortcomings, strength and applicability of existing methods in different environmental scenarios. Furthermore, it overlays the path to device a proficient method of crowd monitoring and classification which can deal with most of the challenges related to this area.

**Keywords** Crowd monitoring · Behaviour analysis · Crowd classification

## 1 Introduction

Crowd phenomenon has been an important research issue in the era of computer vision since past few years. Due to growing population, understanding and monitoring crowd behavior has now become an essential exercise and challenge for the security agencies across the world. Thus, various groups of researchers and intellectuals set their mind to find proper solution in such field so as to control and manage crowd in their specific manner. To monitor a large area of crowd there has been an exponential increase in surveillance cameras installed around the world, limited number of human resources is not sufficient in analyzing these large number

S. Lamba (✉) · N. Nain
Department of Computer Science & Engineering, MNIT, Jaipur, India
e-mail: lamba.sonu5@gmail.com

N. Nain
e-mail: nnain.cse@mnit.ac.in



S.K. Bhatia et al. (eds.), *Advances in Computer and Computational Sciences*, Advances in Intelligent Systems and Computing 553,
DOI 10.1007/978-981-10-3770-2_3

of video frames simultaneously. So there must be an automated way for crowd monitoring and classification. Intelligent surveillance system is one of the extremely important applications of crowd analysis.

Researchers have proposed several methods and techniques to understand crowd behaviors for developing a safe and secure environment in order to avoid crowd congestion, public riots and terror attacks etc. In crowded scenes, standard computer vision techniques are not applicable in first hand manner due to severe occlusion and complex background scenarios. Many computer vision algorithms exist in literature for tracking, detecting and in analyzing behavior of crowded scene. Although, they provide a good result in a low to medium density of population, but it is still a challenge to deal with a dense crowd. There is an adequacy to present a review of crowd monitoring and classification. Most of the recent surveys focused on the activity analysis of a single person or small group of people, rather than focusing on a crowded scenario. The survey papers by Zhan et al. [1] and Teng et al. [2] are the only two focusing on crowd video analysis as per best of our knowledge. Zhan et al. [1] focused on pedestrian detection and tracking in severely occluded scene but crowd behavior understanding and abnormality detection are not at all covered by them. Although, Teng et al. [2] focused on crowd behavior recognition, motion pattern segmentation and anomaly detection but did not provide generally accepted solution for an unseen crowd scene. This survey suggests many open issues for further research in crowd monitoring and classification.

## 2 Crowd Scene Monitoring Applications

Crowd monitoring has wide range of applications in the real-world scenarios.

1. **Intelligent Surveillance**: For security point of view, the very crowded places should be under camera surveillance such as railway stations, stadiums, subways and shopping malls etc. The normal surveillance system should be replaced by intelligent surveillance which can perform crowd behavior analysis and control the crowd by alarming.
2. **Crowd Management**: Crowded scene analysis helps to develop crowd management strategies. In mass gathering situations, these strategies control the crowd motion in order to avoid the overcrowded situations and public stampede.
3. **Public Space Design**: Monitoring and classification of crowd and their relevant dynamics can provide prior instruction in public space designing by ensuring safety measures and comfort level in the construction of railway stations, airport and terminals etc.
4. **Entertainment**: Computer games could be designed to simulate crowd analysis techniques to derive a correct mathematical model of crowds.

### 2.1 *Motivations to Crowd Monitoring and Classification*

In the video scene analysis and understanding, the focus is on object detection, tracking and behavior recognition [3, 4]. The conventional methods are not appropriate or sometimes fail for densely crowded scenes which have severe occlusions, ambiguities and are extremely cluttered, where undetected anomalous activities might lead to adverse situations which are terrible. Some incidents of crowd disaster at mass events are illustrated in Fig. 1. A crowd has both dynamics and psychological characteristics so analysis of behavior is a very complex task. Human crowds are often goal oriented. It is a very difficult to model dynamics of a crowd at an appropriate level. There is a need to detect, count and classify the behavior of crowd in most surveillance scenario. The rest of this paper is structured as follows. In Sect. 2, we introduce the features which are generally used in the literature of crowd scene analysis. Section 3, categorizes crowd analysis and the relevant approaches are detailed. The commonly used database for crowded scene analysis is summarized in Sect. 4. In Sect. 6 we conclude this paper by furnishing some encouraging future directions.

## 3 Feature Categorization in Crowded Scene

A proper feature categorization can benefitted the subsequent tasks. In crowd scene analysis, motion features play a vital role. Representation point of view, motion features can be classified into three levels as: flow-based features, local spatio-temporal features and trajectory/tracklet. In flow based features, each and every pixel is analyzed. In local spatio-temporal features, local information is extracted from 2-D patches or 3-D cubes. On a next stage, being a basic feature of motion representations, trajectory/tracklet evaluates the individual tracks. These feature representations have been used in crowd analysis and perform various tasks as crowd counting, crowd behavior recognition and crowd anomaly detection.

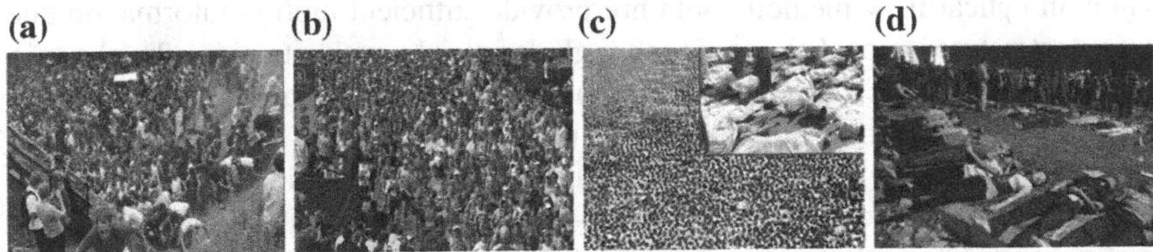


**Fig. 1** Examples of crowd calamities incidents at mass gathering events: **a** Love Parade disaster-Duisburg, Germany (2010). **b** Boston marathon bombing Massachusetts, United States (2013). **c** Mina, Mecca-Saudi Arabia (2015). **d** Khmer water festival-Penh, Cambodia (2010)

### 3.1 Flow-Based Features

When we look at the crowd, our concern is to detect what actions are performed not who is performing it, where a certain set of activities of an individual may visible proportionately random, but to analyze whole crowd scene they have some pattern in their actions [5]. Various flow based features have been presented in recent years [6, 7]. Motion feature can be computed by the conventional optical flow method if the brightness of an image at a time t is represented by I(x, y, t) then

$$\frac{dI}{dt} = \frac{\partial I}{\partial x}\frac{dx}{dt} + \frac{\partial I}{\partial y}\frac{dy}{dt} + \frac{\partial I}{\partial t} = \frac{\partial I}{\partial x}u + \frac{\partial I}{\partial y}v + \frac{\partial I}{\partial t}$$

We calculate motion vector (u, v) for all points in an image. The equation is solved by Optical Flow Constraint (OFC) [6], where image brightness is supposed to be constant with respect to time. Flow-based features are further divided into three categories.

- **Optical Flow**: Optical flow is used for motion detection in video sequences by using flow vector of moving objects. However, long range dependencies are not captured by optical flow.
- **Particle Flow**: The computation of particle flow is done by moving grid with the help of optical flow. The trajectories provided by particle flow are related with an initial and later position of a particle. It has shown very excellent results on abnormal behavior detection and crowd segmentation.
- **Streak Flow**: Mehran et al. [8] introduced streak flow which helps in computation of an accurate motion field in crowded video. It also provides temporal evolution of moving object in a period of time and measures flow in visualization and fluid mechanics.

### 3.2 Local Spatio-Temporal Features

When an optical flow method could not provide sufficient motion information due to an unstructured crowded scene with high density. In such situations, local spatio temporal features are one of the solutions to gain motion flow of crowd. In the estimation of pedestrian movement, the nonuniform motion is produced by any number of objects in each local area. In these kinds of scenarios, only dense local pattern is utilized. To provide the spatio-temporal relationship, two methods are existing such as spatio-temporal gradients [9, 10] and histogram functions [11]. They extract whole motion of crowd video and specify its spatio-temporal distributions based on local 2D patches or 3D cubes. The spatio-temporal gradient, for each pixel i in each patch is calculated as:

$$(I_{ix}, I_{iy}, I_{iz})^T = \left(\frac{\partial I}{\partial x}\frac{\partial I}{\partial y}\frac{\partial I}{\partial z}\right)^T$$

where x, y and t are the video's horizontal, vertical, and temporal dimensions, respectively. Motion histogram plots the motion information defined in local region. In fact, it is not suitable for crowd motion analysis because computing motion is not only time consuming but also erroneous. In contrast of motion histogram, a novel feature descriptor called multi-scale histogram of optical flow (MHOF) is proposed by Cong et al. [11]. Spatio-temporal features are widely used in crowd anomaly detection due to their strong descriptive power.

### *3.3 Trajectory/Tracklet*

Comparing with other features, trajectory is more attractive and semantic than flow based and spatio temporal representation. We can observe crowd scene activities by motion features due to its repetitive pattern. The features such as acceleration, motion energy and relative distance between objects etc. are extracted from trajectories of crowd video. However, as previously mentioned, object detection, feature extraction and tracking are not performed accurately due to severe occlusion and clutter. To overcome these difficulties and obtaining complete trajectory, a motion feature is introduced in [12]. They are a piece of trajectory acquired from the tracker within a short interval termed as tracklet.

When severe occlusion and clutter scene are aroused, tracklets are terminated because they are more stable and rarely change with scenes. In [12–14], they used track-lets to obtain complete trajectories for tracking or activity recognition. In [12] various tracklet based methodologies are proposed to analyzing and clustering semantic regions in unstructured areas. In their work, they extract tracklets from dense crowd video then apply a defined model and it executes the spatial and temporal coherence between tracklets and eventually obtains a behavior pattern to analyze densely crowded scenes. An approach to segment video in the form of trajectories that helps in further analysis of activities present in a video is proposed in [15].

## 4 Crowd Analysis and Monitoring

The research on crowd analysis is comprehensively divided into three parts: People counting/density estimation, people tracking and behavior understanding or anomaly detection. A taxonomy of crowd analysis and monitoring is systematically represented in Fig. 2 with subcategories so the readers can easily understand the

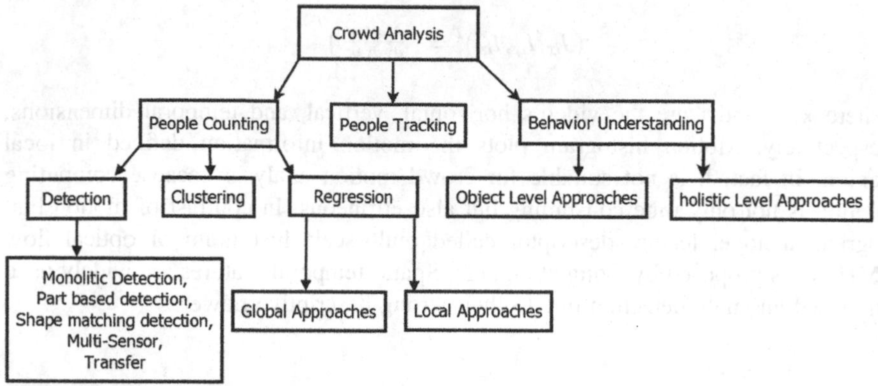


**Fig. 2** Systematic representation of crowd monitoring and classification

crowd phenomenon. Brief explanations of these categories of crowd classification are provided in the following subsections.

## *4.1 People Counting or Density Estimation*

People counting are mainly focused on overcrowded areas for both security and safety purposes [16]. People counting or density estimation is a dominant problem to define the level of a crowd as dense or sparse. People counting can be applied on static images and video sequences in both outdoor and indoor scenarios. In recent years, people counting can be arranged as: counting by detection, counting by clustering and counting by regression [17]. The various crowd counting strategies are illustrated in Table 1 along with its advantages and shortcomings.

## *4.2 People Tracking*

In people tracking, we need to locate the position of an individual in successive frames. The problems of people counting and tracking are correlated, as both have the target of identifying people in crowded scene. However, the problem of counting generally needs to approximate the number of participants present in crowd, instead of their position. On the other side, the tracking problem involves to locate each individual in the frames as a function of time. Mikel et al. [29] combine crowd density and tracking of individual people by optimizing joint energy function. A ground truth density Fo(p) is proposed by them as a kernel density estimation based on the positions of annotated points as

**Table 1** Comparative study of people counting literature with advantages and disadvantages

| Reference and year | Methods | Advantage | Disadvantage | Datasets |
|---|---|---|---|---|
| 2015 [18] | DPM + SIFT + GLCM + Fourier | Combination of multiple features | Perform count only on still images | UCF dataset |
| 2011 [19] | BP-neural network | High accuracy in low density | False detection in dense crowd | Masjid al-Haram |
| 2013 [20] | Kinect + HOG + SVM | Overcome occlusion and overlaps | Constraint to kinect camera | Real time video |
| 2013 [21] | HOG + SIFT + MRF | Count in extremely dense crowd | No explicit density function, count in still images only | 50 crowd images with 64 K annotated Humans |
| 2010 [22] | SURF + SVR | In several cases more accurate than Albiols [23] | Problems due to partial occlusions and perspective | PETS2009 |
| 2014 [24] | Channel state information (CSI) | Device free crowd counting | Cannot work well in a dim or dark environment | Own data set |
| 2015 [25] | Multi-view head shoulder detection | Both static and moving crowd count | Limited to sparse crowd, poor performance on people with strange clothes | Indoor videos: classroom, meeting room |
| 2015 [26] | Part based detection | Detect partially visible humans | Difficult in low resolution, camera position | Internet source: flickers |
| 2013 [27] | CNN + Deep learning | Introduction of a novel loss function | Time consumes for training | Millions of images for a very deep network |
| 2015 [28] | Deep convolution neural network (CNN) | Solve the cross-scene crowd counting problem | Pre-trained CNN model required | UCF CC 50 dataset, UCSD dataset |

$$Fo(p) = \frac{1}{2\pi\vartheta^2}\sum_i \exp\left(\frac{\|\epsilon_i - \mathrm{p}\|}{2\vartheta^2}\right)$$

where $\vartheta$ is size of feature in feature map, $\epsilon$ is ground-truth annotations of feature positions and p is pixel in an image. In [1], object tracking algorithms are comprehensively covered along with taxonomy of approaches with some references on crowd tracking.

## 4.3 Crowd Behavior Analysis

To understand and monitor the crowd behavior is still a challenge despite the various advances in human behavior analysis. Abnormal behavior can be defined in various ways due to its personalized essence. It has been seeding much confusion in the literature. Some researchers describe the abnormality in terms of frequency. The event which occurs infrequently is called as abnormal or which happens rarely. Crowd can be classified as structured or unstructured as shown in Fig. 3. It is easy to analyze structured crowd but an unstructured crowd is very dangerous due to random motion. In behavior understanding we mainly focus on velocities, direction of flow and abnormal events like fighting, running etc.

Validation of crowd behavior is again a challenging problem because ground truth video footage containing specific abnormal behaviors in the typical crowd are not easily accessible and available. To resolve the problem of validation of ground-truth video, Andrade et al. [30], [8] achieved controlled situation with known ground truth data set to test their algorithm by using crowd simulation algorithm. However, [8] discriminate normal and abnormal behavior of crowd by exploring some concepts which are related to crowd simulation. They divided crowd behavior analysis approaches in object-based and holistic based. Object based approach defines crowd as a collection of individual person whereas in a holistic approach, focus on individual difference is ignored. This approach considers the entire individual in a crowd to have similar motion characteristics.

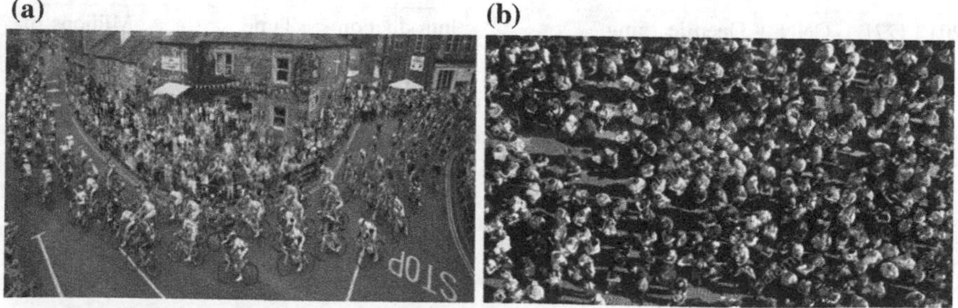


**Fig. 3** Illustration of crowded scenarios: **a** structured event, **b** unstructured event

**Table 2** Crowd video dataset

| Dataset | Description | Size | Label | Accessibility |
|---|---|---|---|---|
| UCF | Videos of crowds, vehicle flows and other high density moving objects | 38 video | Partial | Yes |
| UMN | Scenarios of 3 different indoor and outdoor events | 11 videos | All | Yes |
| UCSD | 34 training video clips and 36 testing videos | 98 video clips | Partial | Yes |
| CUHK | 1 traffic video sequence and 1 crowd video sequence | 2 videos | Partial | Yes |
| QMUL | 3 traffic video sequence and 1 pedestrian's video sequence | 4 videos | Partial | Yes |
| PETS2009 | 8 video sequences of different crowd activities with calibration data | 8 videos | All | Yes |
| Rodriguezs | Large collection of crowd videos with 100 labeled object trajectories | 520 videos | Partial | No |
| UCF | Behavior image sequences from the web videos and PETS2009 | 61 sequences | All | Yes |
| Violent-flows | Real-world video sequences of crowd violence | 246 videos | Partial | Yes |

## 5 Crowd Video Dataset

When we deal with crowd video analysis, we need to validate our results with some real-world database for which some datasets are publically available and accessible. Table 2, illustrates some set of real-data bases with its brief descriptions which includes size, label and accessibility.

## 6 Summary and Future Scope

This paper explores the various aspects related to crowd modeling and crowd analysis which uses various techniques for various applications in the real word. A detailed comparison of the state-of-the-art methods related to people counting has been summarized along with its advantages and shortcomings. This survey presents a futuristic view of the crowd monitoring and classification. An integrated framework is required for crowd management which can deal with any type of crowd analysis ranging from the panic situations to the large scale misbehaved crowds. It is done in order to identify the common lacunae of the existing techniques and to overlay a path for the further research in this area. Though, various researches have been concluded but some issues are still untouched, which demand further research as:

- Multi-sensor Information Fusion.
- Tracking-Learning Detection Framework.
- Deep Learning with neural network for crowd analysis.
- Wireless Sensor Networks for density estimation.
- Real-Time Processing and Generalization.

## References

1. Zhan, Beibei, et al. "Crowd analysis: a survey." Machine Vision and Applications 19.5–6 (2008): 345–357.
2. Li, Teng, et al. "Crowded scene analysis: A survey." Circuits and Systems for Video Technology, IEEE Transactions on 25.3 (2015): 367–386.
3. Hu, Weiming, et al. "A survey on visual surveillance of object motion and behaviors." Systems, Man, and Cybernetics, Part C: Applications and Reviews, IEEE Transactions on 34.3 (2004): 334–352.
4. Badal, Tapas, et al. "An Adaptive Codebook Model for Change Detection with Dynamic Background." 2015 11th International Conference on Signal-Image Technology and Internet Based Systems (SITIS). IEEE, 2015.
5. Leggett, Richard. Real-time crowd simulation: A review. R. Leggett, 2004.
6. Hu, Min, Saad Ali, and Mubarak Shah. "Detecting global motion patterns in complex videos." Pattern Recognition, 2008. ICPR 2008. 19th International Conference on. IEEE, 2008.
7. Wang, Xiaofei, et al. "A high accuracy flow segmentation method in crowded scenes based on streakline." Optik-International Journal for Light and Electron Optics 125.3 (2014): 924–929.
8. Mehran, Ramin, Akira Oyama, and Mubarak Shah. "Abnormal crowd behavior detection using social force model." Computer Vision and Pattern Recognition, 2009. CVPR 2009. IEEE Conference on. IEEE, 2009.
9. Kratz, Louis, and Ko Nishino. "Tracking pedestrians using local spatio-temporal motion patterns in extremely crowded scenes." Pattern Analysis and Machine Intelligence, IEEE Transactions on 34.5 (2012): 987–1002.
10. Kratz, Louis, and Ko Nishino. "Anomaly detection in extremely crowded scenes using spatio-temporal motion pattern models." Computer Vision and Pattern Recognition, 2009. CVPR 2009. IEEE Conference on. IEEE, 2009.
11. Cong, Yang, Junsong Yuan, and Ji Liu. "Abnormal event detection in crowded scenes using sparse representation." Pattern Recognition 46.7 (2013): 1851–1864.
12. Zhou, Bolei, Xiaogang Wang, and Xiaoou Tang. "Random _eld topic model for semantic region analysis in crowded scenes from tracklets." Computer Vision and Pattern Recognition (CVPR), 2011 IEEE Conference on. IEEE, 2011.
13. Bak, Slawomir, et al. "Multi-target tracking by discriminative analysis on Riemannian manifold." Image Processing (ICIP), 2012 19th IEEE International Conference on. IEEE, 2012.
14. Kuo, Cheng-Hao, Chang Huang, and Ramakant Nevatia. "Multi-target tracking by on-line learned discriminative appearance models." Computer Vision and Pattern Recognition (CVPR), 2010 IEEE Conference on. IEEE, 2010.
15. Badal, Tapas, Neeta Nain, and Mushtaq Ahmed. "Video partitioning by segmenting moving object trajectories." Seventh International Conference on Machine Vision (ICMV 2014). International Society for Optics and Photonics, 2015.
16. Fruin, John J. Pedestrian planning and design. No. 206 pp. 1971.
17. Chen, Ke, et al. "Feature Mining for Localised Crowd Counting." BMVC. Vol. 1. No. 2. 2012.

18. Bansal, Ankan, and K. S. Venkatesh. "People Counting in High Density Crowds from Still Images." arXiv preprint arXiv:1507.08445 (2015).
19. Hussain, Norhaida, et al. "CDES: A pixel-based crowd density estimation system for Masjid al-Haram." Safety Science 49.6 (2011): 824–833.
20. Tian, Qing, et al. "Human detection using HOG features of head and shoulder based on depth map." Journal of Software 8.9 (2013): 2223–2230.
21. Idrees, Haroon, et al. "Multi-source multi-scale counting in extremely dense crowd images." Proceedings of the IEEE Conference on Computer Vision and Pattern Recognition. 2013.
22. Conte, Donatello, et al. "A method for counting moving people in video surveillance videos." EURASIP Journal on Advances in Signal Processing 2010 (2010).
23. Albiol, A., Silla, M.J., Albiol, A., Mossi, J.M., 2009. Video analysis using corner motion statistics. In: IEEE International Workshop on Performance Evaluation of Tracking and Surveillance, pp. 31–38.
24. Xi, Wei, et al. "Electronic frog eye: Counting crowd using WiFi." INFOCOM, 2014 Proceedings IEEE. IEEE, 2014.
25. Luo, Jun, et al. "Real-time people counting for indoor scenes." Signal Processing (2015).
26. Fradi, Hajer, and Jean-Luc Dugelay. "Towards crowd density-aware video surveillance applications." Information Fusion 24 (2015): 3–15.
27. Lehman, Logan, and Haroon Idrees. "Counting in Dense Crowds using Deep Learning".
28. Zhang, Cong, et al. "Cross-scene crowd counting via deep convolutional neural networks." Proceedings of the IEEE Conference on Computer Vision and Pattern Recognition. 2015.
29. Rodriguez, Mikel, et al. "Density-aware person detection and tracking in crowds." Computer Vision (ICCV), 2011 IEEE International Conference on. IEEE, 2011.
30. Andrade, Ernesto L., Scott Blunsden, and Robert B. Fisher. "Modeling crowd scenes for event detection." Pattern Recognition, 2006. ICPR 2006. 18th International Conference on. Vol. 1. IEEE, 2006.

# A Roadmap to Identify Complexity Metrics for Measuring Usability of Component-Based Software System

**Jyoti Agarwal, Sanjay Kumar Dubey and Rajdev Tiwari**

**Abstract** Component Based Software System (CBSS) is widely popular in the modern era because of the reduction of development cost, time, and effort. To increase the success rate and acceptability of CBSS among the users, it is important to increase the quality of CBSS. Usability is one of the important quality factors, but challenges exist in measurement of usability. Complexity plays important role in acceptance of usable software system. So, to measure the usability, it is important to measure its complexity by using complexity metrics. Various complexity metrics have been proposed in the literature. The main objective of this research paper is to identify the complexity metrics of traditional and object-oriented software system and to provide a roadmap for the requirement of complexity metrics for CBSS. Present paper may help system designers, developers, and analysts to select the appropriate complexity metrics for CBSS on the basis of provided analytical results. Based on the selected complexity metrics, usability can be measured in easier way.

**Keywords** Component • Complexity • Object oriented • Usability • Metrics • System

## 1 Introduction

Usability is considered as the important quality factor in all quality models and also accepted by mostly all the researchers. It plays major role in success/failure of any software system. If usability of software system is complex, then it will not be the

J. Agarwal (✉) · S.K. Dubey
Amity University Uttar Pradesh, Noida, UP, India
e-mail: itsjyotiagarwal1@gmail.com

S.K. Dubey
e-mail: skdubey1@amity.edu

R. Tiwari
Greater Noida Institute of Engineering & Technology, Greater Noida, India
e-mail: rajdevtiwari@yahoo.com



S.K. Bhatia et al. (eds.), *Advances in Computer and Computational Sciences*,
Advances in Intelligent Systems and Computing 553,
DOI 10.1007/978-981-10-3770-2_4

choice of users. Usability techniques increase the user efficiency and lower the cost of software development. In modern era, software industries are focusing on the Component Based Software System because CBSS is developed by using existing software component which reduces the development time, cost, and effort. For developing efficient CBSS, it is important to improve the usability of CBSS so that users can easily accept it and success rate of the software system can be increased. For Component Based Software System, usability depends on various factors like component interface, complexity, learnability, operability, attractiveness, etc. Complexity has considered as a synonymous with understandability or maintainability [1]. Complexity is also considered as an important sub-factor of usability for CBSS [2]. This paper presents the idea to measure the usability of CBSS by measuring the complexity of CBSS. Different metrics have been proposed by the researchers for traditional and object-oriented software system and usability of such software system has been also evaluated by using the appropriate metrics. In this paper, focus is given for selecting the appropriate complexity metrics for measuring the usability of CBSS and to provide a solution for the need of different software metrics for different software systems. These complexity metrics will be helpful to select the correct metrics to measure the usability of software system as per the research requirement. The ideal software metrics should be simple, precise, objective, obtainable, valid, and robust [3].

The main contribution of this research paper is to systematically analyze the metrics for traditional, object-oriented, and CBSS and to enlighten the reasons for the requirement of different complexity metrics for different software systems. Present paper will be useful for the researchers to find the applicability of complexity metrics in various scenarios.

The paper is structured as follows: first the review process is described, and then literature of the software metrics is done. After literature review, analysis of the framed research question is provided and then conclusion of the paper is given.

## 2 Review Process

### *2.1 Inclusion and Exclusion Criteria*

Various keywords pertaining to component, complexity, object-oriented, usability, metrics, were used to search relevant papers from different digital libraries. Review process includes those research papers and articles which describe the metrics for traditional, object-oriented and CBSS. Research papers which focus on the evaluation of the predefined software metrics and not written in English were excluded. In total, more than 50 research papers and articles were collected, out of them 35 are used in the present paper. Classification of the included research papers is shown in Table 1.

**Table 1** Classification of research papers

| Category | Total no. of papers | % Proportional |
|---|---|---|
| Traditional Software System | 12 | 34.29 |
| Object Oriented Software System (OOSS) | 10 | 28.57 |
| Component Based Software System | 13 | 37.14 |

## 2.2 *Research Questions*

Following research questions are framed based on the studied research papers:

RQ1: Why different complexity metrics are used for different software system?
RQ2: What are the different metrics for traditional, object-oriented, and CBSS?
RQ3: What are the complexity metrics to measure the usability of CBSS?
RQ4: Which and how many journals/conferences include papers on software metrics?
RQ5: What is the research progress for software metrics in successive years?

# 3 Literature Review

The present section provides the complexity metrics for traditional, object-oriented, and component-based software system. This literature will be helpful to select the appropriate metrics for the software system.

(a) Complexity Metrics for Traditional and Object-Oriented Software System

Different authors have been proposed various metrics for traditional software system [4–11]. These software metrics are used for measuring different attributes of traditional software system. Source Line of Code (SLOC), Function Point (FP) and Cyclometric Complexity (CC) are important traditional software metrics.

Tegarden et al. [9] defined that traditional software metrics can be used to measure the complexity of OOSS, however additional metrics are required to measure all aspect of OOSS. For this purpose various metrics have been proposed [12–18]. These metrics are proposed based on the method, complexity, attribute, class, coupling, inheritance, polymorphism, etc. Object-oriented complexity metrics have been further classified into method complexity metrics, class complexity metrics, and system complexity metrics [19]. Chidambar and Kemerer (CK) metrics are the most popular metrics among all the object-oriented software metrics [20–22].

(b) Metrics for Component-Based Software System

The classification and analysis of CBSS metrics are shown in Table 2 and Fig. 1, respectively.

**Table 2** Classification of different metrics for CBSS

| Category | Metrics | Source |
|---|---|---|
| Class | Weighted methods per class | [23] |
| | Lack of cohesion method | |
| | Weighted class per component and number of classes | [24] |
| | Response set for a component | |
| Size | Line of code | [23] |
| | Size critically metric | |
| Coupling | External coupling between objects | [24] |
| | Coupling metric | [25] |
| | Interface count | [26] |
| | Components path | |
| | Architecture coupling | |
| | Coupling between objects | [23] |
| Cohesion | Component cohesion | [24] |
| | Cohesion metric | [25] |
| Interface | Component interface metric | [25] |
| | Architecture component interaction factor | [26] |
| | Architecture interface count | |
| | Component incoming and outgoing interaction density | |
| | Component average interaction density | |
| | Link and bridge critically metric | |
| Complexity | Total number of components | [27] |
| | Average number of methods per component | |
| | Total number of implemented components | |
| | Total number of links | |
| | Average number of links between components | |
| | Average number of links per interface | |
| | Total number of interfaces | |
| | Average number of interfaces per component | |
| | Depth and width of the composition tree | |
| | Component complexity | [28] |
| | Dependency oriented complexity metrics | [29] |
| | Component dependency metrics | [29] |
| | Self-component complexity metrics | [25] |
| | Average interface complexity of a component | [30] |
| | Average incoming and outgoing interaction complexity | |
| | Interface complexity metric | [31] |
| | Interface method complexity metric | [32] |
| | Component coupling complexity metric | |
| | Component plain complexity metric | [33, 34] |
| | Component static complexity metrics | |
| | Component dynamic complexity metrics | |
| | Component cyclometric complexity metrics | |
| | Coupling complexity for black box components | [35] |
| | Component interaction density | [23] |
| | Component packing density | |

(continued)

**Table 2** (continued)

| Category | Metrics | Source |
|---|---|---|
| Inheritance | Maximum of depth inheritance tree and mean of unrelated tree | [24] |
| | Number of children for a component | |
| | Depth of inheritance | [23] |
| | Inheritance critically metric | |
| Interaction | Actual interaction metric | [25] |
| | Component fan-in factor (CFIF) | [26] |

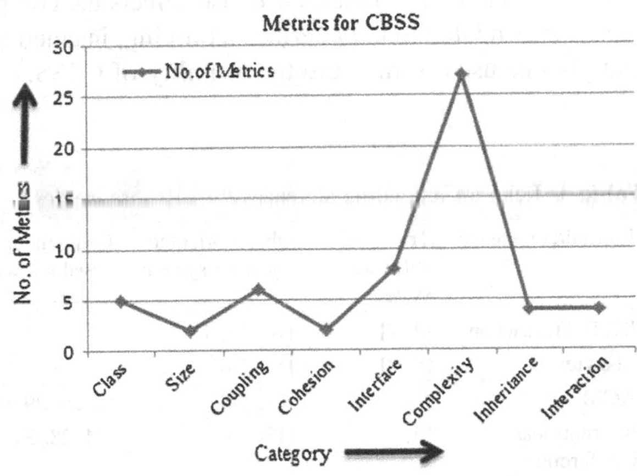


**Fig. 1** Classification of different metrics for CBSS

## 4 Result Analysis

In this section, the result analysis of the framed research questions is given:

(a) Applicability of Complexity Metrics for Different Software System (RQ1)

From the literature review, it is analyzed that we need different complexity metrics for different software systems. Complexity metrics for traditional and object-oriented software system cannot be used to measure the complexity of CBSS. The reasons are given by Rana and Singh [23]:

(a) Traditional metrics are based on the size of code, so these metrics cannot be applied to CBSS because the component size is not known to the developer.
(b) Traditional metrics are based on small programs but CBSS metrics should depend on interoperability aspects of the components.
(c) Traditional cyclometric complexity metrics cannot be used for CBSS because operator, operand, and independent paths cannot be counted for CBSS. Traditional metrics do not focus on interfaces and integration, which is important for CBSS.

(b) Metrics for Different Software System (RQ2)

The different complexity metrics are required for traditional, object-oriented, and component-based software system. These metrics are described in the literature review.

(c) Complexity Metrics to Measure Usability of CBSS (RQ3)

Complexity metrics to measure the usability of CBSS can be selected from Table 2. Complexity of CBSS depends on component and the interaction process among the components. Complexity can be considered as design complexity or interface complexity. Inheritance also affects the complexity of the components so the metrics related to the interface, coupling, interaction, and inheritance in Table 2 can also be used to measure the usability of CBSS.

**Table 3** Relevant journals/conferences for software metrics

| Journal/conference | Traditional software system | Object-oriented software system | Component based software system | Total no. of relevant papers found | % Proportion |
|---|---|---|---|---|---|
| IEEE Transaction | [4, 7] | [14, 17, 18] | – | 5 | 14.29 |
| Elsevier | [6, 8] | [12, 19] | – | 4 | 11.43 |
| ACM | [5] | [21, 22] | [2, 26, 29, 34] | 7 | 20 |
| International Conference | [9] | [13, 16] | [24, 28, 31] | 6 | 17.14 |
| Books, Thesis | [11] | [15, 20] | – | 3 | 8.57 |
| Other International Journals | [3, 10] | [11] | [23, 25, 27, 30, 32, 33, 35] | 10 | 28.57 |

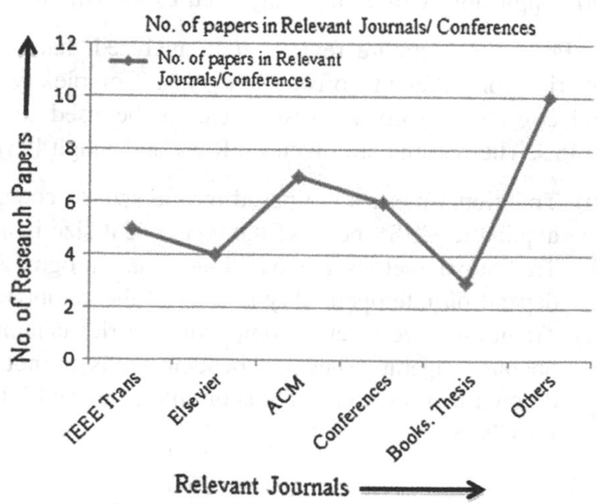


**Fig. 2** Number of research papers in journals/conferences

**Table 4** Research progress for software metrics in successive years

| Year | Total number of papers | % Proportion |
|---|---|---|
| 1976–1986 | 5 | 14.29 |
| 1986–1996 | 8 | 22.85 |
| 1996–2006 | 4 | 11.43 |
| 2006–2016 | 18 | 51.43 |

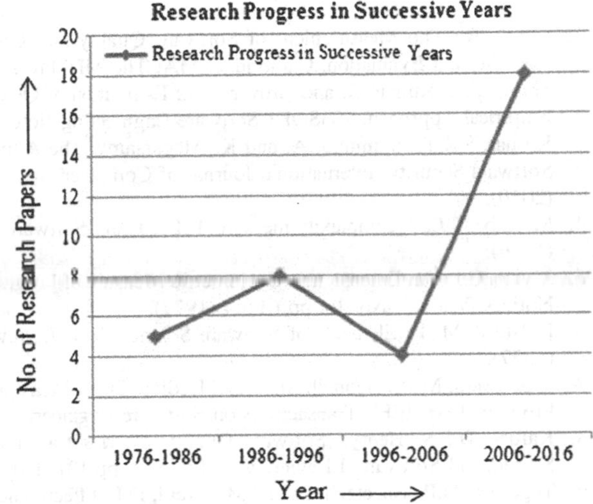


**Fig. 3** Research progress for software metrics in successive years

(d) Relevant Journals/Conferences (RQ4)

The distribution of the research papers in different journals/conferences is given in Table 3 and graphical representation is shown in Fig. 2.

(e) Research Progress for Software Metrics (RQ5)

Research progress for the complexity metrics of software system in the successive year (1976–2016) is shown in Table 4 and its graphical representation is shown in Fig. 3.

## 5 Conclusion

Complexity has been proved as an important sub-factor of usability. Present paper provides a list of various complexity metrics for traditional, object-oriented, and CBSS. It is identified that different complexity metrics are required for different software systems. The reasons for the same have been identified and mentioned in the analysis section. A number of complexity metrics have been proposed by the researchers for CBSS but still there is scope to identify the complexity metrics for black box components, for which information of the source code is not required to the developers. In future, usability of CBSS can be measured in terms of complexity by selecting the complexity metrics from the given list and similar work can also be done for other important usability sub-factors.

## References

1. Curtis, B.: The Measurement of Software Quality and Complexity. Software Metrics: An Analysis and Evaluation. Cambridge, MA: The MIT Press, pp 203–224 (1981).
2. Sharma, A., Kumar, R. and Grover, P.S.: Estimation of Quality for Software Components-an Empirical Approach. SIGSOFT Software Engineering Notes, Vol. 33, No. 6, pp 1–10 (2008).
3. Kumar, S.R.T., Sumithra, A. and K. Alagarsamy: The Applicability of Existing Metrics for Software Security. International Journal of Computer Applications, Vol. 8, No. 2, pp 29–33 (2010).
4. McCabe, T.J.: A complexity measure. IEEE Trans. Software Engineering, Vol. 4, pp 308–320 (1976).
5. Myers, G.J.: An Extension to Cyclometric Measure of Program Complexity. ACM SIGPLAN Notices, Vol. 12, No. 10, pp 61–64 (1977).
6. Halstead, M.H.: Elements of Software Science, Vol. 7. New York: North-Holland, Elsevier (1977).
7. Woodward, M.R., Hennell, M.A. and Hedley, D.: A Measure of Control Flow Complexity in Program Text. IEEE Transactions on Software Engineering, Vol. 1, pp 45–50 (1979).
8. Kafura, D., S. Henry.: Software Quality Metrics Based on Interconnectivity, Journal of System and Software, Elsevier, Vol. 2, No. 2, pp 121–131 (1981).
9. Tegarden, D.P., Sheetz, S. D. and Monarchi, D.E. Effectiveness of traditional software metrics for object-oriented systems. In System Sciences: Proceedings of the Twenty-Fifth Hawaii International Conference, Vol. 4, pp 359–368 (1992).
10. Kandpal, M., Kandpal, A.: Critical Analysis of Traditional Size Estimation Metrics for Object Oriented Programming. International Journal of Computer Applications, Vol. 58, No. 13, pp 38–44 (2012).
11. Patidar, K., Gupta, R.K. and Chandel, G.S.: Coupling & Cohesion Measures in Object Oriented Programming. International Journal of Advanced Research in Computer Science and Software Engineering, Vol. 3, No. 3, pp 517–521 (2013).
12. Chen J.Y., Lu, J.F.: A New Metrics for Object-Oriented Design. Information of Software Technology, Elsevier, Vol. 35, No. 4, pp 232–240 (1993).
13. Li, W., Henry, S.M.: Maintenance metrics for the object oriented paradigm. In Proceedings of the First International Software Metrics Symposium, IEEE, pp 52–60 (1993).
14. Chidamber, S.R., Kemerer, C.F.: A Metrics Suite for Object Oriented Design, IEEE Transactions on Software Engineering, Vol. 20, No. 6, pp 476–493 (1994).
15. Lorenz, M., Kidd, J.: Object-Oriented Software Metrics, NJ, USA, Prentice Hall, Inc (1994).
16. Brito, F., Abreu, E.: The MOOD Metrics Set. In Proceeding of ECOOP'95, Workshop on Metrics, Vol. 95, pp. 267 (1995).
17. Li, W., Sallie, Henry: Metrics for Object Oriented System, Transactions on Software Engineering (1995).
18. Bansiya, J., Davis, C.G.: A Hierarchical Model for Object-Oriented Design Quality Assessment. IEEE Transactions of Software Engineering, Vol. 28, No. 1, pp 4–17 (2002).
19. Abreu, E., Brito, F., Carapuça, R.: Candidate metrics for object-oriented software within a taxonomy framework. Journal of Systems and Software, Elsevier, Vol. 26, No. 1, pp 87–96 (1994).
20. Sarker, M.: An Overview of Object Oriented Design Metrics (Thesis Report). UMEA University, Department of Computer Science, pp 1–53 (2005).

21. Dubey, S.K., Rana, A.: Assessment of Usability Metrics for Object-Oriented Software System. ACM SIGSOFT, Vol. 35, No. 6, pp 1–4 (2010).
22. Dubey, S.K., Rana, A.: Usability Estimation of Software System by using Object-Oriented Metrics. ACM SIGSOFT, Vol. 36, No. 2, pp 1–6 (2011).
23. Rana, P., Singh, R.: A Study of Component Based Complexity Metrics. International Journal of Emerging Research in Management & Technology, Vol. 3, No. 11, pp 159–16 (2014).
24. Vernazza, Tullio, Granatella, G., et.al.: Defining metrics for software components. 5th World Multi-Conference on Systemics, Cybernetics and Informatics, Florida, Vol. 11, pp. 16–23 (2000).
25. Chen, J., Wang, H.: Complexity Metrics for Component-based Software Systems. International Journal of Digital Content Technology and its Application, Vol. 5, No. 3, pp 235–244 (2011).
26. Sengupta, S., Kanjilal, A.: Measuring Complexity of Component Based Architecture: A Graph based Approach. ACM SIGSOFT Software Engineering Notes, Vol. 36, No. 1, pp 1–10 (2011).
27. Salman, N.: Complexity Metrics as Predictors of Maintainability and Integrability of Software components. Cankaya University Journal of Arts and Sciences, Vol. 1, No. 5, pp 39–50 (2006).
28. Sharma, A., Kumar, R., Grover, P.S.: Empirical Evaluation and Critical review of Complexity Metrics for Software Components. WSEAS International Conference on Software Engineering, Parallel and Distributed Systems, pp 24–29 (2007).
29. Gill, N.S., Balkishan.: Dependency and Interaction Oriented Complexity Metrics of Component-Based Systems, ACM SIFSOFT Software Engineering Notes, Vol. 33, No. 2, pp 1–5 (2008).
30. Kumari, U., Upadhyaya, S.: An Interface Complexity Measure for Component-based Software Systems", International Journal of Computer Applications, Vol. 36, No. 1, pp 46–52 (2011).
31. Chillar, R.S., Ahlawat, P., Kumari, U.: Measuring Complexity of Component Based System Using Weighted Assignment Technique. International Conference on Information Communication and Management, Vol. 55, pp 19–27 (2012).
32. Kaur, N., Singh, A.: A Complexity Metric for Black Box Components. International Journal of Soft Computing and Engineering (IJSCE), Vol. 3, No. 2, pp 179–184 (2013).
33. Diwakar, C., Rani, S., Tomar, P.:Metrics Used in Component Based Software Engineering. IJITKM special Issue (ICFTEM), 46–50 (2014).
34. Tiwari, U., Kumar, S.: Cyclometric Complexity Metric for Component Based Software ACM SIGSOFT Engineering Notes, Vol. 36, No. 1, pp 1–6 (2014).
35. Kumar, S., Tomar, P., Nagar, R and Yadav, S.: Coupling Metric to Measure the Complexity of through Interfaces. International Journal of Advanced Research in Computer Science and Software Engineering, Vol. 4, No. 4, pp 157–162 (2014).

# Smart Bike Sharing System to Make the City Even Smarter

**Monika Rani and O.P. Vyas**

**Abstract** In the past few years, the growing population in the smart city demands an efficient transportation sharing (bike sharing) system for its development. The bike sharing as we know is affordable, easily accessible and reliable mode of transportation. But an efficient bike sharing system should be capable not only of sharing bike but also of providing information regarding the availability of bike per station, route business, time/daywise bike schedule. The embedded sensors are able to opportunistically communicate through wireless communication with stations when available, providing real-time data about tours/minutes, speed, effort, rhythm, etc. Based on our study analysis data to predict regarding the bike's available at stations, bike schedule, a location of the nearest hub where a bike is available, etc., reduces the user time and effort.

**Keywords** Smart cities · Bike sharing · NS2 simulator

## 1 Introduction

Smart city demands for energy-efficient transport system also an emphasis on sharing system for utilization of vehicle (e.g., bicycle, motor bikes, etc.). Sharing bike system has various benefits like appropriate resource management, reduces pollution, and leads to improved health. To motivate the bike user first we need to explore various parameters like a number of stations, the number of routes, and the number of available bikes per station using the appropriate statistical approach. Once we obtain the correct status regarding the parameters, then the serving of bikes to the growing population in the smart city become quite an easy job.

M. Rani (✉) · O.P. Vyas
Department of Information Technology, Indian Institute of Information Technology, Allahabad, India
e-mail: monikarani1988@gmail.com

O.P. Vyas
e-mail: dropvyas@gmail.com


S.K. Bhatia et al. (eds.), *Advances in Computer and Computational Sciences*, Advances in Intelligent Systems and Computing 553,
DOI 10.1007/978-981-10-3770-2_5

The motivation of our work is to enhance the efficiency of bike sharing system by making bikes smart by deploying sensors on bikes which will help in collecting real-time data and to forward them to nearby stations. In the recent years popularity of a bike sharing program has shown a vital growth. Another reason is easily accessible and economical in promoting short-term bike rental system. There would be less congestion for commuters in traffic of the car parking. Tourists can also enjoy hassle-free travel without changing multiple busses and taxis. The environment also gets the benefit of less smog after a weekday commute. Cities are managing development and urban living culture is facing major challenges in our daily lives. Based on statistical data of 2007, half of the population of the world was living their lives in cities. UN population fund forecasts that by the end of 2030 nearly 60% of the world population would live their life in cities [1]. Out of all major issues, we can outline air quality, environmental crisis, and transportation issues. Use of a bicycle is an important mode of transportation that could be helpful to many urban transportation issues. As the use of motor vehicles is increasing, problems as cost, congestion, accidents, loss of amenity and space, noise, air pollution, and energy consumption have an adverse effect on the natural environment. In future, use of the bicycle as transport should be the transportation solution for cities as it has no adverse effects [2, 3].

One more critical issue in any modes of transportation is of pod cars to bike; the issue is about picking people from the transport hub as a railway station, bus depots, and to their destination; this problem is called last mile problem [4]. Ever wondered if we could take our vehicle out and then forget it after reaching the destination or if you do not have to plea anyone to drop you down or pick you up. Bike sharing is required where a person can borrow a bicycle from one of our stations in the city and can return to another station. For the same smart bike sharing system to develop, the concept of riding a bike from one point and returning it back at another point can help to solve last mile problem. An efficient bike sharing system collects the real-time data using sensors deployed on them and sends the data to stations on the way. Smart bike sharing not only solves the issue of the last mile, but also a problem of area/acres requires for car parking and reduces the waiting time for a local bus.

Our problem was to increase the efficiency of bike sharing systems and increase the participation of people in these sustainable systems, thus moving towards the concept of the smart city. Due to growing population and change in the transportation usage in urban cities, some people have demanded usage of bike sharing system in the last few years [5, 6]; however, still there are some people who are the reluctance to a combination of the mode of transportation of traditional days [7]. Some explanation was given by people for choosing a traditional transportation, over bike sharing system is that the bike riding is not safe and one has to travel a longer distance also; the weather is another deciding factor. The transportation authorities and local council need to encourage the use of a bike and also need do develop separate infrastructure and alternative routes for bikes to provide safety and provide shorter distance [8, 9]. Also, it might be the case that at any point of time the sharing station can be empty and there is no bike to rent; on the other hand it also be possible that people unable to park the bikes due to the station is full and there is no availability of parking slots. The problem can be extended to use other public transport systems so that whole city transport can be transformed to make the city, even smarter. These problems gave us the motivation to innovate and implement a bike sharing system that is different and better than existing systems.

The paper is structured into five sections; initially, we focus on the introduction of the paper in Sect. 1. In Sect. 2, we describe related works, where we have explained the works that have been done on a bike sharing system and proposed in the relative fields of sensors and smart cities. Section 3 describes a website to enable users of the system to book the bikes from any station, see the station map and check the current availability and see their previous rides, virtually demonstrating the system using NS2 simulator and collecting the simulation data along the way. The section also depicts experimental results that prove the usefulness of the smart bike sharing system. Basically, our analysis of data is from a similar existing system for minimizing the redistribution problem and proposing such analysis when actual data is collected in our system. Section 4 focuses on contribution of work and limitations of work. Finally, Sect. 5 draws the conclusion and future research opportunities in the bike sharing system for smart cities.

## 2 Literature Review of Related Work

Much has been said about how making a city smart and many international conferences have taken place on this issue for a sustainable future. Many researchers have published their research on using a bike sharing system to make the city smarter and how to analyze the bicycle sharing data for generating insights into sustainable transport systems. We have primarily taken guidance from one research paper which deals with how the bikes will send the data on the way and what will be the protocol that will govern this transmission [10] and also took some motivation for better analysis from another research paper which has discussed an analysis of various bike share systems currently existing in various countries [11]. The bike sharing system can use various technologies to make it smart and real-time example, advances sensors, collaborative agents, and ontologies for storing user's and heterogeneous station details [12, 13]. The former research paper has thoroughly covered the topic that was needed at work and some of the terms use like bike availability, busiest station, load factor, etc., and also characteristics of docking stations are described as follows: Aggregate characteristics, spatial characteristics, temporal characteristics, demographic and community detection of data, and the redistribution problem.

Several cities in the United States and Europe have started implementing bike sharing programs in their cities, a few of which we encountered during our research on the project are as follows: Germany has bike sharing programs in many cities, including Aachen, Berlin, Cologne, Düsseldorf, Frankfurt, Hamburg (StadtRAD Hamburg), Karlsruhe, Kassel (Konrad), Mainz (MVGmeinRad), Munich, and

Stuttgart. Station-based system metropolradruhr is located in the Ruhr Area. Bike sharing stations are also located in over 50 ICE railway stations [14, 15]. French cities offering a sharing system include Marseille, Lyon, Bordeaux, Nice, Toulouse, Rennes, Rouen, La Rochelle, Orléans, Montpellier, Nantes, Lille, Strasbourg, Clermont-Ferrand, Avignon, Saint-Étienne, ChalonsurSaône, Belfast, and Aix-en-Provence. Similar work has been done in cities like California and San Francisco [16, 17]. For the simulation, we referenced the standard tutorial for NS2 by MarcGreis [18].

## 3 Methodology

For a website to enable users of the system to book the bikes from any station, see the station map and check the current availability and see their previous rides. For the same a workflow of a bike sharing system for users is shown in Fig. 1 which is

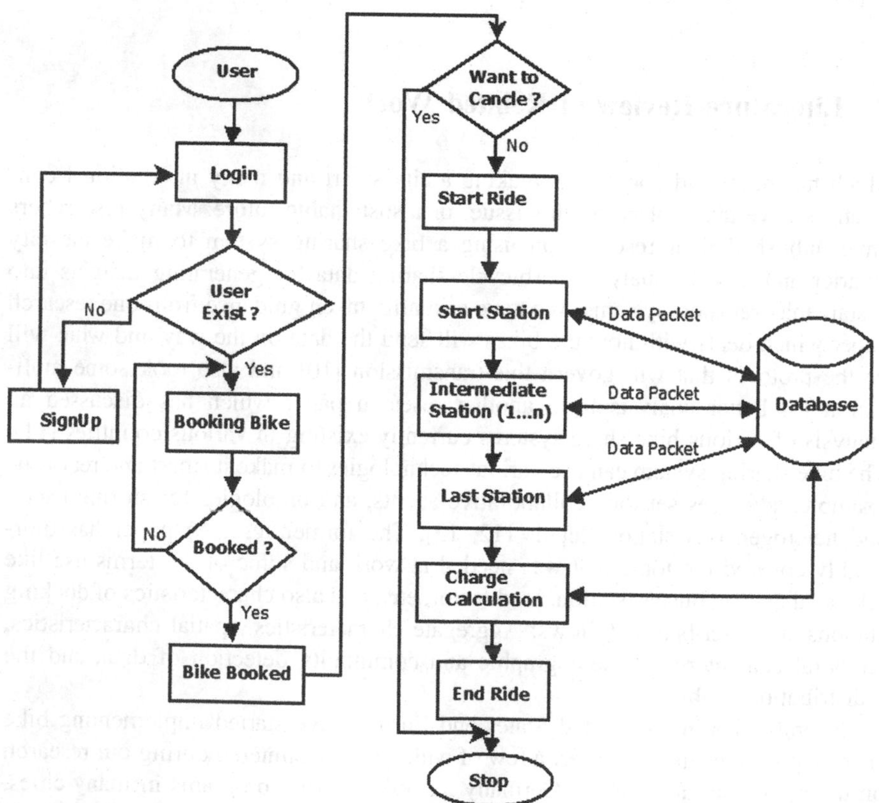


**Fig. 1** Workflow of smart bike sharing system

capable of keeping detail regarding reserve bikes, see their previous rides (as shown in Fig. 2) and get to know about how bike sharing works. The deployed sensor on the bikes which is used to collect the data this data can be further used for predictive user requirement like a number of bikes available at a particular station, route business, and the average time taken between two stations [19]. Simulating the smart bike share system with the help of NS2 simulator and plotting the graphs of data send by bikes and data received at a station as shown in Fig. 3. Analysis of data is from a similar existing system for minimizing the redistribution problem and

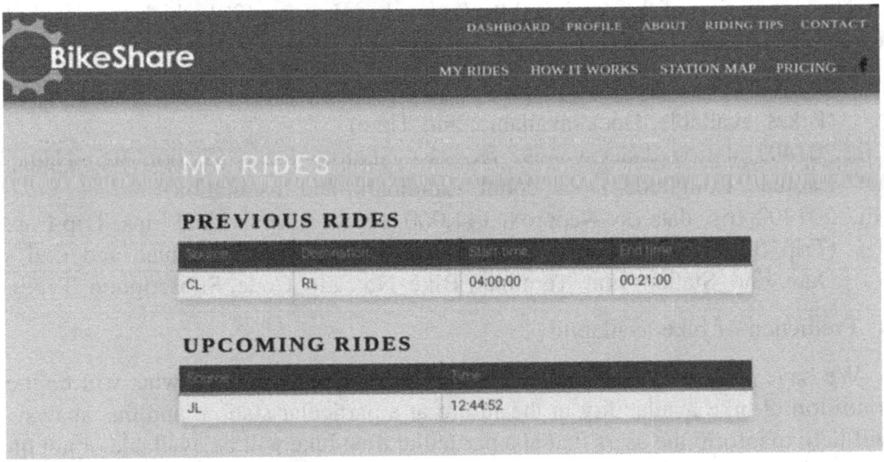

Fig. 2 User's previous and upcoming rides

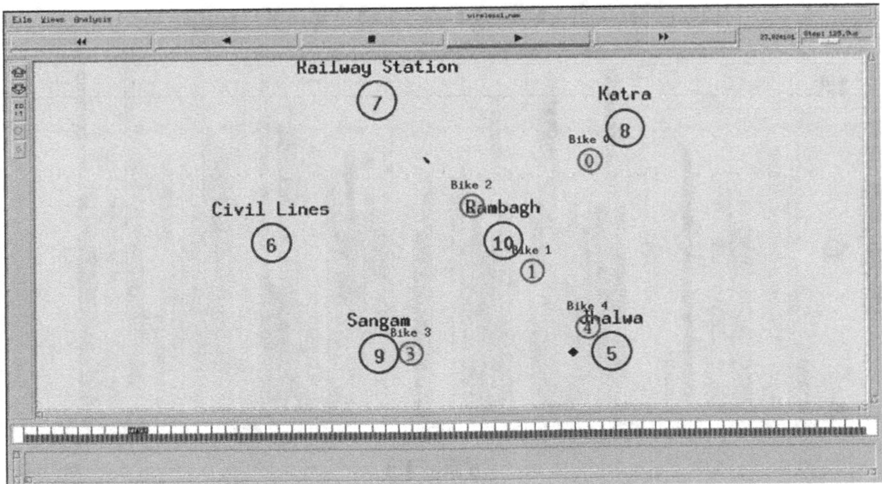


Fig. 3 NS2 simulating the smart bike share system

proposing such analysis when actual data is collected in our system, equipping bikes with sensors to retrieve the information to draw an inference for various aspects like air pollution, noise pollution, and other effects on human and its lifestyle.

## 3.1 Data Analysis

There is the following analysis that has been done on available data in our paper:

The description of the data set (http://www.kaggle.com) [20] which consisted of following files and further described are attributes of each file.

i. 201402_status_data.csv—approx. 17 million records of Status Data (Bikes_available, Dock_available, and Time)
ii. 201402_station_data.csv—69 records—Station Data (Station_ID, Name, Latitude, Longitude, Dock_count, Landmark, and Installation)
iii. 201402_trip_data.csv—approx. 144,000 records of individual trips. Trip Data (Trip_ID, Duration, Start _Date, Start_ Station, Start_Terminal, and End_-Date, End_Station, End_Terminal, Bike_No., Zip_Code, Subscription_Type).

- Prediction of bike availability

We have analyzed our available data to get the prediction that what will be the condition of bike availability in the future at a particular station, and this analysis will help to inform the users that at a particular time bike will be available or not on that station also informs users about the estimated waiting time for a bike, example as a number of bikes that should be available at the station shown in Fig. 4.

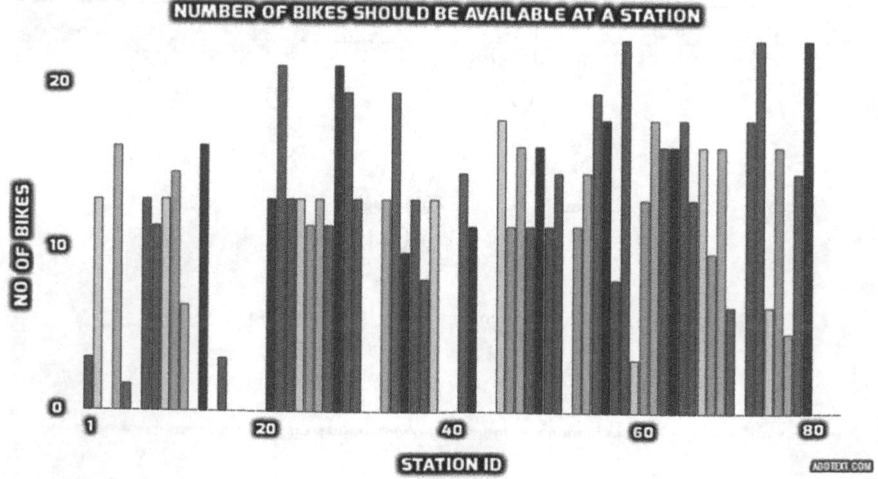


**Fig. 4** Number of bikes that should be available at the station tomorrow

We have used the concept of weighted arithmetic mean [21] on some parameters that are listed below (Suppose today's date is Wednesday, 18th November 2016):

i. No. of bikes at a Day of the Week (DoW) (25% weight)—We analyzed previous data for last 6 Wednesdays.
ii. No. of bikes in Current Week (CW) (50% weight)—Analyzing the trends of the current week and giving them a weight of 50%.
iii. No. of bikes on the Day of the Month (DoM) (25% weight)—Analyzing the data for the current date of last 6 months (last 6 18's in the current year).

No. of bikes that should be available today at stations X:

$$N = 0.25 * \text{DoW} + 0.5 * \text{CW} + 0.25 * \text{DoM}.$$

- Busiest station

We have analyzed the data to find the busiest stations (busiest means where bike incoming and bike outgoing is more) as shown in Fig. 5. Analysis of data of last 6 months revealed the stations that are most busy. For a station X busyness is denoted by K, and the total no. of incoming bikes in a station X (Incoming data = L) and total no. of outgoing bikes from station X (Output data = M). For a station X, the busyness is calculated as

$$K = L + M.$$

The busiest station at a particular time or day: At a particular time, when station has more rush, this information will help the user to know about the availability of bike at a given time. We have analyzed data for last 6 months and calculated density of busyness of all stations at a particular hour. The busyness of a station X at the hour t is defined as the sum of no. of all incoming + outgoing bikes from

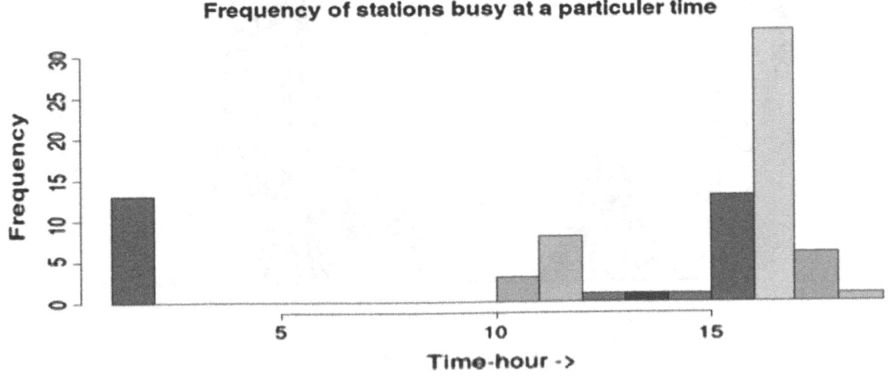


**Fig. 5** Number of stations that are busy at a particular time

station X in the interval t to t + 1. For example, our approach tries to provide useful information like the number of bike trips from each station and the density of busyness of stations at a particular time as shown in Figs. 6 and 7 respectively. Also, we can calculate which station has a maximum number of incoming bikes at a particular time.

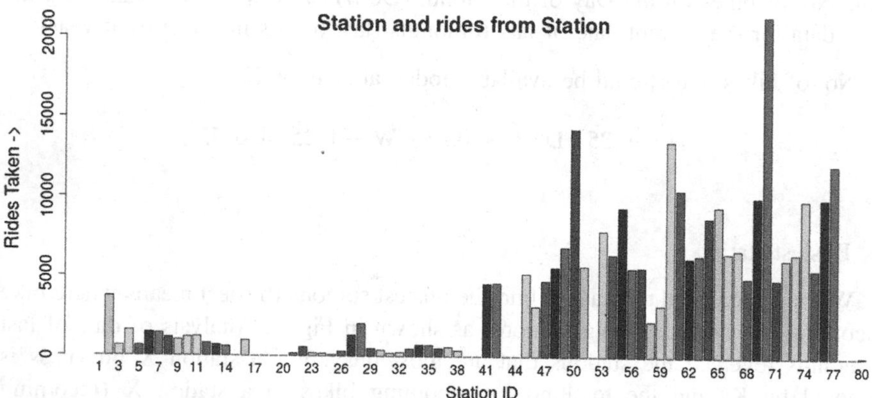


**Fig. 6** Number of bike trips from each station

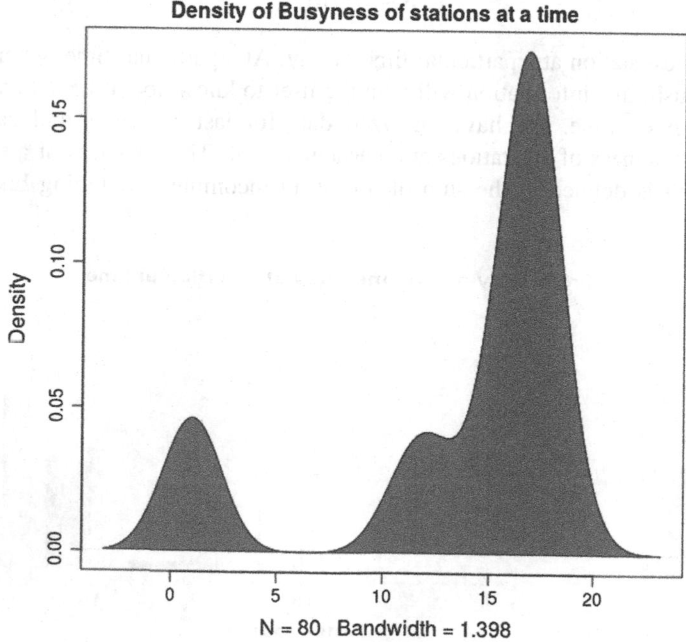


**Fig. 7** Density of busyness of stations at a particular time

- The Average time took between two stations

The average time is taken by the user from a station to reach other stations. We analyzed trip data and took the arithmetic mean of all the trip times for trips between X and Y ($xy_1$, $xy_2$, $xy_3$...$xy_n$) and Y to X ($yx_1$, $yx_2$, $yx_3$...$yx_m$) in the past 6 months, which gave us an average trip time between two stations. For calculating trip times between stations X and Y:

$$((xy_1, xy_2, xy_3 \ldots xy_n) + (yx_1, yx_2, yx_3 \ldots yx_m))/N + M.$$

- The Busyness of all routes

To analyze the busyness of all routes, we analyzed trip data between two stations. This information helps the user to reduce their trip time. The busyness of a route between two stations is defined as the total number of trips between those two stations over the past 6 months.

- Load Factor

The load factor of a particular station is shown in Fig. 8. The calculated load factor (P) of a station represents the load that a station is bearing. Load factor is calculated by the sum of a number of available bikes and a number of empty docks, whereas number of available bikes and number of empty docks are denoted by Q and R, respectively. Load factor increases are directly propositional to load increases on a station. The load factor for a station is calculated as

$$P = Q + R.$$

- Estimated time of availability of a bike

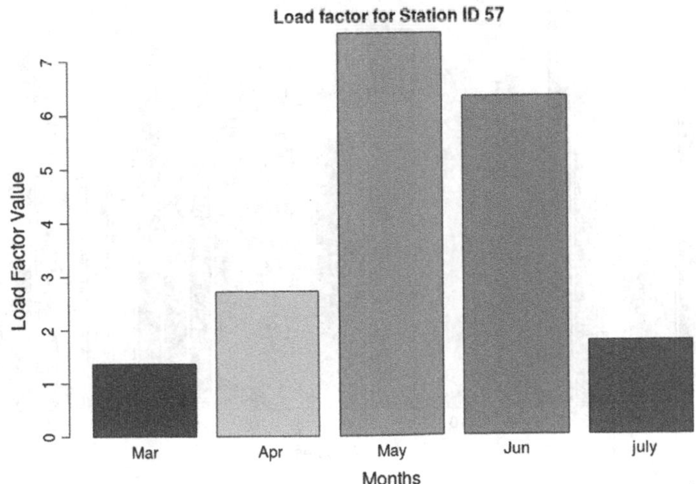


**Fig. 8** Load factor of a particular station

We try to calculate the probability of availability of a bike for each of the next 30 min as shown in Fig. 9. The data set is (status_data describe attributes Bikes_available, Dock_available, and Time). The attribute time is used to calculate the probability of availability of bike. The attribute time is actually timestamp of the form (YYYY-MM-DD HH: MM: SS). Let Max_bikes (U) denote the maximum number of bikes that can be present at a station X for which the probability is to be calculated. Suppose today's date is June 1, 2016 (Wednesday) and the time at which a user arrives at a station and finds zero bikes available is 15:45:00. In our bike sharing system, we attempt to predict and recommend to the user whether he/she should wait for an incoming bike or not. The prediction is based on calculating the probability for an availability of a bike for the next 30 min. To calculate the probability for an availability of a bike for the next 30 min we account three factors of the current timestamp (at which user finds zero bikes available).

i. No. of bikes on the day of a week (Wednesday) at the current time (t) (for the last 6 Wednesdays) denoted by V1.
ii. No. of bikes in the current week at the current time (t)—this includes all the days of the current week denoted by V2.
iii. No. of bikes on the day of a month (1st of every month) for the last 6 months at the current time (t) denoted by V3.

Let t denote the time at which the probability is to be calculated, t ranges from (time of user arrival) to (time of user arrival + 30 min). Let w1, w2, and w3 denote the weights assigned to the factors (V1, V2, and V3) experimentally. Therefore, probability of availability of a bike at t'th minute (T) can be calculated by

$$T = (w1 * V1 + w2 * V2 + w3 * V3)/U.$$

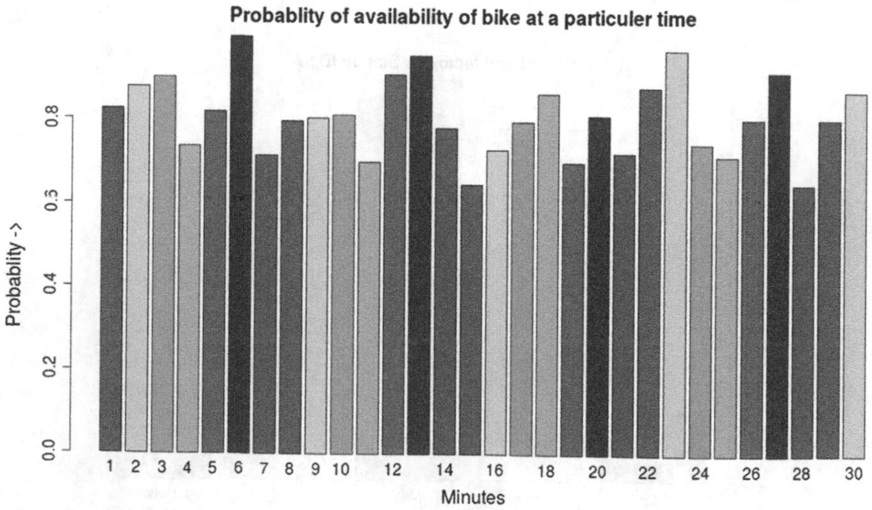


**Fig. 9** Probabilities of availability of a bike for each of next 30 min

### 3.2 Simulating the Smart Bike Sharing System Using the NS2 Simulator

Better understanding of bicycle habits, path, and utilization rate, making bikes intelligent by deploying sensors that send real-time data to base stations, we have simulated the system with the help of network simulator. Key features of simulation are as follows:

- Bike starts from the source and ends at, a destination, both of which are not known a priori. We have tried to show that our system will work under any random scenario which may occur in practice. Simulator starts n (taken as command line argument) bikes from 6 previously defined stations and marks the destinations of the bikes also (randomly). Then each bike starts a trip at a random time (which is the case that will occur in practice).
- On the way to the destination, a bike collects data via sensors and sends the data to stations following AODV (Ad hoc on demand Distance Vector routing) protocol [22].
- When a bike comes in the range of a station, it makes a TCP connection with the station and sends the data to it.
- If a bike comes in the range of multiple stations, it sends equal amounts of data to all of them, thus leading to more network utilization. After simulation, we get the following data in the trace file:
    i. The Amount of data sent by each bike.
    ii. The Amount of data received by each station.

## 4 Discussion

### 4.1 Contribution of Work

We have tried to give extended functionality and eliminate the flaws of some of the existing bike share systems. The facilities provided by our system are as follows:

i. Bikes can be booked sitting at home using our website so that users do not have to wait in case of unavailability.
ii. Previous years' data have analyzed to avoid unavailability.
iii. Hubs will be scattered densely all over the city. Users can take a bike from any hub and return it to any hub in the city.
iv. The whole system will be automatic, and there will be no need of human assistance on each hub.
v. Sensors will be deployed in bikes that will collect data on the way and send to the nearest station.

### 4.2 Limitations of Work

i. The current system does not identify the type of data sent by the bike.
ii. Continuously updating the data received from bikes in the database, and analyzing the data set of about over 17 million records, the availability of data with respect to a particular country is a challenge.
iii. When the bike is not in range of any station, it will not be able to send the data and we will try to use the protocol required for multi-hop communication.

## 5 Conclusion and Future Work

In this paper, we have represented detailed design, implementation plan and evaluation of smart bike sharing system along with sensor networking techniques. The bike sharing system represents the first comprehensive mobile sensing system conveying the cyclist experience. Bike sharing provides the collection and communal environmental sampling. It also supports two modes of operation in support of delay tolerant and real-time sensing. Collected data could be presented both locally to the cyclist and to others as well through back-end services. Bike sharing portal concept promotes social and friendly network among cyclists. Our smart bike sharing system allows the users to easily book a bike using the website at any time without human intervention. There is no need of human for conducting this smart bike sharing system. A user can take a bike from the station using his/her smart card (a smart card that will be given to the user after the SignUp) and start the ride and after completing the trip drop the bike to the station which is near to his/her destination. The simulator in our system is using the sensor to trace the bike and to update the information of the bike position at each time. The sensor will send packets to its nearest station and all these stations will be connected to the website and send the information regarding the bike's status to the app which will update the record.

The data sent by the bike can include Traffic data, Air Quality data, Road Conditions data, etc., which will benefit the operator in solving the redistribution problem as well as users of the system thus saving on operational cost as well as the time of users. In the future when conventional sources of energy would be scarce, bike share system will provide an effective means of transport and within the city, it can be made compulsory to travel through bicycles. In future, our smart bike sharing system can be improved using collaborative software agents on users and station details store on ontologies. Ontologies can easily expand with the addition of users and stations in the system, and provide a secure environment, and machine-readable data for agent's interaction. Software agents can monitor data packets at heterogeneous stations to provide real-time information.

## References


1. Naphade, M., Banavar, G., Harrison, C., Paraszczak, J., & Morris, R.: Smarter cities and their innovation challenges. Computer, 44(6), pp. 32–39 (2011).
2. Rietveld, P., & Daniel, V.: Determinants of bicycle use: do municipal policies matter?. Transportation Research Part A: Policy and Practice, 38(7), pp. 531–550 (2004).
3. Geels, F., & Raven, R.: Non-linearity and expectations in niche-development trajectories: ups and downs in Dutch biogas development (1973–2003). Technology Analysis & Strategic Management, 18(3–4), pp. 375–392 (2006).
4. DeMaio, P.: Bike-sharing: History, impacts, models of provision, and future. Journal of Public Transportation, 12(4), 3 (2009).
5. Quiguer, S.: Acceptabilité, acceptation et appropriation des Systèmes de Transport Intelligents: élaboration d'un canevas de co-conception multidimensionnelle orientée par l'activité (Doctoral dissertation, Université Rennes 2) (2013).
6. Melaina, M., & Bremson, J.: Refueling availability for alternative fuel vehicle markets: sufficient urban station coverage. Energy Policy, 36(8), pp. 3233–3241 (2008).
7. Ballet, J. C., & Clavel, R.: Le covoiturage en France et en Europe: état des lieux et perspectives (2007).
8. Pucher, J., & Buehler, R.: Why Canadians cycle more than Americans: a comparative analysis of bicycling trends and policies. Transport Policy, 13(3), pp. 265–279 (2006).
9. Clavel, R., Legrand, P., & LOXANE, P.: Le covoiturage dynamique: étude préalable avant expérimentation (2009).
10. Perkins, C., Belding-Royer, E., & Das, S.: Ad hoc on-demand distance vector (AODV) routing (No. RFC 3561) (2003).
11. DeMaio, P.: Bike-sharing: History, impacts, models of provision, and future. Journal of Public Transportation, 12(4), 3 (2009).
12. Rani, M., Nayak, R., & Vyas, O. P.: An ontology-based adaptive personalized e-learning system, assisted by software agents on cloud storage. Knowledge-Based Systems, 90, pp. 33–48 (2015).
13. Rani, M., Srivastava, K. V., & Vyas, O. P.: An ontological learning management system. Computer Applications in Engineering Education (2016).
14. Blanke, J. E. N. N. I. F. E. R., Chiesa, T. H. E. A., & Herrera, E. T.: The travel & tourism competitiveness index 2009: Measuring sectoral drivers in a downturn. The Travel & Tourism Competitiveness Report 2009: Managing in a Time of Turbulence, pp. 3–37 (2009).
15. Hoffmann, C.: Erfolgsfaktoren umweltgerechter Mobilitätsdienstleistungen: Einflussfaktoren auf Kundenbindung am Beispiel DB Carsharing und Call a Bike. Osnabrück (2010).
16. Cervero, R., & Duncan, M.: Walking, bicycling, and urban landscapes: evidence from the San Francisco Bay Area. American journal of public health, 93(9), pp. 1478–1483 (2003).
17. Maurer, L. K.: Feasibility study for a bicycle sharing program in Sacramento, California. In Transportation Research Board 91st Annual Meeting (No. 12–4431) (2012).
18. http://www.isi.edu/nsnam/ns/tutorial (as on 29-6-2015).
19. http://www.smartcomp2014.comp.polyu.edu.hk/workshop.html (as on 24-8-2015).
20. https://s3.amazonaws.com/babs-open-data/babs_open_data_year_1.zip (as on 12-7-2015).
21. D. Terr. "Weighted Mean." From MathWorld-A Wolfram Web Resource, created by Eric W. Weisstein. http://mathworld.wolfram.com/WeightedMean.html 58, pp. 13918–13927 (2004).
22. Perkins, C., Belding-Royer, E., & Das, S.: Ad hoc on-demand distance vector (AODV) routing (No. RFC 3561) (2003).

# Effects of Mean Metric Value Over CK Metrics Distribution Towards Improved Software Fault Predictions

**Pooja Kapoor, Deepak Arora and Ashwani Kumar**

**Abstract** Object Oriented software design metrics has already proven capability in assessing the overall quality of any object oriented software system. At the design level it is very much desirable to estimate software reliability, which is one of the major indicators of software quality. The reliability can also be predicted with help of identifying useful patterns and applying that knowledge in constructing the system in a more specified and reliable manner. Prediction of software fault at design level will also be helpful in reducing the overall development and maintenance cost. Authors have classified data on the basis of fault occurrence and identified some of the classification algorithm performance up to 97%. The classification is carried out using different classification techniques available in Waikato Environment for Knowledge Analysis (WEKA). Classifiers were applied over defect dataset collected from NASA promise repository for different versions of four systems namely jedit, tomact, xalan, and lucene. The defect data set consist of six metrics of CK metric suite as input set and fault as class variable. Outputs of different classifiers are discussed using measures produced by data mining tool WEKA. Authors found Naive Bayes classifier as one of the best classifiers in terms of classification accuracy. Results show that if overall distribution of CK metrics is as per proposed Mean Metric Value (MMV), the probability of overall fault occurrence can be predicted under consideration of lower standard deviation values with respect to given metric values.

**Keywords** CK metrics · Classifier · Threshold · WEKA · Naive Bayes

P. Kapoor (✉) · D. Arora
Department of Computer Science & Engineering, Amity University, Lucknow, India
e-mail: pkhanna@lko.amity.edu

D. Arora
e-mail: deepakarorainbox@gmail.com

A. Kumar
Area of IT & Systems, IIM, Lucknow, India
e-mail: ashwani@iiml.ac.in



S.K. Bhatia et al. (eds.), *Advances in Computer and Computational Sciences*,
Advances in Intelligent Systems and Computing 553,
DOI 10.1007/978-981-10-3770-2_6

## 1 Introduction

Since the advent of software designing, metrics has been an integrated part of software development. Metrics are the measure to ensure the complexity level and reliability of a software product. By the time different software development methodologies came into existence, different measures have been involved to ensure better software quality. The software development process has major impact with the advent of Object Orient methodology. In Object Oriented programming paradigm object represents a real world entity. A Class is a template that describes properties and behavior of the entity. An object is an instance of the class. This Object Oriented approach added different dimension to software development process and helped the software industry to meet its current design parameters and user requirements. Object Oriented Metrics are the measures to judge the Object Oriented design level complexity and thus ensure the reliability and robustness of any software product at early stages of its development. Standardization of Object Oriented metrics is in demand as the complexity of software development process is increasing and distributed in nature over time. Various studies have been conducted to ensure the effectiveness in predicting the quality of the software at early development stages [1]. Quality of software design also ensures less occurrence of fault in the final software product [2, 3]. Through massive literature survey it is found that various researchers have proposed their own standardize threshold value for Object Oriented metrics that categorizes good and bad software design.

A NASA study [4] was conducted to find easy-to-use minimum set of software metrics capable of measuring the overall quality of object oriented software. NASA study validates that higher quality systems has Object Oriented metrics values within a specified range for metric set of CK metric suite. Different researchers have established the correlation between Object Oriented and class fault proneness [5–8]. But till now a threshold value is not set that could ensure less occurrence of faults. A threshold value is heuristic value that sets the boundary of Object Oriented metrics like the depicted range of threshold value for Weighted Methods per class (WMC), is beyond 20, hence one can say that probability of fault occurrence is high.

In present research work, experiments have been conducted with supervised techniques present in WEKA, on different data sets from NASA promise defect repository. Supervised techniques are used to when target output is very clear, like, to solve classification and prediction problems. Classification accuracy of these classifiers is compared. The defect set considered contains different CK metrics and output variable as faulty and non-faulty class. Classification accuracy means how well a classifier is in categorizing test data as per the learning from training data sets. If a class A with certain values of CK metrics (WMC, CBO, RFC, NOC, DIT, LCOM) has faulty status (Y), the classifier should keep this class in category of faulty classes. On the basis of correctly classified percentage, author found best performance of Naive Bayes classifier. Threshold range of CK metrics for individual module is proposed by many researchers in literature, but authors suggest that instead of looking for threshold values of CK metrics, for individual module and its probability of fault occurrence, controlling the distribution of metric values across the modules of a system, probability of fault occurrence at system level can be reduced.

## 2 Background

A wide variety of Object Oriented metrics have been proposed to assess the testability of an Object Oriented system. Most of the metrics are focused on the Object Oriented properties like encapsulation, inheritance, class complexity, and polymorphism. Object Oriented metrics can be categorized into two groups: project-based metrics and design-based metrics. Project-based metrics contain process, product, and resources. Design-based metrics contain traditional metrics and object oriented metrics [9]. Abreu et at. [10] defined MOOD (Metrics for Object Oriented Design) metrics. MOOD refers to a basic structural mechanism of the object oriented paradigm as encapsulation (MHF, AHF), inheritance (MIF, AIF), polymorphism (POF), and message passing (COF). Each metrics is expressed as a measure where the numerator represents the actual use of one of those feature for a given design. In MOOD metrics model, two main features are used in every metrics; they are methods and attributes. Methods are used to perform operations of several kinds such as obtaining or modifying the status of objects. Attributes are used to represent the status of each object in the system. CK metrics suite is a set of six metrics proposed by Chidamber and Kemerer [9, 11] which capture different aspects of an Object Oriented design; these metrics mainly focus on the class and class hierarchy. It includes complexity, coupling, and cohesion as well. This metric suite offers informative insight into whether developers are following object oriented principles in their design. Using several metrics available in suit managers and designers can take better design decisions. CK metrics have generated a significant amount of interest among researchers and are currently the most well-known suite of measurements for Object Oriented software.

Many software metrics have been validated theoretically and empirically as good predictors of quality factors using several methodologies, including statistical models and machine learning models [2]. Different supervised and unsupervised learning algorithms have been tested on various software defect data set. Study empirically validates Random Forest Classification technique with highest accuracy rate in classifying the faulty classes [3]. According to Basili et al. [1] Object Oriented metrics are strong predictor of fault proneness rather than the traditional metrics. Qureshi evaluated CK metrics and proposed effectiveness of all CK metrics. Using regression analysis efficiency of CK metric has been evaluated and study reveals insignificant effect of all metrics of CK metric suit except LCOM on total number of defects [12]. Various studies conclude that finding a threshold value for each metric that ensures the highest probability of fault in a module is very difficult. An empirical study done on three releases of eclipse, found that the prediction accuracy varies across the releases [13]. Cognitive theory suggests that as complexity of design increases it starts affecting the fault occurrence. Benlabi et al. [14] concludes there is no value for the studied CK measures where the fault proneness changes from being steady to rapidly increasing. To find out the threshold value a statistical model is validated using Eclipse 2.0 and Eclipse 2.1. Though study [15] suggested threshold values for CBO, RFC, and WMC metrics but results were not consistent across Mozilla and Rhino projects so concluded that these values can not be generalized. Research [16] was done to test two hypothesis and study conclude that practically there were no threshold values of the Object Oriented metrics that neither categorize modules as faulty and non-faulty (binary category) nor for ordinal category. There are many literature entries found on power law distribution. Power law distribution is a phenomenon exhibits in many object oriented properties. A power law is described as "there are few very complex modules while most modules have low complexity. Five metrics, NOC, WMC, NOM, NOV, and SLOC, have shown a potential to follow a power law distribution". The study is a great attempt to use power law characteristics to set threshold values for software metrics or to find the maximum value of a particular metric [17]. Shatnawi concluded that such threshold is not consistent across different projects [18]. The research results show that some object oriented metrics can predict class error probabilities in the three error-severity categories. Research reports that there are significant differences among the metric means in the three error-severity categories and these means can be used to calculate the metrics threshold values. Shatnawi [19] proposed a new dependent variable, and suggested four categories as none, low, medium, and high. The categorization has been done on the basis of fault occurrence in previous and current release of software. The result for classification indicates better predictions. Jin et al. [20] presents a model using artificial neural network (ANN) and support vector machine for software fault prediction using metrics. The study claims better performance for fault prediction. A comparison amongst four techniques: linear, logistic analysis and machine learning techniques: decision tree, neural network has been carried out to find technique that best suited for fault prediction. Observations of study say that all methods come out with similar results [21]. Turnu, validated correlation between the entropy of CBO and RFC and the number of bugs for eclipse and Net beans. Study says that the tendency for software, over time, to become difficult and costly to maintain [22].

## 3 Experiment Design

To conduct experiment authors have chosen the data sets from NASA promise defect data repository with total of seven parameters, six CK metrics as input and fault as out variable. To predict fault, on basis of CK metric values, occurrence of fault can be predicted at design level of software development. Thus it will ensure better software design with less occurrences of faults. To solve the problem, authors have designed three major research objectives.

RO1: To ensure correlation between CK metrics and occurrence of faults. The objective here is check, out of six CK metrics, how many directly or indirectly affect the occurrence of fault.

RO2: To identify the best supervised technique available with better classification accuracy for classification of faulty and non-faulty classes.

RO3: To check the existence of threshold value for these design level metrics that can ensure less occurrences of faults.

## 3.1 Data Set

Object Oriented metric Defect data is collected from Promise repository by NASA [23]. The data set contains total 21 attributes. These attributes are different Object Oriented metrics including six metrics of CK metric suit. For this study authors are concentrating mainly on CK metric suit. Majority of researchers used CK metrics for quality prediction for design purpose. Ck metrics suite is widely accepted for checking the quality of the final product.

The Defect data set includes Bug attribute that specifies the total no of fault occurred in respective instance/class. Data analysis reveals that mostly classes are fault free, less than 20% of instances report bugs. The data collected is noise free so no preprocessing was required, except that we introduced a new categorical attribute fault that reports presence or absence of fault in respective classes on the basis of Bug count. The changes have been incorporated as Naïve Bayes classifier requires a categorical attribute for its functioning. Different versions of four systems were tested separately for the study. For reference mean value of each CK metric as represented by study of NASA, based on three different quality systems were considered, as shown in Table 1. Authors have considered different versions of the system individually to increase prediction efficiency, as each new version is updated with few more modules.

**Table 1** Reference value of CK matrices given by NASA study [4]

| System analyzed | System A developed in Java | System B developed Java | System C developed in C ++ |
|---|---|---|---|
| **Classes** | 46 | 1000 | 1617 |
| **Lines** | 50,000 | 300,000 | 500,000 |
| **Quality** | Low | High | Medium |
| **cbo** | 2.48 | 11.25 | 2.09 |
| **lcom1** | 447.65 | 78.34 | 113.94 |
| **rfc** | 80.39 | 43.84 | 28.6 |
| **noc** | 0.07 | 0.35 | 0.39 |
| **dit** | 0.37 | 0.97 | 1.02 |
| **wmc** | 45.7 | 11.1 | 23.97 |

## 3.2 *Classifiers Used in Experiment*

The experiment is conducted using previous recorded values of CK metrics of each module in a system and corresponding number of faults occurred during implementation. Supervised approach in data mining is used to predict the output category on the basis of input values. Authors applied threshold selector, classification with clustering, Zero R, BFTree, and naive bayes.

### 3.2.1 Naive Bayes

Naïve Bayes classification is a supervised learning approach to get posterior probability of NO fault. Classifier is being used to analyze the accuracy in fault prediction of CK metrics. Authors have chosen Naïve Bayes classifier because it works on strong assumption of conditional independence, it is one of the probabilistic classifier. Different studies have already explored its prediction performance in various domains. Till now researcher analyzed one or two systems using different techniques but for better understanding study we studied four different systems with their versions in total counting thirteen systems.

Bayesian classifier [24] is able to predict class membership probabilities such as the probability that a given tuple belongs to a particular class, based on the bayes theorem. In its general form, Bayesian theorem provides a way of calculating the posterior probability, P(c|x), from P(c), P(x), and P(x|c).

$$p(c|x) = \frac{p(x|c)p(c)}{p(x)} \tag{1}$$

P(c|x) is the posterior probability of class (target) given predictor (attribute). P(c) is the prior probability of class. P(x|c) is the likelihood which is the probability of predictor given class. P(x) is the prior probability of predictor. The above expression for better understanding can be written as:

$$\text{posterior} = \frac{\text{prior} * \text{likelihood}}{\text{evidence}}$$

Bayesian theory is governed by the assumption that the probability distributions of observed data leads to optimal decisions making. Reasoning based on Bayesian theory provides a probabilistic approach to inference knowledge that helps in learning. To get the maximum posterior probability of class (target)

$$\begin{aligned} \text{i.e. CMAP} &= \text{argmax}\ p(c|d) \\ &= \text{argmax}\ \frac{p(d|c)p(c)}{p(d)} \ldots \text{ by applying bayes rule} \end{aligned}$$

as P(d) does not directly affects in getting the maximum likely target class so we can drop this component from the equation, hence forth the

$$C\,MAP = \text{argmax}\; p(d|c)p(c)$$

Naive Bayes classifier works with a strong assumption of Independence among the feature of a class. In this we need to classify a record to a certain class on the basis of training data D, where D is composed of $x_1, x_2, \ldots x_n$ independent attributes then

$$\begin{aligned} &C\,MAP = \text{argmax}\; P(x_1, x_2, \ldots x_n|c)P(c) \text{ using conditional chain rule} \\ &p(c_i|x_1, x_2, \ldots x_n) = p(c_i)\pi_{i=1}^{n} p(x_i|c_i) \end{aligned} \quad (2)$$

Naive Bayes classifier assumes that the effect of these value of a predictor(x) on a given class(c) is independent of the values of other predictors. This assumption is called class conditional independence. In some domains its performance has been shown to be comparable to that of neural network and decision tree learning. CK metric suits of six different metrics show conditional independence for given class fault/bug. In this case $c_i$ has two values (Y/N) Y for faulty class and N for Non-faulty class. $x1$, $x2$, ………$x$ represents six metrics of CK suit, i.e., $x2$ = noc, $x2$ = rfc, $x3$ = wmc, $x4$ = dit, $x5$ = lcom, $x6$ = cbo. Then if we observe that a faulty class is present, the probability of either attribute/metric, e.g., noc, being present is determined (being present means its value lies in a threshold range). Existence of one attribute being present will not alter the probability of the occurrence of the other attribute. In simple words, value of noc metrics does not affect the value of other metrics.

### 3.2.2 Other Classifiers

A threshold selector is a meta classifier that works on F measure. A meta classifier first reduces the data set and then run the base classification technique. Classification via clustering is another meta classifier, this classifier run simple k mean clustering to perform classification. Zero R is rule-based classifier, it is the simplest classifier that classify the data on the basis of the value that is most commonly found for the target attribute in the given dataset. BF Tree classification is a tree-based classifier that performs binary split at the best node chosen. A best node is the attribute selected, that leads to maximum information gain after split. Authors also conducted classification using classifiers like j48 decision tree classifier, K Nearest Neighbor, decision tree naive bayes hybrid classifier (DTNB), and multilayer Perception. On the basis of measures like classification accuracy, ROC curve, total model building time, classification technique used, Naive bayes is considered for further analysis of the research objective of the present work.

## 3.3 Classification Measures Considered for Experiment

For the experiment authors considered certain measures that are produced as output after execution of classifiers in WEKA 3.6 like correctly classified, incorrectly classified, Root Mean Squared Error, Mean Absolute Error, Kappa statistics, TP, FP, F measure, ROC, etc. Few measures are exclusive to classification techniques used like number of rules generated in rule-based classification and number of leaves generated in tree-based classification techniques.

### 3.3.1 Classification Accuracy and Building Time

Figure 1 represents classification accuracy, defined as the percentage of correctly classified instances. The figure represents graph between correctly classified percentage on Y axis and X axis representing total system considered for the experiment. Lines are corresponding to different classifiers. Naive bayes classifier is found to be good in terms of classification accuracy among the classifiers considered for this data set. There are many classifiers that are found as good as naive bayes like DTNB, Multilayer Perception but their building time is found to be far more than the Naive bayes classifier. Model building time is defined as the total time required by a classifier in WEKA to execute and present the output.

### 3.3.2 ROC Curve

A threshold curve or ROC curve is drawn using cost benefit analysis in WEKA. Curve is a two dimensional plot with true positive rate on Y axis and False Positive

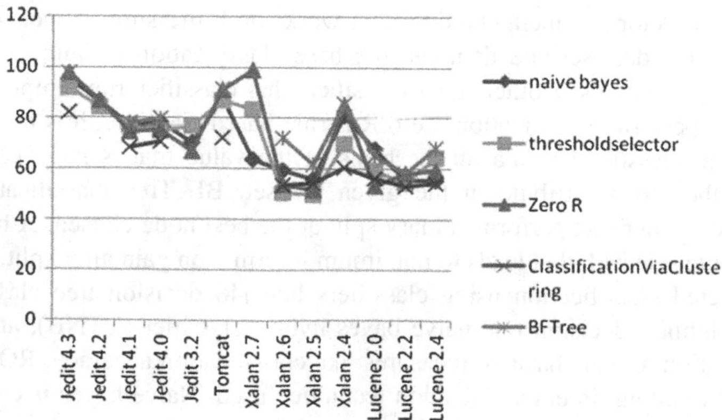


**Fig. 1** Represent classification accuracy of naive bayes, threshold selector, Zero R, classification via clustering, and BFTree classifiers

rate on X axis. The true positive rate (also called hit rate and recall) of a classifier is estimated [25] as:

$$\text{TPR} = (\text{Positives correctly classified})/(\text{Total positives})$$

The false positive rate (also called false alarm rate) of the classifier is:

$$\text{FPR} = (\text{Negatives incorrectly classified})/(\text{Total negatives})$$

Additional terms associated with ROC curves are

$$\begin{aligned}\text{sensitivity} &= \text{recall}\\ \text{specificity} &= (\text{True negatives})/(\text{True negatives} + \text{False positives})\\ &= 1 - \text{FPR}\end{aligned}$$

$$\text{positive predictive value} = \text{precision}$$

ROC are insensitive to changes in class distribution, i.e., if proportion of positive to negative instances changes in a test set, the ROC curves will not change. For the purpose of study, authors have considered threshold curve to check the performance of the classifier. The greater area under the ROC curve in the test indicates that computer-assisted classification increases a predictor's ability to correctly identify faulty classes and control false alarms [25]. Importance of ROC in any study is to use it to compare the results. Here in our study we are trying to check validity of Naive Bayes classifier prediction capabilities. A perfect predictor is one which can identify 100% true as true (sensitivity) and 100% False as False (specificity), however, theoretically any predictor will possess a minimum error.

Figure 2 presents the comparison of ROC values of different classifiers used in experiment. J48 and KNN are found to be better than naive bays in terms of

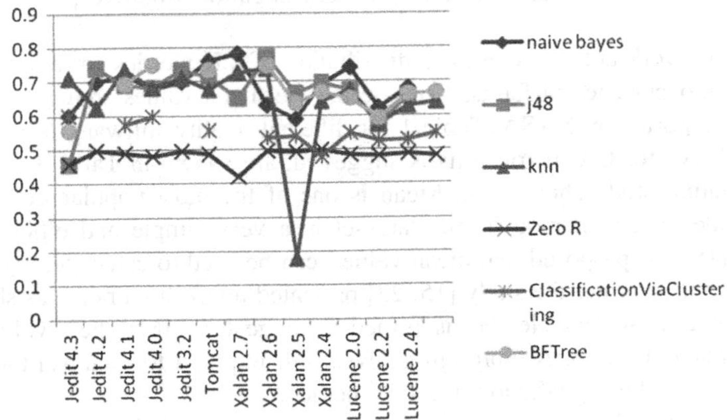


**Fig. 2** Presents the ROC values on Y axis and different systems used in experiment on X axis

classification accuracy but their ROC values are lower than the Naive bayes classifier.

## 4 Result and Discussion

Authors used WEKA3.6 (the Waikato Environment for Knowledge Analysis) tool, which is an open source software by University of Waikato. It is a comprehensive resource, for all techniques of Data mining and Machine Learning. These techniques are implemented in JAVA. In current work, the experiment is conducted, using Naive Bayes Classifier, on different versions of four datasets Jedit, Tomcat, Xalan, and Lucene. This Classifier works on strong assumption of feature independence and classify instances in two different classes (YES/NO). Table 2 shows mean values of different CK metrics, resulting 10 cross fold, Naive Bayes Classifier, on WEKA.

In case of Xalan 2.7 the probability of occurrence of NO fault is 0.01, as shown in Table 2. With the given distribution of CK metrics values in case of Xalan 2.7, there are more chances of fault to occur. Similarly Jedit 4.3 indicates highest probability of getting NO fault (0.97). Table 2 shows a gradual increase in chances of getting NO error across the versions of Jedit. Comparing values of CK metrics for jedit 3.2 and jedit 4.3, it is revealed that probability of occurrence of NO fault (n:p(c)) improved from jedit 3.2 to Jedit 4.3. It shows that design of Jedit 4.3 has distribution of CK metrics values such that mean values for instances with NO fault are, WMC = 11, CBO = 13, RFC = 38. These values of CK metrics given in Table 2 are comparable with the threshold values suggested by NASA study Table 1 for high quality systems and other work Table 3.

It is clear that except CBO and LCOM, nearly all metrics fall in the range of NASA study Table 1. By comparing CK metrics of different versions of Xalan, Jedit, and other systems it is observed that if overall design of system exhibits a distribution of metrics value accuracy of classification improves, as shown in Table 5.

In current work authors suggest a distribution of CK metrics for a system that ensures less occurrence of faults. Considering the mean values shown in Table 2 and values reported by NASA, Table 1 for different quality software, Mean Metric Values (MMV) for CK metric suits is suggested, are shown in Table 3.

As literature study shows that, Mean is one of the most popular measures of central tendency and it models the data set in a very simple and efficient way. Shatnawi [18] has proposed that mean values, can be used to calculate, the metrics threshold value. Further the study [15, 22] presented a threshold range as shown in Table 4 for Eclipse and Net beans projects. Suggested MMV lies within these threshold range Table 4. A thorough analysis still required to establish the strong relation between fault prediction and CK metrics.

n:p(c) indicates that the probability of getting no bug (probability of an instance, being in NO class), results shows that probability of getting NO fault is very high if

**Table 2** Mean values of CK metrics, result after Naive Bayes classification

| | Jedit 4.3 | Jedit 4.2 | Jedit 4.1 | Jedit 4.0 | Jedit 3.2 | Tomcat | Xalan 2.7 | Xalan 2.6 | Xalan 2.5 | Xalan 2.4 | Lucene 2.0 | Lucene 2.2 | Lucene 2.4 |
|---|---|---|---|---|---|---|---|---|---|---|---|---|---|
| n:p(c) | 0.97 | 0.86 | 0.74 | 0.75 | 0.66 | 0.9 | 0.01 | 0.53 | 0.51 | 0.84 | 0.53 | 0.41 | 0.4 |
| wmc (mean) | 11.8 | 10.1 | 9 | 9.8 | 10.1 | 11.3 | 3.5 | 7 | 8.7 | 9.3 | 6.2 | 6.5 | 6.5 |
| dit | 2.3 | 2.5 | 2 | 2.6 | 2.2 | 1.6 | 1.8 | 2.4 | 2.4 | 2.5 | 1.7 | 1.8 | 1.6 |
| noc | 0.45 | 0.5 | 0.5 | 0.5 | 0.5 | 0.3 | 0.5 | 0.35 | 0.42 | 0.5 | 0.89 | 0.52 | 0.4 |
| cbo | 13.5 | 10 | 9.5 | 9.2 | 9.7 | 6.5 | 2.9 | 9.98 | 11.5 | 13 | 7.5 | 8 | 6.7 |
| rfc | 38 | 31 | 25 | 27 | 24 | 27 | 6.7 | 20.9 | 22 | 24.6 | 14 | 16.2 | 14 |
| lcom | 243 | 198 | 41 | 77 | 90 | 117.36 | 2 | 36.9 | 70 | 87.92 | 8.33 | 7.9 | 9.6 |
| Instances | 493 | 367 | 312 | 306 | 273 | 858 | 909 | 885 | 803 | 723 | 64 | 247 | 340 |

**Table 3** Mean metric value identified by the study using Naive Bayes classifier

| Metrics | Mean metric values (MMV) |
|---|---|
| WMC (mean) | 11 |
| DIT | 2 |
| NOC | 0.5 |
| CBO | 13 |
| RFC | 40 |
| LCOM | – |
| Instances | 900 |

**Table 4** Threshold range of 3 CK metrics presented by [15, 22]

| Threshold | T1 | T2 |
|---|---|---|
| CBO | 9 | 29 |
| RFC | 40 | 164 |
| WMC | 20 | 98 |

**Table 5** Statistical values after Naive Bayes classifier

| System | Correctly classified instances (%) | TP | ROC | Kappa |
|---|---|---|---|---|
| Jedit 4.3 | 95.5 | 0.97 | 0.60 | 0.19 |
| Jedit 4.2 | 86.6 | 0.95 | 0.70 | 0.26 |
| Jedit 4.1 | 77.2 | 0.95 | 0.72 | 0.24 |
| Jedit 4.0 | 78.4 | 0.94 | 0.68 | 0.28 |
| Jedit 3.2 | 70.5 | 0.88 | 0.70 | 0.25 |
| Tomcat | 88.1 | 0.94 | 0.76 | 0.22 |
| Xalan 2.7 | 63.9 | 0.63 | 0.78 | 0.02 |
| Xalan 2.6 | 59.7 | 0.20 | 0.63 | 0.15 |
| Lucene 2.0 | 68.2 | 0.92 | 0.74 | 0.34 |
| Lucene 2.2 | 55.8 | 0.29 | 0.62 | 0.19 |
| Lucene 2.4 | 55.8 | 0.31 | 0.68 | 0.2 |

mean values of the CK metrics follow the given distribution. It is very clear from the data shown in Table 1.wmc, cbo, and rfc are the metrics from CK suits which shows major variations, and hence is effecting the final classification. lcom is another metric with varying values, but is independent in overall classification using Naive Bayes and this has been verified by many other research works. Study concludes Mean Metric Value (MMV) of cbo, rfc, and wmc for a system. The system designed, such that the central tendency of metric values follows MMV, ensures better classification accuracy. Resulting values of the current work for all CK metrics, are in reference shown in NASA study except the value of cbo. The kappa statistic is a measure of how closely the instances classified by classifier matched the data labeled as a ground truth, controlling for the accuracy of a random classifier as measured by the expected accuracy. The result shows slight (0.0–0.2) and fair (0.2–0.4) measured values of kappa for each run. ROC value above 0.5 indicates significance of the classification done. It is clear from the table that ROC is above 0.6 Applicability of MMV depends on distribution of metrics values across the system. If values are scattered distributed then MMV is unable to do accurate prediction. Based on above facts and statistics, authors have designed two case studies to ensure that if distribution of values is not scattered or with low standard deviation, the mean can be considered as threshold instead of considering individual metric value of modules.

**Case 1**: MMV for CBO is set as 13, consider a case where there are total of ten modules (c1, c2, c3, c4, c5, c6, c7, c8, c9, c10) in a system with CBO values distribution as (1, 2, 17, 10, 23, 14, 15, 25, 18, 5). This is a metric value distribution with minimum value as 1 and maximum 25, with a standard deviation of 8.3. Average of these values is 13 so we can suggest that design with such distribution of CBO metrics will ensure less occurrences of faults.

**Case 2**: Consider a case where a system with CBO values distribution as (1, 2, 6, 10, 13, 14, 12, 28, 47, 5). This is a metric value distribution with minimum 1 and maximum value 47 with a standard deviation of 13.99, though average of this distribution is also 13. Such a distribution is highly scattered in such case MMV will not ensure better prediction results.

Through this analysis it is found that, value of ROC curve lies in a range of (0.6–0.7), for all datasets, listed in Table 5 and these values reflects statistical significance of the classifier.

Naive Bayes classifier works well at large amount of data to get optimized results. Authors have tested MMV for few other data sets like Velocity, Xerces. Availability of CK defect data set makes the analysis and validation of the results difficult. In this research work authors have tried to set the reference values of metrics and from the analysis of four systems we concluded the results. Other factors that influence the prediction need to be explored. As data is right screwed so instead of mean median will better represent the central tendency. In this research work, value of lcom (lack of cohesion of method) is not considered here as it shows huge variation among the versions and systems.

## 5 Concluding Remarks and Future Work

This research work validates the relation between CK metrics and occurrence of faults. As per classification accuracy and other measures like model building time and ROC it has been observed that the performance of Naive Bayes is found to be better than other classifiers. Further, setting any value as threshold values for each metric of the CK metric suit depends on various factors. The distribution of CK metrics values (Mean Metric values) across the system justify that we can minimize the overall fault/defect fault occurrence. The fault probability estimation on the basis of MMV of the system can help to improve the overall design of the system at early stages of its development. Such methodology will definitely be helpful in ensuring less occurrence of fault. Authors also discussed the cases depending upon largely scattered and less scattered distributed mean values. Further it has been observed that the consideration of standard deviation is an important factor to establish ground that the mean metric values can be considered to improve the overall efficiency of prediction system.

## References

1. V. Basili, L. Briand, and W. Melo, "A Validation of Object-Oriented Design Metrics as Quality Indicators," IEEE Trans. Software Eng., vol. 22, no. 10, pp. 751–761, Oct. 1996.
2. A. Deepak, K. Pooja, T. Alpika, S. Sharma,"Software quality estimation through object oriented design metrics", IJCSNS International journal of computer science and network security, april 2011, pp 100–104.
3. Chug, S. Dhall, "Software Defect Prediction Using Supervised Learning Algorithm and Unsupervised Learning Algorithm", The Next Generation Information Technology Summit (4th International Conference), 2013, pp. 5.01–5.01
4. Laing, Victor & Coleman, Charles: "Principal Components of Orthogonal Object-Oriented Metrics". White Paper Analyzing Results of NASA Object-Oriented Data. SATC, NASA, 2001.
5. S Benlarbi and W. Melo: "Polymorphism Measures for Earl Risk Prediction". In Proceedings of the 2 Lt lntenational Conference on Software Engineering, pages 334–344, 1999.
6. A. Binkle and S. Schach: "Validation of the Coupling be pendency Metric as a Predictor of Run-Time Fauilures and Maintenance Measures". In Proceedings of the 20th lnternational Conference on!$oftware Engineering, pages 452–455, 1998.
7. L. Briand, P. Devanbu, and W. Melo: "An Investigation into Coupling Measures for C++". In Proceedings of the 19th lntemational Conference on Software Engineering, 1997.
8. L. Briand, J. Wuest, S. Ikonomovski, and H. Lounis: A Comprehensive Investigation of Quality Factors in Object-Oriented Designs: An Industrial Case Stud International Software Engineering Researcc'Network technical report.
9. Chidamber, Shyam, Kemerer, Chris F. "A Metrics Suite for Object-Oriented Design." M.I.T. Sloan School of Management E53–315, 1993.
10. Abreu, Fernando B., Carapuca, Rogerio.: "Candidate Metrics for Object-Oriented Software within a Taxonomy Framework.", Journal of systems software 26, 1(July 1994).
11. Lorenz, Mar, K. Jeff: "Object-Oriented Software Metrics", Prentice Hall, 1994.
12. Qureshi, M., and WaseemQureshi. "Evaluation of the Design Metric to Reduce the Number of Defects in Software Development." arXiv preprint arXiv:1204.4909, 2012.
13. R. Shatnawi and W. Li, "The effectiveness of software metrics in identifying error-prone classes in post-release software evolution process," Journal of Systems and Software, vol. 81, no. 11, pp. 1868–1882, 2008.
14. Benlarbi, S., El-Emam, K., Goel, N., and Rai, S., Thresholds for ObjectOriented Measures. NRC/ERB 1073. (National Research Council of Canada), 2000.
15. ShatnawiRaed "A Quantitative Investigation of the Acceptable Risk Levels of Object-Oriented Metrics in Open-Source Systems", IEEE TRANSACTIONS ON SOFTWARE ENGINEERING, VOL. 36, NO. 2, MARCH/APRIL 2010.
16. Shatnawi, Raed, et al. "Finding software metrics threshold values using ROC curves." Journal of software maintenance and evolution: Research and practice 22.1, 2010: 1–16.
17. ShatnawiRaed, and QutaibahAlthebyan. "An Empirical Study of the Effect of Power Law Distribution on the Interpretation of OO Metrics." ISRN Software Engineering 2013.

18. ShatnawiRaed, "The Validation and Threshold Values of Object-Oriented Metrics", Ph.D. Dissertation. University of Alabama in Huntsville, Huntsville, AL, USA. Advisor(s) Wei Li, 2000.
19. RaedShatnawi, "Empirical study of fault prediction for open-source systems using the Chidamber and Kemerer metrics", IET Software, Volume 8, issue 3, 2014, pp. 113–119.
20. C. Jin, S.-W. Jin, J.-M. Ye, "Artificial neural network-based metric selection for softwarefault-prone prediction model", IET Software, Volume 6, issue 6, 2012, pp. 479–487.
21. T. Gyimothy, R. Ferenc, and I. Sik et, "Empirical validation of object-oriented metrics on open source software for fault prediction," IEEE Transactions on Software Engineering, vol. 31, no. 10, pp. 897–910, 2005.
22. I. TURNU, G. CONCAS, M. MARCHESI, R. TONELLI, "Entropy of some CK metrics to assess object oriented software quality", International Journal of Software Engineering and Knowledge Engineering 2013, 173–188.
23. http://promisedata.googlecode.com.
24. T. Mitchell, Machine Learning, Tata McGraw-Hill, 2013. ISBN: 9781259096952.
25. Fawcett, Tom. "An introduction to ROC analysis." Pattern recognition letters 27.8, 2006: 861–874.

# Feedforward and Feedbackward Approach-Based Estimation Model for Agile Software Development

**Saru Dhir, Deepak Kumar and V.B. Singh**

**Abstract** In the software project development software estimation and planning is a crucial process. This paper proposes a framework that outlines the estimation of a software development at the initial stages and consists of feedforward and feedbackward approaches in the whole development cycle. Improved software estimations are also elaborated on an agile project implementation using the proposed framework.

**Keywords** Agile software development · Software estimation · Feedforward · Feedbackward

## 1 Introduction

Estimation of a software project includes ideal time (productive time for the completion of the project), velocity (Measure the total user stories and implementation tasks during the iteration cycles), and total effort (Actual effort taken to complete the project within the total number of days) for the completion of product.

There are different approaches such as empirical and iterative approach where the requirements can be changed according to the user's need and thus deliver a high-quality product on time [1]. Robert Glass acclaims agile methodologies for their importance on continuous unit testing, integration testing and development, client involvement, and programmer friendly environment [2].

S. Dhir (✉) · D. Kumar
Amity University, Noida, UP, India
e-mail: sdhir@amity.edu

D. Kumar
e-mail: deepakgupta_du@rediffmail.com

V.B. Singh
University of Delhi, Delhi, India
e-mail: singh_vb@rediffmail.com



S.K. Bhatia et al. (eds.), *Advances in Computer and Computational Sciences*,
Advances in Intelligent Systems and Computing 553,
DOI 10.1007/978-981-10-3770-2_7

Size estimation methods in agile projects include story points, different methods such as judgements of expert people and techniques like function point analysis (FPA). Iterations are planned at the initial phase, according to the estimated size of user requirements. Effort estimation models are categorized as regression-based models, Bayesian methods, and learning-oriented models. Estimation methods that are based on regression models are easy to understand and develop; for example, least square regression is the simplest model [3]. Learning-oriented models include the different techniques like artificial neural networks, case-based reasoning, fuzzy logic model, machine learning models, knowledge acquisition, and rule induction [4, 5].

The paper is divided into different sections, where Sect. 2 describes the literature review of software estimation in agile development using different software estimation models, such as algorithmic and non-algorithmic process. In Sect. 3, implementation of a web project using SDV process model (model SDV is named on the first letter of author's name Saru, Deepak and V.) is elaborated and estimated with algorithmic way. Section 4 discusses the quantitative estimation of a project implementation, where MRE and MMRE values are calculated.

## 2 Literature Review

Planning of a project includes estimation of the project size, followed by estimation of effort and budget of the project. In the traditional development most of the tasks are finalized in the detailed design, whereas in agile development new tasks are identified at the iteration cycle and each iteration cycle is a tactical development for a specific duration. Traditionally, there were different estimation models that were used to estimate the cost and effort of a project to assess whether a project was good value for money or not [6]. The estimation models can be categorized into algorithmic and non-algorithmic model [7, 6]. Agile estimations are based on non-algorithmic models Wideband Delphi methods (planning poker technique) are also one of the estimation models that are mostly used in agile projects. An analogy technique was compared with predictive models based upon regression analysis for nine datasets and produces a predictive performance in all cases by MMRE [8]. At the bottom-up method, software system components are decomposed and assessed separately; and results are added to generate an estimate of the overall project. Top-down method is opposite to bottom up, where the cost estimation can be set at initial stage and the total cost can be divided into different components [6].

Among all of the non-algorithmic models, expert-based estimation is the most dominant strategy used in the industry, which includes process guidelines, experience, and expert judgement of team members [9]. Lucas et al. [10] elaborated case study using XP evaluation framework to adapt XP model. The XP evaluation framework used feedback loop in the evaluation process. The team for the development of the project was typically agile to deploy XP process model.

C.J. Torrecilla-Salinas et al. [7] proposed a united framework for estimating, planning, and controlling the web projects, including existing agile techniques with web engineering principles. Fuqua, A., and Hammer, J.M. [11] illustrated a report that the velocity varies with a 20% variation, where unstable velocity formulates the estimation difficult. In the agile web development different estimation challenges and comparison with traditional web estimation are demonstrated and also presented in the AgileMOW model [12].

# 3 Implementation Using SDV Estimation Model

## *3.1 SDV Estimation Model*

Dhir S. et al. discussed the ineffectiveness of practices, estimation and tracking problems such as lack of effort estimation, lack of experienced staff, ambiguity, lack of feedback responses, and change of role activity with existing approaches [13]. An estimation model was proposed to estimate the agile project at initial stages, including feedforward and feedbackward approaches. An algorithmic approach was also proposed to estimate the project, where the estimation process starts from the initial stage (EPIS) to the last testing stage by maintaining the functional and non-functional requirements. Case study of a project was also discussed according to the proposed model and algorithmic approach using story points and function point analysis. Figure 1 represents the SDV process model (model name SDV is based on the first letter of author's name Saru, Deepak and V.), which is designed according to proposed estimation model [13].

SDV process model is used to avoid the existing agile estimation issues using the quantitative measurement. The SDV process model includes different phases from the initial phase to testing phase, such as follows:

1. Elicitation and prioritization of requirements
2. Estimate the project
3. Feedforward
4. Division into sub-modules and iterations
5. Implement the sprint cycle
6. Release the sprint cycles
7. Root cause analysis (RCA)
8. Feedbackward
9. Regression testing
10. Generate the final report

## *3.2 Quantitative Estimation of the Model Using an Algorithmic Approach*

An algorithm is implemented to evaluate the effort, velocity and time estimation on the project at the initial stage of the project using the feedforward and feedback approach.

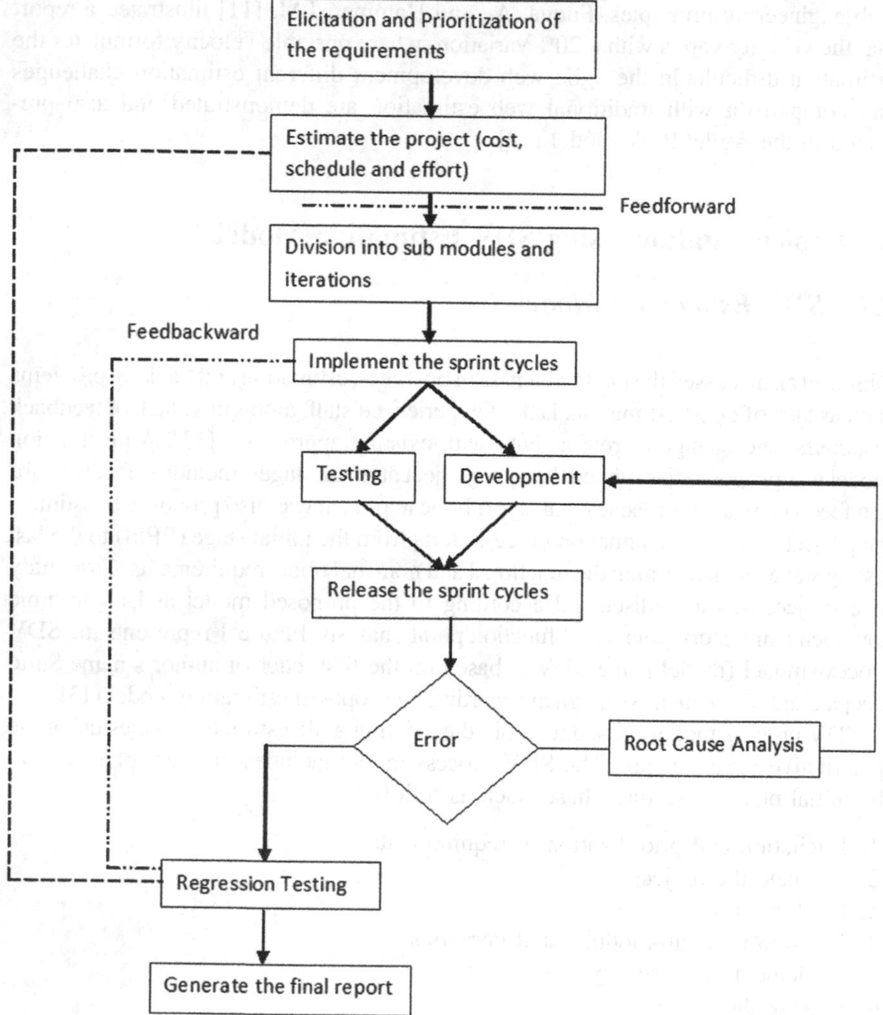


**Fig. 1** SDV process model

In the algorithmic approach of the proposed model [13], estimation of process at the initial stage (EPIS) is maintained in functional and non-functional requirements at the beginning of each iteration cycle. Final estimation (time and effort) is evaluated using an algorithmic approach as follows:

**Effort**:

$$BSP = US * SS, \tag{1}$$

where BSP is the baseline story point, US represents the user story, and SS represents the sprint size. The baseline story point is the total number of user stories and sprint size, and p is people and project-related factors. P = {p1, p2, p3…pi… pn} where $1 < i \leq n$

$$\text{Unadjusted Value (UV)} = \{p1 + p2 + p3 + \cdots + pn\}. \tag{2}$$

Unadjusted values (*UV*) are the additional time, effort or cost estimation that is taken during external review and calculated during the effort, cost, and time estimation process. From Eqs. (1) and (2)

$$\text{ESP} = \text{BSP} + 0.1(\text{UV}). \tag{3}$$

ESP is the estimation according to story point:

$$\text{EstimatingTesting Effort } (ETE) = \text{Testing Effort per Iteration} \times (\text{No. of Iterations} - 1). \tag{4}$$

No. of iterations = TET/(10 − 1).
From Eqs. (3) and (4)

$$\text{Final Estimated Effort } (E_F) = \text{ESP} + \text{ETE}. \tag{5}$$

**Velocity**:

$$\text{In general: Velocity} = \text{Distance/Time}. \tag{6}$$

In the agile development, distance is equivalent to the completion of story point in the iteration and time is equivalent to the length of the sprint. From Eq. (1)

$$\text{EV} = \text{BSP/Total number of completed sprints}. \tag{7}$$

EV is the estimated velocity

$$\text{Initial Velocity } (V_i) = \text{Total number of story points in one iteration/SprintSize}. \tag{8}$$

**Velocity Factors**:
Friction and dynamic factors are considered with some range of values [14], for analyzing the quantitative values of the project.
From Eq. (8),

$$\text{Decelerated Velocity (DV)} = \text{V} * \text{VFR}. \tag{9}$$

**Time:** Estimate the complete duration (per month) of the project as from Eqs. (3) and (8)

$$\text{Estimated Development Time (EDT)} = \text{ESP/V} \quad (10)$$

$$\text{Estimated Testing Time (ETT)} = \sum T_t. \quad (11)$$

$\sum^n T_i$ is the sum of different testing activities based on different factors. From Eqs. (10) and (11)

$$\text{Total Estimated Time (TET)} = \text{EDT} + \text{TET} \quad (12)$$

$$\text{Testing Time (TT)} = \text{Regression testing in one iteration (RTT)}. \quad (13)$$

From Eqs. (12) and (13)

$$\text{Total Time } (T_T) = \text{TET} + \text{RTT}. \quad (14)$$

MRE is the magnitude of relative error (MRE) and MMRE is the mean magnitude of relative error. From Eq. (12)

$$\text{MRE (Time)} = |\text{Actual Time} - \text{Estimated Time}|/(\text{Actual Time}). \quad (15)$$

## 4 Implementation and Experiment Results

**Inputs**:

Total number of user stories (US) = 30
Sprint Size (SS) = 10 SP
Total number of story point complete in 1 iteration = 55 SP
Estimated Unadjusted Value = 100
No. of days in one iteration = 10 days
Regression testing effort per iteration = 10 SP

**Results**:
**Effort**:
From Eq. (1);

$$\text{BSP} = 30 * 10 = 300 \quad (16)$$

From Eq. (16)

$$\text{ESP} = 300 + 0.1(100) = 310 \quad (17)$$

BSP is the baseline story point with units SP and ESP is the estimated story point.

**Velocity**:

From Eq. (16) and sprint size value as inputs,

$$\text{Initial Velocity } (V_i) = 300/55 = 5.45 \quad (18)$$

**Factors**:

$$\text{Velocity Factor Rate (VFR) is average of velocity factor of all factors} = 0.95. \quad (19)$$

From Eqs. (18) and (19)

$$\text{Decelerated Velocity (DV)} = 5.45 * 0.95 = 5.17\ \text{SP/Day}. \quad (20)$$

**Time**:

Estimated testing time (days) is calculated according to the different testing attributes. According to Eq. (10) values are calculated from Eqs. (17) and (20)

$$\text{Estimated Development Time (EDT)} = 310/5.17 = 59.96\ \text{Days}. \quad (21)$$

From Eq. (21) and estimated testing time (ETT) = 15 Days

$$\text{Total Estimated Time (TET)} = 59.96 + 15 = 74.96\ \text{Days}. \quad (22)$$

From Eq. (22), estimated time is

$$\text{MRE (Time)} = |80 - 74.96|/(80) = 0.063 \quad (23)$$

MMRE (Time) = 6.3%

**Effort**:

For Eq. (4), values of TET are taken from Eq. (22) and testing effort per iterations, 10, are given.

Estimated testing effort (ETE) = Testing effort per iteration * (no. of iterations – 1)

No. of iterations = TET/10 = 74.96/10 = 7.496

$$\text{ETE} = 10 * (7.496 - 1) = 64.96\ \text{SP}. \quad (24)$$

For Eq. (5), values are taken from Eqs. (17) and (24)

$$\text{Final Estimated Effort } (E_F) = \text{ESP} + \text{ETE} = 310 + 64.96 = 374.96\ \text{SP}. \quad (25)$$

# 5 Conclusion

The paper concludes with the implementation of the estimation technique by implementation a small project. The SDV process model is based on the estimations at the initial stages by following the feedforward and feedbackward approach. Effort and time of the project are estimated to include the different factors that can affect the project estimation during the implementation. MRE, MMRE (Time) and effort is evaluated.

# References

1. M. Pikkarainen et al.: The Impact of Agile Practices on Communication in Software Development, Empirical Software Engineering, Springer (2008).
2. Glass R.: Extreme Programming: The Good, the Bad, and the Bottom Line, IEEE Software, vol. 18, no. 6, pp. 112–111 (2001).
3. Jørgensen, M. and Shepperd M.: A Systematic Review of Software Development Cost Estimation Studies Document Actions, IEEE Transactions on Software Engineering, 33 (1): pp. 33–53 (2006).
4. Mukhopadhyay, T. and Kekre, S.: Software Effort Models for Early Estimation of Process Control Applications. IEEE Transactions on Software Engineering, pp. 915–924 (1992).
5. Burgess, C. J. and Lefley, M.: Can Genetic Programming Improve Software Effort Estimation? A Comparative Evaluation. Information and Software Technology. 43, (2001).
6. Armario, J., Gutiérrez, J. J., Alba, M., García-García, J.A., Vitorio, J. and Escalona, M. J.: Project estimation with NDT. In: Proceedings of the 7th International Conference on Software Paradigm Trends, Rome, Italy, pp. 120–126 (2012).
7. Torrecilla-Salinas, C. J., Sedeño, J., Escalona, M. J. and Mejías, M.: Estimating, planning and managing Agile Web development projects under a value-based perspective. Information and Software Technology, vol. 61, pp. 124–144 (2015).
8. Shepperd, M. and Schofield, C.: Estimating Software Project Effort Using Analogies. IEEE Transactions on Software Engineering, 23(12):736–743 (1997).
9. Jorgensen, M.: A review of studies on expert estimation of software development effort. The J. of Systems and Software, 70, pp. 37–60 (2004).
10. Lucas, L., Laurie, W., and Lynn, C.: Motivations and measurements in an agile case study. J. of system architecture. 52, (11), pp. 654–667 (2006).
11. Fuqua, A., and Hammer, J.M.: Embracing Change: An XP Experience Report. In: 4th int. conference on Extreme programming and agile processes in software engineering, XP, pp. 298–306 (2003).
12. Litoriya, R. and Kothari, A.: An Efficient Approach for Agile Web Based Project Estimation: AgileMOW. J. of Software Engineering and Applications. pp. 297–303 (2013).
13. Dhir, S., Kumar, D. and Singh, V.B.: An estimation technique in agile archetype using story points and function point analysis. Int. J. Process Management and Benchmarking, (2016).
14. Tipu, Z. S. K. and Zia, S.: An Effort Estimation Model for Agile Software Development. Advances in computer science and its applications (ACSA), Vol. 2, No. 1 (2012).

# Investigation of Effectiveness of Simple Thresholding for Accurate Yawn Detection

**Viswanath K. Reddy and K.S. Swathi**

**Abstract** Drowsiness of a person is major cause for accidents and to avoid accidents alerting person at right time is very necessary. Yawning is one of the signs, which indicates whether the person is drowsy or not. Most of the algorithms in literature detect yawn state considering the region between the lips. Mouth localization is the fundamental step in yawn detection. Region between the lips is segmented using algorithms of different complexities. In this work, a simple segmentation algorithm like thresholding is investigated for its effectiveness. The segmented region with maximum area within the mouth region is considered to classify the frame as yawn frame or otherwise. Yawn video sequences from YawDD dataset are used to test and validate the algorithm. Yawn detection accuracy using the proposed algorithm is 76% which is bit higher than the accuracy obtained with more complex algorithm. Such simple algorithms might be more useful for real-time applications. The time consumption of the implementation is to be verified.

**Keywords** Driver drowsiness · Face and eye detection · Focused mouth region · Image processing based · Yawn detection

V.K. Reddy (✉)
Faculty of Engineering and Technology, Department of ECE,
M. S. Ramaiah University of Applied Sciences, Bengaluru, India
e-mail: viswanath.ec.et@msruas.ac.in

K.S. Swathi
Digital Signal and Image Processing, M. S. Ramaiah University of Applied Sciences, Bengaluru, India
e-mail: swathi.kanle@gmail.com



S.K. Bhatia et al. (eds.), *Advances in Computer and Computational Sciences*,
Advances in Intelligent Systems and Computing 553,
DOI 10.1007/978-981-10-3770-2_8

## 1 Introduction

Most of the vehicle crashes are due to driver poor operating practices which include lack of seat belt use, distraction, fatigue, alcohol, and drug use while driving. Out of all these, driver's inattention due to fatigue is a major issue. As per U.S. National Highway Traffic safety, approximately 10000 accidents each year [1] are due to driver fatigue. Therefore, there is a need for a system to warn the driver in such situations. Reasons for person feeling drowsy may be due to illness, sleep disorders.

Measuring person fatigue through heart rate and EEG is more accurate but not realistic [2], because sensing electrodes cause discomfort to the driver. Hence, it is better to measure fatigueness of a person through facial features such as mouth, eye, and even through head pose. Driver alertness [3] can be determined by eye state or through head nodding angles. When eye is occluded head pose can be considered for obtaining information about lack of attention of a person. If a person is wearing glasses [2], techniques based on detecting eye characteristic may not give good results in detecting fatigue of driver as glasses can cause glare and may be totally opaque to light, making it impossible for a camera to monitor eye movement. Furthermore, the degree of eye openness may vary from person to person.

In the non-intrusive techniques of yawn detection, mouth segmentation is a primary step. Several complex algorithms are being currently used for mouth segmentation. In this work, the effectiveness of a simple segmentation algorithm like thresholding is investigated.

## 2 Related Work

Yawning detection can be done by measuring the rate of mouth opening. Normal mouth opening rate and yawning rate has to be differentiated by algorithms to tell whether the person is drowsy or not. Opening of mouth can be measured either by tracking lip movements or by width of the lip region. Basically, there are three different approaches for yawn detection: feature based, appearance based, and model based.

In featured-based approach, differentiation between normal mouth opening rate and yawning is done using features such as color, edges, and texture. Color properties can be used to segment mouth region. Major differentiation can be done using intensity rather than color itself, and hence transformed color spaces such as normalized RGB space, HSV, and YCbCr are used. YCbCr color space transformation is used [4] to segment the mouth region to detect yawn of person. Initially, face of a person is captured by video camera installed in front of mirror inside car. Based on skin color segmentation and template matching, face region is detected.

Based on color mouth region is segmented. Mouth map equations are applied to face region in order to segment mouth region.

Before segmenting mouth region, to detect mouth region, mouth map equations are applied to face region. To reduce misclassification of mouth region using mouth map equations, active snake contour technique is used. The rate of increase in mouth contour area is used as indication of yawning. Algorithm implemented is tested on videos recorded by authors under various lighting conditions. Initially, face region is classified using SVM and to identify wide mouth regions for detection of yawning. Circular Hough transform [5] is applied on mouth region. Circular Hough transform extract circles from edge images. If mouth is opened for significant number of consecutive frames, algorithm tells the person is yawning and alerts the person by issuing warning signal. Using skin tone detection algorithm [6], yawning of a person is detected. To track mouth region for detecting yawn of a person, mathematical models for LIPMAP in YCbCr and RGB spaces are developed based on characteristic features of color components in mouth region. Lip region is segmented using back projection theory [7] in OpenCV, and yawning of a person is detected. To detect yawn of person, ratio of black pixels and white pixels is compared with a threshold value. If value is greater than particular threshold defined, then algorithm says person is yawning. The performance of their algorithm is tested on YawDD dataset and an accuracy of 75% is obtained.

In appearance-based approach, different machine learning algorithms are used and important features corresponding to region of interest are extracted in order to train the algorithms. Once face is detected, to classify lip and skin region, Fischer linear classifier [6] is used. Kalman filtering is used to track driver's mouth in real time by prediction and localization. Geometric features such as width, height of the mouth, and height between top lip and bottom lip vary greatly when person is talking, yawning, and when mouth is closed. Yawn of person is detected even when there is occlusion [8, 9]. System begins with initialization operation in first frame which consists of face acquisition and region initialization algorithms. During face acquisition, algorithms to detect face, eye, and mouth region are performed sequentially before region of interest algorithms is executed. Region of interest initialization algorithms involves initialization of two regions namely focused mouth region and focused distortion region. To detect yawning of a person in real time these two regions initialized are tracked continuously. Focused mouth region is initialized to measure the openness of mouth region and focused distortion region is initialized to measure the distortions during yawning when mouth is covered. Using adaptive threshold technique, the darkest between lips is identified. When mouth is not occluded, yawning of a person is detected based upon widest area of the darkest region between lips. When mouth is occluded, yawning is detected by identifying distortions in focused distortion region using any edge detector. Distortions in focused distortion region will be more during yawning. Algorithm was tested on SFF dataset. Out of 30 videos yawn was accurately detected in 28 videos.

# 3 Yawn Detection

Process of detecting drowsiness of a person begins with capturing video and converting video into frames for further processing. Once video is converted into frames, each frame of video sequence being captured is processed one by one. Processing consists of face detection, eye detection, mouth localization, segmenting region between lips in each frame, and finally based on area of segmented lip region drowsiness of a person is detected.

## 3.1 *Face Detection and Eye Detection*

First step after converting videos into frames is to detect face in each frame. Face region is detected using Viola Jones algorithm implemented in OpenCV software library. Features that define face region are stored in xml file in OpenCV software library and those features are given as input to the classifier to classify face region. Using cascade classifier, facial and non-facial features are classified and face region is detected in each frame. Similar to face detection, eye region is searched and located within the face region being detected using Viola Jones algorithm. Searching of eye region location starts from the upper left corner of the face frame to lower right corner.

## 3.2 *Mouth Localization*

Once eye region is located, center point of eye region is calculated and coordinates of center point of both left and right eye are stored in a variable. Once the coordinates are obtained, distance formula is applied to calculate the distance between eyes named ED [5]. Midpoint of upper lip region is located. Distance between center point of upper lip region and midpoint between eyes named EMD is obtained using distance formula. Face, eye, and mouth region located is shown in Fig. 1.

With using only xml file available in OpenCV software library mouth region was not properly located for the YawDD dataset. For locating mouth region in a better way new approach mentioned in [5] is used. Coordinates of focused mouth region are obtained using formula mentioned in Eqs. (1) and (2). Formula to obtain coordinates of focused mouth region is slightly changed from the formula mentioned in [5]. The threshold values 0.75 and 0.5 are used based on observation test undertaken for YawDD dataset. X coordinates for mouth region are same as x coordinates of eyes center point. Y coordinates for mouth region are eye center point y coordinate plus 0.75 times the distance between mouth center point and eyes midpoint and 0.5 times the distance between mouth center point and eyes midpoint

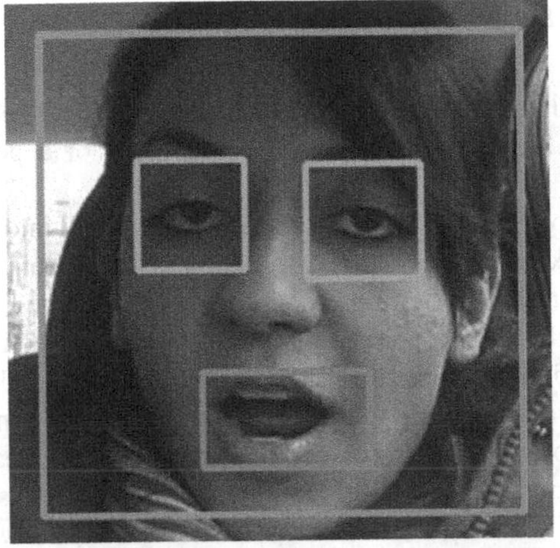

**Fig. 1** Detection of face, eye, and mouth

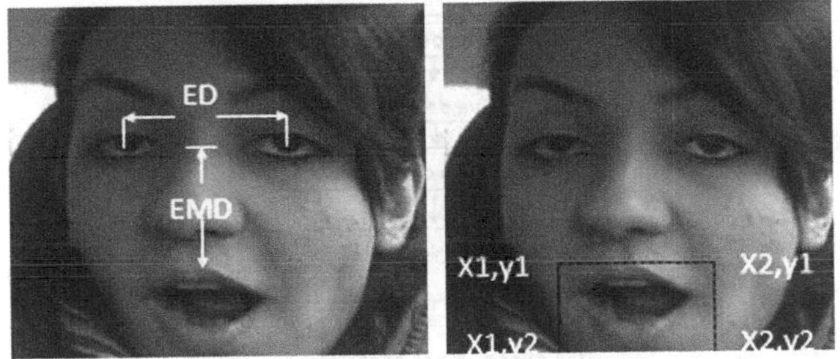


**Fig. 2** Focused mouth region coordinates

plus distance between eyes center point. Focused mouth region with coordinates is as shown in Fig. 2.

$$\begin{bmatrix} x_1 \\ y_1 \end{bmatrix} = \begin{bmatrix} x_r \\ y_r + 0.75EMD \end{bmatrix} \tag{1}$$

$$\begin{bmatrix} x_2 \\ y_2 \end{bmatrix} = \begin{bmatrix} x_l \\ y_l + 0.5EMD + ED \end{bmatrix}. \tag{2}$$

$x_1, x_2, y_1, y_2$ are coordinates of focused mouth region and $x_r, x_l, y_r, y_l$ are x and y coordinates of right and left eyes. Mouth region being detected is stored in vector and rectangular regions of interest are marked for mouth region being detected within face region.

### 3.3 Yawn Analysis

Next step after localizing mouth region is to segment region between lips to detect yawn of a person. Steps involved in analyzing yawn of a person are shown in Fig. 3.

Mouth region localized in RGB format is converted into grayscale. Region between lips is darker when compared to skin and lip region. Intensity of darker region will be nearer to zero. Setting proper threshold to the region will segment the region between the lips. Area corresponding to region between lips is set to white where all other regions are set to black. The number of pixels corresponding to region of interest is counted as it helps in differentiating different states of a person such as yawn and normal state. When person is yawning, the area of region between lips increases and number of pixel count in that region also increases correspondingly. Hence, counting number of pixels is one of the ways to differentiate different states of person. Segmented image after applying thresholding for different states is shown in Fig. 4.

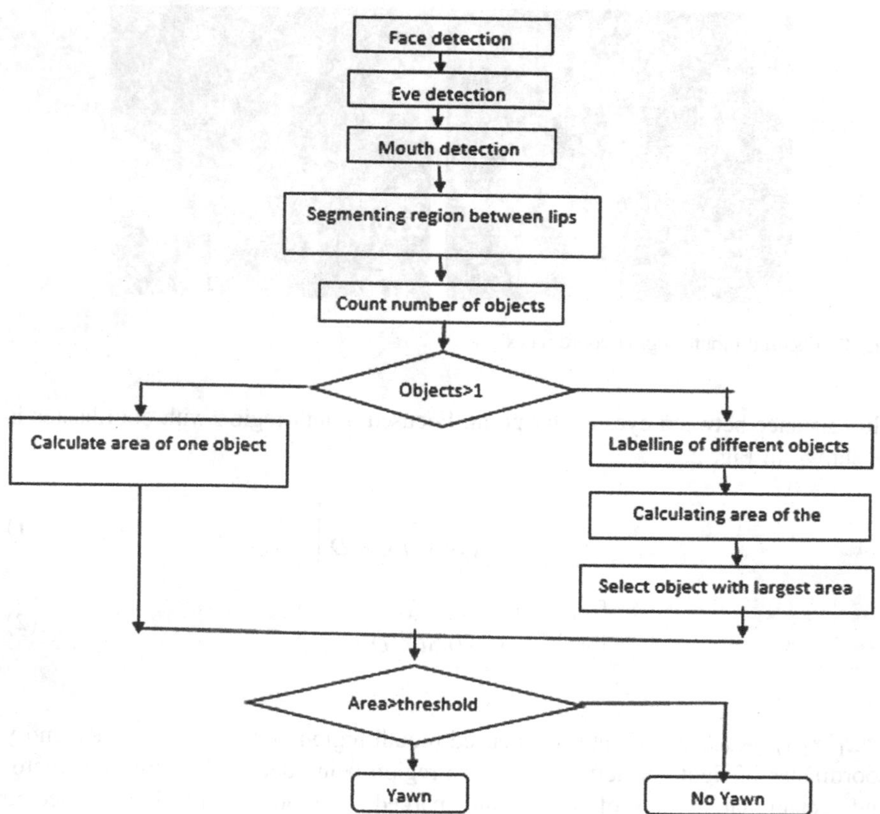


**Fig. 3** Flowchart for yawn analysis

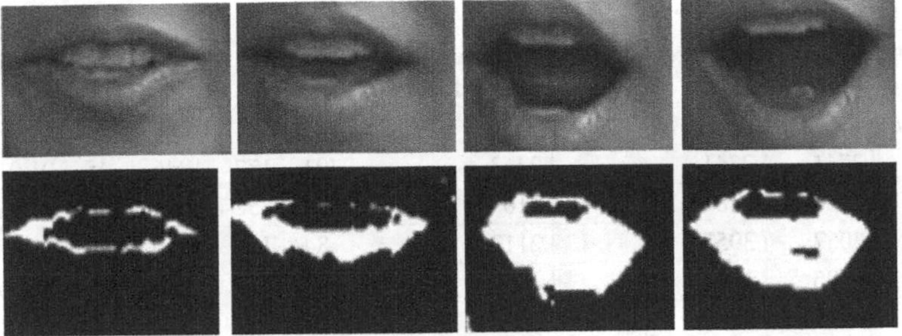

**Fig. 4** Segmentation of mouth regions by thresholding

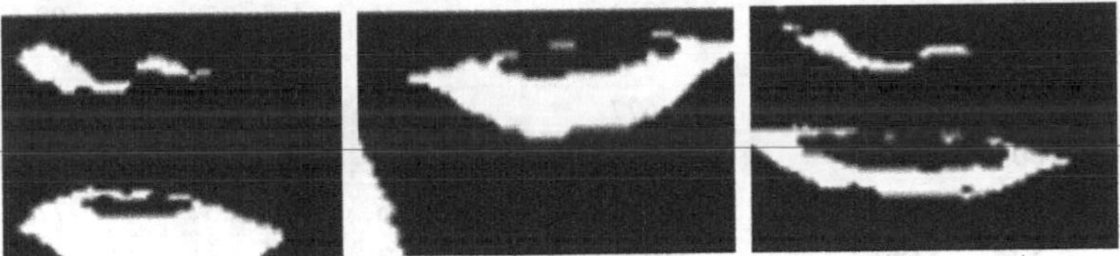

**Fig. 5** Improper segmented region

Regions that are not corresponding to region between lips may also be segmented as shown in Fig. 5, due to same threshold value.

If more than one object is present, labeling is done for different objects and area of each object is calculated. Pixels corresponding to largest area are region between lips. If only one object is present, area of that object is calculated. To detect yawn of a person area of region segmented after thresholding and also the largest area count obtained after labeling process is considered.

## 4 Results

Performance of proposed algorithm is tested for videos sequences in YawDD [10] dataset. By trial and error, a threshold of value 70 is chosen to segment darker region between lips. Fourteen video sequences in the dataset are used for computing the performance of the algorithm in terms of accuracy [7] and the same are tabulated in Table 1.

The accuracy is overall good for female participants (except for Video # 6) when compared to male participants. Average yawn detection accuracy for female participant is 81.7% while that for male participants is 62.6%. Average yawn detection rate for all the participants is 76% using our method.

**Table 1** Tabulation of results

| Video# | # of frames | # of face detected frames | # of eye detected frames | TP | FP | TN | FN | Accuracy |
|---|---|---|---|---|---|---|---|---|
| *Female participants* | | | | | | | | |
| 1 | 2741 | 2741 | 2463 | 401 | 177 | 1846 | 45 | 91.01 |
| 2 | 2180 | 2180 | 1109 | 107 | 652 | 345 | 5 | 40.76 |
| 3 | 3057 | 3057 | 2113 | 84 | 147 | 1750 | 130 | 86.88 |
| 4 | 1496 | 1496 | 856 | 54 | 12 | 788 | 2 | 98.36 |
| 5 | 2149 | 2149 | 1620 | 25 | 24 | 1497 | 73 | 94.01 |
| 6 | 1491 | 1491 | 792 | 162 | 485 | 117 | 27 | 35.28 |
| 7 | 3618 | 3618 | 3341 | 68 | 223 | 2954 | 96 | 90.46 |
| 10 | 1532 | 1532 | 1343 | 0 | 5 | 1327 | 10 | 98.89 |
| 13 | 2488 | 2488 | 321 | 0 | 1 | 320 | 0 | 99.69 |
| *Male participants* | | | | | | | | |
| 5 | 2395 | 2395 | 677 | 21 | 332 | 321 | 1 | 50.67 |
| 11 | 1760 | 1760 | 671 | 7 | 324 | 334 | 4 | 50.98 |
| 12 | 1957 | 1957 | 119 | 26 | 11 | 64 | 16 | 77.00 |
| 13 | 2190 | 2190 | 1094 | 3 | 399 | 681 | 11 | 62.52 |
| 15 | 2639 | 2639 | 1507 | 62 | 421 | 1021 | 2 | 71.92 |

## 5 Conclusions and Future Scope

Yawn detection technique using simple segmentation algorithm like thresholding is developed. Performance of algorithm is tested and evaluated for video sequences in YawDD dataset. Accuracy which is comparable to the published is obtained when tested on 14 of the 30 video sequences. This encourages us to look into possibility of using simple algorithms for real-time implementations. This algorithm works well when mouth is not occluded. Improvements can be done to the algorithm by detecting yawn of a person even when mouth region is occluded.

## References

1. Ji, Q., Zhu, Z. and Lan, P.: Real-time Nonintrusive Monitoring and Prediction of Driver Fatigue. Vehicular Technology, IEEE Transactions on, 53(4), pp. 1052–1068 (2004).
2. Oyini Mbouna, R., Kong, S.G. and Chun, M.G.: Visual Analysis of Eye State and Head pose for Driver Alertness Monitoring. Intelligent Transportation Systems, IEEE Transactions on, *14*(3), pp. 1462–1469 (2013).
3. Picot, A., Charbonnier, S. and Caplier, A.: On-line Detection of Drowsiness using Brain and Visual Information. Systems, Man and Cybernetics, Part A: Systems and Humans, IEEE Transactions on, *42*(3), pp. 764–775 (2012).
4. Reddy, K., Sikandar, A., Savant, P. and Choudhary, A.: Driver drowsiness monitoring based on eye map and mouth contour. IJSTR, 3(5), pp. 147–156 (2014).

5. Alioua, N., Amine, A. and Rziza, M.: Driver's Fatigue Detection Based on Yawning Extraction. International Journal of Vehicular Technology, (2014).
6. Lin, C.T., Wu, R.C., Liang, S.F., Chao, W.H., Chen, Y.J. and Jung, T.P.: EEG-based Drowsiness Estimation for Safety Driving using Independent Component Analysis. Circuits and Systems I: Regular Papers, IEEE Transactions on, *52*(12), pp. 2726–2738 (2005).
7. Omidyeganeh, M., Shirmohammadi, S., Abtahi, S., Khurshid, A., Farhan, M., Scharcanski, J., Hariri, B., Laroche, D. and Martel, L.: Yawning Detection Using Embedded Smart Cameras.
8. Rongben, W., Lie, G., Bingliang, T. and Lisheng, J. Monitoring mouth movement for driver fatigue or distraction with one camera. In Intelligent Transportation Systems, 2004. Proceedings. The 7th International IEEE Conference pp. 314–319. IEEE (2004).
9. Ibrahim, M.M., Soraghan, J.J., Petropoulakis, L. and Di Caterina, G.: Yawn analysis with mouth occlusion detection. Biomedical Signal Processing and Control, 18, pp. 360–369 (2015).
10. Abtahi, Shabnam, Mona Omidyeganeh, Shervin Shirmohammadi, and Behnoosh Hariri.: YawDD: A Yawning Detection Dataset. In: Proceedings of the 5th ACM Multimedia Systems Conference, pp. 24–28. (2014).

# A Survey to Structure of Directories in File System

**Linzhu Wu and Linpeng Huang**

**Abstract** Many applications require high throughput for operating small files, such as picture server. File systems often use tree structure to manage data blocks, which can improve the efficiency of adding, deleting, and searching files under a directory. This article gives an introduction to design of the directory of several popular file systems and shows the influence of structure on the performance by several experiments.

**Keywords** File system • Linux kernel • Directory design

## 1 Introduction

FS is an important part of the operating system (OS). In the OS, it is used to manage storage devices and store user data in an efficient way, which is totally transparent to users. Without a FS, we cannot retrieve meaningful data in storage devices and save data efficiently. Usually, resources in the OS can be described as files, which are components of FS. There are many kinds of files such as regular file, directory, character device file, symbolic link file, pipe file, etc. Data and metadata of these files are managed by FS, which can be obtained by the file system interfaces. As is known to us, regular file and directory are two common types of file objects while the others are special files for specific devices. A directory is a categorical structure which stores references to other files, including regular files, directories and special files. The structure of FS is hierarchical, which is similar to a tree with multiple children. The internal nodes of the tree are directories.

With the increase in the runtime of the OS, numbers of files stored in FS become larger. At the same time, size of the directory is larger. It is a key problem how to search, insert, and delete a file under a huge directory. Without an elegant design of

L. Wu (✉) · L. Huang
Department of Computer Science and Engineering, Shanghai Jiao Tong University, Shanghai, China
e-mail: wulinzhu1954@gmail.com



S.K. Bhatia et al. (eds.), *Advances in Computer and Computational Sciences*, Advances in Intelligent Systems and Computing 553,
DOI 10.1007/978-981-10-3770-2_9

directory structure, it is a great overhead in indexing a file, which will decrease the performance of the OS and the user program. Different FSs with different devices have different structures of the directory. For example, for a block device, a file can be located by sector number of its first sector. The directory could store a pair (*name, sector-num*) in its data area. In the process of searching, FS can get the sector number by comparing with names in all pairs. To accelerate searching, FS can sort all pairs by order for names in inserting and search in binary way. Modern OS, such as Linux, supports a cache for directory lookup [8], which is usually a map from the path name to real addresses of file in storage devices. However, FS is responsible to provide the location of given files when it is found in the cache. Therefore, efficiency of the directory is key to the performance of FS. Design of the directory is relevant to the FS. Tree, array, and linked list are three fundamental data structures in design. FS will choose a proper data structure which is compatible with other modules to obtain a good performance. Furthermore, storage devices also have impact on design of the directory. Different devices can provide different hardware primitives to software [4], which can be used to improve performance. However, most of the FSs are designed for general. They avoid using specific hardware primitives to improve versatility. In other words, FS that takes advantage of devices entirely may work badly on other devices.

In this paper, we will give an introduction to directory structures of several FS in Linux on design and implementation. These FSs, such as EXT4 [1], BTRFS [10], and F2FS [6], adopt different ideas of constructing directories. Some of the structures are good at searching, while some do well in adding and deletion. We also do experiment on the directory in different FS to show the impact of structure on performance. In Sect. 2, we present several concepts related to FS. In Sect. 3, we introduce directory structures of several well-known Linux FS. Finally, we show the results of experiments about the directory and give a conclusion.

## 2 Concepts

### 2.1 Classes of Data

Data in FS can be classified into three types according to their usage [2]:

- Raw data of file is from end users of FS. And FS does not know the meaning of it. But it is important to end users and FS is constructed to provide service to it and responsible to maintain the integrity and consistency of it.
- Metadata of file is information to describe a file. In modern FS, we often use Inode to represent a file, which is a data structure of metadata.
- Metadata of FS is information to describe the FS. The most important part is the super block of FS. Under the help of the super block, we have an idea of identity of FS, storage device managed by FS, space usage, the root directory of FS, etc.

The super block locates in the head of the storage device. The inode table, which records the relationship between inode number and its body, is next to the super block or a fixed position. The remaining area is used to save raw data and metadata of files. The metadata of files can be indexed by inode table, while the raw data can be found under the help of metadata of files.

## 2.2 *Inode Structure*

The inode is a data structure used to gather attributes of a file [12], including file modes, modification time, owner, etc. For example, Fig. 1 is the part of definition of inode structure of F2FS. As we can see, inode includes file mode(i_mode), ID of owner(i_uid, i_gid), size of file(i_size), reference count(i_links), time information (i_atime, i_ctime, i_mtime), name(i_namelen, i_name), and location message of its data(i_addr, i_nid).

## 2.3 *File Lookup*

How does FS finds a file when given a path p? For example, a user wants to read a file with name "/first/foo":

1. OS determine the type of FS of root directory "/" by internal information.
2. The inode number of "/" usually is known to FS and saved in super block. Then FS searches the inode table to get the location of the root inode.

```
struct f2fs_inode {
        __le16 i_mode;
        __le32 i_uid;
        __le32 i_gid;
        __le32 i_links;
        __le64 i_size;
        __le64 i_atime;
        __le64 i_ctime;
        __le64 i_mtime;
        __le32 i_flags;
        __le32 i_pino;
        __le32 i_namelen;
        __u8 i_name[F2FS_NAME_LEN];
        __le32 i_addr[ADDRS_PER_INODE];
        __le32 i_nid[5];
}
```

**Fig. 1** Structure of f2fs_inode

3. FS reads the root inode and gets the location of its data, which consists of pairs of file name and corresponding inode number. Then FS scans all the pairs by name "first" to get its inode number.
4. FS can get the address of "/first" by its inode number from step 3. Similar to the process of locating "/first", FS can retrieve data of "foo" under "/first".

In these steps, there are two kinds of data searching, inode table searching and directory pair *<name, inode-num>* searching. Actually, inode table is a huge array. Inode number is the index and the location of inode is the value of the array. That is, FS can get the location in one step. Therefore, the directory searching becomes a big deal in reading a file.

## 3 Directory Structure

In this section, we introduce directory structure of three widely used FS, FAT, PMFS, and F2FS. In design of FS, optimization for specific storage devices is a key point to improve performance. There are various storage devices in computing system. The most common devices are HDD (Hard Disk Drive), memory, and Flash. Features of these devices are totally different. FS work well in HDD may work badly in memory. FAT is designed for small and embedded computing system, whose storage device is often HDD. Therefore, one of key concepts in FAT is the cluster, which is corresponding to contiguous sectors in HDD. Furthermore, due to design of file allocation table in FAT, maximum size of disk that FAT supports is restricted. And the idea of the block chain works good in HDD, while it is not suitable for memory. Because HDD needs to move the magnetic head to the specific sector before it reads data while the processor can read multiple blocks at the same time. Under the help of MMU, FS maps the separated blocks as a contiguous array to accelerate speed. To make the discussion more credible, we choose three FS for three typical storage devices, FAT (Hard Disk Drive), PMFS (Memory), and F2FS (Flash Storage).

B-Tree is widely used to manage data blocks in HDD because it requires seldom blocks moving to maintain tree balance [10]. It is easy to read and write the contiguous blocks of B-Tree, which is a common operation in FS. Radix tree is often used to manage page caches in DRAM of the inode. FS can find dirty data pages and their addresses in storage device quickly via radix tree. But radix tree is not a good choice of managing data blocks in HDD because it requires updates in byte granularity. However, the minimum size of an update is the size of sector. The usage of array restricts the maximum size that FS supports. For example, FAT uses array to represent block usage of storage device, which limits the maximum size of the device. F2FS uses array to describe ID of inode and internal node block, which limits the maximum number of files to 232. Because array requires contiguous area and a fixed position for potential data, it is hard to resize the array with consistency and efficiency. One of the most common usages of array is used to construct the journal. Journal is a widely used method to provide consistency of FS. It is used in many state-of-art file FS, such as EXT3 [13], EXT4 and PMFS. The potential data will be written along journal before written to real data blocks. Therefore, FS can roll back and roll forward under the help of journals. The journal can be erased once data has been written and FS is in a consistent state. In design of directory structure, FS uses array to construct slots of entries. The size of data block of a FS is fixed, which is preferred by array data structure. Another usage of array is the so-called log-structured FS, such as LFS [11] and NILFS2 [5]. The idea of log-structured FS is append-write and copy-on-write. The storage device is divided into multiple segments. Size of each segment is fixed. The segment is divided into multiple blocks. FS writes data and metadata in block size. Linked list and tree are two important data structure in FS because of good extendibility. FAT uses the linked list to manage data blocks.

### *3.1 FAT*

FAT is a simple and robust system [15], which is widely used in Windows operating system. It is compatible with legacy system [14], such as embedded systems and old Windows 9.x. Therefore, its implementation is very simple but stable. As the first popular FS in Windows, its design has much influence on latter FS, such as allocation table, the cluster, and the block chain. Structure of the directory is as Fig. 2 shows. As we can see, data area of directory is divided into many slots. A slot consists of 5 types of message, including name, extension, attribute, time, and data address. Size of name is 8 bytes, if file name is too long to be saved in one slot, it is saved in the consecutive entries, as blue part shows in Fig. 1. The extension is the extension name of file. Attributes include file types and permission. Data blocks of the directory are constructed as a linked list, and each block consists of several directory entries. For Linux kernel supports 255 bytes file name at most, a long entry needs 64 slots. Therefore, if file name is long, space of data blocks will be wasted due to the meaningless messages (time, data address, etc.). As an old FS, FAT does not have Inode structures. File metadata is saved in the upper directory. We show how to add a new entry to a directory:

1. Calculate number of slots n that an entry need according to length of its name
2. Scan the data block list, and find n consecutive free entries. A free entry is marked by its member attribute
3. n consecutive free entries can be found in internal data blocks or by adding new data blocks
4. FAT fills in n entries with information of new file

Steps of deleting a file are similar to adding, but FAT finds entries by comparing file name instead of free entries. Then FAT marks the entries free. Thus, FAT needs to scan all entries under the directory when searching a file. Time overhead of

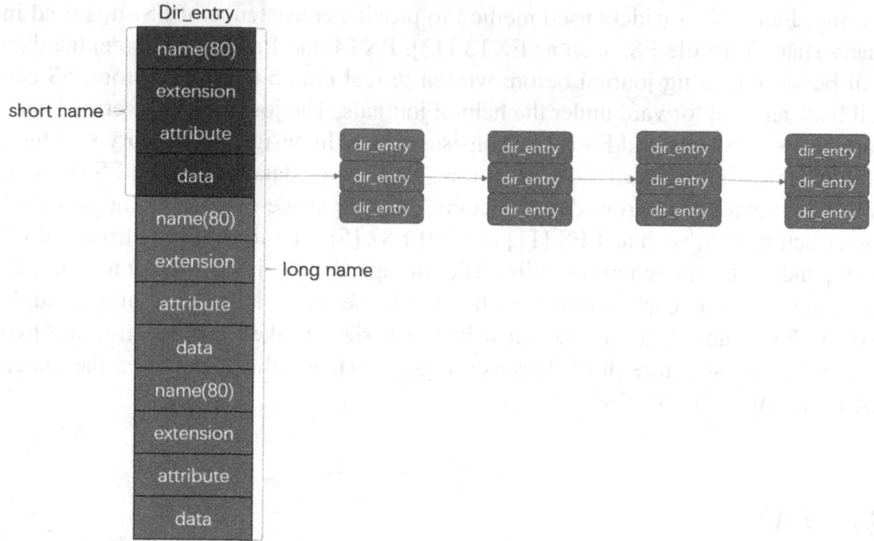


**Fig. 2** Structure of Dir_entry in FAT

searching, adding, and deleting is directly proportional to the number of entries in the directory, i.e., O(# of entries). When size of the directory increases, performance of FAT decreases rapidly. As design and implementation of FAT are simple, it is suitable for small system, which does not need complicated storage.

## *3.2 PMFS*

With the development of NVM, high performance gap between storage and processor become small. To make use of features of NVM, such as byte-addressable, high throughput, and durability, Intel has developed a new Linux FS, PMFS [3]. It is a light-weight system with high performance both in read and write on NVM. Furthermore, it adopts the idea of execute-in-place to accelerate data copying, which also improves directory operations. It is too complicated to explain design of PMFS in short. Now we just show how directory operations work in PMFS. The structure of directory is as in Fig. 3. A file under the directory is described in a directory entry, which consists of 5 attributes, inode number, length of entry, length of name, file types, and buffer of name. Usage of these entries is easy to understand except length of entry and length of name. Why does entry have 2 attributes about length? Actually, length of name is the exact value of length of name, while length of entry is size of area that an entry consumes. For example, the yellow entry in

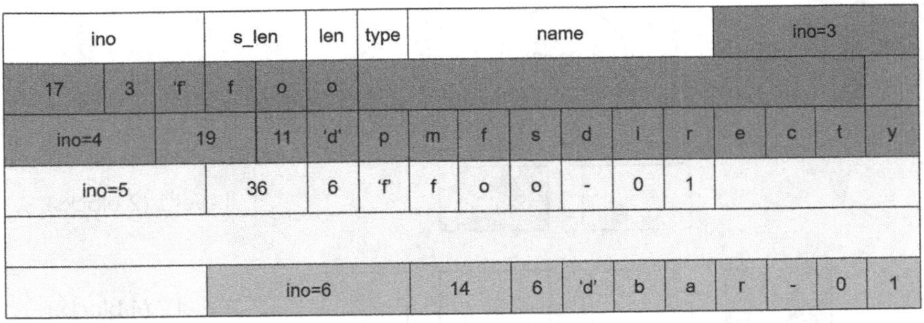

| ino | s_len | len | type | name | | | | | | | ino=3 | | | |
|---|---|---|---|---|---|---|---|---|---|---|---|---|---|---|
| 17 | 3 | 'f' | f | o | o | | | | | | | | | |
| ino=4 | 19 | 11 | 'd' | p | m | f | s | d | i | r | e | c | t | y |
| ino=5 | 36 | 6 | 'f' | f | o | o | - | 0 | 1 | | | | | |
| | | | | | | | | | | | | | | |
| | ino=6 | 14 | 6 | 'd' | b | a | r | - | 0 | 1 | | | | |

**Fig. 3** On disk layout of directory in PMFS

figure, length of name "foo-01" is 6. But it also controls 26 more bytes behind name, which can be added a new proper entry later. Thus, the process to add a new entry is:

1. read an entry in data block of directory
2. length of name is name_len, and length of entry is entry_len. If entry_len - name_len is larger than size of new entry, stop searching; otherwise, skip entry_len bytes and repeat step 1 and 2
3. If PMFS gets a proper slot in last step, sets value of entry length to be its name length. Length of new entry is entry_len(last entry) - name_len(last entry)

## *3.3 F2FS*

F2FS is a flash-friendly file system developed by Samsung, which performs well on modern flash storage device, such as mobile phone and wearables. Key points of designing F2FS include append-only logging, roll back recovery, roll forward recovery, multiple logging, etc. The biggest difference of F2FS is that size of data blocks and metadata blocks are the same, exactly 4096 bytes, which is required by so-called log-structured writing. Therefore, a data block of the directory is able to contain many entries. F2FS uses technologies of hashing and bitmap to manage entries, which will accelerate searching speed. Inode and other internal node block in F2FS are represented by NID, a 32-bit unsigned integer. F2FS also cache NID and real address of node block by a radix tree in DRAM. As Fig. 4 shows, data blocks of a directory can be divided into several levels. Size of level n is at most 2n. When F2FS looks up file in a directory, it first calculates the hash value of its file name. Then it scans data block from level 1 to maximum level. The data block it scans in each level is the remainder of taking hash value from size. For example, suppose hash value is 9, F2FS will scan block 1(=9 mod 2) in level 1, block 1 in

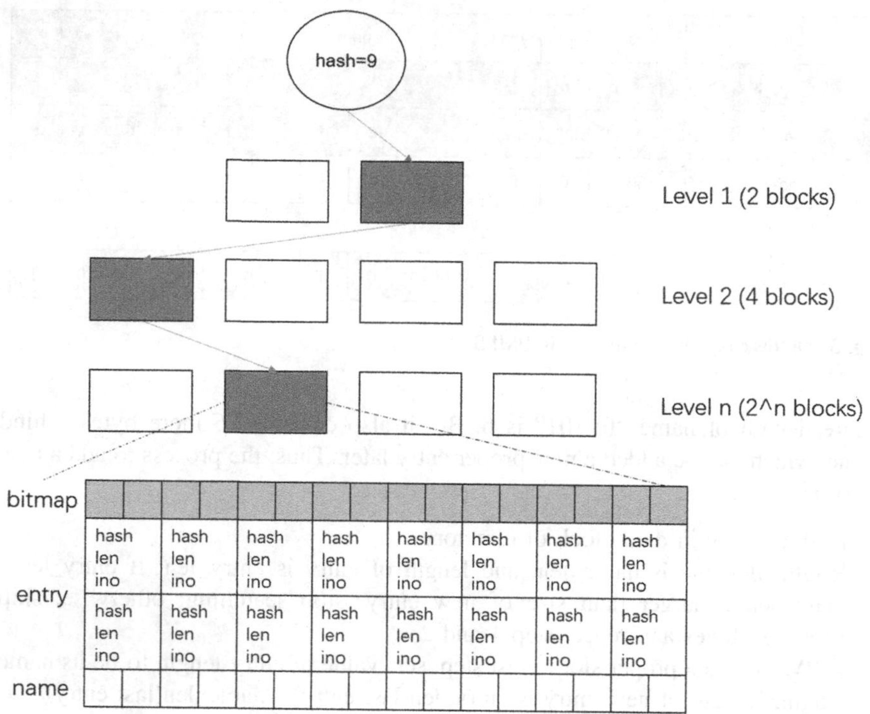


**Fig. 4** Structure of directory in F2FS

level 2(=9 mod 4), block k(=9 mod 2n) in level n until an entry is found. In each level, it scans one block, resulting in an O(log(# of entries)) complexity, which is better than both PMFS and FAT. The process of scanning the block in each level is similar to FAT. A data block consists of three parts, Bitmap Area, Entry Area, and Name Area. But time of searching in a block is less than FAT because of the use of bitmap. The entry in EntryArea is free if its value in bitmap is 0. F2FS will find free slots in bitmap instead of EntryArea. And size of slot in Name Area is 8 bytes. Therefore, directory entry is represented by three isolated slots. But one slot can represent name length of 8 bytes at most. If name length exceeds 8 bytes, it needs (length + 8 − 1)/8 consecutive slots, which can be calculated by name length. When F2FS searches a file, it first compares hash value of given file with value in Entry Area. If equal, compares name; otherwise, skips to next entry. Thus, F2FS also does an optimization in searching.

# 4 Experiments

## 4.1 Experimental Setup

PMFS is a nonvolatile memory system, which runs on memory. But F2FS and FAT run on block device. For speed of memory is much faster than HDD, results from experiments are not credible if PMFS runs on memory but the others run on block device. To address the problem, we do experiment on F2FS and FAT under the help of Ram disk [9]. The Ram disk uses system memory from buddy system to provide block device interfaces to users. Size of Ram disk grows up dynamically and maximum size can be setup in building kernel by default. We run the benchmark on a personal computer with Intel Core processor platform. Each processor runs at 3.60 GHz, has 8 cores. The computer supports up to 20G DDR3 DRAM and 1 TB hard disk drive. And the host operating system is Ubuntu 14.04 LTS 64 bits. We rebuild kernel v3.11 of Ubuntu, setup 2 Ramdisk devices with size 4 GB. Then we rebuild modules of FAT, F2FS, and PMFS. Versions of these FS are 3.0.26, 1.2.0, and 1.0.0 respectively. Filebench is a popular benchmark for storage system [7]. It provides numbers of workloads to test performance of systems. In my experiment, we use several classic workloads, such as listdirs, makedirs, and removedirs.

## 4.2 Add Directory Entries

We first evaluate the performance of adding entries. In Linux, we use command "touch" to create numbers of empty files under a directory. FAT will add a new entry to data blocks of the directory and writes down auxiliary information. PMFS

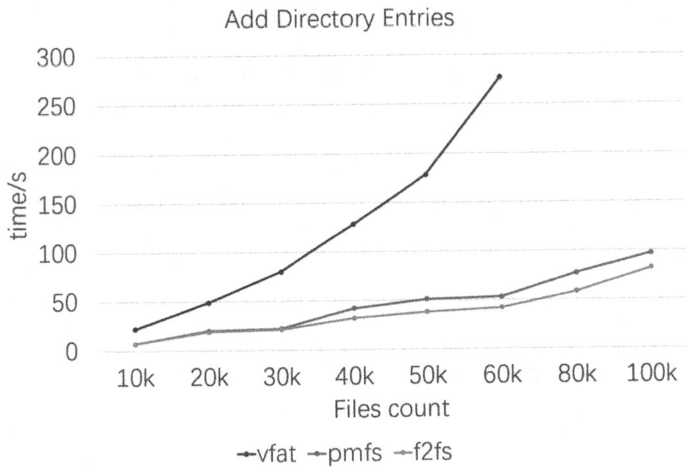


**Fig. 5** Performance of adding directory entries

and F2FS need to allocate space for new inode in addition. Figure 5 is result of experiment. F2FS runs faster than PMFS and FAT. However, performance of PMFS is close to F2FS. Time overhead of them grows steadily as size of entries increases. But time overhead of FAT grows faster than the others. To find an empty slot in data area, FAT has to scan all entries, which slows down speed of adding. Though PMFS also adopts idea of liner searching, it uses execute-in-place to decrease writing overhead. Another reason why performance of F2FS is not significant is that F2FS adopts COW update and provides consistency of both data and metadata. Writing to same inode will result in several writing due to COW. Furthermore, size of inode in F2FS is 4096 bytes, while size of inode in PMFS is 96 bytes. That is, adding same number of entries in F2FS results in more writing than PMFS and FAT.

## 4.3 Lookup Directory Entries

Next we evaluate the performance of searching entries. To show the difference between performance, we evaluate the time to search numbers of entries under directory of 60000 entries. To reduce impact from other issues, benchmark just finds whether specific entry exists or not, which is done by "lookup" system call in VFS. In lookup system call, FS will scan data area of specific directories and return a boolean value as the result. Figure 6 shows the result of lookup. To my surprise, PMFS, FAT, and F2FS perform closely in searching. Time overhead grows steadily as size of queries increase. And Time to search is much less than time to add entries. Why difference of performance is not significant? We think one of the reasons is the existence of DCache and read ahead in VFS of Linux. As we know, DCache(Linux Directory Entry Cache) is a key part of the kernel's filename lookup mechanism,

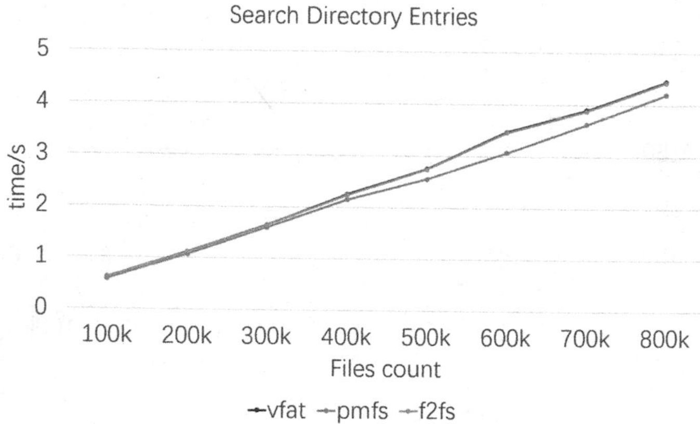


**Fig. 6** Performance of lookup directory entries

which accelerate file lookup. When a filename is lookup before, its inode structure and name are cached in VFS. VFS will search.

DCache for file first [8], so if a file is lookup again, it will be found quickly. Another point is the existence of read ahead. That is, if a data block is read from block device, its neighbor will be read according to space locality. For example of searching file “1”, “1” locates in data block 2 of directory. FS will read block 2 and 3 into memory. There are many entries in block 2 and 3. Thus these entries will be cached in DCache. When benchmark lookup entry close to “1”, it is just cached in DCache and time to read data block in HDD is negligible.

## 4.4 Delete Directory Entries

Next, we evaluate performance of deleting entries. Deleting entries is different from adding entries. Sometimes FS will not really delete entries. It is decided by design of file system to discard and erase entries. For example, F2FS need to set bitmap while PMFS need to set length of entry. In other words, time to allocate and setup new inode is none. Figure 7 shows the result of deleting directory entries. Similar to adding entries, FAT performs the worst among three FSs. But the difference between them is small. Time to delete entries in FAT is appropriately 1.2x larger than PMFS and F2FS. Difference between F2FS and PMFS is negligible. Though design of directory structure of F2FS is better than PMFS, it does a tradeoff in other aspects, which slows down speed of deletion. For example, F2FS make use of multiple logging of Flash storage, which allows F2FS writing and reading more than 1 block at the same time. Metadata area and data area are separated in F2FS. Time to read metadata and data is negligible in Flash storage due to multiple logging. However, reading data and metadata in hard disk simultaneously is not

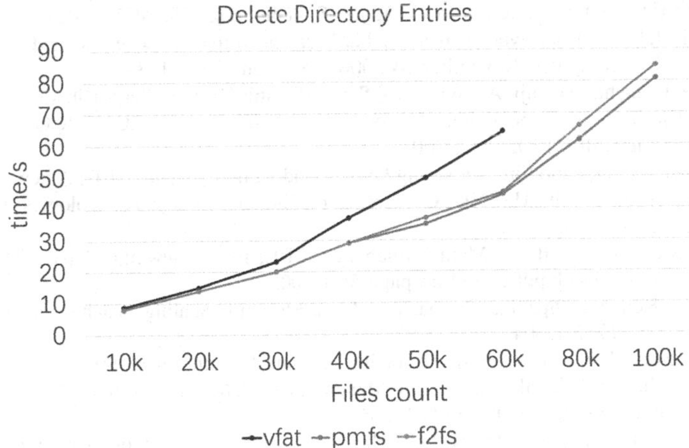


**Fig. 7** Performance of deleting directory entries

allowed because there is only one head for HDD. Optimization done for Flash storage in F2FS is in vain. However, PMFS is designed for memory and makes use of advantage of memory such as byte-addressability and execute-in-place. In this way, advantage of design in F2FS is lost compared with PMFS.

## 5 Conclusion

This paper gives an introduction to directory structure in Linux file system, including the normal layout, structure of inode, and super block. Furthermore, we show directory structure of FAT, PMFS, and F2FS in details. Some experiments have been done to confirm the proposed analysis. Results show that design of the directory has huge impact on efficiency. FS designer has done a trade-off between performance and other features, such as consistency, security, robustness, and light weight.

**Acknowledgements** The work described in this paper is sponsored by the Chinese 863 Program under Grant No. 2015AA015303 and the National Natural Science Foundation of China under Grant No. 61472241.

## References

1. Mingming Cao, Suparna Bhattacharya, and Ted Ts'o. Ext4: The next generation of ext2/3 filesystem. In LSF, 2007.
2. Vijay Chidambaram, Tushar Sharma, Andrea C Arpaci-Dusseau, and Remzi H Arpaci-Dusseau. Consistency without ordering. In FAST, page 9, 2012.
3. Subramanya R Dulloor, Sanjay Kumar, Anil Keshavamurthy, Philip Lantz, Dheeraj Reddy, Rajesh Sankaran, and Jeff Jackson. System software for persistent memory. In Proceedings of the Ninth European Conference on Computer Systems, page 15. ACM, 2014.
4. Steven R Jahnke. Data synchronization hardware primitive in an embedded symmetrical multiprocessor computer, November 14 2006. US Patent 7,137,118.
5. Ryusuke Konishi, Yoshiji Amagai, KojiSato, Hisashi Hifumi, Seiji Kihara, and Satoshi Moriai. The linux implementation of a log-structured file system. ACM SIGOPS Operating Systems Review, 40(3):102–107, 2006.
6. Changman Lee, Dongho Sim, Jooyoung Hwang, and Sangyeun Cho. F2 fs: A new file system for flash storage. In 13th USENIX Conference on File and Storage Technologies (FAST 15), pages 273–286, 2015.
7. Richard McDougall and Jim Mauro. Filebench. URL: http://www.nfsv4bat.org/Documents/nasconf/2004/filebench.pdf (Cited on page 56.), 2005.
8. Paul E McKenney, Dipankar Sarma, and Maneesh Soni. Scaling dcache with rcu. Linux Journal, 2004(117):3, 2004.
9. Mark Nielsen. How to use a ramdisk for linux. Linux Gazette (44), 1999.
10. Ohad Rodeh, Josef Bacik, and Chris Mason. Btrfs: The linux b-tree filesystem. ACM Transactions on Storage (TOS), 9(3):9, 2013.
11. Mendel Rosenblum and John K Ousterhout. The design and implementation of a log-structured file system. ACM Transactions on Computer Systems (TOCS), 10(1):26–52, 1992.

12. Priya Sehgal, Sourav Basu, Kiran Srinivasan, and Kaladhar Voruganti. An empirical study of file systems on nvm. In Mass Storage Systems and Technologies (MSST), 2015 31st Symposium on, pages 1–14. IEEE, 2015.
13. Stephen Tweedie. Ext3, journaling filesystem. In Ottawa Linux Symposium, pages 24–29, 2000.
14. Werner Vogels. File system usage in windows nt 4.0. In ACM SIGOPS Operating Systems Review, volume 33, pages 93–109. ACM, 1999.
15. Zhang Mingliang Zhang Zongjie. Analysis of fat32 file system. Computer & Digital Engineering, 1:014, 2005.

# A User-Mode Scheduling Mechanism for ARINC653 Partitioning in seL4

**Qiao Kang, Cangzhou Yuan, Xin Wei, Yanhua Gao and Lei Wang**

**Abstract** seL4 is formally verified for its functional correctness and provides a trusted code base for ARINC 653 partitioning operating systems. ARINC 653 needs a two-level scheduler to enforce temporal isolation between partitions. We cannot modify the scheduler provided by seL4 to adapt ARINC 653, which may invalidate the formal correctness of seL4. Thus, we propose a user-mode scheduling mechanism, where several user threads serve as the partition scheduler and process schedulers. The execution trace result shows that the temporal partitioning can be enforced by this mechanism. We also elaborate the scheduling overheads.

**Keywords** ARINC 653 · sel4 · Partitioning · Hierarchical scheduling

## 1 Introduction

The ARINC 653 standard [1] provides a powerful partitioning mechanism to protect applications in Integrated Modular Avionics systems [2]. Errors within an ARINC 653 partition cannot be propagated to another. However, mistakes can also happen inside the kernel. Once kernel crashes, everything crashes.

Q. Kang · C. Yuan
School of Software, Beihang University, Beijing 100191, China
e-mail: kangqiao@buaa.edu.cn

C. Yuan
e-mail: yuancz@buaa.edu.cn

X. Wei · Y. Gao
Beijing Institute of Control and Electronic Technology, Beijing 100038, China
e-mail: yaojianni@163.com

Y. Gao
e-mail: gloveg@sina.com

L. Wang (✉)
School of Computer Science and Engineering, Beihang University, Beijing 100191, China
e-mail: wanglei@buaa.edu.cn



S.K. Bhatia et al. (eds.), *Advances in Computer and Computational Sciences*,
Advances in Intelligent Systems and Computing 553,
DOI 10.1007/978-981-10-3770-2_10

Existing ARINC 653 operating systems are usually built on previous kernels, such as the L4-based PikeOS [3], the Linux-based [4] and the Xen-based [5]. But neither L4, Linux, nor Xen hypervisor can guarantee no mistakes within kernel codes.

Now we have seL4 [6], a 3rd generation microkernel. seL4 is first-ever formally proved that the C and compiled binary code both act uniformly with the high level specification [7, 8]. Other work verified different security aspects of seL4 such as integrity [9] and confidentiality [10]. If possible, an ARINC 653 operating system established in seL4 will own a formally verified code base of all time.

ARINC 653 requires both spatial and temporal isolation between partitions. We now focus on the temporal aspect. To build a seL4-based ARINC 653 system, a two-level scheduling mechanism is needed to enforce temporal partitioning. However, we cannot implement an ARINC 653 scheduler in kernel-mode which will invalidate the proved functional correctness of seL4 (will be discussed in Sect. 2). An earlier work by Åsberg [11] tried to establish a user-mode partition scheduler in seL4. But his scheduler is not a hierarchical one, and does not satisfy ARINC 653 standard.

PikeOS [3] implements a kernel-level two-level scheduler based on L4 microkernel. However, it cannot be ported to seL4 because no modification of seL4 is permitted to keep its formal correctness.

Our goal is to implement a hierarchical scheduling mechanism in seL4, enforcing ARINC 653 temporal isolation between partitions. The detailed requirements will be discussed in Sect. 2.1.

ARINC 653 standard also requires limitation of memory and communication authorities of each partition. We just focus on the timing aspect and leave other aspects of ARINC 653 requirements as our future work. Benefiting from the expandability of seL4, memory and communication requirements can be easily handled by adding one or more server threads inside or outside the partitions.

**Contribution** We propose a user-mode approach in seL4 to enforce ARINC 653 scheduling mechanism. Test results show that the execution of partitions is isolated. We propose a compulsive method to switch partitions for partition scheduling and use a kernel-driven method, not relying on any timer hardware, to resume the scheduler periodically for process scheduling.

Note that we do not implement specific scheduling algorithms, and leave the flexibility for users to add different kinds of scheduling policies. Our method can be imitated by other work when designing a user-level scheduler in seL4-like, capability-based microkernels.

**Outline** Section 2 gives a brief description of ARINC 653 scheduling requirements and the kernel scheduler of seL4, explaining the reason that we choose the user-mode approach. Section 3 gives the design of the user-mode scheduling mechanism. The performance is evaluated in Sect. 4 and some conclusions are made in Sect. 5.

# 2 Problem Analysis

## 2.1 ARINC 653 Scheduling Requirements

Two kinds of objects are scheduled in ARINC 653 scheduling: partitions and processes. One process belongs to exactly one partition. Thus, a partition can be viewed as a gather of processes.

Scheduling in ARINC 653 is hierarchical, specifically two-level. The first level is the scheduling of partitions while the second level is the process scheduling within each partition. Partition scheduling policy is a simple one, i.e., fixed time window scheduling. In each partition, processes are scheduled by their priorities.

The main goal of the ARINC 653 scheduler is to guarantee that no process can run beyond the time window of its partition.

## 2.2 Why not Kernel-Mode

seL4 itself implements a two-level scheduler inside the kernel [12]. In a global view, domains are scheduled by fixed time windows. A thread belongs to exactly one domain, and will run only if its domain is active. Within a domain, the kernel scheduler implements a priority-based policy. When more than one thread with the same priority in the active domain are ready, they are scheduled with round robin scheduling. Corresponding to 256 priorities, there are up to 256 ready lists for each domain.

Apparently, domains can be easily mapped into ARINC 653 partitions. However, when building a full ARINC 653 operating system, some standard APIs (called APEX, defined in [1]) must be provided to partitions, such as GET_PARTITION_ STATUS. Such APIs should be dispatched to a server thread by IPC. If the server and the caller belong to different domains, the API call cannot be answered until the server's domain is active, which means a huge delay. Thus, we abandon seL4 domains and adapt the user-mode approach.

# 3 User-Mode Design for ARINC 653 Temporal Partitioning in SeL4

In our design, only one seL4 domain exists and everything belongs to it. ARINC 653 processes are mapped into seL4 threads. We move the scheduling work to several user threads. A Global Scheduler (GS) is used to schedule partitions and each partition needs a Local Scheduler (LS) to schedule other in-partition threads. The affiliation of partitions and processes is managed by the GS.

### 3.1 Preliminaries

seL4 enforces access control by capabilities. A capability is an unforgeable token that references a specific kernel object (such as a thread control block, TCB) and carries access rights that control what method may be invoked. A capability resides in a slot of a thread's CSpace (capability space) and can be granted from a CSpace to another by its owner.

seL4 provides three APIs to help implement user-mode schedulers. A pair of system calls, Resume and Suspend, can be used to setup the status of one thread into ready or blocked. A thread calling Yieldgives up the rest of its time slice and will be appended to the end of its priority list. Note that thread A calling Suspend or Resume on thread B needs a capability to the TCB kernel object (CAP_TCB) of thread B. The Yieldcall needs no capabilities.

seL4 starts the initial thread after the initialization of kernel. The initial thread runs in the highest priority and can create more threads, owning their CAP_TCBs.

### 3.2 Initialization

Consider an ARINC 653 system configuration with 2 partitions, where partition A needs two processes to work and partition B needs one. To meet this requirement, 6 threads in seL4 are needed: the initial thread, a GS, two LSs and 3 other threads, which are shown in Fig. 1a.

The initial thread is responsible for creating all other threads. Since the initial thread runs in the highest priority, the initial process is sequential and cannot be preempted by other threads. When creating threads, the initial thread (1) gives the

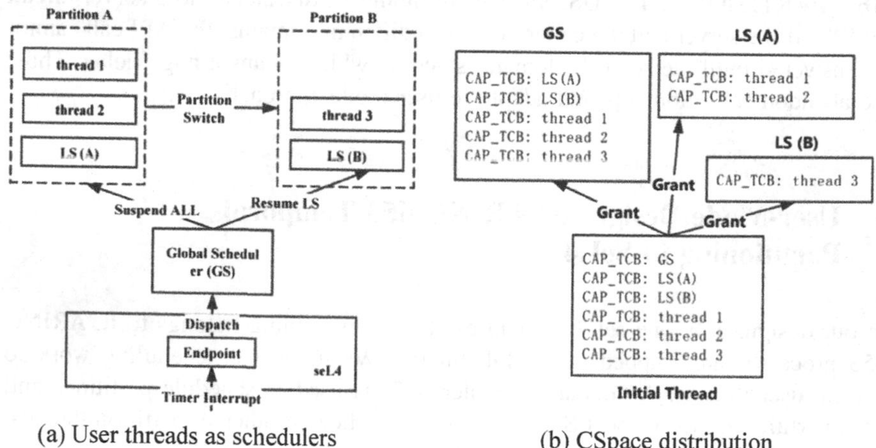


(a) User threads as schedulers (b) CSpace distribution

**Fig. 1** The user-mode solution for two partitions

GS the second highest priority, and (2) gives the LSs and other user threads the same middle priorities.

To use *Suspend* and *Resume*calls, the GS needs CAP_TCBs of all other threads except the initial thread, and LSs need CAP_TCBs of user threads within the partition.

These needed CAP_TCBs are owned by the initial thread and should be granted from the CSpace of the initial thread to the CSpace of the LSs and the GS. The correct final status after the capability transmission is shown in Fig. 1b. Once finishing initialization, the initial thread will stop running and transfer the control to the GS.

## 3.3 Partition Scheduling

The GS registers itself as the handler thread of the timer device via an endpoint (EP). Then the GS can be resumed periodically by the timer device. Since the GS runs in the second highest priority(except the initial thread) and is resumed by timer interrupt periodically, there is no ultimate difference between the user-mode GS and a kernel implementation. That means the GS is qualified a real-time partition scheduler.

The main body of the GS is designed as an infinite loop, which is shown in Fig. 2. Each time the GS runs, it (1) suspends all threads which belong to the previous partition, and (2) resumes the LS of the next partition. The action of the GS to switch partition from A to B is also shown in Fig. 2.

As ARINC 653 standard describes, partition switching must be compulsive and do not rely on the correct execution of partitions. Hence, the GS in our implementation suspends a partition by suspending all threads in it, including the LS.

```
Require: Variables:currPart, ResumeTime.
Initialise partition schedule from ARINC 653 XML file
Set currPart to the first partition in schedule
Resume currPart.LS to run
Set ResumeTime to ResumeTime.timeSlice
Configure timer with ResumeTime
loop
    Suspend currPart.allThreads
    Set currPart to currPart.next
    Resume currPart.LS to run
    Set ResumeTime to currPart.timeSlice
    Configure timer with ResumeTime
end loop
```

**Fig. 2** Algorithm design of the GS

### 3.4 Process Scheduling

The LS should guarantee that only one user thread is active in the current partition and simply suspends the last one and resumes a new one for thread switching. The main body of the LS is designed as an infinite loop, just like the LS. At the end of the loop, the LS gives up its remnant slice by calling *Yield*system call. As described previously, the *Yield*call appends the LS to the priority list (behind the current active user thread) instead of hanging up the LS.

Since only one thread and the LS are active, they are scheduled alternately by seL4 kernel scheduler, which is shown in Fig. 1a. Thus, our LS is triggered periodically just like being resumed by a *virtual* timer periodically. The interval of LS is only decided by user threads time slices.

## 4 Evaluation

We use QEMU on Linux Ubuntu 14.04 to test our schedulers. The QEMU is configured to emulate the Intel core2 CPU running in 2.4 GHz.

### 4.1 Execution Trace

We construct the real system with two partitions, which is shown in Fig. 1, and configure the LSs to be resumed every 5 ms. The length of time windows for partition A and B are both configured to 50 ms.

We trace the execution of the 6 threads (GS, two LSs and three user threads) shown in Fig. 1a. In partition A, the LS simply schedules thread 1 and thread 2 with round robin of their 5 ms time slices. The LS of partition B always schedules thread 3. Part of the trace results are shown in Fig. 3 which indicate no interference between

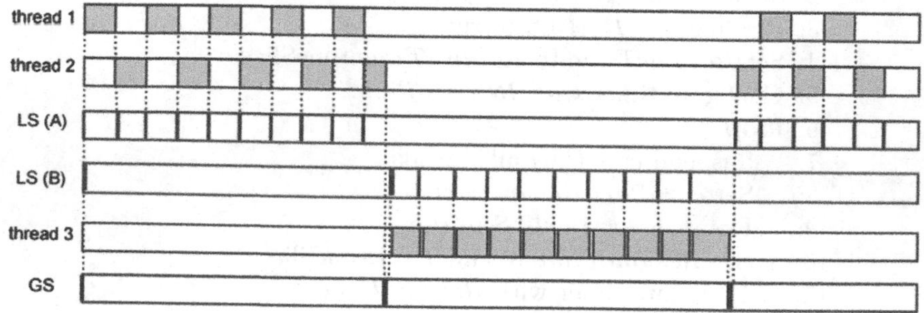


**Fig. 3** Execution trace with two partitions

partitions. For simplicity, the execution of the GS which does not switch partitions is omitted.

## 4.2 Overhead Analysis

Firstly, we focus on the overhead of partition scheduling. When a switch from partition A to B happens, the overhead of the GS is mainly influenced by the number of threads in partition A. The results are given in Fig. 4, which shows that the GS has an O (n) time complexity.

As a comparison, Åsberg uses *Wait* and *Send* IPC primitives in seL4 to switch threads, which need a constant overhead of 213 μs (dotted line in Fig. 4).

For the LS, we implement a simple priority-based algorithm in partition A of Fig. 1a. The overhead of the LS (A) is shown in Table 1, and is compared with an ARINC 653 scheduler design in Linux by Sanghyun Han [4]. Benefiting from the low overhead of context switching in seL4, the process switching overhead in our scheduler is slightly better than Han's.

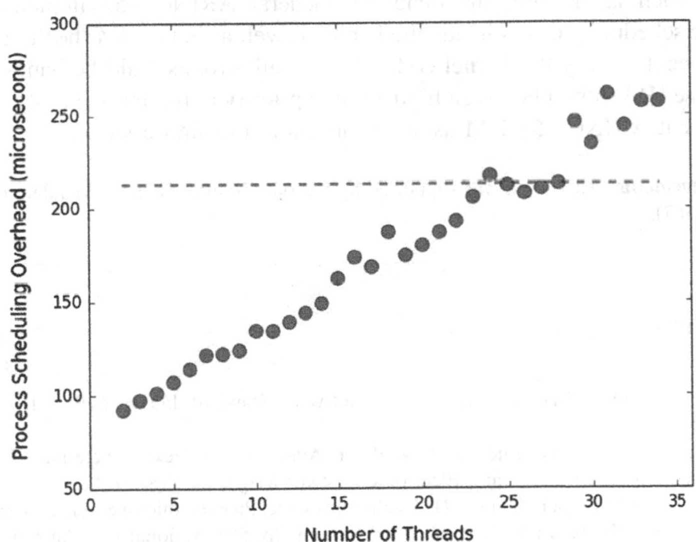


**Fig. 4** Partition scheduling overhead

**Table 1** Test results of process scheduling overheads (μs)

| ARINC 653 Process Schedulers | Platform | Max | Min | Ave |
|---|---|---|---|---|
| User-mode in seL4 | Intel core2 2.4 GHz | 40.0 | 33.9 | 45.1 |
| Kernel-mode in Linux [4] | Intel Mobile 1.4 GHz | 40.4 | 37.2 | 51.1 |
| User-mode in Linux [4] | Intel Mobile 1.4 GHz | 42.5 | 35.9 | 51.4 |

## 5 Conclusion and Future Work

The user-mode approach to establish an ARINC 653 hierarchical scheduling mechanism in seL4 is feasible. Without any modification of the seL4 microkernel, our pure user-mode scheduler can maintain the functional correctness of the kernel. We use the kernel mechanism to execute the user-mode process scheduling and achieve an acceptable overhead.

However, the scheduling of partitions is relatively time-consuming, since the global scheduler actually needs to operate all threads in the previous partition in order to enforce temporal partitioning. A possible solution is to support to *Suspend* a group of threads using one-time system call in seL4, which will be given in the future work.

The missing part of about ARINC 653 scheduling in our implementation is error handling, such as missing the deadline. General ARINC 653 implementations handle the scheduling errors inside the kernel as well as in the HM (health monitor). For we cannot modify the kernel code of seL4, all errors should be handled in the HM and the HM should be given high enough priority to run momentarily. We plan to implement ARINC 653 HM as seL4 threads in our future work.

**Acknowledgements** This work was supported by National Natural Science Foundation of China (No. 61272167).

## References

1. Radio, A. (2008). Avionics Application Software Standard Interface Part 1 - Required Services.
2. Morgan, M. J. (1991). Integrated Modular Avionics for Next-Generation Commercial Airplanes. IEEE Aerospace and Electronic Systems Magazine, 6(8), 9–12.
3. Kaiser, R., and Wagner, S. (2007). Evolution of the PikeOS microkernel. In International Workshop on Microkernels for Embedded Systems (p. 50). National ICT Australia.
4. Han, S., and Jin, H. (2012). Kernel-level ARINC 653 partitioning for Linux. In Proceedings of the 27th Annual ACM Symposium on Applied Computing – SAC12, 1632–1637.
5. VanderLeest, S. H. (2010). ARINC 653 hypervisor. AIAA/IEEE Digital Avionics Systems Conference, 1–20.
6. Klein, G., Elphinstone, K., Heiser, G., Andronick, J., Cock, D., Derrin, P., Winwood, S. (2009). seL4: Formal verification of an OS kernel. In Proceedings of the ACM SIGOPS 22nd Symposium on Operating System Principles, 207–220. ACM.

7. Sewell, T. A. L., Myreen, M. O., & Klein, G. (2013). Translation Validation for a Verified OS Kernel. In Proceedings of the 34th ACM SIGPLAN Conference on Programming Language Design and Implementation (pp. 471–482).
8. Boyton, A., Andronick, J., Bannister, C., Fernandez, M., Gao, X., Greenaway, D., … Sewell, T. (2013). Formally verified system initialisation. Lecture Notes in Computer Science, 8144 LNCS, 70–85.
9. Sewell, T., Winwood, S., Gammie, P., & Murray, T. (2011). seL4 Enforces Integrity. In 2nd ITP (pp. 325–340). LNCS.
10. Murray, T., Matichuk, D., Brassil, M., Gammie, P., Bourke, T., Seefried, S., … Klein, G. (2013). seL4: From general purpose to a proof of information flow enforcement. Proceedings - IEEE Symposium on Security and Privacy, 415–429.
11. Åsberg, M., and Nolte, T. (2013). Towards a User-Mode Approach to Partitioned Scheduling in the seL4 Microkernel. 5th International Workshop on Compositional Theory and Technology for Real-Time Embedded Systems, 15–22.
12. Lyons, A., and Heiser, G. (2014). Mixed-Criticality Support in a High-Assurance, General-Purpose Microkernel. Workshop on Mixed Criticality Systems, 9–14.

# Vectorizable Design and Implementation of Matrix Multiplication on Vector Processor

**Junyang Zhang, Yang Guo and Xiao Hu**

**Abstract** Matrix-vector multiplication is one of the core computing of many algorithms calculation in scientific computing, the vectorization algorithm mapping is a difficult problem to vector processors. In this study, based on the background of BP algorithm for deep learning application, on the basis of in-depth analysis of the BP algorithm, according to the characteristics of vector processor architecture, we proposed an efficient vectorization method of matrix-vector multiplication. The L1D configured into SRAM mode, with double buffer "ping-pong" way to smooth data transmission of multistage storage structure, makes the calculation of the kernel and the DMA data moving overlap, let the kernel run at a peak speed, so as to achieve the best calculation efficiency. Through the way of transpose matrix transmission with DMA to avoid the inefficient access to column of matrix and summation reduction of floating-point calculation between the VPEs, Obtain the optimal kernel computing performance. Experimental result on MATRIX2 shows that the single-core performance of presented double precision matrix multiplication achieves 94.45 GFLOPS, and the efficiency of kernel computation achieves 99.39%.

**Keywords** Matrix-vector multiplication · Vector processor · BP algorithm · Vectorization

J. Zhang (✉) · Y. Guo · X. Hu
College of Computer, National University of Defense Technology,
Changsha 410073, China
e-mail: zhangjunyang11@nudt.edu.cn

Y. Guo
e-mail: guoyang@nudt.edu.cn

X. Hu
e-mail: xiaohu@nudt.edu.cn



S.K. Bhatia et al. (eds.), *Advances in Computer and Computational Sciences*,
Advances in Intelligent Systems and Computing 553,
DOI 10.1007/978-981-10-3770-2_11

## 1 Introduction

With digital image processing, high-definition video, radar signal processing, large dense linear equations solving and deep learning technology which is in today's popular compute-intensive applications demands increasing for high-performance computing, the architecture of microprocessor has the obvious change, some new type of architecture appeared, and vector processor architecture is one kind of new architectures [1]. Generally speaking, it includes scalar and vector processing parts, scalar processing unit is responsible for computing of scalar task and flow control, vector processing component is responsible for vector calculation, it includes a number of vector processing units, each processing unit contains rich arithmetic units, powerful computing capacity, can greatly improve the computing performance of the system. Meanwhile, it also puts forward a new challenge for software development, how to fully develop all levels of parallelism and all kinds of applications efficient vectorization is currently facing major difficulties, according to the architecture characteristics of more processing units, multifunctional components of vector processors [2].

Scientific computing program has a high requirement on the computing speed, many scientific computing program need to invoke various calculation library which is related to matrix and vector. Therefore, exploiting high performance of this kind function library is very important. Basic Linear Algebra Subprograms (BLAS) is the kernel math library of all kinds of necessary scientific calculation [3], widely used in electronic engineering, computer science, physics, and other science and engineering calculation. The optimization of BLAS has always been a hot research topic at home and abroad according to different architecture. Goto adopted manual assembly optimization to realize the efficient GotoBLAS library [4, 5] and adaptive optimization technique to accomplish the ATLAS library [6] for different architectures; Intel realized the optimized basic math library MKL [7] for their own CPU; Marker studied the optimization of matrix multiplication implementation on multithreading architecture platform; Volkov studied the optimization of BLAS library implementation on GPUs; Zhang Xianyi developed high-performance OpenBLAS library [8] based on the multi-core processors of loongson. The matrix-vector multiplication is the most frequently invoked function in BLAS libraries.

Today, the most popular technology in the field of artificial intelligence is deep learning which have lots of matrix-vector operations, especially in the full connection of the neural network, the calculation of the adjacent layer can be abstracted as the matrix-vector multiplication operation. Consequently, this paper mainly puts forward a kind of efficient matrix-vector multiplication algorithm according to the system structure characteristics of vector processor Matrix2, and by the method of software pipelining to fully excavate the instruction level and data level parallelism of Matrix2, thus effectively, improve computing performance of GEMV (General Matrix-Vector Multiplication) to explore the application of deep learning technology in high-performance multi-core vector processors.

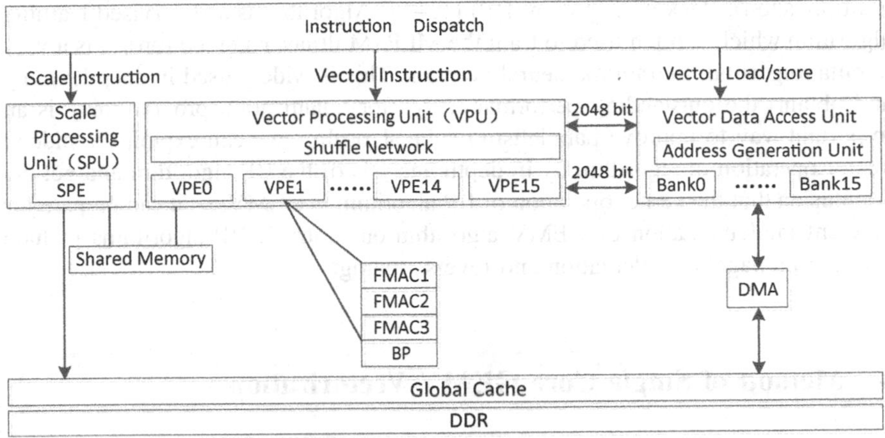


**Fig. 1** Architecture of Matrix2

## 2 The Architecture of Matrix2

Matirx2 is a high-performance floating-point multi-core vector processor for high density calculation which is developed by national university of defense technology, its single chip integrated 12 vector processor cores, 1 GHZ frequency, double precision peak performance up to 1152 GFLOPS. The single-core structure is shown in Fig. 1, each core have vector processing parts (Vector Processing Unit, VPU) and Scalar Processing parts (Scalar Processing Unit, SPU), the SPU is responsible for the scalar calculation and flow control, VPU is responsible for the vector calculation, it includes 16 Vector Processing Elements (VPE), and each VPE contains a local register file and three Floating-Point Multiply accumulation units (FMAC), two load/Store and one BP, a total of six parallel parts. Local register file contains 64 64-bits registers. SPU can support broadcast instructions to broadcast scalar data to vector registers and can exchange data with shared registers between VPUs.

## 3 Bp Algorithm Analysis

In neural networks, most arithmetic operations (e.g., additions, multiplications and activation functions) can be aggregated as vector operations [9, 10], and the ratio can be as high as 99.992% according to our quantitative observations on a state-of-the-art Convolutional Neural Network (GoogLeNet) winning the 2014 ImageNet competition (ILSVRC14) [11]. In the meantime, we also discover that 99.791% of the vector-matrix operations (such as dot product operation) in the GoogLeNet can be aggregated further as matrix operations (such as vector-matrix multiplication). Back propagation (BP) [12–14] Algorithm is a supervised learning algorithm which is often used to train the MLP (Multilayer Perceptron), it is a very popular algorithm to train the neural network. This is widely used in deep learning and obtains the outstanding performance. Vector quantization programming is an important way to improve parallelism of the algorithm, we can exploit the hidden vector operation of algorithm by in-depth analysis of the BP algorithm and get the conclusion that the kernel operation of BP algorithm is GEMV, so it can prepare for efficient implementation of GEMV algorithm on Matrix2. BP algorithms include forward propagation calculation and reverse tuning.

# 4 Method of Single-Core GEMV Vectorization

BLAS2 [3] subroutine is mainly complete relevant operation between matrix and vector, most of the time wasted on memory access, so the way of memory access of BLAS2 library function has great effect on the implementation performance.

## *4.1 GEMV Algorithm Analysis*

By the analysis of GEMM algorithm, we found that there are two kinds of calculation in calculating matrix-vector multiplication, through the analysis of these two kinds of calculation model to determine the most suitable algorithm for Matrix2 vector processor, as $m \times n$ scale matrix an example.

(1) With a line of matrix A and a vector doing dot product operation to carry out one result after shuffle and reduction sum. Although this calculation straightforward and easy to implement, while it has obvious drawback on Matrix2 vector processor, since the inner loop requires reduction to get an element of vector y and it takes larger overhead. Implementation of the algorithm process is shown in Fig. 2.

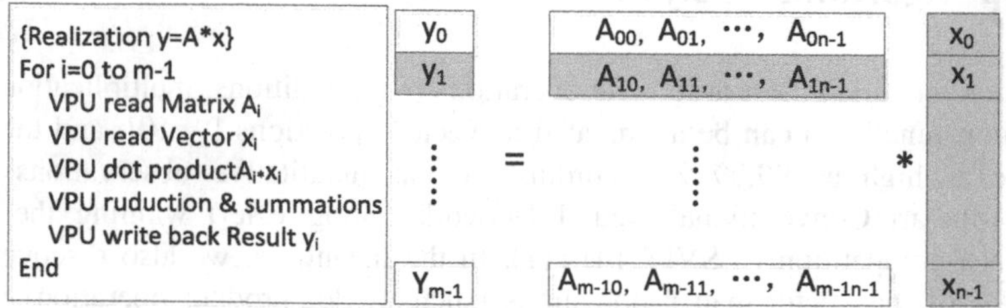


**Fig. 2** Implementation method of traditional GEMV algorithm

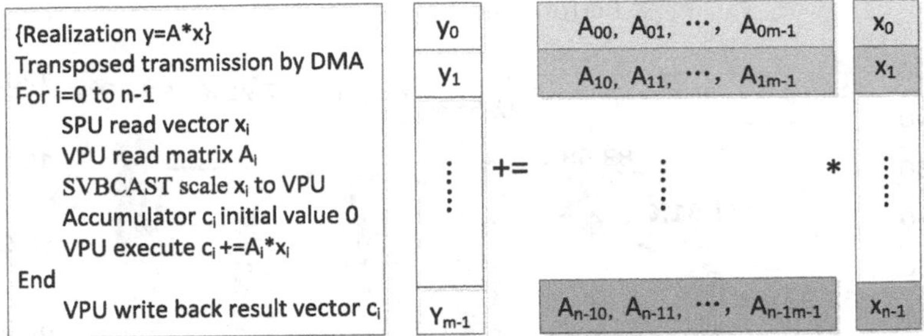


**Fig. 3** Vectorization implementation method of GEMV

(2) In addition to the previous calculation method, we can also use an element of vector x to multiply by the first column elements of matrix A with using broadcast instructions, and use an accumulator which is initialized to zero to accumulate operations, after n cycles we can have a set of result vector y value instead of a value y, the inner loop of this algorithm avoids shuffle operation which is occupy larger overhead can greatly improve the realization efficiency of GEMV. Therefore, this paper uses the second method to achieve vector operation of GEMV. Owing to the vector processor does not support data access in column, so we can use transpose operation when the transmission matrix A by DMA. The implementation of the algorithm process is shown in Fig. 3.

## 5 Performance Measurement and Analysis

We test the performance with different size of GEMV on Matrix2 vector processor platform and compare with Opteron 8387 and Xeon X5472 and X5550 which is very popular models in high-performance computing. The computing performance and implementation efficiency of single-core is shown in Fig. 4.

As can be seen from the Fig. 4, the computing performance and efficiency increase along with the data size in almost linear manner, it is mainly due to use suitable algorithm and optimization methods based on the Matrix2 architecture. We can also find that when the size of the data is relatively small, the computing performance and implementation efficiency is low, which is mainly because the proportion of loop filling and emptying is large when the loop count is small. So it is difficult to exert the advantage of Matrix2 vector processor. While along with the increase of data size, the computing performance, and efficiency of the algorithm is increase in a linear way, when the scale of matrix beyond 3072 × 3072, we can reach a performance of 94.45 GFLOPS under the peak performance 96 GFLOPS and the computation efficiency is 98.39%, which is basically achieve the peak performance of the single-core processor.

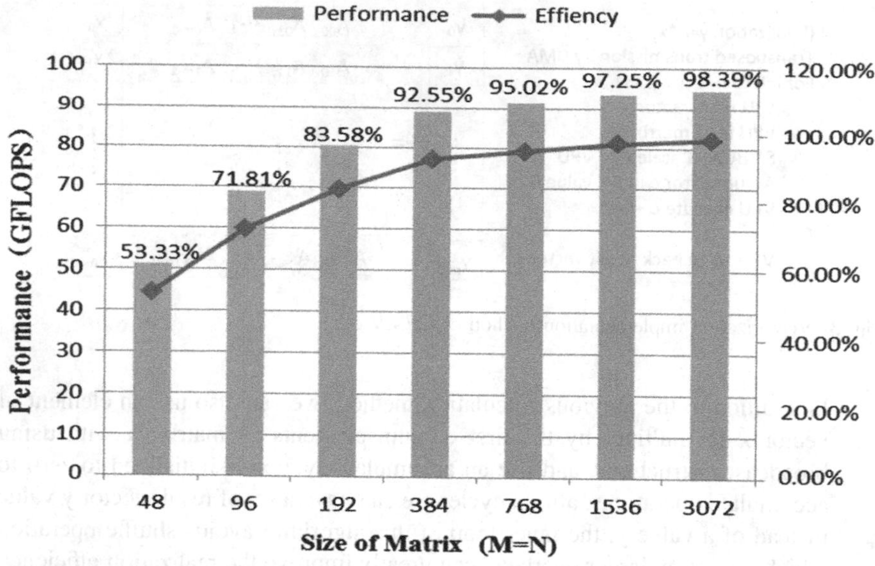


**Fig. 4** Performance and efficiency of single-core

**Table 1** Performance comparison on different platforms

| CPU/DSP | BLAS | GEMV | Peak performance (MFLOPS) |
|---|---|---|---|
| Opteron 8378 | GotoBLAS<br>Atlas [6]<br>MKL [7]<br>ACML | 1000.1<br>637.2<br>727.3<br>727.3 | 9536 |
| Xeon X5472 | GotoBLAS<br>Atlas<br>MKL<br>ACML | 920.8<br>699<br>1000.2<br>1000.1 | 11915 |
| Xeon X5550 | GotoBLAS<br>Atlas<br>MKL<br>ACML | 3200.4<br>3200.4<br>3600.3<br>3600.3 | 10666 |
| Matrix2 | MYBLAS | 80831 | 96000 |

Table 1 lists the average performance and peak performance of the GEMV algorithm based on the BLAS2 library functions in different platforms. From Table 1 we can see the followings: First, different BLAS libraries have different performance on the same platform; Second, the same BLAS library has different performance on different platforms. Third, it can achieve better performance on each platform, which is due to it is optimized through the manual assemble for the mainstream X86 platform. Fourth, the special optimization of MKL is for Xeon processor, so the performance on X5472 and X5550 processor is very close to or

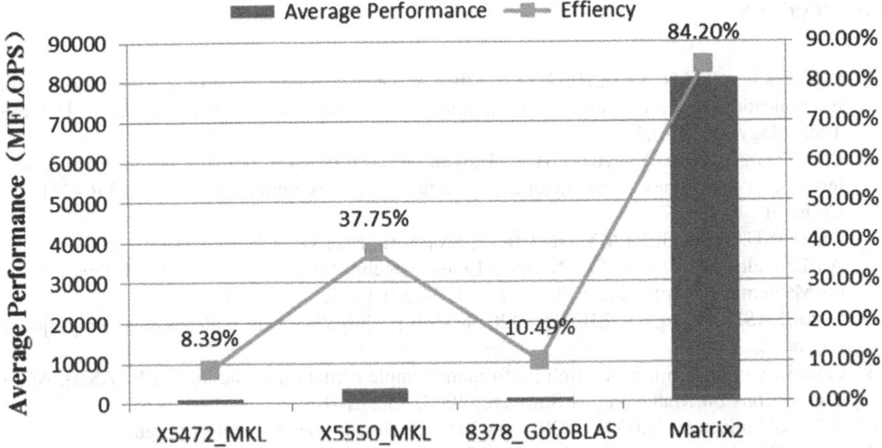


**Fig. 5** Compared with other algorithms library

even more than GotoBLAS [4, 5]. Therefore, we choose GotoBLAS based on X86 and MKL based on Xeon for performance analysis and comparison, the result shown in Fig. 5.

Figure 5 is performance comparison and single-core efficiency of GEMV on CPU or DSP platform, it can be seen that the average performance and implementation efficiency is much higher than the other algorithm library.

## 6 Conclusion and Future Work

According to the structural characteristic of multi-core vector processor Matrix2, combined with the current popular deep learning techniques application background, through in-depth analysis of BP neural network, we proposed an efficient vectorization design method of GEMV, with a reasonable data layout and manual assembly methods to optimize the core assembly code, finally, its performance and efficiency are analyzed. Through the comparison with the GEMV algorithm which is commonly used in other BLAS library functions, we find that our GEMV algorithm can achieve the best performance. Therefore, the study of this algorithm has important theoretical significance and application value for the further study of deep learning technology to the domestic high-performance multi-core vector processor Matrix2. Since Matrix2 is a multi-core processor, based on the efficient implementation of single core, still further, we will focus on exploring the multi-core implementation of GEMV algorithm.

**Acknowledgements** This paper is supported by the National Natural Science Foundation of China (61133007 and 61572025)

## References


1. LIU Zhong, TIAN Xi, CHEN Lei. Efficient vectorization method of triangular matrix multiplication supporting in-place calculation [J]. Journal of National University of Defense Technology, 2014(6):7–11.
2. LIU Zhong, CHEN Yueyue, CHEN Haiyan. A vectorization of FIR filter supporting any length and data types of coefficients [J]. Acta Electronics Sinica, 2013, 41(2):346–351. (in Chinese).
3. J.J. DONGARRA, JEREMY DU CROZ, SVEN HAMMARLING, RICHARD J. HANSON, An Extended Set of FORTRAN Basic Linear Algebra Subprograms [J], ACM Transactions on Mathematical Software, Vol. 14, No. 1, March 1973, Pages 1–17.
4. GotoBLASHomepage. [EB/OI]. [2014-04-24]. http://www.tacc.utexas.edu/tacc-projects/gotoblas2.
5. Goto K, van de Geijn R A. High-performance implementation of the level-3 BLAS[J]. ACM Transactions on Mathematical Software, 2008, 35(1):1–14.
6. ATLASHomepage. [EB/OL]. [2014-04-24]. http://math-atlas.SourceForge.net/.
7. Intel MKL Homepage [EB/OL]. [2014-04-24]. http://software.intel.com/en-us/articles/intel-mkl/.
8. ZHANG Xianyi, WANG Qian, ZHANG Yunquan. OpenBLAS: a high performance BLAS library on loongson 3A CPU [J]. journal of Software, 2011, 22(zk2):208–216. (in Chinese).
9. H. Esmaeilzadeh, P. Saeedi, B.N. Araabi, C. Lucas, and Sied Mehdi Fakhraie. Neural network stream processing core (NnSP) for embedded systems. In Proceedings of the 2006 IEEE International Symposium on Circuits and Systems, 2006.
10. V. Vanhoucke, A. Senior, and M. Z. Mao. Improving the speed of neural networks on CPUs. In Deep Learning and Unsupervised Feature Learning Workshop, NIPS 2011, 2011.
11. Christian Szegedy, Wei Liu, Yangqing Jia, Pierre Sermanet, Scott Reed, Dragomir Anguelov, Dumitru Erhan, Vincent Vanhoucke, and Andrew Rabinovich. Going Deeper with Convolutions. In arXiv:1409.4842, 2014.
12. Zhao Z. Study and Application of BP Neural Network in Intrusion Detection[M] Proceedings of the 2012 International Conference on Cybernetics and Informatics. Springer New York, 2014:379–385.
13. Y.K. Li, “Analysis and Improvement Application of BP Neural Network,” Anhui University of Science and Technology, 2012.
14. Y.M. Li, “The Study of BP Learning Algorithm Improvement and Application in Face Recognition,” Shandong University, 2012.

# Construction of Test Cases for Electronic Controllers Based on Timed Automata

**Xiaojian Liu, Junmin Li and Ting Jiang**

**Abstract** In the verification and testing of electronic embedded controllers, one of the main difficulties is how to generate a collection of consistent and complete test-cases, and how to implement the test process in an automatic means. In this paper, we propose an approach to generating test-cases based on the timed automata–the formal models of controllers. The main contribution of this work is that we present a number of rules, which can be used to guide the generation of test-cases and to reduce their total number. Furthermore, we prove that the presented rules enjoy the completeness property in the sense that the set of the test-cases is able to cover all the functionalities of controllers which formally specified with timed automata.

**Keywords** Software testing · Generation of test cases · Timed automata · Embedded controllers

## 1 Introduction

Electronic Controller Unit (ECU) is an embedded system deployed in a particular system to perform specific control functions. With continuous increasing of system complexity, more and more functions of ECUs tend to be implemented by software, thus the scale and complexity of embedded software grow sharply, which bring great challenges for software testing and verification [1, 2].

This paper is supported by the Fund of Planned Scientific Projects of Education Department of Shaanxi Province (2013JK1188) and the Doctor Fund of Xi'an University of Science and Technology (2013QDJ023).

X. Liu (✉) · J. Li · T. Jiang
Computer School, Xi'an University of Science and Technology, No. 58, Yanta Road, Xi'an 710054, Shaanxi, People's Republic of China
e-mail: 780209965@qq.com





S.K. Bhatia et al. (eds.), *Advances in Computer and Computational Sciences*, Advances in Intelligent Systems and Computing 553,
DOI 10.1007/978-981-10-3770-2_12

The main difficulties in testing and verification of software in ECUs are represented as follows.

- **Real-time** properties. Generally, real-time property mainly concerns about the response time of output against that of input. This property depends on various factors, including the software itself, underlying software platform, hardware, even physical environment. Apparently, it is hard to involve all these factors into consideration to determine the real-time properties for an ECU.
- **Concurrency**. ECUs usually work in physical world, and react to external signals with responses. However, these external signals are usually unordered, unpredicted, and even conflicted, this fact makes the behavior of ECUs to be concurrent and very complicated. To test and verify the correctness of all these behavior is usually a challenge task for testers.
- **Testing environment**. The operations of ECUs usually rely on the right environment to which they cooperate. However, in the early stage of development, this environment is usually unavailable, therefore, we need to construct a real or virtual environment in order to test and validate the functions of ECUs effectively.
- **Regression testing**. Usually, the numbers of test-cases for ECUs are as large as several hundred, thus manual testing is a time-consuming task to perform. Furthermore, if we consider the regression testing very time after the design is changed, the whole testing workload will be a heavy burden. To tackle this problem, automation of testing process is a reasonable solution.

In this paper, we propose an approach to constructing test-cases for ECUs based on the timed automata models.

The testing tools and platforms for embedded controllers are not rare, but the studies of testing methods are relatively less. For example, CANoe [3] is always used to model network topology, simulate network messages sending and receiving, but it cannot resolve the problem of testing functions. Some hardware-in-loop testing tools, such as Labview [4] and dSPACE [5], can be used to model environments and test controllers' functions, but they do not yet concern about testing methods.

## 2 Timed Automata

The behavior of ECUs is usually modeled by timed automata formalism. Timed automata is a traditional automata expanded with time clock variables and time invariants to model real-time properties of systems. In this paper, we use UPPAAL [6, 7] version of timed automata to model the behavior of ECUs.

**Definition 1** (*Timed automata*) Timed automata TA is a tuple:

$$(L, B, C, V, E, I, l_0)$$

where

- $L$ is the set of states, $l_0 \in L$ is the initial state.
- $B$ is the set of channels.
- $C$ is the set of clock variables.
- $V$ is a collection of bounded integer variables. We use $\Phi(C, V)$ as the set of conditional expressions, $R(C, V)$ the set of all clock reset operations and assignment operations of integer variables.
- $E \subseteq L \times B_{?!} \times \Phi(C, V) \times R(C, V) \times L$ is the set of edges, and $B_{?!} = \{a? \mid a \in B\} \cup \{a! \mid a \in B\} \cup \{\tau\}$ the collection of dual ($a!$ represents a sending operation, and $a?$ a receiving operation) and internal operations. An element $(l, \alpha, \varphi, r, l') \in E$ describes a state transition from state $l$ to $l'$, $\alpha$ is an operation, $\varphi$ is a guard condition, $r$ is a reset or assignment operation.
- $I: L \rightarrow \Phi(C, \phi)$ is a timed invariant labeled with a state.

To facilitate the use of timed automata, we need the definition of *path* in timed automata.

**Definition 2** (*Path*) Let $\text{TA} = (L, B, C, V, E, I, l_0)$ be a timed automaton, $l, l' \in L$. A path $p$ is a sequence of edges of TA:

$$p = (l, \alpha_1, \varphi_1, r_1, l_1);\ (l_1, \alpha_2, \varphi_2, r_2, l_2);\ \ldots;(l_{n-1}, \alpha_n, \varphi_n, r_n, l') \quad (1)$$

If there exists a path $p$ from $l$ to $l'$, we say that $l'$ is reachable from $l$ via $p$, or simply $l'$ is reachable from $l$.

We also need the following helper functions on a path.

- ***len(p)***: the length of $p$, i.e., the number of edges contained in $p$.
- ***evseq(p)***: the sequence of events labeled at each edge of $p$. For the above formula (1), $evseq(p) = \alpha_1; \alpha_2; \ldots; \alpha_n$.
- ***opseq(p)***: the sequence of operations labeled at each edge of $p$. For the formula (1), $opseq(p) = r_1; r_2; \ldots; r_n$.

*Example 1* Figure 1 illustrates an example of timed automata for a simplified vehicle lock controller.

In this example, the set of states $L$ = {READY, CRASH UNLOCK, 4DUNLOCK, 4DLOCK}, the set of channels $B$ = {CDUnlock, Key4DUnlock IgnKeyRemove, Crash}. The variable $t$ is a time clock variable, denoting the elapsed time. Other variables, such as "ignState," "speed," and "4DClosed" are bounded integer variables.

The edge (READY, IgnKeyRemove[speed < 5)/4DUNLOCK(), t:=0, 4DUNLOCK) stands for a state transition from READY to 4DUNLOCK, where "speed < 5" is a guard condition, "t:=0" is a reset operation, and 4DUNLOCK() is

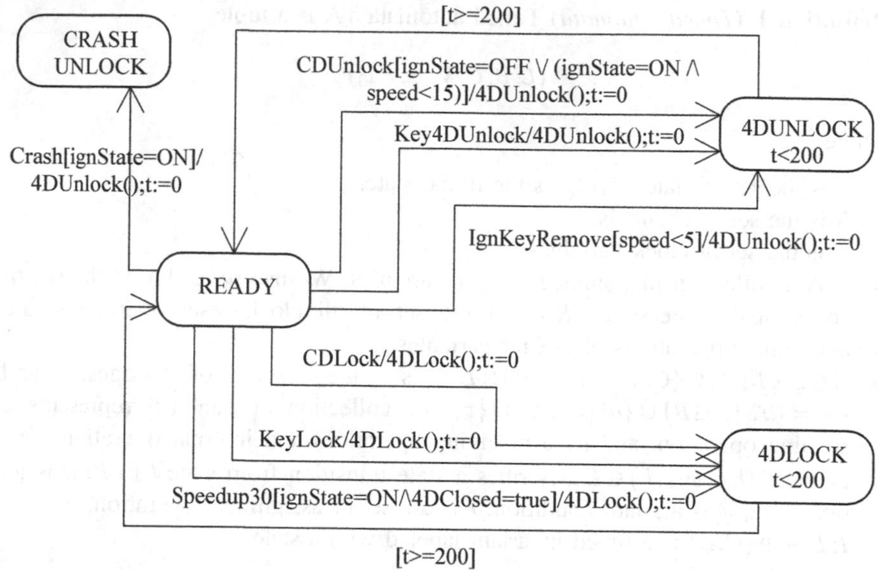


**Fig. 1** Timed automata for lock controller

an assignment operation. This transition describes such a situation "when the ignition key is pulled out in READY state, and the speed is less than 5 km/h, then the controller should take action 4DUNLOCK(), reset timer, and then change the state to 4DUNLOCK".

The expression "t < 200" within the state 4DUNLOCK is a time invariant, meaning that the time staying at 4DUNLOCK state should be no more than 200 ms.

## 3 Test-Case Generation Based on Timed Automata

Timed automata describe the expected behavior of ECUs. By employing them, we can design test-cases for ECUs. First, we should define what is a test-case in our context.

**Definition 3** (*Test-cases*) A test-case for a function is defined as a five-tuple:

$$(S, E, PRE, POST, S').$$

where S and S' are the sets of pre- and post-states of the function respectively, E the set of triggering events. PRE and POST are the sets of pre- and post-conditions of the function.

A test case (*s, e, pre, post, s'*) has the meaning that if the event *e* occurs in a state *s*, and the condition *pre* is met, then the controller should take an action, after then it should meet the condition *post*, and reach the state *s'*. Obviously, a test-case give a contract [8] which should be satisfied by the ECU under testing. If the test-case is passed through successfully, then the controller is correct for this function, otherwise it is incorrect.

The following gives the satisfiability of a test-case respect to timed automata.

**Definition 4** (*Satisfiability of a test-case*) Given the behavioral model TA = $(L, B, C, V, E, I, l_0)$ of a function, and a test case TS = (*s, e, pre, post, s'*). TS is satisfied by TA, denoted as Sat(TA, TS), if and only if one of the following conditions holds:

- $\exists (l, \alpha, \varphi, r, l') \in E \cdot l = s \wedge \alpha = e \wedge pre \Rightarrow \varphi \wedge r \Rightarrow post \wedge l' = s'$
- $\exists (l, \alpha_1, \varphi_1, r_1, l_1), (l_1, \alpha_2, \varphi_2, r_2, l_2), \ldots (l_{n-1}, \alpha_n, \varphi_n, r_n, l') \in E \cdot l = s \wedge \alpha_1;$ $\alpha_2; \ldots; \alpha_n = e \wedge (pre \Rightarrow \varphi_1 \wedge r_1 \Rightarrow \varphi_2 \wedge \ldots \wedge r_n \Rightarrow post) \wedge l' = s'$

In the above definition, we require $pre \Rightarrow \varphi$ and $r \Rightarrow post$. It means that the precondition of TA is more loose than that of TS, and the post-condition of TA is more strict than that of TS. This requirement follows the behavioral refinement theory [9, 10].

According to the Definition 4, we can easily get a collection of test-cases each of which is satisfied by TA.

**Definition 5** Let TA be a timed automaton. A collection of test-cases generated by TA, denoted as TEST, is defined as

$$\text{TEST} = \text{BASIC\_TEST} \cup \text{COMPOSED\_TEST}$$

where

- BASIC_TEST=

$$\{(s, e, pre, post, s') \mid \exists (l, \alpha, \varphi, r, l') \in E \cdot s = l \wedge \alpha = e \wedge pre = \varphi \wedge post = r \wedge s' = l'\}$$

- COMPOSED_TEST = {(*s, e, pre, post, s'*)|

$$\exists (l, \alpha_1, \varphi_1, r_1, l_1), (l_1, \alpha_2, \varphi_2, r_2, l_2), \ldots (l_{n-1}, \alpha_n, \varphi_n, r_n, l') \in E \cdot$$
$$s = l \wedge e = \alpha_1; \alpha_2; \ldots; \alpha_n \wedge (pre = \varphi_1 \wedge l_1 \Rightarrow \varphi_2 \wedge \ldots \wedge post = r_n) \wedge s' = l'\}$$

The basic test-cases in BASIC_TEST are generated by the edges of TA, and the composed ones in COMPOSED_TEST are generated by the paths of TA. Because there may exist cycles in TA, the number of the composed test-cases may be infinite. However, in an actual testing process, we always expect that the testing coverage should be wide while the number of the test-cases remains small. Therefore, we should further explore the approach to reducing the total number of the test-cases.

# 4 Select and Construct Test-Cases

To select and construct typical test-cases, here we mainly consider two properties of the test-cases: **stability** and **history-irrelevance**. For each of the test-cases in TEST, if it satisfies both properties, then it is selected as a candidate test-case; otherwise, it should be replaced by a newly constructed test-case, which has the same semantics to it and enjoy both properties as well.

Since there is a 1-1 correspondence relation between the test-cases and the edges or paths in TA, for the convenience, we use an edge or a path to refer to a test-case and do not distinct between them.

## *4.1 The Stability of the Test Cases and Construction Rules*

**Definition 6** (*Stable and Transient states*) Let $TA = (L, B, C, V, E, I, l_0)$ be a timed automaton. The set $L$ of states can be divided into two mutually disjoint subsets: STABLE and TRANSIENT:

- STABLE is the set of the stable states. A state $l$ is stable if each of its outward transitions is triggered by an external event;
- TRANSIENT is the set of transient states. A state $l$ is transient if each of its outward transitions is controlled by internal logics of ECU, or by short time elapse.

**Definition 7** (*Stable test-cases and Transient test-cases*) Let $(l, \alpha, \varphi, r, l') \in E$ be an edge in TA:

- if $l \in$ STABLE, we call the edge stable one, and its corresponding test-case stable test-case.
- if $l \in$ TRANSIENT, we call the edge transient one, and its corresponding test-case transient test-case.

We denote the set of all stable basic test-cases as BASIC_TEST_STABLE. Because the source state of a stable test-case is stable, testers can conveniently control and observe the test process. However, for a transient test-case, its source state is not stable, thus it is difficult for testers to control the source state and perform testing. Therefore, we prefer to select the stable test-cases.

We propose the following two rules to select and construct stable test-cases from BASIC_TEST.

**Rule 1** Stable test-cases BASIC_TEST_STABLE should be selected from BASIC _TEST.
**Rule 2** For a transient test-case $(l, \alpha, \varphi, r, l')$ in BASIC_TEST, if there is a stable state $l_s$, and a path $p$ such that $l$ is reachable from $l_s$ via $p$, then we use the test-case $p;(l, \alpha, \varphi, r, l')$ to replace the test-case $(l, \alpha, \varphi, r, l')$.

By applying above two rules, all the test-cases we selected or constructed are stable ones.

## 4.2 The History-Irrelevance of Test-Cases and Construction Rules

**Definition 8** (*History-relevance and History-irrelevance*) Let $(l, \alpha, \varphi, r, l')$ be a basic test-case, and $p$ a finite path through which $l$ is reachable.

- if the test-case $(l, \alpha, \varphi, r, l')$ only depends on the source state $l$, and is irrelevant to any path $p$, then we call this test-case history-irrelevant one.
- if the test-case $(l, \alpha, \varphi, r, l')$ depends on $l$ and some path $p$ as well, we call it history-relevant one, where $p$ is called a relevant path.

*Example 2* In Fig. 1, all basic test-cases are history-irrelevant. For example, assuming the controller is in READY state, every operation carried out in READY is irrelevant to all the operations the controller previously performed before going into READY. In this case, the test-cases from READY can be performed independently.

*Example 3* A function called "thermal protection" in the vehicle body controller is history-relevant. Thermal protection means that if locking or unlocking actions are performed repeatedly more than 8 times within 3 s, then the controller enters into a protection state without any response to subsequent locking or unlocking commands. Obviously, thermal protection function is history-relevant, which depends on the number of locking and unlocking actions taken previously.

A history-irrelevant test-case is more attractive to the testers, because when they perform this test-case, they do not need to consider what and how operations have been carried out previously.

The following two rules guide us to select and construct history-irrelevant test-cases from BASIC_TEST.

**Rule 3** The history-irrelevant basic test-cases should be selected from BASIC_ TEST.
**Rule 4** If a test-case $(l, \alpha, \varphi, r, l')$ in BASIC_TEST is history-relevant, and $p$ is the relevant path, then we can construct a new test-case $p;(l, \alpha, \varphi, r, l')$ to replace the original test-case.

By applying **Rule1 ~ Rule4** to BASIC_TEST, we can get a new set of test-cases TEST'. The following lemma gives the properties of TEST'.

**Lemma** *Let TA be a timed automaton. Assume initial state $l_0$ is stable, each edge starting from $l_0$ is history-irrelevant, and all other states are reachable from $l_0$. Then all the test-cases in TEST' are stable and history-irrelevant.*

Obviously, by applying above rules, every test-case in BASIC_TEST is selected or transformed to a stable and history-irrelevant one. Due to space reason, here we omit the proof. According to this lemma, we make a conclusion as follows.

**Theorem** *TEST' is able to cover all the functions of TA.*

*Proof* Let F be a function of TA, then it can be described by a path $P$ in TA:

$$P = (l, l_1); (l_1, l_2); \ldots; (l_i, l_{i+1}); \ldots; (l_n, l').$$

First, we can always assume that the first edge $(l, l_1)$ of $P$ is stable and history-irrelevant, because if this is not the case, according to the above lemma, there must exists a stable and history-irrelevant composed test-case, which can replace it without affecting the test of F. In what follows, we divide the discussion into three cases:

- if each of the edges of $P$ is stable and history-irrelevant, according to the above lemma, the edge must be collected in TEST'. And if all these edges are tested successfully, then $P$ must be successful as well, in other words, $P$ is covered by TEST';
- if there exits an edge $(l_i, l_{i+1})$ in $P$ which is transient and history-irrelevant, according to the above lemma, $(l_i, l_{i+1})$ must be replaced by a stable and history-irrelevant composed test-case in TEST', and if this composed test-case is passed successfully, then $(l, l_1);(l_1, l_2);\ldots;(l_i, l_{i+1})$ is certainly passed successfully as well;
- if an edge $(l_i, l_{i+1})$ in $P$ is history-relevant, and $p$ is relevant path, according to the above Lemma, the composed test-case $p;(l_i, l_{i+1})$ must be included in TEST', and $p$ is actually some prefix of $(l_i, l_{i+1})$ in $P$. For short, if $p;(l_i, l_{i+1})$ is passed successful, you can also guarantee $(l, l_1);(l_1, l_2);\ldots;(l_i, l_{i+1})$ is successful.

Summarily, we can make the conclusion that if every test-case in TEST' is passed successfully, then we can guarantee that every function of TA is correct, in other words, any function F is covered by TEST'.

## 5 Discussion and Case Study

The above theorem resolves the problem of how to construct a complete set of test-cases for controllers. This theorem can guide us to design and develop testing systems. From the theorem, we can make the following conclusions:

- The test cases can be generated through the analysis of formal models of target ECUs, thus it becomes possible to automate the process of the generation of test cases.
- The test cases in the TEST' can cover all the functionalities of ECUs, this fact means that testing only a small number of test cases is enough to ensure all functionalities, which thus can relief our burden of testing task.
- A complex functionality can be tested by decomposing it into several basic test cases, in other words, it can be tested via configuration and assembly of basic cases. This will reduce the complexity of test process.

By applying the theorem, we design and develop a testing system for vehicle body controllers. In this case, the most of the transient states are usually intermediate states, which we usually do not care about when testing; and the most of the functions of the controller are actions, which are intrinsically history-irrelevant, so the number of the test-cases is very limited.

However, in the development of body controllers, the test environment is usually unavailable, because other ECUs which the body controllers will cooperate are either still under development or not available at that time. Therefore, we must build a virtual testing environment, which is illustrated as Fig. 2. We simulate the functionalities of the cooperated ECUs with software, which is running in a desktop computer (PC in Fig. 2), and the input signals are initiated by clicking the buttons, switches in the software interface. The output signals are observed through software oscilloscope, or other software interface elements.

As a case study, Table 1 shows six functional modules and a number of subfunctions for vehicle body controllers. These functional modules closely cooperated to perform the complete functionalities. To facilitate analysis, a timed automaton model is constructed for each functional module, and subfunctions of a module are described as edges or paths of the timed automaton. The test cases are generated by applying the above theorem.

Automatic testing and manual testing can be performed on the testing system. Manual test is helpful, especially in the development of target ECUs. Manual test

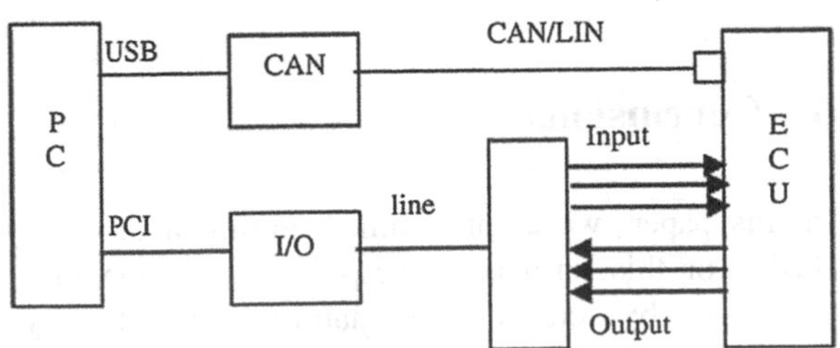


**Fig. 2** Testing system for vehicle body controllers

**Table 1** Functional modules of body controllers

| Functional modules | Subfunctions | Number of test cases |
|---|---|---|
| Central lock control | Checking signals of locks | 31 |
| | Central lock control | |
| | Automatic locking/unlocking | |
| | Remote locking/unlocking | |
| | Front door locking/unlocking | |
| | Hot protection for lock | |
| External lamp control | Checking signals of external lamps | 25 |
| | Small lamps control | |
| | Turn signal lamp control | |
| | Sending home function | |
| | Automatic lamp control | |
| Internal lamp control | Roof lamp control | 16 |
| | Foot lamp control | |
| | Key lamp control | |
| Windscreen wiper control | Wiper control | 15 |
| | Checking signals of wiper switch | |
| Remoter control | Checking signals of remoter | 11 |
| | Remoter control | |
| | Remoter key learning | |
| Windows control | Checking signals of windows switch | 24 |
| | Windows control | |

can be performed through clicking a series of buttons, switches in the software interface, which triggers input signals into the body controller. The output signals or state changes resulting from the body controller is measured via software oscilloscope, or illustrated vividly with the interface elements.

Automatic test is very essential when we perform regression testing. The test cases are constructed through the configuration and assembly of a number of basic test cases. When a test case is initiated, the input and output signals of every basic test case are recorded, which will facilitate postmortem analysis. Obviously, the manual test is more intuitive, which is suitable for a small amount of functional testing and demonstrating the functionalities of ECUs. And automatic test process, once configured, can be tested in a batch style, which can sharply reduce the burden of testing.

## 6 Conclusions

In this paper, we employ timed automata to depict the behavior of controllers. Basing on this formalism, we proposed an approach to selecting and constructing test-cases by considering stability and history-irrelevance properties, and furthermore, we prove the completeness of the constructed test-cases. To apply this approach in practice, we developed a testing system for vehicle body controllers. The case study shows that the proposed approach makes a significant effect in reducing the testing workload and improving the testing process.

## References

1. Lamberg K, Richert J, Rasche R. A New Environment for Integrated Development and Management of ECU Tests. USA: SAE International (2003)
2. Sangiovanni-Vincentelli A, Martin G. Platform-based Design and Software Design Methodology for Embedded Systems. Design & Test of Computers IEEE, 18(6):23–33 (2001)
3. Canoe Test Feature Set Tutorial. http://www.vector.com
4. Wang Z, Shang Y, Liu J, et al. A Labview Based Automatic Test System for Sieving Chips. Measurement, 46(1):402–410 (2012)
5. Luo G, Liu W, Song K, et al. dSPACE Based Permanent Magnet Motor HIL Simulation and Test Bench. Industrial Technology, ICIT2008. IEEE International Conference on. IEEE, 1–4 (2008)
6. T. Amnell, E. Fersman, L. Mokrushin, P. Pettersson and Wang Yi. TIMES - A Tool for Modelling and Implementation of Embedded Systems. Proceeding of TACAS2002, LNCS (2280): 460–464. Springer-Verlag (2002)
7. R. Alur and D. L. Dill. A Theory of Timed Automata. Theoretical Computer Science, 126 (2):183–235, Elsevier Science (1994)
8. Rhanoui M, El Asri B. A Multilevel Contract Model for Dependable Feature-oriented Components Intelligent Systems: Theories and Applications (SITA-14), 9th International Conference on. IEEE, 1–7 (2014)
9. G. Schellhorn. ASM Refinement and Generalizations of Forward Simulation In Data Refinement: A Comparison. Theoretical Computer Science, 336(2–3):403-435, Elsevier Science (2005)
10. Cavalcanti A, Gaudel M C. Testing for Refinement In Circus. Acta Informatica, 48(2):97–147 (2011)

# 'X' Shape Slot-Based Microstrip Fractal Antenna for IEEE 802.11 WLAN

Ram Krishan and Vijay Laxmi

**Abstract** In this paper, a novel fractal microstrip antenna is proposed for IEEE 802.11 wireless local area network (WLAN). The geometry of 'X' shape slots with dissimilar dimensions is used to design the proposed fractal antenna. The proposed fractal antenna is designed with FR4 Glass Epoxy material. The dielectric constant and thickness of antenna are $\varepsilon_r = 4.4$ and 1.6 mm. Radiating patch size of proposed antenna is of 35.4 mm × 27.82 mm with feed width and length 16.4 and 2.6 mm, respectively. Proposed fabricated antenna is analyzed for WLAN frequency band of 2.4 GHz. Ansoft HFSS simulator software is used to obtain and validate the simulation results of proposed antenna.

**Keywords** Microstrip · Fractal antenna · WLAN · HFSS · FR4 glass epoxy

## 1 Introduction

Microstrip antennas [1–3] are commonly used in WLAN applications due to features like light weight, low cost, compactness, easy to manufacture and can be easily incorporated with RF devices [1]. A fractal antenna [4–6] is described as an antenna that makes use of a fractal, self-identical design [7] to increase the boundary (both inside and outside) of the material that is capable of transmitting or receiving electromagnetic radiation [3]. Fractals can be defined as the broken [8] sections of geometry with dissimilar dimensions [7]. Fractal shapes are generally made-up of several replicas of themselves of dissimilar dimensions. Fractal shapes model the miniaturized and wideband antennas by taking two factors into consideration, the unique qualities of self-similarity must be maintained at every reduced scale and it must be capable of accommodating the reduced scale geometries [9].

R. Krishan (✉)
Punjabi University Guru Kashi College, Damdama Sahib (Bathinda), Punjab, India
e-mail: ramkrishan_bansal@yahoo.co.in

V. Laxmi
Guru Kashi University, Talwandi Sabo (Bathinda), Punjab, India



S.K. Bhatia et al. (eds.), *Advances in Computer and Computational Sciences*,
Advances in Intelligent Systems and Computing 553,
DOI 10.1007/978-981-10-3770-2_13

In this paper, a novel design of fractal microstrip antenna with the geometry of 'X' shape slots is presented for WLAN applications [10]. Li [11] presents an antenna design for mobile broadcasting and multiband operation based on the simulation optimization technique. Joshi and Pattnaik [12] presented a rectangular microstrip planar magneto-inductive (MI) patch antenna with 4.16 dBi gain. Immadi et al. [13] presented and developed a microstrip patch antenna through the parallel slot placement on the rectangular patch. The antenna was operated on 4.5 GHz frequency and result shows the regular radiation pattern with improved bandwidth. Sun et al. [10] presents a rectangular microstrip slot antenna designed on a thin substrate with line feed and was accomplish a very high bandwidth. This antenna accomplishes a 36% partial impedance bandwidth at 2.4 GHz resonant frequency. Kurniawan and Mukhlishin [14] design a wideband antenna for wireless technologies, which is capable of working on frequencies from 2.3 to 6.0 GHz. This antenna was designed to the circular patch using FR4 glass epoxy material of 0.8 mm thickness with dielectric constant of $\varepsilon_r = 4.3$ and co-planar waveguide (CPW) feed line. Dorostkar et al. [15] in 2013, proposed a novel T-shaped Fractal antenna for wideband applications. The proposed antenna is made of iterations T-shape around the base shape with 900 rotations in each step. Khanna et al. [7] designed a gap coupled fractal antenna to overcome the limitation of narrow bandwidth with 85.4% impedance bandwidth at 1.88 GHz resonant frequency. Kailas and Kumar [8] proposed a hexagonal microstrip fractal (HMSF) antenna for wireless applications with (UWB) frequency bands from 3.1 to 10.6 GHz.

## 2 Proposed 'X' Shape Slot-Based Fractal Antenna Design

The proposed fractal antenna with 'X' shape slots is designed using FR4 Glass Epoxy material. The dielectric constant and thickness of antenna are $\varepsilon_r = 4.4$ and 1.6 mm. The proposed antenna width and length dimensions are 40 mm × 40 mm used as base geometry and patch width, length is 35.4 mm × 27.82 mm, respectively. Line feed of 16.4 mm length and 2.6 mm width is used to give the input signal to the antenna. The proposed antenna dimensions for its design are shown in Fig. 1 and Table 1.

### 2.1 Design Steps

The design of proposed 'X' shape slots based fractal antenna includes the following steps.

Step 1: Draw base geometry of antenna.

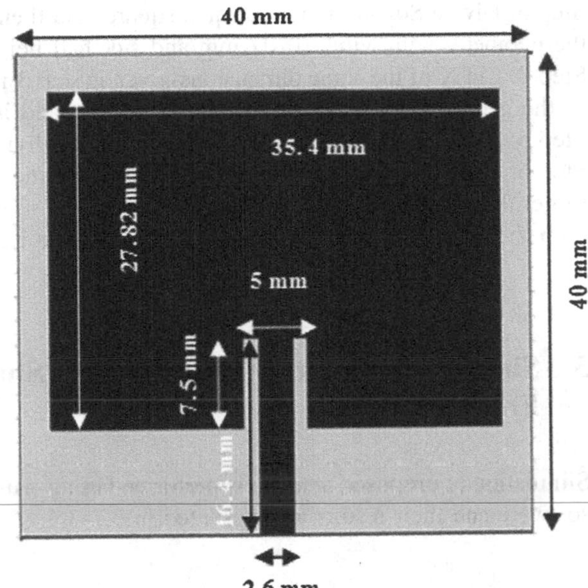


**Fig. 1** Dimensions of antenna

**Table 1** Antenna dimensions

| Sr. no. | Parameter | Value (mm) |
|---|---|---|
| 1 | Width | 40 |
| 2 | Length | 40 |
| 3 | Width of square slot | 35.4 |
| 4 | Length of square slot | 27.82 |
| 5 | Feed length | 16.4 |
| 6 | Feed width | 2.6 |

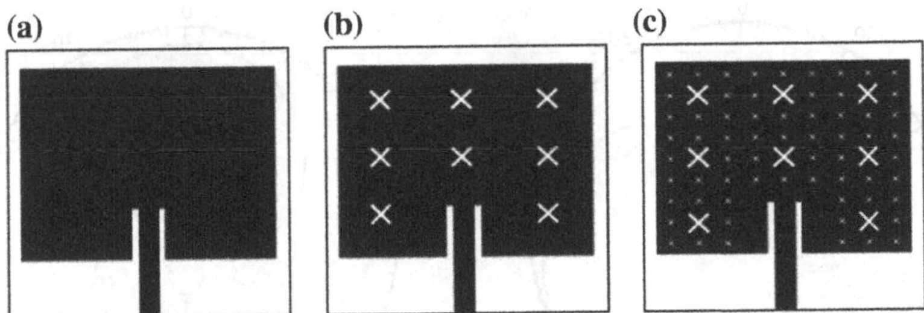


**Fig. 2** **a–c** Antenna iterations 1–3

Step 2: Cut a square of dimensions with width = 35.4 mm and height = 27.8 mm. The feed width and height is given as 2.6 and 23.9 mm to get first Iteration depicted in Fig. 2a.

Step 3: Divide Square into nine equal squares and then cut X from centre square of the geometry with width 0.707 mm and side half height 1.738 mm.
Step 4: Cut X of the same dimensions given in step 3 from remaining eight squares of the geometry to get second Iteration as presented in Fig. 2b.
Step 5: Take one square with X in the centre and divide the square into four parts.
Step 6: Cut eight X shape slots from all corners of the square with width 0.353 mm to get third Iteration.
Step 7: Above steps can be repeated to get infinite iterations of fractal antenna.

## 3 Simulation Results of Proposed 'X' Shape Slot-Based Fractal Antenna

Simulation of proposed antenna is performed using Ansoft HFSS [16, 17] simulator software and their results are presented in 3.1–3.3.

### 3.1 Radiation Pattern

Figure 3 plots the two-dimensional radiation pattern of proposed 'X' shape slots based fractal antenna. The radiation pattern of E-plane and H-plane is plotted at the resonant frequency of 2.49 GHz. Red lines in graphs plots the radiation pattern with $\theta = 90°$ and $\varphi = 90°$.

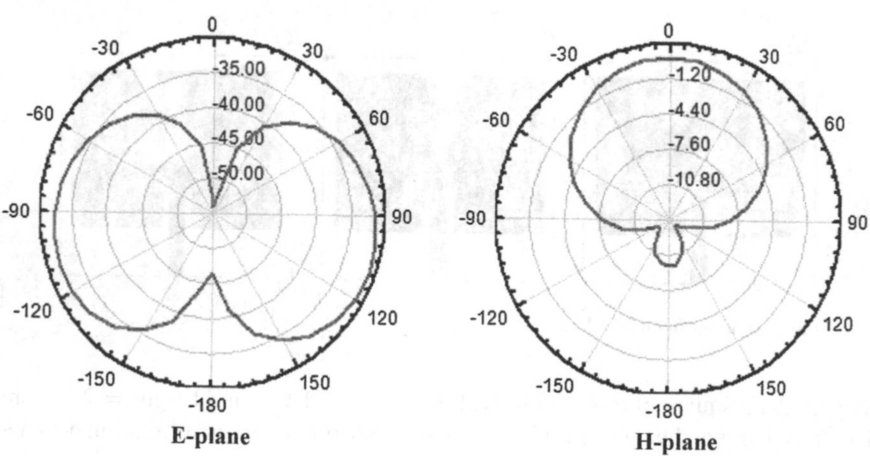


**Fig. 3** Radiation pattern in E-plane and H-plane

### 3.2 Return Loss

Figure 4 shows the Return Loss (S11) versus Frequency (GHz) result. The curve presents that the antenna resonates at 2.49 and 3.98 GHz frequencies.

For practical application of a microstrip antenna, the value of return loss should less than −10 dB. Simulation results of return loss obtained as −29 and −20 at 2.49 and 3.98 GHz resonant frequencies, respectively.

### 3.3 Voltage Standing Wave Ratio (VSWR)

Figure 5 shows the VSWR versus Frequency (GHz) result. The acceptable value of the VSWR lies from 1 to 2 [9]. The curve plots the simulation results of VSWR for proposed antenna are 1.0104 at 2.49 GHz and 1.2208 at 3.98 GHz resonant frequency.

Table 2 shows the simulated performance results of the proposed fractal antenna. The results are presented for resonant frequency (GHz) of 2.49 and 3.98.

## 4 Measured Results of Fabricated Antenna

The final design of proposed 'X' shape slot-based fabricated antenna is depicted in Fig. 6a and experimental setup for testing of the antenna using VNA are depicted in Fig. 6b.

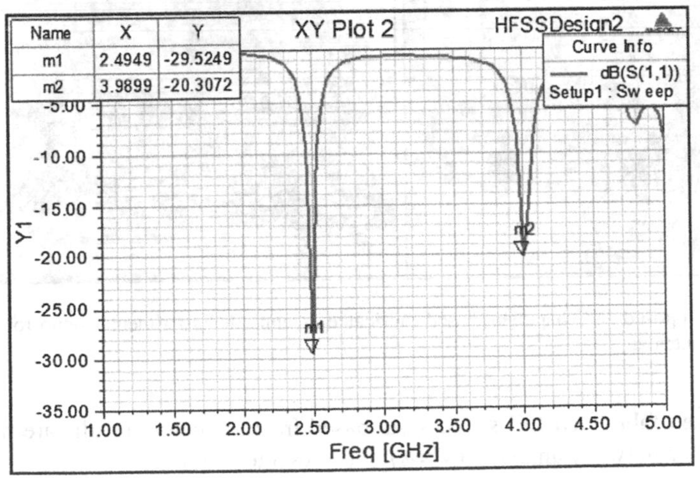


**Fig. 4** Return loss versus frequency plot

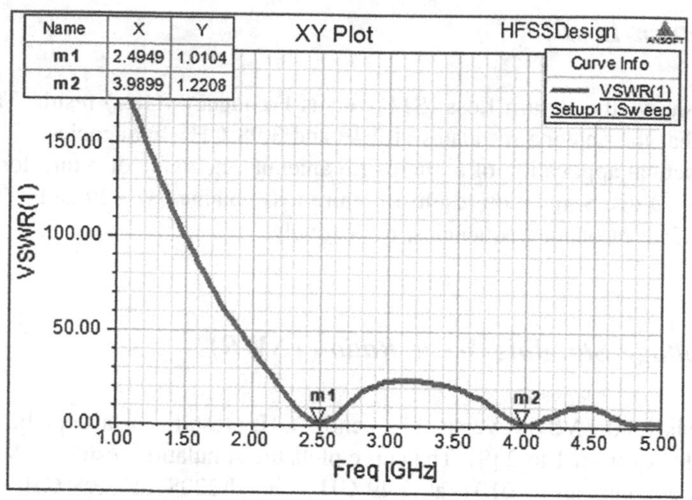


Fig. 5 VSWR versus frequency plot

**Table 2** Simulation results of proposed 'X' shape slot-based fractal antenna

| Sr. no. | Resonant frequency (GHz) | Return loss (S11) | VSWR |
|---|---|---|---|
| 1 | 2.49 | −29 | 1.0104 |
| 2 | 3.98 | −20 | 1.2208 |

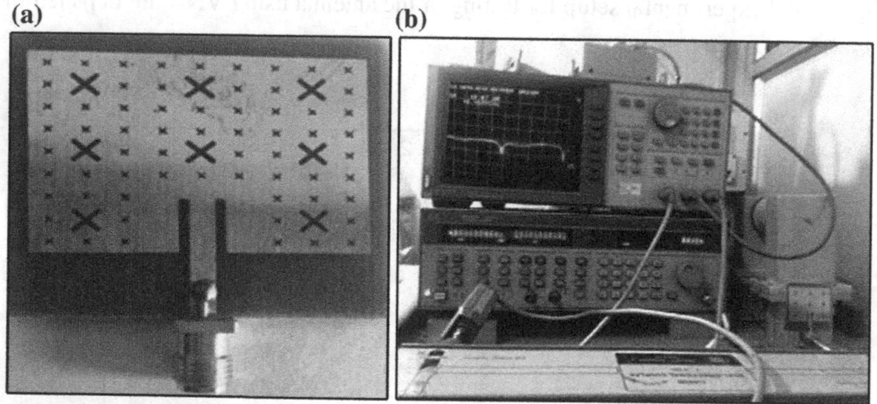

**Fig. 6** **a** Proposed 'X' shape slot based fabricated antenna. **b** Experimental setup for testing of fabricated antenna

Proposed fabricated 'X' shape slots based fractal antenna results are measured using vector network analyzer (VNA) and presented in 4.1.

### 4.1 Return Loss

Figure 7a and b presents the return loss results of the proposed fabricated antenna. Measured return loss results of the proposed antenna are −26, −18 dB at 2.46 and 3.96 GHz, respectively.

## 5 Comparison of Simulated and Measured Antenna Results

Figure 8 shows that the proposed antenna works on two frequency bands at (2.41–2.57) and (3.90–4.11). Comparison of simulated and measured results of proposed antenna is presented in Table 3.

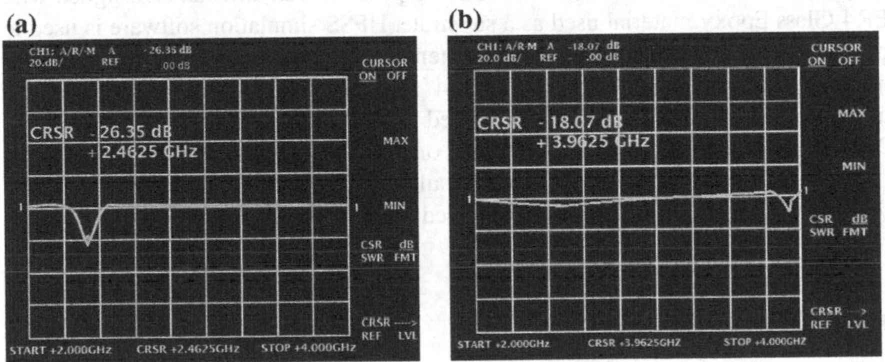


**Fig. 7** **a** Measured return loss at 2.46 GHz frequency. **b** Measured return loss at 3.96 GHz frequency

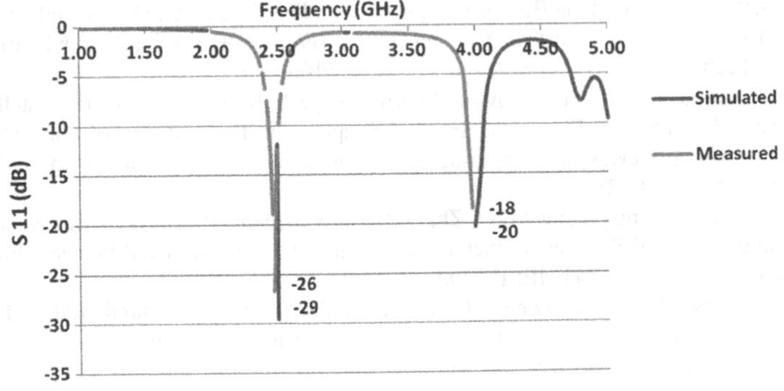


**Fig. 8** Simulated and measured return loss (S11) results of antenna

**Table 3** Comparison between measured and simulated results of proposed antenna

| Results | Resonant frequency (GHz) | Return loss S11 (dB) |
|---|---|---|
| Simulated | 2.49 | −29 |
| | 3.98 | −20 |
| Measured | 2.46 | −26 |
| | 3.96 | −18 |

The simulated and measured results confirm that the proposed 'X' shape slotted fractal antenna is suitable for WLAN applications.

## 6 Conclusions

This paper presents a fractal microstrip antenna designed with the geometry of 'X' shape slots of dissimilar dimensions. The proposed fractal antenna is designed with FR4 Glass Epoxy material used as a substrate. HFSS simulation software is used to perform the simulation of proposed antenna. The antenna results at 2.49 GHz resonant frequency are VSWR = 1.0104 and return loss = −29 dB makes this antenna useful for WLAN. The proposed fractal antenna gives 8.19 dB gain at 2.49 GHz resonant frequency and works on two frequency bands (2.41–2.57) and (3.90–4.11). Simulation and measured results of this antenna confirm its suitability for WLAN applications and give enhanced WLAN coverage.

## References

1. Balanis, C.A., Antenna Theory: Analysis and Design. John Wiley & Sons. (2005)
2. Agarwal Anil Kumar., Pattnaik Shyam Sunder., Devi, S., Joshi, J.G.: Broadband and High Gain Microstrip Patch Antenna for WLAN. In: IJRSP 2011, vol. 40, pp. 282–286. (2011)
3. Sivian Jagtar Singh, Singh Amarpartap, Kamal, T.S.: Design of Sierpinski Carpet Fractal Antenna using Artificial Neural Networks. In: IJCA 0975-8887, vol. 68, pp. 5–10. (2013)
4. Chakraborty Mrinmoy, Rana Biswarup, Sarkar, P.P., Das Achintya.: Design and Analysis of a Compact Rectangular Microstrip Antenna with Slots using Defective Ground Structure. In: Procedia Technology 2012, vol. 4, pp. 411–416. Elsevier (2012)
5. Yogesh Bhomia, Chaturvedi Ashvini, Sharma Yogesh Kumar.: Microstrip Patch Antenna Combining Crown and Sierpinski Fractal Shapes. In: Proceedings of the International Conference on Advances in Computing, Communications and Informatics, pp. 1210–1213. ACM New York (2012)
6. Chen Wen-Ling, Wang Guang-Ming, Zhang Chen-Xin.: Small Size Microstrip Patch Antenna Combining Koch and Sierpinski Fractal-Shapes. In: IEEE Antennas and Wireless Propagation Letters, vol. 7, pp. 738–741. IEEE (2008)
7. Khanna Anshika, Srivastava Dinesh Kumar and Saini Jai Prakash.: Bandwidth Enhancement of Modified Square Fractal Microstrip Patch Antenna using Gap Coupling. In: JESTECH. vol. 18, pp. 286–293. Elsevier B.V. (2015)

8. Kailas Kantilal Sawant, Suthikshn Kumar C.R.: CPW Fed Hexagonal Microstrip Fractal Antenna for UWB Wireless Communications. In: International Journal of Electronics and Communications (AEU). vol. 69, pp. 31–38. Elsevier B.V. (2015)
9. Singh Amandeep, Singh Surinder.: A Modified Coaxial Probe-fed Sierpinski Fractal Wideband and High Gain Antenna. In: AEU-IJEC, vol. 69, pp. 884–889. Elsevier B.V. (2015)
10. Sun Xu-bao, Cao Mao yong, Hao Jian-jun, Guo Yin jing.: A Rectangular Slot Antenna with Improved Bandwidth. In: International Journal of Electronics and Communications (AEU). vol. 66, pp. 465–466. Elsevier, GmbH (2012)
11. Li Yiming.: Simulation-based Evolutionary Method in Antenna Design Optimization. In: Journal of Mathematical and Computer Modelling. vol. 51, pp. 944–955. Elsevier (2010)
12. Joshi Jayant, G., Pattnaik Shyam, S., Devi Swapna, Lohokare Mohan, R.: Bandwidth Enhancement and Size Reduction of Microstrip Patch Antenna by Magneto-inductive Waveguide Loading. In: Wireless Engineering and Technology. vol. 2, pp. 37–44. SciRes (2011)
13. Immadi Govardhani, Swetha, K., Narayana Venkata, M., Sowmya, M., Ranjana, R.: Design of Microstrip Patch Antenna for WLAN Applications using Back to Back Connection of Two E-Shapes. In: International Journal of Engineering Research and Applications, vol. 2, pp. 319–323. (2012)
14. Adit Kurniawan, Salik Mukhlishin.: Wideband Antenna Design and Fabrication for Modern Wireless Communications Systems. In: Procedia Technology 2013, vol. 11, pp. 348–353. Elsevier (2013)
15. Dorostkar Ali, M., Azim R., Islam M.T.: A Novel T shape Fractal Antenna for Wideband Communications. In: Procedia Technology 2013, vol. 11, pp. 1285–1291. Elsevier (2013)
16. Nobrega, L., Clarissa de, da Silva Marcelo, R., Paulo da Silva, H.F., Adaildo Assuncao, D.G. Experimental Characterization of FSS for WLAN Applications with Low-Cost UWB Elliptical Microstrip Monopole Antennas. In: Microwave and Optical Technology Letters, vol. 56, pp. 1331–1333. Wiley (2014)
17. Haung, J.J., Shan, F.Q., She, J.Z.: A Novel Multiband and Broadband Fractal Patch Antenna. In: PIER Symposium, pp. 57–59, USA (2006)

# Performance Enhancement of an E-shaped Microstrip Patch Antenna Loaded with Metamaterial

**Akshit Kalia, Rohit Gupta, Gargi Gupta, Asmita Rajawat, Sindhu Hak Gupta and M.R. Tripathy**

**Abstract** In this paper, an efficient structure of metamaterial is proposed to enhance the performance of an E-shaped micro strip antenna. E-shaped antenna has been considered for the research because of its advantages over other shapes. It is less vulnerable to interference when several other antenna elements are present in the vicinity. The frequency band selected for the design is 6–7 GHz, C-band. The E-shaped antenna proposed is intended to be applied for many satellite communications transmissions. The metamaterial structure is made up of two nested split octagonal rings located on a 10 × 10 FR4_epoxy with 1.6 mm thickness and dielectric constant of 4.4. The patch antenna substrate consists of a 5 × 4 array of such metamaterials. By using this metamaterial in the microstrip antenna, enhancement of performance parameters in terms of return loss is seen.

**Keywords** Metamaterial · E-shaped patch · Microstrip · Miniaturization

A. Kalia (✉) · R. Gupta · G. Gupta · A. Rajawat · S.H. Gupta · M.R. Tripathy
Amity University, Noida, UP, India
e-mail: formalakshitkalia@gmail.com

R. Gupta
e-mail: rohitsempire@gmail.com

G. Gupta
e-mail: gargigupta1305@gmail.com

A. Rajawat
e-mail: arajawat@amity.edu

S.H. Gupta
e-mail: shak@amity.edu

M.R. Tripathy
e-mail: mrtripathy@amity.edu



S.K. Bhatia et al. (eds.), *Advances in Computer and Computational Sciences*,
Advances in Intelligent Systems and Computing 553,
DOI 10.1007/978-981-10-3770-2_14

# 1 Introduction

## *1.1 Related Work*

Antenna is an integral part of communication system. Various parameters which measure the efficiency of an antenna are return loss, VSWR, radiation pattern, gain, resonant frequency, and bandwidth. Metamaterial are materials which are not available in nature, these manmade materials with negative permeability and permittivity enhances the performance parameters of an antenna [1–3]. An E-shaped patch is so-called because of its resemblance to E alphabet, the E-shaped patch has various advantages over the rectangular patch, hence it is used [4, 5]. Metamaterials can be of DNG (Two fold negative-both permittivity and permeability are negative), DPS (Twofold positive–both permittivity and permeability are certain), ENG (permittivity-negative), and MNG (permeability-negative). Certain use of metamaterial incorporates sensor discovery, open security, and remote aviation. Metamaterial has likewise discovered its application in high recurrence front line correspondence and enhancing ultrasonic sensors [6]. DNG metamaterial is exceptionally valuable to build the force which is transmitted by little receiving wire [7].

Microstrip patch reception properties have certain focal points and drawbacks, they have various favorable circumstances like low profile, light weight, and simplicity of creation which accounts as it preferences. A reception antenna is broadly utilized and satisfactory seeing its execution lists, for example high pick up, huge transfer speed, and better proficiency. Diverse work on microstrip patch shows that it is hard to accomplish these files and also keeps up the little size. To defeat the multifaceted nature, distinctive stacking method like opening, winding stacking, shorting pin, and so on; are utilized to inspire these lists and in addition help in size decrease. In any case; utilizing these systems the other execution parameters of the microstrip patch reception properties are disturbed. In this manner to accomplish a superior exchange off between sizes, pick up and transfer speed gets to be troublesome. With the utilization of metamaterial in microstrip patch reception there has been headway of increase, data transmission, directivity, and effectiveness with the decrease of size of antenna [8].

## *1.2 Contribution*

In this paper, we present the possibility of performance enhancement of an E-shaped microstrip patch antenna by using a novel structure of metamaterial, which is placed on the substrate. The software used for designing the antenna is HFSS [9]. The return loss and gain before the addition of array ($5 \times 4$) metamaterial is compared.

(A) Previous research in the field of metamaterial is studied; various related works are effectively read.
(B) An array (5 × 4) of octagonal split rings is designed.
(C) The designed metamaterial structure is placed in the substrate of E-shaped antenna and the result, performance after the insertion of metamaterial are compared.

### *1.3 Organization of Paper*

The research paper has been arranged as follows. Section 2 and the subsequent section present the planned antenna design along with its simulation results, which validates the accuracy of the analysis and demonstrates the strategy for different antenna designs with and without metamaterial and the last section concludes the paper.

## 2 Antenna Design

### *2.1 E-patch*

The shape of the patch used is E shaped. Using an E-shaped antenna instead of rectangular patch antenna; antenna size can be reduced significantly. Antenna equations are used to find the dimensions of the basic patch [10]. The slot length and width of the patch are the various parameters due to which the performance of the patch is influenced. Figure 1 shows the antenna design and Table 1 gives the patch configurations which have been optimized for good results.

### *2.2 Metamaterial Structure*

The configuration of the novel metamaterial structure which has been used in the project is shown in Fig. 2. From the figure, it can be deciphered that the metamaterial structure is actually a nested octagonal split ring which etched on a dielectric substrate. Two concentric metallic ring slots with slits etched in each ring at its opposite sides are used to compose the CSSR; which is loaded on the patch or the ground plane to enhance the performance of the rectangular microstrip patch antenna. The substrate is FR4 Epoxy 5880 with dielectric constant equal to 4.4. Dielectric dimensions ($Ls \times W_s$) are10 mm × 10 mm and the thickness ($t$) is 1.6 mm. The strip width of each octagon is 0.6 mm. The sides of the octagons, from the outer side to the inner side are 4.0, 3.5, 2.8, and 2.3 mm, respectively.

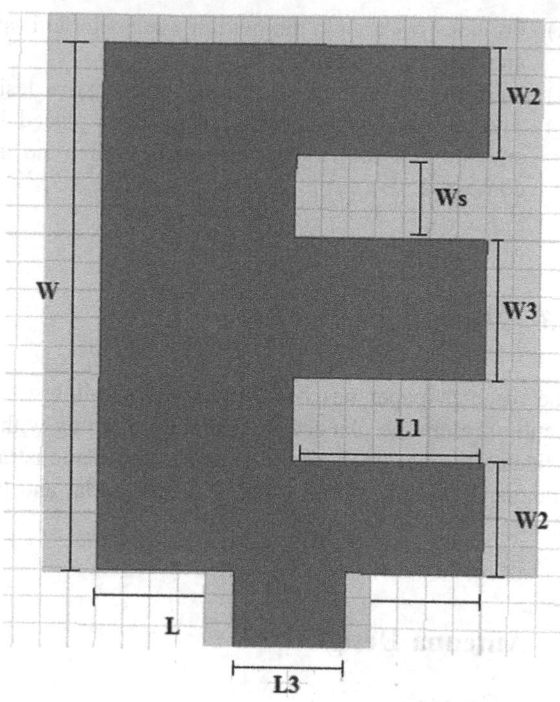


**Fig. 1** E-shaped patch

**Table 1** Patch configurations

| Frequency | 2.4 GHz |
|---|---|
| $W$ | 38 mm |
| $L$ | 28 mm |
| $W_2$ | 8 mm |
| $L_1$ | 12 mm |
| $W_s$ | 6 mm |
| $W_3$ | 10 mm |
| $L_3$ | 5 mm |
| Dielectric ($\varepsilon_r$) | 4.4 |
| Thickness ($h$) | 1.6 mm |

Both the gaps ($g$) in the octagons are 0.3 mm and the distance between octagons ($d$) is 1 mm (Fig. 3).

## 2.3 Antenna Design with Metamaterial

Microstrip line inset feed is fed in this antenna. The width of the feed line ($W_0$) and the depth of the inset ($d$) have been adjusted to match the antenna impedance to

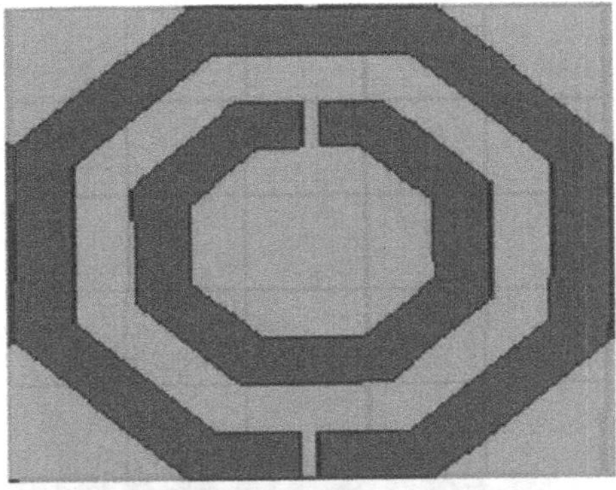

**Fig. 2** Metamaterial split rings. The unit cell structure and its dimensions: $L_s = W_s = 10$ mm, $S_1 = 4$ mm, $S_2 = 3.5$ mm, $S_3 = 2.8$ mm, $S_4 = 2.3$ mm, $g = 0.3$ mm, $d = 1$ mm

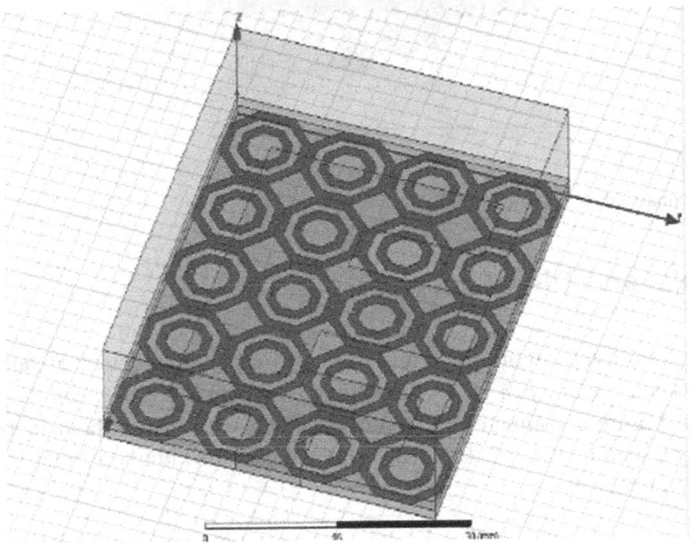

**Fig. 3** Geometry of the proposed antenna. A (5 × 4) array of metamaterial which was described in section *b* is placed in the substrate of a patch antenna

50 Ω. The size of $W_0$ is 4.08 mm and d is 5 mm. The distance between inset line and the patch (*y*) is 1.8 mm (Fig. 4).

When the patch antenna is loaded with the metamaterial, the antenna resonates when the constitutive parameters are negative. The resonant frequency of the antenna is affected with the use of metamaterial. So the antenna resonates with the metamaterial frequency changing the return loss, gain, and other parameters of the antenna.

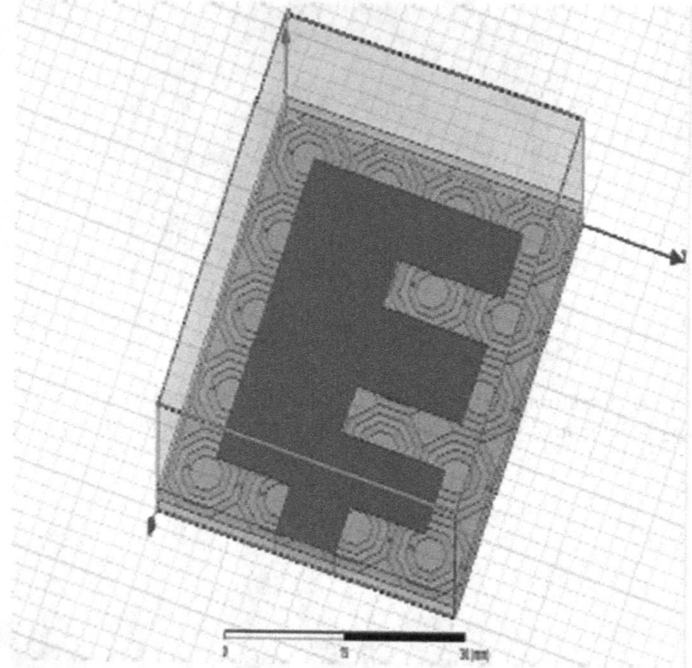

Fig. 4 Antenna with metamaterial array

# 3 Results and Discussion

Significant improvement in the return loss and other parameters of antenna are observed, when the array of (5 × 4) metamaterial is inserted between the substrate of antenna. Figure 5 compares the return loss plot with and without metamaterial which can be seen in Table 2. Figures 6 and 7 shows the radiation patterns with and without metamaterial.

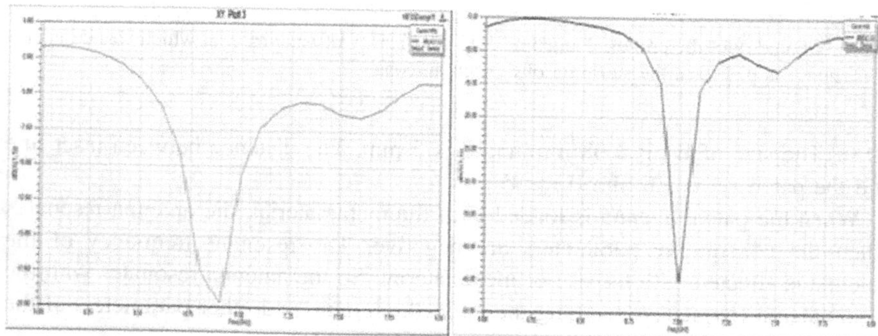

Fig. 5 Return loss comparison without/with metamaterial

**Table 2** Performance comparison

| Properties | Without metamaterial | With metamaterial |
|---|---|---|
| Resonant frequency (GHz) | 6.90 | 7 |
| Return loss (dB) | −20 | −45 |

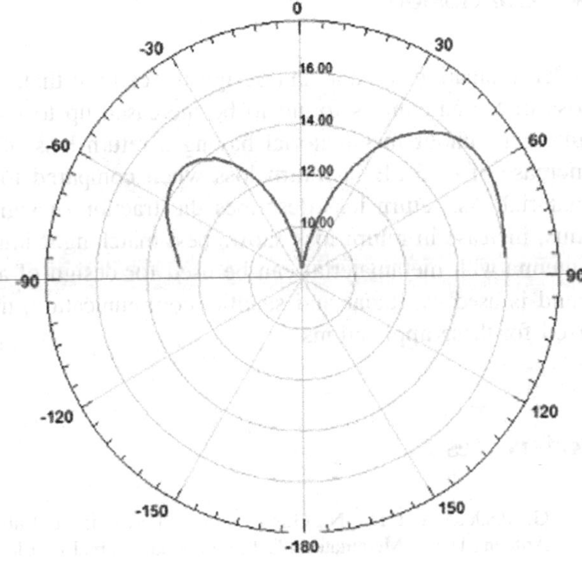


**Fig. 6** Radiation pattern without metamaterial

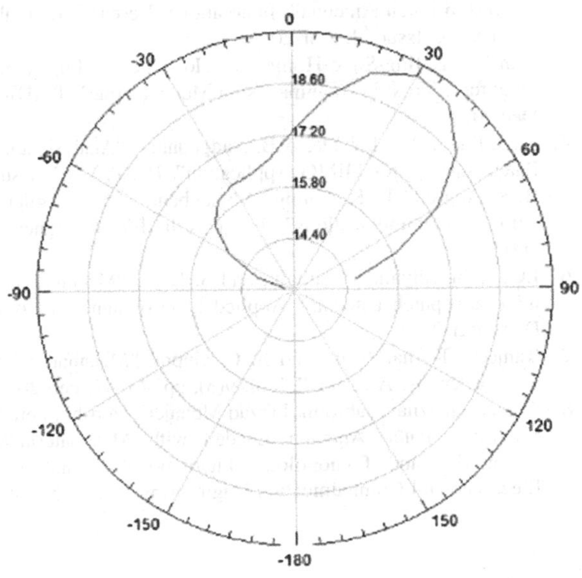


**Fig. 7** Radiation pattern with metamaterial

An E-shaped microstrip antenna inserted with an array structure of (5 × 4) metamaterial was investigated. The performance of antenna after and before the insertion of metamaterial was studied. Significant improvement in the properties of antenna was observed.

## 4 Conclusion

After simulating the antenna design we can see that, with metamaterial the return loss of the antenna is found to be increased up to −45 dB in comparison to the antenna without metamaterial having a return loss of −20 dB. Thus, there is an increase of −25 dB in return loss when compared to the antenna without metamaterial. As, return loss describes the fraction of signal reflected back to source thus, increase in return loss shows best matching. Simulation results show that the antenna with metamaterial can be used for design of antenna in C-band. Since, C band is used by Radar and satellite communication, the proposed antenna can be used for these applications.

## References

1. G. Pradeep and Dr. N. Gunasekaran, "Performance Enhancement of Micro strip Patch Antenna Using Metamaterial", International Journal of Electrical and Computing Engineering, Vol. 1, Issue. 4, June 2015.
2. Aakash Mithari, Uday Patil, "Efficiency and Bandwidth Improvement Using Metamaterial of Microstrip Patch Antenna", International Research Journal of Engineering and Technology, Volume: 03 Issue: 4, Apr-2016.
3. Han Xiong, Jing-Song Hong, Yue-Hong Peng, "Impedance Bandwidth and Gain Improvement for Microstrip Antenna Using Metamaterials", RADIOENGINEERING, Vol. 21, No. 4, Dec 2012.
4. M.B. Kadu, R.P. Labade, A.B. Nadgaonkar, "Analysis and Designing of E-shape Microstrip Patch Antenna for MIMO Application", IJEIT, Vol.1, Issue 2, Feb 2012.
5. B.-K. Ang and B.-K. Chung,"A wideband E-shaped microstrip patch antenna for 5–6 GHz Wireless Communications", Progress In Electromagnetics Research, PIER 75, 397–407, 2007.
6. J.G. Joshi, Shyam S. Patnaik, and S. Devi, "Metamaterial loaded square slotted dual band microstrip patch antenna", Applied Electromagnetics Conference (AEMC), IEEE, pp. 1–4, December 2011.
7. Ranjeeta, Kumar Nitin, and S. C. Gupta, "Metamaterial for performance enhancement of patch antennas: A review", Vol. 9(3), pp. 43–47, Feb 2014.
8. Anahita Ghaznavi Jahromi, Farzad Mohajeri, Nooshin Feiz, "Miniaturization of a Rectangular Microstrip Patch Antenna Loaded with Metamaterial", World Academy of Science, Engineering and Techonology, International Journal of Electrical, Computer, Energetic, Electronic and Communication Engineering Vol: 7, No: 4, 2013.

9. EM simulator, ANSYS HFSS, V.13.
10. Constantine A Balanis, Antenna Theory Analysis and Design, 2nd Edition, Singapore, John Wiley and Sons, 2002.

# Narrow Channel Multiple Frequency Microstrip Antenna with Slits

**Manshi Nisha, Sindhu Hak Gupta, Asmita Rajawat, Monica Kaushik and Devesh Kumar**

**Abstract** In this paper, a narrow channel multiple frequency microstrip patch antenna is proposed. This paper entrusts the trend for not using wider channels, which is even though an increasing trend now a days, but instead we advocate that the radio communication should take place over multiple channels for fair and efficient spectrum utilization. For this purpose, we present a narrow channel multiple frequency rectangular microstrip patch antenna having slits. The proposed design is applicable for commercial frequency bands of 3.20, 5.5, 6.25, and 7.96 GHz which makes it useful for the modern wireless communication purposes (2–8 GHz). HFSS version 13.0 is used for design, evaluation, and analysis of the proposed antenna design. The design is analyzed for radiation pattern, return loss, VSWR, and gain, where the simulated results infer that the planned antenna design shows appreciable performance in terms of VSWR, gain, return loss, and radiation pattern at resonant frequencies.

**Keywords** Probe fed · Narrow channels · Multiple frequency · Microstrip patch · Rectangular microstrip · HFSS

Manshi Nisha (✉) · S.H. Gupta · Asmita Rajawat · M. Kaushik · D. Kumar
Amity University, Noida, UP, India
e-mail: manshinisha@yahoo.com

S.H. Gupta
e-mail: shak@amity.edu

Asmita Rajawat
e-mail: arajawat@amity.edu

M. Kaushik
e-mail: mkaushik@amity.edu

D. Kumar
e-mail: dvshkmr@gmail.com



S.K. Bhatia et al. (eds.), *Advances in Computer and Computational Sciences*,
Advances in Intelligent Systems and Computing 553,
DOI 10.1007/978-981-10-3770-2_15

# 1 Introduction

Wireless Communication has become an important aspect of communication world nowadays. With the advancement in research and technology, the size of devices used in communication world is shrinking appreciably with each passing day. Miniaturization of communication devices as well as components is being welcomed by scientists and researchers at large. This imposes a serious restriction on size of integrated antenna, making miniature, low profile but efficient antennas as prime requirement for such applications [1]. In addition to this, the use of multiple technologies (GSM, LTE, 4G, 5G, Wi-Fi, and GPS) on a single device necessitates the use of multiple antennas which thereby results in the increased size of the device. This undesirable feature can be handled by use of single antenna operating at different frequencies. For this purpose multiband antennas were introduced [1, 2]. In this regard, microstrip antennas have found a very crucial role for themselves in the world of modern wireless communication. Microstrip antennas owe their popularity to benefits like conformable, cost effectiveness, light weighted, small size, and highly flexible [3]. They also exhibit features like low complexity, ease of design as well as fabrication and integration. Microstrip antennas can be designed in many shapes like rectangular, circular, square, triangle, semicircular, etc. [3]. These features make microstrip antenna a popular choice for single multiband antenna for portable devices [4]. They find their application in number of areas including satellite communication, biomedical fields, military, radio communication, navigation, etc.

## *1.1 Related Work*

Various antenna structures are present like horn antenna, parabolic reflectors, etc. which provide good performance characteristics but in terms of planar antennas, micro strip antennas have always had an upper hand because of their features. Microstrip antenna consists of a substrate sandwiched between a patch element and a ground element. The patch element is the main radiating entity [3]. When microstrip antennas are excited, there develops a fringing field between the edge of the patch element and the ground which primarily is the basic reason for antenna radiation.

Various methods were used in designing a multi-frequency antenna that can operate at different frequency bands. Some of the common ones include thickening of patch, use of shorting pins, as well as slots in ground plane [3]. In the year 1993, Huynh and Lee introduced the U slot patch antenna [5]. The main flaw in this type of antenna was its low bandwidth and gain. However, U slot patch can be significantly improved by employing frequency reconfigurable features which provides the designer wide opportunities in terms of antenna designing [6, 7]. An H-type microstrip slot antenna structure is proposed in [7]. This type of antenna operates in Ku-band and employs multilayer substrates. A Swastika slot antenna is proposed in

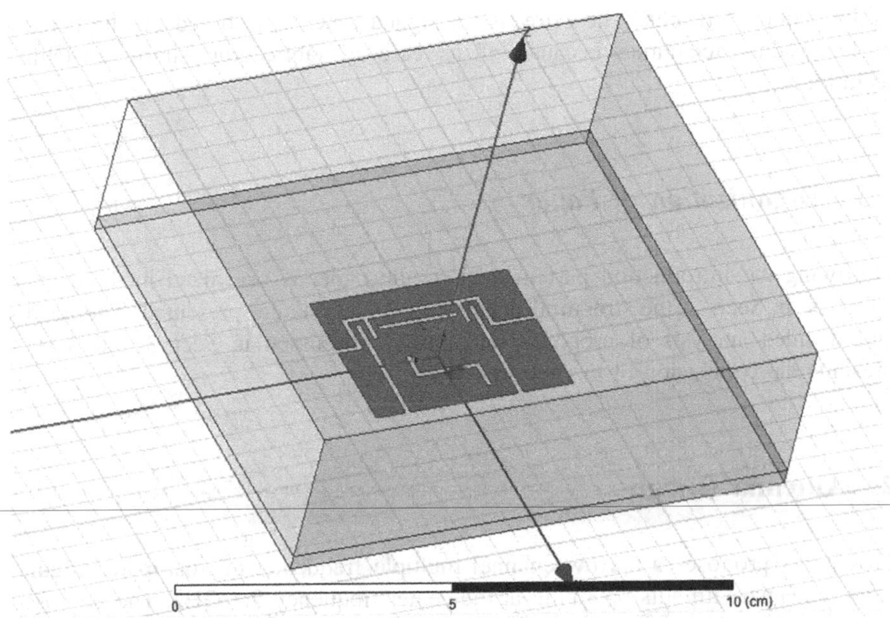


**Fig. 1** Narrow channel multiple frequency rectangular microstrip patch antenna

[8] which manifests a adept impedance bandwidth and operates for GSM AND WLAN applications. A multi frequency planar antenna with multi-frame L slots within a single microstrip line has been discussed in [9] and a Tetra band antenna is studied in [10]. Most of the above discussed antennas though have successfully achieved a multiband frequency operation but they are disadvantageous in terms of design complexity, low gain as well as low efficiency. It thereby puts forth the need for optimum multi-frequency antenna. In this paper narrow channel multi-frequency rectangular Microstrip Patch Antenna design is proposed which operates at 3.20, 5.50, 6.250, and 7.96 GHz frequencies. These frequencies cover different standards of 4G, 5G, and Satcomm communication. In this paper, narrow channel frequencies are proposed for communication purposes instead of wide band so as to facilitate the fair and efficient use of spectrum as well as to avoid interference. Figure 1 depicts the proposed design of to be evaluated patch antenna.

## 1.2 Contribution

The research contribution of the paper is dedicated toward the study and simulation of different configuration of slits introduced in the rectangular microstrip patch antenna for achieving the narrow channel multi frequency results for modern wireless communication systems. The design is modified by employing different slit geometry with different dimensions on the rectangular microstrip patch antenna, for optimization of results. The proposed designed antenna is based on the coaxial probe feeding technique because of its ease of fabrication, simple matching technique.

### 1.3 Organization of Paper

Following the introduction part in Sect. 1, the paper is organized in three main sections. In Sect. 2, the structural details of the antenna are presented. The results and detailed analysis of proposed antenna are discussed in Sect. 3 and Sect. 4 presents the conclusion of the research work.

## 2 Antenna Design

This paper proposes a narrow channel multiple frequency microstrip rectangular patch antenna with slits operating at the center frequency 2.4 GHz. The proposed antenna dimensions are based on the basic antenna design equations [11]. The proposed antenna consists of a patch having dimensions $L_p \times W_p = (3 \times 4)$ cm etched on a dielectric material, Roger RT Duroid 5880 substrate having dimensions $L_s \times W_s = (9 \times 10)$ cm. The height of the substrate employed for the proposed design is 0.32 cm. The rectangular patch has different slots each having different dimensions which are summarized in Table 2. The proposed antenna is excited using probe feeding method and design is analyzed using transmission line model.

The proposed antenna dimensions and the slit dimensions employed in patch of the considered antenna design are summarized in Tables 1 and 2 respectively. Figures 2 and 3 illustrates the geometrical proposal of the patch employed in the planned antenna.

**Table 1** Proposed antenna dimensions (in cm)

| S no. | Antenna dimension | Values |
|---|---|---|
| 1 | Center frequency | 2.4 GHz |
| 2 | Dielectric material used | Rogers RT duroid 5880 ($\varepsilon = 2.2$) |
| 3 | Substrate thickness | 0.32 cm |
| 4 | Substrate length | 9 cm |
| 5 | Substrate width | 10 cm |
| 6 | Patch length | 3 cm |
| 7 | Patch width | 4 cm |
| 8 | Coax pin height | 0.5 cm |
| 9 | Coax pin radius | 0.07 cm |
| 10 | Ground length | 9 cm |
| 11 | Ground width | 10 cm |

**Table 2** Proposed slit dimensions employed in patch (in cm)

| S no. | Slit dimension | Values |
|---|---|---|
| 1 | A | 2.8 cm |
| 2 | B | 0.7 cm |
| 3 | C | 0.5 cm |
| 4 | D | 2.3 cm |
| 5 | E | 2.3 cm |
| 6 | F | 0.6 cm |
| 7 | G | 0.55 cm |
| 8 | H | 1.2 cm |
| 9 | I | 1.2 cm |
| 10 | J | 0.2 cm |
| 11 | K | 0.9 cm |
| 12 | L | 1.6 cm |
| 13 | M | 0.5 cm |

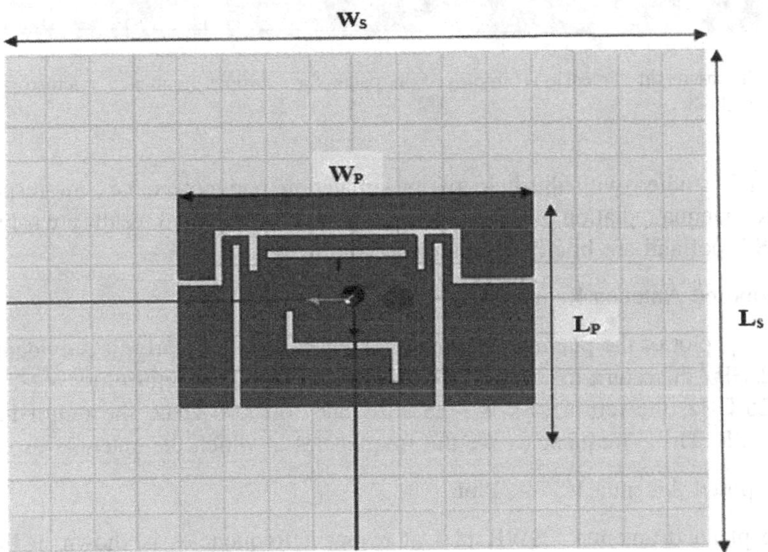


**Fig. 2** Proposed design for multiple frequency micro strip patch antenna with slits

## 3 Simulated Results and Discussion

The conceived antenna can generate multiple frequencies at 3.20, 5.50, 6.250, and 7.96 GHz which makes it useful for the modern wireless communication purposes (2–8 GHz). These frequencies cover different wireless standards including LTE, 4G, 5G, WiMax, Satcomm, and WLAN. The performance analysis of the conceived

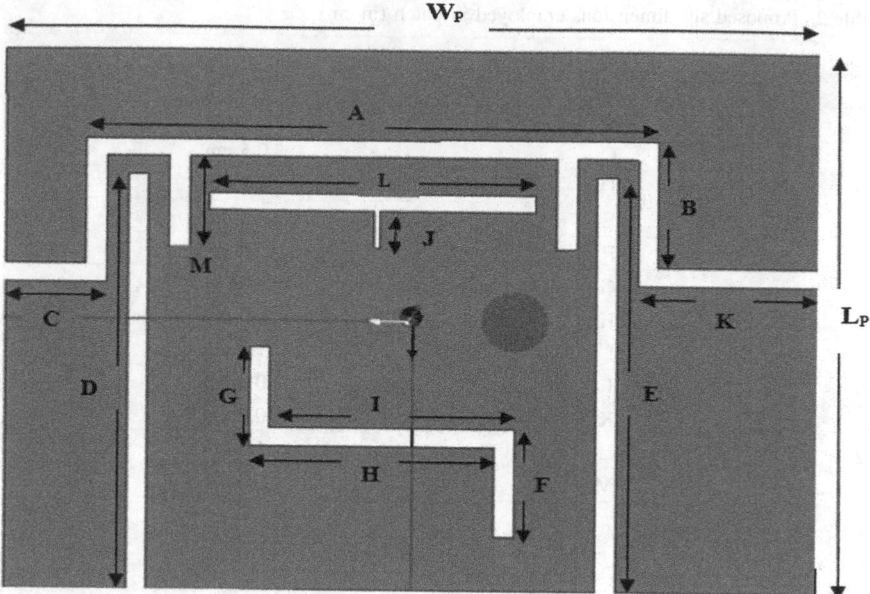


**Fig. 3** Proposed slit dimensions employed in patch for multiple frequency microstrip patch antenna with slits

antenna is studied with the help of its simulation parameters, i.e., $S_{11}$ response, VSWR, antenna radiation patterns, and gain plot. The evaluated results are achieved in HFSS tool and are briefly discussed as following.

(a) Proposed Antenna Return Loss

The $S_{11}$ plot of the planned antenna is given in Fig. 4. Clearly, it is evident that at 3.32 GHz the return loss is −18.14 dB, for 5.5 GHz, the return loss is −22 dB, for 6.25 GHz, the return loss is −28.8 dB and for 7.96 GHz, the return loss is −28.24 dB. These frequencies are the frequencies at which the antenna resonates.

(b) Proposed Antenna VSWR Plot

The planned antenna VSWR plot at resonant frequencies is shown in Fig. 5. From the VSWR plot it can be noted that at the resonant frequencies of 3.32, 5.5, 6.25, and 7.96 GHz the VSWR is quiet reasonable as it is within the required level, i.e., <2.

(c) Proposed Antenna Radiation Pattern

Planned antenna radiation patterns are shown below. As antenna radiates normally upright to its patch, the elevation pattern for phi (∅) = 0° and phi (∅) = 90° would be significant. Figure 6 depicts the radiation pattern for phi (∅) = 0° and Fig. 7 depicts the radiation pattern for phi (∅) = 90°.

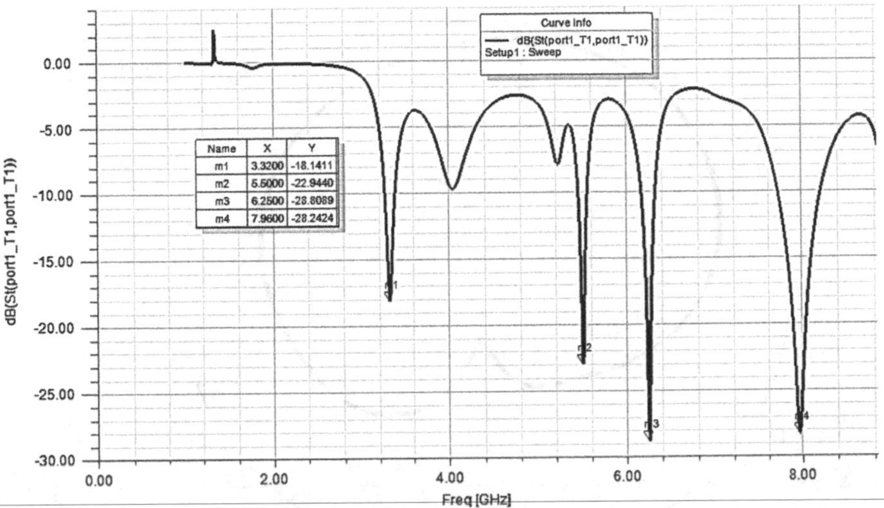


**Fig. 4** Return loss ($S_{11}$)

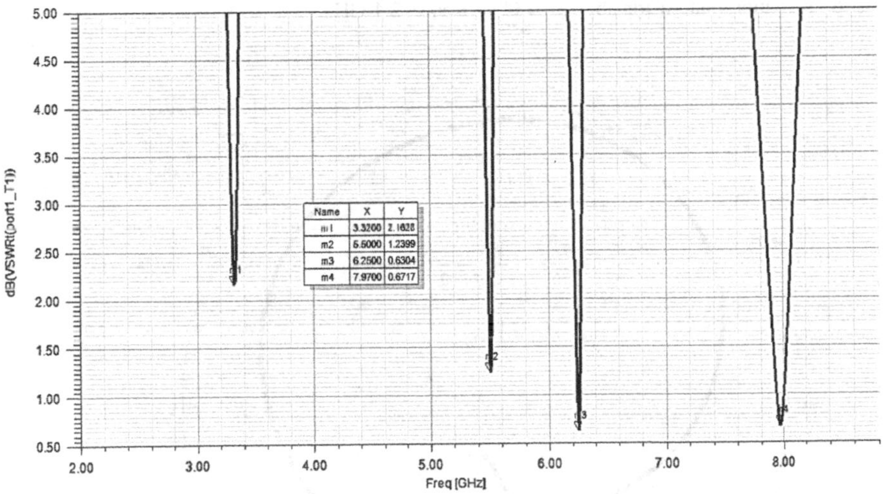


**Fig. 5** VSWR plot

(d) Gain of Proposed Antenna

The proposed antenna plot is shown in Fig. 8. Clearly, it is evident that the antenna gain is 3.81 dB and it surely satisfies the gain criteria for good performance of antenna.

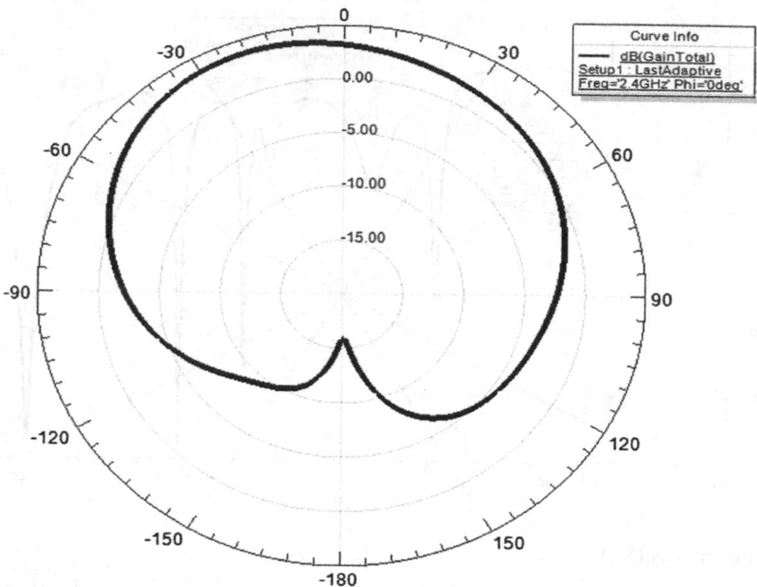


**Fig. 6** Radiation pattern for phi (∅) = 0° at freq = 2.4 GHz

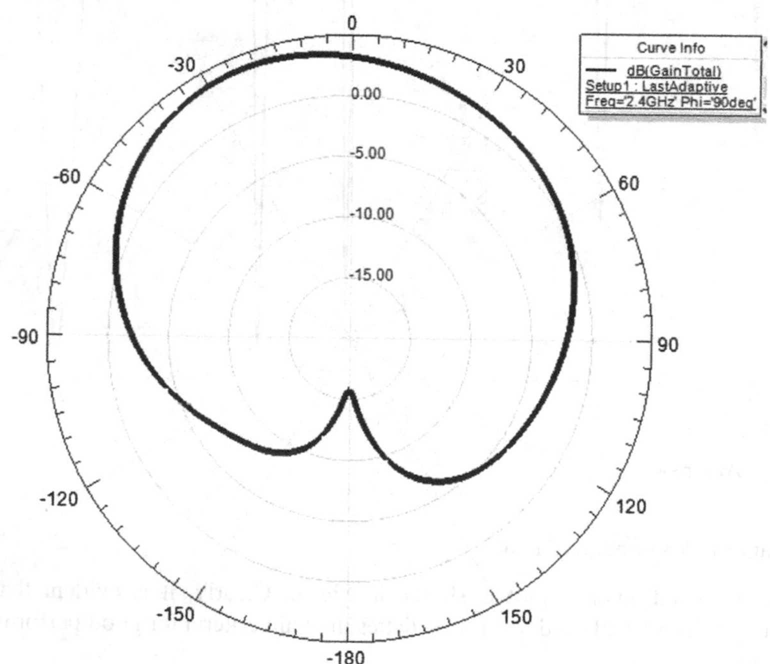


**Fig. 7** Radiation pattern for phi (∅) = 90° at freq = 2.4 GHz

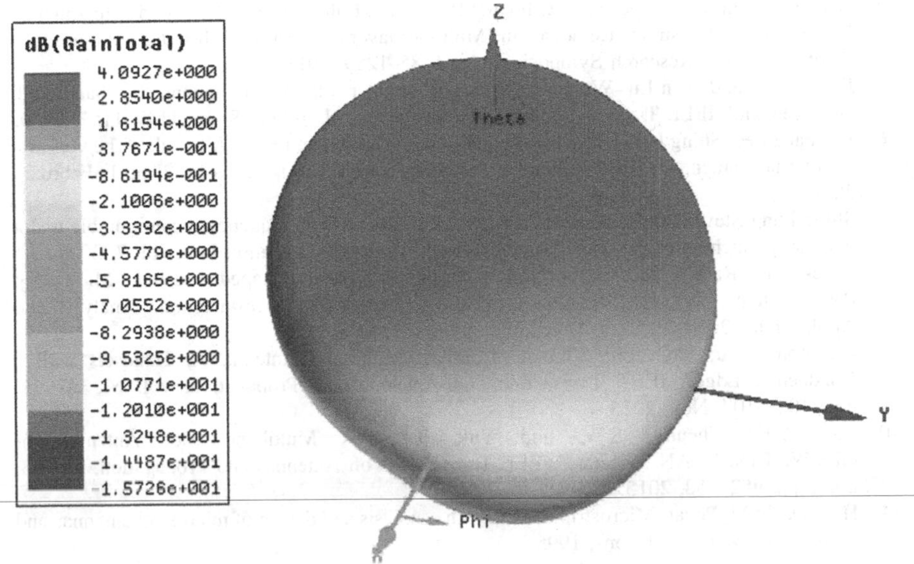


**Fig. 8** Three-dimensional gain plot

# 4 Conclusion

A novel multiple frequency microstrip antenna has been designed, analyzed, and discussed. This approach is novel in sense that it is simpler to design and fabricate. It exhibits promising results in terms of return loss, VSWR and antenna Gain which are certainly better and satisfactory. The proposed multiband antenna can resonate at 3.32, 5.5, 6.25, and 7.96 GHz frequencies to support modern wireless communication purposes including LTE, 4G, 5G, WiMAX, Satcomm as well as WLAN operations. It offers good radiation characteristics with appreciable gain and VSWR. The band channels observed at resonance frequencies are narrow which promise a fair and efficient spectrum utilization with minimum interference. These along with smaller antenna size make the antenna a very suitable choice in portable devices for wireless communication.

# References

1. David A. Sánchez-Hernández, "Multiband Integrated Antennas for 4G Terminals", Artech House, ISBN: 978-1596933989, 2008.
2. Mohammad A. Matin (Ed.), "Wideband, Multiband, and Smart Reconfigurable Antennas for Modern Wireless Communications", Idea Group (US), ISBN: 978-1466686458, 2015.
3. Soumyojit Sinha and AnjumanAraBegum, "Design of probe feed micro strip patch antenna in S-Band,'' International Journal of Electronics and Communication Engineering. Vol. 5, No. 4, 2012.

4. Neeraj Rao and Dinesh Kumar, "Gain and Bandwidth Enhancement of a Micro strip Antenna Using Partial Substrate Removal in Multiple layer Dielectric Substrate," Progress in Electromagnetics Research Symposium, pp. 1285–1289, 2011.
5. Jia-Yi Sze and Kin-Lu Wong, "Slotted rectangular microstrip antenna for bandwidth enhancement," IEEE Trans. on Antennas and Propag. vol. 48, no. 8, pp. 1149–1152, 2000.
6. Kai Fang Lee, Shing Lung Steven Yang, Ahmed A. Kishk, and Kwai Man luk, "The versatile u-slot patch antenna," IEEE Antennas and Propagation Magazine, vol. 52, no.1, February 2010.
7. Shing-Lung Steven Yang, Ahmed A. Kishk, Kai-Fong Lee "Frequency reconfigurable u-slot microstrip patch antenna," IEEE Antennas and Wireless Propagation Lett. vol. 7, 2008.
8. Vivek Singh Rathor, Jai Prakash Saini "A Design of Swastika Shaped Wideband Micro strip Patch Antenna for GSM/WLAN Application" Journal of Electromagnetic Analysis and Applications, 2014.
9. Cho-Kang Hsu; Shyh-Jong Chung, "Compact Multiband Antenna for Handsets with a Conducting Edge," IEEE Transactions on Antennas and Propagation, vol. 63, no. 11, pp. 5102–5107, Nov. 2015.
10. Cao, Y.F.; Cheung, S.W. and Yuk, T.I., "A Multiband Slot Antenna for GPS/WiMAX/WLAN Systems," IEEE Transactions on Antennas and Propagation, vol. 63, no. 3, pp. 952–958, 2015.
11. H. S. David M. Pozar, Microstrip antennas: the analysis and design of microstrip antennas and arrays: John Wiley and Sons, 1995.

# Part II
# Advanced Communications

# An Analysis of Resolution of Deadlock in Mobile Agent System Through Different Techniques

**Rashmi Priya and R. Belwal**

**Abstract** Mobile Agents are the set of processes that can move from one host to another host. The request granted by the client machine is executed by movement of mobile agents from a host machine to another host machine. The processes called by client are executed on a machine. The host machine must have all the resource needed to implement the service. This paper deals with allocation of resources through a proposed technique. In order to implement the tasks offered by the client, the mobile agents resume execution at the new machine (Nelson in Remote Procedure Call, PhD thesis, Computer Science, Carnegie Mellon University, 2002) [1]. The extension of client and server model is followed in the development of mobile agents. This paper also analyses the resolution of deadlock through different methods of study. The client only performs the operations provided by the server. There is scope of network scalable system when a particular server does not provide request desired by the client (Almes et al. in The Eden System: A Technical Review, IEEE Transactions on Software Engineering, 1998) [2].

**Keywords** Mobile agents · Deadlock · Client server · Resource

## 1 Introduction

As both sever and client keeps moving and interacting, there is a call for distributed applications. This is the reason an improvised technique for traditional approach of distributed computing is needed. Movements of agent are limited by imposing traditional distributed solutions into mobile agent systems as traditional solutions limit the movement of agents [3]. There are many assumptions on which traditional

R. Priya (✉)
TMU, Moradabad, India
e-mail: rashmi.slg@gmail.com

R. Belwal
AIT, Haldwani, India
e-mail: r_belwal@rediffmail.com


S.K. Bhatia et al. (eds.), *Advances in Computer and Computational Sciences*,
Advances in Intelligent Systems and Computing 553,
DOI 10.1007/978-981-10-3770-2_16

distributed algorithms depend. Some of these are data location, communication mechanisms or network organization, i.e., agent host location, number of nodes, and connections between hosts [4].

Mobile agents are becoming more popular in areas of application development [5]. Because of movement of both clients and servers freely through the network. Some of the fundamental assumptions of deadlock detection does not hold true. Hence, new techniques and solutions and techniques are to be developed for mobile agents to solve their distributed coordination problems.

It is providing competitive environment to traditional distributed computing techniques.

## 2 Distributed Deadlock Detection ("Traditional Approach")

Deadlock detection properties are similar in case of Single processor Systems and Distributed Systems. The deadlocks are harder to detect, avoid, prevent and resolve because of the nonavailability of centralized information and resource control [6]. These deadlocks are harder to detect. It becomes the responsibility of server to detect and resolve deadlocks in these cases of distributed systems [5]. As the transactions are distributed and a number of servers are interconnected across multiple servers, detection process of deadlock becomes a cumbersome problem. In such situation detection of deadlock becomes complex problem as number of servers and transactions are interconnected and distributed across network through multiple servers [7]. Under such scenario, distributed deadlock detection solutions can be divided into five categories: centralized, path-pushing, edge-chasing, diffusing computation, and global state detection [8]. A number of solutions have been proposed to detect and resolve deadlocks in distributed systems. A number of solutions are achieved by optimizing number of messages or frequency of detection. Distributed deadlock detection is implemented simply using a centralized server to maintain the global wait-for graph. A local copy of the wait-for graph is added which is analyzed on a periodical basis for each server in the network. In this technique, periodically each server in the network adds to the central server its local copy of the wait-for graph, which is analyzed for deadlocks [9]. As the single server can fail to handle many transactions this method does not follow fault tolerance. A local wait-for graph helps to achieve deadlocks and it is extended to the global graph as per the situation in demand. The global graph has no coordination at the central point as it is an extension to the local graph.

Based on the principle of a spanning tree of the global wait-for graph diffusing computation techniques are present. They represent processes and their links.

The root process sends messages to its connected processes and if a deadlock exists it eventually receives a message [10].

Based on the similar concepts to edge chasing global state detection algorithms category are present. It holds little difference in the ways to detect deadlocks. In this technique, the global wait-for graph is constructed from local graphs without blocking the computation [11]. In such systems processes are linked by the communication paths. Root processes can only issue messages; any non-root processes cannot issue messages [12]. This makes the root process special. All the processes, other than the root must wait until all the communication has been received by them. The deadlock detection techniques based on the concept of global graph are path-pushing and edge-chasing.

The techniques of path-pushing show similar aspects. They differ in the manner in which information is sent to neighboring nodes [3].

The other two categories of distributed deadlock detection are diffusing computation and global state detection. The technique developed by [10] is based on this paradigm.

There are two common criteria to evaluate Detection schemes. The two criteria are cited as (1) if an actual deadlock exists, it must be detected in a finite amount of time; (2) phantom deadlocks are not detected, i.e., the scheme must not find a deadlock that does not exist. As per the techniques used for single processor system there is no distributed deadlock detection which is universally detected, there is no universally accepted and deployed distributed deadlock solution, which is consistent with techniques used for single processor deadlock detection. Depending on the properties of an environment the two evaluation criteria must be applied and the most suitable scheme selected. Edge-chasing [13] has an assumption that standard deadlock avoidance techniques are used for deadlock detection. These techniques are two-phase transactions and resource locking. There is a link between servers and the interconnected hosts, for which local wait-for graph is created and then converted to global graph and for each connection a permanent edge is created. In case of edge-chasing, whenever a transaction waits for another transaction which in turn is waiting with a resource on remote server, a probe is initiated. The probe contains the local wait-for graph of the server. The probe is then sent to the remote server which holds the transactions placed on the resource present on remote server. In case a distributed deadlock is present, a cycle is found in the global graph, which needs to be broken.

# 3 Distributed Deadlock Solution

## *3.1 Deadlock Shadow Agent Technique*

The foundation and assumptions of this algorithm are presented first, followed by the detection and resolution phases of the algorithm. Finally, an example deadlock detection process is described in the section below. The following sections deals with topic of the thesis as proposed, i.e., it deals with the solution to distributed deadlock detection (Table 1).

**Table 1** Agents detection table

| Agent | Blocked resource | Primary locks |
|---|---|---|
| A11 | Accounted | User1.db |
| A22 | User1.db | Test1.db, Payroll1.db |
| A33 | Test1.db | Account1.db |
| A66 | Account1.db | Foo1.db |

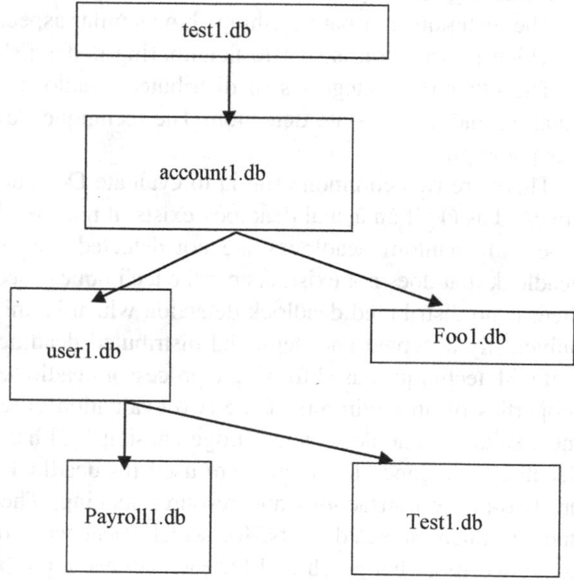


**Fig. 1** Detection graph of agents

**Table 2** Deadlock information

| Agent name | Blocked on resource | Primary locks |
|---|---|---|
| A1 | R2 (R2, A3, E2, 2) | R1 (R1, A1, E1, 1), R4 (R4, A1, E4, 4) |
| A3 | R1 (R1, A1, E1, 1) | R2 (R2, A3, E2, 2) |

The above criteria represents that there a cycle is detected in the graph Fig. 1. The detection table creates the detection graph.

1. The root node is the resource where detecting agent is found.
2. This is account1.db. The primary locks entry from the next nodes in the graph.
3. These nodes create entry to the agents.
4. This is used to locate the root node.

For each child node as it is added to the graph. The above step is repeated (Table 2).

## 3.2 Premise and Assumptions

The solution follows the topology to be static in the beginning of the working of algorithm. A mobile ad hoc networks (MANET) type topology update protocol must execute in the background to keep routing tables synchronized with the actual topology. This movement is accomplished via node or resource-based routing and the agent requests to be routed to a particular environment or resource. It is assumed that the host environment to route the agent on a best effort basis. This technique is similar to the routing scheme used in Internet Protocol (IP) routers. The mobile agents of all types are independent from the network topology. The agents can move independent of nodes with no detailing of number of nodes and their connections. They can move independently through out the network. This property allows host environments to communicate the details of Agent C. The assumption is taken that the Agent C can only lock and unlock the resources. It is assumed that the resource should be at the same environment physically. The agents manipulate the resource requests to the counterparts which detects the deadlock.

This solution assumes that agents must inform the host environment whenever any resource is blocked by them. Moreover, here the properties of a mobile agent solution are presented. These properties are responsible for deadlock detection.

This assumptions provide the environment for the distributed deadlock detection in a mobile agent system [7].

The state of agent is communicated to deadlock detection agents. The additional requests desired by the blocked agents are not granted. The blocked agents during the deadlock detection process are available in the agent system. The agents are assigned identifier preceding the blocking if any done by Agent C. The agent identifiers can be assigned even when a resource is locked.

The host environment approves the agents request to unblock. A unique identifier is needed by the Agent to identify them uniquely. The agents can block other agents only when their request is granted by the host environment. It is assumed that two-phase commit or priority transactions are used. These are standard deadlock avoidance technique. An agent during the detection process will not unlock any resource instantaneously. These properties allow the agent to hold a resource. It helps to avoid any phantom deadlock [11].

According to this property locking of resources can be locked or unlocked when its physical presence is in the same environment. The locking and unlocking of resources are done by Agent C. The property stated above helps in preventing phantom deadlock detection significantly.

There should be strong understanding and cooperation between the agents and shared resources to allow the deadlocks to occur. As the agents perform their tasks resources can be “locked,” which means their exclusivity is made to an individual consumer agent.

In addition the host is the ultimate authority and can allow or deny access to a resource. The assumption that during the locking process the Agent C must communicate with the host environment holds.

The host environment can deny the lock request depending on the tasks granted by an agent. It may block or wait on the resource. It may continue with the processing and its movement through the network. It is assumed that the blocking of agent is not done automatically. As the dynamicity of mobile agent could be disturbed by this.

## 4 Proposed Solution Theoretical Measurements

Due to the asynchronous nature of the distributed deadlock detection technique [4], the events suggested in the example do not execute in a sequential manner. Many of the migrations and processing required during the deadlock detection process occur in parallel and exploit locality of reference. This property means that estimating the total run time of the technique should not be based on the number of migrations, since many occur at the same time [12]. Factors such as host processor speed, deadlock implementation, network congestion, and host environment efficiency have a significant impact on the time required to detect and resolve a deadlock.

Assuming that a deadlock condition exists and the assumed properties are true, the following steps must occur to detect the deadlock:

Detector 1 migrates to agent environment 2 and checks the locks for resource 1. Detector 2 migrates to agent environment 1 and checks its locked resource, resource.

Detector 1 learns that Agent 2 is blocked on resource 1 and returns to its parent shadow agent. Detector 2 learns that Agent 1 is blocked on resource 2 and returns to its parent shadow agent.

Shadow Agent 1 checks the returned information, determines that there is no deadlock and adds Agent 2 (and its info) to its deadlock table. Shadow Agent 2 checks the returned information, determines that there is no deadlock and adds Agent 1 (and its info) to its deadlock table.

Shadow Agent 1 and Shadow Agent 2 re-initialize their detectors and restart the process.

Detector 1 migrates to agent environment 2 and checks the locks for resource 1. Detector 2 migrates to agent environment 1 and checks its locked resource, Detector 1 learns that Agent 2 is blocked on resource 1 and Agent 1 is secondarily blocked on resource 1. Detector 2 that Agent 1 is blocked on resource 2 and Agent 2 is secondarily blocked on resource 2. Both detector returns to their parent shadow agents.

Shadow Agent 1 checks the returned information, notices that Agent 1 was returned in the list of blocked agents and flags the deadlock. Shadow Agent 2 checks the returned information, notices that Agent 2 was returned in the list of blocked agents and flags the deadlock.
Both Shadow Agents used a defined technique to determine that resource 1 must be unlocked.
Shadow I initializes its detector to unlock resource 1 and to notify Agent 2 when the resource is unlocked.
Detector 1 migrates and unlocks the resource. Agent 2 is notified. Detector 1 returns to its parent and notifies it of unlock completion.

Assuming a fault free environment, it is possible to obtain theoretical values for the number of agent migrations and number of lock checks during the deadlock detection process. In-order or sequential execution must be maintained to predict these theoretical values. For the purposes of this discussion, consider a simple distributed deadlock situation. In this situation two agents (Agent 1, Agent 2), resources (Resource I, Resource 2) and environments (environment 1, environment 2) are interacting to create a distributed deadlock.

Before starting an analysis of the preceding sequence of events, several key concepts must be defined. A migration is defined as the movement of a mobile agent from one environment to another. Migrations are unidirectional; therefore at least two migrations are required for an agent to check a lock and return.

A detector agent is defined as the number of messages required to visit all of the resources locked by a consumer agent and return. The number of migrations in a trip is network topology dependent, but is always greater than the number of resources to vislt.

The migration of agent to a resource is named as a visit. In a similar manner to a trip, the number of migrations required to complete a visit is network topology dependent. The minimum number of migrations in a visit is two but normally this value is higher.

The following variables can be defined to represent the various elements of the suggested distributed deadlock detection technique: number of agents in a deadlock (NA), number of repetitions (NR), detection migration volume (MD), resolution migration volume (MR) messages in a trip (MT), and messages in a visit (Mv).

Analysis of the steps required to complete the example deadlock detection case leads to the following general statements.

NR = NA
MD = NAXNRXMT
MR = Mv

In the example deadlock, NA is 2 and MT and M, are 2. Applying these values gives: two repetitions, eight migrations for detection, and two migrations for resolution. By checking the event sequence, these values match the actual numbers found in the example.

Additional analysis shows that since NR equals NA the detection migration volume can be restated as: $N^* \times$ (MT), where N is the number of agents involved in the deadlock. It should be noted that these values are upper bounds for the volume of migrations and hence network load. They are not execution time estimates, since many factors not considered influence the time required to detect and resolve a deadlock using the suggested technique.

**Lemma** *The algorithm detects a deadlock within 2d time units.*

*Proof* When the initiator initiates the algorithm, it sends the CALL messages to all its successors which in turn propagates CALL message to its own successors. Consequently, a CALL message must travel from the initiator to the farthest leaf process of DST that requires at most d time hops in the worst case. Similarly, the leaf processes of DST must propagate the REPORT.

Message that must travel along the edges traversed by CALL messages. As a result, a REPORT message must travel at most d hops before reaching the initiator in the worst case. Thus, the worst case time complexity of the algorithm is 2d, where d is the diameter of the WFG.

*Proof* A process sends at most one CALL message on any of its outgoing edge and one REPORT message on any one of its incoming edges.

Therefore, the message complexity of proposed algorithm in the worst case is 2e, where e is the number of edges in the WFG.

Let us now measure the length of the control messages. The DSA carries the fixed number of process identifiers. Hence, the message length is a constant. Given Table compares the performance of proposed DSA with all distributed algorithms in the literature in terms of deadlock duration, message complexity, message size, and resolution overhead in terms of number of messages. With all performance measures mentioned, it clearly shows that the performance of proposed algorithm is better or equal to the existing algorithms (Tables 3, 4 and 5).

**Table 3** Performance comparison of distributed algorithms for detecting generalized deadlocks

| Algorithms comparison factor | Bracha-Toueg's algorithm (1987) | Wang et al.'s algorithm (1990) | Kshemkalyani and Singhal's algorithm (1994) | Kshemkalyani and Singhal's algorithm (1999) | DSA deadlock detection |
|---|---|---|---|---|---|
| Message complexity | 4e | 6e | 4e − 2n + 2l | 2e | 2e |
| Message size | O(1) | O(1) | O(1) | O(e) | O(1) |
| Deadlock resolution | No scheme message | No scheme message | No scheme message | No scheme message | 1 Message |

**Table 4** Recent research

| Sr. no. | Papers | Objective/research gap |
|---|---|---|
| 1 | Deadlock detection based on resource allocation graph (IEEE2009)<br>Qinqin Ni, Sen Ma | Uses principle of adjacency matrix<br>Random deadlock is difficult to detect |
| 2 | Formal Specification and Verification of the SET/A Protocol with an Integrated<br>Approach (ACM2009)<br>Vitus S.W. Lam and Julian Padget | SET/A protocol which is an agent-based payment protocol for credit card transactions in UML state chart diagrams is used<br>All properties are not satisfied when agent crashes |
| 3 | Deadlock avoidance method for multi-agent robot system<br>Using network of chaotic elements (IEEE2010)<br>Yoichiro Maeda, Takayasu Matsuura: | Deadlock can be avoided by changing singleton values of fuzzy rule<br>The efficiency of proposed algorithm by using multi-agent robot simulator has been shown needs to be tested |

**Table 5** Comparative study

| Sr. no. | Subject/theme/ objective | Findings | Research gap | Reference |
|---|---|---|---|---|
| 1 | Distributed deadlock detection | Survey articles based on deadlock detection by Knapp and Singhal | Distributed deadlock detection algorithms have substantial message overhead, even when there is no deadlock. These factors need to be encountered | "Deadlock detection in distributed database systems," ACM computing surveys, 2006 |
| 2 | Distributed deadlock detection algorithm | Badal has discussed a distributed deadlock detection algorithm that optimizes performance | The impact of a message loss on the various deadlock detection algorithms has not been discussed | Badal [8] |
| 3 | Path-pushing distributed deadlock detection algorithms | Path-pushing algorithms have been discussed by Gligor and Shattuck, Goldman and Menasce and Muntz | False deadlocks can not be detected. When a process detects a deadlock, it does not identify the lowest priority deadlocked process | Gligor and Shattuck [10] |

## 5 Conclusion

The idea of distributed system is much broader and effective than centralized systems. There are many examples of distributed systems, such as distributed databases, computer networks, real-time process control systems and distributed information processing systems. There are many areas expanding the middleware due to the implementation of application-level protocols for communication [7]. In case of distributed systems, the tasks to control the operating systems are cumbersome [10]; this is because database systems deadlock, mutual exclusion, and concurrency control are difficult in case of centralized systems.

As the processes have used resources at various sites to improve throughput distributed systems is more vulnerable to deadlock occurrence. Hence, there is a need to resolve deadlock by proper resource allocation. Further research needs to be initiated on detection of deadlock that requires lower cost and is more time efficient.

## References


1. Nelson, B.J., "Remote Procedure Call", PhD thesis, Computer Science, Carnegie Mellon University, (2002).
2. Almes, G.T., A.P. Black, E.D Lazowska, "The Eden System: A Technical Review", IEEE Transactions on Software Engineering, vol. 11, (1998).
3. Choudhary, A. L., W. H. Kohler, J. A. Stankovic, and D. Towsley, "A Modified Priority Based Probe Algorithm for Distributed Deadlock Detection and Resolution," IEEE Trans. on Software Engineering, 2008.
4. Chandy, K. M., and L. Lamport, "Distributed Snapshots: Determining Global States of Distributed Systems, "ACM Trans. on Computer Systems, 2003.
5. Bacor, J.M., and K.G. Hamilton, Distributed Computing with RPC: The Cambridge Approach, "ACM Transaction on Computer Systems", vol. 5, (2003).
6. Goldman, B., "Deadlock Detection in Computer Networks," Technical Report MIT/LCS/TR185, MIT, 2006.
7. Cheriton, D.R., "The V Distributed System", Communications of the ACM, Vol. 30, (2000).
8. Badal, D. J. "The Distributed Deadlock Detection Algorithm," ACM Trans. on Computer Systems, (1998).
9. Elmagarmid, A.K., N. Soundararajan, and M.T. Liu. "A Distributed Deadlock Detection and Resolution Algorithm and Its Correctness," IEEE Trans. on Software Engineering, 2007.
10. Gligor, V. D. and S. H. Shattuck. "On Deadlock Detection in Distributed System," IEEE Trans. on Software Engineering, 2010.
11. Bracha, G., and S. Toueg, "Distributed Deadlock Detection," Distributed Computing, (2000).
12. Bracha, G., and S. Toueg, "A Distributed Algorithm for Generalized Deadlock Detection," Proc. Of the ACM Symposium on Principles of Distributed Computing, Aug. 1999.
13. Chandy, K. M., J. Misra, and L. M. Haas. "Distributed Deadlock Detection," ACM Trans. on Computer Systems, 2005.
14. Lampson, B., "Remote Procedure Calls", Lecture Notes in Computer science, vol. 105, Springer Verlag, New York, (2004).
15. Chang, E., "Echo Algorithms: Depth Parallel Operations on General Graphs," IEEE Trans. on Software Engineering, July 2007.

# Robustness Analysis of Buffer-Based Routing Algorithms in Wireless Mesh Network

**Kanika Agarwal, Nitin Rakesh and Abha Thakral**

**Abstract** Wireless Mesh Network arises as a promising innovation for providing quick and productive communication for which numerous algorithms have been proposed in networking infrastructure. For routing there are various performance parameters such as throughput, network congestion, resiliency, fairness, robustness, network jitter, delay, stability, optimality, simplicity, completeness, etc. Robustness provides the capability to deal with all the failures that come across during the connection in the network to increase the network performance. In this paper, we have shown and analyzed three algorithms namely robustness parameter Resilient multicasting, Resilient Opportunistic Mesh Routing for Wireless Mesh Network (ROMER) and Buffer-Based Routing (BBR) in Wireless Mesh Networks. We have shown that network performance of BBR is better than these approaches which is analyzed through various parameters such as network congestion, network throughput and resiliency.

**Keywords** Resilient multicasting · ROMER · Buffer-based routing · WMN · Robustness

## 1 Introduction

Wireless Mesh Networks (WMN) proposes a decentralized structural engineering for setting multi-hop wireless communications. The decentralized structural planning brings advantages such as ease of deployment, maintenance, scalability and consistency. However, WMN is deficient in high-level services such as handoff and

K. Agarwal (✉) · N. Rakesh · A. Thakral
Amity School of Engineering and Technology, Amity University, Noida, UP, India
e-mail: kanikaagarwalgtb@gmail.com

N. Rakesh
e-mail: nitin.rakesh@gmail.com

A. Thakral
e-mail: abhareads@gmail.com



S.K. Bhatia et al. (eds.), *Advances in Computer and Computational Sciences*,
Advances in Intelligent Systems and Computing 553,
DOI 10.1007/978-981-10-3770-2_17

mobility management [1]. Routing is process of transferring information across a network from source to destination. It can also be referred to as the process of selecting a path to send the packet. To provide routing services efficiently and appropriately there are many characteristics that need to be analyzed in a routing algorithm which could help in packet transmission in a network. In context to computer network, robustness is the capability of the network to deal with all the failures that occurs during the transmission of message or packet from source to destination. The most appropriate application for robustness is to make routing algorithm so resistant that if error occurs it should not affect the normal functioning of the network.

The issue that exists during communication is management of bundle transmission from source to destination efficiently and demonstrating the algorithm that it is effective in nature. In our previous work [2, 3] Buffer-Based Routing (BBR) was analyzed on three parameters, i.e. system throughput, network congestion and resiliency and it was established while comparing with Resiliency Multicasting [4] and ROMER [5] that BBR is more effective and efficient to previous approaches in terms of Robustness also as the way BBR operates involves definite buffered nodes that support in travelling of packet to its destination even in the face of failures. In this paper, we consider another critical parameter to further upgrade the effectiveness of BBR methodology.

The paper is divided in five sections. In section first we have introduced the problem. Second section discusses the related work in the field of routing protocols. The third section introduced the proposed work by comparing the algorithms on the basis of cost. In section four, the results and analysis are presented. Section five concluded the manuscript, which is followed by references of the manuscript.

## 2 Related Work

Zhao et al. in [4] presented Resilient Multicasting approach which requires two node disjoint paths for every pair of source to destination. These disjoint paths are link disjoint and node disjoint. Link disjoint do not have any link in common and node disjoint do not have any node in common except the source and destination. Yuan et al. [5] proposed Resilient Opportunistic Mesh Routing for mesh Network as a solution and provides with the balance between long-term and short-term performance. It works on R(credit ratio) and T(threshold) value. This mechanism was used to provide differentiated robustness for various categories of data packets. Rathee et al. in [2] projected an approach as Buffer-based Routing Algorithm as to propose effective solution w.r.t resilient multicasting and ROMER. This algorithm is used to maintain buffer at every alternative nodes in the network. These buffered nodes are half the number of nodes present in the network. This approach maintains a routing table keeping all information of the node. Cho et al. in [6] developed an independent directed acyclic graph for resilient multipath routing which follows a path from source to root. This graph is link disjoint in nature. They also develop an algorithm for computation of link-independent and node-independent graphs. Zeng et al. [7] proposed a protocol named as opportunistic multicast protocol for improving throughput of the network. This protocol helped to enhance the unicast throughput in the network. Main concept of the protocol is its tree backbone. The protocol presents the tradeoff between traditional multicast protocol and unstructured protocols. Xin [8] studied the multipoint multicasting for distributed environment in the mesh network targeting to minimize the time slots for exchanging messages among many nodes in the network. The paper presented an algorithm for multicasting algorithm and analyzes its time complexity. The time taken by the algorithm is O(d log n + k). Bruno et al. [9] proposed a routing algorithm called MaxOPP. It takes a localized routing decision for selection of forwarding nodes. The selection of the nodes is on per packet basis and at run time. Fang [10] proposed an opportunistic algorithm for improving the performance of the network. Various problems have been studied for choosing the route for every user so that they can optimize the total profit of various users in the network concerning node constraint. The paper formulated two problems for programming system. By two methods, i.e. Primaldual and Subgradient, an iterative approach named Consort: node constraint opportunistic routing. For every iteration, it updates the LaGrange multiplier in distributed environment according to user and node. Zhang et al. [11] presented an overview of opportunistic routing, the challenges faced in implementing the routing. The paper presented various routing protocol such as ExOR, ROMER, SROR, etc, for achieving increased throughput in comparison with traditional routing Aajami et al. [12] studied various approaches such as wireless interflow network and opportunistic routing for enhancing the throughput. A solution has been proposed by combining these approaches. The paper suggested a technique abbreviated as MRORNC as an integrated cross-layer approach for determining packet, next hop and transmission rate.

## 3 Proposed Work

*Robustness Analysis of Resilient Multicasting*: Zhao et al. [4] proposed that the algorithm works on disjoint paths having no node as common except the source and destination. We have used the path traversed from source to destination for network with varying number of nodes to demonstrate.

For example in the network [3], Fig. 1 when there are 5 nodes, the algorithm chooses two nodes disjoint paths to send the packet from source (node A) to destination (node E) are: $A-B-D-E$ and $A-C-E$. Choosing first path as $A-B-D-E$ having its cost as $1+2+2=5$ units, and in $A-C-E$ as $2+1=3$ units, leads to total of 8 units for the packet to travel. Figure 2 represents network with 10 nodes. Taking two disjoint paths from source (node A) to destination (node J) are: $A-C-H-F-I-G-J$ and $A-D-B-E-J$. Choosing first path as $A-C-H-F-I-G-J$ having its cost as $1+3+4+2+1+1=12$ units and

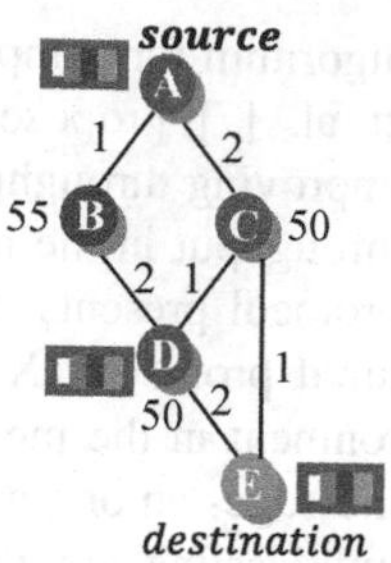

**Fig. 1** Network showing 5 nodes [3]

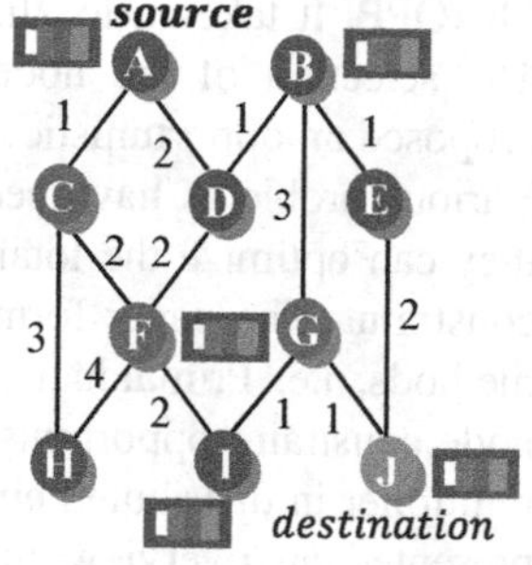

**Fig. 2** Network showing 10 nodes [3]

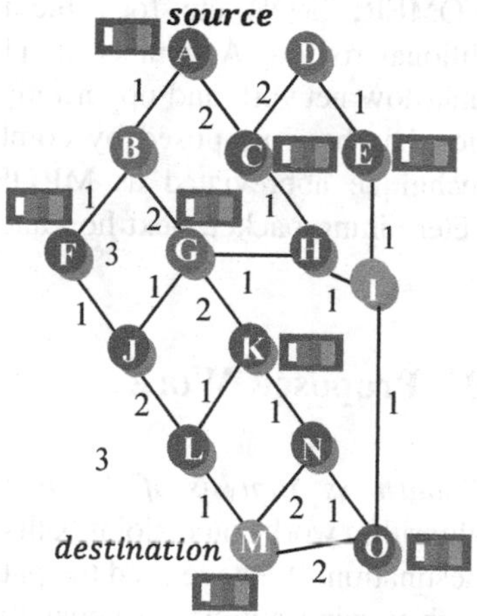

**Fig. 3** Network showing 15 nodes [3]

$A-D-B-E-J$ as $2+1+1+2=6$ units, leads to total of 18 units for the packet to travel. Figure 3 represents network with 15 nodes.

The two disjoint paths from source (node A) to destination (node M) are: $A-C-H-I-O-M$ and $A-B-G-K-N-M$. Choosing first path as $A-C-H-I-O-M$ having its cost as $2+1+1+1+2=7$ and $A-B-G-K-N-M$ as $1+2+2+1+2=8$ units, leads to total of 15 units for the packet to travel. Figure 4 represents network with 20 nodes. Taking two disjoint paths from

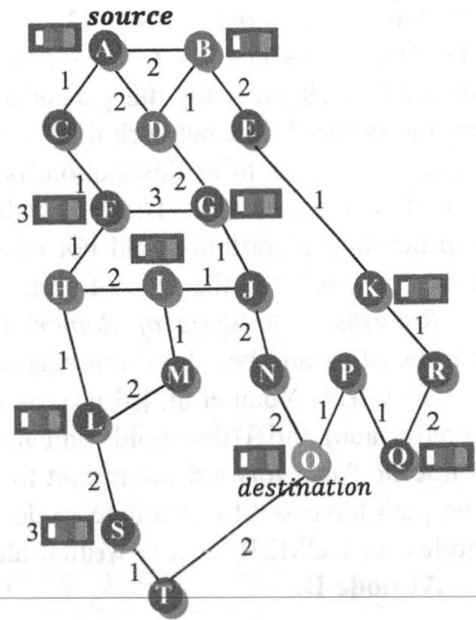


**Fig. 4** Network showing 20 nodes [3]

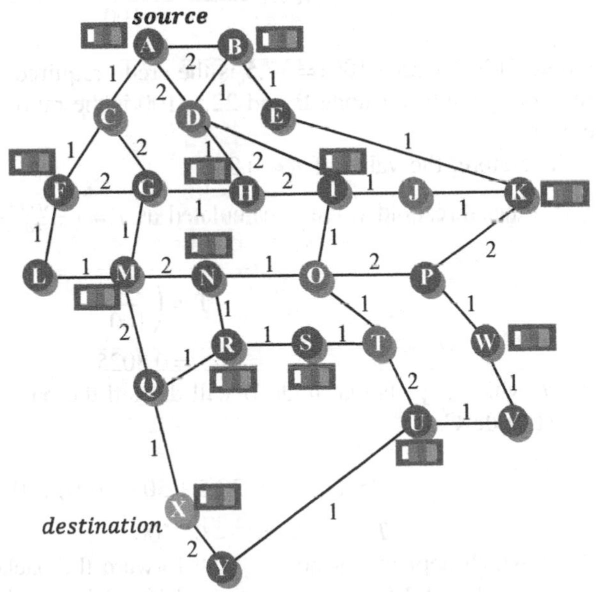


**Fig. 5** Network showing 25 nodes [3]

which packet can travel from source (node A) to destination (node O) are: $A-C-F-H-L-S-T-O$ and $A-B-E-K-R-Q-O$. Choosing first path as $A-C-F-H-L-S-T-O$ having its cost as $1+1+1+1+2+1+2=9$ units and $A-B-E-K-R-Q-O$ as $2+2+1+1+2+1+1=10$ units, leads to total of 19 units for the packet to travel. Figure 5 represents network with 25 nodes.

The two disjoint paths from source (node A) to destination (node X) are: $A-C-G-M-Q-X$ and $A-D-H-I-O-T-U-Y-X$ Choosing first path as having its cost $A-C-G-M-Q-X$ as $1+2+1+2+1=7$ units and $A-D-H-I-O-T-U-Y-X$ as $2+1+2+1+1+2+1+2=12$ units, leads to total of 19 units for the packet to travel. In multicasting algorithm whenever failure occurs in the network the source can prefer another path to route the packet in order to reach to its destination, but when all the routes fail then no packet will travel in the network. Therefore, through all this study we can conclude that multicasting algorithm could not pass this robustness parameter as it is unable to route the packet at the time of node failure [13, 14].

*Robustness Analysis of Romer*: Due to node failure possibility in Resilient Multicasting another algorithm was developed that overcome all its disadvantages, i.e. ROMER. Yuan et al. [5] proposed that the algorithm works on the value of R (credit ratio) and T(threshold) on every node that has its value of R greater than its value of T can forward the packet to the possible route. We will be demonstrating the path traversed from source to destination for network with varying number of nodes for ROMER, cost is written along.

**At node B**:

$$R = \frac{(100-(120.5+55-100)}{100}$$

where $(120.5+55-100)=77.5$ is the credit required, $(100-(77.5))=22.5$ is the remaining credit for node B and $22.5/100$ is the ratio of remaining credit to initial credit.

Therefore, the value of $R=0.245$.

Further threshold value is calculated as $T=\left(\frac{cost\ of\ node\ B}{cost\ of\ source}\right)^2$.

$$T=\left(\frac{55}{100}\right)$$

$$=0.3025$$

$R<T$ which depicts that node B will discard the packet.

**At node C**:

$$R=(100-(120.5+50-100)/100)=0.295$$

$$T=(50/100)^2=0.2500$$

$R>T$ which depicts that node C will forward the packet.

Tables 1 and 2 are representation of Fig. 1 in which there are 5 nodes, the two possible paths (out of three available options) from which packet can travel from source (node A) to destination (node E) are: $A-B-D-E$ and $A-C-E$ and nodes B is having the value of R less than the value of T which shows that this node fails to transmit the packet. Choosing first path as $A-B-D-E$ having its cost as 1 units, and in $A-C-E$ as $2+1=3$ units, leads to total of 4 units for the packet to travel.

**Table 1** Credit unit and Node unit of network of 5 nodes as in Fig. 1

| Node | Credit unit | Node unit |
|---|---|---|
| A | 100 | 100 |
| B | 120.5 | 55 |
| C | 120.5 | 50 |

**Table 2** Showing the value of R (credit ratio) and T (through put) in network of 5 nodes as in Fig. 1

| Node | R | T | Comparison | Function |
|---|---|---|---|---|
| B | 0.245 | 0.3025 | $R < T$ | Discard the packet |
| C | 0.295 | 0.2500 | $R > T$ | Forward the packet |

**Table 3** Credit unit and Node's unit of network of 10 nodes as in Fig. 2

| Node | Credit unit | Node unit |
|---|---|---|
| A | 100 | 100 |
| B | 120.5 | 50 |
| C | 120.5 | 50 |
| D | 120.5 | 55 |
| E | 120.5 | 50 |
| F | 121.5 | 55 |
| H | 122.5 | 51 |

**Table 4** Showing value of R and T in network of 10 nodes as in Fig. 2

| Node | R | T | Comparison | Function |
|---|---|---|---|---|
| B | $R = \left(100 - \frac{122.5 + 50 - 100}{100}\right) = 0.275$ | $T = \left(\frac{50}{100}\right)^2 = 0.2500$ | $R > T$ | Forward the packet |
| C | $R = \left(100 - \frac{(120.5 + 50 - 100)}{100}\right) = 0.295$ | $T = \left(\frac{50}{100}\right)^2 = 0.2500$ | $R > T$ | Forward the packet |
| E | $R = \left(100 - \frac{120.5 + 50 - 100}{100}\right) = 0.295$ | $T = \left(\frac{50}{100}\right)^2 = 0.2500$ | $R > T$ | Forward the packet |
| F | $R = \left(100 - \frac{(122.5 + 51 - 100)}{100}\right) = 0.265$ | $T = \left(\frac{51}{100}\right)^2 2 = 0.2601$ | $R > T$ | Forward the packet |
| H | $R = \left(100 - \frac{121.5 + 55 - 100}{100}\right) = 0.235$ | $T = \left(\frac{55}{100}\right)^2 = 0.3025$ | $R < T$ | Discard the packet |

Similarly, Tables 3 and 4 are representation of Fig. 2 in which there are 10 nodes, the two possible paths (out of all available options) from which packet can travel from source (node A) to destination (node J) are: $A-C-F-I-G-J$ and $A-C-F-I-G-B-E-J$. Choosing first path as having its cost $A-C-F-I-G-J$ as $1+2+2+1+1=7$ units and $A-C-F-I-G-B-E-J$ as $1+2+2+1+3+1+2=12$ units, leads to total of 19 units for the packet to travel.

As previously performed for 5 and 10 nodes, Tables 5 and 6 are representations of Fig. 3 in which there are 15 nodes, the two possible paths (out of all available

**Table 5** Credit unit and Node's unit of network of 15 nodes as in Fig. 3

| Node | Credit unit | Node unit |
|---|---|---|
| A | 100 | 100 |
| B | 120.5 | 51 |
| C | 121.5 | 50 |
| D | 121.5 | 51 |
| E | 122.5 | 50 |
| F | 120.5 | 50 |
| G | 122.5 | 55 |
| H | 121.5 | 51 |

**Table 6** Showing value of *R* and T in network of 15 nodes as in Fig. 3

| Node | R | T | Comparison | Function |
|---|---|---|---|---|
| B | 0.295 | 0.2500 | $R>T$ | Forward the packet |
| C | 0.285 | 0.25 | $R>T$ | Forward the packet |
| E | 0.275 | 0.25 | $R>T$ | Forward the packet |
| F | 0.295 | 0.25 | $R>T$ | Forward the packet |
| G | 0.225 | 0.3025 | $R<T$ | Forward the packet |
| H | 0.275 | 0.2601 | $R>T$ | Forward the packet |

options) from which packet can travel from source (node A) to destination (node M) are: $A-B-F-J-G-K-N-M$ and $A-C-F-J-L-K-N-O-M$. Choosing first path as $A-B-F-J-G-K-N-M$ having its cost as $1+1+1+1=4$ units(till node G) and $A-C-F-J-L-K-N-O-M$ as $1+1+1+2+1+1+1+2=10$ units, leads to total of 14 units for the packet to travel.

As per Tables 7 and 8 for 20 nodes there are two possible paths (out of all available options) from which packet can travel from source (node A) to destination (node O) are: $\boldsymbol{A-B-D-G-F-H-L-S-T-O}$ and $\boldsymbol{A-D-G-J-N-O}$. Choosing first path as $\boldsymbol{A-B-D-G-F-H-L-S-T-O}$ having its cost as $2+1+2+3+1+1+2+1+2=15$ units and $\boldsymbol{A-D-G-J-N-O}$ as $2+2+1=5$ units, leads to total of 20 units for the packet to travel. Therefore, the same calculation can be done in the case of 25 nodes and the paths will be discard are according to incapable nodes.

*Robustness Analysis of Buffer-Based Routing*: Rathee et al. proposed in [2] that the algorithm works on two conditions, i.e. the packet travels through the route that must contain minimum number of buffered node and if more than one path has same number of buffered node than it will select the least cost path from source to destination. We will be demonstrating the path traversed from source to destination for network with varying number of nodes.

Table 9 represents Fig. 1 in which there are 5 nodes, having two possible paths (out of three available options) from which packet can travel from source (node A) to destination (node E) are: A-C-E and $\boldsymbol{A-B-D-E}$. Choosing first path as

**Table 7** Credit unit and Node's unit of network of 20 nodes as in Fig. 4

| Node | Credit unit | Node unit |
|---|---|---|
| A | 100 | 100 |
| B | 121.5 | 50 |
| C | 120.5 | 51 |
| D | 120.5 | 50 |
| E | 120.5 | 55 |
| F | 121.5 | 55 |
| G | 121.5 | 51 |
| H | 122.5 | 50 |
| I | 122.5 | 51 |
| J | 122.5 | 55 |
| L | 123.5 | 50 |
| M | 123.5 | 51 |
| N | 123.5 | 55 |
| O | 124.5 | 55 |
| S | 124.5 | 50 |
| T | 124.5 | 51 |

**Table 8** Showing value of R and T in network of 20 nodes as in Fig. 4

| Node | R | T | Comparison | Function |
|---|---|---|---|---|
| B | 0.285 | 0.25 | $R > T$ | Forward the packet |
| C | 0.285 | 0.2601 | $R > T$ | Forward the packet |
| D | 0.295 | 0.25 | $R > T$ | Forward the packet |
| E | 0.245 | 0.3025 | $R < T$ | Forward the packet |
| F | 0.295 | 0.2500 | $R > T$ | Forward the packet |
| G | 0.285 | 0.2500 | $R > T$ | Forward the packet |
| H | 0.275 | 0.25 | $R > T$ | Forward the packet |
| I | 0.265 | 0.2601 | $R > T$ | Forward the packet |
| J | 0.225 | 0.3020 | $R < T$ | Discard the packet |
| L | 0.265 | 0.25 | $R > T$ | Forward the packet |
| M | 0.255 | 0.2601 | $R < T$ | Discard the packet |

**Table 9** Network of 5 nodes

| Paths | Cost | Buffered nodes |
|---|---|---|
| A-B-D-E | 5 | 3 |
| A-C-D-E | 5 | 3 |
| A-C-E | 3 | 2 |

**Table 10** Network of 10 nodes

| Paths | Cost | Buffered nodes |
|---|---|---|
| A-D-B-E-J | 6 | 3 |
| A-D-F-I-G-J | 8 | 4 |
| A-C-H-F-I-G-J | 12 | 3 |
| A-C-F-I-G-J | 7 | 4 |
| A-D-B-G-J | 7 | 3 |

**Table 11** Network of 15 nodes

| Paths | Cost | Buffered nodes |
|---|---|---|
| A-B-F-J-L-M | 6 | 3 |
| A-B-G-K-N-M | 7 | 3 |
| A-B-G-K-L-M | 7 | 4 |
| A-B-G-J-L-M | 7 | 3 |
| A-B-G-K-N-O-M | 9 | 5 |
| A-C-H-I-O-M | 7 | 3 |

$\boldsymbol{A-C-E}$ having its cost as $2+1=3$ units, and in $\boldsymbol{A-B-D-E}$ as $1+2+2=5$ units, leads to total of 8 units for the packet to travel.

Similarly, Table 10 represents (Fig. 2) two possible paths (out of all available options) from which packet can travel from source (node A) to destination (node J) are: $\boldsymbol{A-D-B-E-J}$ and $\boldsymbol{A-C-F-I-G-J}$ Choosing first path as $\boldsymbol{A-D-B-E-J}$ having its cost as $2+1+1+2=6$ units and $\boldsymbol{A-C-F-I-G-J}$ as $1+2+2+1+1=7$ units, leads to total of 13 units for the packet to travel. Table 11 represents two possible paths (out of all available options) for 15 nodes from which packet can travel from source (node A) to destination (node M) are $\boldsymbol{A-B-F-J-L-M}$ and $\boldsymbol{A-C-H-I-O-M}$. Choosing first path as $\boldsymbol{A-B-F-J-L-M}$ having its cost as $1+1+1+2+1=6$ units and $\boldsymbol{A-C-H-I-O-M}$ as $2+1+1+1+2=7$ units, leads to total of 13 units for the packet to travel.

Table 12 (represents Fig. 4) where there are 20 nodes, the two possible paths (out of all available options) from which packet can travel from source (node A) to destination (node M) are: $\boldsymbol{A-C-F-G-J-N-O}$ and $\boldsymbol{A-B-D-G-J-N-O}$. Choosing first path as A-C-F-G-J-N-O having its cost as $1+1+3+1+2+2=10$ units and $\boldsymbol{A-B-D-G-J-N-O}$ as $2+1+2+1+2+2=10$ units, leads to total of 20 units for the packet to travel. Figure 5 (as detailed in Table 13) when there are 25 nodes, the two possible paths (out of all available options) from which packet can travel from source (node A) to destination (node M) are: $\boldsymbol{A-C-G-M-Q-X}$ and $\boldsymbol{A-B-D-I-O-T-U-Y-X}$. Choosing first

**Table 12** Network of 20 nodes

| Paths | Cost | Buffered nodes |
|---|---|---|
| A-B-E-K-R-Q-P-O | 10 | 5 |
| A-C-F-H-I-J-N-O | 11 | 4 |
| A-B-D-G-J-N-O | 10 | 4 |
| A-C-F-G-J-N-O | 10 | 4 |

**Table 13** Network of 25 nodes

| Paths | Cost | Buffered nodes |
|---|---|---|
| A-C-F-L-M-Q-X | 7 | 4 |
| A-C-G-M-Q-X | 7 | 3 |
| A-D-H-G-M-Q-X | 8 | 4 |
| A-B-E-K-P-W-V-U-Y-X | 12 | 6 |
| A-B-D-I-O-T-U-Y-X | 12 | 5 |

**Table 14** Table showing the units consumed by the packet from source to destination

| Algorithms | 5 Nodes | 10 Nodes | 15 Nodes | 20 Nodes | 25 Nodes |
|---|---|---|---|---|---|
| Resilient multicasting | 8 | 18 | 15 | 19 | 19 |
| ROMER | 4 | 19 | 14 | 20 | 18 |
| BBR | 8 | 13 | 13 | 20 | /19 |

path as $A-C-G-M-Q-X$ having its cost as $1+2+1+2+1=7$ units and $A-B-D-I-O-T-U-Y-X$ as $2+1+2+1+1+2+1+2=12$ units, leads to total of 19 units for the packet to travel.

## 4 Result and Analysis

We are evaluating the robustness of the three algorithms (in terms of cost units). As we can see that robustness is inversely proportional to the cost. Lesser the cost of the packet higher will be the robustness. Cost evaluated on robustness parameter for packet to reach from source to destination in network of different sizes is shown in Table 14 with respect to the three algorithms. In case of multicasting, the cost will be according to two disjoint paths taken in the network. In case of ROMER, the source chooses two paths, if in any path there is a node which is unable to forward the packet further in the network then the cost will be considered up to the node causing failure in the network, in addition to the cost of next path (shown in Fig. 6). In Buffer-Based Routing, the cost depends on the path containing least buffered nodes.

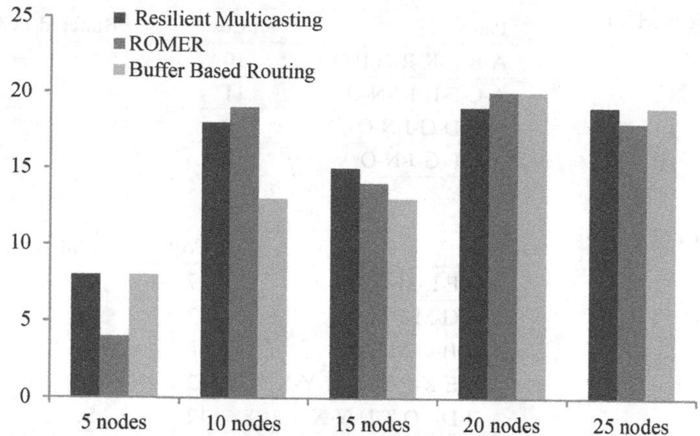


**Fig. 6** Showing the graphical representation of the values of Table 14

# 5 Conclusions and Future Work

This paper explores the robustness parameter in diverse sizes of network that how it deals with errors or failures during the transmission of the packet in all three algorithms. We have shown our results with resilient multicasting algorithm and ROMER using Buffer-based resilient routing approach. We evaluated and compared robustness while transmitting every packet in the network. We have evaluated our results over distinctive size of the networks and we conclude that BBR shows better result when contrasted with resilient multicasting and ROMER. As Resilient Multicasting and ROMER algorithm has more probability of failures whereas in BBR the failure handling capability is more because of buffered nodes are present in the network, which serves to choose another path taken by the previously buffered node whenever the failure occurs.

# References

1. J. Chung, G. González, I. Armuelles, T. Robles, R. Alcarria, A. Morales, "Experiences and Challenges in Deploying Open Flow over a Real Wireless Mesh Network," IEEE Latin America Transactions, vol. 11, 2013.
2. Geetanjali Rathee, AnkitMundra, Nitin Rakesh, S. P. Ghera, "Buffered Based Routing Approach for WMN," IEEE International Conference of Human Computer Interaction, Chennai, India, 2003.
3. Geetanjali Rathee, Nitin Rakesh, "Resilient Packet Transmission (RPT) for Buffer Based Routing (BBR) Protocol," Journal of Information Processing System, October, 2014.
4. Xin Zhao, Jun Guo, Chun Tung Chou, and Sanjay K. Jha, "Resilient multicasting in wireless mesh networks," 13th International Conference on Telecommunication, Polo de Aveiro, Portugal, 2006.

5. Yuan Yuan, Hao Yang, Starsky H. Y.Wong, Songwu Lu, and William Arbaugh, "ROMER: Resilient Opportunistic Mesh Routing for Wireless Mesh Networks," First IEEE Workshop on Wirless Mesh Network (WiMesh), vol. 12, 2005.
6. Cho, Olga, Theodore Elhourani, and Srinivasan Ramasubramanian, "Resilient multipath routing with independent directed acyclic graphs," Communications (ICC), 2010 IEEE International Conference on. IEEE, 2010.
7. Zeng, Guokai, Pei Huang, Matt Mutka, Li Xiao, and Eric Torng, "Efficient Opportunistic Multicast via Tree Backbone for Wireless Mesh Networks," Mobile Adhoc and Sensor Systems (MASS), 8th International Conference on. IEEE, 2011.
8. Qin Xin, and Yanbo J. Wang, "Latency-efficient Distributed M2 M Multicasting in Wireless Mesh Networks Under Physical Interference Model," IEEE 2010.
9. Bruno, Raffaele, Marco Conti, and MaddalenaNurchis, "MaxOPP: A novel Opportunistic Routing for wireless mesh networks, " Computers and Communications (ISCC), Symposium on, IEEE, 2010.
10. Xi fang, Dejun Yang, and Guoliang Xue, "Consort: node-constrained opportunistic routing in wireless mesh networks, " INFOCOM, 2011 Proceedings IEEE, 2011.
11. Zhang, Zhensheng, and Ram Krishnan, "An Overview of Opportunistic Routing in Mobile Ad Hoc Networks," Military Communications Conference, MILCOM, IEEE, 2013.
12. Mojtaba Aajami, Hae-Ryeon Park, and Jung-Bong, "Combining Opportunistic Routing and Network Coding: A Multi Rate Approach, "Wireless Communications and Networking Conference (WCNC): NETWORKS, IEEE 2013.
13. Pawan Kumar Verma, Tarun Gupta, Nitin Rakesh, Nitin, "A Mobile Ad-Hoc Routing Algorithm with Comparative Study of Earlier Proposed Algorithms", Int. J. Communications, Network and System Sciences, 2010, vol. 3, pp 289–293.
14. J. Olsén, "On Packet Loss Rates used for TCP Network Modeling, "Technical Report, Uppsala University, 2003.

# The Security Challenges and Opportunities of New Network Under the Hybrid Cloud Environment

**Yuxiang Dong, Huijun Zhang and Linong Zhao**

**Abstract** A hybrid cloud becomes the preferred solution when enterprises deploy cloud service. Using new technology such as software-defined network (SDN) and network virtualization to form a network will be the trend of the future. This paper introduces the new safety protection opportunities brought by new network technologies, and analyzes the challenges and risks of hybrid cloud system using these technologies. Finally, it puts forward the corresponding solution.

**Keywords** Hybrid cloud · Software-defined network · Network function virtualization · Network security

## 1 Introduction

In the past five years, the Internet has undergone tremendous changes. New IT Infrastructure and applications such as Cloud computing (Cloud computing is a style of computing in which dynamically scalable and often virtualized resources are provided as a service over the Internet [1]), big data, Mobile Internet, and the Internet of things (IOT) are becoming more and more widely used [2]. For enterprise users, 26.1% of users choose Cloud computing as an investment focus, while 27.4% of Small and Medium Enterprises (SMEs) tend to choose SDN center as the focus of investment over the next 12 months [3]. These companies are likely to merge IT systems in physical office environment and virtual private cloud to form a hybrid cloud (hybrid cloud is a cloud computing environment which uses a mix of on-premises, private cloud and third-party, public cloud services with orchestration between the two platforms [4]).

Y. Dong · H. Zhang (✉) · L. Zhao
China Mobile Group Chongqing Co., Ltd, Chongqing 401122, China
e-mail: mms@139.com



S.K. Bhatia et al. (eds.), *Advances in Computer and Computational Sciences*,
Advances in Intelligent Systems and Computing 553,
DOI 10.1007/978-981-10-3770-2_18

Fig. 1 Fusion border of hybrid cloud

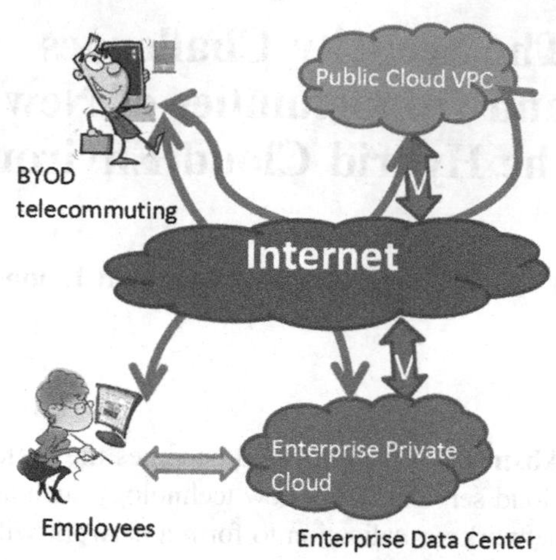


On the other hand, in order to achieve office automation, many enterprises allow the use of mobile devices, so Bring Your Own Device (BYOD) combines with the Cloud computing information system for the office has emerged [5]. Popularization of cloud platform increases the risk of data leakage and network attack. There are all kinds of threats in network, host, virtual resource management, data security, etc. Using mobile office and BYOD hybrid cloud blurs the boundaries between different IT systems of enterprise. As shown in Fig. 1, BYOD blurs the boundaries between fixed assets and mobile assets, so there is not a fixed border. Mobile office enables employees to access internal resources outside of the workplace, so that the entire boundary is beyond the scope of the enterprises physical area. Hybrid cloud is a blend of enterprise of physical devices and virtual assets, and it makes the whole boundary cannot rely on simple physical facilities to control [6]. All safety protective equipment and safety mechanism depend on the boundary will fail, and the attackers can always find the inconsistencies between physical boundaries and logical boundaries to invade. Even if the cloud system builders build a consistent boundary, with the change of virtual resources (such as virtual machines migration), these boundaries may antiquate too, so an attacker can also bypass the boundaries.

At the same time, some new technologies such as SDN and network function virtualization (NFV) appear. On the one hand, these new technologies speed up the Internet industry resource change and flow rate control, and they provide a new technical support for safety protection. On the other hand, these new technologies also bring new risks and challenges for the system.

This paper introduces new challenges of security hybrid cloud system brought by SDN/NFV, and analyzes the new security risks. At last, it gives the appropriate response suggestions.

# 2 New Opportunities of SDN/NFV in the Security Protection

Networks of hybrid Cloud computing information system include Cloud computing physical networks, virtual networks in virtual system, and networks of customer system. In general, security domain is used to isolate networks, so the access control mechanism is deployed on the border. At last, different safety equipment and protective measures are deployed in inner areas. The emergence of SDN and NFV brings a new concept of protection for traditional security system [7]. Mixing a variety of hybrid cloud protective mechanisms will accelerate the speed and efficiency of safety protection greatly [8]. There are three advantages as follows:

## 2.1 Global and Real-Time Flow View

With the help of a centralized architecture, SDN controller has real-time flow information of a global network, including topology, routing, and so on. The flow information is very useful in many protective scenarios. For example, when we take Distributed Denial of Service (DDoS) detection, we can get information of sampled flow (Sflow, it is an industry standard for packet export at Layer 2 of the OSI model. It provides a means for exporting truncated packets, together with interface counters) or OpenFlow (OpenFlow is a communications protocol that gives access to the forwarding plane of a network switch or router over the network). Then, we judge whether there is any malicious attack according to the statistical characteristics of packet stream.

With the aid of flow information based on the global network equipment, we can build real-time and historical knowledge base which based on the flow. Then, we can analyze any access at run time, and confirm whether there is a similar pattern in the history of knowledge base. If answer is no, it may be a malicious attacker existing. So we can deploy virtual safety equipment for deep packet inspection with the aid of NFV technology.

## 2.2 Flexible Security Services

Using NFV technology, security service providers may allocate resources dynamically and it will save resources. For example, they can generate a large number of virtual Web firewalls for the flow of large e-commerce sites in sales season, but reduce number of security virtual machines at other times to get more profit.

### 2.3 Software-Defined Security

Gartner puts forward the Software-Defined Security (SDS) [9] first, and emphasizes the underlying abstract is a kind of resources in safety resource pool. Intelligent and automation business arrangement and management can be realized in the software programming way at the top-level design, to complete the corresponding security functions. Since then, the combination of the definition and security of software become the forefront of the industry development.

NFV can start virtual safety equipment by the software, and with the aid of SDN controller, the appropriate flow can be pulled or mirrored to one or more of the above safety equipment. Then, we can make fine-grained test and form a security service chain. So SDN and NFV naturally become the underlying software-defined security support technology. Security resources and flow control are decided by north security application programming, and thus the entire scheme is flexible and quick.

## 3 Security Challenges of New Network

Adding SDN and NFV into hybrid cloud, the architecture of the whole system will be shown in Fig. 2. The network architecture is divided into enterprise networks, Internet, and cloud system network. The enterprise network contains employee office network, self-built enterprise private cloud, and public wireless network. Cloud system environment contains multiple tenants of the virtual network, and virtual network connects to enterprise network.

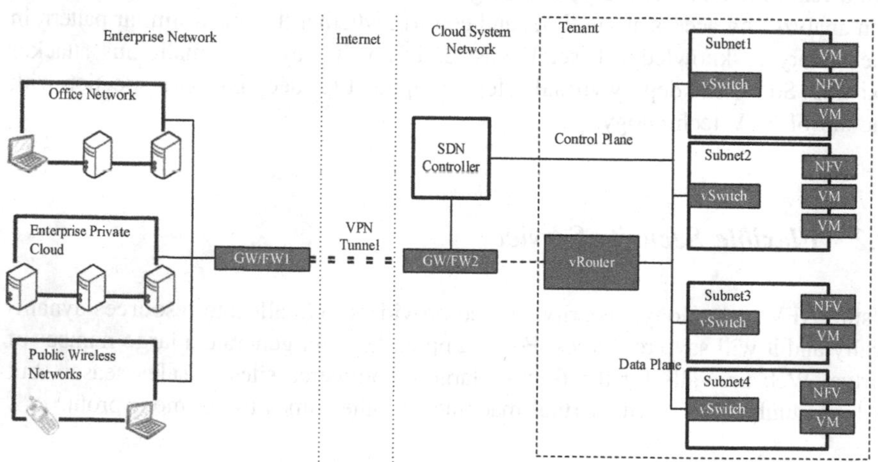


**Fig. 2** Architecture of hybrid cloud

In the cloud environment, SDN controller connects to physical network device or virtual network device in the cloud system. Security management platform deploys virtual safety equipment with the aid of NFV technology, and completes the deep packet inspection, attack prevention and load balancing, and so on.

SDN and NFV technology brought convenience for security protection, but new technology is also associated with new risks and challenges.

## 3.1 Architecture

From an architectural point of view, SDN use centralized control but the cloud system is usually elastic distributed. So when the Cloud computing system scales out, as shown in Fig. 2, the virtual network on the right side is becoming more and more, a single SDN controller may not be able to handle all the network flow. In addition, if SDN controller uses reactive work mode, not issue flow table in advance but do it when received PACKET_IN, it will be easy to be rejected in the larger network services.

## 3.2 System Implementation

SDN architecture includes applications, controllers, switches, and the management system. Virtualization system includes control nodes, management nodes, storage nodes, and the compute nodes. All these components rely on software running. But the software may be fragile, and the attacker can use and gain unauthorized access, thereby blocking of malicious manipulation of the flow.

For example, most of Switches use the open operating system based on Linux, some use unsafe Telnet service, or use the Secure Shell (SSH) service which owns default user name and password, and some use Plaintext communication. Even if the system design and deployment is perfect, some services also have loopholes in themselves. For example, many services use Secure Sockets Layer (SSL, a cryptographic protocol) to encrypt communications, but some use outdated SSL with implementation loopholes, such as Heartbleed (Heartbleed is a security bug disclosed in April 2014, which is a widely used implementation of the Transport Layer Security protocol). It will reveal information of SDN controller or tenant's privacy information in a virtualized system, and it brings great risks to the whole system. In some Cloud computing systems, data network may be shared with management or control network. If tenants attempted successfully, they can access the control node of SDN or virtualized systems, then manipulate the underlying route bypassing safety equipment.

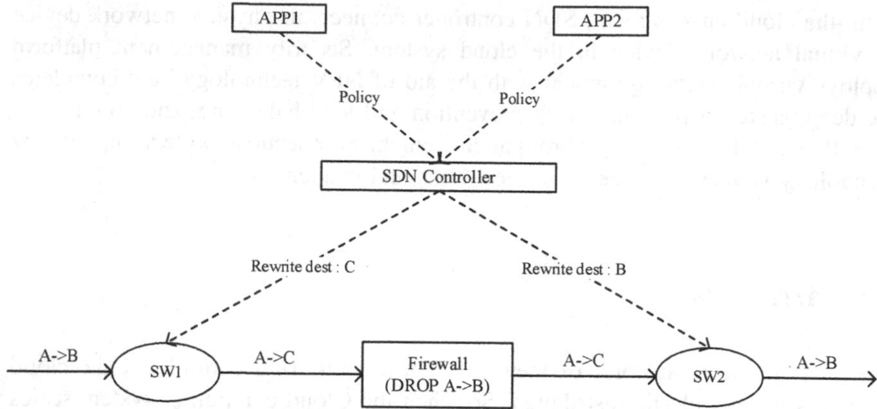


**Fig. 3** Policy inconsistencies cause the firewall be bypassed

## 3.3 *Policy Consistency*

SDN Controller issues upper application policy to the underlying network equipment, and controls all network flow ultimately. But if the SDN controller policy lacks of consistency checking mechanism, it may appear multiple applications issuing policies conflicted and cause network problems. In more severe cases, a malicious attacker can bypass many strategies security deployment in advance.

As shown in Fig. 3, the firewall strategy is to ban all A -> B packets, but the attackers may issue instructions by different applications, and two SDN controllers will issue strategies which rewrite the destination address to SW1 and SW2 respectively. Then strategy 1 rewrites the destination address of packet A -> B for C, strategy 2 rewrites the destination address of packet A -> C packet for B, and manages to bypass the firewall security strategy, implementing unauthorized access control at last.

## 3.4 *Compatibility of SDN and the Virtualization System*

In a typical scene, if the virtualization system use Openstack (Openstack is a free and open-source software platform for cloud computing), and the underlying drivers are OpenVswitch plunging (The main purpose of OpenVswitch is to provide a switching stack for hardware virtualization environments, while supporting multiple protocols and standards used in computer networks), and SDN controller using the Floodlight (an enterprise, Apache license, OpenFlow controller based on Java) or RYU (a kind of controller for OpenFlow) etc., then the Floodlight or RYU cannot distinguish whether flow of different virtual switch port belongs to the same tenant or not. So they can only send packet of tenants' virtual port with no difference, and destroy the tenant isolation which is the fundamental principles of virtualization, and cause a security risk.

### 3.5 Compatibility of NFV and Hypervisor

Migrating the physical security devices to virtualized systems to form a middle box (middle box is a computer networking device that transforms, inspects, filters, or otherwise manipulates traffic for purposes other than packet forwarding) of NFV, is a big challenge for two reasons: First, many devices, such as firewalls and Intrusion prevention system (IPS, a network security appliance), require customized NIC driver and kernel, in order to achieve better performance. But there are compatibility problems existing if you are deploying on the common hypervisor of compute nodes. Second, some virtualization systems offer hypervisor application interface for flow traction. Although it is a solution for SDN controller which is not deployed, but security personnel tend to consume a lot of energy to adapt these nonstandard interface. Even some virtualization systems do not open this interface, resulting in the deployment in L2 network in virtual environment is impossible such as network intrusion prevention system NIPS (Network Intrusion Prevention System, appliances that monitor network and/or system activities for malicious activity).

### 3.6 System Availability

Although the flexible deployment and resource pooling features of virtual safety equipment provide a great convenience for cloud security mechanism, but it is also need to solve the problem of high availability, especially when deploying equipment such as firewall and IPS. In the physical environment, they commonly deploy the active-standby equipment, using the heartbeat message and special line to achieve synchronization, and achieve a high availability. But under the condition of virtualization, network itself is not reliable, so the heartbeat mechanism cannot guarantee the security mechanism of usability.

In addition, virtualization security equipment has performance bottlenecks, so when there are a large number of simultaneous connections in a Website, the virtual safety equipment may deny service.

# 4 Security Advice

Considering the security risks involved in the third chapter, we need to make security design of the whole SDN and NFV architecture, the following aspects should be considered and implemented in system design phase.

## *4.1 The Global System Design*

When deploying SDN system in a cloud environment, you should ensure the SDN controller is not a bottleneck of the whole system. There should be many controllers for collaboration to ensure the availability. Meanwhile, the cloud should have a hierarchical design, such as by reducing the size of each sub-domain in order to reduce the number of broadcast packets and reduce the load of SDN controller. When the network is designing, the scale of virtual network should be reduced to let down network flow of virtual security equipment.

When designing and implementing the SDN controller, we should eliminate safety hazards as far as possible. We can use application authentication, policy checks and control channel encryption to ensure security of application programming interface (API) and third-party libraries.

At the same time, we should give full consideration to combine SDN and virtualized systems, such as OpenDaylight (The OpenDaylight is a collaborative open-source project hosted by The Linux Foundation) which is a SDN controller platform, is a good adaptation to Openstack, it supports the association of tenants and flow, to ensure network resources and flow isolation.

## *4.2 Multiple Inspection Service Chain*

Someone discusses the strategy of how to resolve the SDN application conflict, in order to solve the safety problems mentioned in 3.3. But there are loopholes exist in controllers always, and attackers may use a series of Attack China and avoid the security check to carry out targeted attacks. But defenders can also use the new features of SDN and NFV mentioned in Chap. 2, and compose multiple security mechanism. Security mechanisms can be deployed in any position, such as the proprietary safety hardware, or computing nodes of L2 network, or the gateway on the network nodes. Using SDN to tract flow, we can implement such as access control, Deep packet inspection (DPI, a form of computer network packet filtering) or behavior analysis, etc.

Even if the attacker escape some of the security mechanisms, but also the safety plans are in the dynamic changes, so abnormal phenomenon is found in a very short time. The response process such as depth inspection, forensics, and restore is corresponding at once.

Considering the hypervisor implementation and openness of virtualization system are quite different, so the security resources deployment and flow traction mechanism should be as standardized as possible. For example, the security resources deployment should be deployed on Openvswtich and use Intel DPDK (The Data Plane Development Kit is a set of data plane libraries and network interface controller drivers for fast packet processing) acceleration card, and the flow traction mechanism can use OpenFlow/SDN to carry on traction of traffic, blocking and mirror.

### 4.3 *Failure Recovery Mechanisms*

If the device is not reliable, in order to guarantee the availability of the security mechanism, we should use fast recovery mechanisms after a failure. If the virtual safety equipment cannot response, we should use hot start technology of original image to start a new virtual machine quickly, and guide the network flow, issue the existing security policies and complete online quickly at the same time. This is different from the traditional scheme, and put forward higher requirements for resources generate (Provision) of virtualization system and collaborative scheduling mechanism.

## 5 Conclusion

Hybrid cloud merges Intranet, mobile office, BYOD, and Public clouds to a complete information system, and it improves the efficiency of the office. But at the same time, it changes the traditional security domain division and increases the attack risks. This paper introduces the new safety protection opportunities brought by SDN and NFV, and analyzes the challenges and risks of hybrid cloud system using these technologies. Finally, it puts forward the corresponding solution.

## References

1. J Rhoton, R Haukioja. Cloud Computing Architected: Solution Design Handbook. Recursive Press, 2016
2. Kim, H., Feamster, N.Improving network management with software defined networking. Communications Magazine, IEEE. 2013
3. XL Wang, L Wang, A Bi, YY Li, Y Xu. Cloud computing in human resource management (HRM) system for small and medium enterprises (SMEs).International Journal of Advanced Manufacturing Technology, 2016:1–12

4. A Gordon. The Hybrid Cloud Security Professional. IEEE Cloud Computing, 2016,3(1):82–86
5. M Dhingra. Legal Issues in Secure Implementation of Bring Your Own Device (BYOD). Procedia Computer Science, 2016,78:179–184
6. J Weinman.Hybrid Cloud Economics. IEEE Cloud Computing, 2016,3(1):18–22
7. Zhou T, Xiangyang G, Hu Y, et al. PindSwitch: A SDN-based protocol-independent autonomic flow processing platform. Globecom Workshops (GC Wkshps) 2013 IEEE. 2013
8. Seugwon Shin, Philip Porras, Vinod Yegneswaran, Martin Fong, Guofei Gu, Mabry Tyson. FRESCO: Modular composable security services for software-defined networks. Proceedings of Network and Distributed Security Symposium, 2013
9. Gartner, The Impact of Software-Defined Data Centers on Information Security, https://www.gartner.com/doc/2200415/

# A Linguistic Rule-Based Approach for Aspect-Level Sentiment Analysis of Movie Reviews

**Rajesh Piryani, Vedika Gupta, Vivek Kumar Singh and Udayan Ghose**

**Abstract** Aspect-level sentiment analysis refers to sentiment polarity detection from unstructured text at a fine-grained feature or aspect level. This paper presents our experimental work on aspect-level sentiment analysis of movie reviews. Movie reviews generally contain user opinion about different aspects such as acting, direction, choreography, cinematography, etc. We have devised a linguistic rule-based approach which identifies the aspects from movie reviews, locates opinion about that aspect and computes the sentiment polarity of that opinion using linguistic approaches. The system generates an aspect-level opinion summary. The experimental design is evaluated on datasets of two movies. The results achieved good accuracy and shows promise for deployment in an integrated opinion profiling system.

**Keywords** Aspect-level sentiment analysis · Linguistic approach · Opinion profiling

R. Piryani (✉)
Department of Computer Science, South Asian University (SAU),
New Delhi 110021, India
e-mail: rajesh.piryani@gmail.com

V. Gupta
Computer Science Department, National Institute of Technology Delhi (NITD),
Delhi 110040, India
e-mail: guptavlabs@gmail.com

V.K. Singh
Department of Computer Science, Banaras Hindu University (BHU),
Varanasi 221005, India
e-mail: vivekks12@gmail.com

U. Ghose
School of ICT, Guru Gobind Singh Indraprastha University (GGSIPU),
Delhi 110078, India
e-mail: udayan@ipu.ac.in





S.K. Bhatia et al. (eds.), *Advances in Computer and Computational Sciences*,
Advances in Intelligent Systems and Computing 553,
DOI 10.1007/978-981-10-3770-2_19

## 1 Introduction

Sentiment analysis is a natural language processing task that utilizes computational approaches to classify the opinionated content into either positive or negative or neutral (having no sentiment polarity) classes. Sentiment can be analyzed at three levels: sentence level (assign polarity to each sentence), document level (assign polarity to the complete document), and aspect level (assign polarity to aspect). Sentiment analysis is beneficial across a large variety of domains. Many organizations are using sentiment analysis for identification of customer's attitude/ opinions towards products/services for efficient organizational decision-making. In addition, sentiment analysis is also useful for general purposes. For example, reviews about movies help the audience decide which movie to watch. Earlier studies show that majority of the work in sentiment analysis was centered on document level and sentence level. Recently researchers have been carrying out experiments and analysis on aspect-level sentiment analysis.

The research work carried out in this paper focuses on sentiment analysis at aspect level. We have proposed an approach which identifies the aspects from the movie reviews and computes its sentiment polarity using linguistics approaches. The sentiment polarity of one particular aspect is aggregated from all the reviews and sentiment profile of movie is generated. The remaining paper is organized as follows. Section 2 briefly explains the related work. Section 3 describes the proposed methodology. Section 4 explains about the experiment and evaluation followed by conclusion in Sect. 5.

## 2 Related Work

There are three kinds of approaches used for sentiment analysis namely: document level (polarity of whole document), sentence level (polarity of a sentence) and aspect level (polarity of an aspect). The methods applied to perform the sentiment analysis are categorized into three types: machine learning approaches, lexicon-based approaches, and hybrids of machine learning and lexicon-based approaches.

In the beginning, sentiment analysis has been performed at document level. Pang Lee et al. [1] applied machine learning techniques—Naïve Bayes (NB), Support Vector Machine (SVM), and maximum entropy (ME) classifiers for analyzing sentiment at document level on movie reviews. In their later work [2, 3], they have applied SVM, regression for assigning sentiment to document in the scale of 3 or 4 point. Gamon [4] used SVM for assigning sentiment analysis to document using scale of 4 point. Durant and Smith [5] applied the sentiment analysis on political blog. However, our aim is not to present detail survey of document-level sentiment analysis.

Recently researchers are gaining interest in aspect-level sentiment analysis. Hu and Liu [6] have worked on identifying the features/aspects in product reviews and generating feature-based opinion summary. They have extracted the features about which consumers have recorded their opinions and then classify the each opinion either in positive or negative class. Carenini et al. [7] have merged the algorithms of Hu and Liu [6] and similarity metrics to cluster the aspects into a user-defined hierarchy. Popescu and Etzioni [8] designed a system to detect the semantic orientation of customer reviews related to the extracted product features. Another similar work [9] talks about generating aspect-level summaries for a particular service such as hotel and restaurant reviews, local hair salons, schools, retailers, museums, and so on. Yi et al. [10] devised subject-based sentiment analysis of news articles and other online reviews. The approach mines subject-related aspect tokens and the sentiment orientation about those aspects. Eirinaki et al. [11] proposed aspect-based sentiment analysis of product reviews. Lu el al. [12] devised a structured PLSA method, which utilized the dependency structure of tokens to identify the aspects in the reviews. An LDA-based aspect-level sentiment analysis has been successfully implemented by Titov and McDonald [13].

We have also targeted aspect-level sentiment analysis. Our work is different from the previous works on aspect-level SA in many respects. We have developed an automated approach which does not require any prior training as is required by machine learning approaches. The approach is being developed as an integrated system which does not require any complex calculations. The system is currently in the process of development with the implementation of new linguistic rules.

## 3 Methodology

The input for processing is user review about movies. Our system is a capable of obtaining user reviews for a given movie using a crawler. Each review describes user's opinion on different aspects of a given movie in freeform text. Therefore, the challenge is to (i) identify aspect term; (ii) locate opinion about each aspect, and (iii) compute sentiment polarity of each aspect. Further, it is also expected from any such approach that it generates an aspect-based opinion summary of the movie from its reviews. Figure 1 shows the block diagram for aspect-based sentiment analysis. First and foremost, the review text is preprocessed by replacing the multiple punctuation marks to single for example, punctuation "????" replace to "?." Next, each review is broken into sentences using delimiters— question mark (?), exclamation mark (!), full stop (.). Next, the sentences are parsed with Stanford POS Tagger.[1]

### 3.1 *Identifying Aspect*

As we are dealing with aspect-based sentiment analysis, so, identifying aspect is the prerequirement that is discussed in this subsection. All possible aspects about

[1]http://nlp.stanford.edu/software/tagger.shtml.

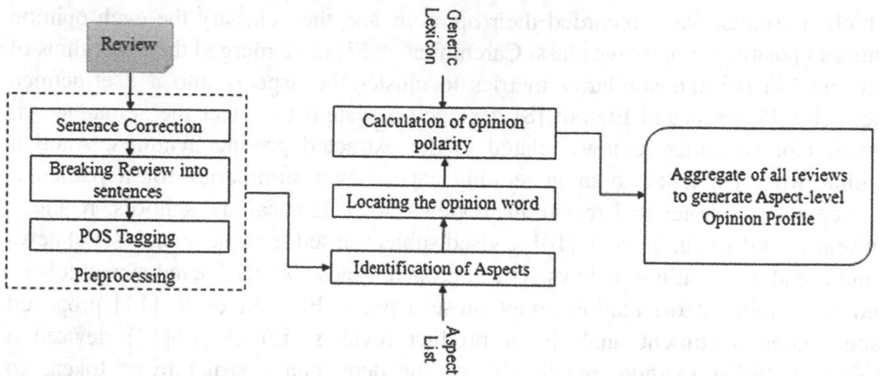


**Fig. 1** Block diagram for aspect-based sentiment analysis

movies have been reviewed by one or more reviewers in multiple reviews. We have assessed all those possible aspects as classified in various movie awards, movie review sites, and movie magazines. Since, specific aspects can be expressed by different words in different reviews. For example acting can be expressed by act, portrayed, etc. We have used the aspect vector list given in [16]. For each sentence we applied the aspect identification algorithm to locate the aspect in sentences using aspect vector list. The algorithm pseudocode is illustrated below.

**Pseudocode for Aspect Detection Algorithm**

```
1. FOR each s ∈ S
      a. Parsed with Stanford POS Tagger yields PS
      b. FOR each w ∈ PS
           i. IF(w exists in aspect vector)THEN
                 1. List(w)
          ii. ELSE search for Noun Phrase (NP)
                 1. IF (NP is actorName)THEN
                      a. List(NP)
                 2. ELSE IF (NP is director_name)THEN
                      a. List(NP)
                 3. ELSE IF (NP is writer_name)THEN
                      a. List(NP)
                 4. ELSE IF (NP is movie_name)THEN
                      a. List(NP)
```

Here S refers to set of sentences, PS refers to parsed sentences and w refers to word, and NP is noun phrase. Only those sentences have been considered which contain an aspect category or aspect term.

## 3.2 Locate Opinion About Each Aspect

After aspect identification, the next task is to locate opinion about those identified aspects. The sentence may contain one aspect or more than one aspect. Figure 2 shows a possible sentence structure, where $S_{Ln}$, $S_{Rn}$ refers to one or more opinion word on L.H.S and R.H.S of aspect $A_n$ respectively. First we located the aspect, then we traversed left and right in search of opinion word. Basically, an opinion word is an adjective as tagged by POS tagger. The whole scheme of assigning which opinion word to which aspect term is described in detail in the Sect. 3.3.

### 3.3 Sentiment Polarity Computation

We have applied sentiment polarity algorithm whose pseudocode is illustrated below. After locating the aspect, the algorithm traverses left and right in search of opinion-bearing word. If it is found to be an adjective, SWN AAC [14, 15] scheme is applied to compute the sentiment score of aspect. Here SWN AAC refers to SentiWordNet adjective adverb combination. We have also used Generic Lexicon [16] as lexical dictionary to compute the sentiment score.

**Pseudocode for Sentiment Polarity Computation**

1. Locate the aspect ($A_n$)
2. FOR each located aspect
   1. Initialize $S_{LA_n} = 0, S_{RA_n} = 0,$
   2. TRAVERSE LEFT in search of polarity feature
   3. IF polarity feature ***adj*** THEN
      i. $S_{LA_n} = SWN\_AAC(adj, Sentence)$
   4. TRAVERSE RIGHT in search of polarity feature
   5. REPEAT step 3, then we will have $S_{RA_n}$
   6. $S_{A_n} = S_{LA_n} + S_{RA_n}$
   7. IF ($S_{LA_n} == 0 \,\&\&S_{RA_n} == 0$) THEN
      i. POL_TO_DET_LIST(An)
3. FOR each $A_n \in$POL_TO_DET_LIST
   1. Locate $A_n$ in LIST
   2. TRAVERSE LEFT to $A_{n-1}$
      i. IF ($S_{RA_{n-1}}! = 0$) THEN
         - $S_{LA_n} = S_{RA_{n-1}}$
      ii. ELSE IF ($S_{LA_{n-1}}! = 0$) THEN
         - $S_{LA_n} = S_{LA_{n-1}}$
   3. IF ($S_{LA_n} == 0 \,\&\&S_{RA_n} == 0$) THEN
      i. TRAVERSE RIGHT to $A_{n+1}$
         - IF ($S_{LA_{n+1}}! = 0$) THEN
           - $S_{RA_n} = S_{LA_{n+1}}$
         - ELSE IF ($S_{RA_{n+1}}! = 0$) THEN
           - $S_{RA_n} = S_{RA_{n+1}}$

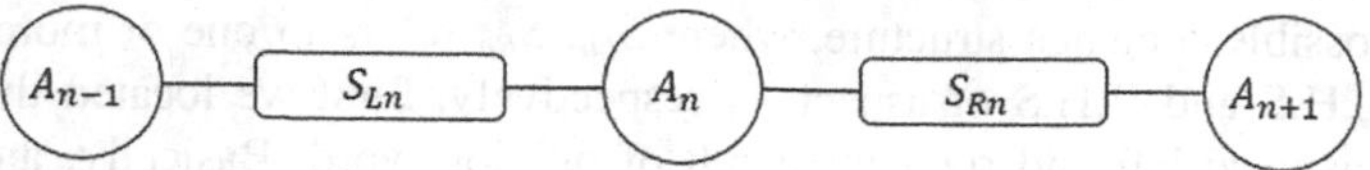


**Fig. 2** Sentence structure

**Table 1** Aspect identified with sentiment polarity

| Aspect | Aspect name | Left-side subjectivity LSS ($S_L$) | Right-side subjectivity RSS ( $S_R$) | Final |
|---|---|---|---|---|
| $A_{n-1}$ | Songs | NIL | Mediocre (−0.778) | Mediocre (−0.778) |
| $A_n$ | Movie | Mediocre (−0.778) | Perfect (+0.75) | (Mediocre + perfect) (−0.028) |
| $A_{n+1}$ | Movie | Perfect (+0.75) | NIL | Perfect (0.75) |

**Table 2** Dataset description

| S. no. | Dataset | Total sentences | Total aspect | IIC (%) | Cohen's Kappa (%) |
|---|---|---|---|---|---|
| 1 | D1-Kick | 219 | 286 | 96.1 | 92.3 |
| 2 | D2-The Dark Knight Rises | 341 | 404 | 92.2 | 82.6 |

For Example the sentence "The **songs** that seemed *mediocre* before watching the **movie**, feel like *perfect* for the **movie.**" The boldface words are aspects, and italic words are subjectivity words. Table 1 shows the aspect with polarity identified by aspect detection and sentiment polarity computation algorithm.

## 4 Experiments and Evaluation

### 4.1 Dataset

In this research experiment, two datasets have been used. These datasets are collected from Internet movie database (IMDb). Each dataset contains sentences. These datasets have been referred as D1 and D2. The datasets D1–D2 are crawled from www.imdb.com and then preprocessed and annotated manually by three different annotators and then reviewed by another two independent annotators for evaluation accuracy. Each annotator annotated each aspect with polarity found in the dataset. The annotation quality has been computed using two inter rater agreement parameters: Inter Indexer Consistency (IIC) [17] and the Cohen's Kappa Coefficient [18]. The analysis obtained suggests high accuracy. Table 2 summarizes description about datasets.

## 4.2 Results and Evaluation

We have obtained the aspect-level sentiment analysis result on two sentence datasets. The results have been evaluated through standard measure of precision and recall. We have used macro-averaging method for the calculation of precision and recall. Table 3 presents the precision and recall results obtained by aspect detection algorithm. The algorithm has achieved nearly 97% recall and 74% precision. Table 4 shows the precision and recall obtained by sentiment computation algorithm. In the table, Z is total number of aspects for which sentiment polarity detected by algorithm. Figure 3 presents the sentiment profile of "Kick" movie. The sentiment profile is generated on different aspects of movie on the basis of aggregation of aspect level sentiment orientation result obtained for each review.

**Table 3** Datasets of aspects identified

| S. no. | Dataset | Manual identified aspects (X) | Observed aspect | Correctly detected aspects (Y) | Precision (%) | Recall (%) |
|---|---|---|---|---|---|---|
| 1 | D1 | 286 | 372 | 277 | 74.46 | 96.85 |
| 2 | D2 | 404 | 540 | 395 | 73.15 | 97.77 |

**Table 4** Sentiment polarity values and accuracy

| S. no. | Dataset name | (Y) | (Z) | Confusion matrix | | Correctly detected aspect | | Total | Precision (%) | Recall (Z/Y) (%) |
|---|---|---|---|---|---|---|---|---|---|---|
| | | | | | | P | N | | | |
| 1 | D1 | 277 | 247 | Actual | P | 112 | 20 | 132 | 79.18 | 89.17 |
| | | | | | N | 32 | 83 | 115 | | |
| 2 | D2 | 395 | 371 | Actual | P | 196 | 28 | 224 | 84.49 | 93.92 |
| | | | | | N | 27 | 120 | 147 | | |

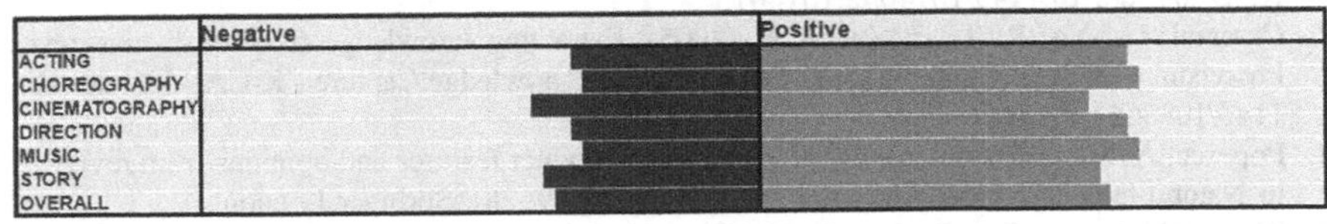


**Fig. 3** Sentiment profile result of a movie ("Kick")

## 5 Conclusion

This paper describes our proposed approach for Aspect-level sentiment analysis of movie reviews. The approach obtains good results that show its usefulness. Overall, this paper shows that a linguistic rule-based approach can be successfully designed and applied for aspect-level sentiment profiling of movies. The approach does not require any prior training as required in machine learning-based approaches. It can be deployed in an automated and integrated opinion profile generation system where reviews are automatically mined and opinion summaries are generated. We are currently working on extending the approach into a web-scale deployable system. The system can also be used for movie recommendation problem as suggested in [19, 20].

## References

1. Pang, B., Lee, L., & Vaithyanathan, S. (2002, July). Thumbs up? Sentiment classification using machine learning techniques. In Proceedings of the ACL-02 conference on Empirical methods in natural language processing-Volume 10 (pp. 79–86). Association for Computational Linguistics.
2. Pang, B., & Lee, L. (2004, July). A sentimental education: Sentiment analysis using subjectivity summarization based on minimum cuts. In Proceedings of the 42nd annual meeting on Association for Computational Linguistics (p. 271). Association for Computational Linguistics.
3. Pang, B., & Lee, L. (2005, June). Seeing stars: Exploiting class relationships for sentiment categorization with respect to rating scales. In Proceedings of the 43rd Annual Meeting on Association for Computational Linguistics (pp. 115–124). Association for Computational Linguistics.
4. Gamon, M. (2004, August). Sentiment classification on customer feedback data: noisy data, large feature vectors, and the role of linguistic analysis. In Proceedings of the 20th international conference on Computational Linguistics (p. 841). Association for Computational Linguistics.
5. Durant, K. T., & Smith, M. D. (2006, August). Mining sentiment classification from political web logs. In Proceedings of Workshop on Web Mining and Web Usage Analysis of the 12th ACM SIGKDD International Conference on Knowledge Discovery and Data Mining (WebKDD-2006), Philadelphia, PA.
6. Hu, M., & Liu, B. (2004). Mining and summarizing customer reviews. Proceedings of the 2004 ACM SIGKDD International Conference on Knowledge Discovery and Data Mining - KDD '04. doi:10.1145/1014052.1014073
7. Carenini, G., Ng, R. T., & Zwart, E. (2005). Extracting knowledge from evaluative text. Proceedings of the 3rd International Conference on Knowledge Capture - K-CAP '05. doi:10.1145/1088622.1088626
8. Popescu, A. M., & Etzioni, O. (2007). Extracting product features and opinions from reviews. In Natural language processing and text mining (pp. 9–28). Springer London.
9. Blair-Goldensohn, S., Hannan, K., McDonald, R., Neylon, T., Reis, G. A., & Reynar, J. (2008, April). Building a sentiment summarizer for local service reviews. In WWW Workshop on NLP in the Information Explosion Era (Vol. 14).
10. Yi, J., Nasukawa, T., Bunescu, R., & Niblack, W. (2003, November). Sentiment analyzer: Extracting sentiments about a given topic using natural language processing techniques.

In Data Mining, 2003. ICDM 2003. Third IEEE International Conference on (pp. 427–434). IEEE.
11. Eirinaki, M., Pisal, S., & Singh, J. (2012). Feature-based opinion mining and ranking. Journal of Computer and System Sciences, 78(4), 1175–1184. doi:10.1016/j.jcss.2011.10.007
12. Lu, Y., Zhai, C., & Sundaresan, N. (2009, April). Rated aspect summarization of short comments. In Proceedings of the 18th international conference on World Wide Web (pp. 131–140). ACM.
13. Titov, I., & McDonald, R. (2008, April). Modeling online reviews with multi-grain topic models. In *Proceedings of the 17th international conference on World Wide Web* (pp. 111–120). ACM.
14. Singh, V. K., Piryani, R., Uddin, A., & Waila, P. (2013, March). Sentiment analysis of movie reviews: A new feature-based heuristic for aspect-level sentiment classification. In Automation, Computing, Communication, Control and Compressed Sensing (iMac4 s), 2013 International Multi-Conference on (pp. 712–717). IEEE.
15. Singh, V. K., Piryani, R., Walia, P., & Devaraj, M. (2014). Computing Sentiment Polarity of Texts at Document and Aspect Levels. ECTI Transaction On computer and Information Technology, 8(1).
16. Thet, T. T., Na, J. C., & Khoo, C. S. (2010). Aspect-based sentiment analysis of movie reviews on discussion boards. Journal of Information Science, 0165551510388123.
17. Rolling, L. (1981). Indexing consistency, quality and efficiency. Information Processing & Management, 17(2), 69–76.
18. Byrt, T. (1996). How Good Is That Agreement? Epidemiology, 7(5), 561.
19. Singh, V. K., Mukherjee, M., & Mehta, G. K. (2011, December). Combining collaborative filtering and sentiment classification for improved movie recommendations. In *International Workshop on Multi-disciplinary Trends in Artificial Intelligence* (pp. 38–50). Springer Berlin Heidelberg.
20. Singh, V. K., Mukherjee, M., & Mehta, G. K. (2011). Combining a content filtering heuristic and sentiment analysis for movie recommendations. In*Computer Networks and Intelligent Computing* (pp. 659–664). Springer Berlin Heidelberg.

# Implementation and Statistical Comparison of Different Edge Detection Techniques

**Deepali Srivastava, Rashi Kohli and Shubhi Gupta**

**Abstract** This paper provides analysis through disparate detection techniques of edge like Prewitt, Sobel, and Robert to detect edges of the image with different analyses. Image segmentation and data extraction are considered process for edge detection. It is an image processing technique for detecting the boundaries within image. The process involves detecting discontinuities in brightness. In this paper, the proposed method shows the performance analysis of edge detection techniques as mentioned above.

**Keywords** Edge thinning · Convolution · Gradient

## 1 Introduction

A set of mathematical techniques which focus at differentiating points where there are sharp distinction in the image intensity or the points at which there are other gaps in a digital image is called as Edge Detection. Edge can be classified as the set of curved lines that are usually constructed by the points where image intensity varies sharply, they are significant changes of depth in an image locally. Within an image, data compression, image segmentation also supports for image renovation

D. Srivastava (✉)
CSE Department, Dr. APJ Abdul Kalam Technical University, Lucknow, UP, India
e-mail: srivastavadeepali15@gmail.com

R. Kohli · S. Gupta
CSE Department, Amity University, Greater Noida, India
e-mail: rashikohli.amity@gmail.com

S. Gupta
e-mail: sr23.shubhi@gmail.com



S.K. Bhatia et al. (eds.), *Advances in Computer and Computational Sciences*,
Advances in Intelligent Systems and Computing 553,
DOI 10.1007/978-981-10-3770-2_20

and considerably more can be done through detecting edge. Detection of edge is an important tool in vision for computers and image processing. This technique is usually used for feature detection and extraction.

Edges are enabled as a outline of image that commonly margin between two distinct areas within an image. The outcome drawn from edge identification is that, by differentiating its data in edges, it conserves the property of an image. The edge is a major local property in a gray-scale image that inside a neighborhood split areas in accordance each of which the greay scale is more or less constant within dissimilar grades on both sides of the edge. Edges detection for a distorted image is tough as the two comprises high frequency satisfying which cannot be extended to any assumption directly as it results in unclear and distorted result.

In image survey or study, edge detection is one of the most frequently used actions. If the edges of images are noticeable easily than better detection and analysis is probable on image.

The blockade in depth from one pixel to another degrades the quality of an image. Thus the key intent is to detect an edge by conserving the key construct assets of an image. A relative analysis of these methods are deliberated in order to analyze various edge detection methods based on definite factors related to the type of edge, localization of the edge, department, cost estimation, execution, etc.

Detection of edge construes a 2D image into curves within a set, removes translate quality of the area, max shortens than pixels.

**Aim**

- Identifying and localization of image edges.
- Indicating significant property from identification of the edge within an image (e.g., corners, lines, curves, sides).
- Higher level computer vision procedures use these features. (e.g., recognition).

**Applications**

The determination of edge detection is to determine data about the shapes and the reflectance or diffusion in an image. In the previous few years numerous edge detection procedures have been established, still there is no particular procedure for edge detection with suitable application (Fig. 1).

**Edge Detection Methods**

To identify variations in the gray-scale gradients edge detection is a type of differential operators. It can be classified into two parts (Fig. 2).

The whole paper is planned in the following consonance. Part-1 introduces objectives, implementation, and approaches of detection of edge. Part-2 clarifies

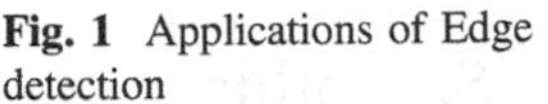

Fig. 1 Applications of Edge detection

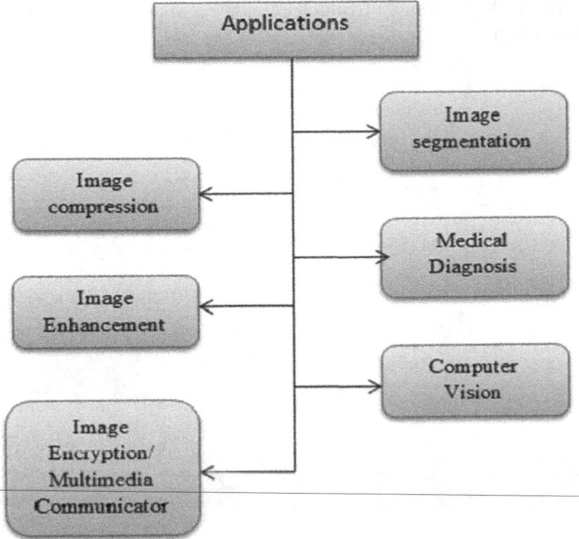


Fig. 2 Categorical methods

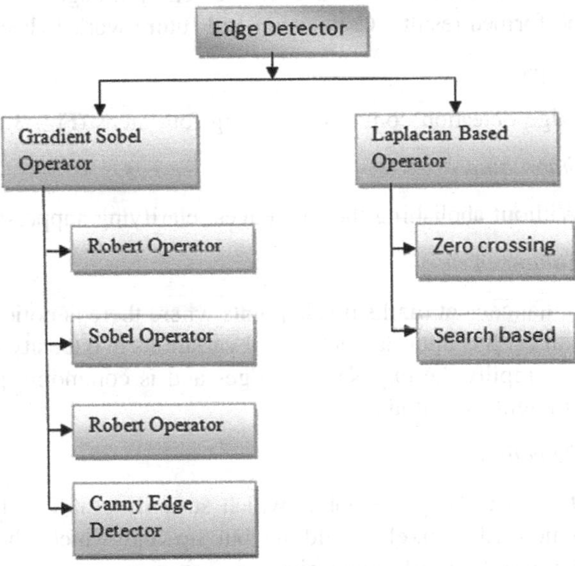

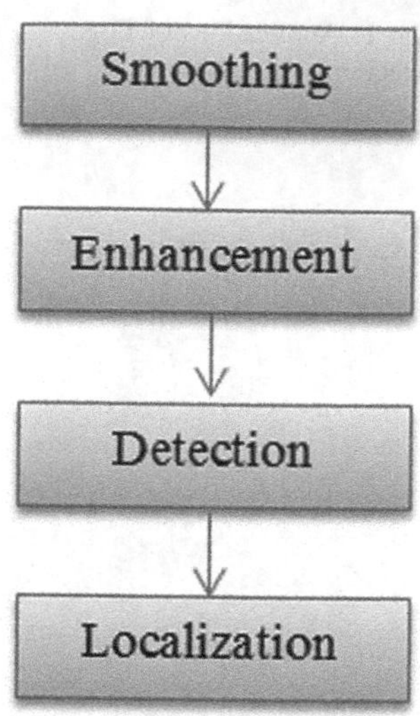


**Fig. 3** Applications of Edge detection

marks of edge detection. Part-3 gives detail on edges and its type. Part-4 Robert, Prewitt, and Sobel methods for detection of edge are specified and Part-5 gives the performed results. Conclusion and future work is lastly ended in Part-6.

Steps

Edge detection contains following four steps (Fig. 3).

*Smoothing*

Without abolishing the true edges, clarifying suppress as much noise as possible.

*Enhancement*

Enhancement marks pixels points where there is noticeable variation in local depth values. It is important to control variations in intensity in the adjacent point, in order to simplify the exposure of edges and is commonly performed by calculating the gradient magnitude.

*Detection*

It aims to keep the points which strongly gratified the edge for which it resolve which edge pixel should be put up and which should be discarded as noise, commonly thresholding gives the criterion.

*Localization*

If requested by the application, with subpixel resolution, the position of the edge can be assessed. The edge direction can also be assessed.

## 2 Types of Edges

Variations in intensity can provide different types of edges profiles. These variations are due to disjointedness, in boundary, surface direction which leads to variation. Edges can be formed with respect to their amplitude varies as follows (Fig. 4):

Step Edge

The image strength abruptly changes from one unlike value of one side on the disjointedness to a dissimilar value on the side.

Ramp Edge

When the intensity variation is not natural and seen over a bounded area then step edges changes into ramp edges. A step edge where the strength varies is not instant but appears over a definite distance.

Ridge/Line Edge

The image intensity suddenly modulates value but then again restoration to the origin value within some abbreviated width.

Roof Edge

A roof edge is a corrugation edge where the intensity modulation is not instant but happens over a definite distance.

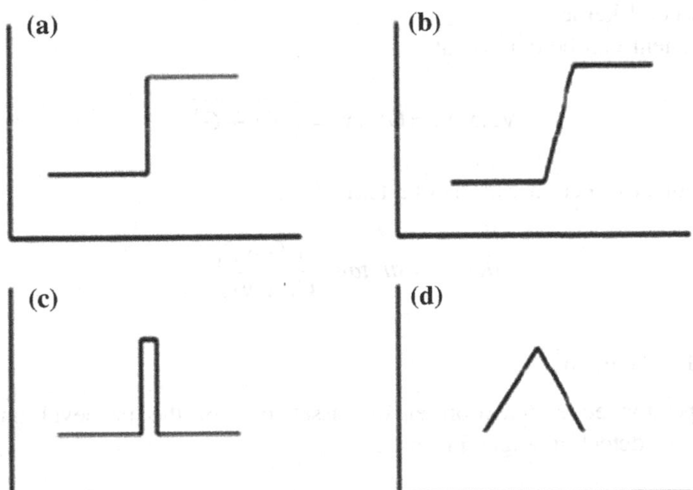


**Fig. 4** Different types of edges

# 3 Techniques for Edge Detection

## *3.1 Robert's Cross Edge Detection*

Robert's edge detection technique is one of the oldest approaches. It is used frequently in applications for hardware where easy processing, simplicity and efficiency are leading features. The idea ahead the Roberts cross works on to imprecise the gradient of an image over discrete variation which is attained by calculating the sum total of the squares of the variances between crosswise adjoining pixels.

Robert's cross operator is used for computer vision and processing of image. First suggested by Lawrence Roberts in 1963 it was one of the first detectors of edge. The Roberts Cross operator highlight areas of huge spatial frequency which frequently co-relates to edges. It implements an easier, efficient, 2-D spatial inclination measurement on an image. In its max regular practice, grayscale image is the input to the operator and the output is the pixel value at all its points specifies the assessed complete significance of the contiguous gradient of the input as an image at that point. It is also known as differential operator and its aim to maximize image gradients through discrete differentiation and is achieved; firstly the original image is convolving with two kernels:

$$\begin{bmatrix} +1 & 0 \\ 0 & -1 \end{bmatrix} and \begin{bmatrix} 0 & +1 \\ -1 & 0 \end{bmatrix}$$

Let $I(x,y)$ be the original image points and $G_x\,(x,y)$ be an image point composed by the first kernel convolving and $G_y(x,y)$ be an image point shaped by convolving with the second kernel.

The gradient can be defined as:

$$\nabla I(x,y) = G(x,y) = \sqrt{G_x^2 + G_y^2}$$

The gradient direction can also be termed as:

$$\theta(x,y) = arctan\left(\frac{G_y(x,y)}{G_x(x,y)}\right)$$

Prewitt Edge Detection

Prewitt operator edge detection encompasses one of the primeval and finest approaches of detecting edges in images.

The Prewitt operator is applied in processing of image, mostly within detection procedures of edge. Theoretically, it is a distinct difference operator, calculating an estimation of the inclination of the image depth function. At every point within the image, the outcome of the operator is the steady gradient vector. The Prewitt operator is low-cost in terms of comparison in calculations since it is depends upon convolving of the image with a small-scaled, divisible, and integer value filter in both horizontal and vertical directions. On the other hand, in specific for higher frequency dissimilarities in an image, the gradient estimate which it produces is comparatively crude.

The Prewitt detection process increases the coordination straight from the kernel through the maximum response obtained. While variance grade identification desires an equitably time overwhelming control to calculate the coordination in the x- and y-orientation by using magnitudes.

Arithmetically, to compute evaluation of the consequences—one for parallel variants and another for perpendicular variants, 3 × 3 kernels which are lapped with the original image are used by operator. Let us describe A as the base image, and $G_x$ and $G_y$ are two pictures which consists of horizontal and vertical obtained estimates, they are planned as:

$$G_x = \begin{bmatrix} -1 & 0 & +1 \\ -1 & 0 & +1 \\ -1 & 0 & +1 \end{bmatrix} *A \; and \; G_y = \begin{bmatrix} -1 & -1 & -1 \\ 0 & 0 & 0 \\ +1 & +1 & +1 \end{bmatrix} *A$$

where * symbolizes the 2-D density procedure. Since Prewitt kernel can be disintegrated as the products or output of an averaging and a separation core, therefore, it is a discrete filter and it calculates grades with smoothing. Let us take an example, $G_x$ can be inscribed as:

$$\begin{bmatrix} -1 & 0 & +1 \\ -1 & 0 & +1 \\ -1 & 0 & +1 \end{bmatrix} = \begin{bmatrix} 1 \\ 1 \\ 1 \end{bmatrix} \begin{bmatrix} -1 & 0 & 1 \end{bmatrix}$$

It is well-defined here that the x-orientation is expanding in the "right"—direction, and the y-orientation is expanding in the "down"—direction. The resulting grade estimates can be joined to give the inclination rule on every point in the image:

$$G = \sqrt{G_x^2 + G_y^2}$$

We can also total the grade's direction by using above information:

$$\theta = atan2(G_y, G_x)$$

where, let Θ is 0 for a upright edge which is ambiguous on the exact side.

Sobel Edge Detection

The Sobel operator, sometimes named as Sobel–Feldman operator or as Sobel filter, are used in techniques like processing of image and computer perception, mostly within edge recognition procedures where it produces an image which shows the highlight edges in the image.

Sobel process is applied to notice an edge. It utilizes two shields with 3 × 3 dimensions, one resembling the grade one in the x-orientation and another resembling the grade in the y-orientation. The shield is fall all over the image, by doubling pixels at a time. At all points in the image, from light to dark, all the resultant directions are delivered to the evolution of the image intensity. Hypothetically, it is a distinct difference operator which is calculating an estimate of the grade of the image strength function. Strong strength differences which are darker or brighter are characterized by edge areas.

The extreme valuation of two convolutions will be mentioned as desirable outcome of the swapping point. It is accessible to achieve Sobel operator in space, it is nearly precious by noise and has a smoothing consequence on the noise. It can deliver more precise edge route info but also distinguish certain improper edges through rough edge span (Fig. 5).

**Fig. 5** Sobel operator

| -1 | -2 | -1 |
|---|---|---|
| 0 | 0 | 0 |
| 1 | 2 | 1 |

| -2 | 0 | 2 |
|---|---|---|
| -2 | 0 | 2 |
| -1 | 2 | 1 |

Arithmetically, to compute approximate of the consequences—one for parallel form (variants), and one for perpendicular form, the operator uses two interactive 3 × 3 kernels with the original image. Let us describe A as the base image, and $G_x$ and $G_y$ are two pictures which contain the vertical and horizontal derivative estimates, they are planned as:

$$G_y = \begin{bmatrix} -1 & -2 & -1 \\ 0 & 0 & 0 \\ +1 & +2 & +1 \end{bmatrix} * A \text{ and } G_x = \begin{bmatrix} -1 & 0 & +1 \\ -2 & 0 & +2 \\ -1 & 0 & +1 \end{bmatrix} * A$$

where * symbolizes the 2-D density operation. Since Sobel kernel can be disintegrated as the output by product of an averaging and a separation kernel therefore it is a discrete filter and it calculates grades with smoothing. Let us take an example, $G_x$ can be inscribed as:

$$\begin{bmatrix} -1 & 0 & +1 \\ -2 & 0 & +2 \\ -1 & 0 & +1 \end{bmatrix} = \begin{bmatrix} 1 \\ 2 \\ 1 \end{bmatrix} \begin{bmatrix} -1 & 0 & +1 \end{bmatrix}$$

It is well-defined here that the x-orientation is expanding in the "right"—direction, and the y-orientation is increasing in the "down"—direction. The culmination grade estimates can be joined to give the gradient rule on every point in the image:

$$G = \sqrt{G_x^2 + G_y^2}$$

We can also total the grade's direction by using above information:

$$\theta = atan2(G_y, G_x)$$

where, let Θ is zero ("0") for a upright edge which is ambiguous on the exact side.

The edge discovery approaches stated primarily change in the styles of leveling screens that are useful and the system of measuring edge vitality are planned. Several detection styles for edge varying permeation used for figuring grade estimations in the x- and y-orientation since they grounded on the computation of picture grades.

## 4 Results

Figure on the right side displays the basis image on which altered methods for detection of edge are practice in MATLAB to observe the edges and estimate the disjointedness in the picture and make comparison between different methods (Fig. 6).

Here, we show the edge detection as result of different methods attained by signal and spectra statistics.

Robert's Cross Edge Detection Implementation

See Figs. 7, 8, and 9.

Edge Detection in Different Axis

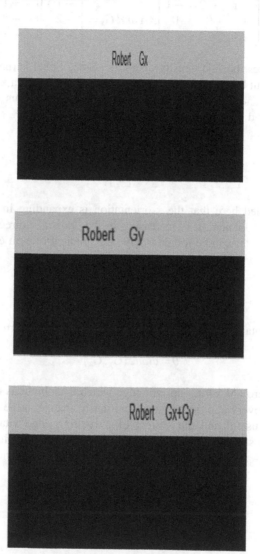


Prewitt Edge Detection

See Figs. 10, 11 and 12.

Edge Detection in Different Axis

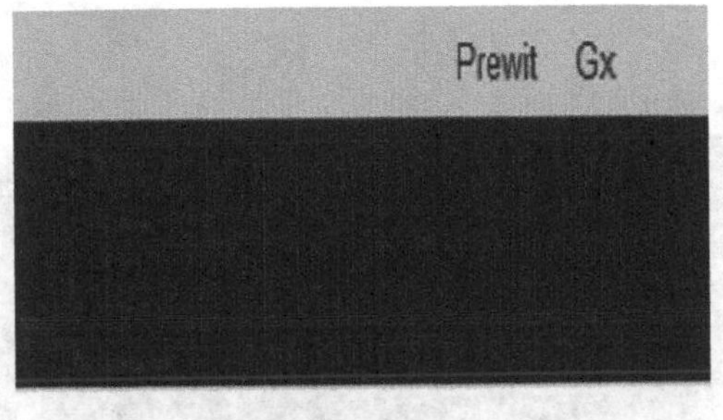


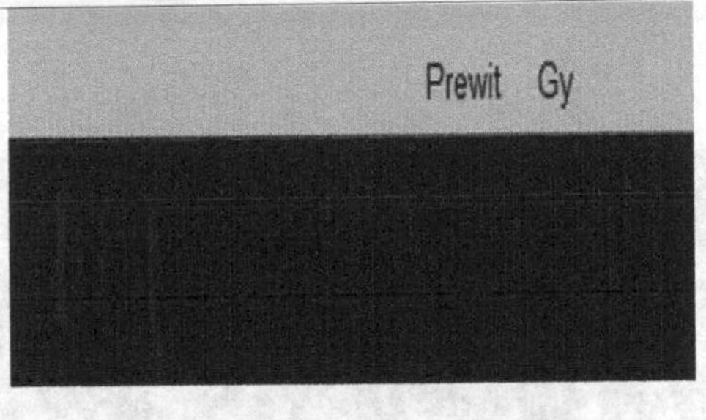

Sobel Edge Detection

See Figs. 13, 14, and 15.

Edge Detected in Different Axis

**Fig. 6** Original image

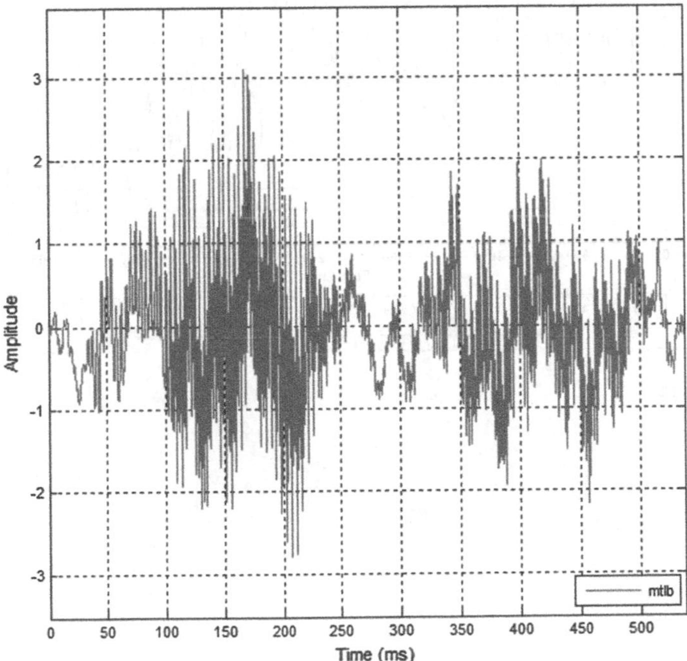


**Fig. 7** Signal

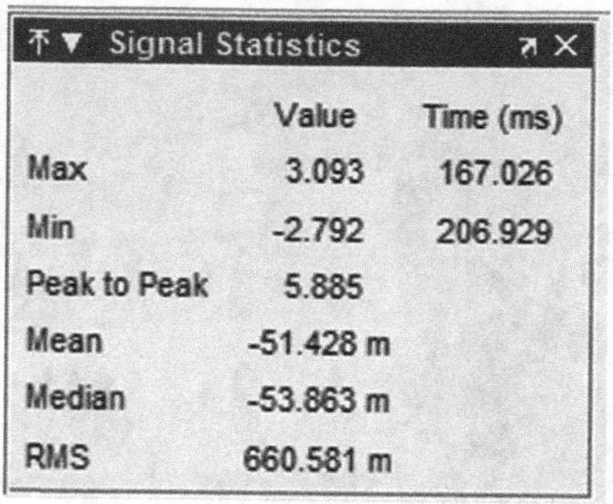

| Signal Statistics | | |
|---|---|---|
| | Value | Time (ms) |
| Max | 3.093 | 167.026 |
| Min | -2.792 | 206.929 |
| Peak to Peak | 5.885 | |
| Mean | -51.428 m | |
| Median | -53.863 m | |
| RMS | 660.581 m | |

**Fig. 8** Data statistic

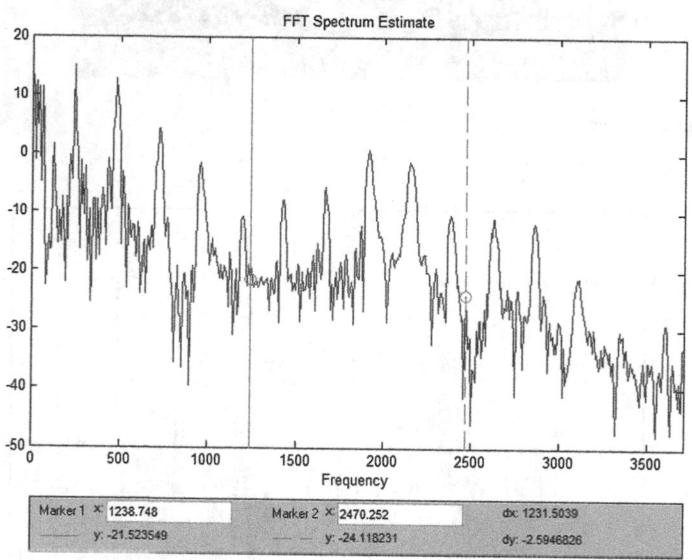


**Fig. 9** Spectra

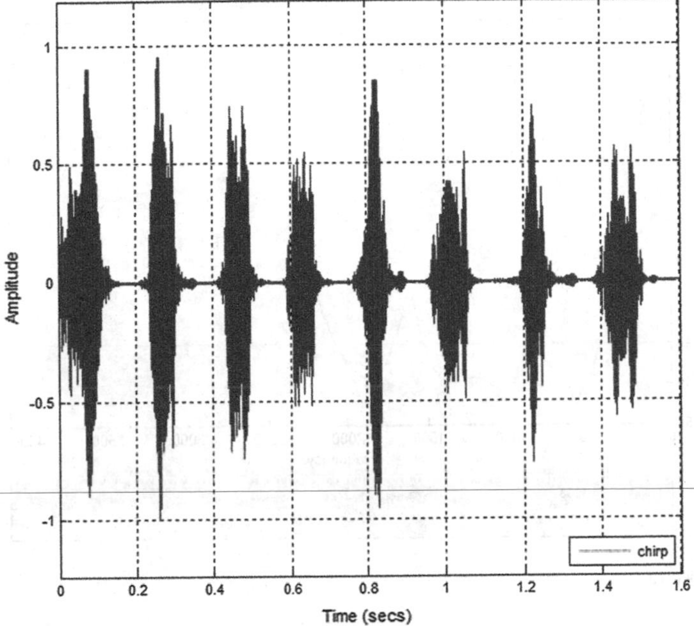


**Fig. 10** Signal

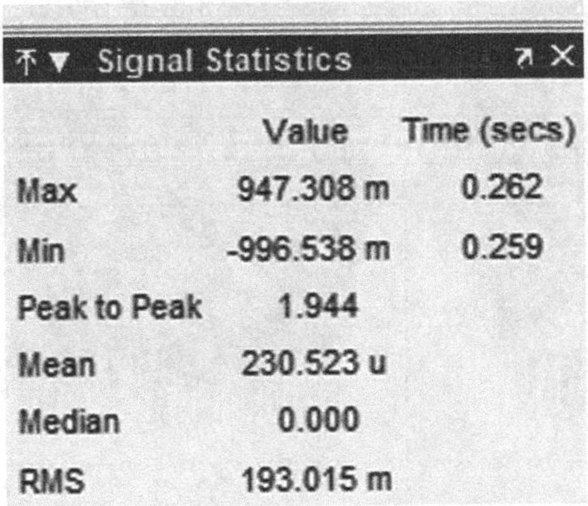

Signal Statistics

| | Value | Time (secs) |
|---|---|---|
| Max | 947.308 m | 0.262 |
| Min | -996.538 m | 0.259 |
| Peak to Peak | 1.944 | |
| Mean | 230.523 u | |
| Median | 0.000 | |
| RMS | 193.015 m | |

**Fig. 11** Data statistic

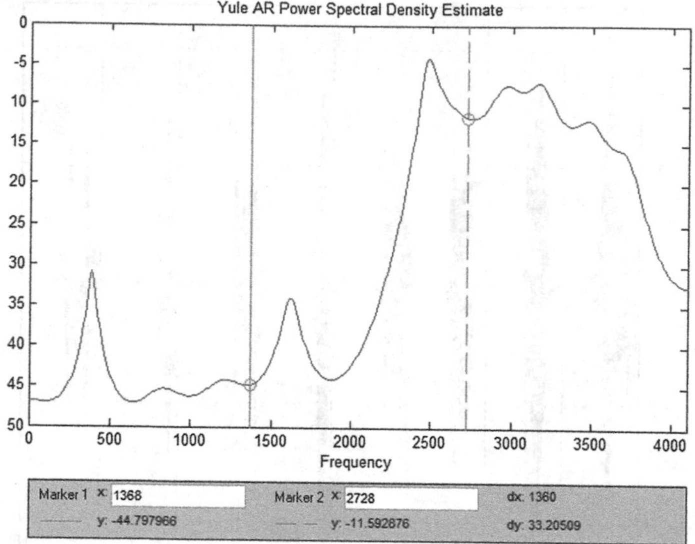


**Fig. 12** Spectra

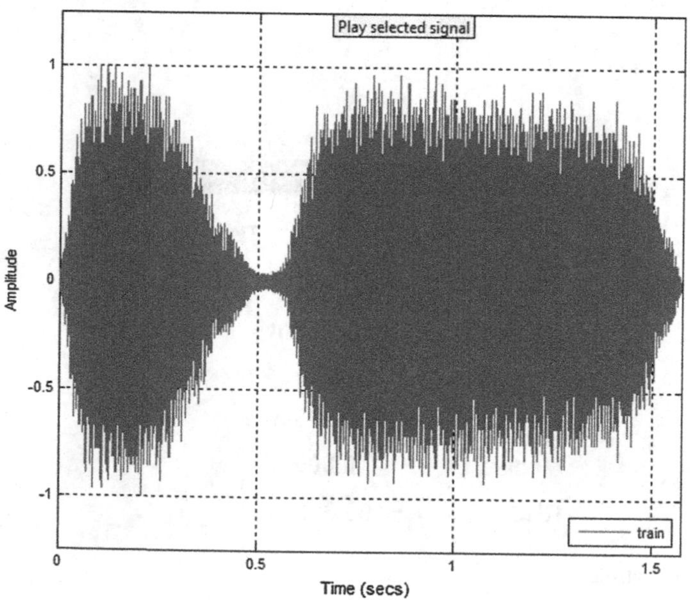


**Fig. 13** Signal

Signal Statistics

| | Value | Time (secs) |
|---|---|---|
| Max | 1.000 | 0.099 |
| Min | -1.000 | 0.210 |
| Peak to Peak | 2.000 | |
| Mean | 572.023 u | |
| Median | -2.294 m | |
| RMS | 338.167 m | |

**Fig. 14** Data statistic

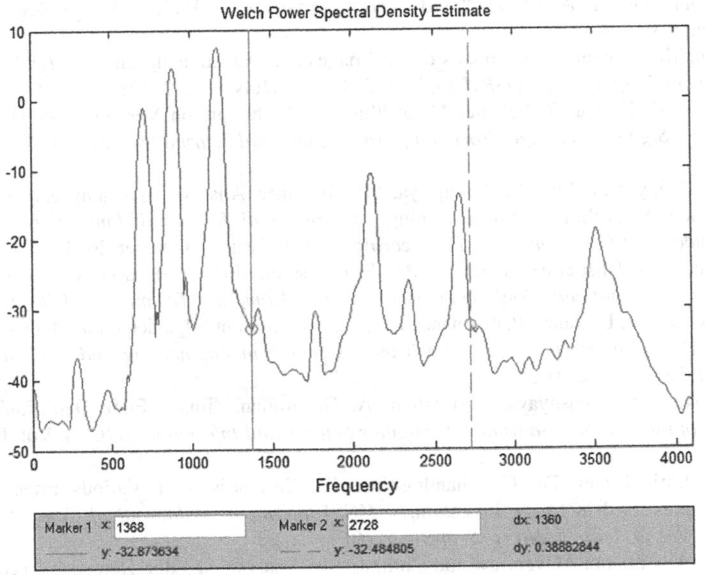


**Fig. 15** Spectra

## 5 Conclusion and Future Work

In Object Recognition the variances between edge detection methods are key to know in prior. In this paper we focused on different methods for edge detection, signifying an image by its edge has the benefit for remembering most of the information of the image as it decreases the quantity of data requisite to be stored. Robert operator, Prewitt operator and Sobel operator are different grade detection methods for edge. From the above results (i.e., Signal Spectra Data Statistics and detected edge in different axis) we detect that Sobel edge detection method provides lesser disjointedness in discovery of edges or outskirt as parallel to Robert and Prewitt detection methods for edge are presented in figures correspondingly. Its equivalent statistics achieved using MATLAB which shows that Robert and Prewitt are having relative equivalent data and Sobel varies from them. In upcoming days, we all can use these uncertain edge detection methods to resolution the gaps molded in Sobel detection process.

## References


1. M. D. Health, S. Sarkar, Member, IEEE Computer Society, T. Sanocki, K. W. Bowyer, Senior Member, IEEE, "A Robust Visual Method for Assessing the Relative Performance of Edge-Detection Algorithms", *IEEE Transaction on Pattern Analysis and Machine Intelligence*, Vol. 19, No. 12, December 1997.
2. R. Maini, Dr. H. Aggarwal, "Study and Comparison of Various Image Edge Detection Techniques".
3. L. Bin, M. S. Yeganeh, "Comparision for Image edge Detection algorithms", *IOSR Journal of Computer Engineering (IOSRJCE)*, Vol. 2, Issue 6 (July-Aug. 2012), PP 01–04.
4. Rashmi, M. Kumar, R. Saxena, "Algorithm and Technique on Various Edge Detection: A Survey", *Signal and Image Processing: An International Journal (SIPIJ)*, Vol. 4, No. 3, June 2013.
5. Dr. S. Vijayarani, Mrs. M. Vinupriya, "Performance Analysis of Canny and Sobel Edge Detection Algorithms in Image Mining", *International Journal of Innovative Research in Computer and Communication Engineering*, Vol. 1, Issue 8, October 2013.
6. A. Halder, N. Chatterjee, A. Kar, S. Pal, S. Pramanik, "Edge Detection: Statical Approach", 2011 3rd *International Conference on Electronics Computer Technology (ICECT 2011).*
7. M. Avlash, Dr. L. Kaur, "Performance Analysis of Different Edge detection Methods on Road Images", *International Journal of Advance Research in Engineering and Applied Sciences*, Vol. 2, No. 6, June 2013.
8. K. Vikram, N. Upashyaya, K. Roshan, A. Govardhan, "Image Edge Detection", *Special Issues of international Journal of Computer Science and Informatics (IJCSI)*, Vol. II, Issue-1, 2.
9. G. T. Shrivakshan, Dr. C. Chandrasekar, "A Comparison of Various Edge Detection Techniques used in Image Processing", *IJCSI International Journal of Computer Science Issues, Vol. 9, Issue 5, No. 1, September 2012.*
10. Chinu, A. Chhabra, "Overview and Comparative Analysis of Edge Detection Techniques in Digital Image Processing", *International Journal of Information and Computation Technology, Vol. 4, No. 10, (2014), PP. 973–980.*

# CALDUEL: Cost And Load overhead reDUction for routE discovery in LOAD ProtocoL

**A. Dalvin Vinoth Kumar, P.D. Sheba Kezia Malarchelvi and L. Arockiam**

**Abstract** Internet of Things (IoT) is a backboneless network. Because of the uni-directional link and mobility nature of the nodes, the network is dynamic. The nodes are self-organized and two nodes can transfer data directly when they are within the transmission range. The nodes in IoT are self-organized and dynamic so MANET routing plays a vital role. The Light weight Ad hoc On-demand Distance Vector (LOAD) is a reactive routing protocol. In LOAD, the routing involves three major processes namely route discovery, path establishment, and route maintenance. The route discovery is carried out by Route REQuest (RREQ), Route REPly (RREP), and Route ERRor (RERR) control packets. The cost of the route discovery is estimated by control packets propagation. The proposed technique Cost And Load overhead reDUction for routE discovery in LOAD ProtocoL (CALDUEL) is proposed to reduce the load overhead, by reducing the cost of route discovery.

**Keywords** IoT · Routing protocol · LOAD · RREQ and overhead

## 1 Introduction

Internet of Things (IoT) is a collection of wireless mobile hosts, wired nodes, Things, and Objects, forming a temporary network without the help of any centralized infrastructure or established administration as shown in Fig. 1a. Thus, the

---

A. Dalvin Vinoth Kumar (✉) · L. Arockiam
Department of Computer Science, St. Joseph's College (Autonomous), Tiruchirapalli 620002, Tamil Nadu, India
e-mail: dal_win@ymail.com

L. Arockiam
e-mail: larcokiam@yahoo.co.in

P.D. Sheba Kezia Malarchelvi
Department of Computer Science and Engineering, J.J. College of Engineering and Technology, Tiruchirapalli 620009, Tamil Nadu, India
e-mail: pdsheba@yahoo.co.in



S.K. Bhatia et al. (eds.), *Advances in Computer and Computational Sciences*, Advances in Intelligent Systems and Computing 553,
DOI 10.1007/978-981-10-3770-2_21

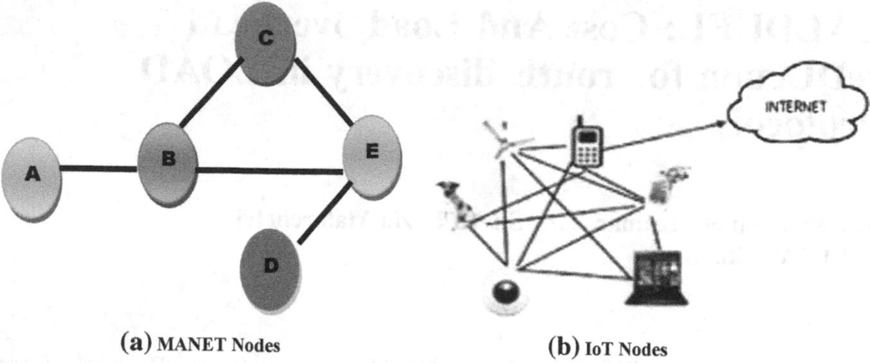


(a) MANET Nodes (b) IoT Nodes

**Fig. 1** **a** MANET nodes, **b** IoT nodes

network's wireless topology may change rapidly and dynamic in nature. Such a network may operate in a standalone fashion, or may be connected to the larger Internet. There are different strategies used in MANET routing [1–4]. The advantage of MANET is low cost, easy to use, and no centralization approach. But, it has some limitations to provide QoS to the user. QoS contains many parameters. Therefore, QoS may vary from one application to another application. In IoT, to achieve QoS mobility is the great challenge. The mobility of IoT nodes are managed by MANET routing protocols.

These networks are fully distributed and might work anywhere without the help of any infrastructure. This property makes these networks extremely robust. In Fig. 1a nodes A and C should discover the route through B so as to communicate. The circles indicate the nominal range of each node's radio transceiver. Nodes A and C are not in direct transmission range of each alternative, since A's circle does not cover C. So if A needs to send data to C it is obvious to involve B to forward the data from A to C. Here A acts as an intermediate node. Without node B, A cannot send the data to C. One of the challenges in MANET is route discovery. As the mobile nodes keep changing their position, the route once discovered from a source to destination cannot be retained forever. Every time a packet has to be transmitted from the source to destination a route has to be discovered. Many algorithms have been proposed for routing in MANET. One such algorithm which is widely used is LOAD. But, the major drawbacks with LOAD are increased cost and overhead involved in finding the paths. In this paper, an attempt has been made to reduce in finding the paths. In Sect. 2, we present the related works. Section 3 elaborates on the proposed work. In Sect. 4, we present the results and discussion and in Sect. 5 we present our conclusions.

## 2 Related Works

A multipath routing protocol proposed by SV. Mallapur et al. [5] is a variant of single-path LOAD routing protocol. The projected methodology established node-disjoint ways with lowest delays and supported interaction of factors from various layers. The authors investigated and compared the proposed protocol's performance to single-path LOAD and multipath AOMDV protocols.

Dalvin Vinoth Kumar et al. [6] proposed a technique to reduce duplication of route request in MANET. The energy consumption is one of the main issues in MANET. By reducing the number of control packets, the author saves energy and increase the lifetime of the network. Although the number of control packets is less, the number of identified path is also less. This is the main disadvantage of this work.

Sheba KeziaMalarchelvi et al. [7] proposed a new routing technique to enhance the QoS in AODV. To achieve the QoS, a technique namely Half Transmission Area toward Sender (HATTS) is employed by which the RREP is forwarded to the only neighboring nodes of destination that are nearer to the source. All the nodes that receive RREP, except the node that forwarded the RREQ to the destination, refrain from forwarding the RREQ to the destination and forward RREP in the reverse path immediately after receiving the RREP from the destination. This drastically reduces the time taken by the source to find all the available paths compared to the LOAD.

The multipath routing protocols proposed by Kapoor et al. [8] to identify the duplication of routes the node does not use disjoint RREQ immediately. Every node has an additional field in RREQ to indicate its first hop. Also, it maintains the permanent hop list of each RREQ from the source to its neighbors. To ensure the link uniqueness in RREP initial hop, destination replies are sent only to RREQ arriving through unique neighbors.

Nismonrio et al. [9] proposed a technique to compute the minimum neighborhood node path that reduces the info transfer delay from supply to destination. Although proposed technique has several benefits, it additionally has weakness like maintenance of tables and routing overhead.

Zhang et al. [10] proposed a dynamic source routing protocol for Mobile circumstantial networks. It uses DSR protocol. In DSR, each node confirms the next hop in the source, till confirmation is received from the next hop. Each packet is simply forwarded once by a node (hop-by hop routing).

Jing Jing Xia et al. [11] proposed neighbor coverage based probabilistic send for reducing overhead in MANETs. It uses the NCPR protocol. It considers the neighbor coverage information. The disadvantage is that the same node can receive identical RREQ packet again and again. It creates routing overhead in Mobile Ad hoc networks.

## 3 CALDUEL: A Proposed Technique

The proposed technique Cost And Load overhead reDUction for routE discovery in LOAD ProtocoL (CALDUEL) is to reduce the number of RREQ propagation and to identify maximum number of available paths. The steps involved in the proposed technique are the following:

Step 1. Identify the network topology.
Step 2. The source node initiates the RREQ to identify the path to the destination node. The RREQ contains the address of source, destination, and also the address of the node where it is forwarded.
Step 3. The Step 2 iterates till the destination is reached or till the TTL expires.
Step 4. Each node in topology maintains two tables, one contains its neighbor information and the other contains information about the nodes which receive RREQ.
Step 5. When the destination receives the RREQ, it constructs the RREP and forward to the nodes if and only if the node is present in, both neighboring table and RREQ table.
Step 6. The intermediate nodes which receive the RREP also forward the RREP like the destination node.
Step 7. Step 5 iterates till the source node is reached or till the TTL expires.
Step 8. The source node receives the RREP from different paths. It finds the optimum path and establishes the path to destination.
Step 9. The data is transferred in the established path.

In CALDUEL technique, the source node constructs the RREQ and sends it to its neighbor nodes. Each and every node in the network maintains two tables namely neighbor table and RREQ table. The neighbor table contains its neighbor information and RREQ table contains the information of the nodes to which the RREQ sent. The source node constructs the RREQ and sends it to its neighbors. The neighboring nodes receive the RREQ and forward it to its own neighbors and also construct the RREQ table. This process continues till the destination is reached or till the TTL expires. When the destination node receives the RREQ, it constructs the RREP and forward it to the nodes, if and only if the node is present in neighboring table and the RREQ table.

## 4 Results and Discussions

In the MANET scenario depicted in Fig. 2, node 1 is the source node, 7 is the destination node, and 2, 3, 4, 5 and 6 are intermediate nodes. The node 1 constructs the RREQ and forward it to its neighbors 2, 3, and 4. The node 2 receives the RREQ which contains the information that the RREQ is already sent to 3 and 4. So the node 2 checks its neighbors list and forward it to its neighbors which are not

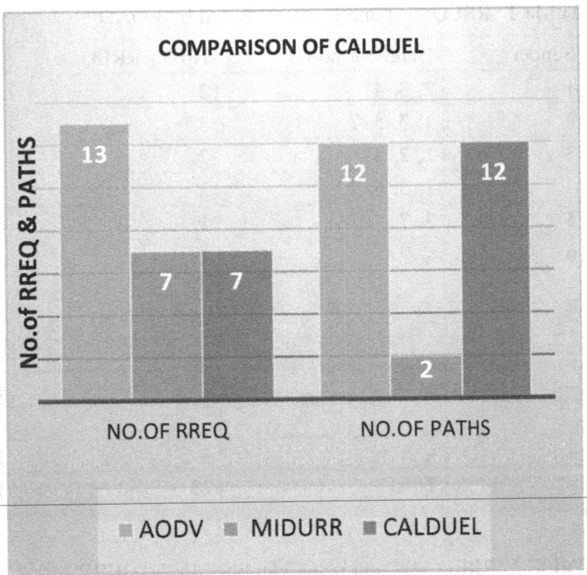


**Fig. 2** Comparison of number of RREQ sent and paths

present in the RREQ table. This process is iterated till the destination is reached or the TTL expires. When the node 7 receives the RREQ, it constructs the RREP and forward it to the nodes 2, 5, and 4. The neighboring table of node 7 contains 2, 4, 5, and 6 as shown in Fig. 2.

The RREQ table formed by the RREQ sent by the node 2 contains 2, 3, 4, and 5. The node 7 compares the neighboring table and RREQ table. The nodes 2, 4, and 5 are present in both tables and hence the RREP is forwarded to these nodes. Even though node 6 is in neighbor list of node 7 it is not present in the RREQ table, so the RREP is not sent to node 6. This process is followed by all the intermediate nodes till the source is reached or TTL expires. The source node establishes connection between the source and destination.

The number of RREQ messages generated and the number of paths identified by LOAD, MIDURR, and CALDUEL are tabulated in Tables 1, 2 and 3 respectively. From the tables, it could be observed that CALDUEL outperforms LOAD in terms of the number of RREQ messages generated, i.e., RREQ for CALDUEL is only 7, whereas it is 13 for LOAD. Similarly, CALDUEL outperforms MIDURR since the number of alternate paths generated by CALDUEL is 12, whereas it is only 2 with MIDURR. From these observations, it is evident that CALDUEL performs better than LOAD and MIDURR.

The proposed work is simulated using Network Simulator2 (NS2). The simulation parameters are tabulated in the below Table 4. The route discovery overhead depicted in Fig. 3. The X-axis denotes the number of nodes and the Y-axis denotes the overhead in bits. The overhead is calculated by number of control packets forwarded in the routing environment. When the control packets increase, the overhead also increases. From the simulation results, it is obvious that the route

**Table 1** RREQ sent and paths identified by LOAD

| Sender | Neighbor nodes | No. of RREQ | No. of available paths |
|---|---|---|---|
| 1 | 2, 3, 4 | 3 | 1 → 2→7 |
| 2 | 1, 3, 5, 7 | 3 | 1 → 2→5 → 7 |
| 3 | 1, 2, 4 | 2 | 1 → 3→2 → 5→7 |
| 4 | 1, 3, 6, 7 | 3 | 1 → 4→3 → 2→5 → 7 |
| 5 | 2, 7 | 1 | 1 → 4→7 |
| 6 | 4, 7 | 1 | 1 → 4→6 → 7 |
| – | – | – | 1 → 3→4 → 6→7 |
| – | – | – | 1 → 2→3 → 4→6 → 7 |
| – | – | – | 1 → 3→4 → 7 |
| – | – | – | 1 → 2→3 → 4→7 |
| – | – | – | 1 → 3→2 → 7 |
| – | – | – | 1 → 4→3 → 2→7 |
| | Total | 13 | 12 |

**Table 2** RREQ sent and paths identified by MIDURR

| Sender | Neighbor nodes | No. of RREQ | No. of available paths |
|---|---|---|---|
| 1 | 2, 3, 4 | | |
| | 3 | – | |
| 2 | 1, 3, 5, 7 | 2 | 1 → 2→7 |
| 3 | 1, 2, 4 | 0 | – |
| 4 | 1, 3, 6, 7 | 2 | 1 → 4→7 |
| 5 | 2, 7 | 0 | – |
| 6 | 4, 7 | 0 | – |
| | Total | 7 | 2 |

**Table 3** RREQ sent and paths identified by CALDUEL

| Sender | Neighbor nodes | No. of RREQ | No. of available paths |
|---|---|---|---|
| 1 | 2, 3, 4 | 3 | 1 → 2→7 |
| 2 | 1, 3, 5, 7 | 2 | 1 → 2→5 → 7 |
| 3 | 1, 2, 4 | 0 | 1 → 3→2 → 5→7 |
| 4 | 1, 3, 6, 7 | 2 | 1 → 4→3 → 2→5 → 7 |
| 5 | 2, 7 | 0 | 1 → 4→7 |
| 6 | 4, 7 | 0 | 1 → 4→6 → 7 |
| – | – | – | 1 → 3→4 → 6→7 |
| – | – | – | 1 → 2→3 → 4→6 → 7 |
| – | – | – | 1 → 3→4 → 7 |
| – | – | – | 1 → 2→3 → 4→7 |
| – | – | – | 1 → 3→2 → 7 |
| – | – | – | 1 → 4→3 → 2→7 |
| | Total | 7 | 12 |

**Table 4** Simulation Environment

| Parameters | Description |
|---|---|
| Simulation area | 500 × 500 m, 1000 × 1000 |
| Number of nodes | 7, 10, 20, 50, 150, 700 |
| Mobility model | Random way point model |
| Nodes mobility | 5 m/s |
| Path loss model | Two ray ground |
| Radio coverage | 250 m |
| Packet size | 512 bytes |
| Routing protocol | LOAD |
| Simulation time | 60,300 s |

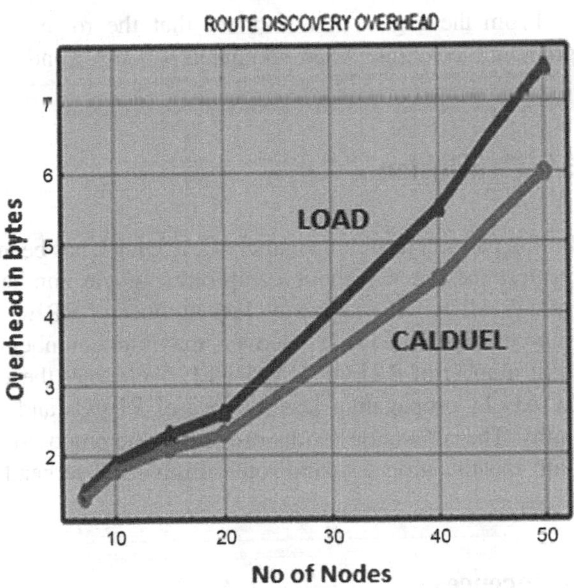


**Fig. 3** Simulation result for overhead in route discovery

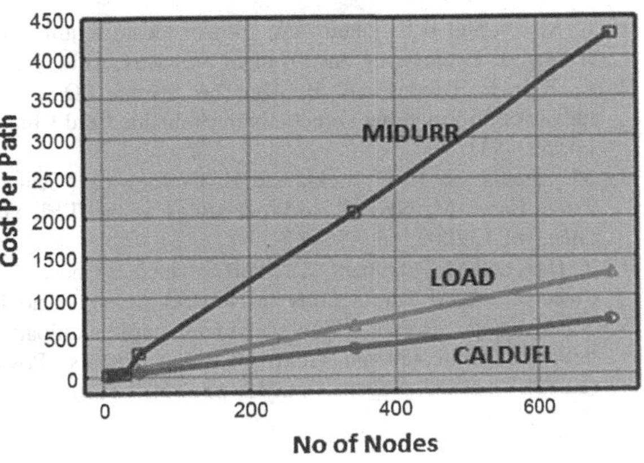


**Fig. 4** Simulation result for path identification cost

discovery overhead is reduced in the proposed technique. Figure 4 explains the route discovery cost. The X-axis denotes the number of nodes and Y-axis denotes the cost. The overhead cost is calculated by number of control packets divided by number of paths identified.

$$RC = (NCP/NP)*C$$

*where,*

*RC* *is Routing overhead Cost*
*NCP* *is Number of Control Packets*
*NP* *is Number of Paths*
*C* *is the Constant*

From the Fig. 4, it is evident that the route discovery cost is least for the proposed technique when compared to LOAD and MIDURR.

## 5 Conclusion

In this paper, a technique namely CALDUEL has been proposed to reduce the route request messages without compromising the number of paths discovered. Even though MIDURR propagates less number of RREQ it fails to identify maximum number of paths. LOAD discovers maximum number of paths but fails to propagate least number of RREQs. CALDUEL overcomes these drawbacks in MIDURR and LOAD by propagating less number of RREQs and finding maximum number of paths. The simulation results prove that the proposed system ultimately reduces the path identification cost and route discovery overhead.

## References

1. S. Mishra and B.K. Pattanayak "power aware routing in mobile Ad-hoc n works: survey", Journal of Engineering and Applied Sciences, APRN vol.9, no.2,2013, pp. 173–189.
2. C. Kim, E. Talipov and B. Ahn. "A reverse LOAD routing protocol in Ad-hoc mobile networks. In Emerging Directions in Embedded and Ubiquitous Computing", Springer, 2006, pp. 522–531.
3. M. Abolhasan, T. Wysocki, and E. Dutkiewicz. "Scalable Routing Strategy for Dynamic Zones-Based MANETs.", In Proceedings of the IEEE Global Telecommunications Conference, vol 1, 2002, pp. 173–177.
4. Z. Haas and M. Pearlman. "The Performance of Query Control Schemes for the ZoneRouting Protocol", ACM Transactions on Networking, vol.9, no.4, 2001, pp. 427–438.
5. SV Mallapur, SR Patil, "Stable backbone based multipath routing protocol for mobile ad-hoc networks", International Conference on Circuits, Power and Computing Technologies (ICCPCT), IEEE, Nagercoil, 2013, pp. 1105–1110.

6. M. Jayakkumar, P. Calduwel Newton, A. Dalvin Vinoth Kumar, "MIDURR: A Technique to Minimize the Duplication of Route Requests in Mobile Ad-Hoc Networks", Soft-Computing and Networks Security (ICSNS), International Conference on IEEE Xplore, 2015, pp. 1–4.
7. A. Dalwin vinoth kumar, P.D. Sheba Kezia Malarchelvi, L. Arockiam, "EQOS-LOAD: A Technique to Enhance Quality of Service using Half Transmission area Towards Sender", Research Journal Medium of Intellectual Search and Beacon of Academic Harmony, vol. 15, 2015, pp. 47–54.
8. RK Kapoor, MA Rizvi, S Sharma, MM Malik, "Exploring Multi Path routing Protocols in Mobile Ad hoc Networks", Journal of Computer and Mathematical Sciences, vol.2, 2011, pp. 693–779.
9. R. Nismon Rio and P. Calduwel Newton, "A Technique to Find Optimum Path for Reducing Data Transfer Delay in Mobile Ad-hoc Networks", Proceedings of the International Conference on Developments in Engineering Research (ICDER) International Association of Engineering, 2015, pp. 50–53.
10. X.M. Zhang, E.B. Wang, J.J. Xia, and D.K. Sung, "An Estimated Distance Based Routing Protocol for Mobile Ad-hoc Networks", IEEE Trans. Vehicular Technology, vol. 60, 2011, pp. 3473–3484.
11. Jing Jing Xia and Xin Ming Zhang, A Neighbor coverage based probabilistic rebroadcast for reducing routing overhead in Mobile Ad-hoc IEEE Transactions on Mo-bile Computing, vol. 12, 2013, pp. 631–636.

# AASOP: An Approach to Select Optimum Path for Minimizing Data Transfer Delay in Mobile Ad-Hoc Networks

**R. Nismon Rio and P. Calduwel Newton**

**Abstract** Mobile Ad-Hoc Networks (MANETs) provides substantial services in the field of network. MANETs is an infrastructure-less network, containing more number of individual wireless mobile nodes (devices) that communicate with one another without any aid of centralized server or base station. Due to dynamic topology, MANETs faces many issues associated with mobile nodes such as link failure, battery power, delay, etc. The main aim of this paper is to reduce data transfer delay by finding optimum path. Ultimately, the outcomes of the proposed work AASOP increase the performance of MANETs by minimizing data transfer delay which in turn increases Packet Delivery Ratio (PDR).

**Keywords** MANETs · Data transfer delay · Wireless mobile nodes · Optimum path

## 1 Introduction

In the present generation of wireless communication systems, there is a need for fast communication of each mobile user. Mobile Ad-Hoc Networks (MANETs) is an infrastructure less network. It has no base station. It sends the data packets through the neighbor nodes to destination node. Mobile nodes can move dynamically and freely self-organize network topologies. MANETs is used to solve challenging real-world issues of military communication like automatic battlefield equipment, sensor networks for sensing the remote weathers and also for other earth activities in case of emergency services such as earthquakes or any disaster recovery. MANETs also has educational applications like setting up a virtual class or conference rooms to share

R. Nismon Rio (✉)
Department of Computer Science, Bishop Heber College (Autonomous),
Tiruchirapalli 620 017, Tamil Nadu, India
e-mail: nismonriocs@gmail.com

P. Calduwel Newton
Department of Computer Science, Government Arts College, Ayyarmalai,
Kulithalai, Karur-Dt 639 120, Tamil Nadu, India
e-mail: calduwel@yahoo.com





S.K. Bhatia et al. (eds.), *Advances in Computer and Computational Sciences*,
Advances in Intelligent Systems and Computing 553,
DOI 10.1007/978-981-10-3770-2_22

their resources among the users. MANETs needs efficient algorithms to determine network topology, link scheduling and routing. It finds the minimum neighborhood node path among multiple paths from source to destination. It calculates the nodes property and stores the information into the path table.

This paper is organized as follows: Sect. 2 presents the related works. Section 3 explains the proposed technique (AASOP), computes the minimum neighborhood nodes that reduces the data transfer delay from source to destination. Section 4 highlights scenarios and discussions. Section 5 is conclusion.

## 2 Related Works

Many researches are being focused on MANETs to provide data service, when there is no infrastructure. Shalini et al. [1] intended to discover neighbor node distance method and key distribution methods to evaluate the performance against masquerading attack to improve network performance and secure data transfer on the network. Kure and Jain [2] developed routing protocol in MANETs to increase the packet delivery ratio and to reduce end-to-end delay. They had set timer to calculate rebroadcast delay.

Roopali Garg and Guneet Kaur [3] concentrated in the Particle Swarm Optimization (PSO) to calculate the fitness value to optimize number of rebroadcasts. In order to control routing overhead, neighbor knowledge and rebroadcast probability method used for rebroadcasting a request. Surya and Anitha [4] implemented a technique called Neighbor Position Verification (NPV) routing protocol designed to protect the network from corrupted nodes by verifying the position of neighbor nodes to improve security and efficiency performance in MANETs routing.

Khaleel Husain and Premala Patil [5] researched in PSO-based neighbor monitoring scheme. It is used to detect and defense from the various attacks. It checks the node status periodically to find out whether the node is valid or malicious. This ultimately improves the network performance. Vijay U. Patil and Tamboli [6] explored Neighbor coverage-based probabilistic rebroadcast protocol for reducing routing overhead in MANETs. It reduces the number of retransmissions and improves the routing performance.

Priyanka Sheela and Sundar Raj [7] proposed Good Node Detection Algorithm (GNDA) to obtain more accurate extra coverage ratio by sensing neighbor node coverage. It selects the trust node to transmit the data by detecting selfish nodes. This technique decreases the number of retransmissions and improves the network performance. Shaik Arif Basha and Preethi Joshna [8] analyzed routing process in the network is more complicated issue due to increase in mobile nodes. Each and every mobile node access services by knowing the location based service. The designed protocol updates the position of nodes in dynamic nature. It adapts the routing position changes when a node movement is frequent.

Nandhini and Malathy [9] discovered an efficient route via suitable path. The neighbor node information and stable path values are considered to reduce the latency and overhead of routing. The proposed protocol mitigates network collision as the result it increases packet delivery ratio and reduces end-to-end delay. Harpreet Kaur and Jasmeet Singh [10] proved that routing overhead goes in peak level because, it provides more number of neighbor discovery messages in the MANETs. To reduce the unnecessary hello messages, they designed random waypoint model and investigated relationship between the hello interval and event intervals.

Kavitha and Sundararajan [11] developed Optimal Link Managed on Demand routing protocol to increase the Quality of Service (QoS) in MANETs. It maintains the available paths and connects to another alternative path, if there is no transmission which reduces the latency by reinitiating route discovery. Sivakumar et al. [12] proposed efficient technique which eliminates the repeat HELLO messages, in order to reduce the unwanted battery power and delay. The nodes have scanned frequently to check whether the routing path got broken.

Directional antenna algorithm to discover the neighbors and assist to reduce number of time slots to discover all neighbors in the network and provide security mechanism to improve cooperation among the neighbor nodes [13, 14].

The ultimate aim of this paper is to select optimum path quickly in case of link failure for minimizing data transfer delay in MANETs. The existing research work considers only hop count as best path for routing the data packets. In existing routing On-Demand Multicast Routing Protocol (ODMRP) algorithm is used to identify the best path when the path is having minimum hop count. ODMRP does not have any novelty except hop count mechanism. And also ODMRP cannot repair the path locally during link failure. No weight factor has given to link quality and it does not support link stability. It causes more delay to setup route from the beginning and achieves less Packet Delivery Ratio. The above literature study shows the various techniques that were developed to select optimum path. To overcome link failure problem AASOP technique is proposed based on the minimum number of neighbor ing nodes and delay.

## 3 AASOP: A Proposed Technique

The proposed technique, AASOP computes the path that has minimum number of neighborhood nodes. It minimizes the data transfer delay from source to destination. The AASOP has the following steps:

***Begin***

/* Determine the path to route data packets from Srce (A) to Destn (H) */

1. Check whether the node moves from initially selected path
   ***If***(Node moves out of range) ***Then***

2. Compute neighboring nodes for each node in the network using this proposed (AASOP) approach and select the path which has minimum number of neighboring nodes
3. ***If***(More than one path equal number of neighbor nodes) ***Then***
   Select the path that has minimum delay
   ***Else***
   Use ODMRP approach to identify the optimum path
   ***Endif***
4. ***If***(Delay is same for more than one alternate path)***Then***
   Select alternate path in which the path is calculated first in routing table
   ***Else***
   Use ODMRP approach to select the optimum path
   ***Endif***
   ***Else***
   Use ODMRP approach to identify the optimum path from Source to Destination
   ***Endif***
5. Repeat steps 1–4 until the data transfer is accomplished
   ***End***
   (Where, Srce(A) = Source Node, Destn(H) = Destination Node)

Initially, Source and Destination nodes are determined to route the data packets. Each and every node monitors and exchanges the status of information in routing table. After a span of time, link failure is likely to be occurring when node moves out from the network topology. Once link gets failure, the overall transmission of data packets are aborted and it is very difficult to find optimum path quickly. This causes high delay and less Packet Delivery Ratio. In this paper, AASOP technique is proposed to select optimum path for reducing data transfer delay in MANETs. According to proposed technique, once if the node moves out of range it computes neighboring nodes for each node in network using AASOP and selects the path which path has minimum number of neighboring nodes. If there is more than one path that has equal number of neighbors then AASOP selects the path that has minimum delay. Even more than one path may have equal number of neighboring nodes and delay. In that case, AASOP selects optimum path which is calculated first in the routing table. Certainly, it resumes data transfer quickly. The steps 1–4 are repeated until the data transfer is accomplished. Probably, forwarding data packets via optimum path minimizes the delay and increase Packet Delivery Ratio.

## 4 Scenarios and Discussions

The AASOP technique is used to select a path that has minimum number of neighborhood nodes among multiple paths. Consider the network topology given in Fig. 1. There are 7 nodes considered in the network topology. They are, A, B, C, D, F, G,

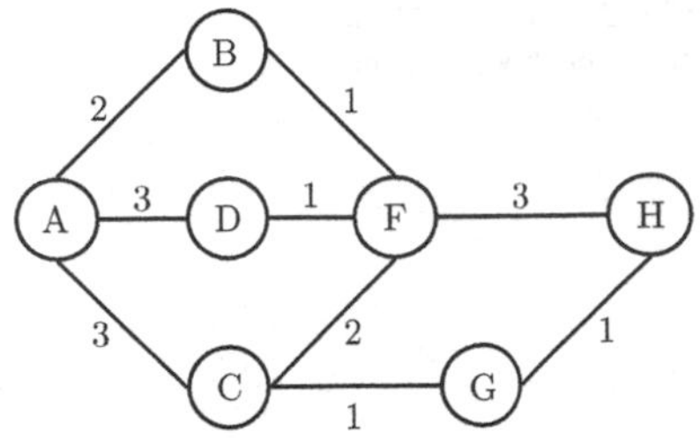


**Fig. 1** Network Scenario 1

**Table 1** Nodes and their Neighborhood Node(s)

| Node(s) | Neighborhood Node(s) | No. of Neighborhood Node(s) |
|---|---|---|
| A | B C D | 3 |
| B | A F | 2 |
| C | A F G | 3 |
| D | A F | 2 |
| F | D B C H | 4 |
| G | C H | 2 |
| H | G F | 2 |

and H. The number on the edges depicts the delay in seconds. Here, A denotes source and H denotes destination, other nodes are intermediate nodes.

**Scenario 1**:

Consider the above network topology in which A has three neighbors (B, C, and D), B has two neighbors (A and F), C has three neighbors (A, F, and G), D has two neighbors (A and F), F has four neighbors (D, B, C, and H), G has two neighbors (C and H) and H contains two neighbors (G and F). For example, the path $A \rightarrow B \rightarrow F \rightarrow H$ has totally 11 neighbors (i.e., $3 + 2 + 4 + 2 = 11$) as given in Table 1. Similarly, all the neighboring nodes of paths are calculated. The initial path is identified based on any existing algorithm (i.e., $A \rightarrow B \rightarrow F \rightarrow H$). If any node leaves from the initially selected path, the AASOP technique identifies alternate path as shown in Table 2 by considering minimum number of neighborhood nodes.

**Table 2** Path(s) with Delay

| Path(s) Number | Routing Path | Total no. of Neighborhood Nodes | Delay (in secs.) |
|---|---|---|---|
| P1 | $A \rightarrow B \rightarrow F \rightarrow H$ | 11 | 6 |
| P2 | $A \rightarrow C \rightarrow G \rightarrow H$ | 10 | 5 |
| P3 | $A \rightarrow C \rightarrow F \rightarrow H$ | 12 | 8 |
| P4 | $A \rightarrow D \rightarrow F \rightarrow H$ | 11 | 7 |

Fig. 2 Nodes and their Neighborhood (Scenario 1)

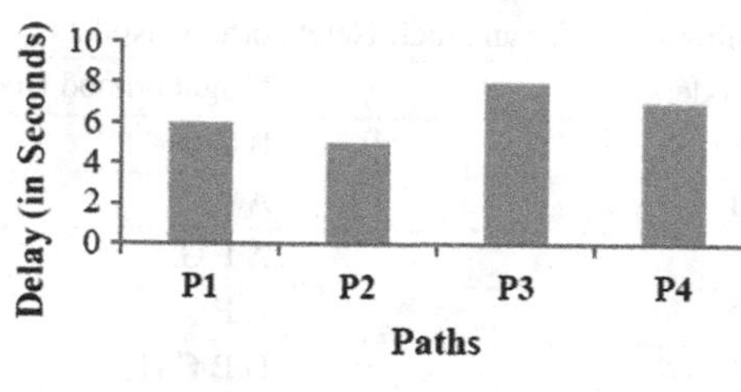


Fig. 3 Nodes with Delay (Scenario 1)

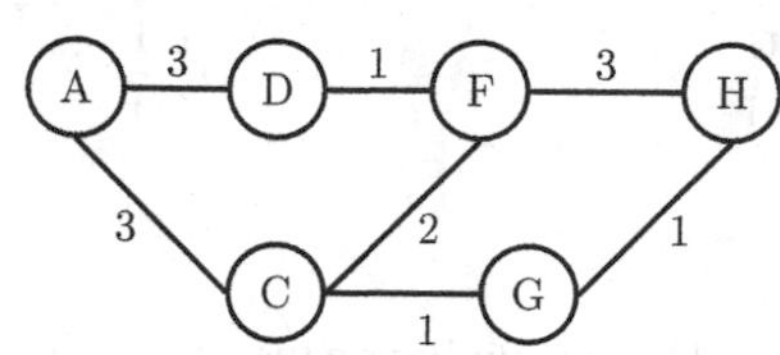


Fig. 4 Network Scenario 2

Here, the alternate path, A→C→G→H is selected since it has minimum number of neighbors, 10 (i.e., 3 + 3 + 2 + 2 = 10).

Figure 2 shows various numbers of nodes and their neighborhood. Figure 3 shows the number of nodes with delay parameter. The AASOP technique selects the Path 2 since, it has minimum of neighboring nodes.

**Scenario 2**

In this scenario, B leaves from the network as shown in Fig. 4. AASOP calculates neighborhood of all the nodes as shown in Table 3. It identifies the alternative paths based on number of neighborhood nodes. Here, there are three alternative paths available A→D→F→H, A→C→G→H and A→C→F→H. In which, A→D→F→H, and A→C→G→H have same number of neighboring nodes. In this case, when neighboring nodes are equal it considers delay parameter to select best optimum path. If the value of neighboring nodes and delay are equal, then the best optimum path is selected based on First Come First Serve (FCFS) as shown in Table 4.

Here, the alternate path, A→C→G→H is selected as it has minimum number of neighborhood, 9 (i.e., 2 + 3 + 2 + 2 = 9) and minimum delay when compared with other paths.

**Table 3** Nodes and their Neighborhood Node(s)

| Node(s) | Neighborhood Node(s) | No. of Neighborhood Node(s) |
|---|---|---|
| A | C D | 2 |
| C | A F G | 3 |
| D | A F | 2 |
| F | C D H | 3 |
| G | C H | 2 |
| H | G F | 2 |

**Table 4** Path(s) with Delay

| Path(s) Number | Routing Path | Total no. of Neighborhood Nodes | Delay (in secs.) |
|---|---|---|---|
| P1 | A → C → G → H | 9 | 5 |
| P2 | A → C → F → H | 10 | 8 |
| P3 | A → D → F → H | 9 | 7 |

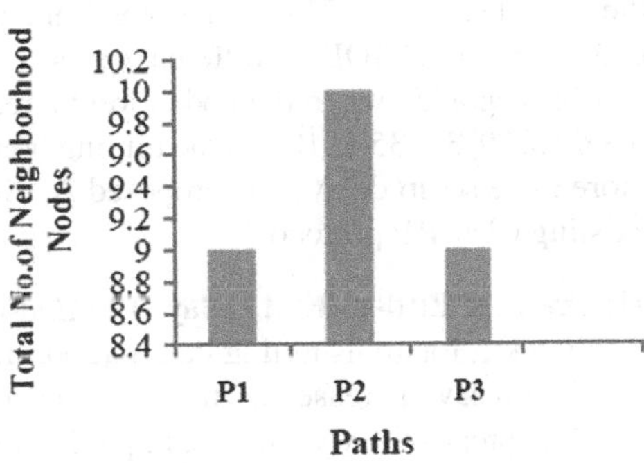


**Fig. 5** Nodes and their Neighborhood (Scenario 2)

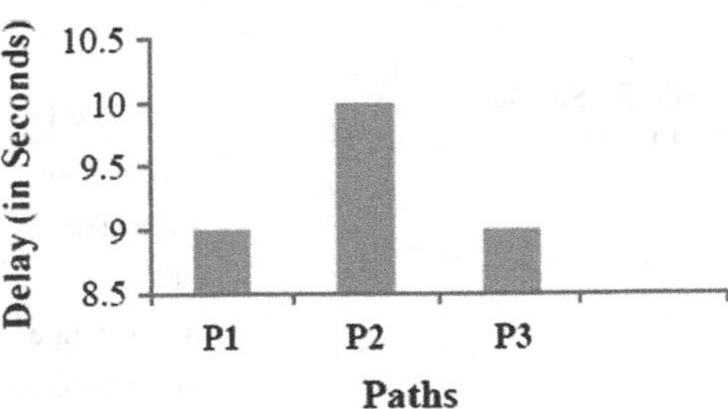


**Fig. 6** Nodes with Delay (Scenario 2)

Figure 5 shows the various numbers of nodes and their neighborhood. Figure 6 shows the various numbers of nodes with delay. The AASOP technique considers the Path 2 as it has equal number of neighborhood nodes and delay.

## 5 Simulation Results

### 5.1 Simulation Parameters and Values

The proposed technique AASOP is implemented using NS-2. The performance of AASOP is compared with the existing On-Demand Multicast Routing Protocol (ODMRP) protocol. It is observed that AASOP outperforms than ODMRP in terms of Packet Delivery Ratio and End-to-End delay. The simulation parameters and values are listed out in Table 5.

### 5.2 Performance Metrics

The AASOP technique is compared with the ODMRP protocol. The performance is evaluated with Quality of Service (QoS) quantitative parameters. In this simulation experiment, the node levels varied as 15, 20, 30, 35 and speed is 10 m/s by keeping the packet size as 512 bytes. Figures 7 and 8 show the comparative performance of ODMRP and AASOP for different cases. The performance of ODMRP routing protocol is degraded when the node level increases. Once if the node level is increased from 15, 20, 30, 35 to 10 m/s the routing breakage will also increase which results in more route setup delay. The proposed AASOP technique incurs less delay than the existing ODMRP protocol.

(i) **Average End-to-End Delay**: The End-to-End delay is referred as the total time is taken for transmitting data packets from source to destination. The End-to-End delay increases as the node level increases which is depicted in Fig. 7. The proposed AASOP technique incurs less delay than the existing ODMRP protocol.

**Table 5** Simulation Parameters

| Parameter(s) | Values (s) |
|---|---|
| No. of nodes | 15, 20, 30, 35 |
| Area size | 500 × 500 |
| Mac | 802.11 |
| Radio range | 250 m |
| Simulation time | 300 s |
| Traffic source | File Transmission Protocol (FTP) |
| Packet size | 512 B |
| Mobility model | Random Way Point |
| Speed | 10 m/s |
| Pause time | 5 s |
| Routing protocol | ODMRP |

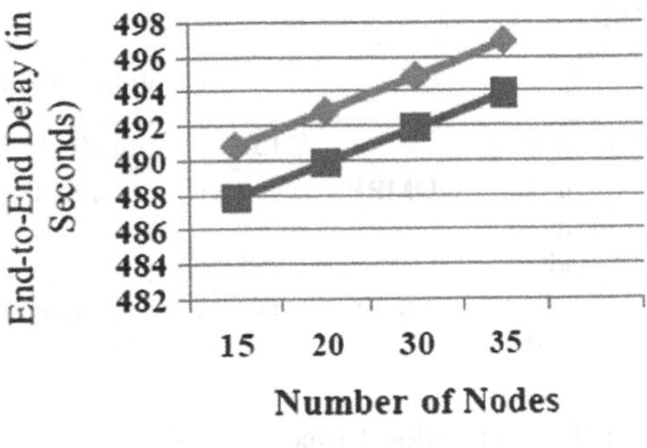


**Fig. 7** Average End-to-End Delay between ODMRP and AASOP

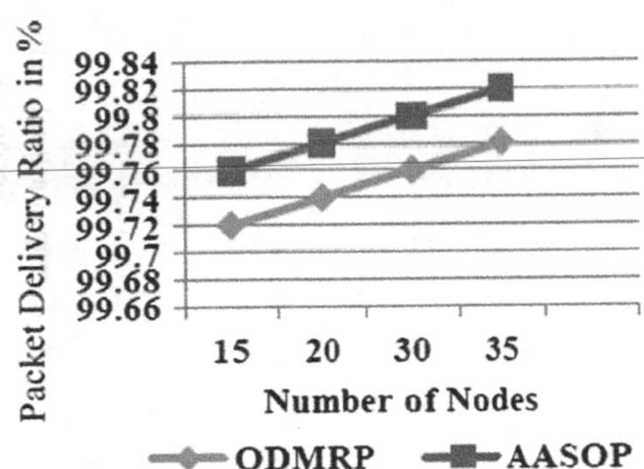


**Fig. 8** Packet Delivery Ratio between ODMRP and AASOP

(ii) **Packet Delivery Ratio (PDR)**: It is the ratio of number of packets received which is divided by number of sent packets. Figure 8 shows the Packet Delivery Ratio between ODMRP and AASOP. The Packet Delivery Ratio decreases when the node level increases. But, AASOP achieves more packet delivery ratio when compared with ODMRP protocol.

$$PDR = \frac{No.\,of\;Received\;Packets * 100}{No.\,of\;Sent\;Packets}$$

Let $n$ be the total number of nodes viz., $x_1, x_2, , x_n$ and the corresponding delay are $d_1, d_2, ..., d_n$, respectively. As shown in Table 6, then the computation or mean is calculated using following equation

$$\text{Computation Time} = \frac{\sum_{i=1}^{n} x_i \,.\, d_i}{\sum_{i=1}^{n} x_i}$$

where, $x_i$—Total Number of Nodes and $d_i$—Delay

**Table 6** Computational Time Calculation

| | | Computational calculation | | | | $\frac{x_i \cdot d_i}{x_i}$ |
|---|---|---|---|---|---|---|
| No. of nodes $x_i$ | | 15 | 20 | 30 | 35 | 100 |
| No. of delay ($d_i$) (in seconds) | ODMRP | 490.81 | 492.83 | 494.85 | 496.86 | 494.54 |
| | AASOP | 487.85 | 489.85 | 491.85 | 493.35 | 494.09 |

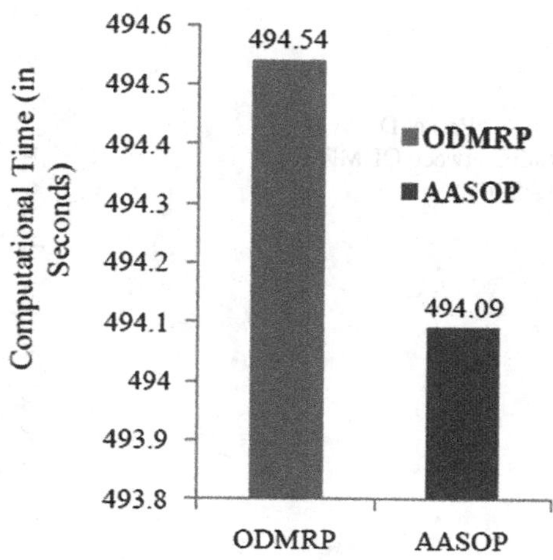


**Fig. 9** Computational Time of ODMRP and AASOP

Figure 9 shows the computation time between ODMRP and AASOP. The existing ODMRP technique computes delay in 494.54 s. Whereas AASOP computes delay in 494.09 s.

## 6 Conclusion

In MANETs, some routing algorithms are used to transfer the data via multiple paths to reduce end-to-end delay between source and destination. Continuous data transfer is not possible since all the nodes change dynamically. The AASOP technique, considers two important parameters such as number of neighborhood nodes calculation and delay. It selects the optimum path from multiple paths. It attempts to transfer the data quickly. Ultimately, it increases the speed of data transfer in MANETs as the path selected by the proposed technique has minimum neighbors. AASOP ensures the performance enhancement in terms of packet delivery ratio and delay. Although AASOP technique has many advantages, it also has weakness like maintenance of tables and routing overhead when the number of node increases. This issue will be addressed in further research.

## References


1. Shalini. A, Arulkumaran. G and Srisathya. K.B, Taming Enactment Using Neighbor Discover Distance Against Masquerading Attack In MANET, International Journal of Computer Engineering & Science, Volume 4, Issue 1, January 2014, pp. 1–6.
2. N. D. Kure, and S.A Jain, Minimum Overhead Routing Protocol in MANET, International Journal of Advance Foundation and Research in Computer, Volume 1, Issue 5, May 2014, pp. 16–23.
3. Roopali Garg and Guneet Kaur, Modified Neighbor Coverage Based Probabilistic Rebroadcast in MANET, International Journal of Advanced Research in Electrical, Electronics and Instrumentation Engineering, Vol. 3, Issue 5, May 2014, pp. 9389–9394.
4. Surya. R.M and Anitha. M, Secured Data Transmissions in MANET Using Neighbor Position Verification Protocol, International Journal of Engineering and Computer Science, Volume 3, Issue3, March 2014, pp. 5067–5071.
5. Khaleel Husain And Premala Patil, Neighbors Monitoring Scheme and Swarm Intelligence Based CrossLayer Attacks Handling In MANET, Proceedings Of 6th IRF International Conference, Bangalore, India, June 2014, pp. 61–66.
6. Vijay U. Patil and A. S. Tamboli, Review of Reducing Routing Overhead in Mobile Ad-Hoc Networks by a Neighbor Coverage-Based Algorithm, International Journal of Technological Exploration and Learning, Volume. 2, No 6, December 2013, pp. 327–330.
7. J. Priyanka Sheela and S. Sundar Raj, A Neighbor Coverage Based Routing By Good Node Detection in MANETs, Proceedings of International Conference On Global Innovations In Computing Technology, Volume 2, Issue 1, March 2014, pp. 3770–3773.
8. Shaik Arif Basha and G. Preethi Joshna, Locating and Verifying of Neighbor Positions in MANETs, International Journal of Computer and Electronics Research, Volume 3, Issue 4, August 2014, pp. 220–222.
9. Nandhini. R and Malathi. K, Power Balancing Approach for Efficient Route Discovery by Selecting Link Stability Neighbors In Mobile Adhoc Networks, American International Journal of Research in Science, Technology, Engineering & Mathematics, Volume 5, Issue 2, February 2014, pp. 196–201.
10. Harpreet Kaur and Jasmeet Singh, Optimization of Hello messaging scheme in MANET On-demand routing protocols using BFOA, International Journal of Application or Innovation in Engineering & Management, Volume 3, Issue 7, July 2014, pp. 333–339.
11. G. Kavitha and Dr. J. Sundararajan, Optimal Link Managed On Demand Routing Protocol in MANET for QoS Improvement, International Journal of Engineering and Technology, Volume 6, No 1, March 2014, pp. 146–154.
12. P. Sivakumar, R. Srinivasan and K. Saranya, An Efficient Neighbor Discovery Through Hello Messaging Scheme in MANET Routing Protocol, International Journal of Emerging Technology and Advanced Engineering, Volume 4, Issue 1, January 2014, pp. 169–173.
13. Suganya Devi. S and Dr. D. Thilagavathy, Neighbor Node Discovery and Trust Prediction in MANETs, International Journal of Science, Engineering and Technology Research, Volume 2, Issue 1, January 2013, pp. 145–149.
14. C. Prasanna Ranjith, Calduwel Newton, Mary Jane, Mobility Prediction in MANET Routing using Genetic Algorithm, International Journal of Innovations & Advancement in Computer Science, Volume 3, 2014.

# Resource Factor-Based Leader Election for Ring Networks

**Tarun Biswas, Anjan Kumar Ray, Pratyay Kuila and Sangram Ray**

**Abstract** A leader election is one of the fundamental problems in distributed systems. A node should have sufficient amount of resources to act as a leader. In this paper, we have proposed a leader election algorithm considering available resources of the nodes. All the nodes compute their resource factor (RF) value and form a process priority status queue (PPSQ) to transmit it to the next connected node. Finally, the node with highest RF value will be elected as a leader. Extensive simulations are performed, and it is shown that the proposed technique is better than the existing random leader election techniques in terms of available resources.

**Keywords** Process priority status queue (PPSQ) · Resource factor (RF) · Leader election · Ring network · Distributed systems

## 1 Introduction

A distributed system is a combination of different computing nodes, interconnected by a communication network. Each node can accept and share information by using communication protocols such as message passing and shared memory [1]. As the

T. Biswas (✉) · P. Kuila · S. Ray
Department of Computer Science and Engineering, National Institute of Technology, Ravangla 737139, Sikkim, India
e-mail: tarun.nitskm@gmail.com

P. Kuila
e-mail: pratyay_kuila@yahoo.com

S. Ray
e-mail: sangram.ism@gmail.com

A.K. Ray
Department of Electrical and Electronics Engineering, National Institute of Technology, Ravangla 737139, Sikkim, India
e-mail: akray.nits@gmail.com


S.K. Bhatia et al. (eds.), *Advances in Computer and Computational Sciences*, Advances in Intelligent Systems and Computing 553,
DOI 10.1007/978-981-10-3770-2_23

nodes may be situated in different geographical regions, it is important as well as hard to maintain the processes coordination among those nodes. Electing a single node as a leader in a distributed system is a critical problem. A leader election algorithm can be explained as a process to determine a central controlling node in a distributed system, to serve as a special node to coordinate among all other live nodes. As soon as a leader node is discovered, it is the responsibility of the leader node to coordinate distributed tasks such as consensus, resource allocation, load balancing, etc. [2–4].

Generally, a leader is elected based on the randomly generated priority value of the nodes [3, 5]. Where, the node with highest priority number is elected as a leader. However, in this process a node with comparably lower resources may be elected as a leader due to the random election. Therefore, the overall Performance of the system may be degraded as the lower resources containing node is being inevitably elected as a leader. In this paper we, propose a leader election algorithm for ring network based on the currently available resources of the nodes. Here, we have considered the available resource parameters like, processing power, memory capacity, shared resources, and residual energy. In order to measure the performance of the algorithm, extensive simulation is done based on the proposed algorithm. We have also simulated few existing algorithms [1, 6–9] and performance of the proposed algorithm is compared in terms of the above-mentioned resource parameters. It is also shown that the proposed algorithm elects a node as a leader having comparably higher set of available resources.

The rest of the paper is organized as follows: In Sect. 2, some existing algorithms are highlighted. The proposed technique is described in Sect. 3. In Sect. 4, an overall discussion on results of this work is presented. Section 5 concludes the paper with possible future directions.

## 2 Related Works

Several leader election algorithms such as the Bully algorithm [1, 6], Ring algorithm [7, 8], Chang and Roberts algorithm [10], and Franklin algorithm [11] have been proposed over the years. The leader is elected based on some randomly system generated priority value. The node which has the highest priority value is elected as a leader. The precondition of election algorithms are based on the following assumptions [1, 6, 12, 13]:

- All nodes in the system are assigned with unique identification numbers.
- All the nodes in the system are organized as a logical ring.
- On recovery, a failed node can take appropriate actions to re-join with the set of active processes.
- When a node wants some service from the leader, the leader is bound to response within the fixed time out period, besides its other tasks.

In all of the above-existing algorithms, the node which is currently holding the highest priority number wins the election and acts as a leader. If a node is failed due to some reasons, remaining nodes must get the updated information accordingly. The following algorithms like Bully [1, 6], traditional ring [8] and Franklin [11], etc. are performed by the system generated random priority numbers.

## 3 Proposed Work

Here, our objective is to elect out the node as a leader which has comparably better resources. We have calculated resource factor (RF) value of a node by considering the following parameters:

1. ***Available Processing Power*** (***$AV_{CPU}$***): The processing capability is the prime factor to execute a process. The processing capability varies machine to machine. Process execution capability of a node depends on its CPU power.
2. ***Available Memory Capacity*** (***$AV_{MC}$***): It signifies the amount of memory space left with a node. The number of tasks that can be executed by a node depends on the remaining memory space of the node.
3. ***Shared Resource (SR)***: According to the application in a distributed system, a node can share its resources with other remote machines. It can also access resources from remote machine. Whenever a node is sharing its resources or accessing some resource, the actual resource parameters need to be updated accordingly.
4. ***Residual Energy*** (***$AV_{RE}$***): The residual energy is also one of the important parameter particularly in energy constrained system like, wireless sensor network. It indicates how much battery power is remaining in any node to handle the processes. The node with marginal residual energy should leave the network.

In this work, we consider four resource parameters to compute the resource factor value of a node for the leader election process, where $\alpha_i$ is the weightage considered for features of all nodes and $f_i$ is different feature of computing nodes as shown in Table 1.

In a distributed system, all the computing nodes have resources in different measures. In this work, we have calculated the resource factor value by considering the above-mentioned four resource features. However, in many applications number of resource features may be different. For an application with $f_n$ number of resource

**Table 1** Parameters for considering resource factor

| Sl. no | Features | $f_i$ | $\alpha_i$ |
|---|---|---|---|
| 1 | $AV_{CPU}$ | $f_1$ | $\alpha_1$ |
| 2 | $AV_{MC}$ | $f_2$ | $\alpha_2$ |
| 3 | $SR$ | $f_3$ | $\alpha_3$ |
| 4 | $AV_{RE}$ | $f_4$ | $\alpha_4$ |

features and $N_n$ number of computing nodes, the Resource Factor (*RF*) value can be calculated by the generalized equation as follow:

$$RF = \frac{\sum_{i=1}^{fn} \alpha_{ij} f_{ij}}{\sum_{j=1}^{Nn} \sum_{i=1}^{fn} \alpha_{ij} f_{ij}}, \quad (1)$$

where $\alpha_{i-j}$ is *i*th weightage of feature *i* of node *j*, $f_{ij}$ is *i*th feature of node *j*, $i = 1, 2, 3, \ldots f_n$. The proposed leader election process is as follows.

At the beginning each node computes its RF value and shares the value along with its node id to its next neighbor node in the ring. After receiving the RF values, the next node adds its own RF value in the list and sorts them in descending order on RF value. Then the sorted list is transmitted to the next node. Thus, the last node of the ring contains RF values of all the available nodes. Therefore, highest RF value is at the beginning of the list. We denote the RF value of a node as its priority number and the sorted list as process priority status queue (PPSQ). It should be noted that all the nodes except the last node contain incomplete PPSQ. Now, the final PPSQ is circulated by the last node throughout the network so that each node gets the complete PPSQ. The first node of the PPSQ will be elected as a leader as it has the highest priority number.

## 4 Results and Discussion

Here, we have considered a ring network consisting of ten computing nodes with randomly generated node ids as follow, 43, 38, 76, 79, 18, 48, 44, 64, 70, 75. The different features values of the nodes are taken as shown in Table 2.

**Table 2** Features values of the nodes

| Sl. no. | $Node_{ID}$ | $AV_{CPU}$ | $AV_{MC}$ | SR | $AV_{RE}$ |
|---|---|---|---|---|---|
| 1 | 43 | 0.8756 | 0.6275 | 0.2024 | 0.2796 |
| 2 | 38 | 0.9455 | 0.9796 | 0.2189 | 0.0554 |
| 3 | 76 | 0.9746 | 0.6288 | 0.3363 | 0.1017 |
| 4 | 79 | 0.9071 | 0.6218 | 0.3717 | 0.1400 |
| 5 | 18 | 0.5983 | 0.6255 | 0.2464 | 0.1893 |
| 6 | 48 | 0.6758 | 0.9154 | 0.2341 | 0.2199 |
| 7 | 44 | 0.4586 | 0.6429 | 0.3029 | 0.3015 |
| 8 | 64 | 0.6902 | 0.7839 | 0.0303 | 0.0216 |
| 9 | 70 | 0.7654 | 0.8896 | 0.3736 | 0.0520 |
| 10 | 75 | 0.7844 | 0.7347 | 0.0048 | 0.1348 |

## 4.1 Case 1 (Each Features of All Nodes Having Same Weightage $\alpha_i$)

Here we have considered fixed weightage value $\alpha_i$ for the features $f_i$ of each node as, $\alpha_1 = 0.4$, $\alpha_2 = 0.3$, $\alpha_3 = 0.2$, and $\alpha_4 = 0.1$. Each node computes their priority (*RF* value) based on Eq. (1). The computed RF values are 0.111, 0.086, 0.118, 0.109, 0.107, 0.095, 0.108, 0.090, 0.072, and 0.098 as shown in Fig. 1a. It is shown in Fig. 1b that the computed priority value of the node 43, i.e., 0.111 is sent to the next node, i.e., node 38. The node 38 will compare its priority with received priority and sort accordingly as described in Sect. 3. Finally, the node 75 has the final PPSQ which circulated again so that each node has the same PPSQ as shown in Fig. 2. It can be observed from Fig. 2 that the node number 76 has the highest priority number 0.118. Therefore, the node 76 is elected as a leader. It is noteworthy that if the current leader node, i.e., node 76 failed due to some catastrophic reasons, the immediate highest priority node (here it is node 43) will be the new leader. We have also compared our proposed leader election process with random election process. It can be observed from Fig. 3a that the elected leaders for the proposed work and the random election process are node 76 and 44, respectively. It can further be observed from Table 2 that the node 76 has richest set of available resources than all other nodes. Though the node 44 has not sufficient amount of resources, it is elected as a leader due to random election process.

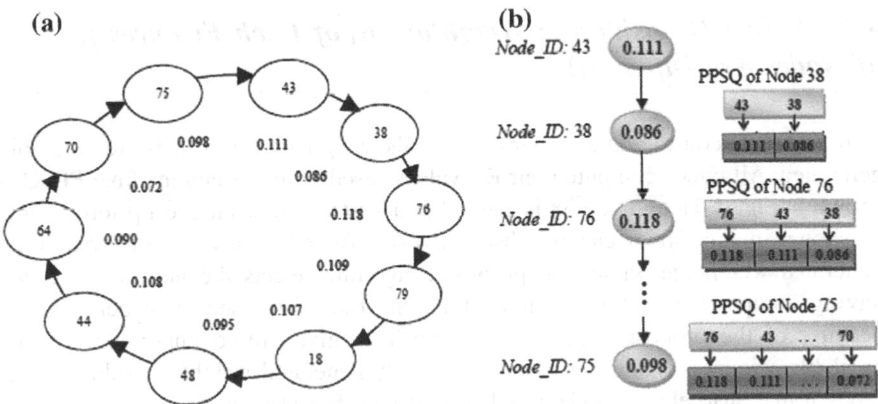


**Fig. 1** **a** Each node computes RF value based on available resources, **b** Formation of PPSQ Table

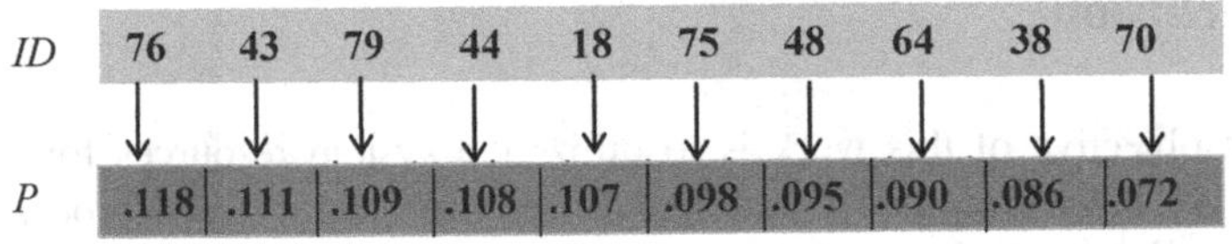


**Fig. 2** PPSQ in Case 1, where ID denotes nodes index and P denotes nodes priority

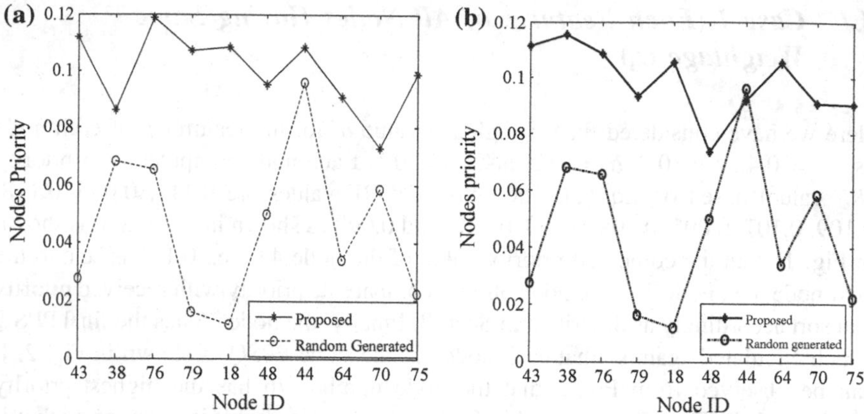


**Fig. 3** Election of leader in **a** case 1 **b** case 2

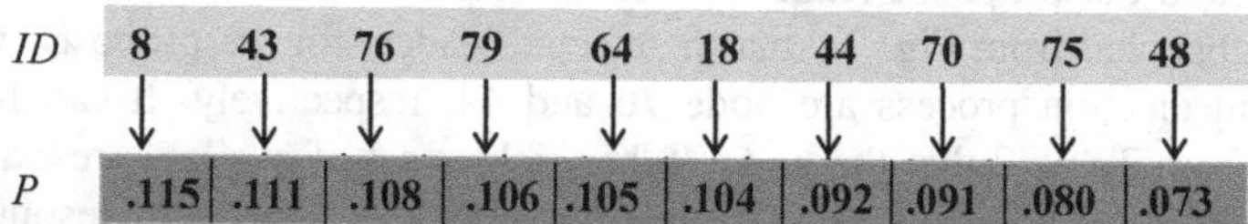


**Fig. 4** PPSQ in case 2, where ID denotes nodes index and P denotes nodes priority

### *4.2 Case 2 (Considering Weightage $\alpha_i$ of Each Features $f_i$ of Nodes Are Different)*

Here, we have considered a scenario where the weightage ($\alpha_i$) values are randomly generated. All nodes compute their RF values based on Eq. 1 and the final PPSQ is shown in Fig. 4. Here, node 38 is elected as a leader as it has highest priority value as shown in Fig. 3b. It can be observed from Table 2 that even after randomly generated weightage value, the proposed algorithm elects the node as a leaded having comparable better resources than the randomly elected leader. This is because of the proposed Eq. 1 where we have given more importance on the available resources. However, due to randomly generated weightage value, sometimes it may also elect a node as a leader having lesser resources.

## 5 Conclusion

The main objective of this work is to utilize the system resources for the leader election. We have presented a strategy for resource factor based process priority generation. We have shown that the node which has the richest set of resource parameters available at the time of election is designated as a leader. A set of case studies is presented to show the effects of system resources on the election of a leader. The results show that the proposed leader election process elects node as the leader which has higher weightage average of available resources. In future, important aspects such as message communication, network overhead, scalability, security features, etc. will be incorporated to enhance the leader election process for real-time deployment.

## References

1. Pradeep K Sinha. Distributed operating systems: concepts and design. *PHI Learning Pvt. Ltd.*, 1998.
2. Ajoy K Datta, Lawrence L Larmore, and Priyanka Vemula. An $o(n)$-time self-stabilizing leader election algorithm. *Journal of Parallel and Distributed Computing*, 71(11): 1532–1544, 2011.
3. Shay Kutten, Gopal Pandurangan, David Peleg, Peter Robinson, and Amitabh Trehan. Sublinear bounds for randomized leader election. *Theoretical Computer Science*, 561: 134–143, 2015.
4. Sung-Hoon Park. A stable election protocol based on an unreliable failure detector in distributed systems. *In Information Technology: New Generations (ITNG)*, 2011 *Eighth International Conference on*, pages 979–984. IEEE, 2011.
5. Seema Balhara and Kavita Khanna. Leader election algorithms in distributed systems. 2014.
6. Ajay D Kshemkalyani and Mukesh Singhal. Distributed computing: principles, algorithms, and systems. *Cambridge University Press*, 2011.
7. Greg N Frederickson and Nancy A Lynch. Electing a leader in a synchronous ring. *Journal of the ACM (JACM)*, 34(1): 98–115, 1987.
8. Hector Garcia-Molina. Elections in a distributed computing system. *IEEE Transactions on Computers*, 100(1):48–59, 1982.
9. Andrew Clark, Basel Alomair, Linda Bushnell, and Radha Poovendran. Minimizing convergence error in multi-agent systems via leader selection: A super modular optimization approach. *IEEE Transactions on Automatic Control*, 59(6):1480–1494, 2014.
10. Ernest Chang and Rosemary Roberts. An improved algorithm for decentralized extrema finding in circular configurations of processes. *Communications of the ACM*, 22(5):281–283, 1979.
11. Randolph Franklin. On an improved algorithm for decentralized extrema finding in circular configurations of processors. *Communications of the ACM*, 25(5):336–337, 1982.
12. Katherine Fitch and Naomi Ehrich Leonard. Information centrality and optimal leader selection in noisy networks. *In Decision and Control (CDC)*, 2013 IEEE 52nd *Annual Conference on*, pages 7510–7515. IEEE, 2013.
13. Fu Lin, Mohammad Fardad, and Mihailo R Jovanovic. Algorithms for leader selection in stochastically forced consensus networks. *Automatic Control, IEEE Transactions on*, 59 (7):1789–1802, 2014.

# TACA: Throughput Aware Call Admission Control Algorithm for VoIP Users in Mobile Networks

**P. Calduwel Newton and K. Ramkumar**

**Abstract** Call Admission Control (CAC) in wireless communication plays major role in deciding the admission of real-time and non real-time mobile users. For non real-time, it does not care for the Quality of Service (QoS) performance. But for real-time services, the CAC cares for the QoS by keeping the sufficient bandwidth throughout the transmission. Bandwidth determines the system capacity and speed. In this paper, a new Analytic Hierarchy Process (AHP)-based CAC algorithm is proposed for increasing the number of admission and reducing the admission of less compressed calls. For this task the various codecs such as G.711, G.729, G.723, G.726, AMR, EVRC, and iLBC have been taken. AHP is applied among them to take the right decision and it produces the ranking order. That rank helps to save more bandwidth and increases the throughput (Packets per second). This task is carried out by taking the criteria like bandwidth, packetization delay, and compression ratio of each individual codecs. The ultimate aim of this TACA is to give better QoS performance for real-time services and increase the system throughput for Voice over Internet Protocol.

**Keywords** Analytic Hierarchy Process · Call Admission Control · Quality of Service · Voice over Internet Protocol

P.C. Newton (✉)
Department of Computer Science, Government Arts College, Ayyarmalai, Kulithalai, Karur-Dt 639 120, Tamil Nadu, India
e-mail: calduwel@yahoo.com

K. Ramkumar
Department of Computer Science, Bishop Heber College (Autonomous), Tiruchirappalli 620 017, Tamil Nadu, India
e-mail: kramdharma@gmail.com



S.K. Bhatia et al. (eds.), *Advances in Computer and Computational Sciences*, Advances in Intelligent Systems and Computing 553,
DOI 10.1007/978-981-10-3770-2_24

## 1 Introduction

Long Term Evolution (LTE) is a technology of fourth generation network. The eventual plan of this technology is to increase the data rate and provide satisfied level of QoS [1]. The term QoS describes the performance seen by the users and it plays a vital role in wireless mobile communications. It has two types: qualitative and quantitative. Quality defines the reliability, scalability, availability, security, etc. These are focusing on the quality as a result for the user. For example, reliability should be greater from 0 and must be lesser than 1, that is error free and scalability works for fairness that is giving equal shares, availability making sure the available resources (time/frequency) at all times, security provides the protection in case of misuse. Similarly, the quantity distributes the good outcome by increasing the parameters like throughput, delay, jitter, packet loss, etc. It is like maximizing the profit while spending limited resources. CAC [2] is an algorithm which accepts all the incoming calls automatically if bandwidth exists or else it rejects when it is not available. This is a usual task of CAC. It does not permit the new subscriber to access it even if the mobile users have high bandwidth. Worldwide Interoperable Microwave Access (WiMAX) [3] is a new technology which provides wider coverage with enriched data rates. According to the WiMAX technology, the subscribers must have enough bandwidth to communicate with base station. Initially, CAC at the base station gives the preference to the user who has high bandwidth, good channel conditions, and application types. Applications can also be differentiated by two methods: real-time and non real-time [4]. This paper only concentrates the real-time service VoIP, and it can be identified by the codec that the users have chosen. Codec is used to convert our analog voice signals into digital signals.

This paper is organized as follows: Section 2 explains the related works and describes the motivations for writing this article, Sect. 3 discusses the TACA, Sect. 4 shows the simulation results, Sect. 5 carries the conclusion and finally references are listed.

## 2 Related Works

Ali et al. [5] explored the data rate which is based on the codec that the user used, type of uplink scheduling algorithm used and header compression techniques used. These factors determine the whole data rate of each VoIP users. Therefore, the silence unsuppressed codec may take high bandwidth and request high data rate when compared with others. So, this stops the others from having an opportunity to make request. Yu et al. [6] proposed a new measurement scale to maximize the call load and calculated the accurate bandwidth with the help of the Erlang B model. They also found the power of call is based on the network devices and not on the physical mediums. ROHC helps to mitigate the number of overheads and not the number of the packets. Saravanaselvi et al. [7] explained the importance of bandwidth and how it only determines the QoS over several mediums. They also stated two types of CAC algorithms: one is Threshold based and other is Queue aware-based CACs. In threshold, the value keeps on comparing with all the incoming calls whether they are matching with that value or not and it is a statistical-based method. If it suits that, it is accepted for communication. Otherwise that should be rejected. Similarly, the Queue aware CAC is fully based on the end user traffic queue status and it notices the number of packets in the queue for considering the process. Hoque et al. [8] discussed the essence of CAC for VoIP and the types of CACs available and also stated the requirements for making VoIP calls, how to decide or select the best CAC also discussed. Mishra et al. [9] elaborated the Resource Reservation Protocol (RSVP) usage of variable bandwidth capacity and their performance, and proved that G.723 codec outperforms other codecs using RSVP.

Ali et al. [10] compared different codecs by taking different network conditions of VoIP application. For this, they took WLAN, UMTS, WiMAX, and wired environments. In WLAN, the G.711 and GSM-FR codecs give better results of QoS to VoIP, the G.729 provides good QoS performance in WiMAX networks, UMTS, and GSM-FR commonly satisfy in all environments and finally G.723 offers good results in WiMAX and UMTS networks on the wired link connection, and they proved all these things through OPNET-based simulations.

Labyd et al. [11] analyzed G.711 codec works well in Voice over LTE technology in various situations and proved in Local Area Network (LAN) connections. This functionality almost equals to VoIP. Their results are not suitable for wireless mobile environments because of the mobility and various bandwidth conditions. Kwon et al. [12] developed a new architecture for mobile VoIP in short form (mVoIP) for improving the QoS by using adaptive codec selection algorithm. They stated that the codec is the most important device for increasing the QoS performance in all situations. Also, they confirmed that the high quality codecs could restrict the admission of low rate codecs in limited resources. Low-bandwidth consumption codec gets first priority among the high bandwidth consumption codecs and they proved using simulation results. Shwetha et al. [13] discussed uplink-scheduling algorithm for increasing the throughput and found the reasons for the greater number of dropping calls and gave some possible solutions to that issue. And, they analyzed the scheduling association with CAC for improving the QoS performances. They also explained that the existing CAC allows the higher priority to the real-time services then the low priority services. Among the real-time services no classifications are made. Bellalta et al. [14] analyzed the features of various AMR codecs. The poor design of CAC could cause blocking large number of VoIP users in unpredictable bandwidths and this may affect other VoIP users.

The main aim of this research is to search how the CAC works in heterogeneous networks for scheduling. The past investigation of this work was only looking at the guaranteeing the bandwidth after the admission of VoIP calls. In the existing CAC algorithms, all the incoming calls are permitted by looking at their requirement and the available resources. CAC did not have any priority or any classification for giving permission to the users. Without this classification, the low-bandwidth compression user with high channel quality may get a chance to access more resources and waste them quickly. Similarly, high compressed user with good channel quality could not get the resources at an appropriate time. So, there must be a mechanism to rank the VoIP users based on the availability of the codecs.

## 3 TACA: A Proposed Technique

The TACA is evaluated using the Multi-Link Point-to-Point protocol based-network condition and adjudged to the CAC algorithm. The following steps list the TACA in detail:

*Algorithm Steps*

1. Let the Available Bandwidth, $B_{avail}$
2. Identify the Required Bandwidth ($B_{tot}$) of Total VoIP incoming calls at the Base Station (eNodeB)

   ***If*** $B_{avail} >= B_{tot}$

   Schedule them as per their request

   ***Else if*** $B_{avail} < B_{tot}$

   1. Use AHP to find ranking among the codecs
   2. Apply Weight Sum Product Algorithm with an alternatives of each codecs
   3. Sort them according their rank in descending order and admit the VoIP users based on the ranking algorithm

   ***Endif***

   ***Endif***
3. Update the available bandwidth using the following formula and update the remaining bandwidth. $B_{avail}$ = Total $B_{ser} - B_{rem}$
4. Repeat step 2 until all the resources has been allocated.

(Where $B_{avail}$ = Available Bandwidth

$B_{tot}$ = Total Bandwidth of all incoming VoIP calls [$B_{tot} = VoIP_1 + VoIP_2 + \ldots + VoIP_n$]

$VoIP_1$ = needed bandwidth to make a call of VoIP user1

$B_{rem}$ = Remaining Bandwidth

$B_{ser}$ = Serviced Bandwidth).

This TACA helps the CAC to take correct decision for adapting more number of incoming calls as well as to increase the overall throughput. Figure 1 show the alternatives which are used in AHP to get the importance metric values. Here the bandwidth has been given higher important than packetization delay. Compression ratio is the second important than packetization delay. Due to the different payload sizes of each codecs, the packetization delay considered the less important factor.

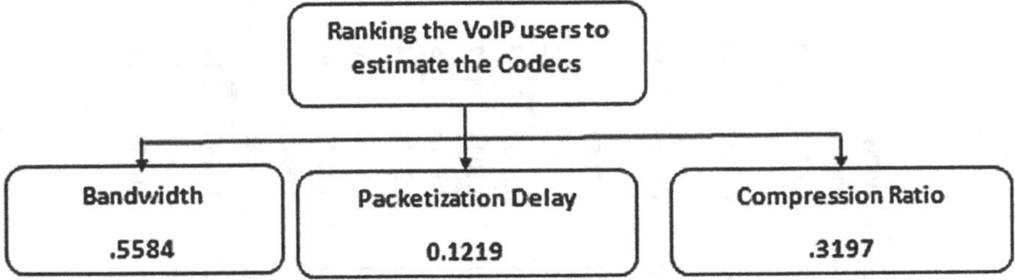


**Fig. 1** Variable alternatives and their priority values using AHP algorithm

Table 1 shows the detailed classifications of various VoIP codecs and their associated rank values. AHP helps to judge the priority among the alternates. After getting the priority, the Weighted Sum Product is applied with these alternates to estimate the rank.

## 4 Simulation Results

TACA is simulated using MATLAB (2014b) software. Let the available bandwidth is 275 Kbps and Number of Users is 8. The users of VoIP codecs are G.711 = 1, G.729 = 1, AMR = 1, G.723 = 2, G.726 = 1, iLBC = 1, and EVRC = 1. Multi-Link Point-to-Point protocol (6 bytes) is taken for wireless communication and the following Eq. (1) is used to calculate the packets per second (PPS). Here codec bit rate defines the size of the bandwidth taken by the codec to sample the voices and payload states the actual packet size. The Bandwidth per call is calculated by using the Eq. (2).

$$\text{Packets Per Second (PPS)} = [\text{codec bit rate}/\text{payload}] \tag{1}$$

$$\text{Bandwidth per call} = \text{Voice packet size} * \text{PPS} \tag{2}$$

### 4.1 *Without Ranking*

Working Methodology of the existing CAC algorithm in 275 Kbps and the number of VoIP calls are summarized in the Table 2. It shows the packets per second of each codecs.

***With Ranking (AHP based)***

Working Methodology of the proposed CAC algorithm in 275 Kbps and the number of VoIP calls are summarized in the Table 3.

Similarly, some more examples are shown in Table 4 that the performance results of the proposed CAC algorithm.

**Table 1** VoIP codecs and their ranking order

| Codec | Bandwidth (Kbps) | Packetization delay (ms) | Compression ratio (%) | Total packet size | Packets Per Second (PPS) | Bandwidth per call (Kbps) | Pay load | WSM values | Rank |
|---|---|---|---|---|---|---|---|---|---|
| G.711 μ law | 64 | 20 | 31.6 | 206(160) | 50 | 82.4 | 160 | 0.0711 | **12** |
| G.711 a law | 64 | 30 | 31.6 | 206(160) | 50 | 82.4 | 160 | 0.0622 | **13** |
| G711 (a & μ) | 64 | 40 | 31.6 | 206(160) | 50 | 82.4 | 160 | 0.0416 | **16** |
| G 729a | 8 | 20 | 76 | 66(20) | 50 | 26.4 | 20 | 0.3852 | **1** |
| G.729b | 8 | 30 | 76 | 66(20) | 50 | 26.4 | 20 | 0.3763 | **2** |
| G 729c | 8 | 40 | 76 | 66(20) | 50 | 26.4 | 20 | 0.3557 | **3** |
| G.723 | 6.3 | 30 | 63.5 | 76(30) | 33 | 20.4 | 30 | 0.1049 | **7** |
| AMR | 4.75 | 20 | 73.4 | 66(20) | 30 | 15.8 | 20 | 0.1815 | **5** |
| AMR | 7.4 | 20 | 73.4 | 66(20) | 46 | 24.2 | 20 | 0.2138 | **4** |
| AMR | 12.2 | 20 | 73.4 | 66(20) | 76 | 40.3 | 20 | 0.1690 | **6** |
| G.726 | 40 | 10 | 70.5 | 66(20) | 250 | 132 | 20 | 0.0844 | **10** |
| G.726 | 32 | 10 | 70.5 | 66(20) | 200 | 105.6 | 20 | 0.0943 | **9** |
| G.726 | 16 | 10 | 70.5 | 66(20) | 100 | 52.8 | 20 | 0.1004 | **8** |
| iLBC | 15 | 20 | 48.8 | 76(30) | 62 | 37.7 | 30 | 0.0830 | **11** |
| iLBC | 13 | 40 | 48.8 | 76(30) | 55 | 33.4 | 30 | 0.0587 | **14** |
| EVRC | 16 | 50 | 61.7 | 66(20) | 100 | 52.8 | 20 | 0.0456 | **15** |

**Table 2** Calculation of 275 Kbps bandwidth and VoIP calls

| Codec type | Required data rate in Kbps | Packets per second (PPS) |
|---|---|---|
| G.726 | 105.6 | 200 |
| G.711 | 82.4 | 50 |
| EVRC | 52.8 | 100 |
| iLBC | 33.4 | 55 |
| **274.2 (Kbps)** | | **405** |

**Table 3** Calculation of 275 Kbps bandwidth and VoIP calls

| Codec type | Required data rate in Kbps | Packets per second (PPS) |
|---|---|---|
| G.729 | 26.4 | 50 |
| AMR | 15.8 | 30 |
| G.723 | 40.8 | 66 |
| G.726 | 105.6 | 200 |
| iLBC | 33.4 | 55 |
| EVRC | 52.8 | 100 |
| **274.8 (Kbps)** | | **501** |

**Table 4** Calculation of 1025 Kbps bandwidth and VoIP calls

| Codec type | Requited data rate in Kbps | Packets per second (PPS) |
|---|---|---|
| G.711 | 11*82.4 = 906.4 | 11*50 = 550 |
| G.729 | 4*26.4 = 105.6 | 4*50 = 200 |
| **1012 (Kbps)** | | **750** |

**Table 5** Calculation of 1025 Kbps bandwidth and VoIP calls

| Codec type | Requited data rate in Kbps | Packets per second (TPS) |
|---|---|---|
| G.729 | 5*26.4 = 132 | 5*50 = 250 |
| AMR | 8*24.2 = 193.6 | 8*46 = 368 |
| G.723 | 6*20.4 = 122.4 | 6*33 = 198 |
| G.711 | 7*82.4 = 576.8 | 7*50 = 350 |
| **1024.8 (Kbps)** | | **1166** |

***Without Ranking***

Let available Bandwidth is 1025 Kbps and Total No. of VoIP calls are 30. They are G.711 = 11, G.729 = 5, G.723 = 6, AMR = 8. Table 4 displays the bandwidth calculation for VoIP calls and their PPS.

***With Ranking (AHP based)***

Working Methodology of the proposed CAC algorithm in 1025 Kbps and the number of VoIP calls are summarized in the Table 5.

**Table 6** Comparisons between Existing and TACA for admitted and neglected users

| Algorithm type | Available bandwidth in Kbps | | Total number of VoIP calls | | Packets per second (PPS) | | Number of admitted users | | Number of neglected users | |
|---|---|---|---|---|---|---|---|---|---|---|
| | Case I | Case II | Case I | Case II | Case I | Case II | Case I | Case II | Case I | Case II |
| Without ranking | 275 | 1025 | 8 | 30 | 405 | 750 | 4 | 15 | 4 | 15 |
| With ranking | 275 | 1025 | 8 | 30 | 501 | 1166 | 7 | 26 | 1 | 4 |

**Table 7** Bandwidth economy and expenditure of each VoIP codecs for 275 and 1025 Kbps

| S. No | Codecs | Compression ratio | 275 Kbps | | 1025 Kbps | |
|---|---|---|---|---|---|---|
| | | | Bandwidth economy | Bandwidth expenditure | Bandwidth economy | Bandwidth expenditure |
| 1 | G.711 | 31.6 | 8 6.9 | 188.1 | 323.9 | 701.1 |
| 2 | G.729 | 76 | 209 | 66 | 779 | 246 |
| 3 | G723 | 63.5 | L74.6 | 101 | 650.3 | 374.2 |
| 4 | AMR | 73.4 | 201.8 | 73 | 752.3 | 272.7 |
| 5 | G.726 | 70.5 | 193.8 | 81.2 | 722.6 | 302.4 |
| 6 | iLBC | 48.8 | 134.2 | 140.5 | 500.2 | 524. 8 |
| 7 | EVRC | 61.7 | 169.6 | 105.4 | 632.4 | 392.6 |

Table 6 depicts the comparison result of the existing CAC algorithm and TACA performances in various bandwidths. And, it clearly shows that the proposed AHP-based algorithm admits greater number of users compared with existing CAC and PPS.

Table 7 the economic level and unused quantity of bandwidth for each codecs which is based on the compression ratio.

Figure 2 shows the PPS values of both existing CAC algorithm and proposed AHP-based algorithm for 275 Kbps bandwidth in 8 VoIP calls (Fig. 3).

CAC algorithm works only for admitted user's purpose and not for the newly arriving users and resource allocations. Figure 4 displays the total packets per second (PPS) values along with the comparison of 275 and 1025 Kbps bandwidths. Obviously, the TACA outperforms in both cases. Figures 4 and 5 show the comparisons of the number of admitted and neglected users by the CAC algorithm and AHP-based algorithm.

This TACA works well even when the total available bandwidth is huge when compare with total bandwidth of all VoIP calls. In this case, it saves much bandwidth to accept more calls. Existing CAC algorithms admit the VoIP users first but without looking at the comparisons among them.

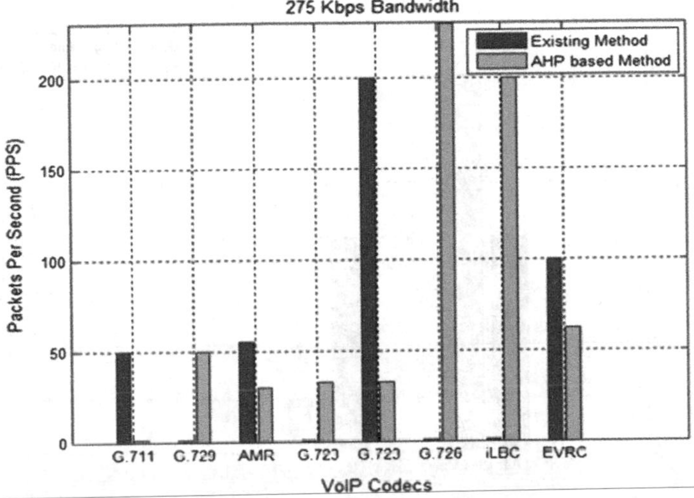


**Fig. 2** Bandwidth 275 Kbps, 8 users

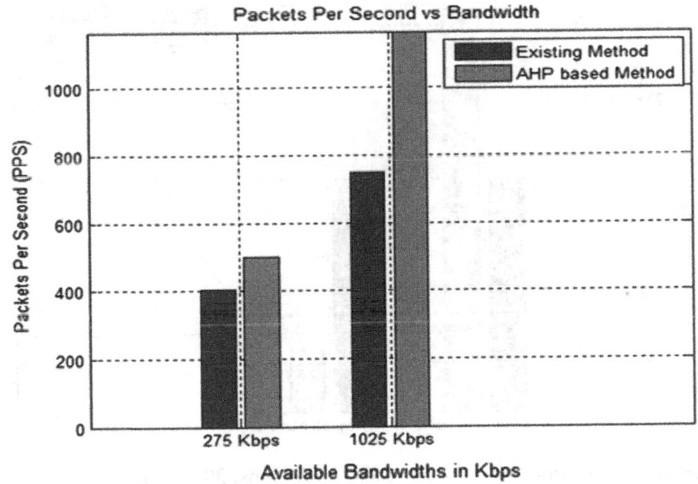


**Fig. 3** Packets per second comparisons with 275 and 1025 Kbps bandwidth

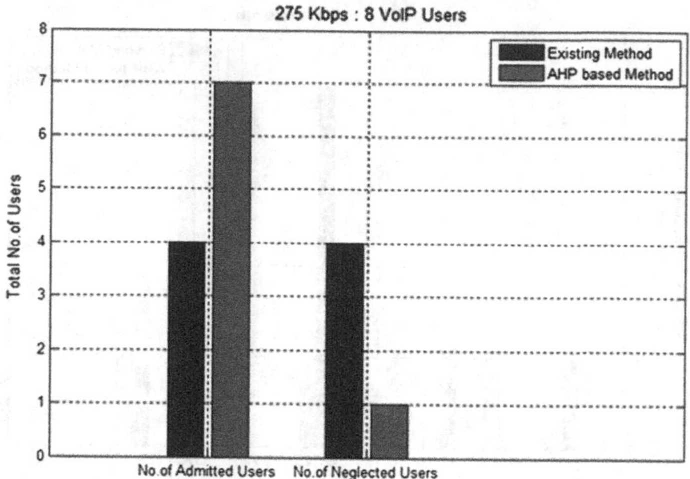


**Fig. 4** Number of admission and rejection calls (275 Kbps, 8 users)

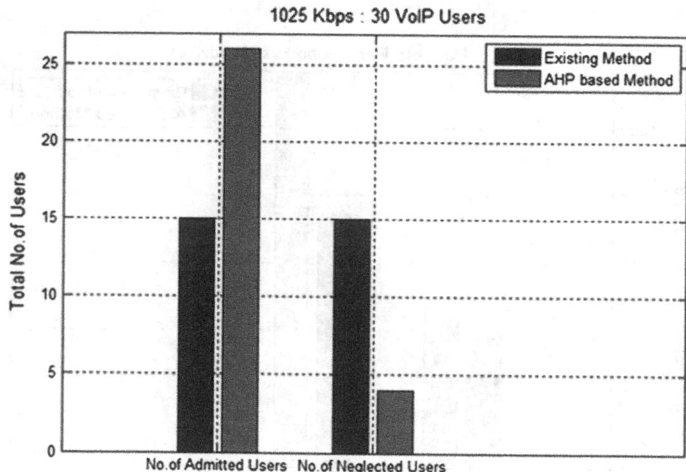


**Fig. 5** Number of admission and rejection calls (1025 Kbps, 30 users)

## 5 Conclusion

In this paper, the Analytic Hierarchy Process-based ranking algorithm is proposed for CAC in mobile networks. TACA classifies and provides the ranking of VoIP calls to the base station that differentiates each codecs. The AHP is used to take the accurate selection from all the available calls by taking required bandwidth, compression ratio and packetization delay criteria. Then, the important parametric value is applied with weight sum method to get the correct ranking order. Finally, based on the sorted ranking, the incoming calls are admitted at the Base station. The major problem of CAC is to give the exact admission for the arriving calls in appropriate times. For VoIP calls, CAC must reserve or guarantee certain bandwidths and it is not possible at all times in limited resources. There is no proper prediction mechanism for categorizing the VoIP users at the base station. So, the proposed work does the rank with correct prediction and permits large number of users and also increases the throughput.

## References

1. P. Calduwel Newton and L. Arockiam, "A Novel Prediction Technique to improve Quality of Service for Heterogeneous data traffic", Springer Science + Business Media, Issue 22, Nov 2009, pp. 867–872.
2. Mumtaz AL-Mukhtar and Huda Abdul wahed, "Optimizing VoIP Using A Cross Layer Call Admission Control Scheme", International Journal of Computer Networks & Communications Vol.5, Issue.4, Jul 2013, pp. 117–130.
3. Anindita Kundu, Iti Saha Misra, Salil K. Sanyal and Suman Bhunia, "VoIP Performance Over Broadband Wireless Networks Under Static and Mobile Environments", International Journal of Wireless & Mobile Networks, Vol. 2, Issue. 4, Nov 2010, pp. 82–93.
4. P. Calduwel Newton and L. Arockiam, "Route Optimization Mechanisms for Internet Applications in Mobile Networks: A Survey", International Journal of Future Generation Communication and Networking, Vol. 3, Issue. 3, Sep 2010, pp. 57–70.
5. Ashraf A. Ali, Spyridon Vassilaras, and Konstantinos Ntagkounakis, "A Comparative Study of Bandwidth Requirements of VoIP Codecs Over WiMAX Access Networks", IEEE Computer Society, pp. 197–203.
6. James Yu and Imad Al Ajarmeh, "Call Admission Control and Traffic Engineering of VoIP", IEEE Conference, Jul 2007, pp. 1–11.
7. P. Saravanaselvi and Dr. P. Latha, "A Survey on Call Admission Control and Bandwidth Allocation for WiMAX", International Journal of Innovative Research in Computer and Communication Engineering, Vol. 1, Issue 9, Nov 2013. pp. 2185–2193.
8. Mohammad Asadul Hoque and Farhana Afroz, "Call admission control QoS: issue for VoIP", IEEE Jan 2008, pp 1–6.
9. B. K. Mishra, S. K. Singh and Kalpana Patel, "Performance Analysis of various Codecs using RSVP on VoIP Quality of Service over Variable Bandwidth", International Journal of Computer Applications (0975 – 8887), Vol. 41, Issue. 7, Mar 2012, pp. 28–36.
10. Malik Ahsan Ali, Imran Rashid and Adnan Ahmed Khan, "Selection of VoIP CODECs for Different Networks based on QoS Analysis", International Journal of Computer Applications (0975–8887), Vol. 84 Issue. 5, Dec 2013, pp. 38–44.
11. Younes Labyd, Mohammed Moughit, Ab derrahim Marzouk and Ab delkrim Ha qiq, "Performance Evaluation for Voice over LTE by using G. 711 as a Codec", International Journal of Engineering Research & Technology, Vol. 3 Issue. 10, Oct 2014, pp. 758–763.
12. Dongwoo Kwon, Rottanakvong Thay, Hyeonwoo Kim and Hongtaek Ju, "QoE-based Adaptive mVoIP Service Architecture in SDN Networks", The Seventh International Conference on Communication Theory, Reliability, and Quality of Service, Feb 2014, pp. 62–67.
13. Shwetha D, Mohan Kumar N M and Devaraju J T, "Modulation Aware Connection Admission Control And Uplink Scheduling Algorithm For WiMAX Networks", International Journal of Wireless & Mobile Networks, Vol. 7, Issue. 1, Feb 2015, pp. 75–90.

14. B. Bellalta, C. Macian, A. Sfairopoulou and C. Cano, "Evaluation of Joint Admission Control and VoIP Codec Selection Policies in Generic Multi rate Wireless Networks", Proceedings of the 7th international conference on Next Generation Tele traffic and Wired/Wireless Advanced Networking, pp. 342–355.

# A Literature Survey on Detection and Prevention Against Vampire Attack in WSN

**Richa Kumari and Pankaj Kumar Sharma**

**Abstract** Wireless sensor network (WSN) is an ad hoc low power wireless network in which sensor nodes cooperatively monitor and gather the information from environment then broadcast that information to other nodes. So security against denial of services (DoS) at routing levels is the most significant area of research. In this paper, denial of service at the routing level are discussed. DoS is cause by resource exhaustion at the network layer, which completely disable the network by consuming node's battery power. This power draining attack is known as "vampire attack" which is not definitive to any routing protocol. A single Vampire can raise network energy usage by a factor of O(N) in worst case, here N is the number of nodes. This paper reviews some methods to detect and mitigate this attack that raise the energy consumption of network and concepts that bounds the damage from Vampires.

**Keywords** Wireless sensor network · Security issue · Routing protocol · Vampire attack

## 1 Introduction

### 1.1 *Wireless Sensor Network*

A wireless sensor network, sometimes called wireless sensor and actuator networks (WSAN). It is a collection of the many low-cost and low-power tiny sensors which cooperatively monitor and gather the information (such as temperature, sound, pressure, etc.) from environment in real time and broadcast the information to other

R. Kumari (✉) · P.K. Sharma
Department of Computer Science and Engineering,
Government Women Engineering College, Ajmer, Rajasthan, India
e-mail: richa.kumari7171@gmail.com

P.K. Sharma
e-mail: pankaj.gmeca@gmail.com



S.K. Bhatia et al. (eds.), *Advances in Computer and Computational Sciences*,
Advances in Intelligent Systems and Computing 553,
DOI 10.1007/978-981-10-3770-2_25

node in the network. The WSN was primarily proposed in domains where infrastructure missing and wired networks are not suitable.

### *1.2 Challenges in Designing WSN*

**Scalability**: Sensor network growing increasingly because sensors are low cost devices and protocol support large network. It is a dare to deploy wireless sensor network to a large scale and work efficiently with huge amount of nodes [1].

**Security**: Security is an important factor of wireless sensor network. It is a big challenge to build WSN with security concerns.

**Synchronization**: It is an important service for Wireless sensor network to synchronize all local clocks of nodes in the network to meet specific requirement.

**Data Confidentiality**: Sensor nodes do not reveal secret information to other nodes.

**Data Integrity**: It assures that data does not change by adversaries during the transmission.

**Authentication**: Data must be accessed by authorized user.

**Data Freshness**: As the name implies that the data must be recent, and it ensures that no intruder replayed previous messages.

## 2 Threat Model of WSN

### *2.1 Passive Attacks*

The passive attack (eavesdropping) is limited to listening and analyzes exchanged traffic. This type of attacks is easier to realize (it is enough to have the adequate receiver), and it is difficult to detect. The target of the intruders to collect the confidential data of significant node that being broadcast into network.

### *2.2 Active Attacks*

An active attack involves monitoring, listening, and modification of the data stream by the malicious nodes/adversaries prevailing inside or outside the network. Active attacks cause direct harm to the network because they can manipulate the data stream.

**Hello Flood Attack**: In most of the protocol, HELLO PACKET is used to show their presence to their neighbour and receiving nodes may assume that it is within the RF range of the sender. Attacker sending a flood of such messages to flood the network and to prevent other messages from being exchanged.

**Black Hole Attack**: A black hole is a malicious node that attracts all the traffic in the network by advertising that it has the shortest path in the network. This black hole drops all the packets it receives from the other nodes.

**Grey Hole Attack**: A grey hole is also a malicious node that selectively drops packets. There are two ways in which a node can drop packets:

- It can drop all UDP packets while transmitting all TCP packets.
- It can drop 50% of the packets or can drop them with probabilistic distribution.

**Wormhole Attack**: Wormhole attack is an attack on the routing protocol in which the packets or individual bits of the packets are captured at one location, tunnelled to another location and then replayed at another location.

**Denial of Service Attack**: In DoS attack [2, 3], attacker try to restrict the authentic user from being use all or part of the network services. DoS attack allows an adversary to disrupt the entire network either by disabling the network or by overloading it with messages, and also diminish a network's capability.

## 3 Vampire Attack

The vampire attack is the class of Denial-of-Service attack. Denial-of-Services in network is caused by consuming the power of the sensor node. It is also called power draining attacks because of this attack consume power of sensor nodes and disable the network. It creates a protocol-compliant message and sends it into network so that the energy used by the network is more than if same message transmitted of identical size to the same destination. Here we can categorize vampire attack into two.

### *3.1 Vampire Attack on Stateless Protocol*

There are basically two attacks on stateless routing protocol:

**Carousel Attack**: In carousel attack, an attacker composes a packet with a path that contain series of loops of that path, so that same of nodes will traverse by the single packet many times as shown in Fig. 1. In this attack, route length of the packet can be greater than the number of node in ad hoc network. This attack can mitigate by limit the number of allowed passage in the source route [4]. Energy uses by this attack is increased by the factor of O ($\lambda$), where $\lambda$ is the maximum route length.

**Fig. 1** Carousel attack

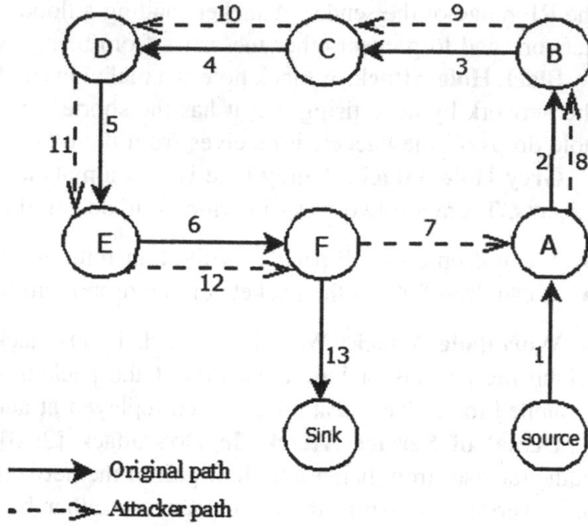


**Fig. 2** Stretch attack

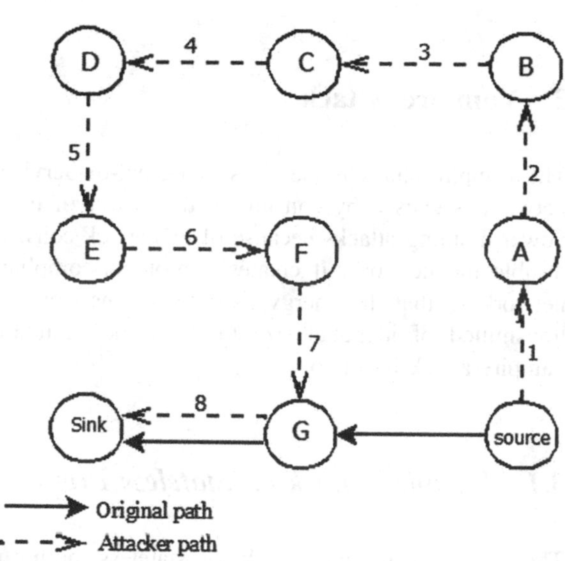


**Stretch Attack**: In stretch attack, intruder builds up fake long source route so that the packets move through large number of node rather than the excellent number of nodes as shown in Fig. 2. The energy consumed by this attack is increased by the factor of O (min (N, λ)), here λ is the maximum path length allowed and N is the number of nodes in the network. These types of attacks are easier to release and difficult to detect.

### 3.2 Vampire Attacks on Stateful Protocol

Stateful routing protocols in which network topology and its state are already known by the nodes so local forwarding decisions at each is done by using stored state. There are two categories of stateful protocols: first one is link-state [5] and other is distance-vector [6].

**Directional Antenna Attack**: In this type of attack, intruder has small control over the packet progress when forwarding decisions are made independently by each node but by using directional antenna, they can still waste energy. In this attack, attackers redirect a packet in any section of the network, that packet is unnecessarily move into network and consume the energy of nodes. Energy use by this attack is increase by the factor of O(D), where D is the diameter of the network. This attack is also called half worm hole attack [7].

**Malicious Discovery Attacks**: Route discovery process in AODV and DSR routing protocol perform on demand. These routing algorithms are vulnerable to vampire, when route discovery initiate by the node at any time, not just during change in a topology. There are many ways for the malicious node to activate topology change: it may wrongly claim in the network that a new link between two fake nodes or a link is fail between two non-existent nodes. This attack is also known as spurious route discovery.

## 4 Related Work

There are some researchers who provide many approaches against vampire.

**Eugene y. Vasserman et al.** [8], Describe PLGP protocol which is developed By Parno, Luk, Gaustad and Perrig to prevent the vampire attack. PLGP have mainly two phases, first is Topology discovery Phase and other is Packet forwarding phase.

In PLGP, adversaries can redirect the packets to any section of the sensor network because the forwarding nodes do not recognize about what actual path a packet took. So at most log (N) logical hops will traverse by packet. Thus, PLGP is vulnerable to vampire attack.

A new concept of "No-backtracking" arrive which can resist vampire attack."-No-backtracking is satisfied if every packet p traverses the same number of hops whether or not an adversary is present in the network". So they have developed a new protocol which fill the property no backtracking, name is PLGPa (PLGP with attestation).

**K. Vanitha and V. Dhivya et al.** [9], They have invented arouting protocol to prevent from Vampire Attacks in WSN, called Valuable Secure protocol. It has mainly three phases: first is network configuration phase, second is key management and last is communication phase. Key management phase is used for the security purpose of the node and data. ECC (Elliptic Curve Cryptography) is most widely used to achieve the security with smaller key size.

In short, PLGPa with ECC is developed to reduce the energy consumption in sensor network as compare to other previous protocol. In VSP, public and private key pair is generated by each node for secure communication.

**B. Umakanth and J. Damodhar et al.** [10], Describe the EWMA (Energy Weighted Monitoring Algorithm) to detect the vampire attack. It has mainly two phases: first is network configuration and other is communication phase. First phase is responsible for creating an excellent routing path from source to sink. Second phase perform aggregation of the packet transmission to avoids the same packets.

In ad hoc sensor network, the death of normal sensor nodes is increased as the number of malicious nodes increases. Rate of nodes alive can be increase by using EWMA. Main goal of EWMA is to increase the overall life-Cycle of network by energy efficient routing.

**Patil and Gaikwad et al.** [11], Describe a method to prevent vampire attack by using EWMA and finding corresponding trust value of each node. For preventing vampire attack first detect carousal and stretch attack. After detection of carousal and stretch attack reduce their impact in wireless sensor networks by using energy weight monitoring algorithm (EWMA). Then finding trust value of each node in the network for performing routing operation.

In proposed system, they use energy consumption and trust value of the node to mitigate vampire attack. The results shows that the impact of this attack reduced in great extent.

**C. Chumble and M. Ghonge et al.** [12], Describe the vulnerability to the vampire attack on exiting routing protocol like AODV even they design to be secure routing. The author also proposed a new routing protocol against vampire known as vampire attack removal protocol (VARP). They basically use clean slate routing protocol with no-backtracking. No-backtracking achieve through Digital signature by using RSA with MD5 so that packet pass in network securely.

**Amee A. Patel and J. Soni et al.** [13], Describe a method for detection and prevention against vampire attack in AODV routing protocol. The main goal of approach is defending the energy draining of nodes in wireless sensor network against vampire attack and increase the lifetime of network.

In this approach, they use energy levels of neighbour node and then find the threshold value for that node. In forwarding phase if the energy of next forwarding node is lower than the threshold value, it will not select that node as next node and broadcast the new RREQ message to other to find new path. According to proposed method, vampire attack in WSN can be detected.

**Shrivastava and Verma et al.** [14], Describe the "Detection of Vampire Attack in Wireless Ad hoc Network" in this scenario they explain a Vampire attacks which alter targeted packets. In AODV routing protocol at the time of route discovery malicious node can change either the destination address or broadcast id and misguide the packets. This action of modification affects the bandwidth and node battery power.

Proposed approach is an efficient and effective approach for detection the malicious packet in the wireless network. But as the number of nodes increases, the performance of proposed approach is decreases frequently. For the small network, the performance of network is much adaptable.

**Besty Haris et al.** [15], Describe wolf routing to detect the vampire in wireless ad hoc network. WSA (wolf searching algorithm) based on wolf victim behaviour for vampire searching. Wolves works in a group with semi-cooperation during

**Table 1** Mitigation technique of vampire attack

| S. no | Technique used against vampire | Feature | Advantages | Disadvantage |
|---|---|---|---|---|
| 1 | PLGP routing protocol | Provably restrict vampire attack | Clean slate secure sensor network routing | Does not satisfy "No-Backtracking" property and vulnerable to vampirc attack |
| 2 | PLGPa routing protocol | Resist vampire attack | Clean slate secure sensor network routing and also satisfy "No-Backtracking" property | Packet forwarding overhead duc to add attestation, it is tractable on 8-bit processors |
| 3 | Valuable secure protocol (VSP) | Secure routing protocol PLGPa with ECC | Reduce the energy consumption in sensor network | Extra overhead due to key management |
| 4 | Energy weighted monitoring algorithm (EWMA) | Detection of vampire attack | Increase the overall life cycle of network | Does not provide fully satisfactory solution during topology discovery phase |
| 5 | Trust value model using EWMA | Prevent from vampire attack | Impact of vampire attack reduced in great extent | It does not provide full solution but avoid some damage |
| 6 | Vampire attack removal protocol (VARP) | Provable mitigate vampire attack | VARP perform better than AODV | No completely mitigation of vampire attack |
| 7 | A novel proposal against vampire attack [13] | Detection and prevention against vampire attack in AODV | Defended against vampire and network lifetime can be maximized | It also does not provide full solution in AODV |
| 8 | A approach to detect the vampire [14] | Detect vampire in AODV routing | Efficient and effective approach for detection | Performance will decrease as no. of node increases in network |
| 9 | Wolf routing | Detect vampire attack | Use heuristic optimization so simultaneously performs autonomous solution | Require additive utility function so increase computational complexity |

hunting. They hide themselves and always move at better position to victim. Each wolf move towards the victim simultaneously and also look for the potential threat (like tiger, loin and human hunter). Here genetic algorithm use for generate better move towards the victim.

**Soni and Pahadiya et al.** [16], Describe a strategy to detect and prevent the vampire attack when energy of nodes drain because of packet flood and RREQ flood in network. They prepare a list of attackable nodes on the basis of broadcast and energy consumption by nodes. If these value are greater than relative variance and energy consumption rate increases then remove susceptive node and declare it as vampire.

## 5 Discussion

In Table 1 author compared the techniques which are used to mitigate the vampire attack. There are some new routing protocols which provably bound the damage from vampire attack in computer networking. There are also some new strategies that applied on existing routing protocol (like LS, DSDV, DSR and AODV) to detect and prevent the vampire attack. Following table shows the features, pros and cons of the routing protocols.

## 6 Conclusion and Future Work

In this paper, various new routing protocol against vampire attack and methods to detect or prevent the vampire in AODV routing protocol are described. Vampire attack is a type of resource depletion attacks which completely disable wireless sensor networks by draining the node's battery power. These attacks do not depend on specific routing protocols so that link state or distance vector is vulnerable to vampire attack.

Strategies which are used in above-discussed papers do not provide complete control over the vampire attack thus reduction in power consumption is not control properly. So in near future increment in the lifetime of sensor node by reducing power consumption at an extent level through some different strategy is aimed.

## References

1. Gowrishankar. S, T. G. Basavaraju, Manjaiah D. H, Subir Kumar Sarkar.: Issues in Wireless Sensor Networks: July 2–4, 2008, London, U.K.
2. I. Aad, J.-P. Hubaux, and E. W. Knightly.: Denial of Service Resilience in Ad Hoc Networks: Proc. ACM MobiCom, 2004.

3. A. D. Wood and J. A. Stankovic.: Denial of Service in Sensor Networks: Computer, vol. 35, no. 10, pp. 54–62, Oct. 2002.
4. P. Rajipriyadharshini and V. Venkatakrishnan, S. Suganya and A. Masanam.: Vampire Attacks Deploying Resources in Wireless Sensor Networks: (IJCSIT) International Journal of Computer Science and Information Technologies, Vol. 5 (3), 2014, ISSN: 0975-9646.
5. T. H. Clausen and P. Jacquet, Optimized Link State Routing Protocol, IETF RFC 3626, 2003.
6. C. E. Perkins and P. Bhagwat.: Highly Dynamic Destination- Sequenced Distance Vector Routing (DSDV) for Mobile Computers: Proc. Conf. Comm. Architectures, Protocols and Applications, 1994.
7. Y.-C. Hu, D. B. Johnson, and A. Perrig.: Packet Leashes: A Defense against Wormhole Attacks in Wireless Ad Hoc Networks: Proc. IEEE INFOCOM, 2003.
8. Eugene Y. Vasserman and Nicholas Hopper.: Vampire Attacks: Draining Life from Wireless Ad Hoc Sensor Network: Ieee Transactions On Mobile Computing, Vol. 12, No. 2, February 2013.
9. K. Vanitha, and V. Dhivya.: A Valuable Secure Protocol to Prevent Vampire Attacks in Wireless Ad Hoc Sensor Networks: IEEE International Conference on Innovations in Engineering and Technology (ICIET'14) Volume 3, Special Issue 3, March 2014 B.
10. Umakant, and J. Damodhar.: Resource Consumption Attacks in Wireless Ad Hoc Sensor Networks: International Journal of Engineering Research ISSN: 2319-6890 (online), 2347-5013 (print) Volume No. 3 Issue No: Special 2, pp: 107-111 22 March 2014.
11. Ashish Patil and Rahul Gaikwad.: Preventing Vampire Attack in Wireless Sensor Network by using Trust.: International Journal of Engineering Research & Technology ISSN: 2278-0181Vol. 4 Issue 06, June-2015.
12. Shrikant C. Chumble1 and M. M. Ghonge.: Simulation of Mitigation of Vampire Attack in Wireless Ad-hoc Sensor Network: International Journal of Advent Research in Computer and Electronics (IJARCE), Vol. 2, No. 9, September 2015 E-ISSN: 2348-5523.
13. Amee A. Patel and Sunil J. Soni.: A Novel Proposal for Defending Against Vampire Attack in WSN: International Conference on Communication Systems and Network Technologies, DOI 10.1109/CSNT in 2015.
14. AnkitaShrivastava, Rakesh Verma.: Detection of Vampire Attack in Wireless Ad-hoc Network: international journal of Software & Hardware Research in engineering volume 1 issue jan-2015.
15. BestyHaris.: Wolf Routing to Detect Vampire Attacks in Wireless Sensor Networks: International Journal of Computer Science and Information Technologies, Vol. 6 (3), ISSN No. 0975-9646 2806-2809, 2015.
16. Manish Soni and Bharat Pahadiya.: Detection and Removal of Vampire Attack in Wireless Sensor Network: International Journal of Computer Applications (0975–8887) Volume 126 – No. 7, September 2015.

# Cloud-Based on Agent Model for Mobile Devices

**Amel Beloudane and Ghalem Belalem**

**Abstract** "Information available for anybody at anywhere and anytime." Ranging from a domestic connection via personal computers toward a mobile access via smart devices using communications technologies, those devices can access to all kinds of information through mobile applications in the cloud. Mobile Cloud Computing (MCC) can be seen as a solution for limitations of cloud computing because all mobile devices are limited by memory capacity, screen, battery, and intermittent connectivity, the MCC exploits the user's information, e.g., localization, memory, power, and bandwidth capacity while running these applications on the cloud. In the aim of addressing the problems of mobile environment which treat mobility of users and services, we propose in this paper a model of cloud-based on agent in mobile environment which ensuring high availability of services by their migration or replication, the aspect of decision-making between mobile devices and mobile applications using top-k algorithm which contribute to find the most appropriate service in the cloud while reducing energy consumption with respecting the SLA.

**Keywords** Cloud computing · Mobility · Availability · Migration · Replication · Energy · Agent · SLA

## 1 Introduction

While the planet becomes more intelligent, the use of mobile devices (smartphones, tablets, laptops, robots, etc.) became an essential part of the human life. The use of services remotely and the representation for the transport of data is named Cloud

A. Beloudane (✉)
Faculty of Exact Sciences and Computer Science, Mostaganem, Algeria
e-mail: amelbeloudane@gmail.com

G. Belalem
Faculty of Exact and Applied Science, University of Oran1,
Ahmed Ben Bella, Oran, Algeria
e-mail: ghalem1dz@gmail.com


S.K. Bhatia et al. (eds.), *Advances in Computer and Computational Sciences*,
Advances in Intelligent Systems and Computing 553,
DOI 10.1007/978-981-10-3770-2_26

Computing. The fast renovations of Cloud Computing become a trend of evolution in the development of techniques relative to business and industry. Cloud Computing groups all disciplines technologies and business models used to deliver computing capacity (software, platforms, hardware) as a service provided by the authors of Cloud (e.g., Google [1], Amazon [2], Microsoft [3], and Salesforce [4], etc.) with a premium and discount factors of quality and cost. Mobile Cloud Computing (MCC) is a new paradigm allowing to sharing resources and applications through a mobile channel while solving the problems of different mobile operating systems by the creation of mobile applications which can be executed on Cloud from any operating system via a web interface. While the application runs on the Cloud, it means that the power of treatment is moved outside the mobile device and the computing is completely supported by the Cloud, which allows to reducing the energy consumption by the mobile devices and economizing the battery life of those devices. The use of cloud computing services is facing to many challenges, for example, limited resources, bandwidth, latency, security, and cost. Furthermore, the mobile devices are confronted to many challenges: the high availability of resources, intermittent connectivity because of the mobility of users and services, the autonomy of battery, storage capacity, and the screen's resolution.

In this paper, we are interested in the mobile cloud computing or rather in his concept of sharing resources via heterogeneous mobile channels. Although this paradigm of distribution of resources is widely adopted by the computer science community, the means of implementation to supply this distribution still possess some limitations, e.g., the problems of resources availability and discovery of the most appropriate service in the cloud; for that, we propose a Cloud Computing approach in a mobile environment based on a mobile agent which is a program installed in a mobile device such smart phone, or tablet PC which can migrate the network bringing its own code and execution state in order to run remotely jobs or to interact with other mobile agents. By integrating the mobile agents into the Clouds, the mobile devices can access to a diversity of applications in Mobile Commerce [5] which aims to save the energy consumption and resources. Mobile agents are able to negotiate with other on behalf of users [6] while addressing: (a) A permanent availability even in the presence of failures, (b) energy conservation, (c) while respecting the SLA (Service Level Agreement). The remainder of this paper is organized as follows: Sect. 2 presents the general concept of Cloud Computing (CC) and Mobile Cloud Computing (MCC), we chain with survey the state of the art of mobile cloud computing in Sect. 3; in Sect. 4, we describe the architecture of our approach and present its modeling. Finally, we conclude and we present the future works in Sect. 5.

## 2 Cloud Computing in Mobile Environment

In this section, we present the differences between CC and MCC.

## 2.1 Cloud Computing

The term "Cloud" has been used historically as a metaphor, used to represent the transport of data and resources. This concept dates backed as early as 1961 when professor John McCarthy suggest that computer time-sharing technology might lead to a future where computing power and even specific applications might be sold. Cloud computing (CC) refers to the use of remote servers to treat or store information. The access is done via a telecommunications network, on-demand and self-service, to shared computing resources, using a web browser.

According to the definition of National Institute of Standards and Technology (NIST) [7]: "CC is a model for enabling ubiquitous, convenient, on-demand network access to a shared pool of configurable computing resources (e.g., networks, servers, storage, applications, and services) that can be rapidly provisioned and released with minimal management effort or service provider interaction." CC is defined as a type of parallel and distributed system consists of a collection of interconnected and virtualized computers, they are dynamically provided and presented as one or many resources of computing based on service level contract established by negotiation between the service provider and the consumer [8]. The computing power in cloud environments is provided by a collection of Data Center, which are typically installed with hundreds of thousands of servers [9].

Cloud computing is typically classified in private cloud, public, hybrid and community based on the cloud location, and in Infrastructure, Platform, Software or, Storage, Database, Information, Process, Application, Integration, Security, Management, and Testing-as-a-service based on a service that the cloud is offering. CloudAgent is concerned with the design and development of software agents for bolstering cloud service discovery, service negotiation, and service composition.

## 2.2 Mobile Cloud Computing

The combination of cloud, portable computing devices, mobile Web, etc., has laid the foundation for a new computing model, called Mobile Cloud Computing (MCC), which allows users an online access to unlimited computing power and storage space. In work [10], authors define MCC as follows: MCC is a model for transparent elastic augmentation of mobile device capabilities via ubiquitous wireless access to cloud storage and computing resources, with context-aware dynamic adjusting of offloading in respect to change in operating conditions, while preserving available sensing and interactivity capabilities of mobile devices. MCC can be represented as "A service that allows resource constrained mobile users to adaptively adjust processing, storage capabilities by transparently partitioning, offloading the computationally intensive and storage demanding jobs on traditional cloud resources by providing ubiquitous wireless access" [11].

From different definitions of MCC, it can be defined as "An architecture where processing and data storage take place outside of the mobile device, the mobile user can access to its information remotely at anytime and anywhere using mobile applications. Those applications move the power of computing and data storage from mobile devices to the cloud, which reduce energy consumption and save battery life of mobile devices."

## 3 Related Works

In paper [12], the author considers actions that buyers can take to resist sellers cheatings, the buyers can adapt their plans with low demands price before migrate to the cloud platform. He uses mathematical model as a Normal-Form Game model in Game Theory called Eavesdropping and Resistance of Negotiation (ERN) Game and analyzes it by the Agent-Based Computational Economic (ACE) approach, which is simulated by a virtual economics for evaluating the co-evolutionary of strategies in ERN game on C++ language, which is extended from the source code developed by McFadzeanin [13].

The authors in paper [5] propose a model named Cloud Agency which offers virtual Clusters on the top of an existing GRID. That can be easily configured with the support of mobile agents based services. They add mobility to a set of features as reactivity, proactivity communication, and social ability which characterize ordinary software agents systems.

Paper [14] proposes a metric called Expected Resource Availability (ERA) and presents preliminary evaluation of the potential of the ERA based on real-world traces to capture the impact of the topology of services and resources. It offers a proxy for the applicability of opportunistic computing schemes to a given network based on the Cambridge iMote traces collected within the context of the Haggle Project [15].

Paper [16] suggest an Intelligent Multi-Agent model for resource Virtualization (IMAV) to automatically allocate service resources suitable for mobile devices in cloud computing environment supporting social media services based on virtualization rules [17] result by multi-agent model. The authors demonstrated a good performance of IMAV model of user correlation between User Context and System Context in terms of accuracy, precision, recall, and F-measure, respectively.

In [18], the authors present the MABOCCF mechanism (Open Mobile Agent-Based Cloud Computing Federation), which combines the advantage of mobile agent and CC in which data and codes are transferred from one device to another by through mobile agents. This last is a program that is executed in a virtual machine called MAP (Mobile Agent Square); it can be moved between MAPs, communicated and negotiated with others, in the aim of realizing portability and interoperability between cloud computing platforms. This approach simulates the performance of 10 CCSPs (cloud computing service providers) for MABOCCF and NMBOCCF (ordinary computing mechanisms that do not support portability between different CCSPs) which has a positive performance in user satisfaction and facility utilization ratio.

The paper [19] proposes a new cloud-based architecture. It addresses the basic elements of the discovery of mobile web services named Discovery as a service (Daas). They provide a comprehensive need analysis for the discovery of mobile Web services in resource-limited environment. In addition, they demonstrate the viability of the proposed architecture take into account user preferences and context. The experimental validation proves that DaaS can effectively respond the best at user's needs.

In paper [20], authors propose a dynamic performance optimization framework for mobile cloud computing using mobile agent-based application partitions. The proposed framework imposes minimal infrastructural requirements on the cloud servers. Experiments were performed with real-world mobile applications to evaluate the performance of the proposed model in terms of application execution time and energy consumption on the mobile device: East1-a region of the Amazon Elastic Compute Cloud (EC2) [2] was used as the cloud hosts using PowerTutor [21], which a free energy measurement tool for Android devices.

In [22], the authors presents MAUI, an architecture that enables fine-grained energy-aware offload of mobile code to the infrastructure in order to demonstrate how to partition the application at maximize energy savings with minimal load on the programmer. MAUI decides at runtime which methods should be remotely executed, driven by an optimization engine that achieves the best energy savings possible under the mobile device's current connectivity constrains. This approach is better in term of resource-intensive face recognition application that consumes an order of magnitude less energy, and latency-sensitive.

Paper [23] presents a system that automatically transforms mobile applications to benefit from the cloud named CloneCloud, which combine static analysis and dynamic profiling to partition applications automatically while optimizing execution time and energy use for a target computation and communication environment. This model is a flexible application partitioned and execution runtime that enables unmodified mobile applications running in an application-level virtual machine to seamlessly offload part of their execution from mobile devices onto device clones operating in a computational cloud.

Paper [24] presents a dynamic computation offloading model for mobile cloud computing, based on autonomous agents. This approach alleviates the management burden of offloaded code by the mobile platform using stateful, autonomous application partitions. The authors present a low-overhead dynamic model integrated into autonomous agents to enhance them with self-performance evaluation in addition to self-cloning capabilities and validate it on two mobile applications.

# 4 Proposed Approach

From the general concept of MCC, the proposed architecture is shown in Fig. 1 to support our contribution which aims to discover the most effective service with a minimum execution cost in the cloud for users.

This architecture is mainly constituted of four main entities which are the following:

- User Entity: It regroups all mobile users with different mobile devices;
- Entity Service: it represents the set of services used in this approach: (i) localization service, (ii) analysis request service, (iii) replication and migration service, (iv) management meta-directory service, (v) quality service and (vi) decision service;
- Database Entity: It is the passive part of the architecture; it includes: (i) Data center existed in the cloud; (ii) local directories for each Data Center, (iii) meta-directory, (iv) historical directory, and network components (routers, BS);
- Entity Agent: It regroups the different agents of the proposed model and their interactions: (i) localization agent, (ii) analysis agent, (iii) replication and migration agent, (iv) maintenance agent, (v) broker agent and (vi) decision agent.

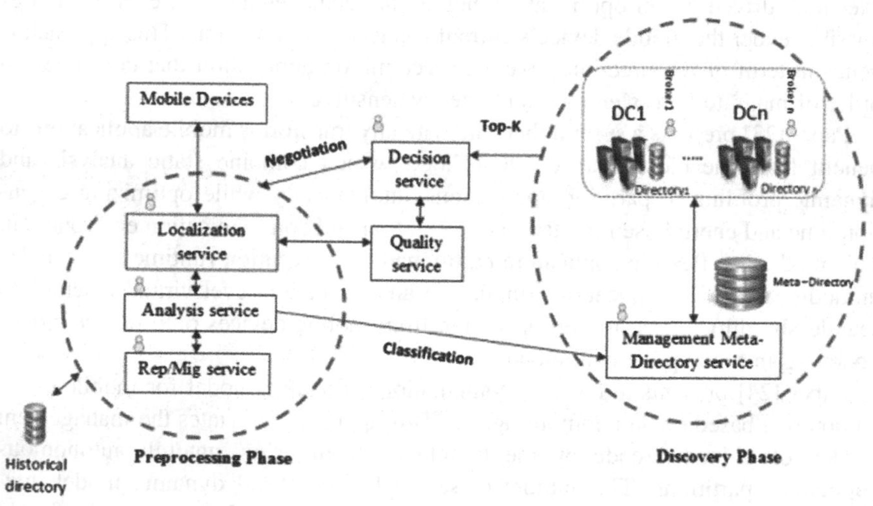


**Fig. 1** Architecture of mobile cloud computing based on agent

### 4.1 Description

This section describes our architecture functioning, the used services and how the different agents interact with them to implement this approach: After the interconnectivity of the mobile device to the network via the base stations (Base Transceiver Station (BTS) or satellites), the request and user's information are transmitted to cloud through Internet. Our system is a continuity of our work in [25]; it is based on the use of mobile agents to discover the best service in the cloud by integrating Top-k and negotiation algorithm. These mobile agents function at an autonomy and asynchrony way, they dynamically adapt to the runtime environment, reduce network traffic; they are also characterized by their mobility and social ability. The interaction between the mobile agents used is described as follows:

- *Localization agent*: It allows to determining user's localization, updates localization according to user's mobility and sends request to analysis service.
- *Analysis agent*: It allows to analyzing user's request, and classifying them according to different categories (using Classification algorithms) registered at Meta-Directory which contains a description of services in local directories (the service's cost and reliability, categories and the associated data center). In interaction with the quality service, it saves the request with SLA, and after the treatment, it returns the best answer to localization service.
- *Replication/Migration agent*: It allows to update the historical directory, which contains the frequency of requests (by category), the categories requested by each user and the number of users according to the categories requested; This service also allows to launching Replication or Migration algorithms of services in order to guarantee availability and discovery of the most appropriate service in the Cloud. For that, this choice proceeds as follows: (i) If the requested service exists in a DC far from the user's localization, the Migration agent will migrate the service in the nearest DC to the user's localization. If the user changes its position (compared to a threshold), then this agent will migrate the service toward the nearest DC to the new localization. (ii) If the service is engaged by other users, or it is about a frequently requested service (by consulting the historical directory), the Replication agent will replicate the service on multiple VM of DC to minimize the overloading on the network. At this stage, the local directories in question and the Meta-directory it will be updated.
- *Maintenance agent*: The Maintenance agent allows to, discover data center's addresses from meta-directory, orient the request to the appropriate data center and manage Brokers agents. The latter allows to updating data Center directories containing lists of services, each Broker agent returns the best K results found using Top-K algorithm to determine the optimal number of virtual machines when running applications in a cloud environment. There are several algorithms in this domain which requiring obtaining the prior intelligent partitioning data clustering for an effective treatment in order to effect as independently as possible the treatments of data fragments with a semantic coherence. The execution of distributed data in the Cloud is a parallel variant of the clustering algorithm

h-means in phase pre-selection of the PMML process (Predictive Model Markup Language) [26]. Many algorithms of parallelism based on k-means are proposed as in [27]. The maintenance agent updates the meta-directory and discovers new categories and their locations. Finally, the coordination agent retrieves and transmits the results to decision service.

- *Decision agent*: It allows to select the best answer corresponding to SLA, otherwise it will launch Negotiation algorithms to decrease some SLA's constraints. Finally, it will return the final answer to the analysis service.

## 4.2 Modeling

This section presents a modeling of the proposed approach which is detailed in [25] while defining the main concepts as follows:

- Data center set: $DC = \{DC_i | 1 \leq i \leq n\}$; where *n* is the number of Data center;
- Services set: Each Datacenter *i* has *j* type of services, where $|S_j|$ is the number of services in $DC_i$: $S_i\{S_{ij} | 1 \leq j \leq |S_{ij}|\}$
- Users set: It represents users who are connected and subscribed to the Cloud, where $U = \{U_i | 1 \leq i \leq |U_i|\}$;
- Requests set: It represents the set of requests submitted to the cloud, where $Req_i$ is the request of $U_i$ and $|Req_i|$ is the number of requests.
  Where $Req = \{Req_i | 1 \leq i \leq |Req_i|\}$. Each request is a triplet represented as follows: ***Req (Id, $S_i$, Cost, Category, reliability)***.
  Where: ***Id***: The user's identifier; ***$S_i$***: The set of services of the *ith* request; ***Cost ($S_i$,$l_{i;t}$)***: The cost of the use of service $S_i$ by $U_i$in location *l* at *t* moment; ***Category ($S_i$)***: the category of services which is defined by the system; ***Reliability***: the reliability of the answer.
- Cloud capacity: Each Data center has a limited number of users:

$$Cap_j = max\,\Sigma_{i=1,n} U_i \cdots;\ j = \{1, \ldots, m\}$$

### 4.2.1 Experimental Study

In order to validate our approach compared in [25], we produce our testes on a simulator platform because of if we do testes in a real cloud computing environment as Amazon EC2 [2], the experiments will be limited because they require a very high cost. For that, we study the behavior of our approach and evaluate its performance on CloudSim Simulator, which is a free platform that generates a cloud environment allowing to modeling heterogeneous environments compared to a basic approach.

We name basic approach a classical approach that allows to researching service in the cloud and migrating it to the adequate localization, the replication algorithm is not made in this approach. We chose three metric for our experimental study: energy consumption while varying the number of VMs, response time and degree of SLA violation while varying the number of services.

### 4.2.2 Energy Consumption

In the aim of analyze the energy consumed by data center, the simulation is shown in Fig. 2. According to the results, we observe a reduction in the energy consumption by applying our approach compared to the classical approach. We estimate the average gain at 0.64% compared to the classical approach.

### 4.2.3 Response Time

To analyze the response time, the results of the simulations are shown schematically in Fig. 3. We observe a significant decrease in response time of our approach compared to the classical approach. We can estimate the average gain at 44.87% compared to the classical approach.

**Fig. 2** Influence of the number of VMs on the energy consumption

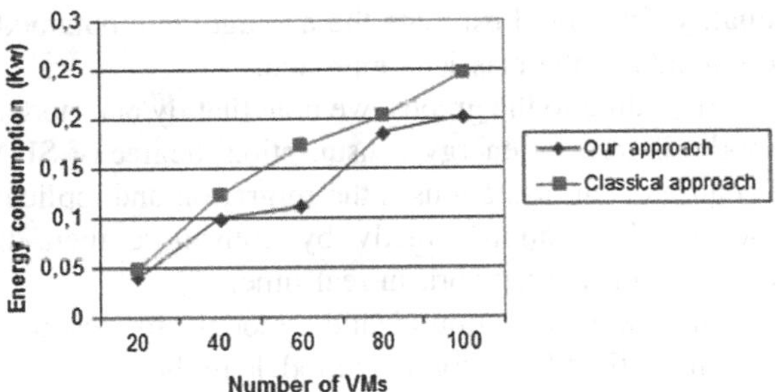


**Fig. 3** Sequence diagram of mobile cloud computing based on agent

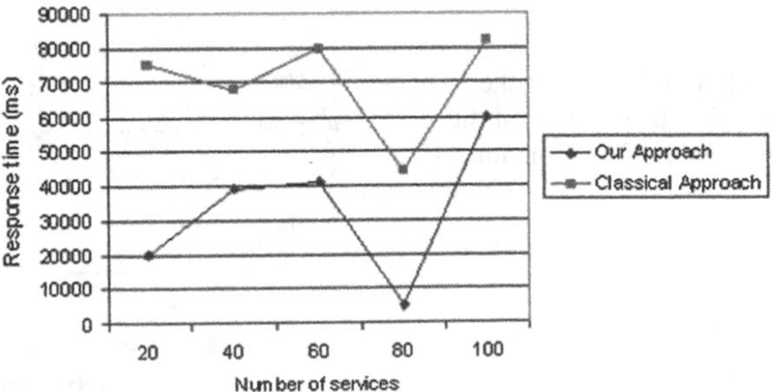

### 4.2.4 Degree of SLA Violation

The requirements of the quality of service requested are extremely important in a mobile cloud computing environment. They are usually formalized in the form of SLA (Service Level Agreement). Since this contract may be violated and not respected by the provider, it is necessary to define a generic metric that can be used in our experiment, which we called degree of SLA violation, in order to estimate the degree of violation of constraints required by the user, we define this metric as follows:

$$DV_{SLA} = |X_r - X_f| + |Reliability_r - Reliability_f|$$

Where:

$X_r$: the cost in % of the requested service;
$X_f$: the cost in % of the most service found;
$Reliability_r$: the reliability of the requested service;
$Reliability_f$: the reliability of the service found

Figure 4 evaluates the results obtained of our experiments which we allow to studying the behavior of our approach in term of SLA violation degree and her position it to the classical approach. We note that the curve of our proposition is below to the curve of the classical approach, which proves that our approach allowing to finding the most effective service, that gives a lowest degree of SLA violation according to the customer's request which qualify our approach oriented quality. We could estimate the average gain obtained by our approach by 31.63% compared to the classical approach.

According to the graphs, we note that by our model, we generate an optimization model in term of energy consumption, degree of SLA violation and response time because of our system uses the migration and replication algorithms depending on the users' situation directly by consulting meta-directory which reducing the overload on the network in real time.

There are other optimization models for resource management in the cloud, which are based on economic models in the market as the work in [28], or based on metaheuristics in [29] which propose workflows-scheduling algorithms.

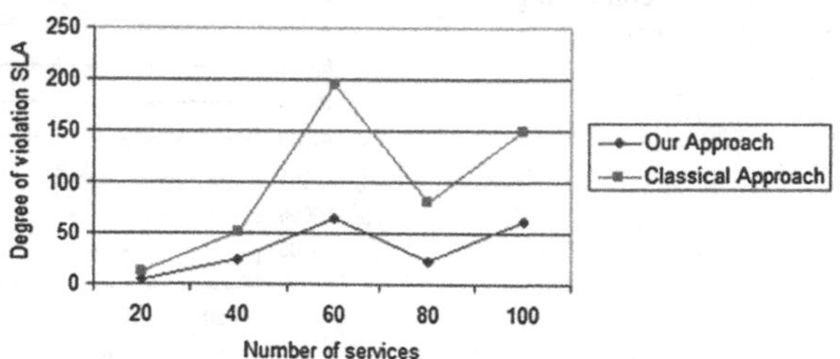


**Fig. 4** Influence of the number of services on the degree of SLA violation

## 5 Conclusion and Future Works

In this paper, we present architecture of cloud-based on Agent in mobile environment for the purpose of scalability and high network availability of services. For that, we propose a model which use the migration or replication of services and classify them to different categories in order to discover the most appropriate service and to optimize the research in the cloud on large-scale while reducing energy consumption. By our contribution, the use of mobile agent provides the appropriate services for users depending on user position with various quality of service. We evaluate our approach by implementing migration and replication approach and compared it to a classical approach according to the influence of the number of VMs on energy consumption, and the number of services on response time and the degree of SLA violation. Having regard to the emerging necessity of using mobile applications, MCC becomes an active research sector, jumping over disciplines such as image processing, robotics, social network, sensors' applications data, etc. In our future works, We will integrate a module allowing to detect the current localization of the user according to his mobility in real time, moreover CloudSim enables to use a broker package that will be served for the selection of top-k services and the negotiation of the latter in terms of quality. Thereafter, we will attend the problem of services' consistency management and fault tolerance by predicting the service's status and we will build our model into a real Cloud Computing such as OpenStack.

## References

1. Google (2012), Google app engine, from http://appengine.google.com.
2. Amazon (2012), Amazon elastic compute cloud (EC2). AWS, from http://www.amazon.com/ec2.
3. Microsoft (2012), Microsoft azure, from http://www.microsoft.com/azure/13.
4. Salesforce (2012), from https://www.salesforce.com/fr.
5. Aversa, R., Di Martino, B., Rak, M., Venticinque, S.; Cloud Agency: A Mobile Agent Based Cloud System. Complex, CISIS'2010, pages: 132–137.
6. Fasli, M. Agent technology for e-commerce. John Wiley and Sons, 2007.
7. Mell, P., and Grance, T. (2011), The NIST definition of cloud computing is available at: http://dx.doi.org/10.6028/NIST.SP.800–145.
8. Buyya, R., Yeo, C. S. and Venugopal, S. Market-oriented cloud computing: Vision, hype, and reality for delivering it services as computing utilities. HPCC'08; pages 5–13.
9. Buyya, R., and Murshed, M. Gridsim, A toolkit for the modeling and simulation of distributed resource management and scheduling for grid computing. Concurrency and computation: practice and experience, 14(13–15): 1175–1220, 2002.
10. Kovachev, D., Cao, Y., and Klamma, R., Mobile cloud computing: a comparison of application models. arXiv preprint arXiv:1107.4940, 2011.
11. Khan, A. N., Mat Kiah, M. L., Khan, S. U., and Madani, S. A. Towards secure mobile cloud comput-ing: A survey. FGCS, 29(5): 1278–1299, 2013.

12. Lai, Y-L., Analyzing Strategies of Mobile Agents on Malicious Cloud Platform with Agent-Based Computational Economic Approach, Expert Systems with Applications 40 (7): 2615–2620, 2012.
13. McFadzean, D., Stewart, D., and Tesfatsion, L. A computational laboratory for evolutionary trade networks. IEEE Trans. on Evolutionary Computation, 5(5): 546–560, 2001.
14. Ferrari, A., Puccinelli, D., and Giordano, S., Characterization of the impact of resource availability on opportunistic computing. MCC'12, Helsinki, Finland, pages: 35–40.
15. Scott, J., Gass, R., Crowcroft, J., Hui, P., Diot, C., and Chaintreau, A. Crawdad Data Set cam-bridge/haggle (v. 2006-01-31). Crawdad wireless network data archive, 2006.
16. Kim, M. J., Yoon, H. G., and Lee, H. K. (2011), An Intelligent Multi-Agent Model for Resource Virtualization: Supporting Social Media Service in Cloud Computing. In Computers, CNSI 2011, Vol. 365, pages: pp. 99–111, 2011.
17. Yoon, H., Lee, H, An Intelligence Virtualization Rule based on multi-layer to support social media-cloud service. In: CNSI 2011.
18. Zhang, Z. and Zhang, X., Realization of open cloud computing federation based on mobile agent, ICIS'2009, vol. 3, pp. 642–646.
19. Elgazzar, K., Hassanein, H. S. and Martin, P., Daas: Cloud-based mobile web service discovery. Journal Pervasive and Mobile Computing. Vol. 13, pp: 67–84, 2014.
20. Angin, P. and Bhargava, B., An Agent-based Optimization Framework for Mobile-Cloud Computing. JoWUA 4(2): 1–17 (2013).
21. M. Gordon, L. Zhang, B. Tiwana, R. Dick, Z. M. Mao, and L. Yang, "PowerTutor: A power monitor for android-based mobile platforms," http://ziyang.eecs.umich.edu/projects/powertutor/, 2013.
22. Cuervo, E., Balasubramanian, A., Cho, D., Wolman, A., Saroiu, S., Chandra, R., and Bahl, P., MAUI: making smartphones last longer with code offload. MobiSys 2010: 49–62.
23. Chun, B., Ihm, S., Maniatis, P., Naik, M., and Patti, A., CloneCloud: elastic execution between mobile device and cloud. EuroSys 2011: 301–314.
24. Angin, P., Bhargava, BK. and Jin, Z., A Self-Cloning Agents Based Model for High-Performance Mobile-Cloud Computing. CLOUD 2015: 301–308.
25. Beloudane, A., and Belalem, G., Toward an Efficient Management of Mobile Cloud Computing Services based on Multi Agent Systems. JITR 8(3): 59–72 (2015).
26. PMML http://www.dmg.org/v4-0-1/GeneralStructure.html.
27. Hartigan, J., Clustering Algorithms. John Wiles and Sons, New York, USA, 1975.
28. Bouamama, S., and Belalem, G., The New Economic Environment to Manage Resources in Cloud Computing. JITR 8(2): 34–49 (2015).
29. Yassa, S., Sublime, J., Chelouah, R., Kadima, H., Jo, G., Granado, B., A genetic algorithm for multi-objective optimisation in workflow scheduling with hard constraints. IJMHeur 2(4): 415–433 (2013).

# Node Mobility Issues in Underwater Wireless Sensor Network

**Kanika Agarwal and Nitin Rakesh**

**Abstract** A fundamental challenge in underwater wireless sensor network (UWSN) is mobility of sensor nodes during the communication held in the network. There is no fixed location of the sensor nodes present under sea level in wireless sensor network. The nodes are mobile which results in improper communication. Although there are various issues in underwater wireless sensor network, some of them have been encountered in this paper. The paper focuses on node mobility in network during the communication among the nodes. An approach has been proposed in this paper so that communication in the network between the nodes can take place even if the node changes its location. The approach is based on the Euclidean distance. The approach is called as an Arc moment and is used to show how nodes placed in the underwater wireless sensor network can communicate with each other based on various assumptions and conditions.

**Keywords** Routing · Underwater wireless sensor network · Communication · Node mobility

## 1 Introduction

A self-configuring network of small sensing nodes that communicates among themselves using radio signals, senses, and monitor, understand the physical world is called wireless sensor network. Each node existing in the network is known as motes that are used to collect all the temporary data. UWSN is also known as subnet for WSNs which is used to answer all those queries that cannot be addressed from

K. Agarwal (✉) · N. Rakesh
Department of Computer Science and Engineering, Amity University, Noida, UP, India
e-mail: kanikaagarwalgtb@gmail.com

N. Rakesh
e-mail: nitin.rakesh@gmail.com



S.K. Bhatia et al. (eds.), *Advances in Computer and Computational Sciences*, Advances in Intelligent Systems and Computing 553, DOI 10.1007/978-981-10-3770-2_27

the surface using remote sensing techniques. Remotely operated vehicles and autonomous underwater vehicles (AUVs) can measure points against the immense backdrop of an evolving ocean environment.

Underwater wireless communication networks (UWCNs) are a network that consist of some sensing devices and autonomous vehicles that are known as acoustic underwater vehicles which interact for performing various applications such as monitoring, data collection from underwater, recording of climate, sampling for ocean, prevention from pollution and disaster, etc. [1]. For making these above application viable, there is a need of communicating protocols among underwater devices.

This acoustic networking is considered as most enable technology for the application of underwater network that helps to acquire a 100 meter distance, where smaller areas are covered by electromagnetic waves. Acoustic network consists of multiple sensing nodes that are deployed for performing monitoring. Large antennae and high transmission power are required at extra-low frequencies for radio frequency (RF) waves for propagating through conductive salty water [2].

Paper is divided into six sections. First section is of introduction. In Sect. 2, various research papers have been discussed. In Sect. 3, the two-dimensional architecture of the wireless sensor network is explained. In Sect. 4, various issues related to wireless sensor network have been encountered. Section 5 presents the proposed approach. Finally concluding the paper and discussing the future scope of the work.

## 2 Literature Review

Akylidiz et al. [3] presented the 2D and 3D architecture of the network, how the communication takes place in acoustic network and issues related to the underwater network and a protocol stack for every layer present in the network. Prasan et al. [4] presented the architecture of the network, challenges, and its applications and identified research direction in acoustic network used for short range, protocol for high latency network, and long duration network sleep. Zhang et al. [5] developed a protocol named cluster-based delay tolerant protocol for addressing the problems and eliminating RTS and acknowledgement handshaking. A self-adaptive algorithm is used by the proposed protocol. The algorithm was used for addressing the consistent delay problem. Han et al. [6] presented localization algorithms for node mobility in underwater wireless sensor network. The algorithm is categorized as stationary, mobile, and hybrid localization. Ayaz et al. [7] presented a model for reliable data delivery. The paper proposed an algorithm for determining the size of data packet delivery. This model used to generate two copies of same data packet. The paper also presented the connection held among the nodes of the network by packet size, throughput, bit error rate and distance. Felemban et al. [8] proposed an optimal positioning strategy for supporting mission critical operations. The paper also presented a strategy for network nodes for attaining maximum coverage and connectivity with minimum transmission loss.

Ayaz et al. [9] proposed the purpose and requirement of dynamically reconfigurable routing and the architecture of underwater wireless sensor network. This routing provides reliable wireless communication. It reconfigures strategy within the protocol that provides alternative paths for efficient delivery of data. The protocol was demonstrated from focused beam routing and sector-based routing protocol. Watfa et al. [10] proposed a mechanism that is used for localizing an event of interest in underwater network. It includes various unique sensor nodes. The approach localizes through a reactive angle. This approach sorts the problem of finding node coverage problem by forming a subnetwork of nodes to represent in original way and is wrapped by few nodes called anchor nodes. These anchor nodes represents as a backbone to this mechanism. Heidemann et al. [11] presented main approaches and challenges in underwater wireless sensor network. The paper elaborates a mechanism regarding the propagation and communicating pattern in the system with the concerning protocol at each layer presented. Naik and Nene [12] proposed an approach called self-organizing localization algorithm that performs its assigned task without any human intervention through U.W.S.N. It eliminates errors when localization process occurs. Effect of speed is studied using the algorithm. It also analyzed anomaly offered because of depth error.

Ali and Hassanein [13] presented underwater hybrid sensor network where radio is used and proposed the turtle Net Architecture and its distance vector routing algorithm, a heuristic algorithm depending upon distance vector protocol for measuring cost surface network. Zhou et al. [14] elaborate a new scheme called DET-based hierarchical localization which incorporates advantages of DNR scheme, such as simplicity and localization ratio which results in decrement of cost, and increment of scalability and its performance and identifies four types of nodes that are useful in network. Anguita et al. [15] described a research work for underwater network related to communication among the nodes in the network. A suitable solution is to have optical communication where radio frequencies are strongly attenuated in water for exploring the communication system in short distance. The paper solely focuses on physical layer using IEEE 802.11 and 802.15.4 protocol properties.

## 3 Architecture of Underwater System

The two-dimensional architecture [3] of the underwater wireless sensor network consist of a group of sensing nodes that are coordinated with deep ocean anchors to the bottom of the sea. Through these acoustic links, the sensing nodes are connected to one or more underwater sinks, these are network devices which are used to

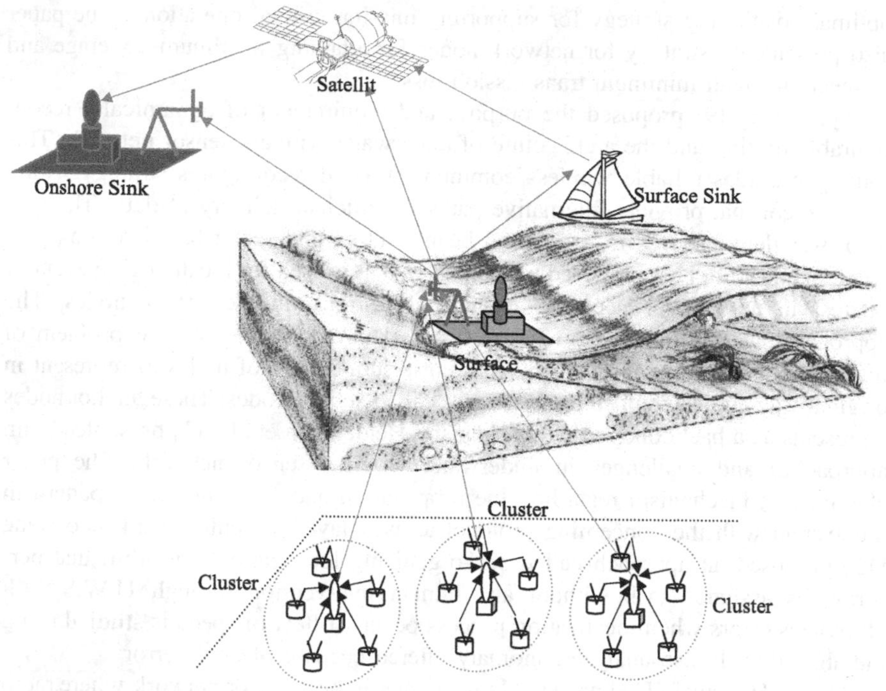


**Fig. 1** Architecture of 2D underwater wireless sensor network [3]

receive/pass on the information from under sea to stations on the surface. To achieve this, sink devices present underwater are having two transceivers, i.e., a vertical and horizontal. The vertical transceivers are long range transceivers, used for applications where depth is 10 km. Underwater sink uses horizontal transceiver for communicating to sensing nodes present in the network (Fig. 1):

- For sending commands and configuring data to sensors (underwater to sensors).
- To collect monitored data (underwater to sinks).

The transceiver is also allotted to the surface station for managing multiple communication lines with underwater sinks. Long range frequency and transmitter for satellite are also provided to the station for communication. Sensors are connected to underwater sink through a direct link or multi-hop path. Every node existing in the network sends the data directly to the underwater sink. This is not an energy efficient way because the distance between the sink and the node is high and power required is also very high. Throughput reduces because of the direct link and interference made by power transmitted. But when it comes to multi-hop links presented in the terrestrial network, sensors are used to pass on the data with the help of intermediate sensors. The resultant of this would be it save energy and the capacity of the network but it would increase the complexity.

## 4 Issues Related to Under Water System

The various issues related to the underwater wireless sensor network are listed as below [4, 5]. These issues are based on various categories such as delay, bandwidth, node mobility, time taken, connection held in the network, cost, power, deployment, distance, etc. The issues are, limited availability of bandwidth, fouling and corrosion in the network is prone to failures, limited power of battery because they cannot get charged as there is a deficiency of solar energy, a regular cleaning mechanism is required because of corrosion and fouling affecting the lifetime of underwater devices, requirement of robust, stable and well-equipped sensors for high temperature due to sensor drift of underwater devices, channel is impaired because of fading and multipath, loss of connectivity and high bit error rates, high magnitude of propagation delay, the network requires a high cost because it needs an extra protective sheaths for sensors, development of robust and less expensive nanosensors, integrated sensors are needed for synoptic sampling for the parameters existing physically, chemically, and biologically for improving its accuracy and precision, constrained amount of computation, energy, communication pattern of nodes in the network it needs more protection as compared to cryptograph, problem of intermittent partitioning because of ships, turbulence, and currents, propagation time is much larger than transmission time, high traffic congestion is also a major problem that occurred in underwater wireless sensor network which is caused by high acoustic propagation delay, a real-time and reliable data transfer is very major issue, node mobility can be considered as an another challenge for data forwarding which then require energy efficient data forwarding protocols, it is a difficult task for underwater nodes to locate themselves, i.e., they cannot find their positions and to synchronize their time with other coordinating nodes, deployment of the network is generally sparse, it requires more power as compared to the terrestrial network, nodes in the network require doing caching because of the noncontinuous nature of the channel, the sensors of the network are not spatially correlated in nature due to the greater distance among sensors, degradation of acoustic communication signal as there exists multipath propagation, the propagation speed is very low as compared radio channel, the network has very low throughput and channel is asymmetric in nature.

## 5 Division of Underwater WSN Based on Node Mobility

The environment under sea level is entirely different from outside world. In case of underwater wireless sensor network, there exist multiple sensing zones where sensor nodes are present (Fig. 2).

The figure shows one of the groups of sensing zone ‘n’ out of ‘m’ zones lying under sea level. So, let us assume this sensing zone ‘n’ is mapped on two-dimensional axis for experiencing their behavior and communication pattern

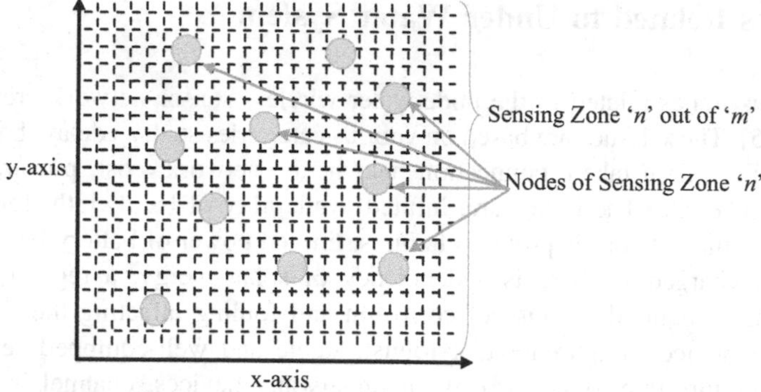


**Fig. 2** The sensing zone of sensor nodes is framed on two axes

among them in the network. The area shared by the zone 'n' is d meter. The distance between the two nodes is calculated by Euclidean distance. The time taken by the node to travel the packet from one node to another is t sec. The rate at which packet is transmitting is r meter per second. The threshold time, i.e., the maximum time taken by the node to travel the packet in the network is $t_0$. Now, further dividing the sensing zone (x-y) axis into four quadrants for the better understanding of communication pattern in the network (Fig. 3).

So, the four quadrants can be called as upper left (UL), upper right (UR), lower left (LL), and lower right (LR) as shown above. This division is made for configuring the mobility of the nodes during the communication among the nodes in the network. Mobility can be defined as changing the original position of the node and locating to another place. But it will not affect the communication pattern, although the communication scenario will surely depend on upon the nature of nodes present in the network. In other words, communication will take place depending upon the changed location of the nodes in the network. The value of threshold $t_0$ will be constant. If the time taken t to transfer a packet between the

**Fig. 3** Division of sensing zone into four quadrants

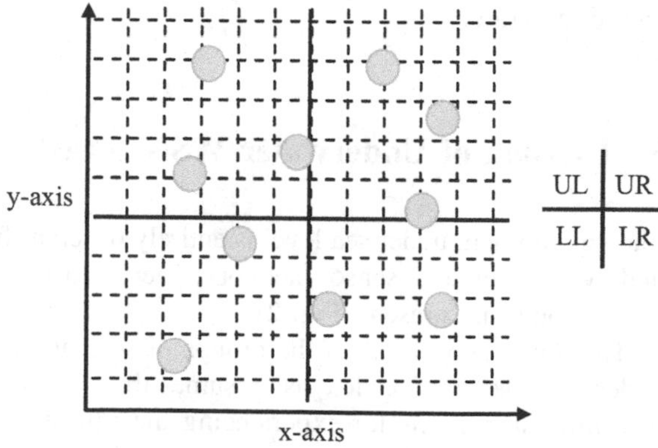

nodes is smaller than the threshold time $t_0$ then only the packet can move to its destination and vice versa. The Euclidean distance is denoted by $d_i$. It is calculated with the help of x and y coordinates of two nodes. For example, a node A having (x1, y1) coordinates and node B having (x2, y2) coordinates in two-dimensional network, then Euclidean distance can be calculated as:

$$d_I = \sqrt{(x2 - x1)^2 + (y2 - y1)^2}.$$

The cases shown below showcase the communication pattern of the system of how the nodes of the network behave and react while communicating with another node. All the possibilities of the node mobility while communication takes place among the nodes have been shown through these cases:

Case 1: If the transferring time t is less than the threshold time $t_0$ then only the packet will be transferred to its destination otherwise not.

| If $t < t_0$ | Packet will transfer to its destination. |
|---|---|
| Otherwise | No packet will be transfer. |

Case 2: A node while transferring the packet to another node in the network will calculate the Euclidean distance to the nearest node, i.e., delivery of the packet is also dependent on the distance between the nodes present in the network.

Case 3: The packet will always travel in Arc movement in the network as mentioned in our previous work. The Arc movement will be in the direction as Table 1:

Case 4: If a node A wants to send a packet through a path to its destination node B, then node A will always sense the intermediary nodes in another quadrant rather than in its quadrant, i.e., communication will always hold in cross quadrants never in a single quadrant. The cross communication is based on the Euclidean distance between the nodes in the network.

Case 5: Not considering the cost parameter, just focusing on the communication pattern in the system.

**Table 1** Directions showing how the communication will take place in quadrants

| Direction | Description |
|---|---|
| | This symbol shows the directions of packet movement or communication that held from U.R quadrant to U.L quadrant and then to L.L. |
| | This symbol shows the directions of packet movement or communication that held from U.L quadrant to U.R quadrant and then to L.R. |
| | This symbol shows the directions of packet movement or communication that held from U.L quadrant to L.L quadrant and then to L.R. |
| | This symbol shows the directions of packet movement or communication that held from L.L quadrant to U.L quadrant and then to U.R. |

The cost of the approach will be very high because the alternative node in the network will maintain buffer with it for keeping the history of the node. The node keeping buffer will also maintain the routing table. The routing table will keep all required information. The information in the table is state of packet sending node, the state of the packet receiving node, the destination of the packet. This approach has been drawn from buffer based algorithm. The simulator used here is ns2 Tracker. The creation of sensing nodes in the wireless network is shown in the Fig. 4. The number of nodes in the wireless sensor network is 10.

These sensing nodes are shown in two-dimensional architecture as explained above. This two-dimensional architecture has x and y-axis in which the nodes are created in the network. The division of the two-dimensional architecture is done for the understanding of the network and the characteristics of the sensing node.

Characteristics of sensing node such as how they behave in the network, how they react in the network, how they move to a different location, at what distance they move, by what time they change their position to a different location.

The network creation is done by connecting all the sensor nodes present in the wireless network [16]. The path is not defined for connecting the nodes in wireless sensor network as shown in Fig. 5.

Figure 6 is shown below, depicting the nodes with their original position in the network. The division of network is done just for the understanding of the network and nodes position. The division creates four different quadrants in which arc movement plays an important because through this arc movement the communication would take place in the network, communication, i.e., from one quadrant to other quadrants.

Figure 7 shows these sensing nodes in wireless sensor network change their location to different location from their previous locations. This is considered as disadvantage in underwater wireless sensor network because it would be very difficult for the sensing nodes to communicate to another node in the network.

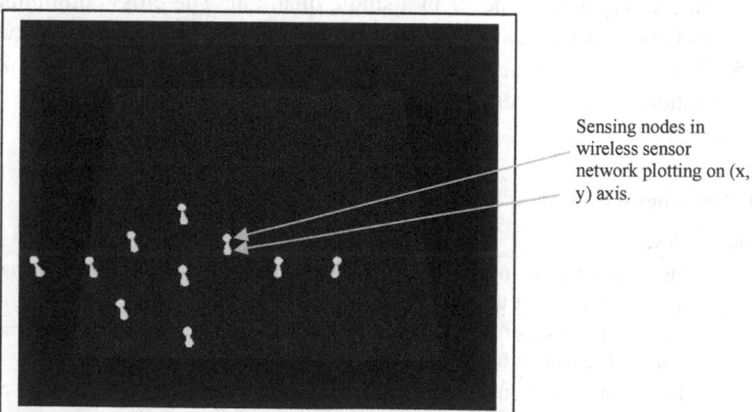


**Fig. 4** Dynamic creation of sensing nodes in wireless sensor network

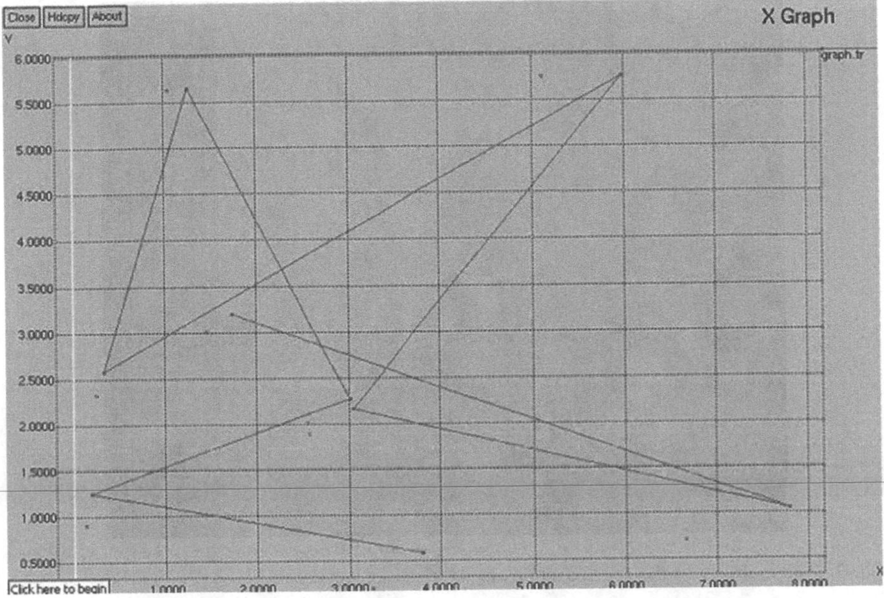


**Fig. 5** The creation of network connecting the nodes present in the network

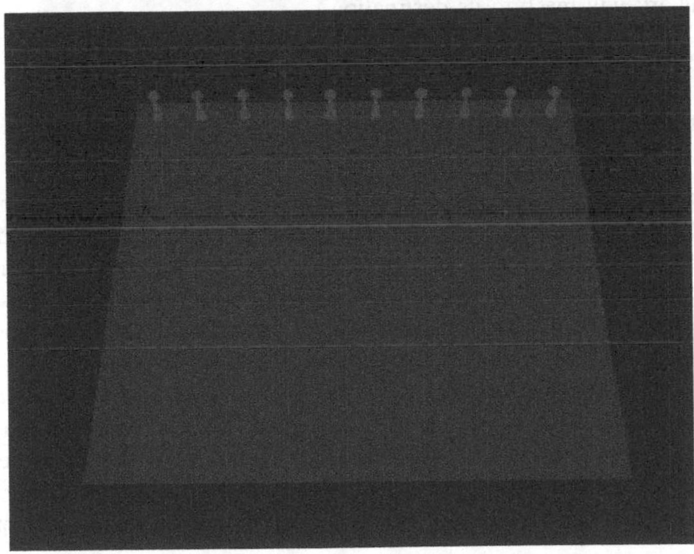

**Fig. 6** The nodes with their original position before the communication take place in the network

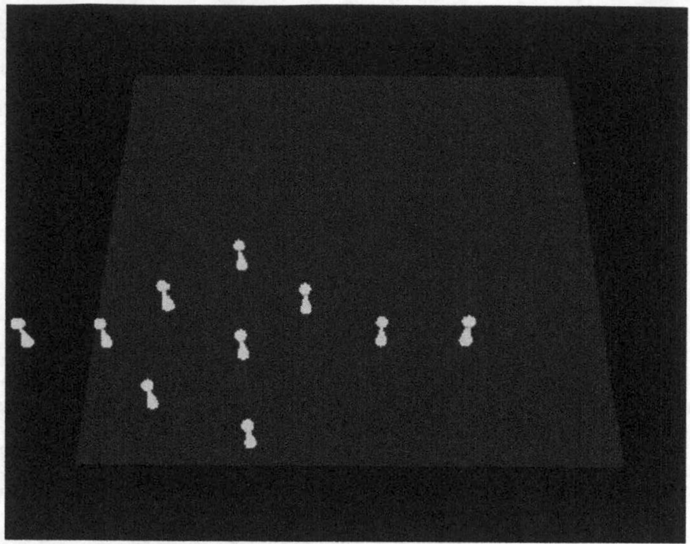

**Fig. 7** The mobility of the sensing nodes in wireless sensor network

For example, if a packet is moving from a source to its destination and in between a node changes its position to another place then how the packet traveling in the network will reach to its destination.

## 6 Conclusion and Future Work

In this paper, routing issues have been encountered related to underwater wireless sensor network and based on one of the categories of issues, i.e., node mobility, the paper proposed an approach for communication among the sensor nodes in underwater wireless sensor network. This approach has been called as an Arc moment. Through this approach, one zone (group of node) located among many sensing zones has been taken and scaled on (x, y) axis. Further the (x, y) axis is divided into four quadrants to see how the nodes of the sensing zone communicate among themselves. The nodes sense the nearest node in another quadrant by calculating Euclidean distance. In future, work can be done on how to achieve more accurate results on how the nodes communicate in real time in underwater wireless sensor network when there is mobility of nodes from one position to another.

## References


1. Ian F. Akylidiz, Dario Pompili, Tommaso Melodia, "Underwater acoustic sensor networks: research challenges", 2005.
2. Dario Pompili, Rutgers and Ian F. Akyildiz, "Overview of Networking Protocols for Underwater Wireless Communications", IEEE Communications Magazine, January, pp 97–102, 2009.
3. Ian F. Akylidiz, Dario Pompili, Tommaso Melodia, "Challenges For Efficient Communication in Underwater Acoustic Sensor Network", ACM, 2004.
4. U. Devee Prasan and Dr. S. Murugappan, "Underwater Sensor Networks: Architecture, Research Challenges and Potential Applications" International Journal of Engineering Research and Applications, Vol. 2, Issue 2, pp. 251–256, Mar-Apr 2012.
5. Zhanyang Zhang and Luciana H. Blanco, "A Self Adaptive MAC Layer Protocol for Delay-Tolerant Underwater Wireless Sensor Networks," IEEE, 2014.
6. Guangjie Han, Jinfanq Jiang, Lei Shu, Yongjun Xu and Feng Wang, "Localization Algorithms' of Underwater Wireless Sensor Network: A Survey", February, 2012.
7. Muhammad Ayaz, Low Tang Jung, Azween Abdullah, Iftikhar Ahmad, "Reliable data deliveries using packet optimization in multi-hop underwater Sensor networks", pp 41–48, vol 24, 2012.
8. Muhamad Felemban, Basem Shihaday and Kamran Jamshaidy, "Optimal Node Placement in Underwater Wireless Sensor Networks", 2013.
9. Beenish Ayaz, Alastair Allen, Marian and Wiercigroch, "Dynamically Reconfigurable Routing Protocol Design for Underwater Wireless Sensor Network", 8th International Conference on Sensing Technology, Sep, pp 2–4, 2014.
10. Mohamed K. Watfa, Tala Nsouli, Maya Al-Ayache and Omar Ayyash "Reactive localization in underwater wireless sensor Networks", University of Wollongong in Dubai – Papers, 2010.
11. John Heidemann, Milica Stojanovic and Michele zorzi "Underwater sensor networks: applications, Advances and challenges", pp 158–175, 2015.
12. S. Naik and Manisha J. Nene, "Realization of 3d underwater wireless sensor networks and influence of ocean parameters on node location estimation", International Journal of Wireless & Mobile Networks (IJWMN) Vol. 4, No. 2, April 2012.
13. Kashif Ali and Hossam Hassanein, "Underwater Wireless Hybrid Sensor Networks", pp. 1166–1171, IEEE,2008.
14. Yi Zhou, Kai Chen, Jianhua He, Jianbo Chen and Alei Liang, "A Hierarchical Localization Scheme for Large Scale Underwater Wireless Sensor Networks" 11th IEEE International Conference on High Performance Computing and Communications, 2009.
15. Davide Anguita, Davide Brizzolara and Giancarlo Parodi, "Building an Underwater Wireless Sensor Network based on Optical Communication: Research Challenges and Current Results", International Conference on Sensor Technologies and Applications, pp. 476–479, 2009.
16. Geetangali Rathee, Ankit Mundra, Nitin Rakesh, and S.P. Ghera, "Buffered Based Routing and Resiliency Approach for WMN", IEEE International Conference of Human Computer Interaction, Chennai, India, 1–7, 2013.

# Bit Error Rate (BER) Performance Enhancement for Wireless Communication System Using Modified Turbo Codes

**Garima Mahendru, Monica Kaushik, Monika Arora, Utkarsh Pandey, Apoorv Agarwal and Jagjot Singh Khokhar**

**Abstract** In wireless communication, turbo and modified turbo codes are used as forward error correction (FEC) codes to improve the performance of the system in terms of bit error rate (BER). This work revisits the FEC using conventional turbo codes and proposes few modifications in them to obtain modified turbo codes. Trivial alterations in the frame length of the codes have been implemented along with puncturing scheme. The decoding complexity of the system is alleviated with the use of Viterbi algorithm for punctured codes. A comparative study between the performance of the system using turbo and modified turbo codes is presented in terms of their BER. Simulation results show that there is significant enhancement in the performance of the system when modified turbo codes are employed.

**Keywords** Efficiency · FEC · Turbo codes · Modified turbo codes · Viterbi algorithm · BER · SNR · Frame length · Punctured codes · MATLAB

## 1 Introduction

Efficiency and reliability are the two vital factors that affect the performance of any communication system. It can be measured and analysed using parameters like bit error rate (BER), signal-to-noise ratio (S/N), etc. For reliable and efficient communication over the channel, it is necessary to minimize the BER and maximize the SNR. So as to accomplish the objective one has to employ error control mechanism. Forward error correction (FEC) is a technique that is used to make communication channel noiseless and efficient. To implement FEC several codes are used, out of which turbo and modified turbo codes are revisited in this paper.

---

G. Mahendru (✉) · M. Kaushik · M. Arora · U. Pandey · A. Agarwal · J.S. Khokhar
Department of Electronics and Communication Engineering, Amity University, Noida, UP, India
e-mail: garimamahendru@gmail.com

M. Kaushik
e-mail: monika4dec@gmail.com



S.K. Bhatia et al. (eds.), *Advances in Computer and Computational Sciences*, Advances in Intelligent Systems and Computing 553,
DOI 10.1007/978-981-10-3770-2_28

Fabrication of turbo codes is basically sourced by the designers which primarily focus on achieving reliable information transfer over bandwidth. These codes [1] are generally formed from parallel concatenation of two or more codes separated by an interleaver. For any reliable data transmission, forward error correction with low BER and low decoder complexity is required. However, the most alarming and disturbing property of turbo codes is that it has large encoding complexity; hence, modified turbo codes have been designed for improving complexity issues and formulating it better than the former. Also, its decoding complexity is almost half than those of turbo codes. Moreover, these modified turbo codes (MTC) not only require less computation but also improve the performance of the system in terms of BER.

Figure 1 presents a brief overview of turbo codes with interleaver where it has been observed that the input bits carrying the information enter the first encoder. These bits are later interleaved by the interleaver; then they enter the second encoder. The code word of the parallel concatenated code entails input bits to the first encoder followed by the parity check bits of both the encoders. The trellis structure of both these codes appears to be similar but the two output sequence does not correspond to the same input sequence. Hence, there occurs a contradiction in the codes generated which needs to be elucidated.

To overcome this incongruity, interleaving is proposed that has an auto-enhance capability of correcting error codes. This technique behaves as one-to-one mapping function. It is a process of re-arranging of data sequence in a deterministic format. This technique helps in improving the performance of forward error correcting codes. By the process of de-interleaving bits, the received sequence can be restored in its original.

In the subsequent sections of this paper, a detailed explanation of modified turbo codes (MTC) is presented, followed by implementation of the codes and simulation results with its interpretation using MATLAB. A comparative study between the standard turbo codes and MTC depicts that MTC outperforms the conventional turbo codes by incorporating modifications in the frame length and puncturing the codes.

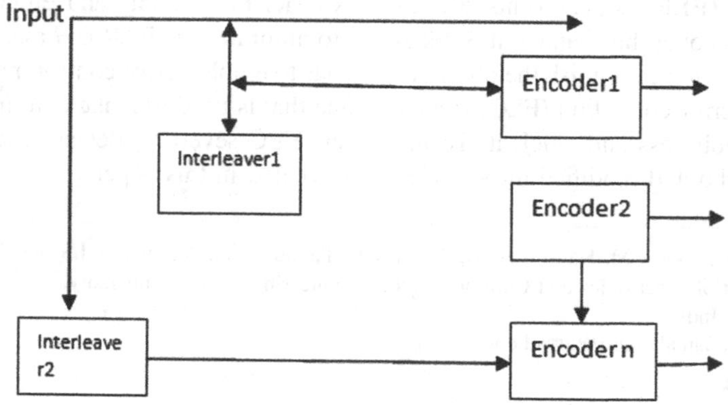


**Fig. 1** Turbo codes with interleaver

## 2 Modified Turbo Codes

Turbo codes possess significant decoding complexity leading to inaccuracy in the outcome. Puncturing technique can be applied for the reduction of complexity but still the reduction is neither significant nor. Moreover, this decoding complexity in turbo codes increases further as these codes require stronger constituent code and low code rates. Therefore, low complexity block turbo codes which is commonly known as modified turbo codes (MTC) have been introduced. By familiarization of MTC, good performance of the system can be achieved which will comparatively have low decoding complexity as compared to turbo codes. MTC generally comprises two types of codes convolution and zigzag code. The zigzag codes [2] have slightly more complexity but shows better performance than the service programming codes (SPC) and that is why they are used in modified turbo codes (Fig. 2).

The convolution code cannot be used in all the components of MTC, so it leads to reduction of error in the channel and provides better decoding complexity than standard turbo codes. In this paper, for simulation of MTC [3] we have used very important parameters—puncturing scheme and the Viterbi algorithm. Puncturing process involves the removal of some parity bits with an error correction code after the encoding process. The punctured bit does not have any effect on the decoder and thus the decoder detects the error bit and removes it. It helps in increasing the flexibility of the system without increasing its complexity. The puncturing technique is also used in Wi-Fi, GPRS, etc. [4].

Viterbi algorithm is the type of algorithm which is mainly used to decode convolution and zigzag codes [5]. Viterbi decoder system used to decode the punctured codes which are used in the simulation of MTC. After decoding different punctured codes, results are obtained which improves bit error rate.

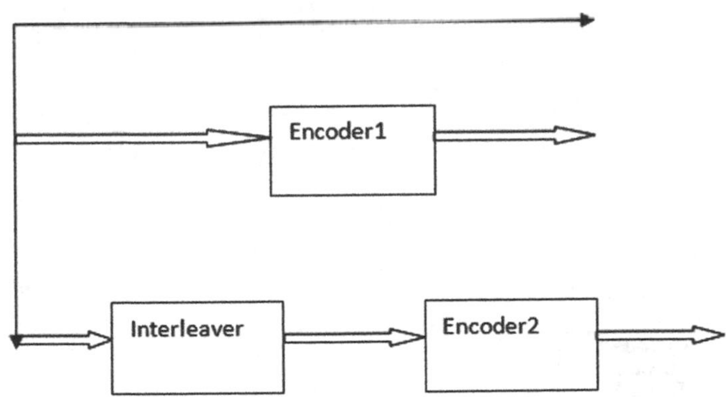


**Fig. 2** General MTC

## 3 Simulation and Result

This section presents the simulated results of turbo codes and modified turbo codes using MATLAB for bit error rate and signal-to-noise ratio.

First, turbo codes are simulated in terms of BER and S/N ratio. The range of S/N ratio is pre-defined for different frame lengths. This simulation has been done using quadrature phase shift keying (QPSK) modulation technique where turbo codes are used to encode the data and then transmitted via AWGN channel [6]. The noisy signal after reception is then demodulated and turbo decoded to calculate the error. The graph in Fig. 3 depicts the plot between BER and S/N ratio.

In MTC simulation, we have used binary phase shift keying (BPSK) technique [7] and Viterbi decoder system for the successful decoding of the punctured codes. Error rate calculator system has been created to calculate the BER, the number of errors observed, and the total number of processed bits. Using different frame lengths graph has been plotted and analysis of bit error rate and signal-to-noise ratio has been done. It can be inferred from Fig. 3 that use of turbo codes enhances the BER performance of the system as compared to a un-coded system for varying frame lengths. The performance gets better for a coded system with increasing frame lengths.

Figure 4 represents the system performance with modified turbo codes where the codes are punctured and decoded using Viterbi algorithm. Simulation results of BER have been obtained for different number of transmissions. The frame length

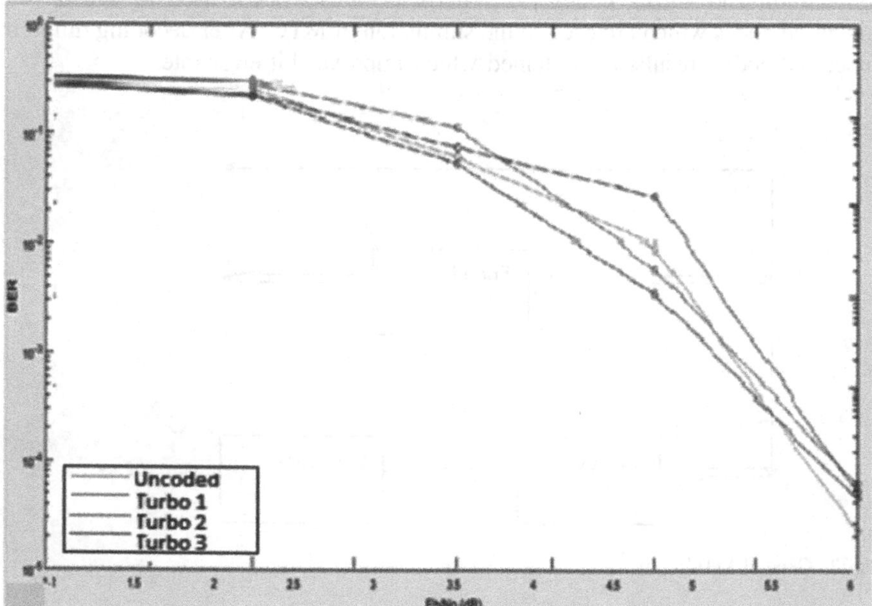


**Fig. 3** BER performance of turbo codes

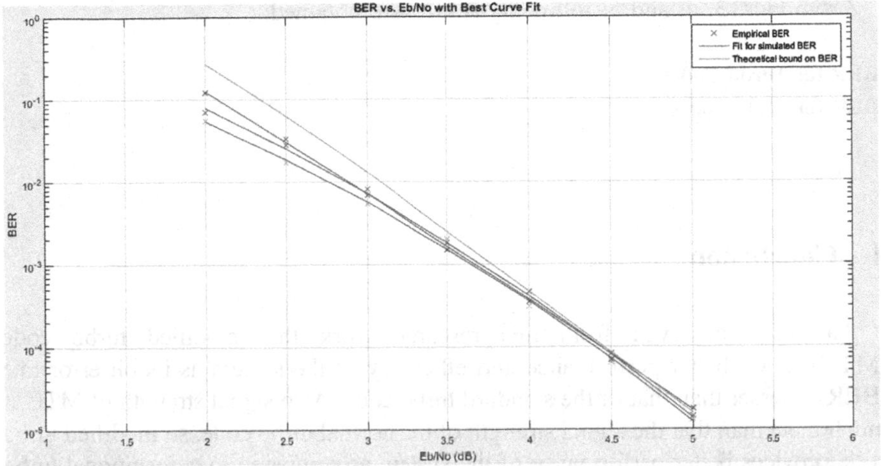


**Fig. 4** BER performance for modified turbo codes

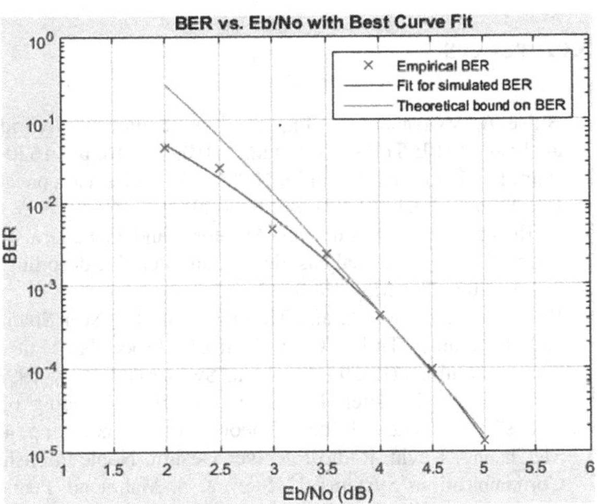


**Fig. 5** BER performance for modified turbo codes

was 3000 and maximum number of transmission taken was 300, 500, and 750, respectively. From the graph, it is shown that as the maximum number of transmissions is increased, BER is reduced and thus the system performance is improved.

The simulation results from Fig. 5 show that when the frame length was 3000 and maximum number of transmissions was 800, the modified turbo codes gives much better system performance for low signal-to-noise ratio. The best simulated result is obtained and after using MTC, system gives a much better performance than standard unmodified turbo codes.

From Figs. 3, 4, and 5, following results are obtained:

| BER for Turbo Codes | $10^{0} - 10^{-1}$ |
|---|---|
| **BER for MTC Codes** | $10^{-1} - 10^{-2}$ |

## 4 Conclusion

It has been observed from this research work that modified turbo code (MTC) shows better performance and efficiency of the system as its bit error rate (BER) is lesser than that of the standard turbo code. Also signal strength of MTC is much more than that the signal strength of the normal turbo code, so modified turbo code provides better performance of the system as compared to conventional turbo codes in terms of BER and SNR of the wireless communication system.

## References


1. Bhise, A., Vyavahare, P. "Improved low complexity hybrid turbo codes and their performance analysis", IEEE Trans. Commun., 2010, 58, (6), pp. 1620–1622
2. Ping, L., Huang, X., Phamdo, N. "Zigzag codes and concatenated zigzag codes", IEEE Trans. Inf. Theory, 2001, 47, (2), pp. 800–807
3. S. Benedetto, D. Divsalar, G. Montorsi and F. Pollara, "Serial concatenation of interleaved codes: Performance analysis, design, and iterative decoding," IEEE Trans. Inform. Theory, vol. 44, pp. 909–926, May 1998
4. Berrou C., Glavieux A. and Thitimajshima P. "Near Shannon Limit Error Correcting Coding and Decoding: Turbo Codes," IEEE Proceedings of the International Conference on Communications (ICC93), Geneva, Switzerland, May 1993, pp. 1064–1070
5. J. Hagenauer, E. Offer, L. Papke, "Iterative Decoding of Binary Block and Convolutional Codes", IEEE Trans. Inform. Theory, vol. 42, no. 2, pp. 429–445, March 1996
6. McClaning, Kevin, Radio Receiver Design, Noble Publishing Corporation
7. Communications Systems, H. Stern & S. Mahmoud, Pearson Prentice Hall, 2004, p. 283

# A Timestamp-Based Strong Designated Verifier Signature Scheme for Next-Generation Network Security Services

**Asif Uddin Khan, Bikram Kesari Ratha and Srikant Mohanty**

**Abstract** Strong designated verifier signature scheme (SDVS) is a unique kind of digital signature scheme that has numerous applications in the networking infrastructure of present and future generations. In recent years, Lee-Chang has put forward a strong designated verifier signature scheme and stated that their scheme is secure. In this paper we show that Lee-Chang's scheme is not secure, and any combatant can forge their scheme unaware of the secret key of the original signer and suggest an improved scheme which solves the fraud attack problem by any adversary. We analyze our scheme and compared with the Lee-Chang's scheme and show that our scheme is secure and meet the security feathers needed by SDVS.

**Keywords** Digital signature • Designated verifier signature scheme • Random oracle model • Forgery attacks • Timestamp

## 1 Introduction

Next-generation wireless network is expected to include many applications and services such as voice, data and multimedia online gaming, and software distribution with very high data rate. Use of new technologies and services such as internet of things cloud computing are growing rapidly day by day where numerous

A.U. Khan (✉) · B.K. Ratha
Department of Computer Science, Utkal University, Vani Vihar, Bhubaneswar, India
e-mail: asifkhan.iiit@gmail.com

B.K. Ratha
e-mail: b_ratha@hotmail.com

S. Mohanty
Department of ECE, CAPGS, BPUT, Rourkela, India
e-mail: srikant.mohanty.2016@ieee.org



S.K. Bhatia et al. (eds.), *Advances in Computer and Computational Sciences*,
Advances in Intelligent Systems and Computing 553,
DOI 10.1007/978-981-10-3770-2_29

network-based services and applications are provided through internet. Further internet of things (IOT) is the big revolution in the future networking technologies where communication for application, data and services can take place anywhere and anytime through any device. Providing guaranteed quality of service (QOS) to these applications and services is an important objective in the design of next-generation network. At the same time various types of security threats in the network-based services are increasing. Providing security such as confidentiality, authenticity, and data integrity is a must in order to achieve guaranteed QOS.

Digital signature is an important technique to provide security benefits such as user verification, data integrity, and non-abrogation in the distributed network environment such as internet. Many changes of digital signature scheme such as proxy signatures, blind signature, and ring signature are there for different applications and different scenario discussed in several papers in [1, 2]. In special scenario like e-voting, software licensing and call for tender ordinary digital signature is not useful. A special kind of signature scheme is required for this new environment. Here the signer does not want the beneficiary of the signature scheme to shift the faith to any third force at his choice.

Addressing the above issue, in [3, 4] authors proposed a new technique known as undeniable signature scheme. This type of signature allows signer of the message to get complete control on the message. In this type of scheme the signature verification needs the involvement of the witness for avoiding undesirable verifiers to get convinced of the potency of the signature. This motivated Jakobsson et al. [5] to propose the idea of designated verifier signature scheme (DVS) to solve the above problem. A DVS is a scheme where it is possible for the signer of the message A to satisfy the designated verifier B that A has witnessed the message in such a manner that B cannot alter the conviction to any third party by sanctioning B the proficiency of simulating a signature effectively which is not distinguishable from A. In [5] Jakobsson et al. also proposed a higher version of DVS to improve the signer privacy. This is called as strong designated verifier signature scheme (SDVS) where any third party cannot even check the efficacy of designated verifier signature, because the verification requires special key of the designated verifier. After that many SDVS designs have been proposed such as in [6]. Huang et al. proposed a variant of designated verifier signature scheme, i.e., short strong designated verifier signature scheme and its variant scheme which is an identity-based scheme. In [7] Yang and Liao proposed a strong designated verifier signature scheme, using key distribution mechanism where both sender and designated receiver share encryption/decryption key to fulfill encryption/decryption algorithm with low cost of communication and computation, respectively, and they proved their security based on Deffi-Helman assumptions. In 2003, Saeednia et al. [8] recommended a strong designated verifier signature theory based on the Schnorr signature theory [9] and Zheng's signcryption scheme [10]. Then Lee and Chang [11] analyzed and found out that Saeednia et al. scheme is not secure and it would disclose the identity of the signer if the secret key of the signer is compromised. Then, they recommended a new strong designated verifier signature scheme that is based on the Schnorr signature scheme [9] and Wang et al. authenticated encryption scheme [12] which can be verified only with the designated verifier's secret key. Lee-Chang claimed their scheme to be efficient and secure but in [13] Hyun et al. proposed several attacks on Lee-Chang's strong designated verifier signature scheme. In this paper we discuss the forgery attack on the Lee-Chang's scheme based on [13]. We then propose a new and improved strong designated verifier signature scheme which is more secure and safe. We enhance the security of the scheme using time stamp value in the signature generation phase. A time stamp [14] is the time that is recorded at which an event occurs; in this context, it is this context that it is the time at which the signer signs the message.

We organize the paper as follows:

Section 2 reviews Lee-Chang's strong designated verifier signature theory. Section 3 shows forgery attack on Lee-Chang's scheme. Section 4 shows the improved scheme, Sect. 5 shows analysis of improved scheme, and finally we conclude in Sect. 6.

*Parameters used throughout the paper are as follows.*

- A: The message signer.
- B: The designated verifier of the signature.
- $p$: A large prime number.
- $q$: A prime factor of $p - 1$.
- $g$: A generator $\in Zq^*$ of order $q - 1$.
- $m$ is the message.
- $H(\cdot)$: a secure one-way hash function such as SHA-2 that outputs values in $Zq^*$.
- $(x_A, y_A)$: A's private and public key pair, respectively, where $x_A$ is a randomly selected secret key in $Zq^*$ and the corresponding.
- Public key $y_A = g^{xA} \bmod p$.
- $(x_B, y_B)$: B's private and public key pair, respectively, where $x_B$ is a randomly selected secret key in $Zq^*$ and the corresponding public key $y_B = g^{xB} \bmod p$.
- ¥: Timestamp value.
- $\sigma$: Signature.

# 2 Lee-Chang's Strong Designated Verifier Signature Scheme

In this section we discuss and analyze Lee-Chang's strong designated verifier signature scheme [11]. The scheme has three phases such as signature generation, signature verification, and signature simulation as follows.

## 2.1 Signature Generation

1. A selects a random value $k \in Z_q^*$
2. Then A computes *r*, *s*, and ***t*** as follows:

$$r = g^k \bmod p \quad (1)$$

$$s = k + x_A r \bmod q \quad (2)$$

$$t = H\,(m,\ y_B^s \bmod p) \quad (3)$$

3. The signature is then σ = (r, t).

## 2.2 Signature Verification

Upon receiving m and σ = (r, t), B verifies the validity of the signature by checking the following equation:

$$t = H\,(m,\ (r y_A^r)^{xB} \bmod p). \quad (4)$$

The signature is valid if Eq. (4) is satisfied.

## 2.3 Signature Simulation

B can simulate the transcript ($r_s$, $t_s$) for the message m by selecting a random number $k_\Psi \in Z_{q^*}$ and computes $r_s$ and $t_s$ as follows:

$$r_s = g^{ks} \bmod p \quad (5)$$

$$t_s = H\,(m,\ (r_s y_A^{rs})^{xB} \bmod p). \quad (6)$$

## 3 Forgery Attack on Lee-Chang's Scheme

In this section we discuss the forgery attack on Lee-Chang's scheme based on [13]. Here it is shown that Lee-Chang's scheme is not secure against forgery attack where any adversary can forge the signature without knowing the secret key of signer or the designated verifier.

Suppose that an adversary E intercepts the signature m, $\sigma = (r, t)$ which is sent from A to B and then E can perform the following forgery attack:

1. $E$ chooses a forged message $m^*$.
2. $E$ finds an integer $r^*$ which satisfies $r^* y_A^{r^*} \bmod p = g$
3. Then $E$ computes $t^* = H(m^*, y_B \bmod p)$.
4. $E$ sends the forged signature $\sigma^* = (r^*, t^*)$ with $m^*$ to B.
5. Upon receiving the forged message $m^*$ and the modified signature $\sigma^* = (r^*, t^*)$, B will verify the validity of the signature by checking the following:

$$t^* = H\left(m^*, \left(r^*, y_A^{r^*}\right)^{xB} \bmod p.\right. \tag{7}$$

If Eq. (7) is satisfied, then we conclude that the signature is verified.
Its correctness can be seen as follows:
Left-hand side is

$$t^* = H\left(m^*, y_B \bmod p\right), \tag{8}$$

and the right-hand side is

$$\begin{aligned} H\left(m^*, \left(r^* y_A^{r^*}\right)^{xB} \bmod p\right) &= H\left(m^*, (g)^{xB} \bmod p\right) \\ &= H\left(m^*, y_B \bmod p\right) \end{aligned} . \tag{9}$$

In the above we can see that Eq. (7) is satisfied and it is confirmed as a legal signature of A by B. Therefore, B will always believe that the real singer is A and the message m* received is A's message. But we know that the signature $\sigma^* = (r^*, t^*)$ is signed by an adversary E and m* is the forged message sent by the adversary. So we conclude that Lee-Chang's scheme is not secure and it is vulnerable to the forgery attack.

Example of the above forgery attack [13] is as follows:

Let the parameters $p = 23$, $q = 11$, $g = 2$, $x = 8$,
$y_A = 28 \bmod 23 = 3$, then it is easy to find an integer r*

which satisfies $r^* y_A^{r^*} \bmod p = g$.

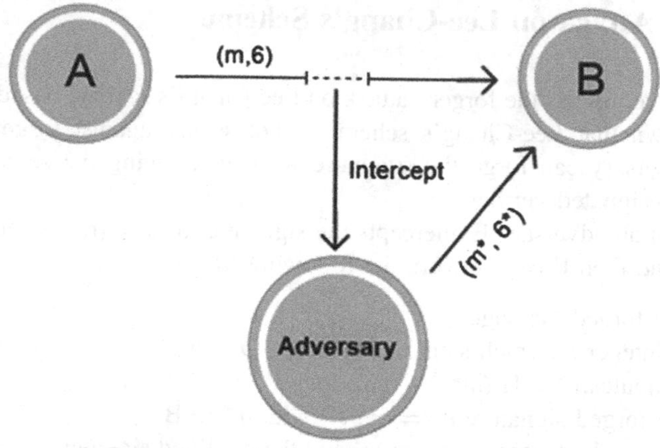


1. Receive (m, 6)
2. Choose m*
3. Find r* which satisfies $r^* y_A^{r^*}$ modp=g
4. Compute t*= H(m*, $y_B^{modp}$)
5. 'A' sends a forged signature σ* =(r*,t*) with m* to 'B'

**Fig. 1** Forgery attack on Lee-Chang's scheme

For example,

If $r = 4$, then $r^* y_A^{r^*} \bmod p = 4.3^4 \bmod 23 = 324 \bmod 23 = 2 = g$.
If $r = 8$, then $r^* y_A^{r^*} \bmod p = 8.3^8 \bmod 23 = 52488 \bmod 23 = 2 = g$.
If $r = 13$, then $r^* y_A^{r^*} \bmod p = 13.3^{13} \bmod 23 = 20726199 \bmod 23 = 2 = g$.

The details of the forgery attack are shown in Fig. 1.

## 4 Improved Scheme

Since Lee-Chang's scheme is not secure against forgery attack, in this section we propose a new scheme which is more secure than the previous one. We use same set of parameters as Lee-Chang's scheme. We replace H by a random oracle. Security of our scheme in the ROM can be proved under the assumption that the discrete logarithm (DL) problem is hard.

There are three phases in the improved scheme. The security is enhanced by adding the timestamp value Ŧ in signature generation phase during calculating r and s as follows.

## 4.1 Signature Generation

1. A selects a random value $k \in Z_q^*$.
2. get the timestamp value ¥
3. now find ₮ = $(¥)^{xA}$
4. A computes $r$, $s$, and $t$ as follows:

$$r = gk + ₮ \bmod p$$

$$s = (k + ₮) + x_A r \bmod q$$

$$t = H(m, y_B s \bmod p)$$

5. The signature is then $\sigma = (r, t)$.

## 4.2 Signature Verification

Upon receiving m and σ = (r, t), B can verify the validity of the signature by checking whether

$$t = H\left(m, \left(ry_A^r\right)^{xB} \bmod p\right). \tag{10}$$

## 4.3 Correctness Proof

We have to prove that

$$\begin{aligned} t &= H\left(m, \left(ry_A^r\right)^{xB} \bmod p\right) \\ \text{But}\, t &= H(m, y_B^S \bmod p) \end{aligned} \tag{11}$$

So if we can prove that

$$y_B^s = \left(ry_A^r\right)^{xB} \bmod p.$$

Then Eq. (11) is proved.

$$\left(ry_A^r\right)^{xB} = \left(r\left(g^{xA}\right)^r\right)^{xB} \bmod p. \tag{12}$$

$$y_B{}^s = y_B{}^{(k+Ŧ)+xAr} \text{ mod p}$$

$$= (g^{xB})^{(k+Ŧ))+xAr} \text{ mod p}$$

$$= (g^{xB})^{(k+Ŧ)} (g^{xB})^{xAr} \text{mod p}$$

$$= ( g^{(k+Ŧ)})^{xB}. ( (g^{xA})^{r})^{xB} \text{mod p}$$

$$= r^{xB}. (( y_A)^{r})^{xB} \text{mod p}$$

$$= (ry_A{}^{r})^{xB} \text{ mod p}$$

So from the above Eq. (11) is proved

## 4.4 Signature Simulation

B can simulate the transcript ($r_\Psi$, $t_\Psi$) for the message $m$ by selecting a random number $k_\Psi \in Z_q^*$ and computes $r_\Psi$ and $t_\Psi$ as follows:

$$r_\Psi = g^{k\Psi} \text{ mod p} \tag{13}$$

$$t_\Psi = H\left(m, \ \left(r_\Psi y_A^{r\Psi}\right)^{xB} \text{ mod p}\right). \tag{14}$$

# 5 Analysis of the Improved Scheme

## 5.1 Security Analysis

**Basic Unforgeability**:

Because the problem of getting private key $x_A$ from the public key $y_A$ equals to solving DLP, no one else can create normal digital signature of the original signer.

**Unforgeability of our scheme**:

Because the timestamp value of the signer A is added that is $r = g^{k+Ŧ}$ mod p and $s = (k + Ŧ) + x_A^r$ mod q and no one except the original signer A knows "Ŧ" and $x_A$, so any adversary cannot forge the signature and also finding $x_A$ from $y_A$ is same as solving DLP which is a hard problem. Again the adversary cannot link t with t* as because H is replaced by Random Oracle. Therefore, our scheme is provably secure. In the previous example, we have seen that adversary can only forge the signature if he/she find r* such that $r^* y_A^{r^*}$ mod p $= g$. In the improved scheme a

**Table 1** Performance comparison between our scheme and Lee-Chang's scheme

| Computational cost | Lee-Chang's scheme | Our improved scheme |
|---|---|---|
| Signature generation | $2T_E + T_M + T_H$ | $3T_E + T_M + T_H$ |
| Signature verification | $2T_E + T_M + T_H$ | $2T_E + T_M + T_H$ |
| Signature simulation | $3T_E + T_M + T_H$ | $3T_E + T_M + T_H$ |

*Note* In the above Table 1 $T_M$, $T_E$, and $T_H$ are the computation times for modular multiplication, exponential operation, and one-way function, respectively

timestamp value, i.e., ₮ = $(¥)^{xA}$ is added in the step 3 and step 4. So adversary cannot guess r* because it does not know the value of ₮ and xA which is used in signature generation (r = gk + ₮) mod p; only the original signer knows the value of xA and timestamp value ¥. Adversary can only forge the signature if and only if he/she get xA from yA and have the knowledge of ¥ which same as DLP a hard problem with large prime. Hence, this proves that our improved scheme is safe and secure.

### 5.2 *Performance Comparison of the Improved Scheme with Lee-Chang's Scheme*

This section shows the comparison of computational cost our scheme with Lee-Chang's scheme. Here we compare the cost incurred in different phases of the schemes in Table 1. Table 1 shows that the computational cost of our improved scheme is same as Lee-Chang's scheme in all phases except one more TE is used in signature generation phase.

## 6 Conclusion

In this paper, we discuss how important is the requirement of designing a secure and safe strong designated verifier signature scheme (SDVS) for the next-generation network services and applications, analyzed the weaknesses of Lee-Chang's strong designated verifier signature scheme against forgery attack, and proposed a new strong designated verifier signature scheme based on timestamp value. We analyzed our new scheme and compared with Lee-Chang's scheme and conclude that our improved scheme can remedy the weaknesses of Lee-Chang's scheme and meets the security aspects needed by the strong designated verifier signature scheme. In other words, the new scheme is more secure than the existing scheme.

## References


1. Aki, S.G., "Digital signatures: A tutorial survey," Computer, vol. 16, no. 2, pp. 15, 24, Feb. 1983 doi:10.1109/MC.1983.1654294 (http://ieeexplore.ieee.org/stamp/stamp.jsp?tp=&arnumber=1654294&isnumber=34678).
2. Manik Lal Das, Ashutosh Saxena2, and Deepak B Phatak, "Algorithms and Approaches of Proxy Signature: A Survey", International Journal of Network Security, Vol.9, No.3, PP.264284, Nov.2009 (http://ijns.jalaxy.com.tw/contents/ijns-v9-n3/ijns-2009-v9-n3-p264-284.pdf).
3. D. Chaum, Zero-knowledge undeniable signa- tures. In: Advances in Cryptology—Eurocrypt'90, LNCS, vol. 473, Springer-Verlag, 1990, pp. 458–464.
4. Chaum, D., Van Antwerpen, H., 1990. Undeniable signature. In: Advance in Crypto'89, LNCS, vol. 435. Springer-Verlag, pp. 212–216.
5. M. Jakobsson, K. Sako, R. Impagliazzo, Designated verifier proofs and their applications. In: Advances in Cryptology-Eurocrypt'96, LNCS, vol. 1070, Springer-Verlag, 1996, pp. 143–154.
6. Huang, Xinyi, et al. "Short designated verifier signature scheme and its identity-based variant." (2008).
7. Yang, Fuw-Yi, and Cai-Ming Liao. "A provably secure and efficient strong designated verifier signature scheme." *International Journal of Network Security* 10.3 (2010): 220–224.
8. S. Saeednia, S. Kremer, O. Markowitch O, An efficient strong designated verifier signature scheme, in: ICISC'03, Lecture Notes in Computer Science, vol.2971 (Springer, Berlin, 2004) 40–54.
9. C.P. Schnorr, Efficient signature generation for smart cards, Journal of Cryptology 3 (3) (1991), 161–174.
10. Y. Zheng, Digital signcryption or how to achieve cost (signature & encryption) ¡¡ cost (signature) +cost (encryption), in: Advances in Cryptology -Crypto'97, Lecture Notes in Computer Science, vol.1294 (Springer, Berlin, 1997) 165–179.
11. Ji-Seon Lee, Jik Hyun Chang, Comment on Saeednia et al.'s strong designated verifier signature scheme, Computer Standards & Interfaces (2008), doi:10.1016/j.csi.2008.02.003.
12. G. Wang, F. Bao, C. Ma and K. Chen, Efficient authenticated encryption schemes with public verifiability, in The 60th IEEE Vehicular Technology Conference (VTC 2004)—Wireless Technologies for Global Security, vol. 5 (IEEE Computer Society, 2004) 3258–3261.
13. Hyun, Suhng-Ill, Eun-Jun Yoon, and Kee-Young Yoo. "Forgery attacks on Lee-Chang's strong designated verifier signature scheme." Future Generation Communication and Networking Symposia, 2008. FGCNS'08. Second International Conference on. Vol. 2. IEEE, 2008.
14. Krishna A., "A New Non Linear Model Based Encryption Scheme with Time Stamp & Acknowledgement Support," the International Journal of Network Security, vol. 14, no. 1, pp. 27–32, 2012.

# Motion Estimation Enhancement and Data Transmission Issue Over WiMAX Network

**K. Sai Shivankita and Nitin Rakesh**

**Abstract** High Resolution (HR) image is obtained utilizing set of Low Resolution images using the Super Resolution processes. For motion vector within the frames in Super resolution, the motion estimation algorithms are used. Motion estimation gives a fully designed algorithm in programmable platforms. Highest spatial accuracy is used by the algorithm to adapt the resolution of the image content where it is necessary, that is the border of the moving objects. Binocular vision is the most focused work in reconstruction of image and helps in collecting motion and depth image from defocused images. Shape from Focus (SFF) method is a sequence of images, which is used in the application where high resolution of focused images is given priority. It helps in removing blurred frames and helps in reconstruction. A framework model is proposed in this paper to derive HR images from LR images using motion estimation algorithms and further reconstruction algorithms are implemented using SFF. In this paper, we discuss how to increase the resolution of any image and to reconstruct image and then transferring data to WiMAX network and through WiMAX network this data can be viewed in mobile area as well. The paper proposes with a framework where the model deals with the sequence where we can recreate high resolution image using set of low resolution images and the transferring of data to WiMAX network.

**Keywords** IEEE 802.16e • Mobile WiMAX • Motion estimation • Super resolution • High • Resolution

K. Sai Shivankita (✉) · N. Rakesh
Department of Computer Science and Engineering, Amity University, Noida, UP, India
e-mail: shivankita27@gmail.com

N. Rakesh
e-mail: nitin.rakesh@gmail.com



S.K. Bhatia et al. (eds.), *Advances in Computer and Computational Sciences*,
Advances in Intelligent Systems and Computing 553,
DOI 10.1007/978-981-10-3770-2_30

## 1 Introduction

The effective categorization of broadband wireless access networks, like Worldwide Interoperability for Microwave Access (WiMAX), which enables video service offer greater bandwidth services to remote users. The most popularly used service is video streaming, in which the user can play with the video without the effort of downloading the entire content [1–3]. The Quality of Experience (QoE) which works on video streaming using the concept of Quality of Service (QoS) at transmission network uses downloading schemes which determine packet loss and delay. The network works effectively with QoE when its working and application are known. The data is well analyzed when previously loaded data is viewed instead of the new streaming data.

Giving demanding QoS ensures upon remote diverts as a rule, in any case, the administration stream particularly QoS utilization in WiMAX delivers it in manageable manner for video streaming. WiMAX depends upon IEEE 802.16 standard for remote communication, and likewise gives a design to the radio access system and the availability of IP center system [4]. Administration stream deals with the QoS parameters which work on MAC layer of WiMAX network.

The WiMAX Base Station plans to parcels upon the remote channels while guaranteeing the QoS sureties of every administration stream. This change in QoS may happen because of the variety in the peak at the BS, for example, when the quantity of clients associated within the base station changes, or because of the variety in the got signal quality at the MS, and when the client travels far from the base station, all can possibly adjust the QoE at the client side.

The sensors capture optical images efficiently which are stored as digital information and created a new are of imaging by optical sensors. The resolution is dependent on the size and the number of sensors used to capture an image. It is not possible to improve the sensors resolution always to improve image resolution. Sensor developing techniques are used which decrease the pixel size for improving spatial resolution. Even the amount of light decreases while the size of the pixel decreases. But it effects the quality of the image by generating shot noises [5]. Another way of increasing spatial resolution is to increase the capacitance while increasing the chip size. Also, this approach has a disadvantage by making the speed slow for charge transfer rate. For preventing these issues, we provide with super resolution process where high resolution [6] images are generated using sequence of low resolution images (Fig. 1).

For required HR image, a magnification factor, $L$ is constructed for the image, where $L = \frac{HR\ image}{LR\ image}$, where $L$ depends on the nonredundant $LR$ images. At earlier times, this process was only feasible with hardware solutions and software solutions required computing power for making it user friendly. Now, it is easy to use these models efficiently in software implementation.

In this paper, motion estimation algorithms and reconstruction algorithms are used for an image that are incorporated in stereo system. These algorithms help in

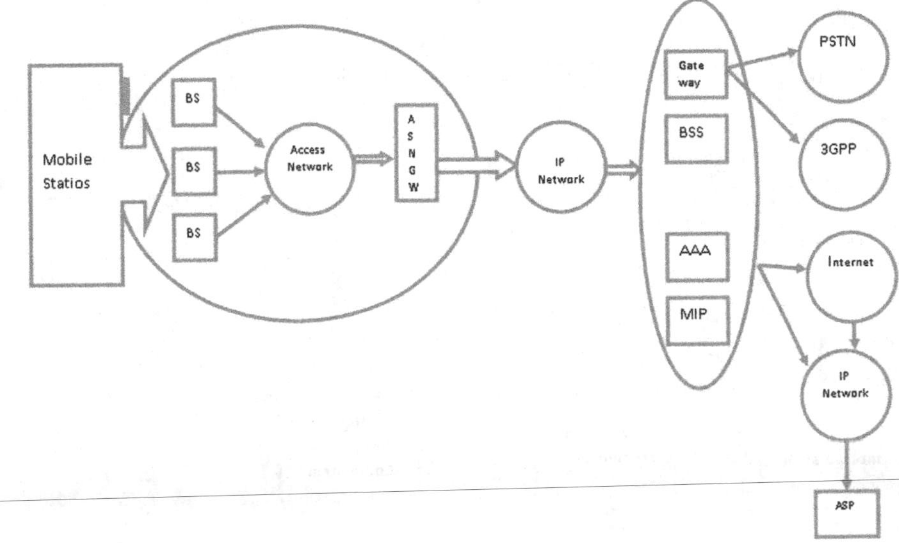


**Fig. 1** WiMAX architecture

analyzing the image and reconstructing low resolution image into a high resolution image.

The structure of an image is estimated using the method shape from focus [5, 6]. Real aperture camera is used to capture the set of the images of an object. The depth of the image is not focused and is prone to noise and blurring of the image. This method helps in estimating high resolution image by de blurring and de-aliasing of sequence of multiple low resolution image frames. The various techniques used for this method are motion free, motion based, and learning based [7]. Motion less super resolution is applied when no motion is there in the low resolution image frames.

Motion dependent super resolution is based upon the relative motion within the camera with the scenario which helps in raising the sub pixel displacement between the LR images for obtaining HR image. Learning based super resolution gives statistical relationship between the image space of low resolution and high resolution images while the training phase [8, 9]. These relationships provide with finer details of the low resolution images. Our goal is to exploit defocus cue to reconstruct high resolution images by analyzing the multiple given low resolution image frames (Fig. 2).

This paper is organized as follows: in Sect. 1, we have presented the brief introduction about the evolution of WiMAX technology over the motion estimation and the super resolution process. In Sect. 2, we have given the literature review. In Sect. 3, we have shown the model used for super resolution along with the algorithm used that is motion estimation algorithms and reconstruction algorithms.

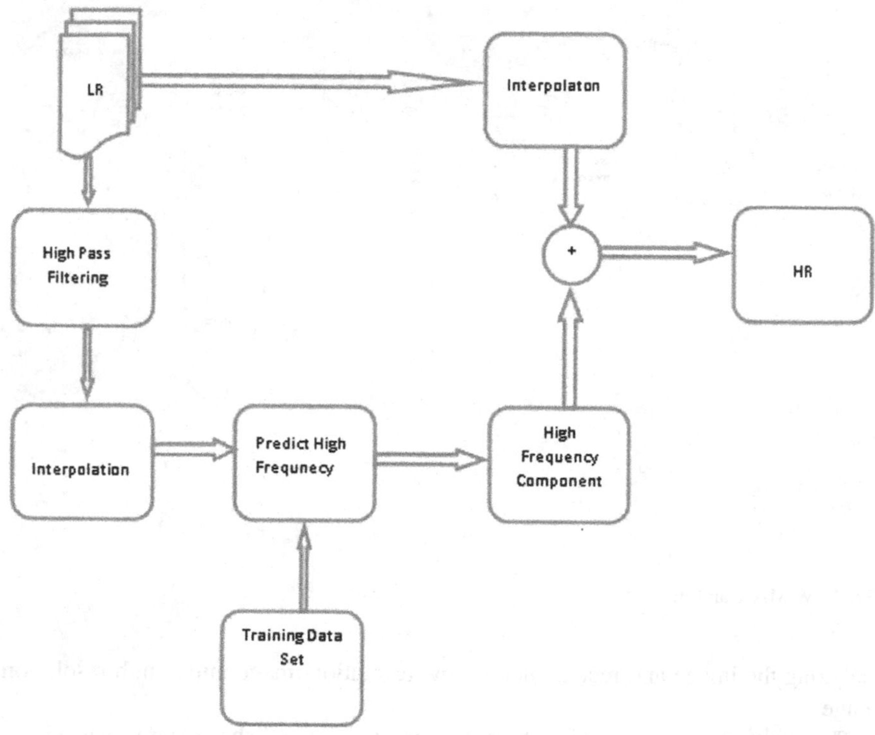


**Fig. 2** Super resolution architecture

Section 4 provides with the proposed framework to reconstruct HR images from LR images in WiMAX network. Section 5 has the conclusion and the future scope.

## 2 Related Work

Motion estimation has been traditionally utilized as a part of the utilization of video encoding, yet these days specialists from different fields rather than using video encoding are moving toward motion estimation for tackling different genuine issues in their individual fields. Additionally the agenda of this concept is to furnish the persuader with a vibe of the execution of the calculations, with specific thoughtfulness regarding the imperative exchange off between calculation unpredictability, expectation quality, output quality, and different applications.

Motion repaid change coding frames the premise of the current streaming pressure Standards H.261 along with H.262 and MPEG-1 along with MPEG-2, where the pressure calculation can misuse the transient and spatial duplicities by utilizing any type of motion pay, took after a change coding, separately. The main stride in evacuating fleeting excess is the motion estimation in which a motion vector can anticipate within the present casing and a reference outline. Taking after the motion estimation, a motion pay level is connected to get the lingering picture that is the pixel contrasts within the present casing and a reference outline.

Repeatedly, the motion estimation along with the term optical stream is used together. It is likewise related in idea to Image enlistment and stereo correspondence. Indeed these terms allude to the procedure of discovering relating focuses between two pictures or video outlines. The focuses that compare to each other in two perspectives (pictures or edges) of a genuine scene or article are generally the same point in that scene or on that protest [4, 10]. Before we do motion estimation, we should characterize our estimation of correspondence, i.e., the coordinating metric, which is an estimation of how comparable two picture focuses are. There is no correct here; the decision of coordinating metric is typically identified with what the last assessed motion is utilized for and in addition the enhancement system in the estimation process.

## 3 Methods Used: S-R Model, Motion Estimation Algorithms and Reconstruction Algorithm

The observation model shows low resolution image analyzed from the original and high resolution image is shown in Fig. 3 [9].

The $X$ denotes continuous image and the $Xs$ be the high resolution image which is sampled using Nyquist rate from the continuous image. The $Yk$ is the $k$th observed low resolution image from image sensor. Model is represented as

$$Yk = D \times B_k \times M_k \times X \times N_k, \quad k = 1, 2, 3 \ldots$$

Here, $D$ represented as down sampling operator and $B_k$ represents the blur on the $k$th low resolution image [8]. $M_k$ contains the motion information which helps in transforming the low resolution image onto the high resolution image. Also $N_k$ is the noise represented at $k$th low resolution image. Here motion estimation plays an important role in super resolution [9, 11]. Here we use block-matching algorithms which provide with motion vector quality and flexible trade off complexity. It is

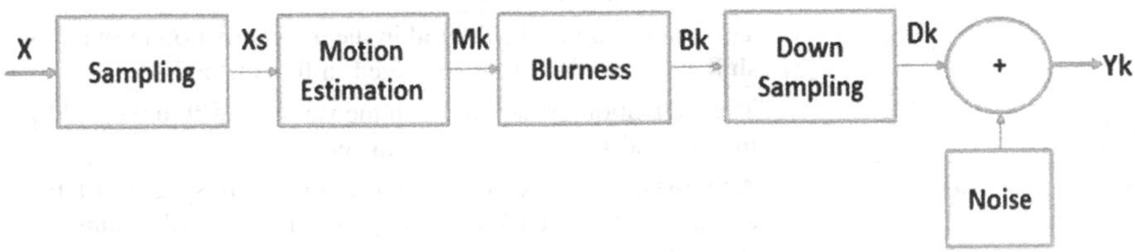


**Fig. 3** Observation model of super resolution

important to provide correct motion vectors in order to increase spatial resolution and able to detect these motion vectors for high resolution [12].

Motion estimation is the process of dealing with motion vectors that produces difference within one 2D image and then to the next; more often than not from adjoining outlines in a video succession. It is an ill-posed issue where the motion works in three manners; also the images are projected at the 3D image on a 2D plane. The motion vectors can recognize the whole image or image parts, like rectangular, squares, or per pixel. The motion vectors might be used in a translational technique or numerous different techniques which can be surmised to motion of a genuine camera, for example, pivot and interpretation in each of the three manners and focusing [13–15]. The motion estimation algorithm used in this papers are Lucchese et al., Marcel et al., Vandewalle et al. and Keren et al. which are explained in the following tables (Table 1).

Image Reconstruction Algorithm is used as mathematical process which helps in reconstructing HR images from LR images. These algorithms are used to increases the quality of the reconstructed image. These algorithms works on the scaling factor and rotating factor which work on the LR images [16–18]. The reconstruction

**Table 1** Motion estimation algorithms

| S.no. | Motion estimation algorithm | Functionality of these algorithms |
|---|---|---|
| 1. | Lucchese et al. | Works within frequency domain. |
| | | The angle of the lines along the axes is identical with half of the rotated angle within the images |
| | | This algorithm implements a three-stage coarse for rotation angle exploitation along the degree accuracy. The standard phase correlation method is used to estimate shifting |
| | | The two images vary in frequency domain through phase shift and is found in correlation |
| 2. | Marcel et al. | Log-polar transform is utilized by the value of the frequency spectra, conversion of horizontal as well as vertical shift sare done by rotation along with scaling image |
| | | A high-pass filter is applied for making the high frequencies strong |
| | | Phase correlation is used to estimate planer shifts. Frequency spectrum is used to reduce the errors in this method |
| 3. | Vandewalle et al. | High-frequency components are discarded which is an advantage for this method |
| | | Translation to linear shift within the transformation through the shift in space domain is exploited in this method |
| | | The estimation updates through the variation within the real LR images and the computed LR images |
| 4. | Keren et al. | A two-stage approach is used for restoration of super resolution along with the spatial domain approach for globally translated shifting and rotation |

**Table 2** Reconstruction algorithm used

| S.no. | Reconstruction algorithm | Usage of the algorithm |
|---|---|---|
| 1. | Bi-Cubic Interpolation | Interpolation technique identifies the grey level value of the 16 nearest pixels of the specific input values and then provide with the final value |
| | | For 1-D bi-cubic interpolation, there are four grid points which compute the function for interpolation. On one way, two grid points of that point of consideration and on the other way, two grid points |
| 2. | Populis-Gerchberg | Until the convergence process is obtained it tries to place the pixel on HR grid and cuts the high frequencies in frequency domain and repeat it |
| | | At low magnification garbled results may occur which suffer from degradation and periodic corruption and also the motion parameters are not taken into account |
| 3. | POCS | The shift estimation parameters are only considered in this method and not the rotation parameters. The image is passed from a low-pass filter which computes approximation of the camera's point spread function (PSF) |
| | | Even at high magnification, this method provides with good output when the LR images are properly reconstructed |
| 4. | Robust Super Resolution | The resolution is increased which results in high accuracy estimates of HR images mainly when outliers are present. Since regions at higher frequencies aliasing is more prone, so it treats every pixel independently in the resultant |
| 5. | Iterated-Back-Projection | The registration procedure is the start of this method and the displacement estimation is refined iteratively |
| | | By decimating, the pixels HR image can be recreated from LR image |
| | | The examined LR images are then down sampled |
| 6. | Structured Adaptive Normalized Convolution | The local channel is the approximation that takes along a projection on a sub-space spanning a set of basic functionalities |
| | | This procedure tries to increases signal-to-noise ratio which results in reducing the diffusion at discontinuity |

algorithms applied are Bi-Cubic Interpolation, POCS, Iterated-Back-Projection, Populis-Gerchberg, Robust Super Resolution, and Structured Adaptive Normalized Convolution (Table 2).

## 4 Proposed Framework

This framework helps in getting focused image in a scene. The S-R Model is used to obtain HR image from various LR images. Motion Estimation and Reconstruction algorithms are used to reconstruct HR images. This algorithm incorporates the resolution on the image content by utilizing the high spatial accuracy on those section of the image where it is necessary, i.e., at the boundaries of moving objects. These concepts are finally used in SFF for getting the focused resolution image.

LR images are obtained using S-R Model. These images are used to convert into HR images by deblurring the image and sharpening the boundaries of the images.

The motion estimation algorithms and reconstruction algorithm are utilized and work in intermediate stage after providing scaling and rotation parameters by the user's necessity. It deals with LR image and works on it. In the final stage, the HR image is processed and reconstructed according to the user's need and provides the result.

These stages are further explained in given bellow figures.

Further this model is connected with WiMAX network for accessing the data in remote areas. This is possible for transferring real time data over the PHY layer of the network (Fig. 4).

In Fig. 5a, b different LR images are obtained using the S-R Model which is used for reconstruction of HR images. Figure 6a, b is the intermediate stage between LR and HR images which takes the parameters from the user according to

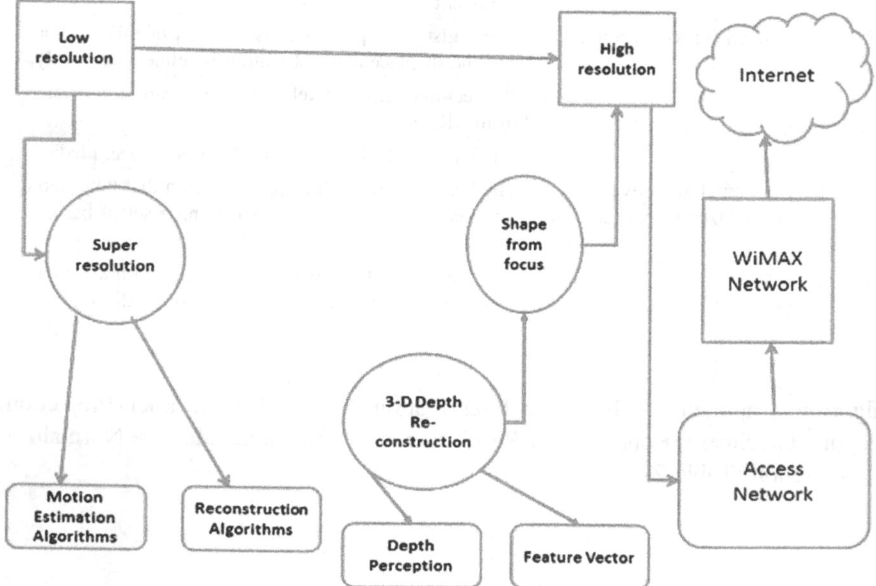


**Fig. 4** Proposed framework to reconstruct HR images from LR images in WiMAX network

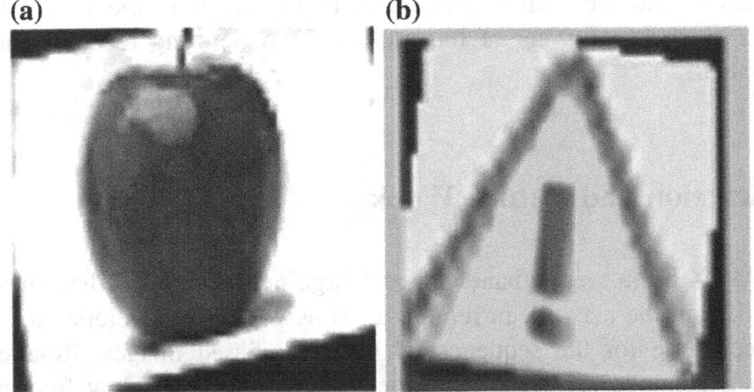


**Fig. 5** **a** and **b** are the LR images

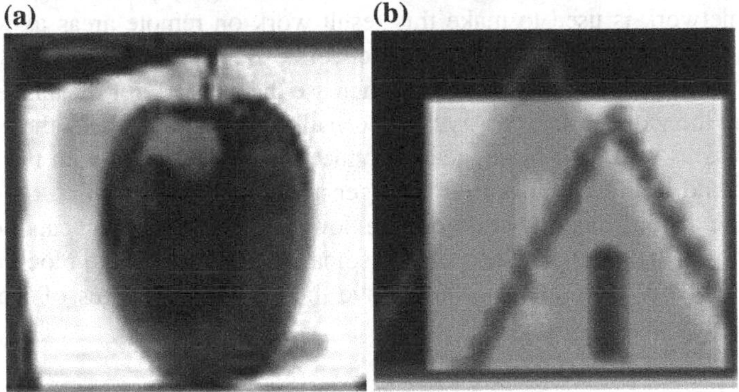


**Fig. 6** **a** and **b** are the rotated and shifted images according to the user parameters

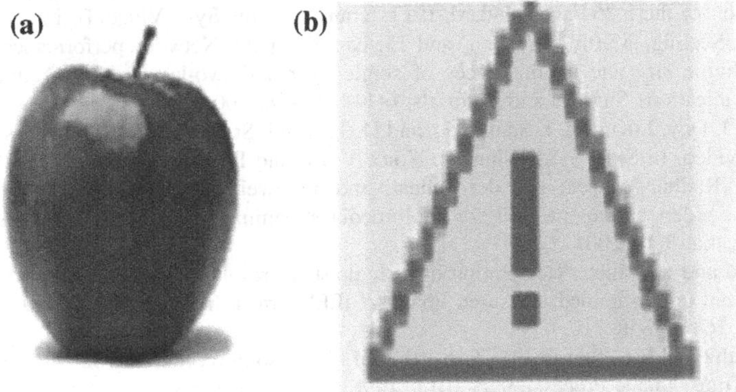


**Fig. 7** **a** and **b** are the corresponding HR images

their necessity and are further processed. Figure 7a, b is the final HR image obtained after the processing of the motion estimation and reconstruction algorithms applied.

## 5 Conclusion and Future Work

Since WiMAX has a greater bandwith and large network area, it becomes easy to view the data in the network in remote areas as well. We developed a frequency domain technique for the sequence of LR and aliased images. Rotational and translational methodologies are performed depending on the low frequency and aliasing less portion of the image. This process is used on super resolution process to develop a HR image from the sequence of aliasing and LR images. The algorithms used are compared with rotation and scaling methods in computational and practical enhancement. Both prove to be valid and highly precise algorithm. The WiMAX network is used to make this result work on remote areas as well. This result can be accessed anywhere in the network.

The future scope for this project is to increase the network for the analysis of the images and the video in such a way that it can allow the user to check the data in the remote areas as well. The basic idea of connection of the model with the WiMAX network could be further increased to larger network.

Also the reconstruction work on the low resolution images can be further improved by the applying more motion estimation techniques like block matching algorithms where different algorithms have their own advantages of making the image a highly resolved image.

## References

1. Heiko Schwarz, Detlev Marpe, and Thomas Wiegand. Overview of the scalable video coding extension of the h.264/avc standard. IEEE Trans. Circuits Syst. Video Techn., 17(9), 2007.
2. Patrick Seeling, Martin Reisslein, and Beshan Kulapala. Network performance evaluation using frame size and quality traces of single-layer and two-layer video: A tutorial. IEEE Communications Surveys and Tutorials, 6(1–4):58–78, 2004.
3. S. Sen, J. Dey, J. Kurose, J. Stankovic, and D. Towsley. Streaming CBR transmission of VBR stored video. In SPIE Symposium on Voice Video and Data Communications, 1997.
4. Hayder Radha, Mihaela van der Schaar, and Yingwei Chen. The MPEG-4 fine-grained scalable video coding method for multimedia streaming over IP. IEEE Transactions on Multimedia, 3(1), 2001.
5. M. Elad and A. Feuer, "Restoration of a single super resolution image from several blurred, noisy, and undersampled measured images," IEEE Trans. Image Process., 1997, vol. 6, no. 12, pp. 1646–1658,.
6. R. Schultz and R.L. Stevenson, "Extraction of high resolution frames from video sequences," IEEE Trans. Image Process., 996–1011, 1996, vol. 5, no. 6, pp.

7. D. Rajan and S. Chaudhuri, "Simultaneous estimation of super-resolved scene and depth map from low resolution defocused observations," IEEE Trans. Pattern Anal. Mach. Intell., vol. 25, no. 9, pp. 1102–1117, 2003.
8. D. Scharstein and R. Szeliski. A taxonomy and evaluation of dense two-frame stereo correspondence algorithms. International Journal of Computer Vision (IJCV), 47(1):7–42, 2002.
9. D.A. Forsyth and J. Ponce. Computer Vision: A Modern Approach. Prentice Hall, 2003.
10. O. Nemethova, M. Ries, M. Zavodsky, and M. Rupp. PSNR-based estimation of subjective time-variant video quality for mobiles. In MESAQIN, 2006.
11. S. Das and N. Ahuja. Performance analysis of stereo, vergence, and focus as depth cues for active vision. IEEE Trans Pattern Analysis and Machine Intelligence (PAMI), 17(12):1213–1219, 1995.
12. Ping Li, W.S. Lin, S. Rahardja, X. Lin, X. K. Yang, and Z. G. Li. Geometrically determining leaky bucket parameters for video streaming over constant bit-rate channels. In IEEE International Conference on Acoustics, Speech, and Signal Processing (ICASSP), 2004.
13. M. Quartulli and M. Datcu. Bayesian model based city reconstruction from high resolution ISAR data. In IEEE/ISPRS Joint Workshop Remote Sensing and Data Fusion over Urban Areas, 2001.
14. N. Cornelis, B. Leibe, K. Cornelis, and L. Van Gool. 3d city modeling using cognitive loops. In Video Proceedings of CVPR (VPCVPR), 2006.
15. C. Hentschel, et al., Scalable video algorithms and quality-of-service resource management for consumer terminals, in International Conference on Consumer Electronics (ICCE), Los Angeles, CA, June 2001, pp. 338–339.
16. C. Hentschel, R. Braspenning and M. Gabrani, Scalable algorithms for media processing, in International Conference on Image Processing (ICIP), Thessaloniki, Greece, October 2001, pp. 342–345.
17. Mark Kalman, Eckehard G. Steinbach, and Bernd Girod. Adaptive media playout for low-delay video streaming over error-prone channels. IEEE Trans. Circuits Syst. Video Techn., 14(6), 2004.
18. J. Klaue, B. Rathke, and A. Wolisz. Evalvid - A framework for video transmission and quality evaluation". In International Conference on Modelling Techniques and Tools for Computer Performance Evaluation, 2003.

# A Reliable Tactic for Detecting Black Hole Attack in Vehicular Ad Hoc Networks

**Isha Dhyani, Neha Goel, Gaurav Sharma and Bhawna Mallick**

**Abstract** Nowadays, Vehicular Ad Hoc networks is one of the most emerging and favorable technology which succor in governing day-to-day road traffic in roads. The main disquiet of VANETs is focused on providing security to moving nodes in the vehicular network, so that possibility of accidents, traffic jams or any other hindrance in communication among different vehicles will get reduced. Black hole is most popular security attack that sends false reply message to the source by advertising itself as having optimal route toward destination. This paper focused on black hole problem. Here, a reliable mechanism is proposed for averting black hole attack in VANET by unicasting data packet to vehicles. To enhance security, trust factor technique is used which detect routing misbehavior and gives surety of relaying data packets to the destination. This elucidation defends against black hole attack and simulation on Ns-2 will sustain its efficiency and reliability.

**Keywords** Vehicular Ad hoc networks · Black hole attack · Security · Attack

## 1 Introduction

The wide advancement of wireless communication technologies have transformed human standards of living by providing the most flexibility and accessibility ever in handling Internet services and numerous kinds of personal communication

I. Dhyani (✉) · N. Goel · B. Mallick
Galgotia College of Engineering and Technology, Greater Noida, India
e-mail: isha.dhyani25@gmail.com

N. Goel
e-mail: goelneha3012@gmail.com

B. Mallick
e-mail: bhavna.mallick@galgotiacollege.edu

G. Sharma
Amity University, Noida, India
e-mail: sharmagaurav@ieee.org



S.K. Bhatia et al. (eds.), *Advances in Computer and Computational Sciences*,
Advances in Intelligent Systems and Computing 553,
DOI 10.1007/978-981-10-3770-2_31

applications. In recent times, car developers and telecommunication companies have unified cars with the technology that allows car drivers or passengers to communicate with each other as well as with the roadside units (RSU), situated at certain critical division of the road. The RSU can be at any intersection point, traffic light or any existing stop sign placed on road. Vehicular Ad Hoc network (VANET) consists of two type of applications. The first one is known as security application; it gives real-time report of road conditions to the driver. The real-time report entails of information about congestion, collision warning or emergency information. The second one comprises of nonsafety application. This application makes the journey on road more enjoyable by providing information about hotels, music, video streaming, etc.

Vehicular Ad hoc network uses wireless communication channel [1] for data transfer. Therefore, privacy, reliability, and security are major concerns needed to be taken care. As information is transmitted through wireless medium, one must be aware that whatsoever message or information that is being broadcasted by sender will be received by all the nodes present in its communication range. Vehicular Ad hoc networks is a part of mobile Ad hoc networks, that means each and every node can move freely inside the network and can communicate with each other or with roadside units. The vehicles communicating among themselves are known as On-Board unit (OBU) and units that are stationary nodes or fixed at roadside regions are known as Roadside unit (RSU). Intelligent vehicular Ad hoc Networking or VANET carries an erudite approach of using vehicular networking. The main delinquent in vehicular networks is the fast movement of nodes as well as recurrent changes in network topology. Due to its highly dynamic nature of nodes, it leads to provide various challenges in network security and privacy.

The enthralling features of VANETs will surely experience higher risks if such networks do not possess security into positioning nodes. For instance if the security messages between the nodes is altered or tampered due to any circumstances, serious consequences might be possible which can lead to small injuries or even deaths. To improve the efficiency of VANET, many different routing protocols have been proposed. For securing communication, one should understand different security attacks and their consequences in VANETs. Sybil attack, Gray hole attack, Wormhole attack, Black hole attack, Flooding attack, Denial of service (DoS), Impersonation attacks, Bogus Information attack, etc. [2, 3], are kind of attacks which reduces the performance as well as efficiency of the network.

Due to dynamic behavior of nodes in vehicular networks, security issues become the vital part of VANET. Several attacks take place on vehicular networks, but the main disquiet in VANET is Black hole attack. A black hole is a vindictive node that replies to the route requests messages, averring it has the shortest path to the destination node and dropping all the incoming packets from the source node or any other intermediate nodes [4, 5]. Cooperative black hole attack includes more than one black hole nodes. These nodes work in cooperation with each other to launch the attack [6].

The rest of the paper is organized as follows: In Sect. 2, the overview of existing related work for detecting black hole attack is covered. Section 3 explains proposed mechanism to ensure secure communication of data between sender and receiver. Section 4, introduces the impairment caused by black hole attack and also presents the amendment in network performance by proposed technique in the form of simulation results and graphs. Section 5 concludes the paper and provides future work.

## 2 Related Work

This section describes several solutions provided by various researchers. Almutairi et al. [7] presented a solution to detect black hole node in a vehicular network. Every node is provided with a trust routing table (TRT) [8] which includes all the reliable nodes in the network. Each sender node periodically updates its trust value according to the packets it receives from destination node. The trust value is joined with progress to select the best neighbor. Here the author uses greedy geographic routing protocol, i.e., position-based routing for detecting single black hole node. Tamilslvan and Sankaranarayanan [9] has enhanced original AODV by setting timer in the RimerExpiredTable for collecting request from other nodes after getting the first RREQ. It also stores destination number and received time in CCRT (Collect Route Reply Table), by counting timeout value on the basis of first RREQ and judges the validity of route through threshold value. Here End-to-End delay may increase when malicious node goes away from source node.

Marti et al. [10] also presented a solution for detecting malicious node by watchdog technique. In this technique, neighboring nodes keep records of overhearing transmission of data packets sent and received by their neighboring nodes. If number of failure increases outside threshold value, message will be sent to source node. This approach is unable to find malicious node at time of collisions. Dhingra and Arora [11] offered a solution to detect and eliminate the black hole attack. In this approach, the header of AODV is modified by adding parent vehicle. This parent vehicle field consist address of earlier source of packet. The route redundancy approach is used for eliminating route recovery that is link failure sessions. It reduces E2E delay but routing overhead increases.

Aware and Bhandari [12] proposed a solution to prevent black hole attack on AODV by using hash function. In this, the first optimal path is rejected and the second optimal path is preferred. To maintain data integrity, here hash function is used which is used to detect the malicious node in the second optimal path. The hash value of the message is calculated at the source node as (SHA-ONE) and the received message will be calculated separately at the destination node as (SHA-TWO). If both hash values will be same, then the message received is not altered otherwise data packet error (DPE) message will be broadcast in the network. All the information will be saved in routing table. Here it takes additional processing time and E2E delay increased as additional time is required to find secure route.

Rani and Kumar [13] presented a solution to remove black hole attack by using AOMDV routing protocol. By calculating legitimacy of the node, this approach finds node disjoint multipath which detect black hole node and minimize the overhead of the specific node. Yang et al. [14] put forward a new approach namely, Anti-Blackhole Mechanism (ABM), in which, the number of RREQs and RREPs are compared. It consists of two tables: RQ and SN table. The RQ table stores RREQ message like sequence number, source, and destination ID whereas SN table calculate suspicious value suggested by ABM. If the value exceeded beyond threshold, it will mark it as blackhole. Here packet loss rate increased upto 92.40 and 10.05% (when threshold is 5) for 9 IDS nodes.

A mechanism is proposed by Kozma and Lazos [8] in which trust-based multipath AOMDV routing is merged with a soft-encryption methodology. This approach is based on message encryption using XOR operations, second is message routing which uses multiple path node disjoint AOMDV routing protocol with different trust-based values. Last step is message decryption in which message is decrypted at destination end. Huang et al. [15] presented a Resource Efficient Accountability (REAct) scheme which involves audit, search, and identification process. Whenever a packet is dropped, a feedback is sent to source node that chooses an audit node. The audit node uses bloom filter to get a behavior proof. This scheme reduces routing overhead although identification delay is increased. The major limitation of this approach is lack of cooperative black hole detection.

## 3 Our Contribution

Through this paper, we would like to expound a reliable tactic for detecting black hole attack in VANET. Here we proposed a reliable defense method to protect against black hole attack which will help in finding more secure path toward destination. We mentioned existing related work in the above section that defines black hole problems and their solution. The main problem while averting black hole attack from network is due to increase in E2E delay and routing overhead. Alert messages cause additional overhead in network. Therefore, our reliable proposed technique utilizes control message of AOMDV to notify any unsuspected behavior to other nodes which will help in minimizing routing overhead. As well as the algorithm used helps in reducing E2E delay and gives surety of data delivery to the destination.

## 4 Proposed Work

In our proposed work, Data packets handling technique of normal AOMDV [14] is enhanced for detecting black hole in vehicular network. Each node has its own black_list table which contains all information of black hole nodes. Source node will broadcast RREQ message to all nodes in networks that are not in black_list table. On receiving RREQ message, nodes will unicast [16] RREP to the source node. The node having first highest and second highest sequence destination number will be chosen for sending data packets. If the first node is black hole node [17] then data will be transmitted through second node. For reducing routing overhead in AOMDV flare packets are used which will notify all other nodes when any mischief is detected and also have last hop information. This method eliminates extra control packets overhead making algorithm more efficient. For trust factor, the threshold value [9, 18] is calculated and compared with trace value which is based on number of packets forward and dropped by the node. Although, we have used single malicious node in the proposed algorithm to analyze network performance but multiple malicious nodes can also be detected by adding little modification (Fig. 1).

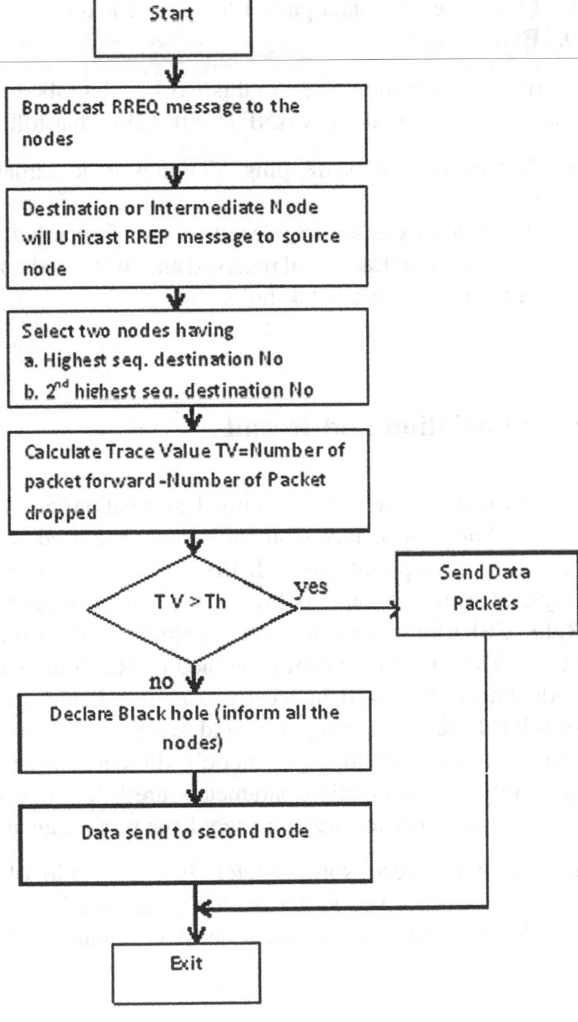


**Fig. 1** Proposed algorithm flow chart

### *4.1 Algorithm*

1. Broadcast RREQ message to those node that are not in black_list table. If node sending RREP is in the black_list Table then block that path.
2. Destination or Intermediate node unicast RREP message to the source node.
3. Select two nodes that having highest sequence destination number and second highest sequence destination number.
4. For trust factor
5. Calculate Trace Value (TV).
6. TV = Number of packet Forward—Number of packet dropped.
7. Assume the threshold value (Th), if Trace value < Threshold value then declare that node as black hole node, store it black_list and flare packets will be used to notify this malicious node to all other nodes.
8. Otherwise send data packet to second node.
9. Exit

Basic Assumption taken in this work is described below. Here it is assumed that each and every node in VANET will follow the following items.

- At the time of bootstrapping, all nodes are legitimate nodes (that is no black hole present).
- All nodes have already received and forwarded few packets in the network. (otherwise subtraction of received and forwarded nodes will be zero, thus TV = 0).
- There is only one black hole.

## 5 Simulation and Results

In this section, we have described simulation scenario and report of simulation results. For performance evaluation, we have used Network Simulator 2 (Ns-2.35) [19] in our proposed work. IEEE 802.11p and wireless channel [20] is used at physical layer and data link layer. To set up simulation traffic source Continuous Bit Rate (CBR) for generating node randomly and ensures its connectivity with other nodes. The node movement is defined by Random Waypoint Model [16], here each node lingers for fixed interval pause time before proceeding to new location of simulated area of rectangular field within (1000 × 1000 m2). Due to frequent change in network topology node's direction, speeds, and locations are chosen randomly. The simulation parameters are listed in Table 1.

Following metrics are considered for performance evaluation.

- Packet Delivery Ratio (PDR): It is the ratio of number of data packets or messages received at destination to the number of data packet send by source. That is, PDR define as the successful number of data delivered at the endpoint.

**Table 1** Simulation parameters

| Parameters | Values |
|---|---|
| Routing protocol | AOMDV, BlackholeAOMDV |
| Simulation time (sec) | 100 |
| Simulation area (m) | 1000 × 1000 |
| Simulation model | TwoRayGround |
| Mac type | 802_11p |
| Traffic type | CBR |
| Link layer type | LL |
| Antenna | OmniAntenna |
| Packet size | 512 bytes |
| Pause time (sec) | 1–12 |
| Queue length | 50 |
| Number of nodes | 25, 50, 75, 100 |

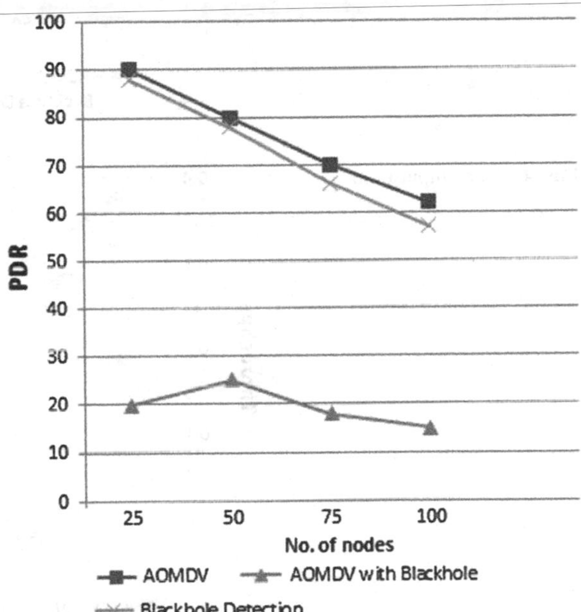


**Fig. 2** Packet delivery ratio

$$\text{PDR} = \frac{\sum (\textit{number of data packets received at destination})}{\sum (\textit{number of data packets send by source})}$$

The result of PDR is given in Fig. 2, here as we see number of data delivered at the endpoint or destination is greater than AOMDV with Black hole node. So, our method delivers high rates of data packets by eluding black hole node.

- Aggregate End-to-End Delay (E2E Delay): It is the ratio of aggregate delay obtain while transmitting data packets/messages to the destination node to the

Fig. 3 End-to-End delay

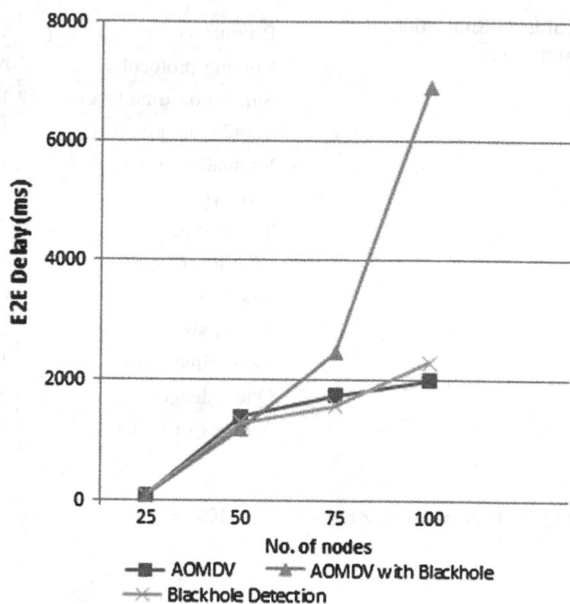


Fig. 4 Throughput ratio

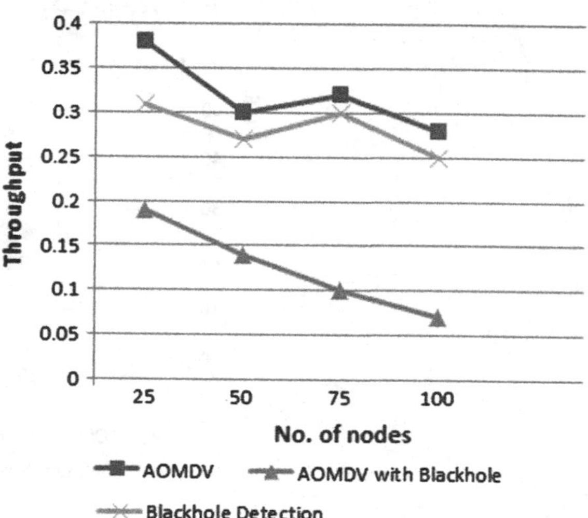


total number of data transmitted in the network. The average delay is generally measured in seconds. In given Fig. 3, the graph is illustrated between number of nodes and end–to-end delay.

- Routing overhead: It is defined as the routing packets transmitted during entire session.

**Fig. 5** Routing overhead

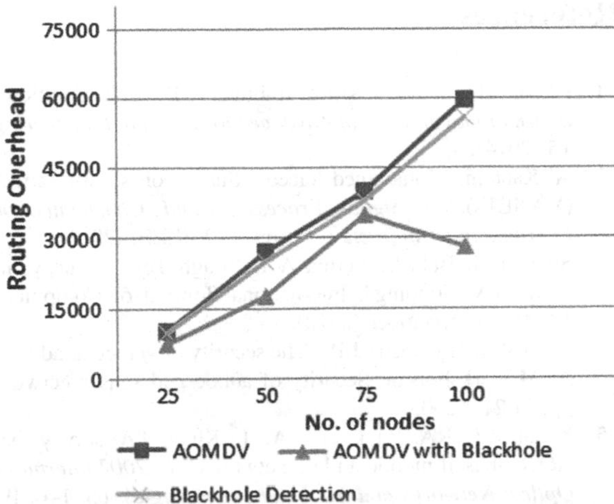


- Throughput: It is the ratio of successful data transferred to its final location per unit time. It is represented as T and its unit is packets/sec. The value of throughput depends upon two main factors that are limited bandwidth and limited power (Figs. 4 and 5).

$$T = \frac{Number\ of\ data\ trasmitted\ at\ destination}{total\ time\ taken}$$

## 6 Conclusion and Future Scope

VANET is emerging research topic. Different security algorithms have been evaluated but there is a strong need of reliable and efficient security measures. The performance of our proposed scheme for detecting black hole in AOMDV has been evaluated and compared with simple AOMDV and AOMDV with black hole detection under various numbers of nodes and different parameters. It is observed that by detecting black hole in AOMDV gives favorable result in the network. In future, existing algorithm can be applied in other VANET routing protocols as well as can be enhanced for cooperative Black hole attacks. This work can be extended to prevent black hole attacks in vehicular network using various encryption techniques.

## References

1. Khan, Mohammad Arifin Rahman. "Possible solution for traffic in roaming system." *International Journal of Advanced Research in Engineering and Applied Sciences* 3, no. 8: 1–15 (2014).
2. Al-Kahtani, Mohammed Saeed. "Survey on security attacks in Vehicular Ad hoc Networks (VANETs)." In *Signal Processing and Communication Systems (ICSPCS), 2012 6th International Conference on*, pp. 1–9. IEEE, (2012).
3. Sharma G, Bala S, Verma A K, Singh Tej, "Security in Wireless Sensor Networks using Frequency Hopping", International Journal of Computer Application(0975-8887), Volume 12- No. 6 December (2010).
4. Raya M. and Hubaux J-P, "The security of vehicular ad hoc networks", Proceedings of the 3rd ACM workshop on Security of adhoc and sensor networks. New York, NY, USA: ACM,, pp. 11.21, (2005).
5. Mahmood, RA Raja, and A. I. Khan. "A survey on detecting black hole attack in AODV-based mobile ad hoc networks." In *2007 International Symposium on High Capacity Optical Networks and Enabling Technologies*, pp. 1–6. IEEE, (2007).
6. Ramaswamy S.,Huirong Fu, Manohar S. "Prevention of Cooperative Black hole attack in wireless Ad-Hoc Network", *International Conference on Wireless Networks*, pp. 570–575, June (2003).
7. Almutairi H, Chelloug S, Alqarni H, Aljaber R, Alshehri A, Alotaish D, "A new Black Hole Detection Scheme for Vanets" ACM, September (2014).
8. Kozma W, Lazos L (2009) REAct: Resource-Efficient Accountability for Node Misbehavior in Ad Hoc Networks based on Random Audits. Paper presented at *the Second ACM conference on Wireless Network Security*, Zurich Switzerland, 16–18 March (2009).
9. Tamilselvan, L., Sankaranarayanan, V., Prevention of Blackhole Attack in MANET. Paper presented at the 2nd International Conference on Wireless Broadband and Ultra Wideband Communications, Sydney, Australia, 27–30 August(2007).
10. Marti S, Giuli TJ, Lai K, Baker M: Mitigating Routing Misbehavior in Mobile Ad Hoc Networks. Paper presented at the 6th *annual International Conference on Mobile Computing and Networking*, Boston, Massachusetts, 6–11 August (2000).
11. Dhingra S, Arora K, "Detection and Prevention of Black Hole Attack in VANET" International Journal of Computer Science & Management studies, Vol. 13, June (2013).
12. Aware A.A, Bhandari K, "Prevention of Black hole Attack on AODV in MANET using Hash Function", 3rd IEEE International Conference on Reliability, Infocom Technologies and Optimization (ICRITO 2014), pp 388–393, October (2014).
13. Rani J, Kumar N "Improving AOMDV Protocol for Black Hole Detection in Mobile Ad hoc Network" *IEEE International Conference on Control, Computing, Communication and Materials (ICCCCM)*, (2013).
14. Su, M.-Y. Prevention of selective black hole attacks on mobile ad hoc networks through intrusion detection systems. Computer Communications, 34(1) 107–117, January (2011).
15. Huang J-W, Woungang I, Chao H-C, Obaidat M-S, Chi T-Y, Dhurandher S, "Multi-Path Trust-Based Secure AOMDV Routing in Ad Hoc Networks" IEEE Globecom, (2011).
16. Ros, Francisco J., and. Ruiz M. P. "Implement ing a new manet unicast rout ing protocol in ns2." Sun Microsystems Inc (2004).
17. Khamayseh, Y., Bader, A., Mardini, W., BaniYasein, M.,"A New Protocol for Detecting Black hole Nodes in Ad Hoc Networks", "*International journal communication network & information security*" April (2011).
18. Chauhan, R. K. "An assessment based approach to detect black hole attack in MANET." In *Computing, Communication & Automation (ICCCA), 2015 International Conference on*, pp. 552–557. IEEE, (2015).

19. Beraldi, R., Querzoni, L., and Baldoni, R., "A Hint-Based Probabilistic Protocol for Unicast Communications in MANETs", in *Proc. of IEEE 11th International Conference on Parallel and Distributed Systems (ICPADS'05)*, (2005).
20. Khan, Mohammad Arifin Rahman. "Analysis the channel allocation for removing the traffic problems from the roaming systems." *International Journal of Advanced Research in Engineering and Applied Sciences* 3.7,74–86 (2014).

# On Solutions to Vehicle Routing Problems Using Swarm Optimization Techniques: A Review

**Ashima Gupta and Sanjay Saini**

**Abstract** Vehicle Routing Problem (VRP) is among the intensively studied problem in the field of operations research. The literature of VRP has spread to dozens of variants that are studied till now, which makes the problem more complex. Due to its complexity and several real-time constraints, it is difficult to find optimal solutions for VRP models. In recent decades, swarm optimization techniques have emerged as promising solution to solve these problems optimally. The purpose of this research is to develop structural classification of different domains and attributes of VRP solved using swarm techniques. The findings of the study show the most studied attributes, capacitated VRP, time windows VRP, objective function with cost minimization and the least studied attributes, maximization objective function. The VRP literature is summarized in a manner that provides a clear view to identify future research directions.

**Keywords** Vehicle routing problem · Meta-heuristic · Swarm optimization

## 1 Introduction

Vehicle routing problem is a class of combinatorial optimization problems in which a homogeneous fleet of vehicles must service a set of customers while satisfying certain constraints at minimum transit cost. Its relevance stems from the real-world applications in domains like travel, transportation, logistics and distribution [38]. All these domains are backbone of today's business and have direct impact on the cost of goods and so on modern economy. The importance of this field has

---

A. Gupta (✉) · S. Saini
Department of Physics & Computer Science,
Dayalbagh Educational Institute, Agra 282005, India
e-mail: ashimagupta17@gmail.com

S. Saini
e-mail: sanjay.s.saini@gmail.com



S.K. Bhatia et al. (eds.), *Advances in Computer and Computational Sciences*,
Advances in Intelligent Systems and Computing 553,
DOI 10.1007/978-981-10-3770-2_32

motivated us to look for all possible alternatives to optimize the overall length of transportation.

Because of its complexity as being NP-hard problems, VRPs are solved by many heuristics, such as exact methods, tabu search, simulated annealing and nature inspired techniques, etc. In the recent year's swarm techniques are gaining much attention of the researchers to find optimal solutions for VRP variants. Over the last three decades, we can find large number of academic publications, case studies, taxonomies and surveys on different domains of VRP solved by swarm optimization.

According to the VRP literatures, CVRP, VRPTW, VRPPD are the most studied attributes and have practical importance in real life applications. Hence, this paper mainly focuses on these three variants of VRP for both single and multi depot problems. The remainder article is organized as follows: Sect. 2 highlights some of the variants of VRP. Section 3 gives the formulation for CVRP, VRPTW and VRPPD. Section 4 discusses the swarm techniques that are used to solve complex problems. Section 5 gives the literature on VRP variants, solved using swarm algorithms. Section 6 is the conclusion and future research directions.

## 2 Variants of VRP

According to literature there exist numerous variants of VRP with the aim to find minimum travel cost of a set of routes. Some of the variants are described as follows:

*Capacitated VRP (CVRP)* is the most elementary version of VRP, where the vehicles have limited capacity (weight, no. of pallets, which they cannot exceed) [38]. Hence, in CVRP the sum of demands of clients should not exceed the vehicle capacity.

*Distance Constrained VRP (DVRP)* is the variant where length is the main constraint. A non-negative length is associated to each arc and it is important that for each route the total length of the arcs should be less than the maximum tour length.

*Multi Depot VRP (MDVRP)* is the case of more than one depot, like an organization has multiple service points. To fulfil customers' demands, grouping is performed and customers are clustered around depots on the basis of distance between them.

*Vehicle Routing Problem with Time Window (VRPTW)* adds an additional time interval constraint which is associated with each customer [33] and the service has to be done within that time slot.

*VRP with Pickup and Delivery (VRPPD)* is the variant in which the products are delivered both ways, i.e. from depot to customers and from customers to depot [29].

*Vehicle Routing Problem with Backhauls (VRPB)* [31] deals with the services of two types of customers: (a) Linehaul customers, requiring delivery of products. (b) Backhaul customers, for which products have to be picked up.

*Periodic Vehicle Routing Problem (PVRP)* permits to extend the basic service period to N-days. Thus, the vehicle might not return to depot on the same day it departs. Also, customers may be visited more than once in this extending period.

*Stochastic VRP (SVRP)* [17] considers one or more variables that are unknown during tour, such as customers' presence or demand and the service or travel time [15].

In addition to the above variants, many other side constraints are also added to the problem structure; hence, the VRP class of problems is very large when considering real time constraints and objectives.

## 3 CVRP, VRPTW and VRPPD Formulation

The terms and notations for CVRP, VRPTW and VRPPD can be formally defined as:

*Graph:* an undirected graph $G = (V, E)$, $V = \{v_0, v1, \ldots, v_n\}$ is the set of vertices and $E$ is the set of all possible connections [38].
*Depot*: in the graph, vertex $v_0$ is the depot from where routes starts and terminates.
*Customers*: the problem is defined for $n$ customers represented by vertices $v_1, v_2 \ldots v_n$. Each customer has a non-negative deterministic demand $q$.
*Vehicles*: vehicles can serve many customers provided the sum of demands for each customer should not exceed the vehicle capacity $Q$. Also, the vehicle must start and end at the same depot after servicing the customers.
*Travelling Cost*: $C_{ij}$ represents the travelling cost between customers $i$ and $j$. It is generally calculated using Euclidian distance between the customers.
*Route*: is a sequence of nodes, starting and ending at the depot. The length of the each route $r$ depends upon the number of customers.
*CVRP*: it ensures that the sum of customers' demand $q_i$ cannot exceed the vehicle capacity $Q$. Similarly, the total route distance $d_{ij}$ of a vehicle cannot exceed its route length limit. It also ensures that each customer can be served by only one vehicle and the continuity at each node for every vehicle should be maintained.
*Time Window*: a time interval $[e_i, l_i]$, is associated to each client, where $e_i$ is the earliest arrival or opening time, $l_i$ is the latest arrival or closing time. A vehicle must arrive at the customer before the closing time and if it reaches before opening time then it will wait until the service starts. Hence, the time instance for a route includes: (i) start time from the depot, (ii) total travel time and (iii) the service time for each client.
*Pick up and Delivery*: a customer has two properties $d$ and $p$, i.e. required demand of goods to be delivered and to be picked up. It is always assumed that the delivery must be performed before pickup so that capacity constraint should not violate. Therefore, the total load on the vehicle before arriving to next location can be calculated as: [*Initial load – Demands that are delivered + Demands that are already picked up*].

**Table 1** Some frequently encountered attributes of VRP

| | | | |
|---|---|---|---|
| 1. Stochastic | 3. Operation | 4.1.2. Distance | 6. Applied Methods |
| 1.1. Customer | 3.1. CVRP | 4.1.3. Vehicle | 6.1 Exact methods |
| 1.2. Demand | 3.2. VRPTW | 4.1.4. Time | 6.2 Heuristics |
| 1.3. Service time | 3.3. VRPPD | 4.2. Maximize | 6.3 Meta-heuristics |
| 1.4. Travel time | 3.4. Others | 4.3. Others | 6.4 Hybrid methods |
| 2. Depots | 4. Objective | 5. Objective Function | |
| 2.1. Single | 4.1. Minimize | 5.1. Single objective | |
| 2.2. Multiple | 4.1.1. Cost | 5.2. Multi objective | |

- *Objectives*:
  - Minimize the total cost of travelling.
  - Minimize the total number of vehicles.
  - Minimize the distance travelled by all the vehicles.
- *Constraints*:
  - Each customer should be visited only once.
  - Each vehicle must start and end at the depot.
  - Total demand of clients of any route does not exceed the vehicle capacity.
  - Customers should be serviced within the specified time interval.
  - Customers that have delivery demands must service before the customers with pickup demands.

In addition to these constraints several other real life constraints: fixed number of vehicles, customer priority, cleaning [20], working hours, drivers' regulations, road restrictions, production and distribution planning, etc., which have to be satisfied. Table 1 below shows some of the attributes repeatedly encountered in VRP literature.

## 4 Swarm Intelligence Techniques

Beni and Wang were first to introduce swarm intelligence in the context of cellular robotics. On the set of probabilistic meta-heuristics swarm intelligence (SI) is attracting much attentions [13]. The SI algorithms can be interpreted as "distributed problem optimization methods" inspired by the collective behaviour of decentralized, and self organized agents like ants, bees, birds, etc. Swarm intelligence works on two basic principles: self organization and stigmergy (simulation by work) [26]. The table below shows some of the popular swarming algorithms mentioned in the literature.

Algorithms like ACO, ABC, PSO, BFO, BATCO are inspired from animals' collective behaviour, whereas, IWD imitate natural rivers to find optimal paths to their destination. Though there are number of swarm based algorithms are developed, but the first four algorithms in the Table 2 are quite mature, so we concentrate

**Table 2** Swarm inspired algorithms

| SI algorithm | Pioneer | Year | Motivation |
|---|---|---|---|
| ACO (ant colony optimization) | M. Dorigo | 1992 | Ant colonies |
| PSO (particle swarm optimization) | J. Kennedy | 1995 | Group of birds |
| ABC (artificial bee colony) | D. Karaboga | 2005 | Honey bees |
| IWD (intelligent water drops) | Shah Hosseini | 2007 | Water drops |
| BFO (bacteria foraging optimization) | K.M. Passino | 2002 | E. coli, M. xanthus |
| WCO (wasp colony optimization) | P. Pinto | 2005 | Wasp |
| TCO (termite colony optimization) | M. Roth, S. Wicker | 2003 | Termite |
| BATCO (bat colony optimization) | Xin-She Yang | 2010 | Bat |

our study on these algorithms. Rest of the paper will review these four algorithms for their applications in solving vehicle routing problems.

## 5 Literature Review

Scientists and researchers all over the world have shown great interest in VRPs. Several exact methods and meta-heuristic approaches have been used till date, to get better solutions for the problem. Therefore, the VRP domain and its solution methods has very huge background. In this paper the literature is categorized by ACO, PSO, ABC and IWD algorithms used for solving CVRP, VRPTW and VRPPD variants.

Gambardella et al. [14] proposed a Multi Ant Colony System *(MACS-VRPTW)* algorithm to solve VRP with hard time windows constraints. In their approach two artificial ant colonies works for multiple objectives. One colony deals with tour optimization and the other minimizes the distance. Sandhaya and Katiyar [33] presented an Ant Colony System *(ACS)* to solve VRPTW problem. Experiments are performed on Solomon benchmarks and the results are compared with the best known results available. MACS-VRPTW and ACS algorithms are the best known ACO algorithms, in the literature.

Iredi et al. [18], proposed two methods to solve bi-criterion optimization problems, one is for cooperation between the colonies and the other for comparing them with a multi-start ant. Reimann et al. [29], presented an ant system to the VRP with backhauls and time windows (VRPBTW). They use an insertion procedure to construct solutions and compare the results with heuristic approach made for the problem. Rizzoli et al. [30] applied ACO to VRPTW, Dynamic VRP, VRPPD and to two real-world, industrial scale applications. Manfrin [23] applied ACO on CVRP and compared the results of five meta-heuristics: ACO, evolutionary computation, Iterated local search, tabu search, and simulated annealing. Bell and McMullen [7], applied ACO to a set of VRPs and experimental results finds the solutions within 1% of known optimal solutions and also found that the multiple ant colonies are competitive solution approach for large and complex problems. Favaretto et al. [12] implemented an ACS based algorithm for the VRP with multiple time windows on a set of benchmark problems. They [27] presented a multiple-ACO on VRP constraints: heterogeneous fleet of vehicles, periodic constraint and limited duration subtours. Their algorithm performed better than nearest neighbour and tabu search algorithms.

Donati et al. [11] presented a time dependent model based on MACS-VRPTW with an enhanced local search method to perform efficiently in terms of optimality of solution and computational times. Zhen and Zhang [42], presented a hybrid ant colony system *(DSACA-VRPTW)* to solve VRPs and tested it on Solomons' instances. The results show that the DSACA is efficient in finding solutions for VRPTW. Ting and Chen [36] presented a hybrid approach, by combining the strengths of MACS and simulated annealing in order to improve the solution quality. Their algorithm is tested on several benchmark problems and results show that the algorithm is effective in solving the MDVRPTW as compared to other algorithms.

Zhang and Tang [41], presented *SS_ACO,* a novel hybrid approach with the solution construction method of ACO and scatter search (SS). In which, ACO and greedy method generates initial solutions and then in the SS framework, new solutions are generated by updating the pheromone. Neighbourhood search is also applied to improve the solution. The results found the algorithm competitive as compared to the best existing algorithms. Saravanan and Sundararama [32] proposed an ACO based time constraint VRP, where services have to be completed within the permissible time. They compared the results with shortest path heuristic and found it more efficient.

Yu and Zhen-Yang [40] proposed an improved ACO for periodic VRPTW and tested it on the benchmark instances. Obtained solutions are found better than the existing solutions and prove IACO as an effective tool to solve PVRPTW. Bin et al. applied [8], a new ant-weight strategy to update the pheromone and used mutation operator (GA) to explore the solution space. Bouhafs et al. [9] combined ACS with savings algorithm and local search to get better results for CVRP. Ruinelli [31] combines an ACS, with column generation method and a general purpose linear problem solver, to solve VRPPD. Balseiro et al. [6] suggested a hybrid ACO with an insertion heuristic algorithm for time dependent VRPTW (TDVRPTW).

Ting and Chen [37], developed a multiple ant colony optimization algorithm (MACO) to solve the location routing problem with capacity constraints. The MACO is applied to optimize different subproblems: location selection, customer assignment and VRP. Proposed technique is tested on benchmark sets and MACO is found competitive with existing well-known algorithms. Doerner et al. [10] implemented a parallel D-Ant algorithm to speed up the operation. In this approach, ants are used to solve only subproblems rather than solving the whole problem. Some of the researchers [7, 40, 41] have made major contributions to the development of ACO to solve VRP variants.

PSO has been used in many application domains and is successful to obtain good results in a fast and cheaper way as compared to others. It has wide literature in the field of VRP also. Ai and Kachitvichyanukul developed a PSO algorithm for VRPTW [1], VRPPD [2], CVRP [4] and for other logistics problems [3]. The experimental results are compared with other existing meta-heuristics and found efficient in solving VRPs. Kanthavel and Prasad [19] proposed a *Nested_PSO* for VRP; they focused on maximum utilization of loading capacity and determining the optimum set of vehicle routes for CVRP. Zhu et al. [43] presented an improved PSO to solve VRPTW, which they tested on numerical examples and after comparison conclude that the PSO algorithm gets best solution much faster than the genetic algorithm (GA).

Yin and Liu [39] hybridized PSO with GA (*GPSO)* to solve a single-depot complex VRP (SCVRP). This hybrid combination is able to avoid being trapped in local optimum. In Ponce [28] PSO method each particle represents a set of feasible routes, and a modified crossover operator (GA) move particles towards better solutions. Ai-ling et al. [5] proposed a hybrid discrete PSO *(DPSO)* with simulated annealing. The DPSO searches for the optimal results globally and locally and SA avoids the convergence to local optima. The results show that the DPSO is feasible and efficient to solve large CVRP problems. Liu et al. [21] proposed a hybrid PSO *(HPSO)* with the crossover operator to improve the speed of the algorithm. Results are compared with the results of PSO, GA, and parallel PSO algorithms and found HPSO is superior for solving VRPTW. Masrom et al. [25] proposed PSO with variable neighbourhood descent and named it as *h_PSO*. The *h_PSO* is investigated on benchmark problems and the results indicate that the *h_PSO* competes with the existing heuristics.

Researchers also used artificial bee colony algorithm to solve VRP variants. Marinakis et al. [24] proposed honey bees mating algorithm for efficiently solving the VRP. The proposed *HBMOVRP* algorithm combined with the multiple phase neighbourhood search and greedy randomized adaptive search process *(MPNS-GRASP)*. Gomez and Salhi [16] proposed a new ABC algorithm for CVRP. The novel approach relies upon the two specializations named diversification and intensification. Benchmark datasets are used to test the performance of the algorithm. Szeto and Ho [34] proposed an enhanced ABC, to improve the solution quality. The performance is evaluated on benchmark sets and compared with the original ABC approach and other heuristics. The results show that the enhanced version performs better than both.

IWD is a newly developed method, but due to good optimization results it has drawn continuous attention to solve complex combinatorial problems such as Job shop scheduling, VRP, etc., using IWDs. Li and Zhao [22] developed an efficient IWD algorithm for solving VRPTW. Simulation results show that the IWD algorithm can found optimal solution of VRPTW with high probability. The average of solutions found by IWD algorithm is better than the solutions of other algorithms. Teymourian et al. [35] solved CVRP by developed four state-of-the-art algorithms: an improved IWD *(IIWD)*; an advanced cuckoo search *(ACS)*; and two effective hybrid meta-heuristics, local search hybrid algorithm *(LSHA)* and post-optimization hybrid algorithm *(POHA)*. Results of the experiments are compared with the best obtained results reported in literature. Specifically, for 92.9% of Christofides and

50% of Golden instances, the best solutions are obtained. It is found that the LSHA and POHA methods can effectively solve such problems.

## 6 Conclusion

Over the years many SI techniques have been developed to solve complex single, multiple, and many objective optimization problems. However, this paper restricts its discussion with ACO, PSO, ABC and IWD to solve, three variants of VRP, i.e. CVRP, VRPTW and VRPPD. As per the given literature, ACO and PSO algorithms are widely used and found to be highly effective and adaptable for various domains of VRP with considerable potential for hybridization and integration with other intelligent systems. Though, very few researchers implemented ABC and IWD algorithms on VRP variants, but these algorithms have potential to find optimal solutions to large and complex problems. The findings of the study also show that the most studied attributes of VRP are: CVRP, VRPTW, time dependent VRP and objective function with cost minimization and the least studied attributes are: VRPPD domains, maximization objective function, etc.

Hence, the future research will focus on solving the least studied attributes and other variants of VRP such as VRPB, MDVRP, and VRPPD, etc., using highly efficient swarm algorithms. Also, the least implemented algorithms like ABC and IWD can be used to obtain optimal solutions by combining them with other intelligent systems.

## References

1. Ai, J. and Kachitvichyanukul, V. A Particle Swarm Optimisation for Vehicle Routing Problem with Time Windows. *International Journal of Operational Research*, 56, 1 (2009), 519–537.
2. Ai, J. and Kachitvichyanukul, V. A particle swarm optimization for the vehicle routing problem with simultaneous pickup and delivery. *Computers & Operations Research*, 36, 5 (2009), 1693–1702.
3. Ai, J. and Kachitvichyanukul, V. A Study on Adaptive Particle Swarm Optimization for Solving Vehicle Routing Problems. In *9th Asia Pacific Industrial Engineering and Management Systems Conference* (Bali, Indonesia 2008).
4. Ai, J. and Kachitvichyanukul, V. Particle swarm optimization and two solution representations for solving the capacitated VRP. *Computers & Industrial Engineering*, 56, 1 (2009), 380–387.
5. Ai-ling, Gen-ke, YANG, and Zhi-ming, WU. Hybrid discrete particle swarm optimization algorithm for capacitated vehicle routing problem. *Journal of Zhejiang University SCIENCE A*, 7, 4 (2006), 607–614.
6. Balseiro, S. R., Loiseau, I., and Ramone, J. An ant colony algorithm hybridized with insertion heuristics for the time dependent vehicle routing problem with time windows. *Computers & Operations Research*, 38 (2011), 954–966.

7. Bell, J E and McMullen, P R. Ant colony optimization techniques for the vehicle routing problem. *Advanced Engineering Informatics* (2004), 41–48.
8. Bin, Yu, Zhong-Zhen, Yang, and Baozhen, Yao. An improved ant colony optimization for vehicle routing problem. *European Journal of Operational Research* (2009), 171–176.
9. Bouhafs, Lyamine, Amir and Koukam, A. Hybrid Heuristic Approach to Solve the Capacitated Vehicle Routing Problem. *Journal of Artificial Intelligence: Theory and Application*, 1, 1 (2010), 31–34.
10. Doerner, K F, Hartl, R F, and Lucka, M. A parallel version of the D-Ant algorithm for the Vehicle Routing Problem, Parallel Numerics' 05, (2005), 109–118.
11. Donati, V., Montemanni, R., Rizzoli, E., and Gambardella, M. Time Dependent VRP with Multi Ant Colony System. *European Journal of Operational Research*, 185, 3 (2008), 1174–1191.
12. Favaretto, D., Moretti, E., and Pellegrini, P. Ant colony system for a vrp with multiple time windows and multiple visits. *Journal of Interdisciplinary Mathematics*, 10, 2 (2007), 263–284.
13. Gambardella, L M, Rizzoli, A E, Oliverio, F, Donati, A V, Montemanni, R, and Lucibello, E. Ant Colony Optimization for vehicle routing in advanced logistics systems. In *Proceedings of MAS 2003 – International Workshop on Modelling and Applied Simulation* (Bergeggi, Italy 2003), 3–9.
14. Gambardella, L M, Taillard, E, and Agazzi, G. MACS-VRPTW: A Multiple Ant Colony System for Vehicle Routing Problems with Time Windows. In D. Corne and M. Dorigo, ed., *New Ideas in Optimization*. McGraw-Hill, UK, 1999.
15. Gendreau, M., Laporte, G., and Eguin, R. S. An exact algorithm for the vehicle routing problem with stochastic demands and customers. *Transport. Sci.*, 29, 2 (1995), 143–155.
16. Gomez, A. and Salhi, S. Solving capacitated vehicle routing problem by artificial bee colony algorithm. *Computational intelligence in Production and Logistics Systems (CIPLS)* (2014), 48–52.
17. Hadjiconstantinou, E. and Roberts, D. Routing under uncertainity: an application in the scheduling of field service engineers. In Toth, P. and Vigo, D., eds., *The vehicle routing problem*. SIAM, (2001).
18. Iredi, S., Merkle, and Middendrof, M. Bi-Criterion Optimization with Multi Colony Ant System. *Proc. International Conference on Evolutionary Multi-Criterion Optimization (EMO'01)*, (2001), 359–372.
19. Kanthavel, K. and Prasad, P. Optimization of Capacitated Vehicle Routing Problem by Nested Particle Swarm Optimization. *American Journal of Applied Sciences*, 8, 2 (2011), 107–112.
20. Lahyani, R., Khemakhan, M., and Semet, F. Rich vehicle routing problems: from a taxonomy to a definition. *European Journal of Operation Research*, 241, 1 (February 16 2015), 1–14.
21. Liu, X., Jiang, W., and Xie, J. Vehicle routing problem with time windows: a hybrid particle swarm optimization. Proc. *Natural Computation, ICNC'09*, 5th International Conference, (2009) 502–506.
22. Li, Z., and Zhao, F., Intelligent water drops algorithm for vehicle routing problem with time windows. In Proc. *Service Systems & Service Management (ICSSSM), 11th International Conference*, (2014), 1–6.
23. Manfrin, M. *Ant Colony Optimization for the Vehicle Routing Problem*. 2004.
24. Marinakis, Y., Mariniki, M., and Dounias, G. Honey Bee Mating Optimization Algorithm for the Vehicle Routing Problem. *Studies in Computational Intelligence (SCI)*, 129 (2008), 139–148.
25. Masrom, S., Abidin, S. Z. Z., Nasir, A., and Rahman, A. Hybrid particle swarm optimization for vehicle routing problem with time windows. In *Proceedings of the International Conference on Recent Researches in Computational Techniques, Non-Linear Systems and Control* (2011), 142–147.
26. Parunak, H. Dyke and Brueckner, S. Engineering swarming systems. *Methodologies and Software Engineering for Agent Systems* (2004), 341–376.

27. Pellegrini, P., Favaretto, D., and Moretti, E. Multiple ant colony optimization for a rich VRP: a case study. *In Knowledge-Based Intelligent Information and Engineering Systems*, 4693 (2007), 627–634.
28. Ponce, Daniela. Bio-inspired Metaheuristics for the Vehicle Routing Problem. In *Proceedings of the 9th WSEAS International Conference on APPLIED COMPUTER SCIENCE* (), 80–84.
29. Reimann, M., Doerner, K., and Hartl, R. F. Insertion based ants for vehicle routing problems with backhauls and time windows. *Ant Algorithms: Third International Working,* (2002), 135–148.
30. Rizzoli, A E, Oliverio, F, Montemanni, R, and Gambardella, L M. *Ant Colony Optimisation for vehicle routing problems: from theory to applications*. Istituto Dalle Molle di Studi sull'Intelligenza, 2004.
31. Ruinelli, L. *Column generation for a rich vrp: vrp with simultaneous distribution, collection and pickup-and-delivery*. University of Applied Sciences and Arts, Southern Switzerland, 2011.
32. Saravanan, M and Sundararama, A. Ant colony optimization for one-sided time constraint vehicle routing problem. *International Journal of Services, Economics & Management*, 23–4 (2010), 332–349.
33. Snadhaya and Katiyar, V. An Enhanced Ant Colony System for Solving Vehicle Routing Problem with Time Window. *international Journal of Computer Applications*, 73, 12 (2013), 27–31.
34. Szeto, Y., and Ho, C. An artificial bee colony algorithm for the capacitated vehicle routing problem, (2011), 126–135.
35. Teymourian, E., Komaki, M., and Zandieh, M. Enhanced intelligent water drop and cuckoo search algorithms for solving the capacitated vehicle routing problem. *Information Sciences*, (2016), 354–378.
36. Ting., C. J. and Chen, H. Combination of multi ant colony and simulated annealing for the multi depot vehicle routing problem with time windows. *Journal of Transportation Research Board* (2009), 85–92.
37. Ting, C. J. and Chen, C. H. A multiple ant colony optimization algorithm for the capacitated location routing problem. *International Journal of production Economics*, 141, 1 (2013), 34–44.
38. Toth, P. and Vigo, D. The Vehicle Routing Problem. *SIAM*, 2001.
39. Yin, L. and Liu, X. A Single depot Complex Vehicle Routing Problem and its PSO Solution. *Proc. Symposium on International Computer Science & Computational Technology (ICSCT)* (2009),266–269.
40. Yu, B. and Yang, Z. Zhen. An Ant Colony Optimization: The Periodic Vehicle Routing Problem with Time Windows. *Transportation Research: Logistics and Transportation Review*, 47, 2 (2011), 166–181.
41. Zhang, X. and Tang, L. A new hybrid ant colony optimization algorithm for the vehicle routing problem. *Pattern Recognition Letters*, 30, 9 (2009), 848–855.
42. Zhen, T., and Zhang, Q., Hybrid Ant Colony Algorithm for the Vehicle Routing with Time Windows. *Computing Communication, Control and Management, ISECS International Colloquium,* (2008), 8–12.
43. Zhu, Q., Li, Y., and Zhu, S. An Improved Particle Swarm Optimization Algorithm for Vehicle Routing Problem with Time Windows. *IEEE Conference on Evolutionary computation* (Vancouver, 2006).

# Scrutiny of VANET Protocols on the Basis of Communication Scenario and Implementation of WAVE 802.11p/1609.4 with NS3 Using SUMO

**Arjun Arora, Nitin Rakesh and Krishn K. Mishra**

**Abstract** Around the world governments have made many policies and laws in order to avoid accidents that occur on road, however, they seem to be unavoidable. Therefore, it becomes important to take care of road safety. This is where VANETs have helped in introducing various applications primarily based on V2V and V2I communication which have helped in increasing safety of commuters on road. Although to deploy VANET successfully the following things have to be kept in mind: First range of wireless networks, second secure communication and lastly stable network performance. When we look into VANETs we find that there are primarily two unique forms of communication available. The environments where VANETs are deployed are mainly the highways and cities/towns. VANET communication on highways is fairly simple to implement but when it is to be deployed in cities communication becomes complex. This is because cities have congested outlook mainly because of buildings, trees and houses, etc. And it is not possible to always have direct communication between nodes in the intended direction. This paper analyses the different protocols in VANET and make a comparison on the basis of communication. Also this paper shows a scenario based implementation of Wireless Access in Vehicular Environments (WAVE) 802.11p/1609.4.

A. Arora (✉)
University of Petroleum & Energy Studies, Dehradun, India
e-mail: arjunarora06dit@gmail.com

N. Rakesh
Amity University, Noida, UP, India
e-mail: nitin.rakesh@gmail.com

K.K. Mishra
Motilal Nehru National Institute of Technology, Allahabad, India
e-mail: kkm@mnnit.ac.in



S.K. Bhatia et al. (eds.), *Advances in Computer and Computational Sciences*,
Advances in Intelligent Systems and Computing 553,
DOI 10.1007/978-981-10-3770-2_33

## 1 Introduction

In order to inform the commuters about road conditions, traffic influx, etc. It is very crucial to send such information quickly and accurately. To achieve this VANETs are deployed and too much extent VANETs are able to do so. Also in case of an emergency such as car malfunction, accident, health issues, etc., advantage can be taken of facilities provided by VANET technologies.

The limitations of VANETs in coverage and channel capacity, high degree of mobility among nodes, presence of obstacles lead to frequent data packet losses, changes in topology of the network and fragmentation. Therefore there is a need for a routing strategy which is able to make better use of resources which leads to an efficient deployment of VANETs. Packet routing in VANETs is a challenging task as there lacks a controller entity, which manages the routing paths centrally among the given set of nodes.

VANETs have a limited list of protocols, which can be directly implemented in comparison to MANETs which have a larger list of protocols that support it directly. After rigorous study the primary reasons that came out for this deprived list of protocols is due to (a) high speed movement of nodes (b) amount of information that needs to be communicated at high speeds for timely delivery. Thus identification and administration of correct routes in a quick and timely manner involves lot of processing overhead [1, 2]. This in turn has led to an array of research challenges related to security and privacy, which need to be considered in order to design and develop suitable protocols for VANETs.

This paper is divided into sections the first section provides the introduction the second talks of the communication architecture with some discussion on applications and challenges. The third section gives a brief explanation of all protocols

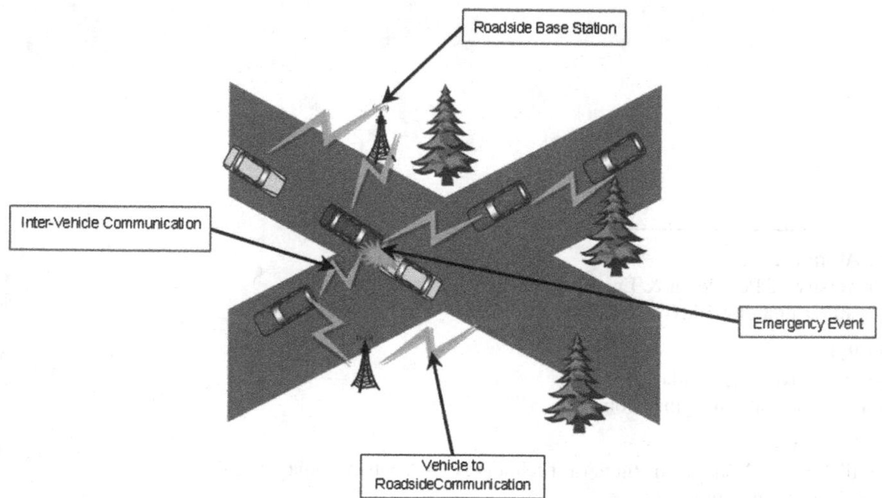


**Fig. 1** VANET

analysed. Fourth section illustrates a tabular comparison of different VANET routing protocols. The fifth section weighs the various protocols in comparison to WAVE 802.11p/1609.4 and shows the implementation of the protocol using a simplistic road map. The sixth section concludes the analysis made in the paper with discussion on future scope. (Fig.1).

## 2 VANET Architecture, Applications and Challenges

A successful deployment of VANET enables communication among smart vehicles in the area or with the infrastructure. VANET has been considered as the new way of providing an array of some very interesting services in case of intelligent transportation. Figure 2 shows some very interesting forms of VANET architecture. In a VANET deployment there could be use of cellular gateways and access points especially at traffic junctions in order to get updated information on traffic flow which can then be used for routing of traffic [3]. In such a case the VANET architecture would look like as shown in Fig. 2a below. This category of communication under VANETs is termed as V2I communication and is helpful in scenarios where exist a varied type of wireless technologies such as LTE, 3G, WiFi, WiMax, etc. Figure 2b shows that in some scenarios the nodes may be engaged in line of sight communication because of economic constraints and or it may be difficult to erect cellular towers and access points.

In such a scenario the sensors fixed inside a vehicle will be used as a primary source for communication and alert other vehicles of road conditions or other emergencies. Such an environment where infrastructure is non-existent and communication is mainly ad hoc is termed as V2V communication. A third scenario is also possible as shown in Fig. 2c which is called as hybrid architecture. In this architecture communication among networked devices may be with fixed infrastructure on the roadside like cellular towers or access points and also among themselves. A variety of applications are possible for such a network related to traffic monitoring like driver support, security and entertainment. In such a network communicating nodes access active information much beyond the range of network they are part of and are able to send this information through a peer to peer communication network. The hybrid architecture makes use of both V2I and V2V communication thus providing more variety of content and enhanced flexibility. VANET provides a variety of applications which can be broadly classified into two categories mainly:

Safety Applications: These applications are generally aimed at increasing the safety of drivers/co-passengers by continuous transmission of information among vehicles. Such applications generally send alerts directly to drivers or an active safety system. Example of such applications could be accident management, emergency streaming of live video, warning of lane changes, etc.

Comfort Applications: These applications are generally aimed at on enhancing the comfort level of the driver/co-passenger within the vehicle. Example of such

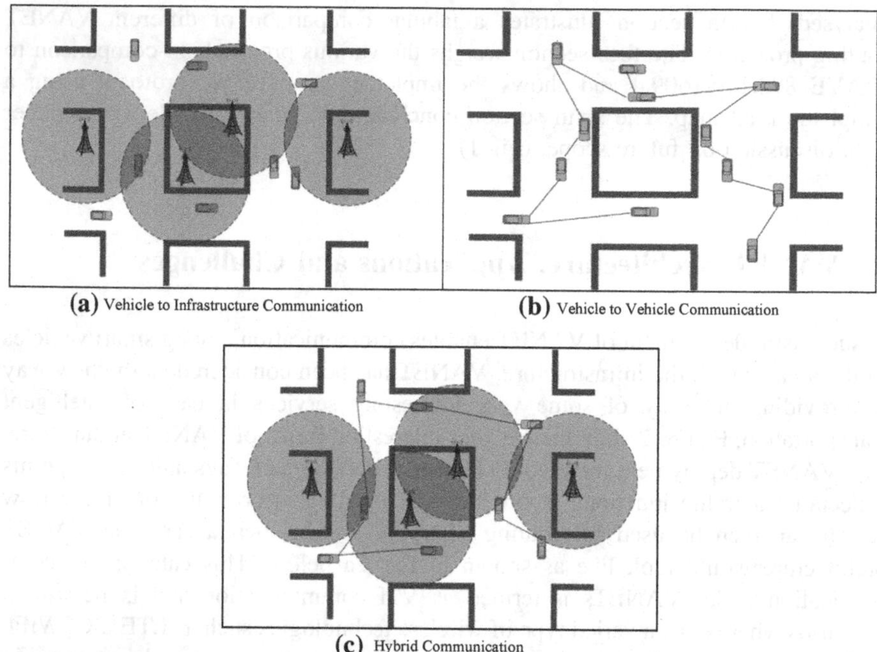


**Fig. 2** VANET communication architecture of V2I, V2V and hybrid type respectively

applications could be automatic toll collection, electronic parking payment, management of traffic, etc. Similar to MANET information within VANET can be arranged and managed by nodes without the need of any central authority. In case of VANET as nodes are in motion data transmission is hampered thus making it less reliable. Some of the more complex VANET challenges are:

(a) *Network Downtime*: In VANETs constant connectivity is difficult due to fast movement of vehicles. Especially where vehicle density is low there are high chances of the communication link being disconnected. This is mainly because of high mobility, the network topology and channel condition which change rapidly because of which structures like trees cannot be used and maintained for the rapidly changing topology. Many applications which are internet based may be hampered [4]. A probable solution could be installation of a large number of access points and relay nodes however such a solution may not be financially feasible.
(b) *Constantly changing Topology*: In VANETs the topology is not constant they keep changing and change is based directly on the speed of vehicles. The traffic conditions are different in rural and urban areas because of which network partitions occur frequently resulting in congestion and collision. Assume a scenario where wireless transmission range is 300 m. In such a case two vehicles can only communicate until they are not more than 300 m apart

from each other. Also point of concern is that such a link will only remain active for roughly 18 s if the vehicles are moving in opposite direction at a speed of 60 km/hr.

(c) *Different communication environments*: In case of VANETs the traffic flow varies in cities and highways, thus making it more difficult and complex to maintain seamless connectivity. This is because VANETs use electromagnetic waves for communication. These waves get affected by environment, thus while deploying VANET the impact of environment must also be considered. Also point of concern is obstruction from trees, buildings, and other such objects which may hamper the direction of intended communication.

(d) *Modelling and Prediction of vehicle location*: In case of VANETs modelling and prediction of vehicle location is a crucial part considering the fact that in case of VANETs there is high movement in nodes and the topology is of a dynamic nature [5]. Also the nodes prediction is limited by streets, roads and highways which make it possible to predict position of the vehicular nodes. As VANETs are also used for road safety applications security of position based messages which are being sent over the network are also a point of concern. VANETs generally use shared medium to communicate hence MAC designing is crucial.

## 3 Analysis of VANET Routing Protocols

In this paper we classify routing protocols based on communications scenario. The VANET communication scenario can broadly be classified into two main categories Routing Protocols in Congested Scenario (RPCS) and Routing Protocols in Non-Congested Scenario (RPNCS). In case of RPCS the condition of roads are more complex as in congested scenario the number of obstructions are more, number of streets, roads are more which form junctions where traffic flow is more. Therefore, in such scenarios a large network is formed because of high traffic density. Also implementing VANET in congested scenarios will also have concerns such as low strength of signals because of obstructions. The network scalability needs to be high for deployed VANET and conditions change in topology and network density based on time of the day and area within the network. Hence it becomes a challenging task to develop and implement routing protocols in congested scenarios. VANET deployment in congested scenario even though has problems it comprises of benefits as well such as the vehicular nodes may make full utilization of all resources provided which in turn would result in making better routing base decisions. The main objective of routing protocols in congested scenarios is to prevent accidents and provide support instantaneously in case of emergency. In case of RPNCS the VANET routing protocols are implemented in low-density scenarios with high to very enormous node mobility. Thus, it becomes difficult to develop and design routing protocols for non-congested scenarios. Protocols which fit in the above-described scenarios are:

i. *Geographic Source Routing Protocol (GSR)*: This protocol combines topological information with position based routing and is primarily aimed at routing in congested scenarios. The location information of destination is fetched by making use of Reactive location service. It makes use of digital maps which helps in understanding the location of all cross-roads from source to destination. It makes use of Dijkstra algorithm to calculate the shortest path through source to the destination. The junction points act as anchors to GSR are also included in the shortest path calculation. The entire bundle of packets in GSR are tagged with the source location; destination and anchors which as a result are not comprehended by the GSR [6] and the existence of transports on road are offered on the basis of intensity of traffic in lanes before the path is chosen between the source and destination.

When this protocol is compared to other protocols it gives a better packet delivery ratio than others. The disadvantage of this protocol is that it ignores situations when there are very few forwarding packets. It can also lead to a dead-end, i.e. when there are no neighbours close enough to destination. Then is such a case in order to recover face routing may be used to find another alternative path from where forwarding can be resumed. Also this protocol generates high overhead by making use of control messages in routing.

ii. *GpsrJ+ Routing Protocol*: This protocol removes unnecessary details at an intersection and only keeps details necessary for competent planning of topographical maps. It makes use of a technique termed as twin leap neighbour beaconing to capture those segments of road that needs to be taken up by the cross-road node. If the prediction defines that its neighbouring cross-roads will forward the packet along an opposite direction of the road then it sends the packet to junction node [7] else it takes a diversion and sends the packets to the outermost neighbour node as illustrated in Fig. 3.

Node which is closest to target node becomes the next best forwarder. This is done by setting the timer $T_m$ as:

$$T_m = C_m \times \left(\frac{D_{RT}}{D_{ST}}\right) \tag{1}$$

Where $C_m$ is the delay in forwarding which varies with processing time and transmission rate. While setting timer $T_m$ we need to keep in mind the problem of packet duplication which can be removed by selecting the forwarder in such a manner that it does not include a node say $N_m$ and the following inequality is checked:

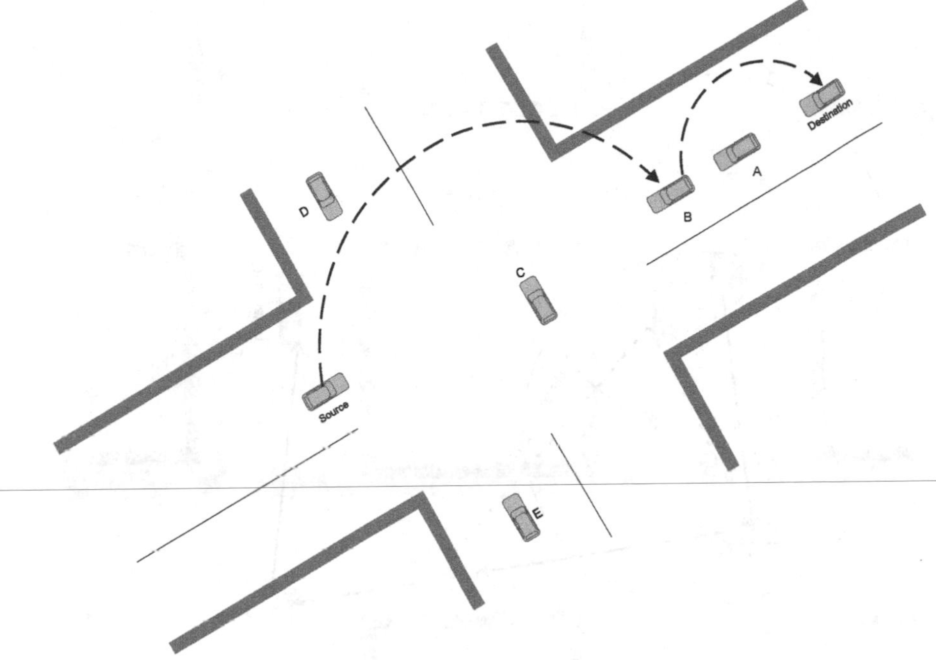


**Fig. 3** Predictions in GpsrJ+

$$|T_{Nm} - T_{Ni}| < \partial \tag{2}$$

Where $\partial$ is minimum time for suppression.

The main advantage with this protocol is that it does not require expensive planarization strategy. The numbers of hops are reduced in comparison to other protocols and the packet delivery ratio [8] is also increased. However, the main disadvantage of this protocol is that it is not suitable for sensitive applications, i.e. it follows complex path even if it has simple path.

iii. *Greedy Perimeter Coordinator Routing Protocol (GPCR)*: The GPCR is not dependent on a digital map. The digital map is fundamentally used by GSR protocol. The path from source to destination based on these digital maps is computed by Dijkstra Shortest Path and Dijkstra least weight path algorithm. Instead of transmitting data packets through the junction GPCR protocol intermittently sends data packets [9] to a node present on the junction. As seen in Fig. 4, node A simply sends data packets to only the junction node present, i.e. node B even though the range of node A covers node C. The term co-ordinator is used to define a node present near the junction. Restricted greedy approach is used for packet forwarding. The coordinator becomes the featured node till it is positioned near the junction over a non-coordinator. A repair strategy is developed as soon as local occurs.

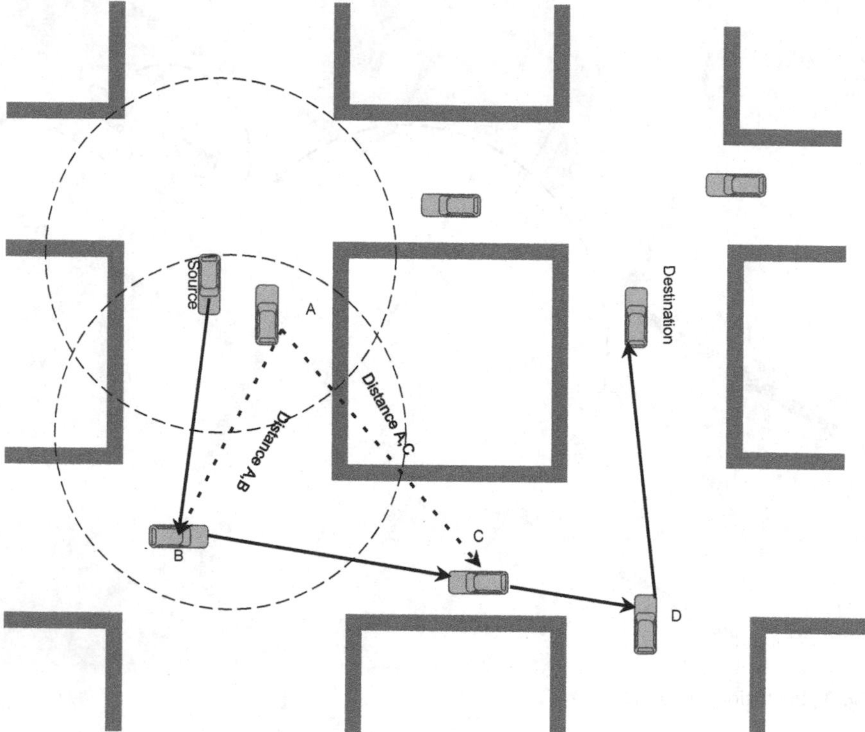


**Fig. 4** GPCR routing

This protocol provides the advantage of that it has no planarization problem like that in case of planar sub graphs also it does not require usage of any external or global information. However, the disadvantage with this protocol is that it is dependent on nodes present at the junction and if there is a problem in node detection at the junction [10] like in case of curved roads or sparse roads then the protocol fails to give efficient results.

iv. *Movement-Based Routing Protocol (MORA)*: The basic idea of MORA comes from the developments in the communication of intervehicular and networking which have been used in deployment of VANETs as an ad hoc and infrastructure free solution for automobiles. Acting as a dependable that offered with competent routing schemes with incidences of comparative movement simulations. As this algorithm requires the nodes to communicate directly with their immediate neighbours placed within transmission range, we can say that this protocol is completely disseminated. The essential characteristic of MORA is the usage of routing metric, which helps in utilization of positioning information and the direction of movement of the vehicles. The dynamic changes in the network are unambiguously reflected by MORA with the information available of the topology [11]. The advantages when broadly assessed show that MORA especially in high-congested scenario and in continuous topological changes proves efficient. The link layer of the protocol is integrated with application layer and thus provides with high quality performances. Every vehicle in this algorithm makes estimate on the stability of link ($L_S$) for every neighbouring vehicle. The $L_S$ denotes relation between a constant value ($\sigma$) and communication lifetime. Lifetime of a link can be given as ($L_T[i,j]$) with an estimated time given as:

$$\Delta t = t_1 - t_0 \tag{3}$$

where $t_1$ is the time when $j$ is out of range of communication of $i$ which may be denoted as $C_1$. $C_1$ and $\Delta t$ are approximated using initial location of $i$ and $j$ $[(X_{io}, Y_{jo})]$ and $[(X_{jo}, Y_{jo})]$. Consider the initial speeds to be $\vec{Vi}$ and $\vec{Vj}$ then

$$C_1^2 = \left((X_{io} + V_{xi}\Delta t) - (X_{jo} + V_{xj}\Delta t)\right)^2 \tag{4}$$

$$C_1^2 = A\Delta t^2 + \mathrm{B}\Delta t + D \tag{5}$$

Now $L_S$ can be computed as

$$L_s[i,j] = \frac{L_T[i,j]}{\sigma} \tag{6}$$

where

$$L_s[i,j] = 1$$

when

$$L_T[i,j] \geq \sigma \tag{7}$$

v. *Position-Based Routing with Distance Vector Recovery Protocol (PBRDV)*: In this protocol as the packets lie in the range of local maximum, this protocol deploys AODV-style recovery. The request packet used in this protocol generally consists of the destination location. Once the request packet is acquired the node ensures its proximity at the local maximum less than that with the target node. The other way round situation in case of this protocol is that it takes into consideration the node which receives the packet request and re-transmits the request. If it is not able to then it simply transmits a response of the request to the same node from where it receives it [12]. The transitional nodes simply register with the node placed before it, i.e. the replay packet is sent, the node keeps a near-by node path at local maximum than itself all this is because the reply packet moves near to the local maximum node. The weakness of this protocol lies with its need to overflow in order to detect the non-greedy parts within route.

Extended Greedy Routing Algorithm used in PBRDV [12].

```
metricCal ← d(E,D)
    forK ∈ KG_{E,x} do
    distance ← d(K,D)* sph(K,E)*scharge
    if distance <metricCal then
    metrical ←distance
return metricCal
```

vi. *Ad hoc on demand distance Vector Routing Protocol (AODV)*: In this protocol a routing table is maintained and on receiving of a broadcast query (RREQ) the address of the node which sends the query is stored in the nodes routing table. This protocol is illustrated in Fig. 5a. Backward learning is defined as the method of registering the preceding hops. Once the reply packet reaches is target it is transmitted back traversing the entire path which is acquired by backward learning from destination to source Fig. 5b. The forward path from the source is instituted this is done by the node registering its preceding hops. A full duplex path is formed which is overflowing with query and reply [13] distribution. If the source uses the path that has been formed, it is then maintained throughout. Breakdown in links are accounted for in a recursive manner back to the source and ultimately generate a different query response process in order to identify the best possible path.

This protocol suffers from time management, i.e. it requires more time to set up connection and initial communication for a given route when compared to other protocols. This protocol also involves a heavy control overhead in case of multiple

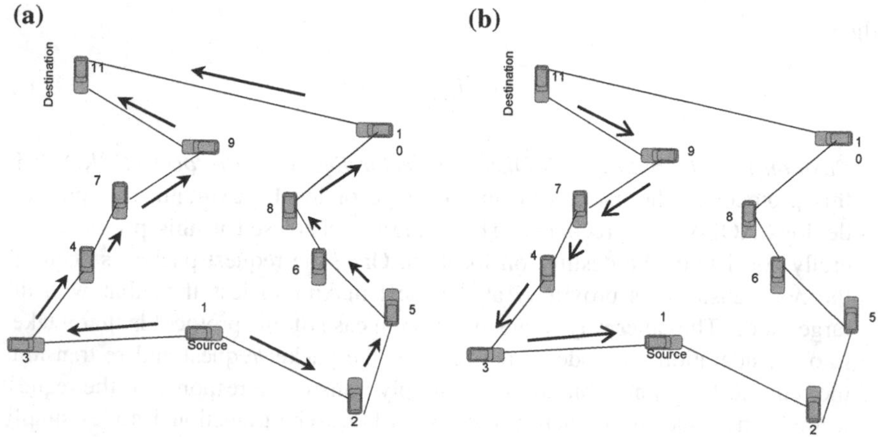


**Fig. 5** Path discovery in AODV, **a** Shows the message being sent from source to destination via vehicles on both the possible paths. **b** Shows the reply to message from destination to source via vehicles on the shortest path

reply packets, which are sent through a simple route of reply. One of the characterizing aspects of this protocol is that it follows an updated path to its destination by using a sequence number. It also helps in reducing excessive memory used and redundancy in the route.

vii. *Vehicle-Assisted Data Delivery Routing Protocol (VADD)*: This protocol is specifically intended for VANETs and enhances the routing in vehicular networks by carrying forward disengagement based on conventional mobility of vehicle. From an intersection the vehicle moves to a road that it chooses which starts at the intersection. The best path for forwarding of packets is chosen by simply switching between packet modes namely intersection, destination [14] and straight way modes. This protocol is suitable in a congested scenario where data delivery is done through multiple hops. The delivery ratio of this protocol is higher in comparison to GPCR and DSR [14]. On the other hand the routing protocol has a disadvantage that it's efficiency is affected by large delays mainly due to change in topology and traffic flow.

viii. *Geographical Opportunistic Routing (GeOpps)*: This protocol is benefitted from the fact that it takes up recommended routes from vehicle navigation in order to choose vehicles close to travel to from source to destination of the packet. It simply computes straight distances from destination packets to the concerned nearest point in the path of the vehicle and then time from packet to destination is computed approximately. Figure 6 shows the computation of nearest point of neighbours N1 and N2. As N2 is closer to destination the vehicular node will select N1 to send packets [15]. The packet in turn will be sent through another vehicle if it is present such that it has shorter time to

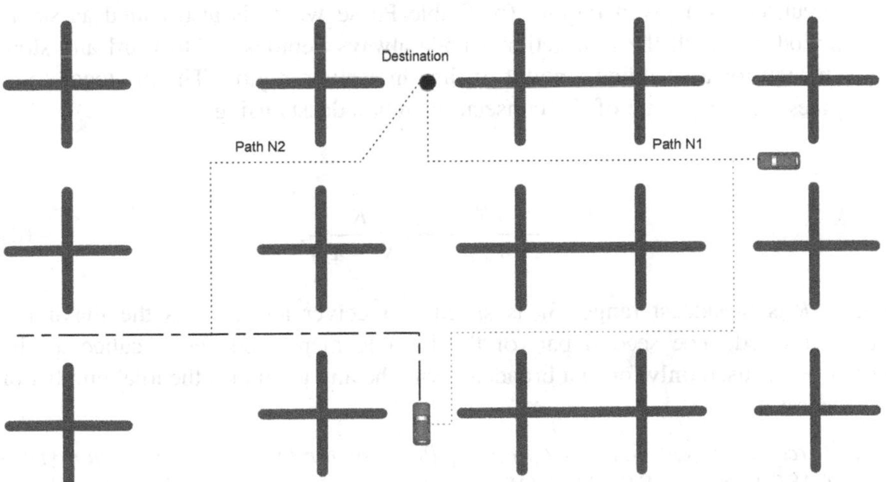


**Fig. 6** Computation of nearest point from destination

travel approximately. This process recursively continues till the packet reaches its destination.

The primary advantage of this protocol is that it requires few parameters to find vehicle destination when vehicle is moving towards it [16]. The ratio of delivery for this protocol is not dependent on density of vehicles which heavily relies on traffic flow and topology. One of the disadvantages with this protocol is that the privacy maintenance in this protocol is an issue this is because of the information regarding navigation which is disclosed in the network of the vehicle.

ix. *Greedy Perimeter Stateless Routing Protocol (GPSR)*: This protocol is based on position routing aimed at non-congested scenarios with mobile nodes. The GPSR protocol provides best results in non-congested scenarios such as highways where obstruction is low and vehicles are distributed almost uniformly. This protocol follows two approaches for routing (a) Greedy Approach (b) Recovery Approach. In case of greedy approach the mobile simply forwards the packet to its nearest direct neighbour which is closest to destination node. In case of recovery approach as soon as a packet reaches local maximum the recovery mode forwards packets to the node which is closer to destination in comparison to the node which faced local maximum. GPSR is not suitable for congested scenarios because greedy approach fails in case of impediments because of lack in direct communication. Also GPSR toggles routing so in case of obstacles efficiency and packet ratio delivery reduces resulting in higher amount of delays.
x. *Dynamic Time Stable Geo-cast Routing (DTSG)*: The main objective of this protocol is to give high efficiency in non-congested scenarios with low obstructions. It works efficiently based on the density of network mobile nodes and also takes into account the speed of vehicles for performance enhancement [17]. It has two phases (a) Pre-Stable Phase which assists messages to be circulated in a given region. (b) Stable Phase which is also called as stable period in which the transactional node always deploys the forward and store strategy for a specified amount of time in a given region. This protocol computes the sleep time of the consecutive rebroadcasts using:

$$\tau_k = \frac{2R}{S_k + S_{max}} + \frac{R}{(X_s - X_k)}, \quad (8)$$

where $R$ is broadcast range, $S_k$ is speed of receiver and $S_{max}$ is the maximum allowed speed. The second part of the formula mentioned above called as the coefficient is used only for first broadcast with the aim to control the total number of rebroadcasts.

xi. *Direction-Based Geo Cast Routing Protocol for Query Dissemination (DG-CASTOR)*: The DG-CASTOR [18] protocol is a new version of geo-cast routing. This protocol has been specifically customized for entertainment as well as information based applications in a vehicular network. Its main purpose is to develop a virtual platform, which depends upon locations of potential forecast of the mobile nodes in a network. Such a platform is generally termed as Rendezvous. It is at this platform that all nodes combine a few years later. On the contrary the query being generated is simply distributed among nodes having the same Rendezvous.

xii. *Distributed Vehicular Broadcast Protocol (DVCAST)*: In this protocol all mobile vehicles make use of a flag variable in order to give confirmation to redundancy in packets. It makes utilization of restricted topology information by developing recursive messages to send information [19]. This protocol segments the vehicles into three different types (a) strongly connected (b) loosely connected (c) disconnected. The first category uses a weighted perseverance scheme for various slots. In the second category, the different broadcast messages of the vehicle instantaneously re-broadcast messages of vehicles which are moving along in the direction same as that of the messages [20]. In the last category, the packet gets completely discarded until no other vehicle gets into the transmission range till the expiry of time.

## 4 Communication Scenario-Based Comparison of VANET Protocols and Scenario Based Implementation of Wave 802.11p/1609.4

Different VANET protocols are compared qualitatively based on their crucial requirements and tabulated in the Table 1. Basic simulations are made of WAVE 802.11p/1609.4 over a scenario showing a crossroad and cars traversing from source to destination. When WAVE is compared to other protocols mentioned in previous sections we can see the impact on performance that is performs faster and more efficiently [21]. In case of VANET simulations one of the important characteristic is the mobility model that is it could be either congested or non-congested. In this evaluation NS-3 with SUMO is used to map a crossroads. Vehicles travel with an average speed of 50 km/h. In order to evaluate the algorithm the speed is recorded with simulation time. The actual speeds of the vehicles may vary with real world traffic conditions. The monitoring is done in order to determine if vehicles are in congestion. As soon as the speed of the vehicles drops below a threshold value the algorithm starts to re-route vehicles on the next best possible path to its destination. Figure 7 shows screen shots of the simulation in SUMO installed with NS-3 on which results are complied, it can be seen how the algorithm is able to reduce congestion at the crossroad. There are several other research aspects in which this is also implemented [22–29].

Figure 8 shows the graph of speed vs. time. As it can be clearly seen that expected graph should have flattened out at the speed of 50 km/hr which in our case is the average speed of the vehicles. However with time we can see a gradual dip in

**Table 1** Comparison of different VANET routing protocols

| | Protocols | Forwarding approach | Recovery approach | Digital map | Location services | Predictive nature |
|---|---|---|---|---|---|---|
| Congested scenario based | GSR | Greedy | Store and Forwarding | No | Yes | Yes |
| | GpsrJ+ | Greedy | Store and Forwarding | No | Yes | Yes |
| | GPCR | Greedy | Store and Forwarding | Yes | Yes | No |
| | MORA | Greedy | Store and Forwarding | Yes | Yes | No |
| | PBRDV | Greedy | Store and Forwarding | Yes | Yes | No |
| | AODV | Multi-hop | Store and Forwarding | No | No | No |
| | VADD | Greedy | Store and Forwarding | No | Yes | No |
| | GeOpps | Greedy | Store and Forwarding | Yes | Yes | No |
| Non-congested scenario based | GPSR | Greedy | Store and Forwarding | No | Yes | No |
| | DTSG | Multi-hop | Flooding | No | No | No |
| | DVCAST | Multi-hop | Store and Forwarding | No | No | No |
| | DG-CASTOR | Multi-hop | Flooding | No | No | Yes |

**Fig. 7** Congestion starts to reduce re-routing begins

the graph this is so because after some time into the simulation there occurs congestion and at this point the algorithm starts to re-route the vehicles to the next best path possible in order to remove congestion and we can see that the graph gradually again moves up towards 50 km/hr mark.

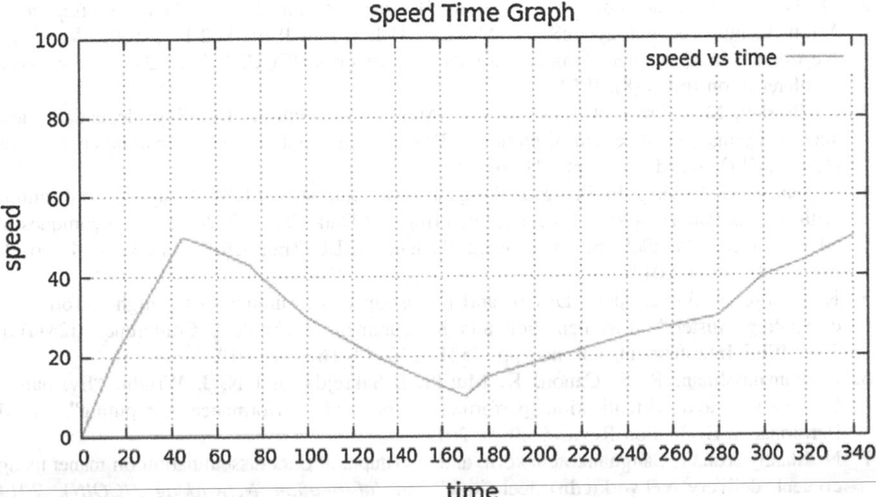


**Fig. 8** Efficient re-routing with 802.11p for VANETs

## 5 Conclusion and Future Scope

Re-routing and routing are important aspects of VANET communication be it of V2V type or V2I type. This paper provides an overview of different VANET protocols. It is a difficult task to develop an efficient and reliable protocol for all application in VANETs. VANET protocols rely on various parameters out of which few are mentioned in the Table 1 the others could be like model support, obstruction and environmental support, etc. In this paper an elaborate study has been carried in order to bring about the various VANET protocols and have been classified on the basis of their communication scenario.

Routing of VANETs over the past few years have gained lot of weight and are now in consideration for a large number of future application. One of such applications could be security which affects the life and death of the commuter and depends upon accurate and precise decisions which due to any destructive interference may cause severe consequences. Also to consider the fact those VANETs are a more demanding communication network when it comes to secure routing, distribution of data efficiently and sharing of data, these are among some of the most crucial challenges that need to be overcome for a successful VANET deployment.

## References

1. Chen, Q., Jiang, D., and Delgrossi, L. 2009. "IEEE 1609.4 DSRC multi-channel operations and its implications on vehicle safety communications". In Vehicular Networking Conference (VNC), 2009 IEEE (pp. 1–8). IEEE.

2. Di Felice, M., Ghandour, A. J., Artail, H., and Bononi, L.2012. "On the Impact of Multi-channel Technology onSafety-Message Delivery in IEEE 802.11 p/1609.4Vehicular Networks". In Computer Communications andNetworks (ICCCN), 2012 21st International Conference on (pp. 1–8). IEEE.
3. A. Bessani, M. Correia, B. Quaresma, F. Andr, and P. Sousa, "Dep-Sky: dependable and secure storage in a cloud-of-clouds," Proc. Sixth conference on computer systems (EuroSys'11), ACM, 2011, pp. 31–46.
4. S. Sehrish; G. Mackey; P. Shang; J. Wang; J. Bent, "Supporting HPC Analytics Applications with Access Patterns Using Data Restructuring and Data-Centric Scheduling Techniques in Map Reduce," Parallel and Distributed Systems, IEEE Transactions on, vol. 24, no. 1, pp. 158,169, Jan. 2013.
5. K. George; V. Venugopal, "Design and performance measurement of a high-performance computing cluster," Instrumentation and Measurement Technology Conference (I2MTC), 2012 IEEE International, vol. no., pp. 2531,2536, 13–16 May 2012.
6. L. Ramakrishnan, R. S. Canon, K. Muriki, I. Sakrejda, and N. J. Wright. "Evaluating Interconnect and virtualization performance for high performance computing", ACM Performance Evaluation Review, 40(2), 2012.
7. N. Khaitiyakun, T. Sanguankotchakorn, and A. Tunpan. "Data dissemination on manet using content delivery network(cdn) technique". In *Information Networking (ICOIN), 2014 International Conferenceon*, pages 502–506, Feb2014.
8. Jinwoo Nam, Seong-Mun Kim and Sung-Gi Min, "Extended wireless mesh network for VANET with geographical routing protocol," *11th International Conference on Wireless Communications, Networking and Mobile Computing (WiCOM 2015)*, Shanghai, 2015, pp. 1–6.
9. Y. Bai, D. Xie, S. Wang and M. Zhong, "Multi-path transmission protocol in VANET," *2015 International Conference on Connected Vehicles and Expo (ICCVE)*, Shenzhen, 2015, pp. 204–209.
10. D. Jiang, & L. Delgrossi, (2008). "IEEE 802.11p: towards an international standard for wireless access in vehicular environments". In Proceedings of 67th IEEE vehicular technology conference on vehicular technology (pp. 2036–2040), May 2008.
11. J. Zhao, and G. Cao,2008. "VADD: Vehicle-Assisted Data Delivery in Vehicular". *Vehicular Technology, IEEE Transactions on*. 57.3.1910–1922.
12. I. O. Chelha and S. Rakrak, "Best nodes approach for alert message dissemination in VANET (BNAMDV)," *RFID And Adaptive Wireless Sensor Networks (RAWSN), 2015 Third International Workshop on*, Agadir, 2015, pp. 82–85.
13. K.C. Lee, U. Lee, and M. Gerla, Oct. 2009. "Survey of routing protocols in vehicular ad hoc networks". *Advances in Vehicular Ad-Hoc Networks: Developments and Challenges, IGI Global.*
14. G. Yan, D. B. Rawat and B. B. Bista, "Towards Secure Vehicular Clouds," Proc. Complex, Intelligent and Software Intensive Systems (CISIS), 2012 Sixth International Conference on, 2012, pp. 370–375.
15. F. Granelli, G. Boato, and D. Kliazovich, 2006. "MORA:A movement-based routing algorithm for vehicle adhoc networks". *1st IEEE Workshop on Automotive Networking and Applications (AutoNet 2006)* 2006. San Francisco, USA.
16. S. Pearson and A. Benameur, Privacy, "Security and Trust Issues Arising from Cloud Computing". Proc. Cloud Computing Technology and Science(CloudCom), 2012 IEEE Second International Conference on, 2012, pp. 693–702.
17. T. Atéchian, L. Brunie, J. Roth, and J. Gutiérrez. "DGcastoR: direction-based geo-cast routing protocol for query dissemination in VANET". *IADIS International Telecommunications, Networks and Systems*. 2008.
18. A. Festag, (2009). "Global standardization of network and transport protocols for ITS with 5 GHz radio technologies". In Proceedings of the ETSI TC ITS workshop, Sophia Antipolis, France, February 2009.

19. R. K. Shrestha, S. Moh, I. Chung, and D. Choi 2010. "Vertex-based multi-hop vehicle-to infrastructure routing for vehicular ad hoc networks". In *IEEE proceedings of 43rd Hawaii International Conference on System Sciences (HICSS).*
20. L. Kristiana, C. Schmitt and B. Stiller, "Survey of angle-based forwarding methods in VANET communications," *2016 Wireless Days (WD)*, Toulouse, 2016, pp. 1–3.
21. S. Grafling, P. Mahonen, & J. Riihijarvi. "Performance evaluation of IEEE 1609 WAVE and IEEE802.11 p for vehicular communications". In Ubiquitous and Future Networks (ICUFN), 2010 Second International Conference on (pp. 344–348). IEEE.
22. Praveen. K. Gupta and Nitin Rakesh, "Different Job Scheduling Methodologies for Web Application and Web Server in a Cloud Computing Environment", 2010 3rd International Conference on Emerging Trends in Engineering and Technology, IEEE, Pages 569–572, Goa, India, Nov. 19–21.
23. Nitin Rakesh and Vipin Tyagi, "Linear-code multicast on parallel architectures," Elsevier Advances in Engineering Software, vol. 42, pp. 1074–1088, 2011.
24. Nitin Rakesh and Nitin, "Analysis of Multi-Sort Algorithm on Multi-Mesh of Trees (MMT) Architecture", Springer Journal of Supercomputing, vol 57, no 3, pp. 276–313, 2011.
25. Nitin Rakesh and Vipin Tyagi, "Linear Network Coding on Multi-Mesh of Trees using All to All Broadcast," International Journal of Computer Science Issues, vol 8, no 3, pp. 462–471, 2011.
26. Nitin Rakesh and Vipin Tyagi, "Parallel architecture coding: link failure–recovery mechanism (PAC: LF–RM)," Springer International Journal of System Assurance Engineering and Management, vol 4, no 4, pp. 386–396, 2013.
27. Sandeep Pratap Singh, Shiv Shankar P Shukla, Nitin Rakesh, Vipin Tyagi, "Problem reduction in online payment system using hybrid model", International Journal of Managing Information Technology (IJMIT) Vol.3, No.3, pp. 62–71, August 2011.
28. Praveen K Gupta, Nitin Rakesh, "Different job scheduling methodologies for web application and web server in a cloud computing environment", 2010 3rd International Conference on Emerging Trends in Engineering and Technology (ICETET), pp. 569–572, 2010, India.
29. I. B. Jemaa, O. Shagdar, F. J. Martinez, P. Garrido and F. Nashashibi, "Extended mobility management and routing protocols for internet-to-VANET multicasting," *2015 12th Annual IEEE Consumer Communications and Networking Conference (CCNC)*, Las Vegas, NV, 2015, pp. 904–909.

# Efficient Module for OHM (Online Hybrid Model)

**Akash Agarwal, Nitin Rakesh and Nitin Agarwal**

**Abstract** The Internet-based businesses are increasing day by day and even the new concept of digital India is developing, through which every government and private companies are switching to Internet and cloud services. Internet users are increasing day by day at State Bank of India and about 69% of daily transactions happen through alternative channels, including Internet, ATM, and mobile banking. This figure is rising every year and more young generation is using online services. But on the other hand, there are many security concerns as well; presently we have many secured transaction channels. In this paper, we have proposed the new algorithm to prevent frauds and track the transactions location if any fraud occurs in the Internet. Using Online Hybrid Model algorithm, we will generate unique Internet id of the user. This algorithm also supports in the prevention of fraudulent activities, for example, if terrorist do any transaction online, we can track easily.

**Keywords** Online phantom transactions · Online fraud detection · OHM · E-Commerce

A. Agarwal · N. Rakesh (✉)
Amity School of Engineering and Technology, Noida, UP, India
e-mail: nitin.rakesh@gmail.com

A. Agarwal
e-mail: akashgrwl79@gmail.com

A. Agarwal · N. Rakesh
Amity University Uttar Pradesh, Noida, UP, India

N. Agarwal
KPMG, Bengaluru, Karnataka, India
e-mail: nitinagarwal1@kpmg.com



S.K. Bhatia et al. (eds.), *Advances in Computer and Computational Sciences*,
Advances in Intelligent Systems and Computing 553,
DOI 10.1007/978-981-10-3770-2_34

## 1 Introduction

With the advent of digital technology, e-commerce [1] has become very popular for online shopping, online bidding, merchandise, etc. It has resulted in commercialization of e-shopping and thus lead to brand awareness and a source of revenue generation, as well as making shopping easy for an individual [2]. At the same time, it resulted in malpractices and misuse of the technology. Fraudsters attempt to get some illegal benefits through misinterpretation of facts and figures [3]. One of the major parts of online frauds is Internet auction fraud [4]. Here the illegitimate users try performing fraudulent activities to gain illegal profit [5]. An online auction in general terms is bidding which takes place in the Internet. [1]. It comprises of two groups of people [6]:

1. Seller and
2. Buyer

Malicious sellers post their products on the Internet auction sites and when people purchase those goods, it does not get delivered after payment [7]. That item may be posted for bidding with false specifications to deceive the buyer about its actual price. In this paper, we are proposing different techniques concerning Online Hybrid Model which was discussed earlier [8]. In the earlier module, we find how the OHM registers the user verifying its credentials and authenticating the user as buyer or seller.

This model will generate the Unique Internet ID of all users who use the Internet as discussed in the algorithm [9]. In this algorithm, we are taking certain ID proofs of users and validate all ID proofs with the government database and with reference with that database, we are generating the Unique Internet ID. The overall process will take 15 days' time to approve UID and if all documents are legal, then UID will be generated and will be sent over mail and to the physical address [10]. This UID will keep track of all the online transactions; through this model it is very easy to track all fraudulent and unethical activates on Internet [11].

## 2 Previous Work

OHM model can be broadly classified in three categories (shown in Fig. 1).

1. **Module 1**: It comprises of Fraud [5, 6] Avoidance Techniques in online auction process,
2. **Module 2**: It includes the techniques of fraud detection [12].
3. **Module 3**: In this we are proposing the techniques of fraud prevention

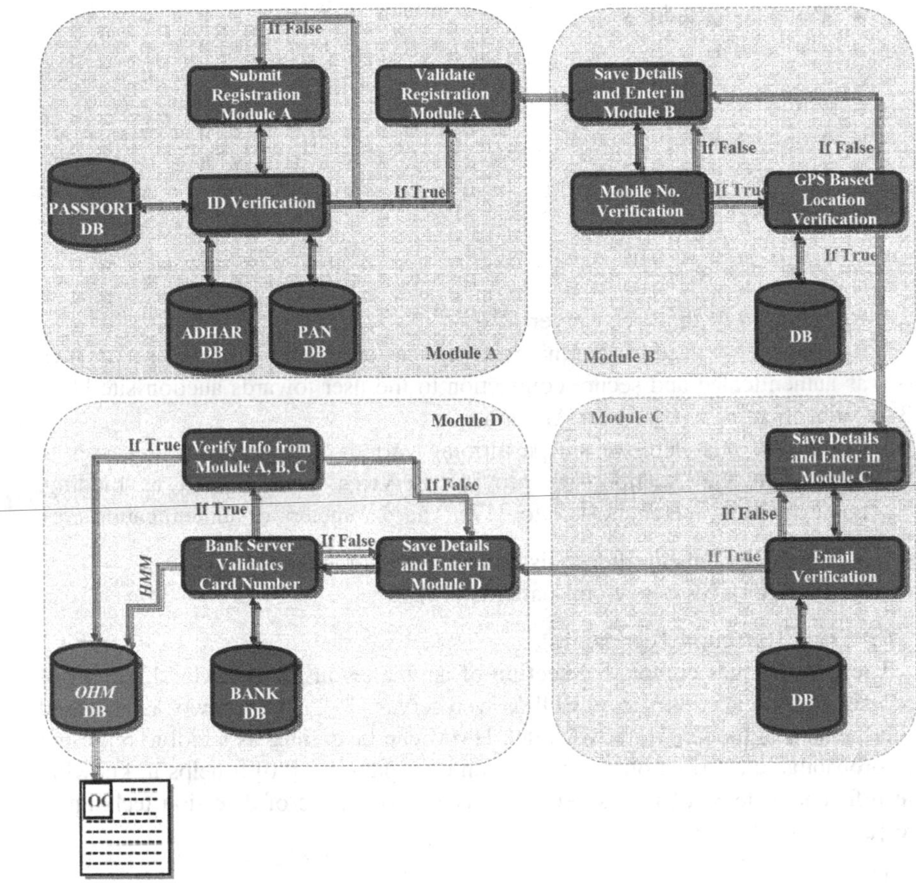


**Fig. 1** OHM Logical Design for USER Registration

(a) Fraud Avoidance Techniques [13]:
With reference to the previous paper, these can be classified as.

(1) Registration of the Seller and Buyer [14]:
The users of the auction site can be authenticated with the help of Modules A, B, and C discussed previously. In this, the parameters needed are:
**Module A**

- Name of the buyer/seller
- Date of Birth
- Sex

- Permanent Address
- Nationality

**Module B**

- Mobile Number
- Current Address

**Module C**

- Email Address

(2) Registration of the Web Server [15]:
In this part, we need to scrutinize the services provided by the web server such as authenticated and secure connection to the user towards auction site [16].

(3) Authenticating the Internet Auction Site
Coming to this step, we are performing various authentication processes to verify that the Auction site provides services to user such as bidding, buying/selling, merchandise, etc. [17]. The parameters of authentication are

- Policy agreement after registration of auction site.
- Web Services behavioral patterns [18].

(b) Fraud Detection Techniques:

These techniques comprise detection of anomalies in the behavioral pattern of the user and the auction site as well as web server [19]. The behavior is analyzed with the help of hidden Markov Model. HMM can be defined as a tool to represent the probabilistic distribution of observation over patterns. [20] It helps in knowing the different patterns of user's behavior. The components of detection techniques are [21]:

- User
- Mobile Phone
- Auction Site
- Authentication Center
- Data Collector
- Authentication Decision Maker
- OHM to provide OHM Certificate
- Web Server

From the previous paper, we know that there are certain threshold values chosen for the authentication of user's identity and mobile phone [22]. By comparing the entries made by user and the threshold values, we can conclude whether the user can proceed to get the OHM certificate and participate in auction process [23].

## 3 Proposed Model

In the registration module we saw that A, B, and C are responsible for user in terms of address, mobile number, and email address [24]. We can optimize the three modules by unique random number generator through retrieving authentic data of the user(seller/buyer) from various domains like the passport, pan card, voter id card, etc. [3].

The unique number generation technique follows following set of steps:

1. First of all it takes the input parameters from the user like pan card details, passport details, Aadhar card details, etc. [25].
2. These details are then sent to the authentication center for the validation of data entered by the user by comparing with the threshold values.
3. The threshold values are taken from previous paper on "online fraud prevention" [26].
4. Suppose the user's passport number is XXX00000YYYY and pan number is XYZAB0000 and Aadhar card number is @22#333$555*.
5. The generator will perform the operation by taking the values from these parameters in the subsequent steps
   - Selection of keys like "X0Y" for passport, "XA0" for pan Card, and @#$ for Adhar Card.
   - These keys are modulated again by generator so as get another robust unique number like "X@0Y#A0X$" and so on.

Initial Condition: User Registration [27].

Local Variable: Aadhar card, driving license, Pan card, passport, voter id, bank pass book [12];

Final Condition: User Authentication and Issue of Internet Unique ID.

Algorithm for Identity Authentication [28].

(a) Input: Upload photo using webcam, Aadhar card details, passport details, driving license details, pan card details, bank passbook details, voter ID details, upload soft copy of all documents.

(b) Details should be in the set of
Set (1)–Passport, Pan, Driving License, Voter ID
Set (2)–Aadhar, Pan, Bank Account, DL
Set (3)–Aadhar, Passport, Voter ID, Bank Account

(c) Verification: Details are sent to authentication center connection with the database of respective departments and are compares the threshold value complete process take time of 15 days after submission of all details successfully [29].

```
Input:
Data Center Value(s)
If
{If Input value == Data center value then authenticate user
Else
{Registration Failed}
If
{Data == Passport
Passport == A9900905444CZE
Generate: First two digit, Fifth digit from left, Third digit from
right}
If
{
Data == PAN CARD
PAN NO == BIIPA990
Generate: Third digit from right, second digit from left
}
If
{
Data == Driving license
Driving license No == RJ1220050000188
Generate: Ninth digit from left, seventh digit from right, Two
alphabets from starting
}
If
{
Data == Voter ID
Voter ID == YCV0164798
Generate: Fifth digit from left, fifth digit from right, first two digit
}
If
{
Data == Bank account passbook
Bank Account Number == 32796878470
Generate: Second digit from right, Fifth digit from left, First three
digit from right
}
If
{
Data == Aadhar card
Aadhar card == 12346789011
Generate: Fifth, Six, Seventh, eight digit from left
}
}
If
{
User authenticate
Generate: Internet Digital ID
}
If
{
```

(continued)

(continued)

```
Input:
Data Center Value(s)
User Submit == Set 1
Then IUD == Set 1
}
If
{
User Submit == Set 2
Then IUD == Set 2
}
If
{
User Submit == Set 3
Then IUD == Set 3
}
{
Upload Photo == Photo ID
Verified UID;
}
Output:
SET 1: UID = A906I900RJ16YC
SET 2: UID = 4678I926832700RJ
SET 3: UID = 4678A90616VC268327
```

## 4 Conclusion

In this paper, we have proposed the algorithm for assigning a Unique Internet ID to the Internet users when the user does any Internet transaction online. Then all transaction of users are traceable with its respective details and if any user is found to be involved in any fraud case, we can track his address and all the details. This algorithm will prevent all frauds on Internet and by this algorithm, we are also validating the physical address of person. When the implementation of this algorithm takes place, the companies will keep it mandatory to fill Internet Unique ID on the products which they want and customer cannot buy the products online without filling UID. This will help to avoid any type of frauds in online transaction; so it is clear in the previous paper that we are not validating the physical address of the user and the transactions are not traceable in the previous model of OHM.

## References

1. U.S. Commerce Department, Forrester Research, Internet Retailer, ComScore., www.statisticbrain.com/total-online-sales/.
2. Prasad, B.: Intelligent techniques for E-Commerce. J. Electron. Commer. Res. 4(2), 65–71 (2003).

3. Kristin M. Finklea, "Identity Theft: Trends and Issues," Congressional Research Service, A CRS research report for congress, 2012.
4. National Crime Prevention Council, http://www.ncpc.org/cmsupload/ncpc/File/aucfraud.pdf.
5. Understanding Credit Card Frauds, www.popcenter.org/problems/credit_card_fraud/PDFs/Bhatla.pdf.
6. Fei Donga, Sol M. Shatza and HaipingXub, "Combating Online In-Auction Fraud: Clues, Techniques and Challenges," Computer Science Review, 3 (4) 245–258, 2009.
7. W.L. Wang, Z. Hidvègi, and A. B. Whinston, "Shill Bidding in English Auctions," Technical report, Emory University, 2001, http://oz.stern.nyu.edu/seminar/fa01/1108.pdf.
8. National White Collar crime center: Report on Internet fraud, www.nw3c.org/docs/whitepapers/internet_fraud.pdf?sfvrsn=7, June 2008.
9. J. Trevathan and W. Read, "Detecting Collusive Shill Bidding," Proc. of International Conference on Information Technology: New Generations, 799–808, 2007.
10. Porter, R., Shoham, Y., On cheating in sealed-bid auctions. J. Decis. Support Syst. Special issue of the fourth ACM Conference on Electronic Commerce, 39(1), 41–54 (2005).
11. U.S. Commerce Department: Forrester Research, Internet Retailer, ComScore., http://www.statisticbrain.com/total-online-sales/.
12. B. Prasad: "Intelligent Techniques for E-Commerce," Journal of Electronic Commerce Research, 4 (2) 65–71, 2003.
13. W. L. Wang, Z. Hidvègi, and A.B. Whinston, "Shill Bidding in Multi-Round Online Auctions," Proc. of the 35th Annual Hawaii International Conference on System Sciences, January 2002.
14. Fault tolerance based routing approach using WMN http://ieeexplore.ieee.org/xpls/abs_all.jsp?arnumber=7361345.
15. AnkitMundra, Nitin Rakesh, (2013) "Online Hybrid Model for Online Fraud Prevention and Detection," International Conference on Advance Computing, Networking, and Informatics–ICACNI-2013, Springer, pp. 805–815.
16. Wang, W.L., Hidvègi, Z., Whinston, A.B.: Shill Bidding in Multi-Round Online Auctions. In: Proceedings of the 35th Annual Hawaii International Conference on System Sciences, Jan 2002.
17. Trevathan, J., Read, W.: Detecting Collusive Shill Bidding. In: Proceedings of InternationalConference on Information Technology: New Generations, pp. 799–808 (2007).
18. Liang Zhang Jie Yang Belle Tseng, "Online Modeling of Proactive Moderation System for Auction Fraud Detection," World Wide Web Conference (WWW), 669–678, 2012.
19. Srivastava, A., Kundu, A., Sural, S., Majumdar, A.K.: Credit card Fraud Detection using Hidden Markov model. IEEE Trans. Dependable Secure Computer5(1), 1062–1066 (2008).
20. SandeepPratap Singh, Shiv Shankar P. Shukla, Nitin Rakesh and VipinTyagi, "Problem reduction in online payment system using hybrid model," International Journal of Managing Information Technology, 3(3) 62–71, August 2011.
21. Chui, K., Xwick, R.: Auction on the Internet: A Preliminary Study, http://repository.ust.hk/dspace/handle/1783.1/1035, July 2008.
22. S.O. Falki, B.K. Alese, O.S. Adewale, J.O. Ayeni, G.A. Aderounmu and W.O. Islamia, "Probablistic Credi Card Fraud Detection System in Online Transactions," Interantional Journal of Software Engineering and Its Applications, Vol.6, No.4, 69–78, 2012.
23. S. P. Singh, S. S. P. Shukla, Nitin Rakesh,VipinTyagi., Problem reduction in online payment system using hybrid model. Int. J. Manag. Inf. Technol. 3(3), 62–71 (2011).
24. AbhinavSrivastava, AmlanKundu, S. Sural, A.K. Majumdar, "Credit Card Fraud Detection Using Hidden Markov Model," IEEE Transactions on Dependable And Secure Computing, 5 (1) 1062–1066,2008.
25. Stephan Kovach, Wilson Vicente Ruggiero, "Online Banking Fraud Detection Based on Local and Global Behavior," Proc. Of ICDS: The Fifth International Conference on Digital Society, 166–171, 2011.

26. Yungchang Ku, Yuchi Chen, Chaochang Chiu, “A Proposed Data Mining Approach for Internet Auction Fraud Detection,” Intelligence and Security Informatics Lecture Notes in Computer Science Volume 4430, 238–243, 2007.
27. Wang, W.L., Hidvègi, Z., Whinston, A.B.: Shill Bidding in English Auctions, Technical report, Emory University, http://oz.stern.nyu.edu/seminar/fa01/1108.pdf (2001).
28. Internet Crime Complain Center: Internet Crime Report, 2004–2011, http://www.ic3.gov/media/annualreports.aspx.
29. National White Collar crime center, Report on Internet fraud, June, 2008, www.nw3c.org/docs/whitepapers/internet_fraud.pdf?sfvrsn=7.

# SANet: An Approach for Prediction in Music Trends

**Fei Hongxiao, Chen Li, He Jiabao, Xiao Yanru and Liu Han**

**Abstract** The precise prediction of the popular trend in music can contribute to the exploration of the potential entertainment market. According to surveys, the technical difficulties of such prediction contain the difference between computer simulation and the real human emotions, as well as the comprehensive factors and data that are processed. Therefore, this thesis will present SANet which can forecast the popular trend in songs by self-accommodating and nonlinear mapping. It will be demonstrated by focusing on the discussion in the areas on data preprocessing, model constructing, and accommodating of hidden columns, as well as the test of partial data by random sampling and the analysis of the experiment result.

**Keywords** Self-adapting · Prediction · Trend of music

F. Hongxiao (✉) · C. Li · H. Jiabao · L. Han
Software Engineering Department of Central South University,
Changsha 410000, Hunan, China
e-mail: hxfei@csu.edu.cn

C. Li
e-mail: vchenli@csu.edu.cn

H. Jiabao
e-mail: hejiabao@csu.edu.cn

L. Han
e-mail: liuhan@csu.edu.cn

X. Yanru
Information Science and Engineering Department of Central South University,
Changsha 410000, Hunan, China
e-mail: sugarray@csu.edu.cn



S.K. Bhatia et al. (eds.), *Advances in Computer and Computational Sciences*,
Advances in Intelligent Systems and Computing 553,
DOI 10.1007/978-981-10-3770-2_35

# 1 Introduction

The preference for people to music mainly depends on the emotion. In addition, some other subjective factors like the background of creation, propaganda, language, and so on will also bring a certain influence [1]. The analyses of music focus more on the user's recommendation and only a few institutions work on this field now. A series of solutions have been raised with the use of neural networks [2], Gaussian Mixture Model [3], Support Vector Machine [4], and other methods to forecast music trends in domestic and foreign. Foreign scholars C. Laurier used a combination of audio and lyrics to classify emotions expressed by the music, and then study music trends [5]. But, mostly, the expression of language is ambiguous.

During the process of analyzing trends of song, we find the main difficulties as follows: (1) the trends of pop music reflect the particular trend of public aesthetic while a group of people being famous will popularize their main element; (2) a lot of factors, such as characteristics of music and times, people's emotions, are combined together to predict music popular. And these factors are of great change and contingency; (3) taking too much into consideration results in an enormous data size. Low efficiency of common computing model for data processing will lead the loss of timeliness.

The thesis put forwards SANet, based on back propagation, to improve its ability of forecasting the popular trend in music. The contributions of the thesis are as follows:

(1) The whole network structure reduces the computational burden which makes it more straightforward to achieve with powerful expansibility.
(2) The hidden layer nodes in the network can adjust automatically based on the data size, and the thesis can make a more accurate prediction by making more elaborate adjustments on the study rate.
(3) After the farther optimizing, this model can predict the data of the following M day(s) depending on the data of the former N day(s).

# 2 SANet

## *2.1 Adjustment of Node Number in Hidden Layers*

Although the selection of nodes number in hidden layers will exert a tremendous influence, there is no complete theoretical direction on selecting the number of neuron in the hidden layer at present. If the number of nodes is not enough, the network will fail to get enough useful information which can cause bad learning effect and low fault tolerance; if the number of nodes is too large, it would increase the training time and store many irregular contents from sample which can cause an over-fitting problem and a loss of generalization ability. Therefore, selecting reasonable hidden layer nodes have a significant effect to the network performance.

In order to find out the best number of hidden layer nodes quickly, the processes are as follows:

- Determine the selecting range $n_1 \leq n \leq n_2$ according to the empirical formula.
- Make $n = n_1$, calculate the back propagation when the number of the hidden layer nodes is $n_1$, get the network sum of squared errors $M_1$.
- Make $n = (n_1 + n_2)/2$, n is integer. Calculate sum of squared errors $M_2$ of n hidden nodes in the back propagation.
- If $M_1 < M$, then make $n_1 = n$, $M_1 = M_2$; otherwise, make $n_2 = n$.
- If $n_1 < n_2$, then return to the step 2, follow this cycle. Otherwise, exit program and end the algorithm.

The n is the better number of the hidden layer nodes at the end of the algorithm.

## 2.2 Adjustment of Studying Rate

In the BP neural network, training rate is the important factor which can influence the training speed and accuracy of training. If the training rate is too small, accuracy of training can be guaranteed, but the convergence speed and cycle would be too long. If the training rate is too big, it can shorten the training cycle, improve the convergence speed, but can lead to oscillation or divergent. Thus a fixed training rate could not be suitable for the whole network training process.

The back propagation algorithm provides that the steepest descent method has been got a kind of approximate in the calculation of weight space [6]. The smaller the training rate parameter n is, the smaller the variation of the network synaptic weight from one iteration to the next iteration and the smoother the locus in the weight space will be.

We can use the following methods to solve this problem:

$$\Delta w_{ij}(n) = \alpha \Delta w_{ij}(n-1) + \eta \, \delta j(n) yi(n)$$

In the formula, $\alpha$ is a momentum constant [7], which is usually an integer.

# 3 The Construction of Model

## 3.1 Prediction Model

The design of music popular trend prediction model in this paper is divided into two periods: training and prediction. The training period is a process to use the sample to study and predict the trend in popular music. The prediction stage is a process to

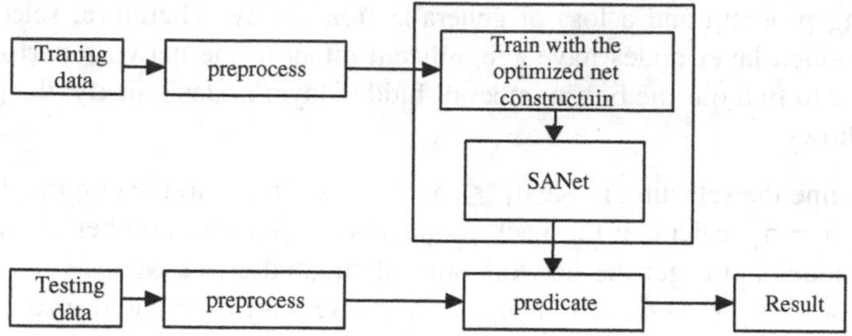


**Fig. 1** Songs popular trend model

use the neural network which has been studied during the last stage to predict the popular trend in music. To improve the generalization ability of this model and reinforce the veracity to the popular trend in music, we choose to use the neural network integration technology. In this paper, the integration of the neural network reflects on data preprocessing and results fusion these two aspects. The details show in the Fig. 1.

## 3.2 Data Preprocessing

Through the analysis of the data, we find that plays are much higher than the collections and downloads; so it is obvious that the music plays have a greater influence on the future data. However, collections and downloads, having different characteristics, cannot be ignored easily because they are small on the numerical. So we want to find a ratio between them instead of getting average of them.

Assuming pi, $c_i$, $d_i$, $t_i$ ($i \geq 1$) respectively represent the plays, collection, downloads, and the output in the ith day. $w_1$, $w_2$, $w_3$ respectively represent the weight value of inputs from $p_i$, $c_i$, $d_i$ to $t_i$. Because the relationships among the three properties of each song are the same, w1, w2, w3 of each song should be the same. Moreover, the back propagation algorithm is a self-study network that w1, w2, w3 can be operated by the structure itself.

## 3.3 Determine the Model

As Gerald Tesauro [8] put forward in his paper that the proportionate relationships are determined by the size of training set, the BP neural network and the 80% of the performance level should be learned. Therefore, we can divide the data with the ratio of 4 to 1 to get a better result.

As for choosing the number of hidden layers, the theory suggests that too much variables can get better fitting effect and even over-fitting. Hornik has proved that if a linear conversion function is used in the input layer and output layer, with hidden layer using sigmoid function, the neural network with one hidden layer can approach any rational function with arbitrary-precision [9]. Therefore, the model selects a three-layer BP neural network with an input layer, a hidden layer, and a output layer.

As for hidden layer nodes and learning rate, apply what the essay mentioned about the number of hidden layer nodes and adjustment earlier.

## 4 Experiment Design and Results

With the data from Ali, music users from March 1, 2015 to August 30, 2015, the downloads, plays, and collections are processed and integrated as the data pre-processing mentioned. The SANet taken in this thesis is a three-tier structure with 30 nodes in the input layer and six nodes in output layer, so the plays of music in the next 6 days can be estimated by the data during the former 30 days. And then, the impossible appropriate number of hidden layers may be 5, 9, 12, 16, 19. As there are 147 groups of data, the first 120 groups of data can be used as training data and the remaining 27 groups of data can be used as the validation data. In the model, weights are initialized by Gaussian random.

According to the above analysis of the number of nodes in hidden layers, we need to compare the number of hidden nodes' deviation which are 5, 12, 19 first. And the following Table 1 shows result when the hidden layer nodes are 5, 12, 19. Finally, we know when the hidden nodes are 16, it has the best fitting results.

After we determine the model structure, we use the same principles to come up with an analysis diagram which is the n 30 day songs popular trends. As shown in Fig. 2, Red is the actual amount of playing, green for predicting the amount of playing.

From the result, we find that the prediction of song trend is more accurate in the prediction stage; after a day's relative to the amount of songs played, the day before the situation is almost right. Although it is not completely close in value, it is proved to succeed in avoiding over-fitting phenomena.

**Table 1** During the experiment, we find the sum of squared errors is large when the nodes number is 5 than 12 and 19. So we can limited the hidden nodes between 12 and 19

| Sum of squared error | 2000 epochs | 4000 epochs | 8000 epochs |
|---|---|---|---|
| 5 nodes in hidden layer | 0.0130 | 0.0121 | 0.0118 |
| 12 nodes in hidden layer | 0.0094 | 0.0072 | 0.0068 |
| 19 nodes in hidden layer | 0.0112 | 0.0070 | 0.0042 |

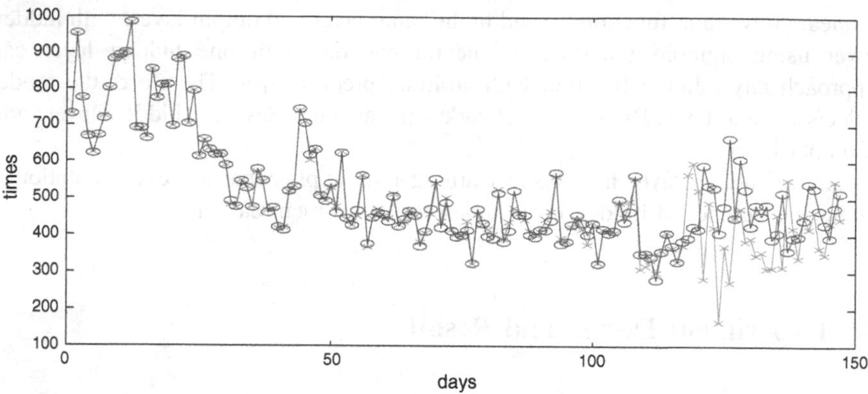


**Fig. 2** Prediction by SANet when the hidden layer nodes is 16

# 5 Conclusion

This paper proposes SANet to analyze the pop music trend, which helps to predict the sum of play in the future. While building the model, a concrete analysis of study rate and spots in hidden layers is focused on to improve the correctness of prediction. This model has a great expansibility and can be applied to other predictions because this construction learns by itself instead of relying on a special functional relationship.

# References

1. Meredith D. Music analysis and point-set compression [J]. Journal of New Music Research, 2015, 44(3): 245–270.
2. Burr G W, Shelby R M, Sidler S, et al. Experimental demonstration and tolerancing of a large-scale neural network (165 000 Synapses) using phase-change memory as the synaptic weight element[J]. Electron Devices, IEEE Transactions on, 2015, 62(11): 3498–3507.
3. GuanghuShen; Quang Nguyen; JongSuk Choi IFAC Proceedings Volumes 2012–6
4. Samsudin R, Shabri A, Saad P.A Comparison of Time Series Forecasting using Support Vector Machine and Artificial Neural Network Model [J]. Journal of Applied Sciences, 2010, 10(11).
5. Chathuranga Y, Jayaratne K L. Automatic Music Genre Classification of Audio Signals with Machine Learning Approaches [J]. Gstf Journal on Computing, 2013, 3(2):1–12.
6. Jenab A, Sarraf I S, Green D E, et al. The Use of genetic algorithm and neural network to predict rate-dependent tensile flow behaviour of AA5182-O sheets [J]. Materials and Design, 2016, 94:262–273.
7. S. K. Lenka, A. G. Mohapatra. Gradient Descent with Momentum based Neural Network Pattern Classification for the Prediction of Soil Moisture Content in Precision Agriculture[C] IEEE International Symposium on Nano electronic and Information Systems. IEEE, 2015:63–66.

8. Gerald Tesauro, Scaling relationships in back-propagation learning: dependence on training set size". Complex Systems 1 (1987) 367–372.
9. Ye Weijun, NingShan, Jiang Yanqiu. Convolution neural network optimization algorithm research [J]. Journal of qiqihar university (natural science edition), 2016 (2).

# Multichannel Dual Clocks Two-Dimensional Probability Random Access Protocol with Three-way Handshake Mechanism

**Hongwei Ding, Shengjie Zhou, Kun Yue, Chunfen Li, Yifan Zhao, Zhijun Yang and Qianlin Liu**

**Abstract** Although two-dimensional probability random multiple access protocol can have a better control effect on the system, it features a single agreement, the practical application of the network is not strong. To solve the above problem, the model of multichannel dual clocks two-dimensional probability random access protocol with three-way handshake mechanism (DMTPTH protocol) is proposed. DMTPTH is based on the two-dimensional probability random access model, adding dual-clock, multichannel and three-way handshake mechanism, respectively, to achieve lower system idle rate, confirm the multiservice access and information transmission states. Theoretical analysis and simulation experiments show that the model can not only improve system throughput, but also achieve throughput by changing the two-dimensional probability adapt to different network environments, enhanced agreement practicality, and reliability.

**Keywords** Random multiple access · Dual-clock · Two-dimensional probability · Multi-channel · Three-way handshake

## 1 Introduction

Nowadays, people live in a sea of data, in the face of such a large amount of data [1]. How to effectively use these data, especially how to better transmit such large amounts of data, friendly and easy to realize data exchange, it is the great challenge that we face [2]. Therefore, it is of practical significance to study the random access MAC protocol in big data environment. On the basis of the study of the dual clocks two-dimensional probability random access protocol, the model of multichannel dual clocks two-dimensional probability random access protocol is studied,it can achieve a variety of types of data to achieve differentiated priority transfer, the data of different priorities are transmitted at different priorities, these data include video

H. Ding (✉) · S. Zhou · K. Yue · C. Li · Y. Zhao · Z. Yang · Q. Liu
School of Information, Yunnan University, Kunming 650091, China
e-mail: dhw1964@163.com



S.K. Bhatia et al. (eds.), *Advances in Computer and Computational Sciences*, Advances in Intelligent Systems and Computing 553,
DOI 10.1007/978-981-10-3770-2_36

data, voice data, text data, and so on [3]. In order to better ensure the reliability of transmission data, three-way handshake mechanism is introduced into the model. Through introducing the three-way handshake, the system can transmit data more reliably. Of course, this will lose a small amount of channel resources, in order to achieve a higher reliability; a small amount of loss of system resources is worth [4].

In wireless communication network, the MAC layer protocol plays a key role on the performance of the whole system. Therefore, MAC layer protocol is designed in consideration of factors such as: throughput, security, scalability, and robustness and so on. For different tasks and network environment, experts and scholars at home and abroad proposed a variety of protocols [5].

The basic idea of two-dimensional probability random access protocol is: when the channel is idle, the site sends the information packet with the probability P1 if there are new information packets arriving, abandoning the action of sending with range of probability 1-P1 [6]; when the channel is busy, the site continue to detecting the channel with probability P2, abandoning the action of detecting with range of probability 1-P2 [7].

From the basic idea, we can see that when the channel is free, selecting a high probability of sending a message packet, the throughput of the system will increase; and when the channel is busy [8], selecting the high probability of detecting the channel, it will cause information high probability of collisions; therefore, if we adjust two probability reasonably, then the two-dimensional random probability agreement will achieve a higher throughput rate [9].

## 2 Protocol Model

First, solve the average length $E(U)$ of packet successfully sent in the event of packet sent successfully.

Packet successfully sent into the following two cases:

(1) If packets arrive during the last slot of idle period, namely packet arrives at the continuous clock control, and in the next slot time, no one but it adhere to send it, then it is sent successfully, the record for the event is $U_1$. Its average length of $U_1$ is:

$$E(U_1) = E(N_{U_1}) \times 1 = \frac{ap_1 G e^{-ap_1 G}}{1 - e^{-ap_1 G}}. \tag{1}$$

(2) If the packet arrives at the busy period, and the packet is the only packet adhere to sent at the current TP period, then the packet will be successfully transmitted within the next TP period, referred to as an event of $U_2$. Its possibility is

$$q_0 = e^{-p_1 p_2 G \frac{32}{23}(1+3a+\tau_R+\tau_C)}. \tag{2}$$

In the transmission period $\frac{32}{23}(1+3a+\tau_R+\tau_C)$, if there is only one information packet to be sent, its possibility is

$$q_1 = p_1 p_2 G \frac{32}{23}(1+3a+\tau_R+\tau_C) e^{-p_1 p_2 G \frac{32}{23}(1+3a+\tau_R+\tau_C)}. \tag{3}$$

In a cycle, the average length of information packets transmitted successfully at the $U_2$ is:

$$E(U_2) = \frac{q_1}{q_0} = p_1 p_2 G \frac{32}{23}(1+3a+\tau_R+\tau_C). \tag{4}$$

Then the average length $E(U)$ is

$$E(U) = E(U_1) + E(U_2) = \frac{p_1 G a e^{-p_1 G a}}{1-e^{-p_1 G a}} + p_1 p_2 G \frac{32}{23}(1+3a+\tau_R+\tau_C). \tag{5}$$

Second, the average length $E(B)$ during the busy period

$$E(B) = E(N_B) \times \frac{32}{23}(1+3a+\tau_R+\tau_C) = \frac{\frac{32}{23}(1+3a+\tau_R+\tau_C)}{e^{-p_1 p_2 G \frac{32}{23}(1+3a+\tau_R+\tau_C)}}. \tag{6}$$

Finally, the average length $E(I)$ during the idle period.

Since the number of idle slots I within the geometric distribution with the mean: $E(N) = \frac{1}{1-e^{-Gp_1 a}}$, an information packet arrive in a time slot with normalized probability: $p_{I1} = \frac{Gp_1 a e^{-Gp_1 a}}{1-e^{-Gp_1 a}}$, more than an information packet arrives in a time slot with the normalized probability [10] $p_{I2} = \frac{1-Gp_1 a e^{-Gp_1 a} - e^{-Gp_1 a}}{1-e^{-Gp_1 a}}$.

Then we get

$$E(I) = \left(\frac{1}{1-e^{-Gp_1 a}} - 1\right)a + \frac{Gp_1 a^2 e^{-Gp_1 a}}{2(1-e^{-Gp_1 a})} + \frac{(1-Gp_1 a e^{-Gp_1 a} - e^{-Gp_1 a})a}{1-e^{-Gp_1 a}}. \tag{7}$$

Besides, we know that the average length $E(I')$ during the idle period under the traditional two-dimensional probability is:

$$E(I') = \frac{a}{1-e^{-Gp_1 a}}. \tag{8}$$

The throughput of the DTPTH protocol is

$$
\begin{aligned}
S &= \frac{E(U)}{E(B)+E(I)} \\
&= \frac{\frac{p_1 Gae^{-p_1 Ga}}{1-e^{-p_1 Ga}} + p_1 p_2 G \frac{32}{23}(1+3a+\tau_R+\tau_C)}{\frac{\frac{32}{23}(1+3a+\tau_R+\tau_C)}{e^{-p_1 p_2 G \frac{32}{23}(1+3a+\tau_R+\tau_C)}} + \left(\frac{1}{1-e^{-Gp_1 a}} - 1\right)a + \frac{Gp_1 a^2 e^{-Gp_1 a}}{2(1-e^{-G_i p_1 a})} + \frac{(1-Gp_1 ae^{-Gp_1 a} - e^{-Gp_1 a})a}{1-e^{-Gp_1 a}}}.
\end{aligned}
\tag{9}
$$

On the basis of the DTPTH protocol model, in order to achieve multiservice access capabilities, we introduce multichannel mechanism.

In the case of load balancing, the arrival rate with N channels is balanced. The system throughput of the N channels is

$$S' = NS. \tag{10}$$

For the channel $i$, its arrival rate

$$\lambda_i = \frac{G_i}{N-i+1}(i \le N). \tag{11}$$

So the throughput of priority $l$ in DMTPTH protocol is

$$S_{pl} = \left(\sum_{i=1}^{l} \frac{1}{N-i+1}\right) S. \tag{12}$$

## 3 Simulation

Based on the above analysis, with the use of simulation tool MATLAB R2010a, the simulation results are shown as following:

Through the above simulation results, we can draw the following conclusions:

(1) From Figs. 1, 2, 3, 4, 5 and 6, the system simulation value consistent with the theoretical value at high degree proved the theoretical derivation, and proves the accuracy of the simulation.
(2) In Fig. 2, the throughput of DMTPTH is higher than the throughput of 1-persistent CSMA/CA, less than the throughput of two-dimensional probability random access system. The reason is that the increase of the three-way handshake mechanism, improve the system stability, take up some of the system resources.
(3) In Fig. 3, through the P1 and P2, system throughput achieves regulation. Therefore, we can make a combination of two probabilities in the case of the optimum throughput.

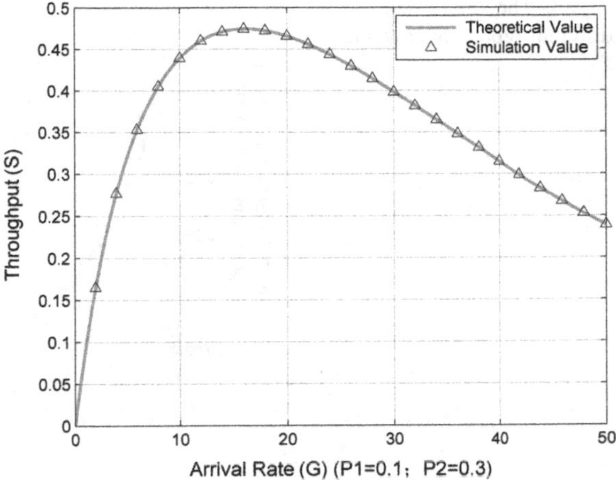


**Fig. 1** Throughput of DMTPTH protocol with 1 channel

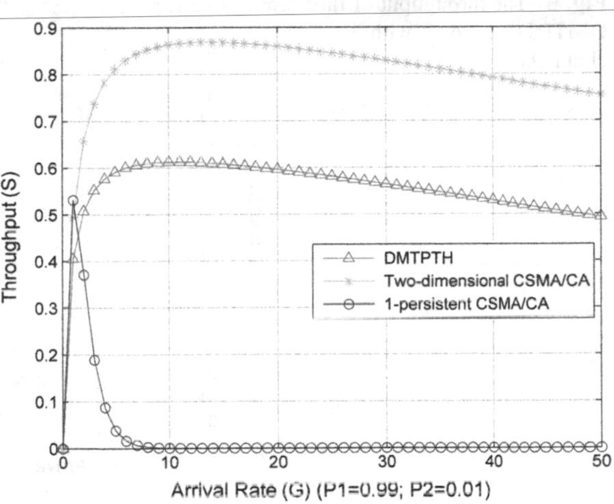


**Fig. 2** The throughput of different protocols

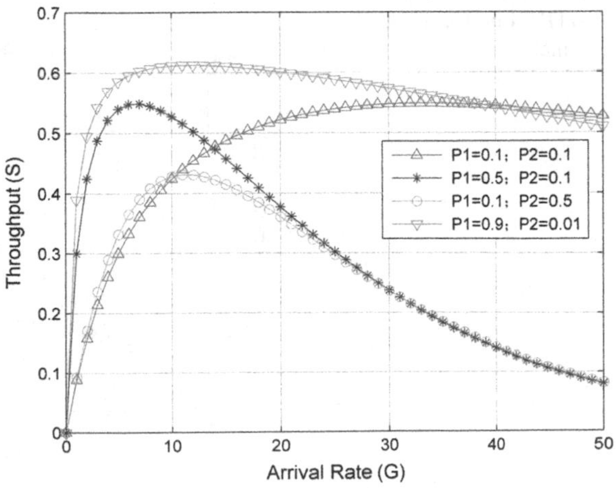


**Fig. 3** Throughput of DMTPTH protocol with different probabilities

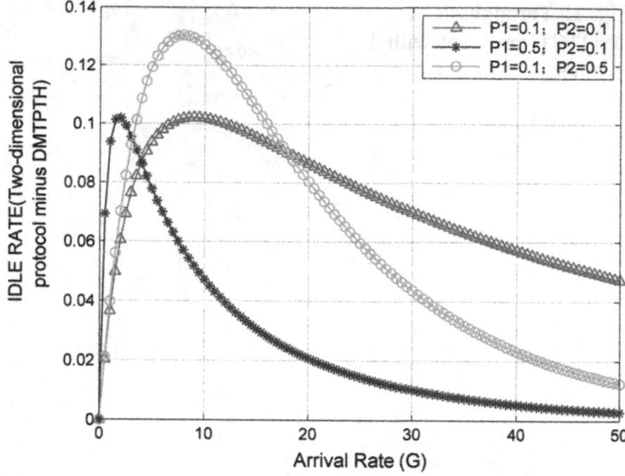


**Fig. 4** The idle rate of different probabilities value

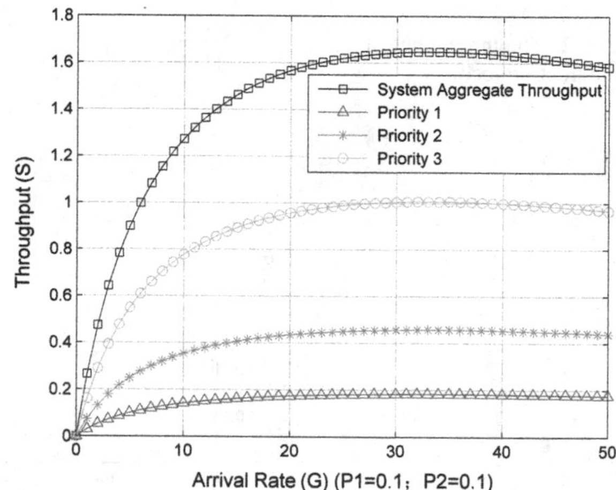


**Fig. 5** The throughput of the DMTPTH protocol with 3 channels

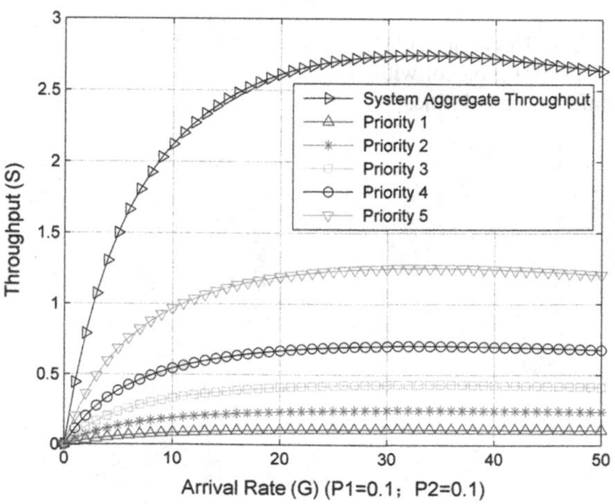


**Fig. 6** The throughput of the DMTPTH protocol with 5 channels

(4) In Fig. 4, by introducing a dual clock mechanism, it can indeed reduce the idle rate of the system; especially at low load, idle rate of the system is reduced by a big margin.
(5) From Figs. 5 to 6, the multichannel mechanism, not only improves the aggregate throughput of the system, but also to achieve the different priorities of different throughput, providing different throughput access for multiple services.

## 4 Conclusions

On the basis of the two-dimensional probability, multiple access protocols with a variety of functions, the paper proposes DMTPTH protocol. The agreement, by the addition of dual-clock, multichannel and three-way handshake mechanism, respectively, achieves lower system idle rate, confirms the multiservice access and status information transmission. Using the average cycle method gets accurate expressions of system throughput and idle rate. Finally, through theoretical analysis and simulation experiments proved the accuracy and validity of the model.

**Acknowledgements** This work was supported by the National Natural Science Foundation of China (No.61461053, No.61461054, No.61072079); Natural Science Foundation of Yunnan Province (No.2010CD023); The Financial Support of Yunnan University (No.XT412004).

## References

1. Zhang Ping, Cui Qimei. Big Data Driven Green Communication Network. Journal of Shenzhen University Science and Engineering, 2013, 30(06):557–564.
2. Wang, H.; Chen, C.; Cui, L.; Tan, S.; Leung, V.C.M. Flow-driven media access control protocols over mobile long-distance ad hoc networks[J]. IET Networks, 2012, 1(1):34–45.
3. Yang O, Member S, Heinzelman W, et al. 0 Modeling and Performance Analysis for Duty-cycled MAC Protocols with Applications to S-MAC and X-MAC[J]. Mobile Computing IEEE Transactions on, 2011, 11(6):905–921.
4. Jian Zhao, Hongtao Bai. Performance Analyses and Improvements for the IEEE 802.15.4 CSMA/CA Scheme for Heterogeneous Sensor Networks Based Adaption and RTS/CTS Mechanism. Sensors, 2012, 12(4):5067–5104.
5. Conti M, Giordano S. Mobile ad hoc networking: milestones, challenges, and new research directions [J]. Communications Magazine, IEEE, 2014, 52(1): 85–96.
6. Hongwei Ding, Yingying Guo, Yifan Zhao, Shengjie Zhou, and Qianlin Liu. Research on the Multi-channel Probability Detection CSMA Protocol with Sensor Monitoring Function. Sensor Letters, 2015, 13(2):143–146.
7. Shengjie Zhou et al., Research on the Discrete time Three-Dimensional Probability Csma Protocol In ad-hoc Network. *International Journal of Recent Scientific Research*, 2015, 6(5):4257–4262.

8. Yingying Guo, Jing Nan, Hongwei Ding, Yifan Zhao and Shengjie Zhou. Research on the multi-channel p-persistent csma protocol with monitoring function [J]. International Journal of Future Generation Communication and Networking, 2015, 8(5): 115–124.
9. Hongwei Ding, Yingying Guo, Qianlin Liu and Shengjie Zhou. The multichannel pd-csma with 3-way handshake based on conflict resolution algorithm in wsn [J]. International Journal of Recent Scientific Research, 2015, 6(4): 3714–3718.
10. Wang L, Wu K, Hamdi M. Attached-RTS: Eliminating an Exposed Terminal Problem in Wireless Networks [J]. IEEE Transactions on Parallel and Distributed Systems, 2013, 24(7): 1289–1299.

# Labeling and Encoding Hierarchical Addressing for Scalable Internet Routing

**Feng Wang, Xiaozhe Shao, Lixin Gao, Hiroaki Harai and Kenji Fujikawa**

**Abstract** Hierarchical addressing and locator/ID separation solutions have been proposed to address the scalability issue of the Internet. However, how to combine the two addressing schemes has not been received much attention. In this paper, we present an address encoding method to integrate hierarchical addressing and locator/ID separation. Our analysis and evaluation results show that the proposed encoding method could guarantee the scalability property, and alleviate the inefficiency of address space.

**Keywords** Variable-length encoding · Hierarchical addressing · Internet routing

## 1 Introduction

The Internet is facing the accelerating growth of routing table size. Since the Internet has become integrated into our daily lives, the scalability of the Internet has become extremely important. There are two trends in improving the scalability of inter-domain routing: locator/ID separation solutions and hierarchical addressing scheme [1–5]. Combining these two trends—the address format follows locator/ID separation technology, and the location allocation is based on a hierarchical routing topology—can significantly reduce routing table size.

F. Wang (✉)
School of Engineering and Computational Science, Liberty University,
Lynchburg, VA 24515, USA
e-mail: fwang@liberty.edu

X. Shao · L. Gao
Department of Electrical and Computer Engineering, University of Massachusetts,
Amherst, MA 01003, USA

H. Harai · K. Fujikawa
National Institute of Information and Communications Technology,
Tokyo 184-8795, Japan





S.K. Bhatia et al. (eds.), *Advances in Computer and Computational Sciences*,
Advances in Intelligent Systems and Computing 553,
DOI 10.1007/978-981-10-3770-2_37

However, integrating hierarchical addressing and locator/ID separation is challenging [3–5]. A common way to encode hierarchical addressing is to use fixed-length addresses. However, fixed-length addresses do not scale well because it is very difficult to preassign address length to accommodate the future growth of the Internet. In addition, it is difficult to automatically allocate locators with minimal human intervention.

In this paper, we employ a prefix-based labeling scheme to label hierarchical addresses, and use variable-length encoding to implement the proposed addressing scheme. We investigate the performance of the encoding method by understanding the degree to which the Internet can benefit from hierarchical addressing. Based on the real routing data, our study shows that it is difficult to use fixed-length encoding to implement hierarchical addressing for today's Internet. Furthermore, fixed-length encoding uses address space inefficiently. Our analysis and evaluation results show that variable-length encoding could resolve both the scalability problem and the inefficiency of address spaces.

The rest of paper is organized as follows. We present a variable-length encoding scheme in Sect. 2. We evaluate the performance of the proposed encoding scheme in Sect. 3. We conclude the paper in Sect. 4 with a summary.

## 2 Variable-Length Encoding Hierarchical Addressing

We employ a locator/ID address format by separating the location and the identity of a host. An address of each host consists of two parts: a *host ID* and a *locator*. Since the location of a node may be dynamic and change with node movement, a locator is used to describe the network attachment point(s) to which a host connects. Each host is assigned a globally unique number, which would not change even if the host moved to another location. Each host may have multiple locators representing different routes toward it. Each AS obtains a locator from a provider in the upper layer. Locators are labeled by a prefix-based labeling scheme. Each locator consists of a *provider label* followed by a *self-label*. A provider label identifies the provider, while a self-label identifies a customer. We use delimiter '.' to join a set of provider labels and self-labels. Because Tier-1 ISPs do not have any provider, their provider labels are empty. A centralized authority, such as IANA, can assign them a unique self-label. So, their self-labels are their locators. For other non-Tier-1 ASes, they first determine a self-label for each customer. Then, each provider concatenates its own provider label and the self-label to generate a customer's locator. With the prefix-based labeling, customers that have the same providers should share a common prefix. To avoid the allocation loop, a provider is prohibited from advertising a locator back to a customer from which the locator was assigned. In addition, a provider can detect a locator containing a loop by examining whether the locator and any one of its locators share the same prefix.

The proposed addressing scheme is used on a loose underlying hierarchical topology, where the hierarchy is not strictly enforced. Even though providers are

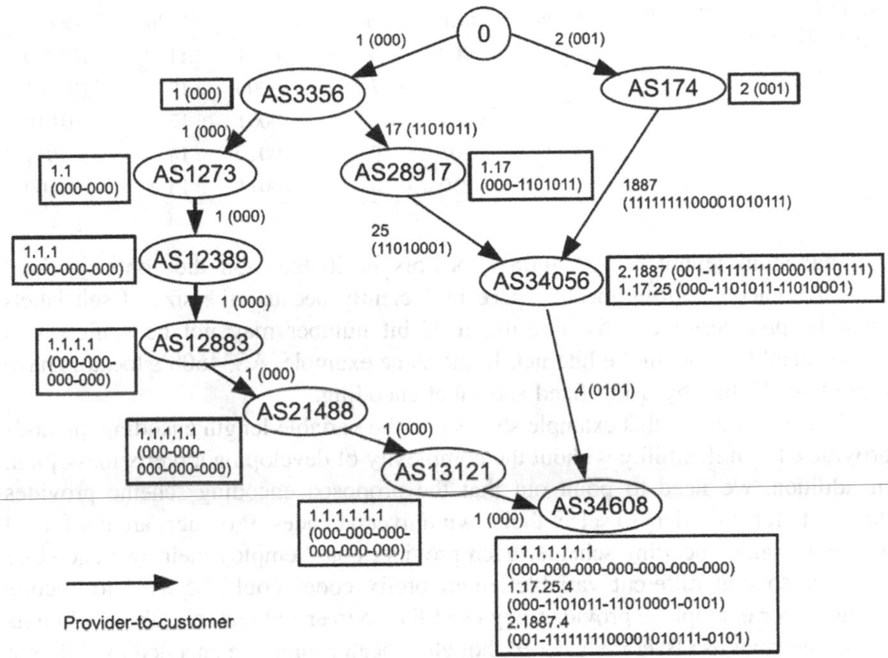


**Fig. 1** An example of Baer encoded locators. The number in parentheses represents an encoded locator or self-label. Dashes are used to make encoded locators more readable

organized as a tiered hierarchy, it only provides a loose characterization of the AS hierarchy. There are interconnections between providers at the same tier, and a non-Tier-2 AS may obtain locators directly from a Tier-1 provider. For example, in Fig. 1, AS28917 (non-Tier-2) obtains a locator from Tier-1 AS3356.

The idea of using a variable-length representation is to assign each self-label a variable-length number. The size of each self-label is essentially determined by a specific variable-length encoding scheme. Using a variable-length encoding method, we can use an infinite number of integers to represent self-labels. That means, it does not have a restriction on the size of self-labels.

Universal codes [6] are particularly suitable for our purpose because each universal code assumes an implied probability distribution. The implied distribution is an estimation of power law or Zipf distribution. Previous work has shown that the number of customers in today's Internet can be described as a power law distribution [7]. We believe that the distribution of self-labels would follow a power law as well. There are several possible universal codes for self-labels: Elias' γ, Elias' δ, Elias' ω codes [8], and Baer code [9]. Table 1 shows some part of Baer codewords. Note that we do not claim that this code is the optimal code for self-labels. For example, based on each AS's rank, which is determined by the number of descendants, we show the encoded locators in Fig. 1. From this Figure, we see that

**Table 1** An example of Baer codeword tables

| Value | Baer | Value | Baer | Value | Baer |
|---|---|---|---|---|---|
| 1 | 000 | 6 | 0111 | 11 | 101000 |
| 2 | 001 | 7 | 10000 | 12 | 101001 |
| 3 | 0100 | 8 | 10001 | 13 | 101010 |
| 4 | 0101 | 9 | 10010 | 14 | 101011 |
| 5 | 0110 | 10 | 10011 | 15 | 101100 |

the maximum length of AS34608's locators is 26 bits. On the contrary, fixed self-label encoding uses locator space inefficiently because the size of self-labels must be predetermined. As a result, a 32-bit number may not be sufficient to represent all locators in the Internet. In the same example, AS34608's locators have more than 32 bits by using fixed self-label encoding.

Most importantly, this example shows that the variable-length encoding methods provide a lot of flexibility without the complexity of developing a prefix mask plan. In addition, we need to point out that the proposed encoding scheme provides flexibility for providers to select their own universal codes. Providers are not forced to use the same encoding scheme. Each provider could employ their own encoding schemes so that different variable-length prefix codes could be used to encode locators. For example, a provider may use Elias' δ to encode its self-labels, while its customers may use Baer code. Even though a locator might be encoded by different universal codes, it is still a unique locator. Note that to maintain this flexibility, Tier-1 ASes' locators must be prefix-free if they are encoded by different encoding schemes. This is not difficult to implement because a centralized authority, such as IANA, can guarantee the uniqueness of the locators.

## 3 Performance Evaluation

We use the AS relationships dataset [10] from CAIDA to build an AS level graph, which can be used to represent today's Internet hierarchy. Then, based on the graph, we infer the distribution of locators encoded by different codes. We use the three Elias' codes and Baer code to encode self-labels. We use two ways to encode locators: *random assignment* and *ranking-based assignment*. Random assignment means that we randomly assign a self-label to a customer, and encode the number using one of the codes. Ranking-based assignment means that we assign a self-label according to the rank of a customer. The ranking is determined by the number of descendants of each customer. Then, we sort the ranking of the customers in non-increasing order. According to its order, a self-label code word is assigned to each customer. Meanwhile, we compare the length of encoded locators with that of fixed-length locators. We use the total number of customers for each provider to determine the number of bits to represent self-labels.

We present the results in Fig. 2. From Fig. 2a, we find that the locator length will be almost same if we randomly select a code word. The maximum length of

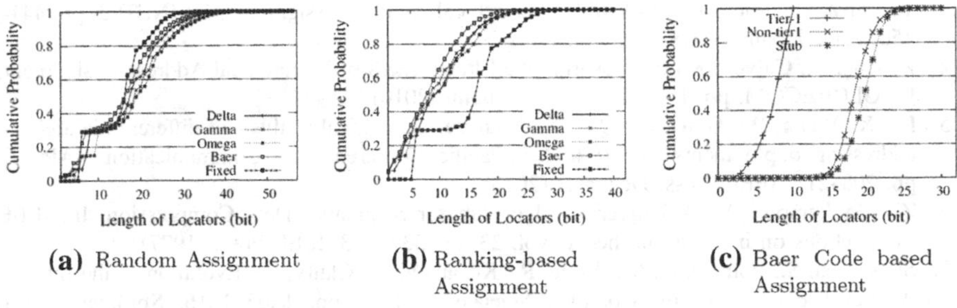

(a) Random Assignment (b) Ranking-based Assignment (c) Baer Code based Assignment

**Fig. 2** Length distribution of locators encoded in different codes

fixed-length locators is 56 bits. Over 80% of the locators encoded in Baer code have no more than 24 bits. As shown in Fig. 2b, if we employ ranking-based assignment, the locators encoded by Elias' δ and Baer codes are always shorter than the others, including fixed-length method. In Fig. 2c, we redraw the length distribution of locators encoded by Baer code. It shows the length at different hierarchical layers. We can see locators at different layers might have different length distribution, and the majority of the locators are 10–25 bits long. In addition, the average length of variable-length locators is about 19 bits, while the average length of the fixed-length locators is about 28 bits. Hence, the variable-length locators are much shorter than fixed-length locators.

## 4 Conclusion

In this paper, we propose a variable-length addressing encoding scheme to mitigate the scalability issue of the current Internet. To support this new addressing architecture, we present an address encoding method to implement the addressing scheme. Our analysis and evaluation results show that the proposed encoding method could guarantee the scalability property of hierarchical addressing, and alleviate the inefficiency of address space due to fixed-length addresses.

**Acknowledgements** This work was supported by National Science Foundation grant CNS-1402857 and CNS-1402594 and under NSF-NICT Collaborative Research JUNO (Japan-U.S. Network Opportunity) Program.

## References

1. Atkinson, RJ., Bhatti, SN., Andrews, U.St.: Identifier-Locator Network Protocol (ILNP) Architectural Description. Request for Comments 6740. (2012)
2. Kafle, Ved P., Otsuki, H., Inoue, M.: An ID/locator Split Architecture for Future Networks. In: Comm. Mag. vol. 48, pp. 138–144. (2010)

3. Tsuchiya, P.: Efficient and Flexible Hierarchical Address Assignment. In: INET92. pp. 441–450. Kobe Japan (1992)
4. Zhuang, Y., Calvert, K.L.: Measuring the Effectiveness of Hierarchical Address Assignment. In: GLOBECOM. pp. 1–6. IEEE Press, Miami (2010)
5. Lu, X., Wang, W., Gong, X., Que, X., Wang, B.: Empirical analysis of different hierarchical addressing deployments. In: 19th Asia-Pacific Conference on Communications (APCC). pp. 208–213. IEEE Press, Denpasar (2013)
6. Ziv, J., Lempel, A.: A Universal Algorithm for Sequential Data Compression. In: IEEE Transactions on information theory. vol. 23, pp. 337–343. IEEE Press (1977)
7. Shakkottai, S., Fomenkov, M., Koga, R., Krioukov, D., Claffy, kc: Evolution of the Internet AS-Level Ecosystem. In: Complex Sciences. vol. 5, pp. 1605–1616. Springer, Berlin Heidelberg (2009)
8. Elias, P.: Universal Codeword Sets and Representations of the Integers. In: IEEE Transactions on Information Theory. vol. 21, pp. 194–203. IEEE Press (1975)
9. Baer, M.B.: Prefix Codes for Power Laws. In: 2008 IEEE International Symposium on Information Theory (ISIT 2008). pp. 2464–2468. IEEE Press, Toronto (2008)
10. The IPv4 Routed/24 AS Links Dataset, http://www.caida.org/data/active/ipv4_routed_topology_aslinks_dataset.xml

# Link Utilization in Hybrid WiMAX-Wi-Fi Video Surveillance Systems

**Smart C. Lubobya, Mqhele E. Dlodlo, Gerhard De Jager and Ackim Zulu**

**Abstract** This paper presents a link utilization performance analysis of the hybrid WiMAX-Wi-Fi and WiMAX video surveillance systems. Link utilization is ratio of throughput to capacity expressed as a percentage. Therefore, Link utilization measurement must take into account throughput, packet losses and signal-to-noise ratio. A WiMAX-Wi-Fi video surveillance system consists of a Base Station (BS) which is connected in a point-to-point configuration with the Customer Premises Equipment (CPE). The base station then connects to the Internet, routers and video servers. The CPE has a Wi-Fi and WiMAX wireless interfaces. Results show that the hybrid WiMAX-Wi-Fi video surveillance system outperforms the WiMAX in throughput, packet loss and signal-to-noise ratio terms. The hybrid WiMAX-Wi-Fi is 4.5 times better in utilizing the WiMAX link than the WiMAX system.

**Keywords** Hybrid WiMAX-Wi-Fi · Link utilization · Packet loss · Signal-to-noise ratio · Throughput · Video surveillance

S.C. Lubobya (✉) · M.E. Dlodlo · G. De Jager
Department of Electrical Engineering, University of Cape Town, Cape Town, South Africa
e-mail: lbbsma001@myuct.ac.za

M.E. Dlodlo
e-mail: mqhele.dlodlo@uct.ac.za

G. De Jager
e-mail: gerhard.dejager@uct.ac.za

A. Zulu
Department of Electrical and Electronics, University of Zambia, Lusaka, Zambia
e-mail: ackim.zulu@unza.zm



S.K. Bhatia et al. (eds.), *Advances in Computer and Computational Sciences*, Advances in Intelligent Systems and Computing 553,
DOI 10.1007/978-981-10-3770-2_38

## 1 Introduction

WiMAX is a wireless Metropolitan Access Networks (MAN) technology that promises broadband Internet service at higher data rate [1]. It also guarantees quality of service [2] through the various qualities of service classes for both real and non real time signal types. WiMAX stems from the IEEE 803.16 family of standards. These standards specify the licensed operating frequency band, modulation scheme and data rate. The standards also specify WiMAX use in line of sight and non line of sight environments.

Wi-Fi is a wireless technology that provides Internet services and operates in the unlicensed ISM frequency bands and is guided by the IEEE 803.11 family of standards. The hybrid WiMAX Wi-Fi systems combines the WiMAX wireless technologies with the Wi-Fi wireless technologies.

In this paper we carry out a comparative analysis of the WiMAX and hybrid WiMAX-Wi-Fi video surveillance systems. The two systems are compared and analyzed in terms of throughput, packet loss, link utilization and signal-to-noise ratio on the WiMAX link. The rest of the sections are divided as follows: the next section gives related work on performance of WiMAX and WiMAX-Wi-Fi video surveillance systems. Section 3 describes the literature review on WiMAX and WiMAX-Wi-Fi networks. Section 4 gives a detailed mathematical model of traffic flows and description of related performance metrics. Sections 5 and 6 gives the simulation methodology and results and discussions, respectively. The conclusion is given in Sect. 7.

## 2 Related Work

Charitos and kalivas [3] deployed a hybrid vehicular wireless network consisting of IEEE 802.11b/g/e and IEEE802.16e inside a tunnel environment for surveillance purposes. They then analysed the performance of such a network during handovers after an emergency situation (fire or explosion). Link quality metrics such as end-to-end delay, Signal-to-Noise Ratio (SNR) and throughput were considered. In the work of Dagar and Sharma [4] concerning video streaming on a WiMAX network, it was established that End-to-End delay and throughput increases with increase in the coverage area of the base station. This effectively means that the closer the SS or MS is to the base station the higher the throughput would be. Ahmad and Habibi proposed a scheme that estimates the utility of different cameras mounted on a public transport and, based on the estimated utility, decides which camera(s) to put offline so that overall utility of the whole video surveillance system improves. Oyman et al. in [5] provided an overview of technology options for enabling multicast and unicast video services over WiMAX and LTE networks, quantified and compared the video capacities of these networks in realistic environments, and discussed new techniques that could be exploited in the future to further enhance the video capacity and quality of user experience [5]. Juan et al. [6] investigate the performance of scalable video streaming services in mobile WiMAX systems. Both CIF and QCIF were used in the simulations.

Kafhali et al. [7] presented a performance analysis for the bandwidth allocation in IEEE 802.16 Broadband Wireless Access (BWA), in which the throughput was measured against traffic intensity (packets/frame). They noted that an increase in traffic intensity had a corresponding increase in throughput—until saturation point. Yousaf et al. [8] conducted, among others, TCP and UDP throughput tests for the downlink and uplink channels of WiMAX network. The throughput tests were carried out under varying modulation types and at varying distances. The throughput results on the stressed WiMAX link were satisfactory, even at the lowest transmission power (13 dBm) and for a distance of up to 9.4 km.

In this paper, we analyze the link utilization, which is a function of throughput and network bandwidth, of hybrid WiMAX-Wi-Fi video surveillance and compare it to the WiMAX network.

## 3 WiMAX and Hybrid WiMAX-Wi-Fi Networks

In general there are two types of WiMAX networks: the Fixed and Mobile WiMAX networks. In a fixed WiMAX fixed end-devices connects to the base station via the Customer Premises Equipment (CPE) or Subscriber Station (SS). The CPE has two wireless interfaces and therefore two ratio links, for connecting to the WiMAX BS and the Wi-Fi end-devices. This creates a hybrid WiMAX-Wi-Fi network. In the mobile WiMAX, mobile end-devices, such as mobile phones, mobile cameras, connects directly to the BS and transmits data, voice or video.

Regardless of the WiMAX network type, the networks architecture for these networks consists of two basic units: the Access Service Network (ASN) and Connectivity Service Network (CSN) [9]. The hybrid network has the additional Subscriber Station (SS), also called the CPE. The mobile WiMAX operate under standard 802.16e while fixed mobile used 802.16d on the WiMAX link.

The Access Service Network (ASN) consists of one or more BSs and the gateway routers to the connectivity service network. It provides service access to the mobile stations and fixed end-devices via the SS or CPE. The CSN is responsible for providing Internet connectivity to the WiMAX radio equipment [10]. It also provides subscriber billing and inter-operator payment, supports communication between Network Access Providers and controls policies for Internet access among others [11]. The role of the BS is to manage radio resources for the WiMAX network, and to provide connectivity between the network service provider equipment and the subscriber stations (Fig. 1).

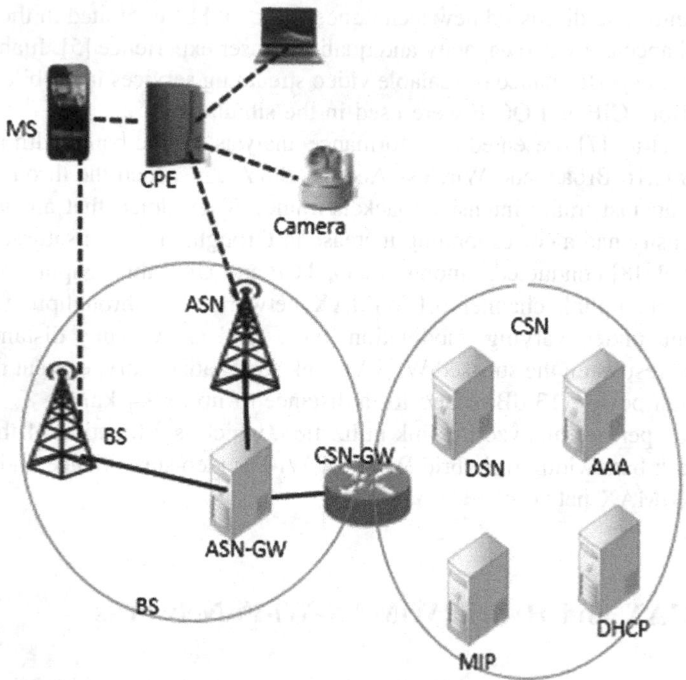


**Fig. 1** WiMAX and hybrid WiMAX-Wi-Fi network architecture

# 4 Theory: Modelling Traffic Flows

This section gives and derives the mathematical model of the performance metrics used to test and compare the performance of the surveillance systems.

## 4.1 Bandwidth

Bandwidth (*B*) requirements increase with increase in the number of cameras. The network bandwidth can then be computed from:

$$B = \frac{FS \times FPS \times n_p \times 8}{1024} \tag{1}$$

where FS is the frame size in kilobytes, FPS is the frames per second and $n_p$ is the planned connected number of cameras.

### 4.2 Actual Throughput at the CPE

We can also model the flow throughput for any number of IP cameras using graph theory [12]. Figure 3 shows a traffic flow diagram of a WiMAX-Wi-Fi network. The traffic model consists of: the supply devices which in this case are wireless IP cameras denoted as node $i$. Node 5 represents the customer premises devices. Let I be a set of directed links to the CPE and $k$ be the set of possible routes to CPE and/or BS. Then we can describe the links and routes with tables or matrices as:

$$A_{ik} = \begin{cases} 1 & \text{If the link K lies on route } k \\ 0 & \text{Otherwise} \end{cases} \tag{2}$$

This defines matrices $A_1$ called the link route incident matrices, which are such that

$$A_1 = A_{\substack{ik \\ i \in I, k \in K}} \tag{3}$$

Now each column of matrix $A_1$ corresponds to one of the route $k$ while each row in $A_1$ corresponds to the link $i$. The columns of route $k$ consist of 1s and 0s. The ones signify which links are on route $k$ and zeros indicates which ones are not. For the rows, ones for link $k$ indicate the routes which pass through that link. Then:

$$x_i = x_{k_{k \in K}} \tag{4}$$

The flow throughput is then the summation of individual flows

$$S_{flow} = \sum_{n \in N} A_{ik} x_k \tag{5}$$

The number $(x_n, n \in N)$ can be thought of as forming a vector.

The above equations can then be represented succinctly in matrix form. To represent the flow throughput at the CPE and/or the BS, Eq. (5) simplifies to:

$$S_{flow} = (A_1 x_1) \tag{6}$$

### 4.3 Link Utilisation

Link utilization ($LU$) is defined as the ratio of the amount of data carried on the link or current data transfer rate to the link's capacity [13, 14]. Mathematically the link utilization can be written as:

$$LU = \frac{S_{flow}}{B} \quad (7)$$

where $S_{flow}$ is the number of packets or bits transmitted in a unit interval of time and is equal to (5). Utilization factor can also be expressed as a percentage in which high percentage indicate busy link while low link utilization percentage indicates that the link is idle (poor link utilization). Ideally, during congestion, $LU = 1$ or 100% [14].

## 5 Simulation Methodology

The OPNET Modeller simulation set-up for the WiMAX and hybrid WiMAX–Wi-Fi video surveillance system was arranged as shown in Fig. 2a and b. In the first scenario, Wi-Fi cameras are transmitting to the WiMAX BS via the CPE. This scenario is compared to a baseline WiMAX system in which WiMAX cameras are transmitting directly to the BS. Several simulations are run for 4, 8, 12, 16, …, 32 number of cameras. In each scenario an average measurement of WiMAX throughput, packet loss is recorded and plotted against the number of nodes or cameras. The Wi-Fi interfaces are configured with IEEE802.11b/g, transmit power of 0.05 W and sensitivity of −95 dB (Fig. 2).

A constant bit rate, 1420 byte, 352 × 288 resolutions at 15 FPS video was used in simulations. Since the source video from IP cameras are unidirectional, transmitting from the camera to the WiMAX BS, the incoming stream inter-arrival time was configured to 'none' in the applications definitions. The CPE was configured with 14 dBi and transmit power of 3 W while the BS had 15 dBi and 5 W power. A Modulation type of QPSK was used. we assume that the Wireless link is the bottleneck and that an End-to-End delay of between 150 and 200 ms [15] and Jitter of below 60 ms [16] recommended for video transmission is maintained.

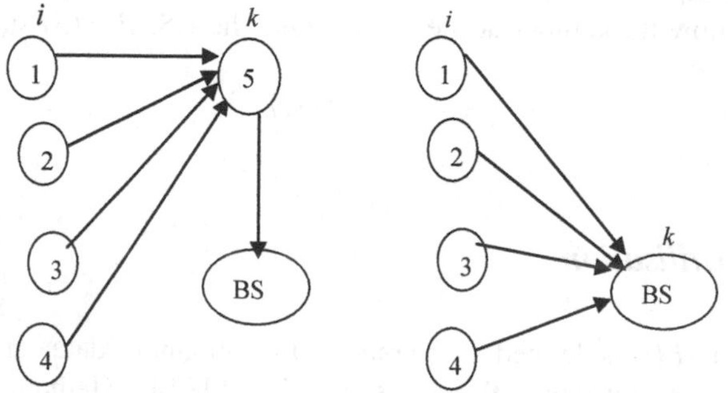


**(a)** Hybrid WiMAX-Wi-Fi Traffic Flows **(b)** WiMAX traffic flows, without CPE

**Fig. 2** Illustration of WiMAX and WiMAX-Wi-Fi traffic flows

## 6 Results and Discussion

Figure 3 shows the measured throughput results of the WiMAX and hybrid WiMAX-Wi-Fi video surveillance system. Throughput increases with increase in the number of cameras for all the two systems. However, beyond sixteen cameras the throughput for the hybrid system saturates with any increase in number of cameras having null increase in throughput. This is in contrast to the WiMAX whose throughput continues to increase with increase in number of cameras and is only limited by the capacity of the WiMAX network. A further look at the hybrid system shows that for the first twelve cameras, the hybrid WiMAX-Wi-Fi throughputs is marginally higher than that for the WiMAX system. The implication of the hybrid WiMAX-Wi-Fi system results is that the CPE can accommodate up to sixteen cameras in a single and simultaneous uplink transmission for the 1420 byte, 15 fps Video. If cameras, above sixteen (16), are to be loaded on a CPE, they should be of low payload and/or frame rate.

In Fig. 4, a performance comparison of packet loss of the WiMAX uplink for the WiMAX and hybrid WiMAX-Wi-Fi is made. From the results, the hybrid WiMAX-Wi-Fi performed relatively better during the first 12 cameras than the WiMAX system, recording an average packets loss of between 0 and 0.80. How-

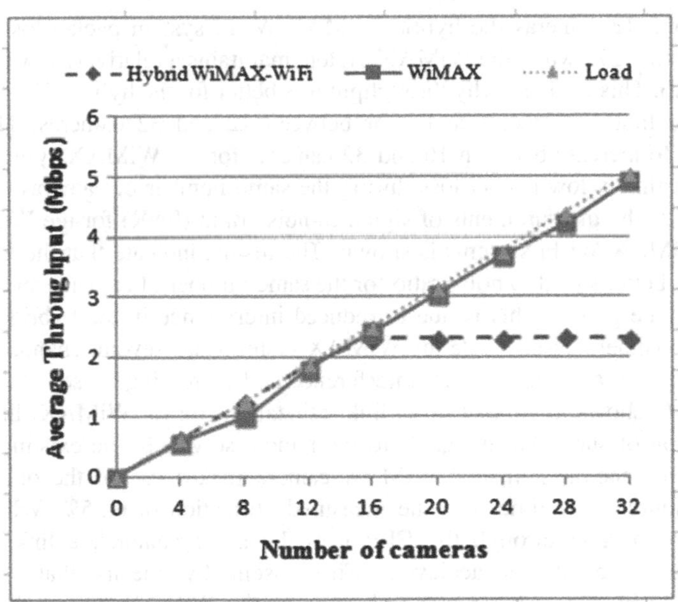


**Fig. 3** Throughput measurements

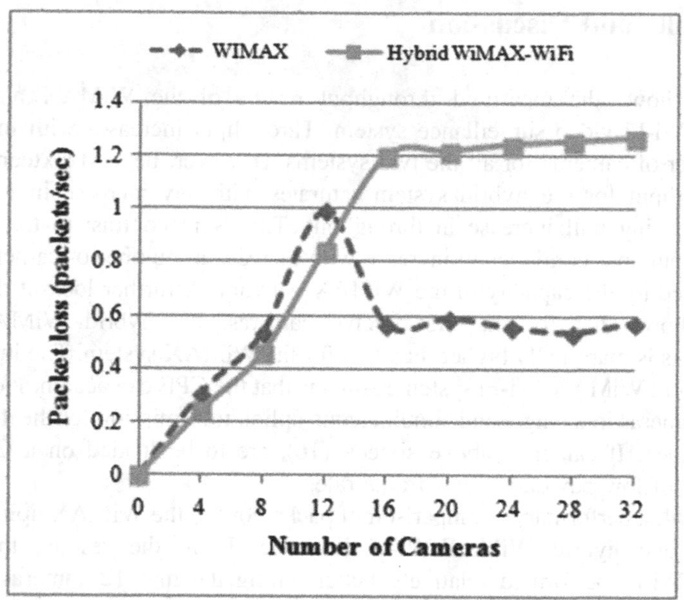


**Fig. 4** Average packet loss measurements

ever beyond 12 cameras the hybrid WiMAX-Wi-Fi system packet loss increases from 0.80 to 1.275 while the WiMAX system maintains a relatively low packet loss of 0.55–0.6. This explains why throughput was better for the hybrid WiMAX-Wi-Fi during the first 12 cameras and poor between 12 and 32 cameras. Throughput continued to increase between 16 and 32 cameras for the WiMAX system because of the marginally low packet loss during the same number of cameras.

In Fig. 5, the measurements of signal-to-noise ratio (SNR) for the WiMAX and hybrid WiMAX-Wi-Fi systems is shown. The results indicate that the hybrid systems has a better signal to noise ratio for the same number of cameras and measured over the same period. This is due to reduced interference in the hybrid system as only one channel is used while the WiMAX systems has several channels for each camera creating adjacent channel interference and increasing noise.

Figure 6 shows the results of link utilization for a WiMAX link during transmission of surveillance signals to the remote server. In the existing WiMAX video surveillance system one WiMAX camera makes use of the one WiMAX uplink resources at a time with the measured utilization of 0.35%. When sixteen cameras connecting through the CPE uses the same channel, a link utilisation percentage of 5.34% is achieved. This essentially means that the hybrid WiMAX-Wi-Fi system has better utilisation of the WiMAX uplink channel than the WiMAX system.

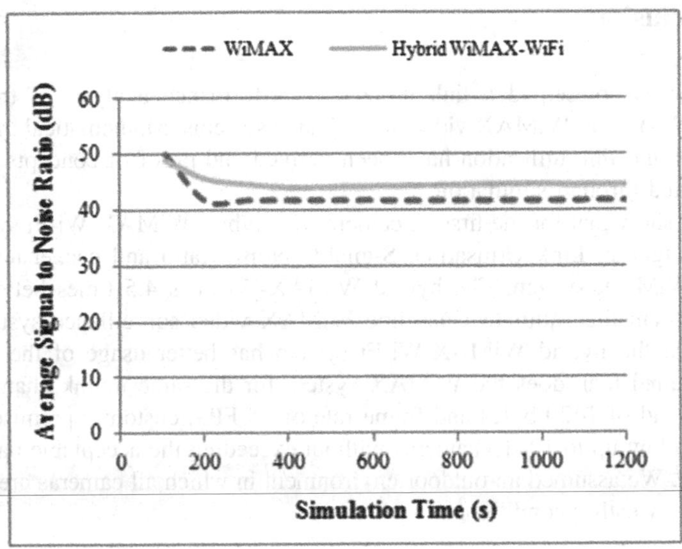


**Fig. 5** Signal-to-noise ratio measurements

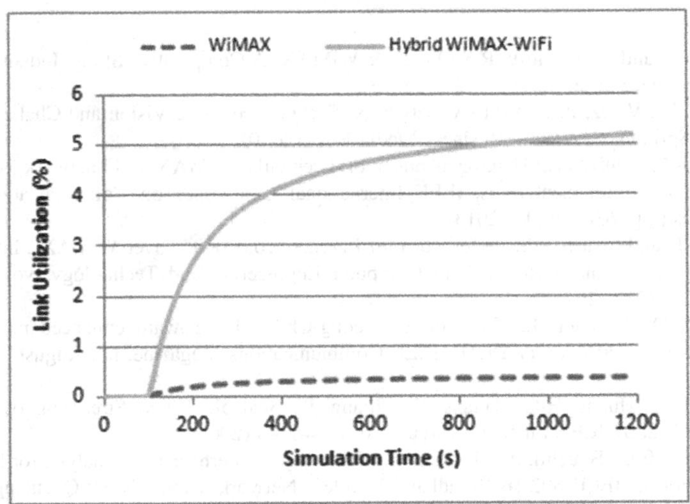


**Fig. 6** Link utilization measurements

# 7 Conclusion

This paper has presented a link utilization performance analysis of the hybrid WiMAX-Wi-Fi and WiMAX video surveillance systems. Mathematical models for throughput and link utilisation have been derived and proof of concepts has been demonstrated through simulation.

Results show that for the first 12 cameras the hybrid WiMAX-Wi-Fi system has better throughput, Link Utilisation, Signal-to-Noise Ratio and packet loss values than the WiMAX system. The hybrid WiMAX-Wi-Fi is 4.5 times better in link utilisation than the equivalent baseline WiMAX video surveillance system. This implies that the hybrid WiMAX-Wi-Fi system has better usage of the WiMAX uplink channel than does the WiMAX system for the same uplink channel. At a video payload of 1420 bytes and frame rate of 15 FPS, customer premises equipment can allow up to 12–16 cameras without exceeding the acceptable packet loss of 1% [16]. We assumed an outdoor environment in which all cameras are exposed to the same weather conditions.

**Acknowledgements** This work is supported by the University Science, Humanities and Engineering Partnerships in Africa (USHEPiA) Fund.

# References

1. Banerji, S and Chowdhury, R.S.: Wi-Fi & WiMAX: A Comparative Study, Indian Journal of Engineering, vol. 2, no. 5, pp. 1–5 (2013).
2. Prasad, R., Velez, F.J.: WiMAX Networks: Techno-Economic Vision and Challenges. New York: Springer Science + Business Media B.V. (2010).
3. Charitos M, Kalivas G.: Heterogeneous hybrid vehicular WiMAX-WiFi network for in-tunnel surveillance implementations. IEEE International Conference on Communications (ICC). Budapest; pp. 6386–6390 (2013).
4. Dagar, K and Sharma, P.: Performance of Internet Protocol TV over WiMAX', International Journal of Advanced Research in Computer Engineering and Technology, vol. 4, no. 6, pp. 2507–2513 (2015).
5. Oyman, O., Foerster, J., Yong-joo T., Seong-Choon L.: Toward enhanced mobile video services over WiMAX and LTE, IEEE Communications Magazine, no. August, pp. 68–76, (2010).
6. Juan H-H., Huang H-C., Huang C, Chiang T.: Scalable Video Streaming over Mobile WiMAX. 2007 IEEE Int Symp Circuits Syst. 3463–6 (2007).
7. Kafhali, S.E.L., Bouchti, A.E.L., Hanini, M., Haqiq, A.: Performance Analysis for Bandwidth Allocation in IEEE 802.16 Broadband Wireless Network using BMAP Queueing, International Journal of Wireless and Mobile Networks, vol. 4, no. 1, pp. 139–154 (2012).
8. Yousaf F.Z., Daniel, K., Wietfeld C.: Analyzing the Throughput and QoS Performance of WiMAX Link in an Urban Environment. In: WIMAX New Development [Internet]. Shanghai, China: Unit 405, Office Block, Hotel Equatorial Shanghai No. 65, Yan An Road (West), INTECH; 2009. pp. 308–20.
9. Andrews, J.G., Ghosh, A., Muhamed, R.: WiMAX Network Architecture, in Fundamentals of WiMAX Understanding Broadband Wireless Networking, Upper Saddle River, NJ: Pearson Education, Inc. One Lake Street Upper Saddle River, NJ 07458, 2007, pp. 335–362.

10. Farrukh, E., Panaousis, E.A., Politis, C.: Performance Evaluation of secure video transmission over WiMAX, International Journal of Computer Networks and Communication, vol. 3, no. 6, pp. 131–144 (2011).
11. Dagar, K.: Performance Evaluation of Video on Demand (VoD) over WiMAX, International Journal of Advanced Research in Computer Engineering and Technology, vol. 4, no. 5, pp. 2039–2043 (2015).
12. Kelly, F.: 'The Mathematics of Traffic in Networks', Princeton companion to Mathematics, vol. 1, no. 1, pp. 862–870 (2008).
13. Adarshpal S.S., Vasil, YH.: The Practical OPNET User Guide for Computer Network Simulation Title, vol. 53, no. 9. CRC Press, Taylor & Francis Group (2013).
14. Camlibel, C.K., Julius, A.A., R. Pasumarthy, R., M. Jacquelien, M.A.S.: Mathematical Control Theory I Nonlinear and Hybrid Control Systems. New York: Springer US, 2015.
15. Hamodi J.M., Thool, RC.: Investigate the Performance Evaluation of IPTV Over WiMAX Networks, International Journal of Computer Networks and Communications(IJCNC), vol. 5, no. 1, pp. 81–95 (2013).
16. Chen, Y., T. Farley, T., Ye, N.: QoS Requirements of Network Applications on the Internet, Information-Knowledge-Systems Manag., vol. 4, pp. 55–76 (2004).

# Part III
# Power and Energy Optimization

# Short Term Price Forecasting Using Adaptive Generalized Neuron Model

**Nitin Singh and S.R. Mohanty**

**Abstract** Deregulation in the electricity industry has made price forecasting the basis for maximizing profit of the different market players in the competitive market. The profit of market player depends on the bidding strategy and the successful bidding strategy requires accurate price forecasting of electricity price. The existing methods of price forecasting can be broadly classified into (i) statistical methods (ii) simulation-based methods and (iii) soft computing methods. The conventional neural networks were used for price forecasting due to their ability to find an accurate relation between the historical data and the forecasted price without any system knowledge. They suffer from major drawbacks like training time dependency on complexity of the system, huge data requirement, ANN structure is not fixed, hidden neurons requirement is large relatively, local minima. In the proposed work, the problems associated with conventional ANN trained using back-propagation are solved using improved generalized neuron model. The genetic algorithm along with fuzzy tuning is used for training the free parameters of the proposed forecasting model.

**Keywords** Generalized neural network · Genetic algorithm · Wavelet transforms · Fuzzy systems electricity price forecasting

## 1 Introduction

The deregulation of electricity market has ensured the reliable electricity supply at reasonable cost for economic and industrial growth. Market players use price forecasting as a tool to manage the risk in deregulated electricity market [1, 2]. Bulk electricity consumers can maximize their load schedules while generation companies maximize their profit using accurate price forecasting. In case of vertically

N. Singh (✉) · S.R. Mohanty
Department of Electrical Engineering, MNNIT Allahabad, Allahabad, India
e-mail: nitins@mnnit.ac.in

S.R. Mohanty
e-mail: soumya@mnnit.ac.in



S.K. Bhatia et al. (eds.), *Advances in Computer and Computational Sciences*,
Advances in Intelligent Systems and Computing 553,
DOI 10.1007/978-981-10-3770-2_39

integrated electric industry, the electricity prices reflect the government policy and price forecasting was based on average costs [3] whereas, deregulated market being a customer-driven market, the price is set by the supply–demand relationship. In deregulated market, the supply–demand have to be balanced in real time which requires that the electricity price must be pre-estimated before real-time operation for maximizing profit. Price forecasting in addition to helping independent generators in setting up optimal bidding patterns also helps in designing physical bilateral contracts, market prices strongly affect the decision on investing a new generation facilities in the long run [4].

Various models for forecasting the electricity prices have been developed based on hard and soft computing techniques out of which the models based on artificial neural network gained popularity due to their capacity to map input–output relation without the knowledge of system. The models based on regression analysis or hard computing techniques require complete information about the system before prediction. Although the results are very accurate, but a lot of information is required for forecasting the electricity prices which make these models computationally very expensive [5, 6].

Several authors have used back propagation algorithm with conventional artificial neural network for training ANN for forecasting time series [7, 8]. The conventional neural networks have some serious drawbacks such as large training time, large data requirement, etc., to overcome these drawbacks a generalized neural model (GNM) has been developed [9] which was successfully implemented for modeling [10] and forecasting [11]. In the proposed work, the generalized neuron model trained using genetic algorithm and fuzzy rules is used for forecasting the price of New South Wales market, the results obtained are compared with conventional artificial neural network, generalized neuron model, auto regressive (AR) integrated (I) moving average (MA), and generalized auto-regressive conditional heteroskedasticity model.

## 2 Drawbacks of Conventional ANN

Conventional artificial neural network due to its inherent black box nature have gained popularity in solving nonlinear problems quite easily but there are some serious limitations with conventional artificial neural network trained using back-propagation learning algorithm [11].

1. For large data sets the training time is large resulting in slow response.
2. No rule exists for predicting number of hidden neurons or hidden layer.
3. Only summation function is used for giving the output of neuron.
4. The famous back-propagation learning algorithm suffers from the following drawbacks:
   a. Learning rate is slow.
   b. Problem of local minima.

c. Training time is dependent on the pattern presented to the network, for complex patterns it may take large time to train.

# 3 Proposed Methodology

## 3.1 Generalized Neuron Model

To overcome the shortcomings of the conventional neural network a new generalized neuron model has been developed [11, 12]. The generalized neuron model due to its features is capable of incorporating the non-linearities present in the system.

1. It possesses the characteristic of high-order as well as low-order neuron.
2. No hidden layer is required, which reduces the complexity of the system.
3. No effect on input output mapping or normalization on the training of neuron.

Unlike conventional neural network which uses only sigmoid activation function and summation as aggregation function, the generalized neuron incorporates both summation as well as product as aggregation function. The structure of generalized neuron is capable of coping up with the nonlinearity involved in the type of application dealt with. The generalized neuron model due to use of product and summation as aggregation function has shown drastic reduction in the training time.

In generalized neuron model, both the aggregated output of the summation and product function is passed through a nonlinear squashing function. The structure of the generalized neuron is shown in Fig. 1.

The output of the $\sum_1$ part is the summation of weighted input with the sigmoid activation function $f_1$ whereas the output of the $\prod$ part is the summation of the weighted input with Gaussian activation function $f_2$. The final output of the neuron is the aggregation of output of sigma part $\sum_1$ and pi part $\prod$, i.e., given as (1)

$$O_i = W \times O_{\Sigma} + (1 - W) \times O_{\Pi} \tag{1}$$

where output of the sigma part is given as (2)

$$O_{\Sigma} = f_1 (W_{\Sigma i} \times X_i) + X_{O\Sigma} \tag{2}$$

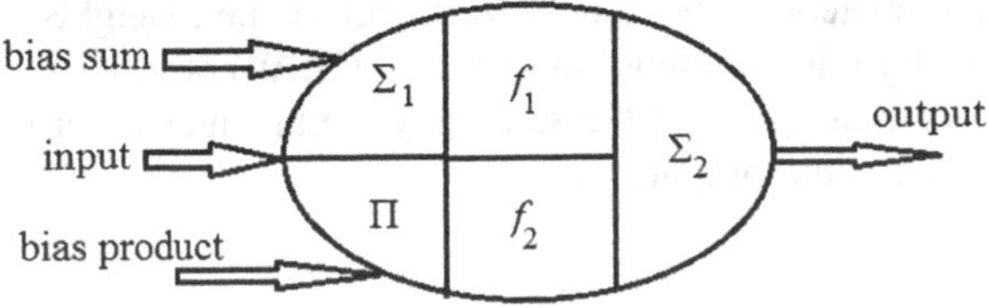


**Fig. 1** Architecture of generalized neuron model

the output of the pi part is given as (3)

$$O_{\prod} = f_2(W_{\prod i} \times X_i) + X_{o\prod} \qquad (3)$$

In the proposed work both the activation functions $f_1, f_2$ used are linear, i.e., ramp function with unity slope.

## 3.2 *Adaptive Genetic Algorithm*

To overcome the shortcomings of the conventional back propagation algorithm, the genetic algorithm is used for training the free parameters of the generalized neuron model. Genetic algorithm being random search method is more robust than the directed search methods. The genetic algorithm is a good optimization tool for providing near-optimal solutions. In the proposed work, to improve the performance of genetic algorithm, fuzzy rules are used.

The weights of the generalized neuron model input are considered as chromosomes of the genetic algorithm, which are modified using GA operators, i.e., crossover and mutation to get the new population. The convergence of the genetic algorithm depends upon the number of variables.

In genetic algorithm, the value of the parameters such as, crossover probability ($P_c$), mutation probability ($P_m$) are initialized in the beginning and kept constant during the optimization process [13]. In adaptive genetic algorithm, the operators of genetic algorithm, i.e., crossover probability ($P_c$), mutation probability ($P_m$), and population size are modified dynamically using fuzzy rule base.

The optimization of genetic algorithm is affected by the variation of parameters. At the starting, the high crossover probability and low mutation probability produce better chromosomes with high fitness value but after some generation the fitness value tends to become constant showing no effect of crossover in the population. At this time, the increase in mutation probability can diversify the population and chromosomes can inculcate new characteristics. Methods for deciding optimal setting of the operators were discussed by many researchers in the past. Dependency of mutation and crossover probability on time was discussed in [14] and optimal settings for all the GA parameters are found by experimentation in [15].

In the proposed work utilizes the fuzzy rule base system for controlling the values of operators $P_c$ and $P_m$ for which the GA parameters are defined using five linguistic terms which are very low (VL), low (L), medium (M), high (H) and very high (VH). The proposed adaptive genetic algorithm is used for training the free parameters of the generalized neuron (i.e., weights) for predicting the future electricity price of New South Wales (NSW) electricity market. The past five values of electricity price of the same day are taken as input values and next hour electricity price is the output.

## 3.3 Wavelet Transform

The price series is highly volatile and nonstationary in nature which makes it difficult to predict. In order to improve the generalization ability of the generalized neuron, input price series is improved by taking the wavelet transform of the series to convert it into a set of constitutive series.

Wavelet transform decomposes the price series data into a set of constitutive series, which has better behavior in terms of data variance, than original price series [16]. The wavelet transform begins with the selection of a proper mother wavelet and then analyzing the translated and dilated version of the signal. A wavelet can be defined as a function (t) with a zero mean as given by (4) [16].

$$\int_{-\infty}^{+\infty} \psi(t)\, dt = 0 \tag{4}$$

A signal can be decomposed into many series of wavelets with different scales a and translation b as shown in (5)

$$\psi(t)_{a,b} = \frac{1}{\sqrt{a}}\, \psi\left(\frac{t-b}{a}\right) \tag{5}$$

Thus, the wavelet transform of a signal $f$(t) at translation $b$ and scale a is defined by (6)

$$wf(b,a) = \frac{1}{\sqrt{a}} \int_{+\infty}^{-\infty} f(t)\, \psi\left(\frac{t-b}{a}\right) dt \tag{6}$$

The original signal $f$(t) can be reconstructed by inverse wavelet transform as shown in (7)

$$f(t) = \int_{0}^{+\infty} \int_{-\infty}^{+\infty} \frac{1}{a^2} wf(b,a)\psi(t)_{a,b}(t)\, db\, da \tag{7}$$

The discrete wavelet transform is implemented using multi-resolution analysis algorithm developed by Mallat. The algorithm has two stages, decomposition and reconstruction [17, 18] as shown in Fig. 2.

In the proposed work three-level decomposition is used, and consequently three detail series, i.e., D1, D2, and D3 and one approximate series, i.e., A3 are obtained. The mother wavelet is taken as daubechies of order 3 (db3). The decomposition level is decided based on the skewness and kurtosis analysis of the transformed price signal.

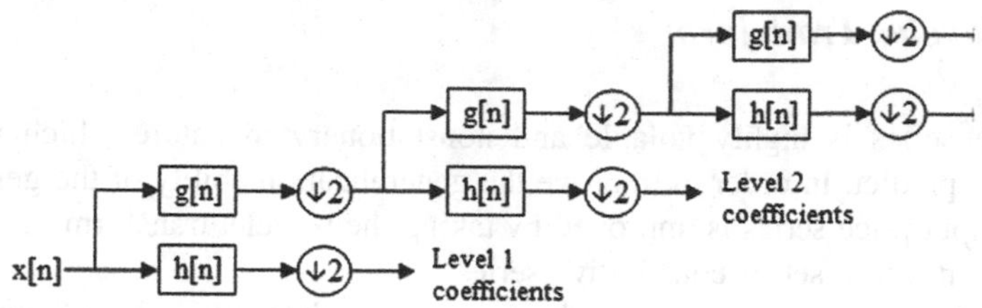


**Fig. 2** Daubechies wavelet of order 3 (db 4) & three level decomposition

# 4 Proposed Model Accuracy

In order to assess the prediction accuracy of the proposed generalized neuron model trained using adaptive genetic algorithm, the mean absolute percentage error (MAPE) of the week and day is calculated as shown in (8) and (9). The weekly absolute error is expressed as (8)

$$MAPE_{week} = \frac{1}{168} \sum_{t=1}^{168} \frac{|f_{pt} - \hat{f}_{pt}|}{P_{week}}, \tag{8}$$

where $f_{pt}$ and $\hat{f}_{pt}$ are the actual and forecasted price of hour $t$, respectively.

$$P_{week} = \frac{1}{168} \sum_{t=1}^{168} f_{pt} \tag{9}$$

Variance of the error term is another important factor which measures the precision of the model. The smaller is the variance, the more precise is the prediction of prices. The weekly error variance is given as (10)

$$EV_{week} = \frac{1}{168} \sum_{t=1}^{168} \left( \frac{|f_{pt} - \hat{f}_{pt}|}{P_{week}} - MAPE_{week} \right)^2 \tag{10}$$

The propose model is compared with other existing model based on the above two accuracy parameters, i.e., mean absolute percentage error (MAPE) and error variance (EV).

# 5 Results and Discussion

The generalized neuron model is used to forecast the electricity price of New South Wales electricity market (NSW). The proposed model is trained using adaptive genetic algorithm and the results obtained are compared with conventional neural network, generalized neuron model trained using back propagation.

The yearly data of the New South Wales market is divided into four seasons, i.e., winter (January to March), spring (April to June), summer (July to September), and fall (October to December). Second week of first month from all the seasons are selected for forecasting.

The curves are plotted between the actual and forecasted price using conventional artificial neural network, generalized neuron model, and proposed approach for weekday (i.e., Tuesday) of the winter and spring season, Sunday of the summer season, and first week complete of fall season. The corresponding results were shown in Figs. 3, 4, 5 and 6. The proposed model is compared with the existing models in terms of mean absolute percentage error (MAPE) and error variances, the results obtained are tabulated in Tables 1 and 2.

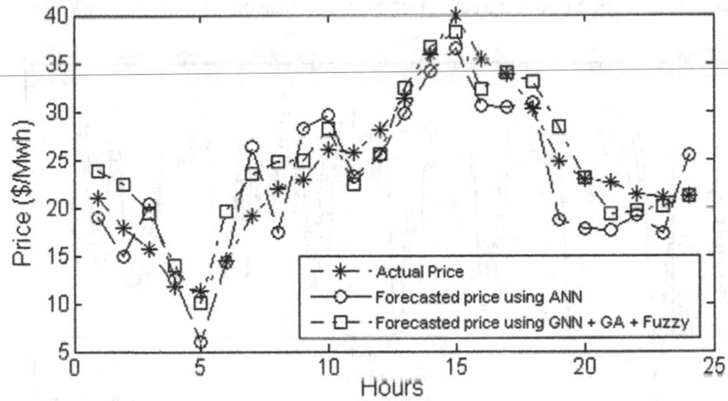


**Fig. 3** Price forecast for Tuesday (week 1) of Winter Season of New South Wales

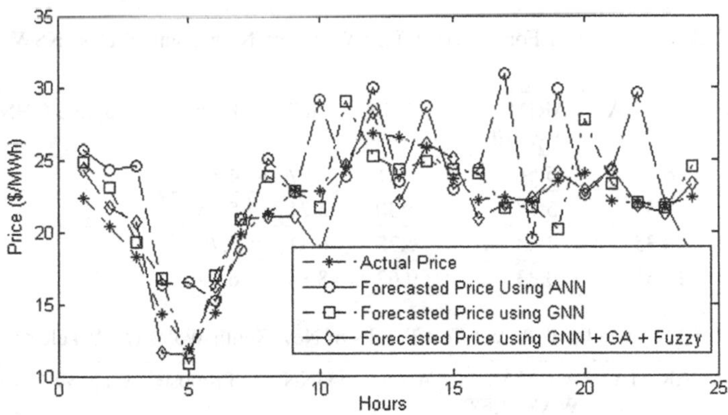


**Fig. 4** Price forecast for Tuesday (week 1) of Spring Season of New South Wales

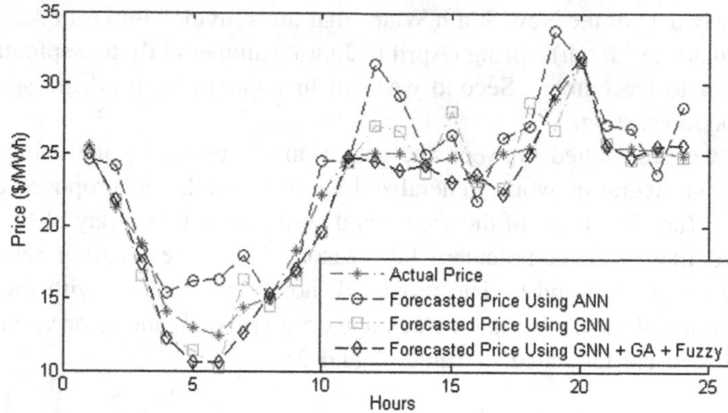


**Fig. 5** Price forecast for Sunday (week 2) of Summer Season of New South Wales

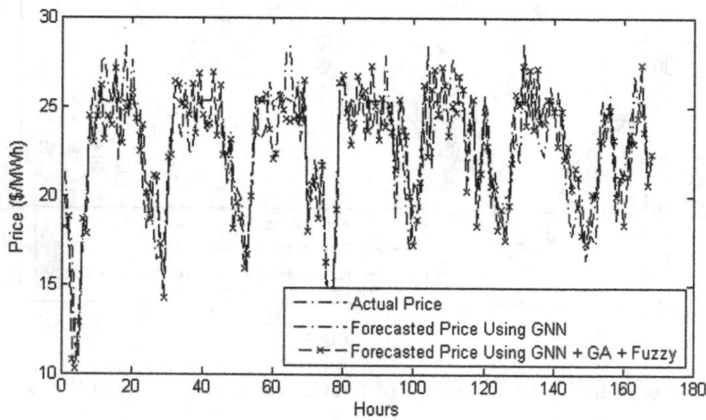


**Fig. 6** Price forecast for second week of Fall Season of New South Wales

**Table 1** Weekly MAPE (%) For Different Test Weeks of New South Wales (NSW) electricity market

| Test Week | ARIMA | ARIMA + WAVELET | ANN | GNN | Proposed Adaptive GNN Model |
|---|---|---|---|---|---|
| Winter | 6.30 | 5.66 | 5.21 | 4.97 | 4.35 |
| Spring | 6.58 | 5.17 | 5.30 | 5.21 | 5.06 |
| Summer | 8.35 | 7.23 | 6.78 | 6.31 | 5.78 |
| Fall | 11.56 | 10.33 | 9.62 | 8.88 | 6.39 |

**Table 2** Error Variance for Different Test Weeks of New South Wales (NSW) electricity market

| Test Week | ARIMA | ARIMA + WAVELET | ANN | GNN | Proposed Adaptive GNN Model |
|---|---|---|---|---|---|
| Winter | 0.0033 | 0.0018 | 0.0017 | 0.0015 | 0.0012 |
| Spring | 0.0025 | 0.0018 | 0.0016 | 0.0013 | 0.0012 |
| Summer | 0.0092 | 0.0082 | 0.0079 | 0.0043 | 0.0039 |
| Fall | 0.0139 | 0.0105 | 0.0088 | 0.0059 | 0.0041 |

## 6 Conclusion

In the proposed work, a hybrid model combining the features of generalized neuron model and adaptive genetic algorithm is proposed. The proposed model is used to predict the electricity prices of the New South Wales (NSW) electricity market, the performance is compared with the conventional neural network, regression based models, i.e., ARIMA and wavelet ARIMA and generalized neuron model. It is observed that the proposed adaptive generalized neuron model gives better prediction and is able to track the variation in electricity price more closely. The obtained results were also validated by using the mean absolute percentage error and error variance as the performance indices and the proposed model shows considerable reduction in the forecasting error.

## References

1. M. Shahidehpour, H. Yamin, and Z. Li, *Market Operations in Electric Power Systems*. New York, USA: John Wiley & Sons, Inc., 2002.
2. S. K. Aggarwal, L. M. Saini, and A. Kumar, "Electricity price forecasting in deregulated markets: A review and evaluation," *Int. J. Electr. Power Energy Syst.*, vol. 31, no. 1, pp. 13–22, Jan. 2009.
3. J. P. S. Catalão, S. J. P. S. Mariano, V. M. F. Mendes, and L. A. F. M. Ferreira, "Short-term electricity prices forecasting in a competitive market: A neural network approach," *Electr. Power Syst. Res.*, vol. 77, no. 10, pp. 1297–1304, Aug. 2007.
4. L. Hu, G. Taylor, H. B. Wan, and M. Irving, "A review of short-term electricity price forecasting techniques in deregulated electricity markets," in *Universities Power Engineering Conference (UPEC), 2009 Proceedings of the 44th International*, 2009, pp. 1–5.
5. S. K. Aggarwal, L. M. Saini, and A. Kumar, "Electricity price forecasting in Ontario electricity market using wavelet transform in artificial neural network based model," *Int. J. Control Autom. Syst.*, vol. 6, no. 5, pp. 639–650, 2008.
6. L. Abdullah, "ARIMA Model for Gold Bullion Coin Selling Prices Forecasting," *Int. J. Adv. Appl. Sci.*, vol. 1, no. 4, Dec. 2012.
7. Z. Quan-yin, Y. Yong-hu, Y. Yun-yang, and G. Tian-feng, "A Novel Efficient Adaptive Sliding Window Model for Week-ahead Price Forecasting," *TELKOMNIKA Indones. J. Electr. Eng.*, vol. 12, no. 3, Mar. 2014.
8. D. E. Rumelhart, G. E. Hinton, and R. J. Williams, "Learning Internal Representations by Error Propagation," Sep. 1985.
9. N. Singh, D. K. Chaturvedi, and R. K. Singh, "A Modified Error Function GNN For Load Frequency Control of Multi-area Power System," in *Proceedings of the 2010 International Conference on Artificial Intelligence, ICAI 2010, July 12–15, 2010, Las Vegas Nevada, USA, 2 Volumes*, 2010, pp. 353–359.
10. D. K. Chaturvedi, P. S. Satsangi, and P. K. Kalra, "New neuron models for simulating rotating electrical machines and load forecasting problems," *Electr. Power Syst. Res.*, vol. 52, no. 2, pp. 123–131, Nov. 1999.
11. D. K. Chaturvedi, M. Mohan, R. K. Singh, and P. K. Kalra, "Improved generalized neuron model for short-term load forecasting," *Soft Comput. - Fusion Found. Methodol. Appl.*, vol. 8, no. 5, pp. 370–379, Apr. 2004.

12. D. K. Chaturvedi, A. P. Sinha, and O. P. Malik, "Short term load forecast using fuzzy logic and wavelet transform integrated generalized neural network," *Int. J. Electr. Power Energy Syst.*, vol. 67, pp. 230–237, May 2015.
13. J. D. Schaffer, R. A. Caruana, L. J. Eshelman, and R. Das, "A Study of Control Parameters Affecting Online Performance of Genetic Algorithms for Function Optimization," in *Proceedings of the Third International Conference on Genetic Algorithms*, San Francisco, CA, USA, 1989, pp. 51–60.
14. T. C. Fogarty, "Varying the Probability of Mutation in the Genetic Algorithm," in *Proceedings of the Third International Conference on Genetic Algorithms*, San Francisco, CA, USA, 1989, pp. 104–109.
15. J. Grefenstette, "Optimization of Control Parameters for Genetic Algorithms," *IEEE Trans. Syst. Man Cybern.*, vol. 16, no. 1, pp. 122–128, Jan. 1986.
16. A. J. Conejo, M. A. Plazas, R. Espinola, and A. B. Molina, "Day-Ahead Electricity Price Forecasting Using the Wavelet Transform and ARIMA Models," *IEEE Trans. Power Syst.*, vol. 20, no. 2, pp. 1035–1042, May 2005.
17. J. P. S. Catalao, H. M. I. Pousinho, and V. M. F. Mendes, "Hybrid Wavelet-PSO-ANFIS Approach for Short-Term Electricity Prices Forecasting," *IEEE Trans. Power Syst.*, vol. 26, no. 1, pp. 137–144, Feb. 2011.
18. S. G. Mallat, "A theory for multi resolution signal decomposition: the wavelet representation," *IEEE Trans. Pattern Anal. Mach. Intell.*, vol. 11, no. 7, pp. 674–693, Jul. 1989.

# Design and Performance Comparison of CNTFET-Based Binary and Ternary Logic Inverter and Decoder With 32 nm CMOS Technology

**Mayuri Khandelwal and Neha Sharan**

**Abstract** This paper attempts to compare ternary and binary logic gate design using CMOS and carbon nanotube (CNT)-FETs technology. Ternary logic is an effective approach over the default binary logic design technique because it allows to define one more voltage level which is VDD/2 and it also allows a circuit to be simple in design and energy efficient due to its property of reduction in circuit overhead such as interconnects and chip area. A CMOS and CNTFET-based ternary logic gates and arithmetic circuit design has been proposed to implement and compare binary and ternary logic design based on CMOS and CNTFET. The main objective is to compare the CMOS and CNTFET results and verify the advantages of CNTFET technology. The proposed CNTFET technique combined with ternary logic provides an usable performance, improved speed and reduces propagation delay characteristics in circuit such as inverter and decoder. Simulation results of proposed designs using H-SPICE are observed and shown that the proposed ternary logic gates consume significant less delay than the CMOS gates implementations. In realistic digital application, the proposed design of ternary logic compared with binary logic results in over 95% reductions in terms of the consumption of propagation delay.

**Keywords** CNT-FET · Ternary · Inverter · Decoder · Propagation delay time

---

M. Khandelwal (✉) · N. Sharan
GLA University, Mathura, UP, India
e-mail: mayurikh12@gmail.com

N. Sharan
e-mail: neha.sharan@yahoo.co.in



S.K. Bhatia et al. (eds.), *Advances in Computer and Computational Sciences*,
Advances in Intelligent Systems and Computing 553,
DOI 10.1007/978-981-10-3770-2_40

# 1 Introduction

Digital (Binary) logic computational technique is generally used to apply over two value logic, i.e., (0 or 1), (true or false) according to Boolean format. Multiple-value logic used in place of traditional boolean characterization of boolean values or variables with either infinite or finite values such as ternary logic for three-valued logic [1]. Ternary logic (three-valued logic) has an advantage over binary that it can define one more logic level so that it become feasible for it to come with simple and more over an energy efficient design as this logic design is able to decrease the effects of complexity in circuit design and reduces chip area [2]. Moreover, logic circuit operations like Inverter, NAND, NOR operations can be results as faster in ternary as compare to binary. Intense research in the designing and implementation of binary and ternary logic designs using CMOS and CNTFET can be found in the literature [2, 3]. Propagation time delay can be reduced up to 90% using an efficient ternary implementation for a ternary decoder compared to its CMOS design [4]. MVL designing methodology has been used in place of binary logic designs to enhance the performance parameters of CMOS and CNTFET technologies [5].

The carbon nanotube field effect transistor (CNT) FET is an effective technology compare to the silicon-based transistor which is CMOS for reducing power and better performance device design due to its ballistic transport and low OFF—current properties [6–9]. The ternary logic based arithmetic design is effective in terms of power and propagation delay. In CNTFET technique, the threshold voltage (VTH) of the transistor is inversely proportional to the diameter (D) of the carbon–nanotube. A design based on multiple threshold can be designed by using CNTs with different diameters in the CNTFETs where as much tubes are used as required. In this research, the binary and ternary logic circuits based on CNTFETs are designed and compared with existing counterpart designs based on CMOSs.

The CNTFET based design of binary and ternary logic gate (inverter) is implemented and analysed in detail and compared with its counterpart CMOS based designs. This analysis has also been done for ternary decoder design and presented as an example of the application and use of basic ternary logic gates design technique. For the arithmetic design, a ternary logic in most efficient form is used to increase the speed and reduce propagation delay time of the circuit. The design of ternary decoder uses NTI, STI, and ternary NOR for its results. The ternary logic gates are useful in designing because it reduces number of transistors and propagation delay and also useful for fast computational speed and efficient performance. In this research, the ternary and binary logic gate and arithmetic circuit design methodology is proposed with intense simulation results. H-SPICE simulation results show advantages of using CNTFET technology in terms of speed and power consumption.

Table 1 Logic symbols

| Voltage level (in volts) | Logic value |
|---|---|
| 0 | 0 |
| ½ $V_{DD}$ | 1 |
| $V_{DD}$ | 2 |

## 2 Review of Ternary Logic

Computer systems are designed on the basis of binary logic circuit in which a signal consists of only two values, 0 and 1, where, 0 stands for ground level and 1 stands for VDD. Then, It also become possible to design ordinary computer system using analog circuit, where the voltage value is directly proportional to a particular value of that signal at that time. A major issue with this approach is that of the circuit which may include noise, which changes the voltage value of the signal. In a digital design, since the value of noise is comparatively small as to the difference between the voltage values which represent as logic 0 and logic 1. The realistic goal of multi-valued logic (MVL) research is to direct the design and fabrication of multivalued circuit. Multivalued circuits should have the following advantages over their binary circuits: A MVL based wired signal that can transmit more information than binary is able to reduce the circuit connections. A MVL design can deal to accumulate information than that of binary one so that complexity can also get reduced. The connections on-off chip connections can be decreased to reduce the pin-out difficulties that accumulate with larger chips. The speed the transmission of serial information becomes faster as the transmitted information per unit time is increased.

The basic operations of ternary logic design are defined in Table 1.

## 3 Carbon Nanotube Field Effect Transistor (CNTFET)

With the continuous increasing trend to reduce the feature sizes, and to employ continuous smaller components on the integrated circuits and new challenges also arose on the way of silicon CMOS circuits and devices. New and Emerging technology of "nano devices" shows the possibility of increased density of integration and reduced power consumption. The new devices due to their small dimensions, shows large variations in their characteristics and behavior. The variation observed in these devices affects the reliability issues and the performance of circuits. The Carbon Nanotube (CNT) is one such device which is used as a major device in this work.

CNTs exhibit properties because of which electronic and mechanical characteristics are observed and due to which extraordinary strength of the carbon–carbon bond is formed which are described in Table 2. It also have small atomic diameter in the atom of carbon. The free π-electrons which are available in the graphitic configuration make it much stronger to replace CMOS technology.

**Table 2** Properties of CNT-FET

| Electrical conductivity | Metallic or semiconducting |
|---|---|
| Electrical transport | Ballistic, no scattering |
| Energy gap (semicod.) | $E_g$[eV] = 1/d [nm] |
| Maximum current density | ~ $10^{10}$ A/$cm^2$ |
| Maximum strain | 0.11% at 1 V |
| Thermal conductivity | 6000 W/Km |
| Diameter | 1–100 nm |
| Length | Up to millimeters |
| Gravimetric surface | >1500 $m^2$/g |
| E-modulus | 1000 Gpa |

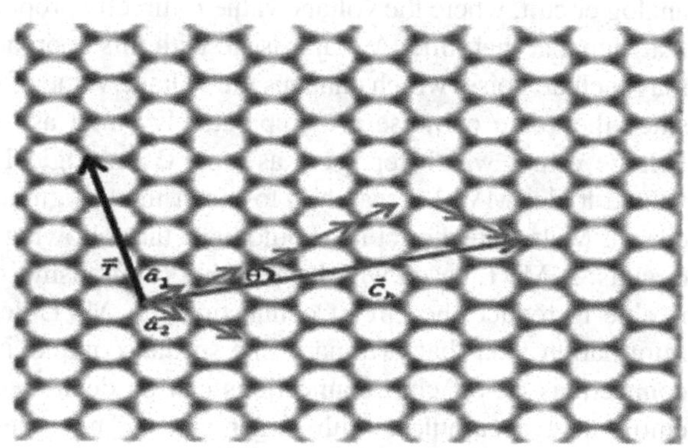

**Fig. 1** Chilarity vector

The unbeatable properties of electrical and mechanical characteristics in carbon nanotubes come from its electronic structure of graphene that can roll up to form a hollow cylinder shape like structure (Fig. 1). For such carbon nanotube, its circumference can be expressed in terms of its chiral vector which is Ĉh = nâ1 + mâ2 in which two crystallographically equivalent sites of the 2-dimensional graphene sheet are joined. In the below structure, n and m represents as integers and â1 and â2 represents as the unit vectors. Therefore, any carbon nanotube or its structure can be described (Fig. 2) as an index with a pair of integers (n, m) that is its chiral vector.

The nanotube diameter $d_t$ and the chiral angle θ in terms of the integers (n, m) [1] are given as

$$d_t = \frac{\sqrt{3}a_{C-C}\sqrt{m^2 + mn + n^2}}{\pi} \tag{1}$$

$$\theta = \tan^{-1}\left(\frac{\sqrt{3}n}{2m+n}\right) \tag{2}$$

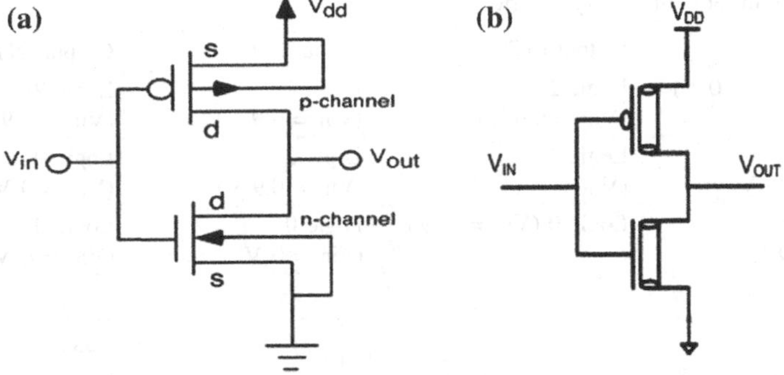


**Fig. 2** Circuit design of **a** CMOS based **b** CNTFET based binary inverter

**Table 3** Truth table of binary inverter

| Input | Output |
|---|---|
| 0 (0 V) | 1 (0.9 V) |
| 1 (1 V) | 0 (0 V) |

# 4 Circuit Design and Techniques

## *4.1 Binary Inverter*

An inverter circuit gives output as a voltage which represents the logic level which is opposite to its input. For low input, output will be high and vice versa as in Table 3.

For CMOS design, construction of inverter can be done by using a single NMOS or PMOS transistor coupled with a resistor or both which is called as CMOS. Since current flows through the resistor in one of the two states so the resistive drain configuration becomes the disadvantage for power consumption and speed. The design of this circuit as shown in Fig. 2 is done on H-spice using $V_{DD}$ is 0.9 V at 32 nm CMOS.

For CNTFET design, CNTs are used as gate material and CNTFET parameter file is used to design circuit Fig. 2 on H-spice at $V_{DD}$ equals to 0.9 V.

## *4.2 Ternary Inverter*

Ternary logic design can be done in 3 ways which are STI, NTI, and PTI. Most popular and efficient ternary logic is STI or simple ternary inverter. For the inputs {0, 1, 2} it yields the output {2, 1, 0}. Its truth table is shown in Table 4.

A CNTFET based network can be design over ternary logic pattern to reduce power consumption, improve speed and propagation time delay and the main advantage is to nullify the use of resistors that reduces the area overhead. Since

**Table 4** Truth table of ternary inverter

| Input | Output (STI) | Output (PTI) | Output (NTI) |
|---|---|---|---|
| Logic 0 (Vin = 0 V) | Logic 2 (Vin = 0.9 V) | Logic 2 (Vin = 0.9 V) | Logic 2 (Vin = 0.9 V) |
| Logic 1 (Vin = 0.45 V) | Logic 1 (Vin = 0.45 V) | Logic 2 (Vin = 0.9 V) | Logic 0 (Vin = 0 V) |
| Logic 2 (Vin = 0.9 V) | Logic 0 (Vin = 0 V) | Logic 0 (Vin = 0 V) | Logic 0 (Vin = 0 V) |

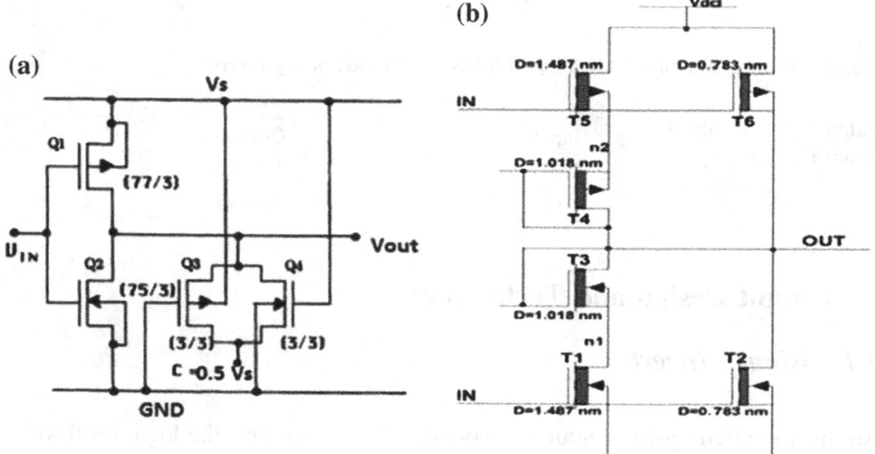


**Fig. 3** Circuit design of **a** CMOS based **b** CNTFET based ternary inverter

diameter of CNTs depends upon the threshold voltage so threshold values are defined for each transistor to specify their performance (Fig. 3).

## 4.3 Ternary Decoder

The ternary decoder is a circuit design which has one-input, three-output combinational logic that can generate unary functions for an input x. The response of the ternary decoder to the input x is given as,

$$X_K = 2, \text{if } x = k$$
$$0, \text{if } x \,!= k$$

where, k can be any logic values from 0, 1, or 2.

In Fig. 4, '–'denotes NTI, '+' denotes PTI and '.' denotes ternary NOR logic gates.

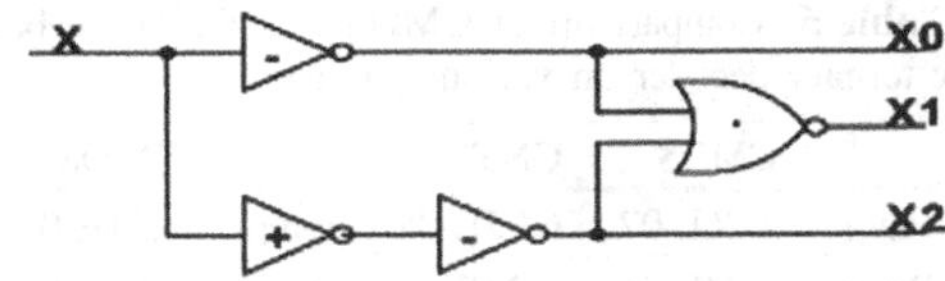


**Fig. 4** Circuit design of ternary decoder

# 5 Simulation Results

## *5.1 CMOS Based*

See Fig. 5.

## *5.2 CNTFET Based*

See Fig. 6.

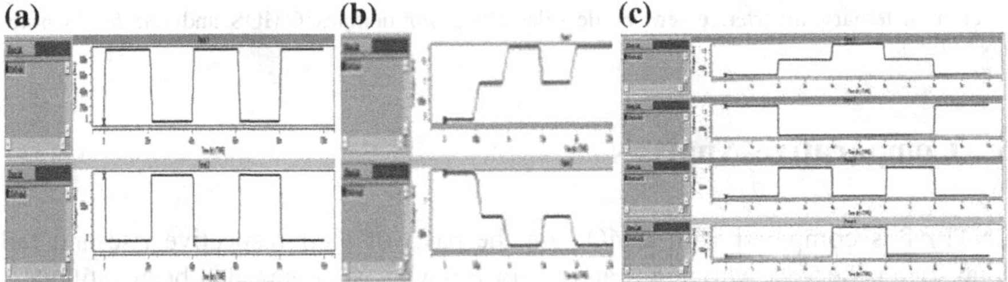


**Fig. 5** Simulation results (Topmost pulse indicates input and bottommost pulse indicates output) of CMOS based **a** binary inverter **b** ternary inverter and **c** ternary decoder (Result follow a sequence (top to down) as input, x0, x1, x2.)

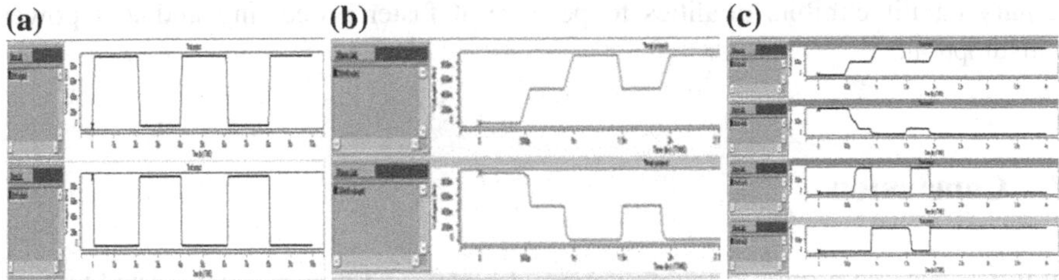


**Fig. 6** Simulation results of CNTFET based **a** binary inverter **b** ternary inverter and **c** ternary decoder

**Table 5** Comparison of CMOS and CNTFET based **a** binary inverter, **b** ternary inverter and **c** ternary decoder on various parameters

| | CMOS | CNTFET | | CMOS | CNTFET | | CMOS | CNTFET |
|---|---|---|---|---|---|---|---|---|
| Pavg | 1.26E-07 | 6.97E-08 | Pavg | 6.31E-04 | 1.28E-06 | Pavg | 5.70E-04 | 1.78E-06 |
| Pmax | 6.74E-06 | 4.24E-06 | Pmax | 6.89E-04 | 6.24E-06 | Pmax | 9.03E-04 | 1.54E-05 |
| TDR | 5.69E-11 | 5.52E-12 | TDR | 1.24E-11 | 2.37E-11 | TDR | 5.98E-06 | 8.46E-10 |
| TDF | 8.75E-11 | 6.29E-12 | TDF | 3.45E-11 | 3.07E-11 | TDF | 2.01E-06 | 4.09E-10 |
| TR | 2.86E-10 | 2.45E-10 | TR | 8.98E-11 | 5.27E-11 | TR | 2.36E-09 | 1.57E-10 |
| TF | 3.98E-10 | 2.44E-10 | TF | 8.04E-11 | 8.50E-11 | TF | 1.12E-08 | 2.63E-10 |

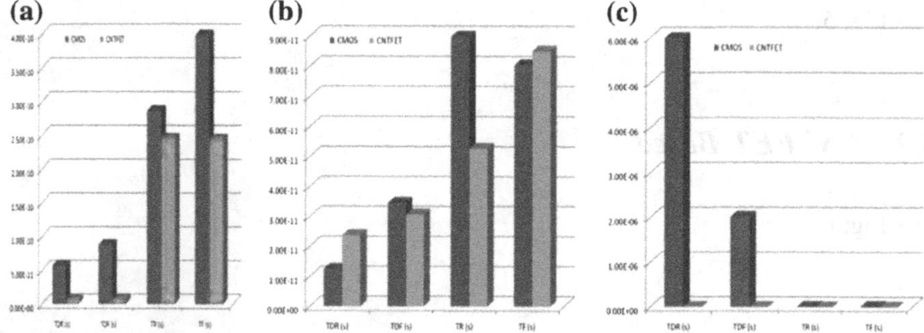


**Fig. 7** Graph of comparison (Parameters of comparison are TDR, TDF, TR, and TF): **a** binary inverter, **b** ternary inverter, **c** ternary decoder (*Blue bar* denotes CMOS and *red bar* denotes CNTFET.)

## 6 Comparative Analysis

CNTFET is compared with CMOS on the basis of their respective rise and fall delays are observed. Maximum and average power value has also been estimated. Comparison of CMOS with CNTFET has been done to find out the best between them on the basis of propagation delay time of rise (TDR) and fall (TDF), rise time (TR), fall time (TF), average power (Pavg), maximum power (Pmax).

From Table 5 and Fig. 7, we can conclude that most of the parameters of CNTFET have very less values as compare to CMOS. So we can say that CNTFET ternary circuit exhibits qualities to perform at faster processing and less power consumption.

## 7 Conclusion

In this paper, we proposed the designing of binary and ternary Inverter and ternary decoder which are based on CMOS and CNTFET technologies. We compared the circuits of different technologies on the basis of propagation delay time, peak and maximum power consumption. Since CNTFET structure and its properties depends upon its construction as diameter of CNT is inversely proportional of the threshold voltage of transistor and number of tubes of CNT are also controllable. Therefore, CNTFET proves as a reliable technique for nanoscale era since it allows better performance, reduces delay and power consumption, improves speed and act as a controllable device. Ternary logic method also act as an efficient methodology over binary since it allows to define three logic state so it increases information content capabilities. Simulation and analysis have been performed using H-SPICE over 32 nm CMOS and CNTFET model schemes. The results have been justified the advantages of using CNTFET as an technology and ternary logic system in designing for future digital demands.

## References


1. S. Kang and Y. Leblebici (2003), CMOS Digital Integrated Circuits - Analysis and Design, McGraw-Hill.
2. Wanjari, Ms Nilmani P., and Ms Shweta P. Hajare. "VLSI Design and Implementation of Ternary Logic Gates and Ternary SRAM Cell."
3. Anand, Aparna, and S. R. P. Sinha. "Performance Evaluation of Logic Gates Based On Carbon Nanotube Field Effect Transistor." International Journal of Recent Technology and Engineering (IJRTE) ISSN: 2277 3878 (2013).
4. Moaiyeri, Mohammad Hossein, et al. "Efficient CNTFET-based ternary full adder cells for nanoelectronics." Nano-Micro Letters 3.1.
5. J. Rabaey, A. Chandrakasan, B. Nikolic, "Digital Integrated Circuits," Pearson Education International, New Jersey, 2003.
6. Radha Tapiawala, Rahul Kashyap "Design of Universal Logic Gates based on CNTFET for Binary and Ternary Logic." International Journal of Engineering Research & Technology (IJERT) Vol. 3 Issue 6, June – 2014.
7. Doostaregan, Akbar, et al. "On the design of new low-power CMOS standard ternary logic gates." Computer Architecture and Digital Systems (CADS), 2010 15th CSI International Symposium on. IEEE, 2010.
8. Ghorbani, Ali, and Ghazaleh Ghorbani. "Energy Efficient Full Adder Cell Design With Using Carbon Nanotube Field Effect Transistors In 32 Nanometer Technology." arXiv preprint arXiv:1411.2088 (2014).
9. Kim, Yong-Bin. "Integrated circuit design based on carbon nanotube field effect transistor." Transactions on Electrical and Electronic Materials 12.5 (2011): 175–18

# Optimized Route Selection on the Basis of Discontinuity and Energy Consumption in Delay-Tolerant Networks

**Lokesh Pawar, Rohit Kumar, Swinky Arora and Amit K. Manocha**

**Abstract** Delay-Tolerant Networks (DTNs) are intermittently connected mobile wireless networks in which the connectivity between nodes changes frequently due to nodes movement. A delay-tolerant network suffers from nodes energy efficiency issues due to frequent mobility and reliable transmission of data packet from source to destination. Due to the mobility of the node there are frequent changes in the path. Due to which enormous amount of energy is wasted. So to conserve the energy of the node energy efficient transmission from one node to another node can be achieved by using the proposed algorithm EADBNHS. Connectivity of the network can be determined by Synchronism, Simultaneousness and Discontinuity using EADBNHS. EADBNHS is the algorithm devised for energy efficiency in this article. To achieve energy efficiency of a node the algorithm is designed in a way, which includes transmission energy, reception energy and the health of the node. This algorithm supports energy efficient transfer of data packet; it selects a node on the basis of minimum energy available on the node and on the basis of its connectivity to other nodes. With the help of Synchronicity, Simultaneousness and Discontinuity we can easily judge a node's condition to become the participant node in the path from source to destination.

**Keywords** Quality of service · Delay tolerant networks · Energy efficiency · Network lifetime · Source node · Destination node · Mandatory storage

L. Pawar · R. Kumar (✉)
Department of CSE, Chandigarh University, Gharaun, Mohali, Punjab, India
e-mail: rohitbhullar@gmail.com

L. Pawar
e-mail: lokesh.pawar@gmail.com

S. Arora
Department of CSE, LLRIET, Moga, Punjab, India
e-mail: swinky.cgc@gmail.com

A.K. Manocha
Department of Electrical Engineering, MAU, Baddi, Himachal Pradesh, India
e-mail: manocha82@gmail.com



S.K. Bhatia et al. (eds.), *Advances in Computer and Computational Sciences*,
Advances in Intelligent Systems and Computing 553,
DOI 10.1007/978-981-10-3770-2_41

## 1 Introduction

Over the long time, the evolution of energy efficient communication and networking algorithms and protocols for DTNs (Delay-Tolerant Networks) has been a challenge for the researchers. Data packets in DTN follow store-carry and forward approach by taking advantage of node's mobility to initiate the communication. With the ubiquitous wireless communication and the era of new wireless technologies, the network communication has evolved and revolutionized. It enabled new technologies or techniques to deal with low power and energy efficiency in DTN [1–4]. Messages are not delivered properly in DTN because in such kind of sparse network one can never find a stable classical network where synchronicity/simultaneous occurrence of events alone affect the whole network. DTN faces battery problem and reliable transmission of the data packet [5, 6]. Several experiments and researches has been made in the field of Delay Tolerant Networks but such kind of work is the first approach toward these factors which deal with the nodes transmission power and receiving power as the prime constraint and next follows the synchronicity, simultaneousness and discontinuity as a prime factor for next hop selection and transferring data packet to a healthy node. Synchronism means simultaneous occurrence of events which seems to be related but cannot be distinguished between the intermediate nodes simultaneousness which means operating at the same time and discontinuity which means irregularity in the intermediate nodes. Discontinuity can be taken as a prime factor for next hop selection since it is concerned with the irregularities in the intermediate nodes from SN (Source Node) to DN (Destination Node). The challenges for DTN are connection and storage.

## 2 Route Selection

Route selection on the basis of below-mentioned techniques might have better path with energy efficiency and reliable transmission.

### *2.1 Synchronism in DTN*

Synchronicity is a concept, first sparked by Carl Jung, which holds that events are "meaningful coincidences" if they occur with no causal relationship, yet seem to be meaningfully related. Synchronism in DTN generally measures how different hops in the network coincidence each other on an end-to-end basis. The Store-and-forward approach is in-built in the DTN network [5, 6]. The Prophet Routing protocol helps the node to store the previous path followed by the data packet to reach the destination which helps in finding the path. If a path exists the value of synchronicity will be '1'

**Fig. 1** Overlapping of hops

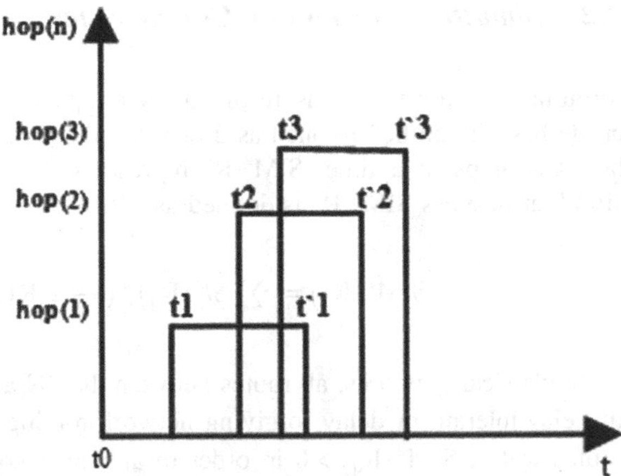


and packet will be delivered else it will be '0' and the node is forced to store the data packet till the synchronicity becomes '1'. At times in DTN where complete connectivity is available, a standard or classical communication is realizable because the data packet coincidently can be sent directly without store-and-forward approach. Route synchronicity SYN N(R) is characterized as [7]:

Synchronicity = Time of window overlap/Route Lifetime$_{End-to-End}$. This time of window overlap depends upon the hops available in the network as shown in Fig. 1.

$$\text{SYNN}(R) = \text{TWO}(R)/\text{RLF}_{E-t-E}(R)\ \text{With}\ \text{SYN}(R)\ 2[0,1] \tag{1}$$

$\text{TWO}(R_q)$ is the period of time window overlap, and $p_i$ represents the propagation delay for all hops [8–10]:

$$\text{TWO}(R_q) = \{t'_1 - t_n + \sum_{j=1}^{n-1} p_j \ \text{if}\ t_n + \sum_{j=1}^{n-1} p_j < t'_1\} \tag{2}$$

$$\{0 \qquad \text{otherwise}$$

Following synchronicity by issuing messages during $[t_n, t'1]$ it assure that:

1. No compulsory storage takes place. //if synchronicity is 1.
2. The route may be "revocable" up to r round trips if $(t_n + 2 * r * \sum_{i=1}^{n} p_i <= t'1)$.

Standard or classical networks have a synchronicity ever set to 1. If it is not then the route is reasoned to be broken [7]. SYN(R) value lies between (0,1) because either there will be a connection to further process the data packet or there will be no connection to further process the data packet.

Figure 1 clearly depicts that hop (1) started at time t1 and was alive till t1 and in between hop (2) overlapped hop (1) which will impact the synchronicity of the network.

### 2.2 Simultaneousness or Concurrency

Simultaneousness is a measure of how subsequently hops overlap each other. On crude basis it can be reasoned as downgraded synchronism because it covers each hop two hops at a time. $SIM^2(R)$ represents the measure of two hops. Route simultaneousness $SIM^2(R)$ is defined as [7]:

$$SIM^2(R_q) = \sum_{i=1}^{n-1} SI_i^2(R_q) / (n-1)RLF_{E-t-E}(R_q) \tag{3}$$

In classical networks, all routes between the SN and DN have a $SIM^2(R_q) = 1$. In delay tolerant or delay forgiving network routing algorithm might only choice routes with a $SIM^2(R_q) > 0$ in order to aid some hop by hop dependable bundle transfer (subjected to propagation delay ($p_i$)). Propagation delay in case of DTN is equal to Distance/Speed. Speed in case of Wireless network is considered as the speed of light (i.e., $3.00 \times 10^8$ m/s) and the distance as per the scenario for simulation experiment. $SIM^2(R) > 0$ or DTN only chooses the node with $SIM^2(R) > 0$ because to transfer a data packet we need to deliver the packet to the next closest node to the destination or to a node which can deliver the data packet on a route which has a connectivity to the destination.

### 2.3 Discontinuity or Irregularity

Discontinuity also sometimes called Irregularity in lay man language is the twofold or dual metric of simultaneousness/concurrency and represents the normalized time period of compulsory storage. Figure 2 presents two situations where compulsory

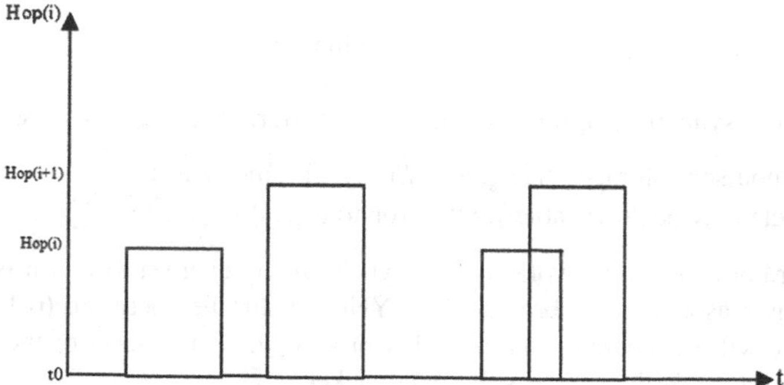


**Fig. 2** Fully and partially disjoint hops

storage occurs, among them is: fully and partially disjoint hops. Likewise end-to-end delay, the time period of compulsory storage depends on the time of issue of messages. The minimal $DSC^m(R_q)$ and maximal $DSC^M(R_q)$ route discontinuity are defined as [7]:

$$DSC^m(R_q) = \sum_{i=1}^{n-1} MSTO_i^m(R_q)/RLF_{E-t-E}(R_q) \tag{4}$$

$$MSTO_i^M(R_q) = t_{i+1} - (t_i + p_i) \tag{5}$$

$MSTO_i^m$ means mandatory storage for the data packets in case of minimal discontinuity. $MSTO_i^M$ means mandatory storage for the data packets in case of maximal discontinuity. $p_i$ in case of fully connected, fully disconnected and partially disconnected is considered to be negligible [11–14]. In standard/classical networks, all legitimate routes have a $DIS^{m,M}(R_q) = 0$, it means that hops are connected and data packet can be sent directly without saving and waiting for a connection to establish.

## 2.4 *Relationship Between (DIS) Discontinuity, (SIM) Simultaneousness and (SYN) Synchronicity in DTN*

Discontinuity, Simultaneousness and Synchronicity are tightly interconnected or interlaced metrics. Classical networks have total or fully connected routes where $SYN(R_q) = 1$ and no storage space is necessary or required. In DTN there are three types of network connectivity: Total or Fully disconnected, Absolute/Partially Disconnected and Partially Connected [15]. Table 1 shows the relationship between $SYN(R_q)$, $SIM^2(R_q)$, $DIS(R_q)$.

Table 1 represents the route conditions and the role of SYN, SIM and DIS. If the value of discontinuity is lower it reduces the amount of mandatory storage on the node and helps reduce memory requirements. If message is passed from 1st node to the 5th node than custody is required only in 1st and 5th node (case should be of simultaneous path.) The custodian node nearest to the node will work as a coordination center.

**Table 1** Routes relationship with synchronicity, simultaneousness and discontinuity

| S.no. | Routes | SYN(R) | $SIM^2(R)$ | DIS(R) |
|---|---|---|---|---|
| 1 | Fully disconnected routes | 0 | 0 | DIS > 0 $DIS^m > 0$ |
| 2 | Partially disconnected | 0 | $SIM^2(R) > 0$ | $DIS^m(R) > 0$ |
| 3 | Partially connected | SYN(R) > 0 | $SIM^2(R) > 0$ | $DIS^m(R) = 0$ |

# 3 Methodologies to Measure Energy Consumption of a Node

## *3.1 Power Consumption Models and Modes*

Mobile nodes in DTN are attached to other mobile nodes via wireless media. Nodes in such an environment are set free to receive and transfer data packet to the path which leads to the destination node and for transferring and receiving data packets the energy of nodes get diluted. Energy efficiency of a node enhances the quality of service which inculcates the service time, the success rate of the node, number of reliable transmissions done through the node. The energy consumption in a node is divided into 3 parts which consumes the energy. There are three modes of energy consumption in a node: (i) Transmission Mode (ii) Reception Mode (iii) Idle Mode. These modes of power intake are described as:

### 3.1.1 Transmission Mode

When a node transmits a data packet it requires transmission power which is a factor calculated with the help of transmission energy and transmission time. A given node is in transmission mode when it is ready to send the data packet with all minimum requirements for power. The transfer or transmission energy can be formulated as:

$$TE_e = ((330) * \text{Paylength}) / (2 \times 10^6) \tag{6}$$

$$TD_t = \text{Paylength} / \text{Bandwidth} \tag{7}$$

$$TPp = TEe / TDt, \tag{8}$$

where TEe is the transmission energy, $TP_p$ is the transmission power and $TD_t$ is time taken or consumed to transfer data packet and Paylength is length of data packet in number of Bits.

### 3.1.2 Receiving or Reception Mode

When a given node get a data packet from neighbor nodes or from other nodes in the networks such mode is known as reception mode/receiving mode and the energy consumed to get or receive packet is called reception Energy (Re). So, the reception energy can be formulated as:

$$REe = (230 * Paylength) / (2 \times 10^6) \quad (9)$$

$$RPp = REe / RDr, \quad (10)$$

where REe is known as reception energy, RPp is known reception power, RDr is a time duration taken to receive data packet.

### 3.1.3 Passive or Idle Mode

In this particular mode, a given node is neither transmitting nor receiving any data packets. But such mode also consumes node power of some predictable magnitude because the given nodes have to respond to the wireless medium incessantly in order to discover a packet that it should receive to communicate further.

## 4 Proposed Algorithm

In the present work an efficient algorithm called **EADBNHS** (Energy and Discontinuity Based Next Hop Selection) has been proposed. In EADBNHS (Energy and Discontinuity Based Next Hop Selection) selection of nodes takes place on the basis of transmission and receiving power $TP_p$, $RP_p$, respectively. The algorithm then follows the Minimum Discontinuity ($DIS^m$) approach, which substantiate the node energy and its next hop selection. Two major events occur during this course, i.e. (a) Route Request and (b) Route Reply. The Packet format for RREQ and RREP is TCPCL. If the transmission power of the next hop node is lower than the minimum threshold, the data packets will be dropped and will not be sent to such nodes whose transmission power is lower than the minimum threshold. ROUTE REQUEST trigger will select those nodes as next hop whose value is greater than the minimum threshold. REQUEST PATH will update the path details and will update the path metric. The list of the nodes which are on the path from source to destination is listed and put together. DIS min is also checked further if the node has a minimum threshold of transmission and receiving power with DIS min = 0 the node is selected as next hop. The second phase takes place when ROUTE REPLY event is triggered. ROUTE REPLY has a minimum requirement of threshold energy on the receiver node the algorithm checks residual energy on the receiver node. The receiver node enlists the path followed and sends back the acknowledgment to the SENDER node via RREP. Algorithm is followed until we reach the destination node.

**Pseudo code for the algorithm is:**

*EADBNHS* (Energy and Discontinuity Based Next Hop Selection)
*Event (RRQ_SENDER) triggered*
*if RRQ.NEIGHBOUR NODE* $TP_p \geq 131.83_{(minimum\ threshold)}$
*SELECT AS NEXT HOP*
*NODE.TPp* $\geq 131.83_{(minimum\ threshold)}$
*// Minimum threshold for Transmission Power is calculated using eq(vi,vii,viii) where bandwidth is 100KBps.*
*RRQ.PTH[Pth.length]=NODE.Tpp*$_{(min}$ *//PTH is Path length vector RreqList.add(RRQ)*
*entryTimeDIS*$_{min}$*.begin()*
*Event(RRQ_SENDER) initiation*
*if RRQ. NEIGHBOUR NODE DIS*$_{min}$*== 0*
*RRQ.PTH[Pth.length]= NODE.DIS*$_{min}$
*RreqList.add(RRQ) with DIS*$_{min}$*==0*
*else*
*RRQ.drop() // if minimum threshold not available*
*// if DIS*$_{min}$ *!==0*
*Else*
*RRQ.Path[Path.length] = NODE.TPt(min).NODE.DISmin*
*For z=0 to route_table.length*
*if (pair(source, destination) = pair[z])*
*ispresent = 1 // here 1 represent "true"*
*else*
***ispresent = 0 // here 0 represents "false"***
*if (ispresent)*
*Event (RRP_RECEIVER) triggered*
*if RRP. NODE RPp* $\geq 92_{(minimum\ threshold)}$
***// Minimum threshold for Reception power is calculated using eq.(ix,x)where bandwidth is 100KBps.***
*create or make new RRP*
*RRP.Path = path[z]*
*RRP.Source = RRQ.Source*
*RRP.Destination = RRQ.Destination*
*RRP.hopCount = path[z].hopCount*
*sendPacket(RRP,source)*
*else*
*RRP.drop() //if minimum threshold is not available*

## 5 Simulation Results

In the proposed work GloMoSim simulation tool has been used to validate our projected *EADBNHS* strategy and compare it with normally distributed delay distribution and node selection process to start the communication between the nodes with less consumption of power of the node. The performance metric used for the proposed scheme are as follows:

$$TE_e = (330 * Paylength)/(2 \times 10^6)\ TD_t = Paylength/Bandwidth\ TPp = TEe/TDt$$

$$REe = (230 * Paylength)/(2 \times 10^6)\ RPp = REe/RDr\ SYN(R) > 0,\ SIM^2\,(R) > 0, DIS^m = = 0$$

Table 2 defines simulation parameters. In Fig. 2, it can be observed that EADBNHS consumes less energy than normally distributed delay distribution based hop selection on nodes in DTN when the time taken for simulation is increased.

**Table 2** Simulation parameters

| | |
|---|---|
| Dimensions of the network | 1000 m × 1000 m |
| Number of nodes taken | 20–100 |
| Bundle layer routing | Epidemic and prophet |
| Simulation time | 500 s |
| Mobility model of nodes | Random way point |
| Max movement speed of nodes | 10 m/s |
| Packet size of the datagram | 512 bytes |

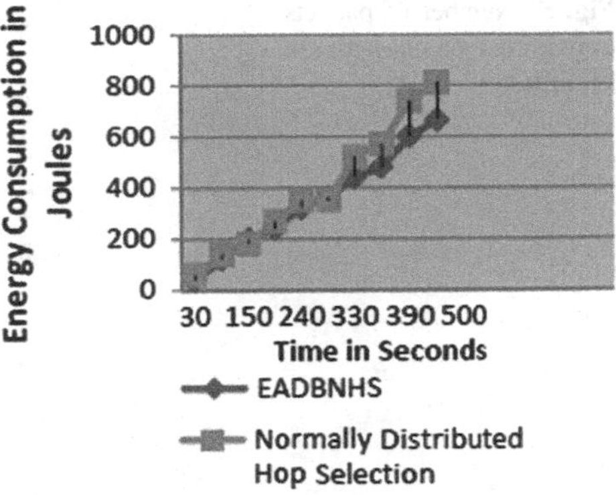


**Fig. 3** Energy used-up or consumed in path selection versus simulation time

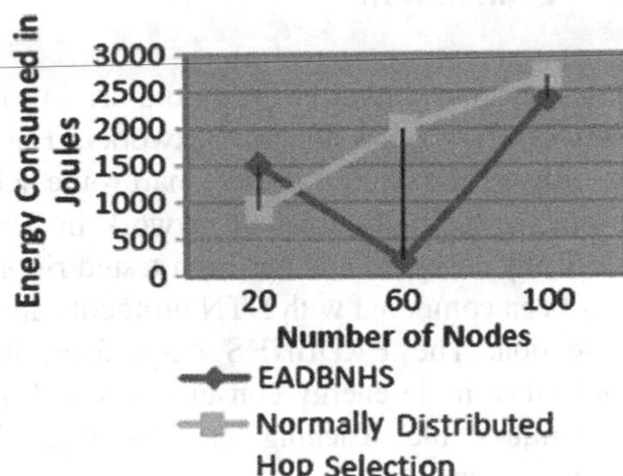


**Fig. 4** Energy used-up or consumed in path selection versus number of nodes

Next, energy used up in path selection has been analyzed when the number of nodes in the network increases. The results are given in Fig. 3. From this it can be comprehended that in its starting phase, EADBNHS used-up or consumed more energy as compared to normally distributed delay based hop selection on 20 nodes. This is due to the fact that EADBNHS is in a path learning stage and may consume higher amount of energy in starting phase as compared to normally distributed delay on the nodes in Fig. 4.

The network lifetime is measured or calculated using the residual energy of the nodes in the path from SN to DN. The health of the nodes is represented by $\mathrm{HTH}_{(x,y)}$.

$$\mathrm{HTH}_{(x,y)} = \mathrm{E}_{res} / \mathrm{E}_{(x,y)} \tag{11}$$

here (x, y) denotes the path from origin or source node to destination/sink node. Network lifetime is determined by $\mathrm{Min}\left(\mathrm{HTH}_{(x,y)}\right)$ amongst the intermediate nodes.

As represented in Fig. 5. Number of packets transmitted using EADBNHS are greater as compared to normally distributed delay in the DTN.

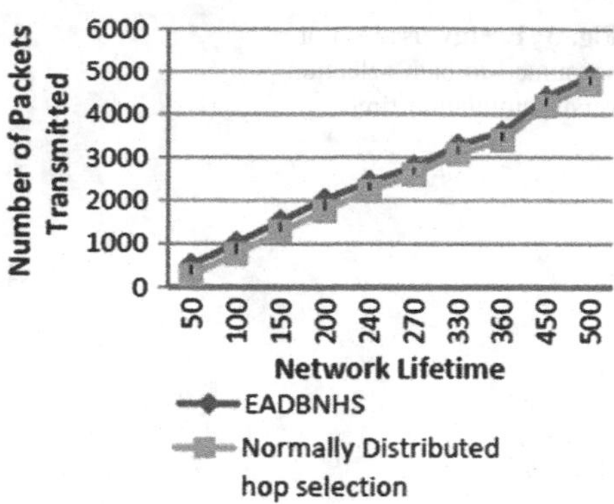


**Fig. 5** Number of packets transmitted versus network lifetime

# 6 Conclusion

Energy conservation has been the driving force when working with delay-tolerant networks or other mobile networks. Energy consumption can be reduced by intelligent and skillful use of smart route selection techniques EADBNHS has been proposed for delay tolerant network in the present work. This protocol has been skillfully designed and has been tested rigorously. The performance of the protocol has been compared with DTN protocols approaches like Normally Distributed Hop Selection. The EADBNHS outperform the competitor algorithms in terms of individual node energy consumption and offers improved network lifetime. Other techniques like caching can be used in addition for further performance improvement.

# References

1. M. Kaur et al., Stochastic Approach for Energy-Efficient Clustering in WSN, Global Journal of Computer Science and Technology, Vol. 14, Issue 7, pp. 1–8, (2014).
2. R. Kumar et al., Cross-Platform Application Development for Smartphones: Approaches and Implications, Proceedings of IEEE Conference, Conference ID: 37465, pp. 2571–2577, (2016).
3. K. Prabhu et. al., Performance Analysis of Modified OLSR Protocol for MANET using ESPR Algorithm, Proceedings of IEEE Conference ICICES, pp. 1–5.
4. P. Jain et al., Energy Efficient Local Route Repair Multicast AODV Routing Scheme in Wireless Ad hoc Networks, Proceedings of IEEE Conference ICACCCT (2014), pp. 1168–1173, (2014).
5. L. Pawar et al., Design of Simulator for Finding the Delay Distribution in Delay Tolerant Networks, Global Journal of Computer Science and Technology, Vol. 12, Issue 14, pp. 32–36, (2012).
6. L. Pawar et al., Stochastic Simulator for Estimating Delay in DTN Environment, IJETAE, Vol.2, Issue 8, pp. 183–189, (2012).
7. H. C. Sanchez et al., Routing Metrics in Delay Tolerant Networks, IRIT/RR-2007 22-FR, (2010).

8. R. Kumar et al., SPF: Segmented Processor Framework for Energy Efficient Proactive Routing Based Applications in MANET, Proceedings of IEEE Conference RAECS, pp. 1–6, (2015).
9. S. Arora et al., Novel Stress Calculation in Parallel Processor Systems Using Buddy Approach with Enhanced Short Term CPU Scheduling, In CRC Press ICCCS, Taylor And Francis September 2016.
10. R.K. Bhullar et.al., Intelligent Stress Calculation and Scheduling in Segmented Processor System Using Buddy Approach, Journal of Intelligent & Fuzzy Systems, IOS Press, doi:10.3233/JIFS-169256, ISSN:1064–1246, In Press.
11. R.D. Joshi et al., Distributed Energy Efficient Routing in Ad hoc Networks, IEEE Transactions, pp. 16–21 (2008).
12. Yi-Chao Wu et. al., Power Saving Routing Protocols in Ad hoc Wireless Networks, IEEE Wireless Communications, pp. 69–81 (2009).
13. Shivshankar et. al., Implementing a Power Aware QoS Constraints Routing Protocols in MANET", IEEE Conference on Communications (ICC), pp. 3166–3170 (2009).
14. S. Sisodia et. al., Performance Evaluation of a Table Driven and On Demand Routing protocol in Energy Constrained MANETs, Proceedings of IEEE Conference ICCCI, pp. 76–80, (2013).
15. K. Prabhu et. al., Performance Analysis of Modified OLSR Protocol for MANET using ESPR Algorithm, Proceedings of IEEE Conference ICICES, pp. 1–5, (2014).

# Remote Fuel Measurement

**Garv Modwel, Nitin Rakesh and Krishn K. Mishra**

**Abstract** This paper introduces detailed study of remote interaction of vehicle with user so that user can have value parameters of parameters like fuel tire pressure, etc., while sitting at remote location. Here we have focused specifically on vehicle fuel measurement and its feasibility. This study is about achieving wireless measurement of vehicle's fuel using wireless protocols, and add the comfort level for user for any kind of automobile like Car, Bike. This kind of feature will not only help private vehicle owners but will help public vehicle owner to know the state of vehicle with parameters like fuel, Tire pressure while they are sitting remotely.

**Keywords** Wireless · Fuel · Android · Remote measurement

## 1 Introduction

Fuel management [1, 2] is one of the most important parts of any automobile like cars, buses. Fuel measurement is always useful and that informs about the performance of the engine. From the end user point of view it is directly related to the pocket of a customer. It is always important to keep eye on fuel consumption to know the health of your vehicle and moreover to predict your budget on fuel expenses. There are lot of ways to know about fuel consumption and respected parameters. This paper is focusing on measurement of fuel consumption remotely, that means user can have all the details of fuel consumption even when user is out

G. Modwel (✉) · N. Rakesh
Department of Computer Science and Engineering,
ASET, Amity University, Noida, UP, India
e-mail: garv.modwel@gmail.com

N. Rakesh
e-mail: nitin.rakesh@gmail.com

K.K. Mishra
MNNIT, Allahabad, India
e-mail: kkm@mnnit.ac.in

 
S.K. Bhatia et al. (eds.), *Advances in Computer and Computational Sciences*,
Advances in Intelligent Systems and Computing 553,
DOI 10.1007/978-981-10-3770-2_42

of reach of vehicle. Fuel management [1, 2] is not only important for the private vehicle but is important for public vehicle like bus truck and tempo. For private vehicle, it is important considering the mileage of vehicle but for public vehicle it is important to track the fuel theft and to track overall running. For example if we consider a taxi operator who is totally dependent on his drivers, may loss big chunk in his profit if driver is trying to play with fuel, similarly a bus and truck operator may not trace exact fuel usage and can loss good chunk in profit.

Now fuel management [1, 2] can be done with various parameters, like using odometer and having a regular check on the consumption, another way to do that is set trip on vehicle and keep the track on that. In both the cases we may not get the real time data and moreover owner of vehicle should be present inside the vehicle. So it is quiet difficult for a transporter of a vehicle owner to know the fuel status. This paper is about the tracking of fuel remotely, that means user can find out how much fuel is left or how much was last or previous run for a vehicle, while sitting at home. This is a win-win situation for private vehicle owner as well as for transporter, they can easily get real-time fuel usage and their fuel need on daily, weekly, and monthly basis (Fig. 1).

Above figure shows typical fuel management system where there is interaction of different layers. Ideally, there is presence of base, control, Presentation and application layer along with the support of system and IPC Hardware. In over all process there is value read via analog device and then there is conversion to digital

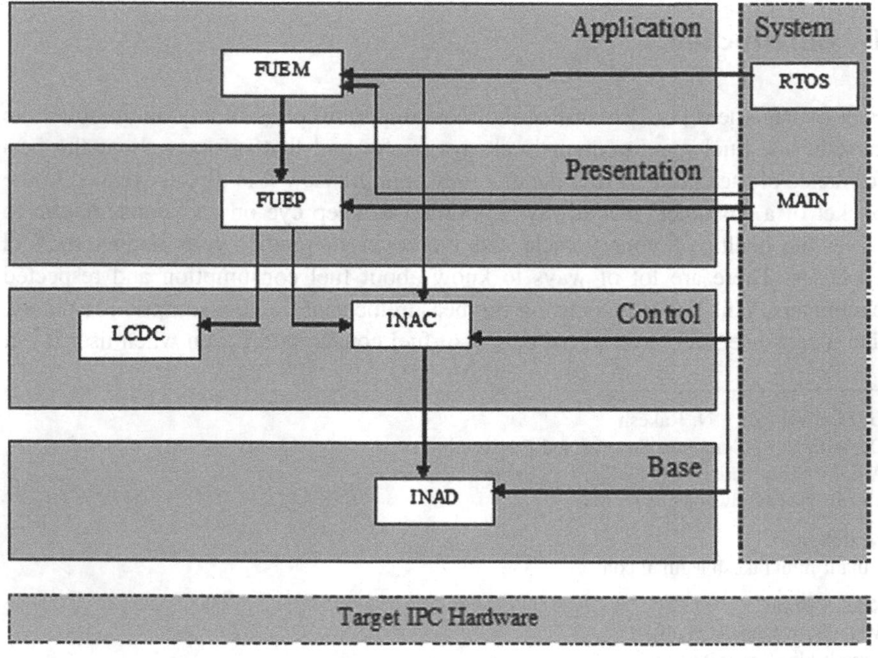


**Fig. 1** Typical fuel management system

value using ADC followed by calibration and display. So overall process involves certain read/write operations and certain calculation and then it get displayed on meter either digitally or with analog signals. These particular processing limits the accessibility of information for user within or near the range of the vehicle and user cannot access during his presence at workplace or home. In above figure FUELM represents fuel management system [1, 2], FUELP represents fuel presentation system, LCDC represents LCD control system, INAC is for in and out control, INAD is for in and out drivers, apart from that RTOS stands for real-time operating systems.

Hoffman et al. and Dinnawi et al. [1, 2] this is typical implementation of fuel management system and that can vary as per the requirements. Fuel processing starts from the reading the value from some analog gauge and then using analog to digital converter value get converted into digital data format. So finally fuel value is some voltage variation. Now there is calculation of fuels past value and current value in order to avoid any kind of error that is called calibration, various calibration techniques are used as such depending on the requirements.

If fuel value is changed from previous value saved in NVM then there is execution of all the calibration algorithms or else it skips after minor checks. Calibration is needed when fuel value is less than the last saved fuel value. Of course logic varies from vehicle to vehicle, for example there are different logics for 2 wheelers and 4 wheelers. After calibration phase values is displayed on display device like analog and digital meter. So this is all about the measurement of the fuel. Now coming to remote measurement we need to enhance the same system and connect the same with cloud or with some connectivity media (Fig. 2).

## 2 Related Work

### *2.1 Communication Protocols Wi-Fi for Fuel Measurement*

There is various ways to measure the fuel remotely. Target is to achieve value of fuel and its usage remotely even we have value stored in vehicle without affecting car body [3]. Consider the scenario if value of fuel can be transmitted using some wireless [4, 5] protocol and at the receiving end it can be captured and shown in any display device without wires for connectivity and the device is there on 802.11 [6]. Considering the approach of wireless transreceiver in the vehicle cluster, it will be an expensive way to do so. Moreover the usage of the Wi-Fi will be very limited so it will be kind of useless investment and will be wastage of money. Functionality wise it is easily achievable. In general all the data flows in automotive [7] system using wired protocol CAN (Controller Area Network) and LIN (Local Interconnect Network). CAN is most efficient wired protocol available for automotive industry.

A Controller Area Network bus is an automotive bus standard designed for microcontrollers and devices to maintain communication with each other basically

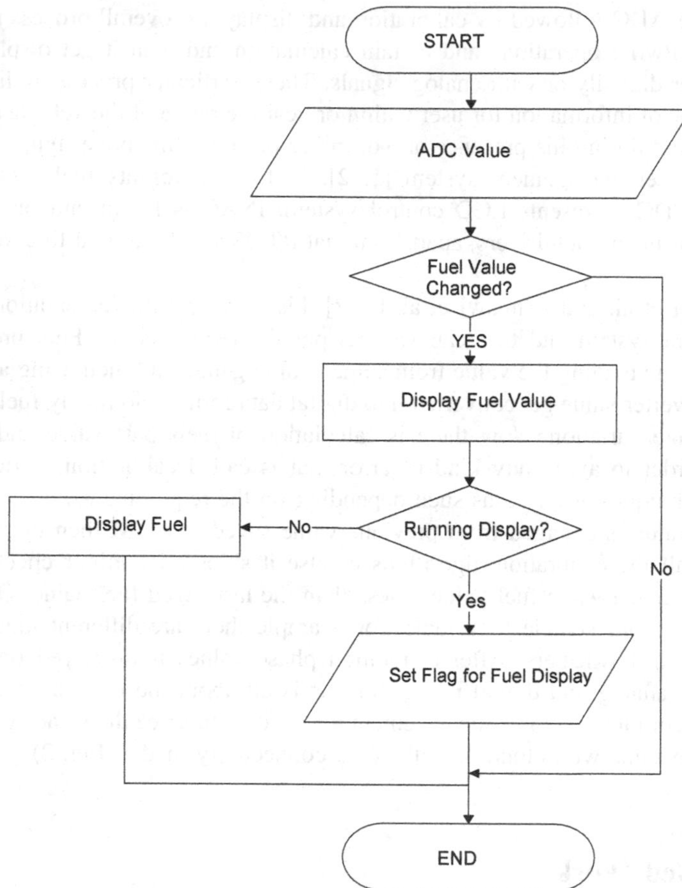


**Fig. 2** Logic for fuel management system

in apps even not having host PC. This protocol is message based protocol; it is designed specifically for the multiplexed automotive wiring, it is used in other context as well. This is one of the most important protocols used for diagnostics of automotive. There is a standard for the diagnostics of automotive called On Board Diagnostics (OBD). There are various versions of OBD and currently OBD-2 is followed by automotive car makers. Moreover OBD-2 is mandatory for all the car makers for now. Now a day's automotive vehicle have 50–80 electronic control units with different functionalities and in general Engine control unit is most important one, so engine control unit have most efficient and powerful processor. Others are used for antilock braking/ABS, transmission, airbags, audio systems, cruise control, electric power steering, power windows, doors, mirror adjustment, battery and recharging systems for hybrid/electric cars, etc. This is subsystem based communication and some subsystems plays independently. Since some of the subsystems are independent that does not means that communication is not needed, that means communication may be needed but can play a role of standalone ECU. ECU in general works as a gateway for the entire component and also called as BCM Gateway [8, 9] as well. In general a subsystem may need command over some hardware like actuators and sensors to have feedback or to send signal. So this kind of need is fulfilled by controller area network or CAN. CAN is a communication that is lossless and it uses bitwise arbitration to manage conflict in the messages.

Main requirement of arbitration is that network connected with CAN network should be synchronized properly so that every bit is there at place properly at same point of any given time. This arbitration method requires all nodes on the CAN network to be synchronized to sample every bit on the CAN network at the same time. Hence sometime we call CAN synchronous. Sometime the term synchronous is irrelevant since the data is transmitted without a clock signal in an asynchronous format. As per the specifications the CAN use the terms like "dominant" bits and "recessive" bits where dominant is a logical 0 and recessive is a logical 1. The idle state is represented by the recessive level. Consider a scenario when one of the node transmits the dominant bit and on another side other node transmits the recessive bit, so in case of collision dominant bit will win and get and respective node will get control and there will not be any delay in high priority message, moreover node transmitting lower priority message will automatically re attempt or re transmit six bit clock as soon as dominant message is over. Hence CAN is very good real time system with prioritizing mechanism. Voltage for 0 and 1 is depending on layer used but as per the basic principle CAN need that all the nodes should listen the data and it should include the transmitting node and transmitting data. Considering the scenario if a logical 1 is transmitted by all the nodes at same time, so as a result logical 1 will be seen all the nodes either transmitting or receiving side. Similarly in case of logical 0, it will be all logical 0 at all the nodes. Consider another scenario If there is a logical 0 is being transmitted by one or more nodes, and there is a logical 1 is being transmitted by one or more nodes, then as a result logical 0 is seen by all nodes and that includes the node(s) transmitting the logical 1. Similarly when a node transmits a logical 1 but sees a logical 0, it is because that there is a contention and it quits transmitting. So by using this kind of process, any node that transmits a logical 1 when another node transmits a logical 0 "drops out" or loses the arbitration and this is concept of arbitration and key feature of the CAN protocol. Any node which loses in the process of arbitration, re-queue its message for later transmission and CAN frame bit streaming continues without any kind of error until there is only one node transmitting. Hence the node that transmits the first 1 loses arbitration. Since the 11 (or 29 for Extended CAN) bit identifier is transmitted by all nodes at the beginning of the CAN data frame, the node having lowest identifier transmits more zeros at the start of the frame, so that node wins the arbitration or has the highest priority.

Coming to value of fuel it can be given to Wi-Fi protocol directly and then it can be used with any mobility device (Fig. 3).

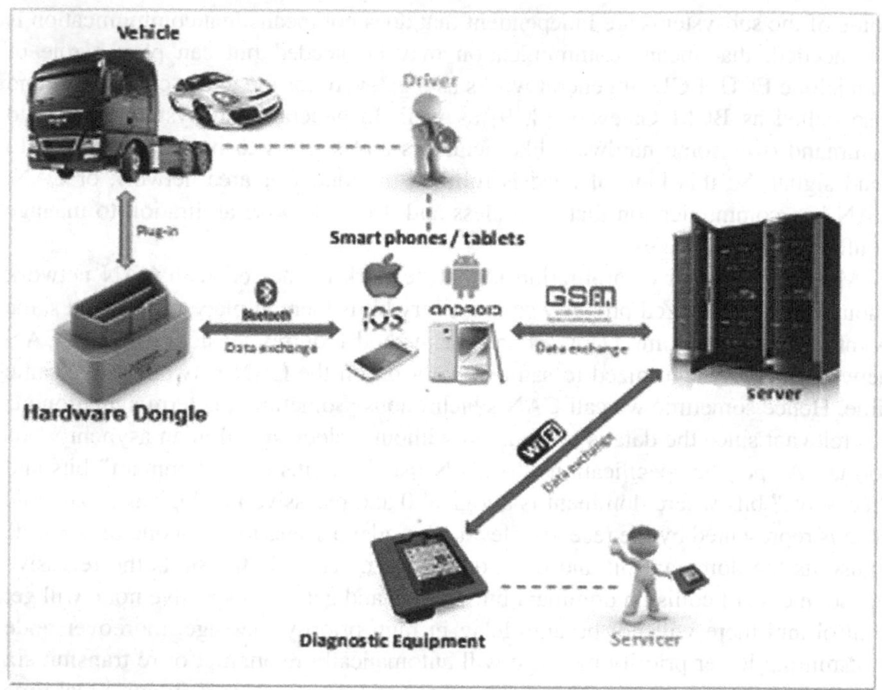


**Fig. 3** Remote fuel management

## 2.2 *Communication Protocols BlueTooth for Fuel Measurement*

Wi-Fi is very important protocol considering any kind of wireless communication, In general transmission from Wi-Fi station happens only in case of clear channel. All the transmissions get acknowledgements, in case there is any collision, there is retry mechanism that comes into picture for random intervals and it keeps trying. Wi-Fi systems configuration is half duplex kind of shared media configurations, in this configuration every station transmits and receives data with one radio channel. Main problem in this environment of radio system is that it is difficult for station to hear during send operation, such situation makes it difficult to detect collision. Moreover to keep data available everywhere we need to put it on cloud by either using Wi-Fi or Bluetooth.

Here Bluetooth is very much preferred protocol as it is available on many automobiles and specifically in cars. Since now a day's lot of vehicles comes with connectivity option and that makes our target easy, in general we are using Bluetooth for telephonic connection or Audio/video connectivity but in this case we can very well use it for data connectivity between cluster and mobile device like tablet of smart phone. In order achieve this functionality we need to have to communication channel. One needs to communicate from vehicle to mobile device and another need to communicate from mobile device to cloud storage. Cloud storage is needed for storing all the data and moreover considers the volume of users and respective communication it is very important to have cloud communication.

## 3 Analysis

If we consider wireless prospective, a car is very complex system, it is important to understand the car communication so that we can manipulate that what features we need to keep in mind while designing such system.

In Fig. 4 down side block diagram is showing the typical hardware for the dongle that can be connected to the vehicle to retrieve the data. There are various parts in the dongle playing different roles. There are five major component in the design of Hardware dongle like microcontroller, CAN Transreceiver, Bluetooth Controller, antenna and power supply. Power supply is typical power supply that will be using power from vehicle and can go up to 12 V DC. Microcontroller can be any controller supporting CA and it can be either of 16 bit or 32 bits. CAN transreceiver is component that is responsible for requesting and receiving CAN message.

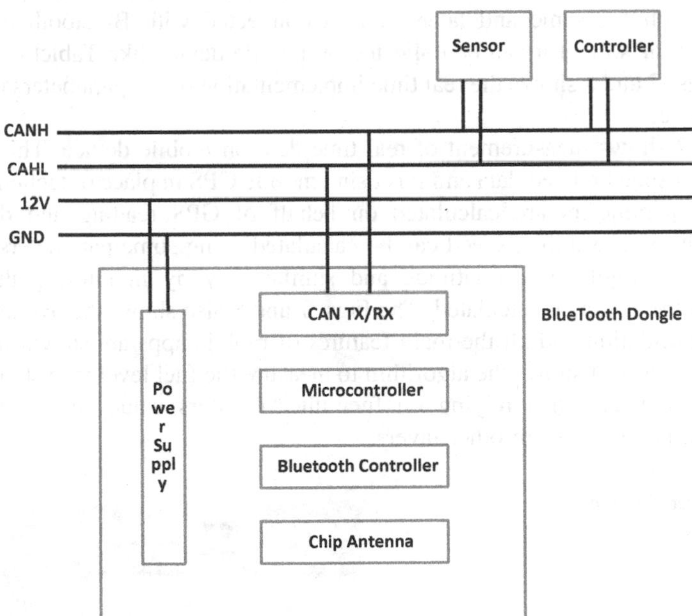


**Fig. 4** Connection diagram remote fuel management

Coming to overall functionalities of dongle, it is dongle having CAN connector and it is connected to CANH and CANL buses in the vehicle, and in a typical vehicle, generally CANH and CANL is connected with CAN bus slave and CAN bus master. So for communication CAN transreceiver receives request from the data from CAR network using OBD-2 standards and by sending some DTC value, DTC values are standard values defined by OBD-2. In OBD-2 certain values are implementation specific and certain are must and fuel value is one of them, so transreceiver gets the value from the CAN network and now we can use the value as per our need.

Now since we want achieve the functionalities of remote measurement that means we need to connect the dongle with some wireless protocol and it can be easily achieved using Bluetooth. Dongle and mobile device can be easily connected using the Bluetooth protocol and all the required values can be updated on cloud using mobile devices. To implement this functionalities on cloud it need server side support as mentioned below.

Figure 5 shows the block diagram for the server side implementation. Mobile device is playing role of data collector and then it need to pass it to Tom cat services which placing and storing the data on the web cloud server, after that same data can be requested from various registered users and all the required data can be given to registered users after validating the authentications of users.

Figure 6 shows the connection of the setup with two wheeler, here using CAN data of bike is retrieved and then connected with Bluetooth dongle. Since main purpose is to get the data and then transmit it using some wireless media. All the main data of vehicle can be collected from CAN with the help of OBD hence it is collected with the same and later on it is connected with Bluetooth dongle to transmit on air so that it can be collected on mobile device like Tablet or Mobile. Below Figs. 7 and 8 shows the real time implementation of the parameters using the mobile device.

Figure 7 shows measurement of real time data on mobile device. This is measurement of speed related data and it is using mobile GPS in place of tachometer, all the speed parameters are calculated on behalf of GPS reading and displayed accordingly. For example, speed can be calculated using some parameters of GPS reading like longitude and latitude and similar way by monitoring the speed maximum speed can be calculated. The Figs. 7 and 8 also shows the overall mobile app implementation and all the main features of mobile apps are shown in Fig. 8. Below code snippet shows the algorithm to measure the fuel level in system. In this code we are very much relying on fuel filter resistors value and values what application is getting from other layers.

**Fig. 5** Server side fuel management

**Fig. 6** Connection with engine

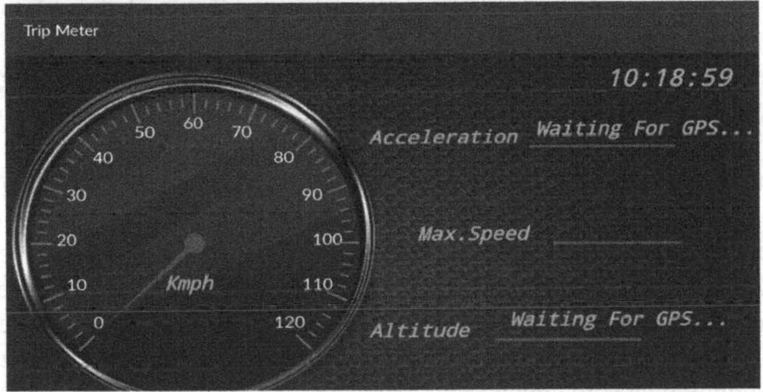


**Fig. 7** Real time measurement

**Fig. 8** Overall mobile application

```
static void CalcMyFuelLevel(void)
{
 Fuel_TimeScheduler;// Fuel Timer
 UpdateFuelFilteredResiVal(); Check for fuel Filter Resistor
 if ( GET_MY_APL_ON )
 {
  Get_APL_CalcFuelLevel); //Get Application level details
 }
 if (Fuel_TimeScheduler == 15)
 {
  if ( APL_M_IS_IN_CHIAVE_OFF )// Check Condition
  {
   Get_APL_CalcFuelLevel_DINA();
  }
  Get_APL_CalcFuelLevel_plus(); //Fuel interpolation lower layer
  Fuel_TimeSchedulers = 0;
 }
}
```

In this application various features are implemented that are helpful for end user, workshop, mechanic and for OEM.

First feature is read data that can read various data like fuel reading, Speed, distance and all and that may be useful for all the stake holders like user, mechanic workshop and OEM at any given point of time. Second important feature is actuator control and that can be useful for mechanic and workshop, user can also use this feature to have control over the vehicle but chances are very less. In general user will be using mechanical part like key of the vehicle to do the same. Another feature is DTC mode that will be used to diagnostic parameters check, as mentioned earlier these parameters are based on the OBD standard. These parameters are helpful for workshop and technician. In general when technician want to know what is the status of vehicle then some coded value are sent to vehicle and in response vehicle send back the response value, based on the response values; vehicle's health can be predicted. In general it is done using wired connection but using new implementation it can be done with wireless setup.

Next feature is configuring data and that is typical configuration part of any mobile application like user Id, Password, setup profile and all. This is standard mobile app configuration. It is End of Line (EOL) action, EOL is the setup that is used to ECU on the production like, this application is helpful in production line also. In general EOL is used to test that after production ECUs are satisfying the basic criteria or not. For that there are requirements of some specific SW and HW, this application eliminates the need of the same, as it can turn the ECU in EOL mode and all the testing can be done wirelessly. Another important mode is monitor modes, monitor modes means various mode of vehicle like economy mode, normal mode, etc., this application supports different type of modes based on the requirement of vehicle, internally it simulates data accordingly. This application is also equipped with Dashboard that is shown in Fig. 7 which is the medium to display needed parameters.

## 4 Conclusions and Recommendations

There is big demand of connecting the cars on cloud and retrieving the data remotely by the private cars users and by public car operators to keep an eye on activities of the vehicles, as mentioned earlier this is not only demand of small vehicle like cars but also it is demand of truck and bus operators. Fuel is one of the parameters what has big impact on vehicle performance in the terms of mileage and t can be read remotely using this approach, apart from that we can do many useful activities. Consider the scenario for the countries where temperature is really high or very low in that case vehicle start adjusting its temperature after user enters into the car and start its engine, driver is forced to stay in some terrible temperature condition for time being till vehicle is adjusting that, so this is big discomfort situation for drivers and can be avoided by using the concept developed for remote measurement of fuel. That means driver can start the engine sometime before entering into the vehicle and set the temperature of vehicle as per the need using mobile or tablets.

By the use of same concept user can do real time tracking of vehicle. Implementation wise we need to implement application with use of Google map API and owner of the vehicle can trace the exact position of vehicle using mobile app. This is win-win kind of situation for transporters and bus operator to avoid any kind of security and fuel usage. Considering the tracking of vehicle by security or by police this implementation can be a milestone and that can help police to control the traffic and as well to take care of any kind of theft. Another important application of this implementation is car care, that means by reading some parameters of the car user can predict when car need service or any kind of repair. This kind of approach is very useful for all the stake holders in automotive industry like, user, technician, workshop and OEMs. All the stake holders will have different type of advantages as mentioned below.

First one and most important one is owner that means owner of the vehicle, For owner such type of the system can help owner in various ways. Owner can make sure that vehicle is not having any kind of issue before ride and in case any issue is pretended it can be rectified before vehicle goes into bad state. Another important benefit what user can avail using such service is fair performance measurement that means user will have all the performance data like mileage, trip and fuel value accurately. This factor not only will help to know fair values of performance but also will help to maintain the expense on vehicle and user can save a lot.

Another important stake holder is technician, who is fixing issues in automotive vehicle, this kind of approach will help him to find out issue immediately with the help of mobile application and technician can save lot of time, so at a give time frame technician can repair more cars and chances of missing any kind of issues are very less. Since system will have all the past history of vehicle so this approach will also help technician to fix current issue. On top of that there will be skill enhancements of technician in the terms of information technologies usage. Third important stake holder is workshop, this system is very helpful for workshops, since

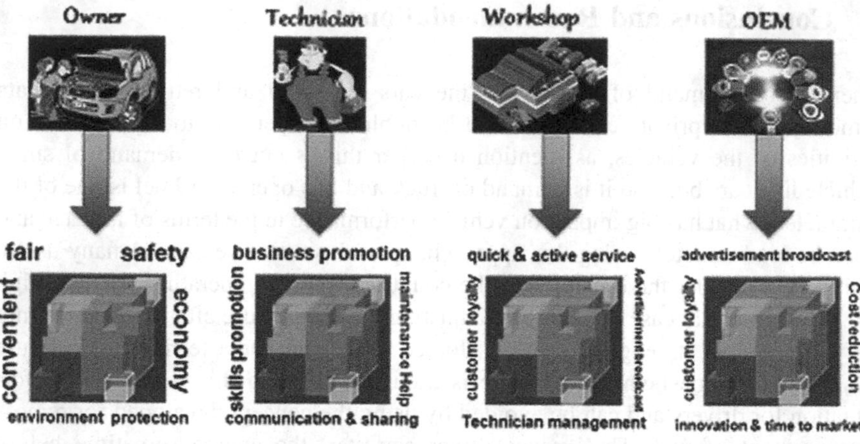


**Fig. 9** Internet age usage for cars

workshops will get reliable data so workshop can fix the issues in efficient way and in sorter time, this will attract more and more customer, moreover since workshop will have all the present and past data so it will help workshop to suggest vehicle owner proper rectification, on top of all this workshop can know when a particular owner's vehicle need service and workshop can inform the owner about the correct timing of service or any kind of malfunctioning in vehicle. Last and most important stake holder is OEMs, they can use this technology very well and monitor the performance of their vehicle over a period of time, this type of technology will be an added feature for their vehicle and can help lot in sale, and moreover such system can help in lot of advertising as well (Fig. 9).

So remotely this very much possible to control car features and develop some remotely controlled vehicle, along with measuring some important parameters of vehicle like fuel.

## References

1. Alwyn J. Hoffman, *Marius van der Westhuizen*; Jitesh Naidoo.
2. Rafika Dinnawi; Electr. & Comput. Eng. Dept., American Univ. of Beirut, Beirut, Lebanon, *Dima Fares,* Riad Chedid; Sami Karaki.
3. S.-C. Kim, H. Bertoni, and M. Stern, "Pulse propagation characteristics at 2.4 GHz inside buildings," *Vehicular Technology, IEEE Transactions on*, vol. 45, no. 3, pp. 579–592, Aug 1996.
4. IEEE Std. 802.11-2007, "Wireless LAN medium access control (MAC) and physical layer (PHY) specifications," in http://standards.ieee.org/getieee802/download/802.11-2007.pdf.
5. IEEE Std. 802.15.1-2005, "Wireless medium access control (MAC) and physical layer (PHY) specifications for wireless personal area networks (WPANs)," in http://standards.ieee.org/getieee802/download/802.15.1-2005.pdf.

6. M. Drienberg and F.Zheng, *Centralized channel assignments for IEEE802.11 WLANs: utilization minimax-sum*, Proc. IEEE WPMC'12, pp 633–637, Taipei, Taiwan, Sept 2012.
7. M. Petrova, J. Riihijarvi, P. Mahonen, and S. Labella, "Performance study of IEEE 802.15.4 using measurements and simulations," in *Proc. IEEE Wireless Commun. and Networking Conf. (WCNC'06)*, Las Vegas, USA, Apr. 2006, pp. 487–492.
8. H. Yun and S. Lee, *vehicle Mobile Gateway for provisioning and support of ITS service on nomadic devices,*17th world congress on intelligent transport system, 2010.
9. ISO 22901 (all the parts), *road vehicles-Open Diagnostic data Exchange (ODX).*
10. IEEE Std. 802.15.4-2006, "Wireless medium access control (MAC) and physical layer (PHY) specifications for low-rate wireless personal area networks (WPANs)," in http://standards.ieee.org/getieee802/download/802.15.4-2006.pdf.
11. H.-M. Tsai, C. Saraydar, T. Talty, M. Ames, A. Macdonald, and O. Tonguz, "Zigbee-based intra-car wireless sensor network," in *Proc. IEEE Int. Conf. on Communications (ICC'07)*, June 2007, pp. 3965–3971.
12. "CC2520 data sheet," in http://focus.ti.com/lit/ds/symlink/cc2520.pdf.
13. Bluetooth SIG, "Bluetooth specification version 2.0 + EDR," in http://www.bluetooth.com.
14. E. Walker, H.-J. Zepernick, and T. Wysocki, "Fading measurements at 2.4 GHz for the indoor radio propagation channel," in *Proc. Int. Zurich Sem. Broadband Commun.*, Zurich, Switzerland, Feb. 1998, pp. 171–176.
15. T. S. Rappaport, *Wireless communications: principle & practice*. Upper Saddle River, NJ: Prentice Hall, 1996. 2704.
16. M. Kim and C. Choi, *Hidden node detection in IEEE 802.11n wireless LANs*, IEEE trans. on vehicle Technology, vol 62, no. 6, pp. 2724, 2734, Jul 2013.

# Design and Implementation of "Reassurance Broiler Project" Traceability Platform

**Xiao-hua Xu and Chang-xi Chen**

**Abstract** "Reassurance Broiler Project" is one of the "Ten Secure Food Projects" in Tianjin, China. The platform, a third-party authentication and management platform, contains eight subsystems and a website for query and information release. It applies object-oriented system analysis and design on the basis of use cases under RUP.".NET" framework is applied, SQL Server is used as DBMS while a mode combining C/S and B/S structure is adopted. A 25-digit unique identification is used as industry chain transmission identification, with RFID as the carrier. Besides, 10-digit one-dimensional tracing code and two-dimensional QR code are used on tracing label. The platform can trace information for processes from broiler production, inspection and quarantine, slaughtering, processing, as well as packaging and selling, which can help improve food safety, reduce management cost, increase brand advantages and help government authorities provide early warnings against animal epidemic and problems related to production quality and safety.

**Keywords** Traceability · broiler · Traceability platform · RFID

## 1 Introduction

In recent years, mad cow disease, tonyred, clenbuterol, and poisonous ham and other major meat food safety crisis has aroused widespread concern in the whole world [1]. These meat food safety incidents, seriously affected the healthy diet of people, lead to lack of consumer confidence in meat products, has affected the development of livestock and poultry breeding and meat processing industry, and restricted the national economic development [2]. Therefore, whether it is a

X. Xu · C. Chen (✉)
TianJin Agricultural University, Tianjin, China
e-mail: chenchangxi@tjau.edu.cn

X. Xu
e-mail: xuxiaohua@tjau.edu.cn



S.K. Bhatia et al. (eds.), *Advances in Computer and Computational Sciences*,
Advances in Intelligent Systems and Computing 553,
DOI 10.1007/978-981-10-3770-2_43

government department, livestock and poultry breeding enterprises, meat processing enterprises, or related research personnel have the responsibility to find a proper solution to the problem. For the government, many governments raise "food safety" to a strategic level of national safety, establish food traceability system through legislation, such as during the 2008 Olympic Games in China the "Olympic meat food traceability system" was established [3]. In 2008, the "Shanghai pork circulation safety information traceability system" was established [4] and so on. Researchers conducted further research and improvements in the field of traceability system, applied the latest research of information technology to the field of food safety traceability, including RFID technology [5], SMS query technology [6], barcode technology [7], QR code technology [8], WSN technology [9], etc.

After getting a thorough understanding of the whole production process from taking care of breeding hen and chickens, incubating, broiler chicken producing, butchering, processing, storing to selling, researchers, applying the newest information technologies, researched and developed "Reassurance Broiler Project" traceability platform in view of the actual demand. The platform covers Good Agricultural Practices(GAP) management system implemented in the production process, in the processing link, Good Manufacturing Practice (GMP) food safety and quality assurance system, Sanitation Standard Operation Procedure (SSOP), Hazard Analysis Critical Control Point (HACCP), ISO9001 quality management system, ISO14001 environment management system, QS food quality and safety market access certification, and other standard systems are executed for broiler product quality security control [10].

## 2 Requirement Analysis

"Reassurance Broiler Project" is one of the top ten morale project of reassurance food series initiated by Tianjin municipal government. After several discussions with Tianjin Animal Husbandry and Veterinary Bureau, the requirements are determined as follows:

Three key control points shall be created and policed rigorously for "Reassurance Broiler Project" in the three stages of breeding, slaughtering, and marketing aspects: in the stage of breeding, standardized practice shall be promoted, the environmental monitoring on the production site shall be strengthened, and contraband goods and animal medicine residue shall be controlled and monitored; quality and safety inspection shall be carried out during slaughtering stage, and the products shall bear labels, this being the second rigorous control point; a market admission system shall be practiced, a complete set of required tickets and certificates shall be present, and a system of daily operation records shall be kept such that complete process quality control is maintained from production to market.

Electronic information archives shall be created to record important information of broilers, covering from chicks taking-in, use of fodder and medicine, epidemic prevention, quality monitoring, place of origin, slaughtering quarantine, to product package labeling, so as to establish quality, safety, and information traceability of marketed broiler products.

Such information shall be interfaced with the inspection and quarantine system of the General Administration of Quality Supervision, Inspection, and Quarantine of the People's Republic of China.

In consideration of the limit in Internet coverage, it is required to provide traceability with respect to livestock farmers without Internet service.

It shall provide broiler industry chain enterprises statistical analysis, query, search, and reporting features and can save worker management cost.

## 3 System Analysis

### 3.1 Workflow

A broiler industry chain enterprise (a breeding, slaughtering, or processing enterprise) intending to establish broiler product quality and safety traceability shall make an application via traceability certification platform. The information needed in application includes the 13-digit code assigned by China Society of Article Coding to the enterprise's major products, the enterprise name, the representative of the legal person, the means of contact, etc. When the check is done satisfactorily, the traceability platform will create automatically an enterprise database for the enterprise which includes enterprise information management, broiler breeding, inspection and quarantine, slaughtering, processing, and traceability. A platform administrator account No. will be allocated to the enterprise, which is sent to the applicant using the email address. The enterprise administrator is authorized to do department management, post management, and employee management (including enterprise administrators, broiler breeder users, slaughtering workers, inspectors and quarantine personnel, and processing workers), enterprise product management, and the management of other internal basic information of the enterprise. The feature of employee management makes it possible to issue the employee FRID card individually or in batch. Each user may log in and access their respective subsystem appropriate to their authority using the 5-digit A/C number and password. The administrator of the broiler enterprise in charge of chick releasing shall enter into a breeding contract with the broiler breeding farmers before releasing the chicks. The broiler breeder user (Whether C/S or B/S architecture) shall maintain a file of broiler breeding information, including daily breeding information, weekly breeding information, medication information, epidemic prevention, and any anomaly observations. A broiler is delivered for slaughtering after 42–45 days of breeding, during which the inspection and quarantine authority needs to manage epidemic prevention and medication. The broilers delivered for slaughtering shall undergo such management steps as pre-slaughtering quarantine inspection,

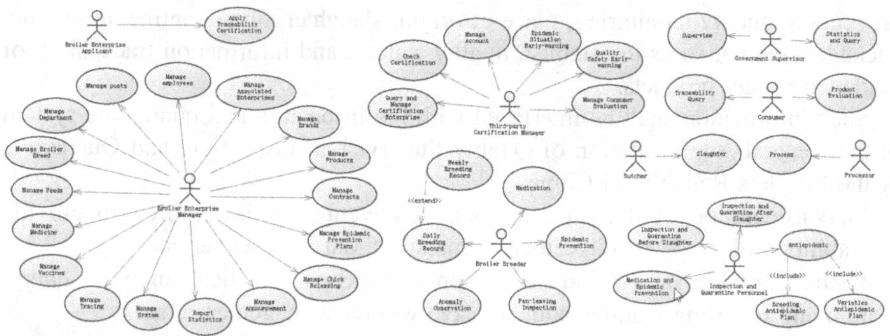


**Fig. 1** Use case diagram

slaughtering, post-slaughtering quarantine inspection, processing, and traceability labeling, these steps being synchronized with the platform database via the database trigger. The government's supervision authority makes query and statistics analysis of broiler production, and the consumers may query, trace, and comment on the management of broilers with the aid of the tracing code. The third-party certification administrator manages epidemic prevention, quality, and safety evaluation in the production process.

## 3.2 Use Case Diagram

Traceability Platform Participants. Participants mainly include broiler enterprise applicants, third-party certification managers, enterprise managers, broiler breeders, inspection and quarantine personnel, butchers, processors, enterprises anti-epidemic managers, government regulators, consumers, and so on.

Use Case Diagram. The main use cases are shown in Fig. 1.

# 4 The Overall Design

## 4.1 Technology Framework Specification

Traceability platform is developed using C#.NET and ASP.NET in Microsoft ".net" framework. Back-end database system uses SQL Server platform, supports smooth transition to Oracle without a code. This system adopts the thinking of module development to reduce the coupling between different sub modules. It uses the service oriented XML data exchange platform to realize the application integration and data exchange.

## 4.2 Platform Architecture

Platform architecture adopts the mixed architecture model of C/S and B/S. The B/S architecture is used for the under featuring internet connection [11]. C/S architecture is mainly used for broiler breeding households without internet. When delivering chicks, broiler breeding households would be distributed with a XML document used for the storage of related structure of database and the content information. Broiler breeding households without internet all install with the Client procedure which would read such XML documents and record the broiler breeding archives like the B/S architecture cultivators. When the broiler are marketed and submitted to the supervisor, a XML document would generate automatically and users boasting Internet connection would read such document and upload it to the remote database.

## 4.3 Composition of the Platform

"Reassurance Broiler Project" traceability platform is a third-party certification and management platform, there are eight subsystems, including information management subsystem, broiler production subsystem, broiler breeding environment monitoring subsystem, broilers slaughtering subsystem, broilers processing subsystem, traceability evaluation subsystem, inspection and quarantine subsystem and government regulation subsystem, the third-party authentication management platform and a network platform for each subsystem user login, consumer query, and the third-party platform information release.

## 4.4 Traceability Identification and Its Transfer

**Tracing Method**. Since the way of chick release and slaughter delivery is "all in all out," large granularity tracing is adopted to trace the entire batches of broiler breeding shed, not each individual broiler.

**Traceability Identification**. Considering the huge number of broiler breeding enterprises in Tianjin, China, and the fact that it might be spread to other provinces and cities vastly for the convenience of consumption query, two types of identifications are used. One is the identification for industry chain delivery, the other is the identification for consumer traceability.

**Industry Chain Delivery Identification**. Combined with international EAN/UCC, GS1 standards and China's "livestock and poultry identification and breeding archives management approach," 25-digit encoding of the industry chain transfer identification is adopted: Top 10 digits in EAN-13 code + employee number inside each enterprise (5 digits) + broiler pen number (2 digits) + date of chick release (8 digits).

**Consumer Traceability Identification**. A 25-digit encoding of the industry chain transfer identification is not easy for consumers to enter and query, so the Hash mapping algorithm is adopted to transfer the 25-digit encoding to 10-digit broiler product traceability code.

**Tracing Carrier**. Two-dimensional QR code, which is mainly used in mobile phones to scan and trace; One-dimensional barcode for consumer tracing and query in network, SMS, and WeChat; RFID card, which carries 25-digit encoding of tracing transfer identification, is used for transferring the only identification of the whole industry chain from production to sales.

# 5 Detailed Design

## 5.1 Database Design

**Database**. Taking into account the connecting number of the database and network speed when "Reassurance Broiler Project" traceability platform is implemented, the design of the platform database uses the multi-database mode, including: network platform database, government database, and broiler enterprise database. Each broiler enterprise has its own broiler enterprise database and its structure is the same.

**Stored Procedure, Trigger, and Function**. Traceability platform adopts stored procedure to implement database's operations of insert, delete, update, and select for high efficiency. Triggers and functions are used to achieve data synchronization and update within the database.

**Reassurance Broiler Traceability Platform Website**. Reassurance broiler traceability platform mainly include several functions: login of each subsystem; consumer traceability queries and product evaluation; all kinds of information such as the recommended broiler products, policy, market trends, breeding technology, including video release; query summary tracing and certificated enterprises, etc. [12].

"Reassurance Broiler Project" traceability platform's website URL is http://www.rjzspt.cn.

## 5.2 User Management and Login

Users of "Reassurance Broiler Traceability Platform" is divided into three types:

**Enterprise Users**. Enterprise user account is set to 5 digit. The 5-digit's meaning is as follows: Enterprise department number(1 digit) + post number (3 digit) + employee number (1 digit).

**Regulatory Users**. Regulatory users include third-party supervision platform users and the government management user.

**Ordinary Users**. Ordinary users mainly are consumers registered as a platform members, involve in query and evaluation of products.

## 5.3 Enterprise Basic Information Management

Basic information management of enterprise include application of information management, department management, post management, staff management, associated enterprise information management, broiler breed management, feed management, medicine management, vaccine management, broiler brand management, enterprise product management, etc.

## 5.4 Chick Releasing Management

The broiler breeding enterprise of the "Company + Farmer" type must first sign the contract with farmers, and then release the chicks.

After signing the breeding contract, chick releasing management is needed in each breeding stage. Chick releasing management will automatically generate the unique 25-digit identification encoding of broiler industry chain and write this code into RFID card. Breeding farmer is in charge of this card in the whole breeding process, no matter C/S or B/S architecture. In slaughter process, the RFID card is needed to transfer the unique broiler identification encoding. Figure 2 shows the writing card process of RFID read and write device at the time of chick releasing process.

**Fig. 2** RFID read and write device

## 5.5 *Broiler Breeding Management*

Broiler breeding management record daily breeding information, weekly breeding information, medication information, epidemic prevention information, anomaly observations, and other information.

Broiler inspection and quarantine management, slaughtering and processing management will not be described in detail; broiler traceability management is described below.

## 5.6 *Broiler Traceability Management*

Broiler traceability management include traceability information management and traceability code printing management, in this process, 25-digit unique identification code of industrial chain is mapped into the 10-digit traceability code. Product traceability label can be generated and printed in page of B/S architecture.

## 5.7 *Third-Party Platform User Early Warning Management*

The third-party platform early warning including epidemic situation early warning and quality safety early warning. Epidemic situation early warning is based mainly on the mortality of broiler and quality and safety; early warning is based on consumer's safety evaluation on the quality and safety of broiler products.

## 5.8 *Government Supervision*

Government supervision management includes broiler epidemic situation management, quality and safety early warning management, and broiler origin quarantine, slaughtering quarantine, nonlocal broiler (including gross broiler and killed broiler query) management within the scope of the city.

# 6 Application and Demonstration

"Reassurance Broiler" platform has been applied and demonstrated in Tianjin Xinnong Broiler Professional Cooperative, the cooperative has more than 300 broiler breeding household, has an annual output of 6 million chickens. The Cooperative's broiler production has won the national "Green Broiler" certification.

**Fig. 3** Traceable tag

The cooperative's tracing label is shown in Fig. 3, the label includes a QR code, bar code, enterprise information, product information, address, and telephone information.

After consumers buy broiler products, they can scan the QR code of broiler products tags to trace through mobile phone, or query the tracing information through internet, SMS, or phone call according to bar code [13]. Detailed information query include daily breeding record, weekly breeding record, medication situation, epidemic prevention situation, pen-leaving information, inspection and quarantine information before slaughter, slaughter information, inspection and quarantine information after slaughter, processing information, and other detailed situation. Through the third-party management background, administrators can determine what kind of information displays.

## 7 Conclusions

"Reassurance Broiler Project" traceability platform covers the standard system of production, slaughter, and processing of Broilers. IOT core technology of RFID is used to bear the unique identification code of broiler industry chain and realize identification transfer. Integrating the tracing of information from broiler production to inspection and quarantine, slaughter, processing, and packaging, the platform has given consumers access to the right to know the product and will ensure the food safety. Practices show that the platform will enhance informatization of enterprise management, cut down management costs, and sharpen brand competitive edge. What is more, the platform will help give the governmental management authorities an access to real-time warning of broiler epidemic and the quality safety of the broiler products.

**Acknowledgements** This work was financially supported by Science & Technology Pillar Program of Tianjin Municipal Science and Technology Commission (13ZCZDNC01100, 14ZCDGNC00099), special fund program Of National System for Broiler Production Technology (CARS-42-17), Tianjin science and technology achievements transformation program (201302060), the Science & Technology Pillar Program sub-project for Agriculture of Ministry of science and technology during the Twelfth Five-year Plan Period (2012BAD39B01, 2012BAD39B0406) and scientific & Research Plan Program of Tianjin Municipal Education Commission (20140811).

## References

1. Chen Changxi, Zhang Hongfu, Feixie Jingwei. Traceability Platform Design of Production Monitoring and Products Quality for Broilers Industry Technology System [J]. Transactions of the Chinese Society for Agricultural Machinery,2010,41(8):100 ~ 105. (in Chinese)
2. Chen Changxi, Zhang Hongfu, FeiXiejingwei, et al. Design of whole process tracking and traceability platform of broilers safety production [J]. Transactions of the CSAE, 2010, 26(9): 263 ~ 269. (in Chinese)
3. Lei Fanghua. Focus on Food Safety of Beijing Olympic Games 2008 [J]. China Food. 2008, 4:50 ~ 51. (in Chinese and English)
4. JIANG Lihong, Yan Shaoqing, XIE Jing, et al. Design of traceability system for the safety production entire processes o f pork [J]. Science and Technology of Food Industry, 2008, 29 (6): 265 ~ 268. (in Chinese)
5. Ahad E, Palacio F, Nuin M, et al. RFID smart tag for traceability and cold chain monitoring of foods: demonstration in an intercontinental fresh fish logistic chain [J]. J. Food. Eng., 2009, 93(4):394 ~ 399
6. HU Dong, XIE Jufang. Application of Short Message Technology in the Remote Control and Traceability System for Pork[J]. Journal of Agricultural Mechanization, 2011, 8:142 ~ 145. (in Chinese)
7. YE Chun-Ling, ZHANG Bing, GU Song-Hao, et al. Design and Implement of Traceable Label of Vegetable Produce Applied in Vegetable Quality and Safety Traceability System [J]. Food Science, 2007(07):572 ~ 574. (in Chinese)
8. ZHANG Jingjing, HE Dong. Development of Warehouse Management System Base on RFID and 2D Barcode [J]. Agriculture network information, 2011, (05): 66 ~ 68. (in Chinese)
9. Darr M J, Zhao L. A wireless data acquisition system for monitoring temperature variations in swine barns [C]//Livestock EnvironmentVIII. ASABE Eighth International Symposium. Iguassu Falls, Brazil, 2008.
10. Chen Changxi, Zhang Hongfu, Wang Yiding, et al. Study on Architecture of Farm Produce Authentication and Traceability System Based on Object-Oriented Style [J]. Journal of Jilin Agricultural University, 2010, 32(5):560 ~ 567. (in Chinese)
11. WU Jianhua, FU Zhongliang, WANG Li, et a1.Drainage geographic information system based on C/S and B/S hybrid architecture [J]. Computer Engineering and Applications, 2007, 43(7):230 ~ 232,235. (in Chinese)
12. Website: http://www.rjzspt.cn. (in Chinese)
13. Zhao Li, Xing Bin, Li Wenyong, et al. Agricultural Products Quality and Safety Traceability System Based on Two-dimension Barcode Recognition of Mobile Phones. Transactions of the Chinese Society for Agricultural Machinery,2012, 43(7): 124 ~ 129. (in Chinese)

# Implementation of Low-Power 6T SRAM Cell Using MTCMOS Technique

**Tripti Tripathi, D.S. Chauhan, S.K. Singh and S.V. Singh**

**Abstract** Electronics industry in present-day scenario is facing the major problem of standby leakage current in most of the electronic devices. As the speed of processor is increasing, the demand for high-speed cache memory is ever increasing. SRAM being mainly used for cache memory design, several low-power techniques are being used to reduce its leakage current. Full CMOS 6T SRAM cell is the most preferred choice for most of the digital circuits. This paper implements 6T CMOS SRAM cell using MTCMOS technique and simulation results show significant reduction in leakage during standby mode. The simulations are done on Cadence Virtuoso Tool using 45 nm technology.

**Keywords** CMOS · SRAM · Sub-threshold leakage

## 1 Introduction

Power consumption particularly off-state leakage current is the major technical problem being faced by present-day electronic industry. As the chip densities increase to a billion of transistors or more, power is the major limiter of design performance or integration. According to International Technology Roadmap for

T. Tripathi (✉)
Inderprastha Engineering College, Ghaziabad, India
e-mail: tripti.chunmun@gmail.com

D.S. Chauhan
GLA University, Mathura, India
e-mail: pdschauhan@gmail.com

S.K. Singh
VIET, Ghaziabad, India
e-mail: sanjaysinghraj@rediffmail.com

S.V. Singh
JIIT, Noida, India
e-mail: sajaivir@rediffmail.com



S.K. Bhatia et al. (eds.), *Advances in Computer and Computational Sciences*,
Advances in Intelligent Systems and Computing 553,
DOI 10.1007/978-981-10-3770-2_44

Semiconductors (ITRS) projections, the number of transistors per chip and the local clock frequencies for high-performance microprocessors will continue to grow exponentially in next 10 years too. As the speed of microprocessor-based electronic equipment increases, there is requirement of large quantity of data at very high speed, which is difficult accomplish. This has led to design of cache memory as major concern. Mostly SRAM is used for cache memory design and full CMOS 6T SRAM cell is preferred choice mostly. Static (or leakage) power affects all kinds of Complementary Metal Oxide Semiconductor (CMOS) circuits but is particularly critical for Static Random Access Memories (SRAMs) since memories have been designed as performance being the primary figure of merit and also memories are accessed in small portions, thereby leaving vast majority of memory cells unaccessed for large fraction of time [1]. As reported in International Technical Roadmap for Semiconductors (ITRS), transistors devoted to memory structures in microprocessor-based system is about 70% today and is expected to increase to 80% in near future [2]. In recent years, the demand for low power has led to development of different techniques for leakage reduction in SRAM cells at various levels of abstraction, architecture, device and circuit level. With continuous technology scaling, power dissipation due to leakage currents has become biggest challenges of VLSI industry in designing of these high-speed and low-power devices. Sub-threshold leakage is the major contributor towards leakage in standby mode of SRAM cell and various techniques have been proposed to minimize it. This paper uses Multi-Threshold CMOS (MTCMOS) technique in 6T SRAM cell at 45 nm technology and compares its delay and power dissipation values with conventional 6T SRAM cell.

## 2 Leakage Current Components in CMOS

Some of the major sources of leakage currents: sub-threshold current due to low threshold voltage, gate leakage current due to very thin gate oxide and band to band tunneling due to heavily doped halo doping profile. The leakage current of a deep submicron CMOS transistor has three major components, junction tunneling current, sub-threshold current and gate tunneling current.

### *2.1 Junction Leakage Current*

The reverse biased junction leakage current has two main components. (1) Caused by EHP generation in the depletion region of reverse biased junction. (2) Due to minority carrier diffusion near the edge of depletion region.

The junction leakage current is quite small and mainly exhibits in access transistors of SRAM memory cell.

## 2.2 Gate Leakage Current

The electric field across the oxide increases if the thickness of gate oxide is reduced. This high electric field results in exponential increase of gate oxide tunneling current due to increase in tunneling probability of electron through the gate oxide (Fig. 1).

## 2.3 Sub-threshold Leakage Current

It is the drain to source current of a transistor when the gate to source voltage is lower than threshold voltage ($V_T$). It is due to carrier diffusion between the drain and source regions of transistor in weak inversion. The behavior of MOS transistor in sub-threshold operating region is similar to that of bipolar device and sub-threshold current exhibits exponential dependence on gate voltage. In present-day technology, it mainly contributes to leakage current. For a 6T CMOS SRAM cell shown in Fig. 2, leakage current can occur either inside the bit cell or in access transistors paths. The sub-threshold leakage has dominant paths: (a) $V_{DD}$ to ground (b) bitlines to ground through access transistors.

In Fig. 1, when nv0 stores '0' then significant leakage current can flow through transistors M1 and M4 and M5. The sub-threshold leakage current in a MOS Transistor varies exponentially with gate to source voltage ($V_{GS}$).

Device material technology has managed to keep gate leakage under control by use of high k-dielectrics. Also the junction leakage current is critical mainly for strong reverse biased junctions and reverse biasing is used selectively and with moderate amount of bias for performance reasons, so the major relevant component of leakage current to be considered is sub-threshold leakage.

It can be reduced by in increasing $V_T$ of all or some of the transistors in the cell. If $V_T$ of p-channel Metal Oxide Semiconductor (PMOS) is increased then write

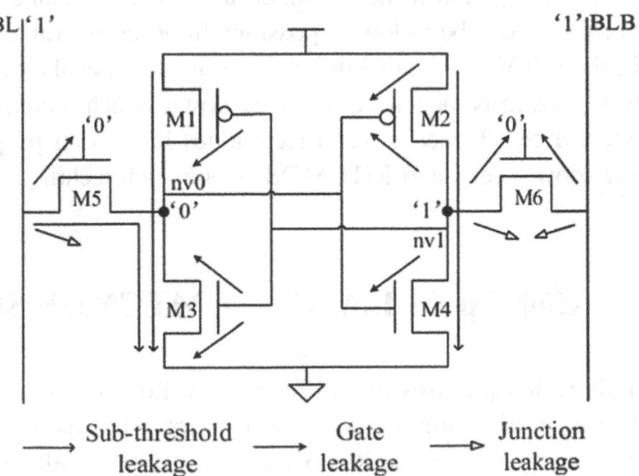


**Fig. 1** Leakage currents in 6T SRAM cell [3]

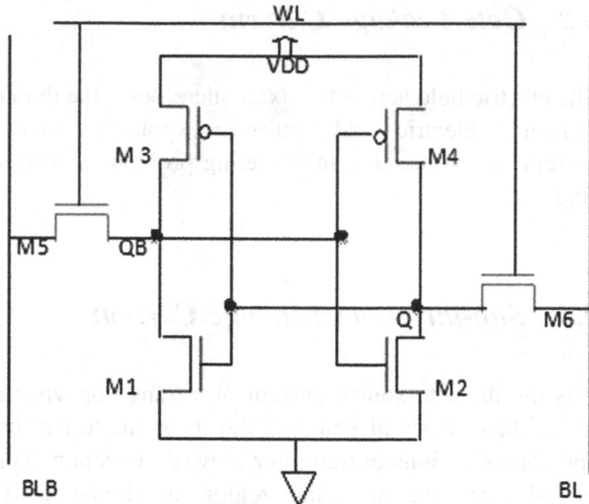


**Fig. 2** Conventional 6T SRAM Cell [4]

delay for the cell is increased while the read delay is negligible and if $V_T$ of n-channel Metal Oxide Semiconductor (NMOS) is increased then read delay in increased while the write delay is negligible. However, if $V_T$ of pass transistors is increased then it leads to increase in both read and write delays of the cell.

## 3 Conventional 6T CMOS SRAM Cell

6T CMOS SRAM cell has two cross-coupled inverters connected back to back. They are coupled through NMOS access transistors, the gate of which are connected to word line (WL) such that data can be written into or read from the cell through bitlines (BL and BLB) which are complementary to improve the noise margin.

It can operate in active mode as well as in standby mode during which leakage current should be as low as possible. In order to reduce the switching power dissipation in the cell $V_{DD}$ is scaled down; but this increases the propagation delay and to improve it $V_T$ has to be accordingly scaled down which increases the sub-threshold leakage in SRAM cell. Hence, several techniques have been proposed to implement high-speed low-power cells and MTCMOS is one such technique [5–7].

## 4 High-Speed Low-Power MTCMOS SRAM Cell

This technique uses two different threshold voltages in the circuit. The SRAM cell is designed using low $V_T$ transistors whereas high $V_T$ transistors are used to effectively isolate the low $V_T$ cell to prevent leakage dissipation in standby mode.

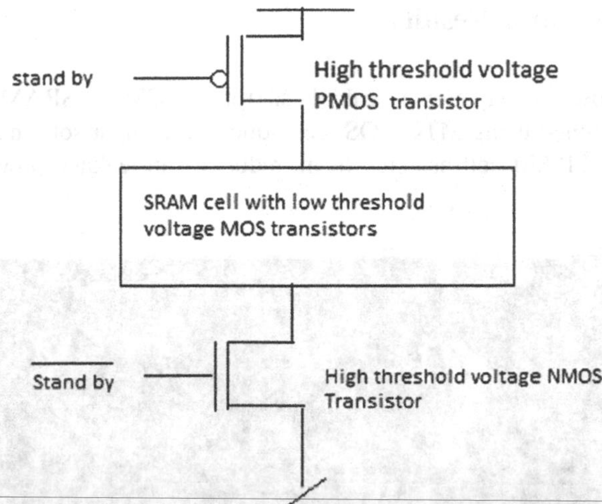


**Fig. 3** MTCMOS technique

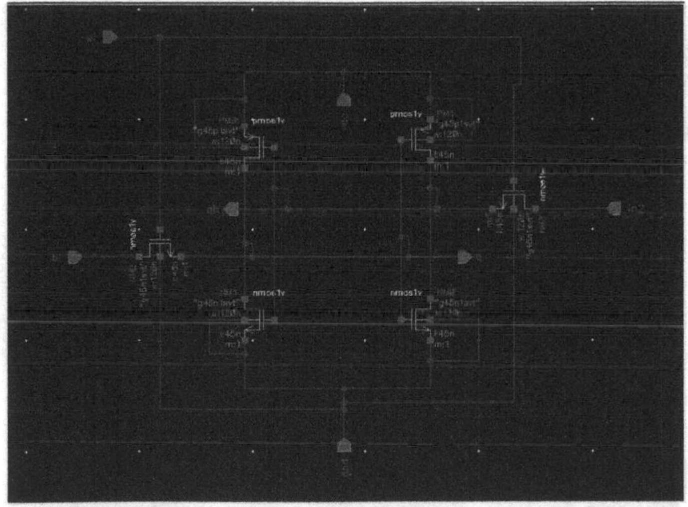

**Fig. 4** Schematic of 6T SRAM cell

In active mode, high $V_T$ transistors are turned on and low $V_T$ SRAM cell operates such that it has low switching power dissipation and small propagation delay. During standby mode the high $V_T$ transistors (also called as sleep transistors) are turned off and the conduction path for any sub-threshold leakage currents that may originate from low $V_T$ cell is effectively cut off. In this configuration, leakage power dissipation is reduced for the cell; however, two extra transistors increase the area and complexity of the circuit (Fig. 3).

## 5 Simulation and Results

The leakage power dissipation is calculated for 6T CMOS SRAM cell and for SRAM cell designed using MTCMOS technique. On comparison, the results show that MTCMOS SRAM cell has significant reduction in leakage power for supply

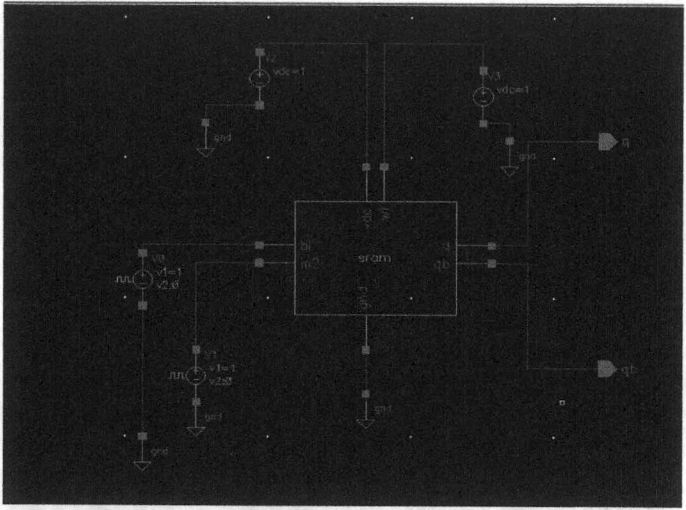

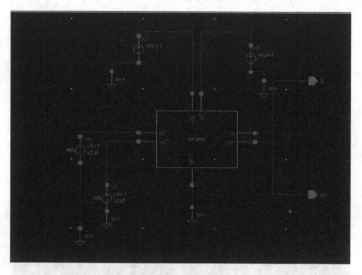

**Fig. 5** Schematic creation from symbol for 6T SRAM cell

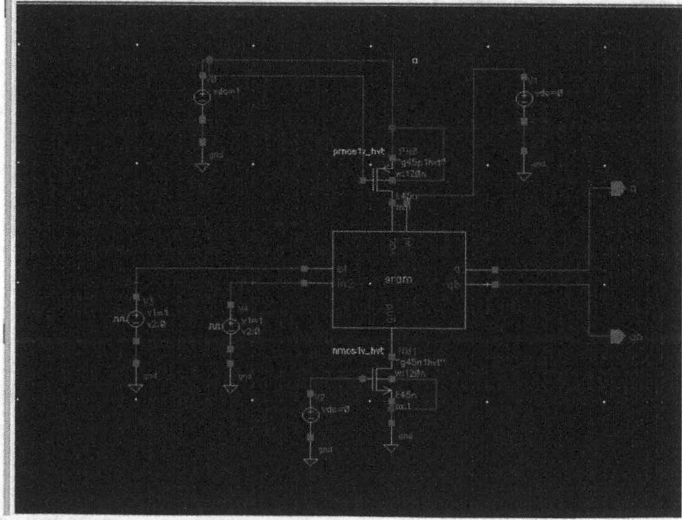


**Fig. 6** Schematic for MTCMOS 6T SRAM cell

**Table 1** Comparison of leakage power in conventional 6T SRAM and MTCMOS SRAM cell

| Supply voltage ($V_{DD}$) = 1 V | | Supply voltage ($V_{DD}$) = 0.8 V |
|---|---|---|
| | Leakage power (in μW) | Leakage power (in μW) |
| 6T CMOS SRAM cell | 1.094 | 0.1353 |
| MTCMOS SRAM cell | $20.02 \times 10^{-6}$ | $9.56 \times 10^{-6}$ |

voltage of 1.0 and 0.8 V using 45 nm technology. Figure 4 shows the schematic of 6T SRAM cell, Fig. 5 shows the symbol of SRAM cell and Fig. 6 shows the schematic of MTCMOS SRAM cell designed using cadence virtuoso tool and 45 nm technology. The circuits are simulated and power dissipation is calculated at supply voltage of 1, 0.8 and 0.5 V respectively.

Table 1 shows the comparison for leakage power dissipation in conventional 6T CMOS SRAM and MTCMOS SRAM cell.

The above figure shows the schematic for 6T SRAM cell from which the symbol can be created and inputs applied to perform simulations. The schematic created from symbol is shown in Fig. 5.

The above table shows that for 45 nm technology, significant reduction in leakage power can be obtained by use of MTCMOS technique.

## 6 Conclusion

With continuous scaling of technology, leakage power is becoming a major concern for VLSI industry and various techniques at various levels of abstraction are being used and proposed to minimize it. This paper shows that for SRAM cell design at 45 nm technology MTCMOS technique can be used to reduce leakage power in standby mode of operation; however, it uses two extra transistors increasing the area.

## References

1. Andrea Calimera et al., 'Design techniques and architectures for low leakage SRAMs', IEEE transactions on circuits and systems, vol. 59, No. 9, pp. 1992–2007, Sept. 2012.
2. 'International technical roadmap for semiconductors', 2009 available online at http:// www.itrs.net/links/ 2009ITRS/home2009.htm
3. Li-Jun Zhang et al., 'Leakage power reduction techniques of 55 nm SRAM cells', IETE Technical Review, Vol. 22, issue 2, pp. 135–145, 2001.
4. Debasis Mukherjee et al., 'Static noise margin analysis of SRAM cell for high speed application', International Journal of Computer Science Issues, Vol. 7, Issue 5, 2010
5. G. Razavipour et al., 'Design and Analysis of Two Low-Power SRAM Cell Structures', IEEE Transaction on Very Large Scale Integration (VLSI) Systems, Vol. 17, No. 10, pp. 1551–1555, Oct. 2009.

6. S. Shigematsu et al., 'A 1-V High-Speed MTCMOS Circuit Scheme for Power-Down Application Circuits', IEEE Journal of Solid-State Circuits, Vol. 32, No. 6, June 1997
7. B. Amelifard et al., 'Leakage minimization of SRAM cells in a dual-Vt and dual-Tox technology', IEEE Transactions on Very Large Scale Integration (VLSI) Systems, vol. 16, pp. 851–859, July 2008.

# Short-Term Electricity Price Forecasting Using Wavelet Transform Integrated Generalized Neuron

**Nitin Singh and Soumya R. Mohanty**

**Abstract** With the advent of deregulation, electricity has become a commodity which is capable of being traded in the deregulated electricity market. In the deregulated environment, accurate electricity price forecasting has become necessity for the generating companies in order to maximize their profits. The existing forecasting models can be broadly classified into statistical models, simulation models, and soft computing models. The soft computing based models have gained popularity among other existing models because of their nonlinear mapping capabilities and ease of implementation. In the presented work, a generalized neuron based electricity price forecasting model has been proposed to forecast the electricity price of New South Wales electricity market. The de-noising capability of the wavelet transform is explored for decomposing the ill-behaved price signal into low- and high-frequency signals for better representation. The low- and high-frequency signals were given as input to the generalized neuron model individually for improving the forecasting accuracy of the model.

**Keywords** Electricity price forecasting · Generalized neural network (GNN) · Wavelet transform · Wavelet analysis · Daubechies wavelet · New South Wales (NSW) electricity market

## 1 Introduction

The deregulation of the electricity market has made the research in the field of electricity price forecasting an important area. The accurate electricity price forecast serves as tool, which is used by various market players that trade in electricity markets. The market operator collects the bids containing price and quantity of the

N. Singh (✉) · S.R. Mohanty
Department of Electrical Engineering, MNNIT Allahabad, Allahabad, India
e-mail: nitins@mnnit.ac.in

S.R. Mohanty
e-mail: soumya@mnnit.ac.in



S.K. Bhatia et al. (eds.), *Advances in Computer and Computational Sciences*, Advances in Intelligent Systems and Computing 553,
DOI 10.1007/978-981-10-3770-2_45

electricity to be traded from the generators and consumers. The bids are accepted in terms of the increasing price until the total desired demand is met. A company that is able to forecast the price accurately can regulate its own price/production schedule accordingly. The market is cleared based on the submitted bids and the supply and demand curve at a price known as market clearing price, accurate MCP prediction along with its confidence interval estimation can be helpful to services and independent power producers in submitting effective bids with low risks.

The knowledge of the supply–demand balance ahead of time is extremely important for all the market players and specifically for generating companies. Hence, electricity price must be predicted before real-time operation to maximize the profit of generating companies [1]. Price forecasts are not only used by the generation companies for optimal bidding but also for designing physical bilateral contracts. Based on the knowledge of the future demand and price, the generation companies invest in new generation facilities for meeting the future demand at competitive price. It can be said that the electricity price forecasts are needed by different market players for serving different purposes [2].

There exist various electricity price-forecasting models in the literature proposed by the researchers across the globe. These models can be broadly classified into hard and soft computing based models. Hard computing based models such as regression-based models requires the model parameters to be defined exactly, although the results are comparable, but a lot of information is needed for prediction, which makes these models computationally expensive and difficult to implement. On the other hand, the soft computing models based on artificial neural networks have gained much popularity in the recent past among researchers due to their nonlinear mapping capability, less information requirement, and ease of implementation.

The training of the artificial neural network based models used for electricity price forecasting is generally done using the popular backpropagation algorithm. The classical backpropagation algorithm has some limitations which does affect the performance of the artificial neural network model. In order to improve the performance of classical artificial neural network model using backpropagation algorithm, the generalized neuron model has been developed, which has been already applied for various applications, such as, forecasting [3], modeling [4] and control [5]. In the proposed work the generalized neuron model along with the wavelet transform is used for forecasting the future electricity price of the New South Wales electricity market, the results obtained are compared with the regression-based and ANN based models.

## 2 Drawbacks of Artificial Neural Network

The classical artificial neural network model trained using backpropagation algorithm has got the following limitations which affect its performance [4]:

1. The number of hidden neurons required is decided on the basis of hit and trial method; there is no technique or method to decide the number of hidden neurons.
2. Slow learning, especially when used for training large networks.
3. Problem of local minima.
4. For complex functions a large number of unknowns to be determined in existing neural network. Because of this, the requirement of the minimum number of input–output pairs increases.

## 3 Generalized Neuron Model

In order to improve the performance of the classical artificial neural network by overcoming its limitations, the generalized neuron model was developed. The generalized neuron is basically a single higher order neuron as shown in Fig. 1. The generalized neuron model uses two aggregation functions, i.e., summation ($\sum$-part) and product ($\prod$-part), the individual output of $\sum$, and $\prod$ goes through sigmoidal and Gaussian basis functions, respectively. Finally, the output of the sigmoidal and Gaussian basis function is aggregated in order to get the neuron output [5].

In the proposed work the generalized neuron model uses $(X+W)^2$ in $\sum$-part with sigmoidal activation function and $(X+W)$ with Gaussian activation function in $\prod$-part, where $X$ and $W$ represent input and weight, respectively. The final output of the neuron can be mathematically represented as (1). In generalized neuron model as shown in Fig. 1, the output of the sigmoidal and gaussian basis function is aggregated and is also called as summation type compensatory neuron model.

$$O_i = O_{\Sigma} * W_{\Sigma} + O_{\Pi} * (1 - W_{\Sigma}). \tag{1}$$

## 4 Generalized Neuron Model Learning Algorithm

The generalized neuron model used in the proposed work is trained using back-propagation through time algorithm. The following steps are involved in the training of the proposed model [5, 6]

**Fig. 1** Generalized neuron model

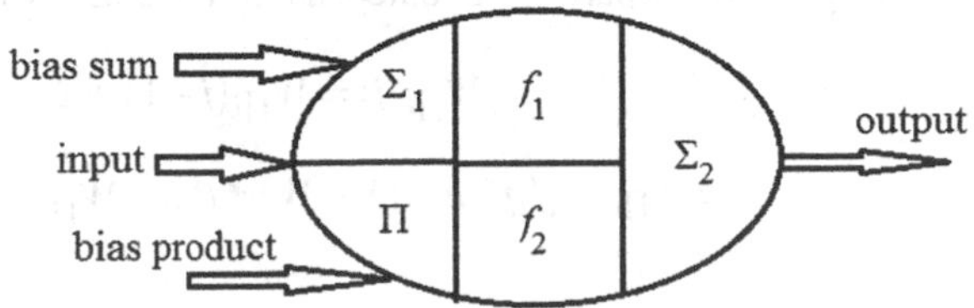

Step 1 The output of $\sum$-part of generalized neuron is calculated using (2)

$$O_{\Sigma} = f_1\left[\left(W_{\Sigma i} + X_i\right)^2 + X_{o\Sigma}\right]. \tag{2}$$

Step 2 The output of $\prod$-part of generalized neuron is calculated using (3)

$$O_{\Pi} = f_2\left[\left(W_{\Pi i} + X_i\right) + X_{o\Pi}\right]. \tag{3}$$

Step 3 The final output of the generalized neuron model can now be calculated using (1).

Step 4 Error is calculated using (4) after final output of the neuron is calculated in the forward pass.

$$E = Y_i - O_i. \tag{4}$$

In (4), $Y_i$ is the desired output and $O_i$ is the actual output of the generalized neuron. The error shown in (4) is minimized by updating the weights of the sigma ($\sum$) and product ($\prod$) parts of the neuron, respectively. Sum squared error (SSE) function is used as the objective function for minimizing the error the SSE function is given as (5).

$$SSE = 1/2 \sum E_i^2. \tag{5}$$

Step 5 After the error is calculated the weight of the neuron inputs or connection strength is modified in the reverse pass. Now weights are adjusted with respect to error. The change in weight $\Delta W$ is calculated using (6)

$$\Delta W = \eta * E * \left\{O_{\Sigma} - O_{\Pi}\right\} + \alpha W(j-1). \tag{6}$$

Weights for the $\Sigma$-part are updated using (7) and (8)

$$W_{\Sigma i}(j) = W_{\Sigma i}(j-1) + \Delta W_{\Sigma i}, \tag{7}$$

$$\Delta W_{\Sigma i} = \eta * E * W * X_i + \alpha W_{\Sigma i}(j-1), \tag{8}$$

where $i$ is the number of input, $j$ is the number of iterations, $\alpha$ is momentum factor, and $\eta$ is learning rate.

Weights for $\Pi$-part are updated using (9) and (10)

$$W_{\Pi i}(j) = W_{\Pi i}(j-1) + \Delta W_{\Pi i} \tag{9}$$

$$\Delta W_{\Pi i} = \{(\eta * E * (1 - W) * P_{net})/W_{\Pi i}\} + \alpha W_{\Pi i}(j-1). \tag{10}$$

The input data given to the generalized neuron is normalized in the range of 0–1 using (11).

$$p_a = \left[(r_a - r_b) * \left(\frac{a-e}{b-e}\right)\right] + (r_b), \tag{11}$$

where $r_a$ and $r_b$ is chosen appropriately between 0 and 1, $a$ = value of the input variable, $e$ = minimum value of the input variable, $b$ = maximum value of the input variable.

# 5 Wavelet Transform

Price signal is a nonstationary signal and these signals can be analyzed using wavelet transform properly. A good local representation can be produced with the help of wavelet transform in both time and frequency domain; hence, it is utilized to solve problem of nonstationary price signal [7–9].

The continuous wavelet transform of any function $x(t)$ with respect to a wavelet $\psi(t)$ can be written as (12) and (13):

$$W(s,b) = \frac{1}{\sqrt{|s|}}\int \Psi^*(t)x(t)dt, \tag{12}$$

$$\psi_{s,\tau}(t) = \frac{1}{\sqrt{s}}\psi\left(\frac{t-\tau}{s}\right), \tag{13}$$

where s and $\tau$ are known as dilation parameter (scale parameter) and translation parameter (time shift parameter), respectively and $*$ denotes complex conjugation. The non-detailed part of the signal is provided by expanding the signal using low frequencies (large scale) and the detailed part of the signal is provided by compressing the signal using high frequencies (low scales). In the discrete wavelet transform, the scale and position values based on power of two, called dyadic dilation and translations, are obtained by the scaling and translation parameters, as shown in (14)

$$DWT_x(m,n) = 2^{-(m/2)}\sum_{t=0}^{T-1} x(t)\psi\left(\frac{t-n\cdot 2^m}{2^m}\right), \tag{14}$$

where the length of the signal $x(t)$ is denoted as $T$. The scaling and translation parameters are functions of the integer variables m and n, where $s = 2^m$ and $\tau = n2^m$, $t$ is the discrete time index. Three level decomposition using wavelet transform is shown in Fig. 2.

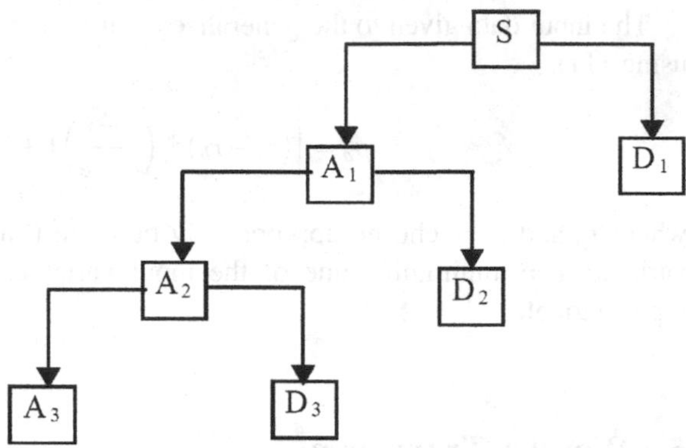

**Fig. 2** Three level decomposition

## 6 Selection of Wavelet Transform

The electricity price is highly volatile in nature and it contains a lot of spikes which making it difficult to predict. Wavelet transform is the most suited technique for converting it into a set of detailed and approximate series which possess better characteristics. The conversion of nonstationary time series to constitutive series requires the proper selection of mother wavelet for the series under consideration and selection of optimal level of decomposition which represents the series correctly with all the hidden features. Out of the several mother wavelets present in the history Daubechies function is most suited for the price forecasting [10].

In the Daubechies family of mother wavelet, the order of the function affects the smoothness as well as the support interval, higher order function may lead to deterioration in prediction therefore the lower order functions are advised [11]. The selection of decomposition level is also important, as this will decide the better classification of data into low- and high-frequency components. The approximate series ($A_i$) is the low-frequency component which essentially contains the information regarding the price signal, and detailed component ($D_i$) is the high-frequency component of the price series which represents information regarding noise in the data, spikes due to weather condition, network congestion, etc.

In the proposed work, the comparison of the Daubechies mother wavelet function for NSW price series is done for two orders, i.e., third and fourth order function and seven decomposition level, in order to select proper order and decomposition level for the NSW series. It has been observed that wavelet of order 3 is more similar to the original series than the other higher order functions. The wavelet of order 3 is chosen for decomposing the original price series of NSW electricity market.

The skewness and kurtosis of the decomposed signal using Daubechies (db) order 3 (db3) with seven decomposition levels is shown in Table 1. Figure 3 shows original New South Wales (NSW) price signal's approximate series from $A_1$ to $A_7$ and detailed series from $D_1$ to $D_7$ using db3. Close observation of Fig. 3

**Table 1** Skewness and kurtosis of db3 decomposed series

| Time series | Skewness | Kurtosis | Time series | Skewness | Kurtosis |
|---|---|---|---|---|---|
| NSW price series | 0.7324 | 4.5895 | NSW price series | 0.7324 | 4.5895 |
| Approx $A_7$ | 0.984 | 1.8232 | Detail $D_7$ | 0.4171 | 1.6777 |
| Approx $A_6$ | 0.6234 | 1.6047 | Detail $D_6$ | −0.3443 | 2.2356 |
| Approx $A_5$ | 0.3269 | 1.5597 | Detail $D_5$ | −0.4072 | 4.1198 |
| Approx $A_4$ | 0.9073 | 2.7613 | Detail $D_4$ | 0.2802 | 3.2924 |
| Approx $A_3$ | 1.3331 | 4.3621 | Detail $D_3$ | −0.0134 | 2.1900 |
| Approx $A_2$ | 0.9861 | 4.4122 | Detail $D_2$ | −0.1467 | 2.7594 |
| Approx $A_1$ | 0.7606 | 4.5905 | Detail $D_1$ | 0.1679 | 3.2087 |

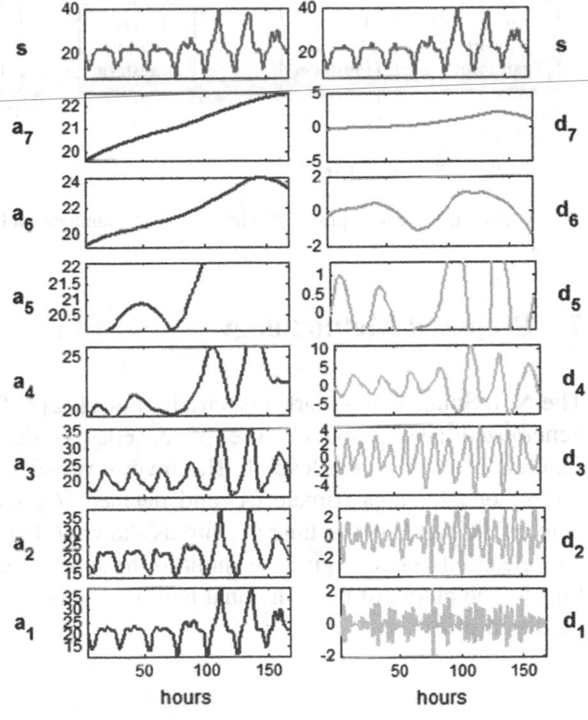


**Fig. 3** Seven level decomposition of price series using db3

reveals that approximate series $A_1$, $A_2$, $A_3$, and $A_4$ are similar in shape to the original price series. The $A_1$ series will have similar characteristics as that of original series and is therefore not used for forecasting.

Now examining the corresponding detailed series it can be observed that detailed series from $D_5$ to $D_7$ do not contain any useful information about the signal and only represents noise contained in the series, while detailed series $D_1$–$D_4$ contain useful information related to the original price signal. In the proposed work mother wavelet of Daubechies family of order 3 (db3) with 4 decomposition levels is used for decomposing the price series of NSW electricity market.

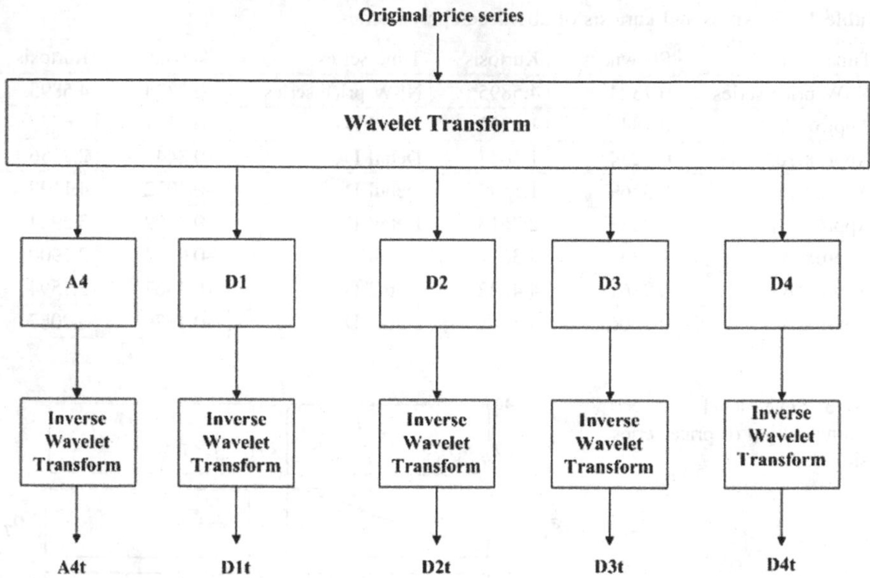


**Fig. 4** Average weekly price of winter, spring, summer and fall seasons in NSW electricity market

## 7 Proposed Methodology

The New South Wales market electricity price [12] is forecasted using the proposed generalized neuron model. The price series is decomposed using Daubechies function of order 3 and level 4. The inverse wavelet transform of the constitutive series done for approximate (*A*) and detailed (*D*) series individually in order to convert the signal to the time domain as shown in Fig. 4. The constitutive series of past electricity price is given as input to the generalized neuron model as shown in Fig. 5. The output of the individual neuron is aggregated to in order to get the final predicted price.

## 8 Results and Discussion

The performance of the proposed forecasting model is verified by using mean absolute percentage error (MAPE) as the performance indices. The mean absolute percentage error MAPE [13] is given by (15).

$$MAPE = \frac{1}{N}\sum_{t=1}^{N}\frac{|F(P(t)) - P(t)|}{P(t)} \times 100, \tag{15}$$

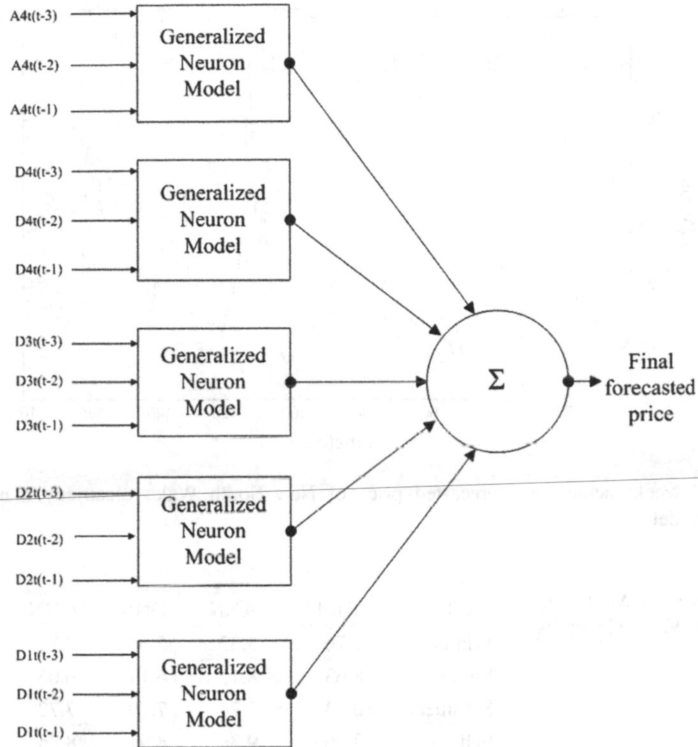


**Fig. 5** Proposed forecasting technique using generalized neuron model

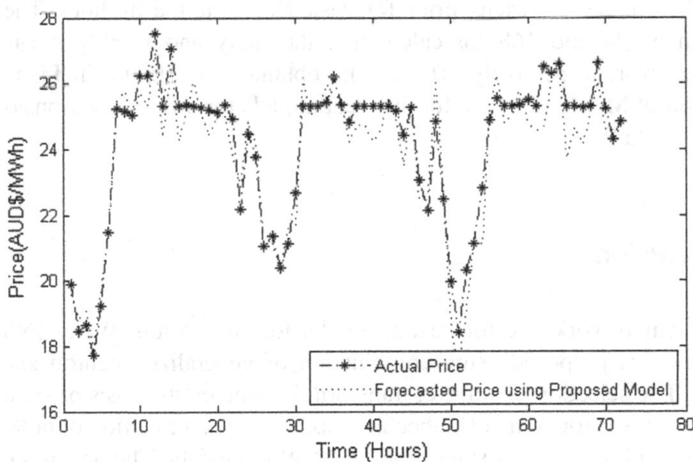


**Fig. 6** Seventy two hour forecast of New South Wales electricity market using proposed model

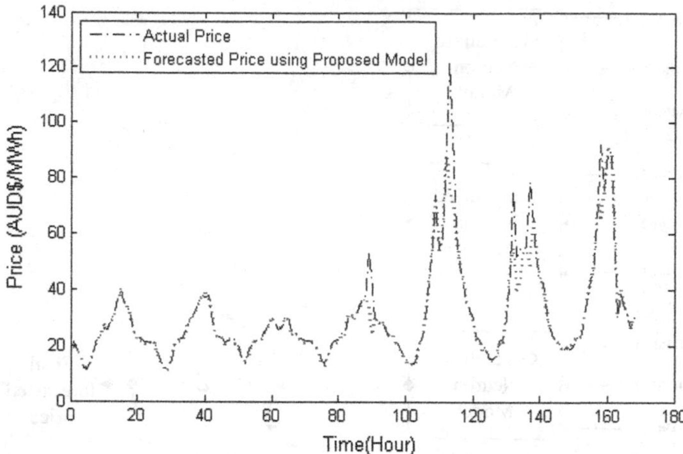


**Fig. 7** Fall week: actual and forecasted price of New South Wales electricity market using proposed model

**Table 2** Weekly MAPE of New South Wales electricity market

| Week | ARIMA | ANN | GNN | GNN + wavelet |
|---|---|---|---|---|
| Winter | 7.76 | 5.33 | 5.66 | 4.74 |
| Spring | 8.63 | 8.15 | 6.14 | 6.05 |
| Summer | 10.53 | 7.71 | 7.79 | 7.72 |
| Fall | 12.36 | 9.36 | 8.91 | 8.86 |

where $N$ represents the forecasting period number; $P(t)$ is the price value at the $t$th hour, $F(P(t))$ is the electricity price forecasted value at the $t$th hour. The value of $N$ is taken as 24 and 168 for calculating the daily and weekly mean absolute percentage error, respectively. The results obtained are shown in Figs. 6 and 7. Comparison of MAPE obtained for different models for different seasons of NSW is shown in Table 2.

## 9 Conclusion

In the presented work, the forecasting model for New South Wales (NSW) electricity market is proposed using combination of generalized neuron and wavelet transform. The selection of wavelet transform is done on the basis of skewness and kurtosis and it was found that Daubechies mother wavelet function of order 3 with 4 decomposition levels is best suited for the NSW price data. The decomposed series is given as input to the generalized neuron model for predicting the future prices and it has been found that generalized neuron model along with wavelet gives better result as compared to regression-based models, generalized neuron model alone and artificial neural network model.

## References


1. M. Shahidehpour, H. Yamin, and Z. Li, "Market operations in electric power systems," John Wiley and Sons, 2002.
2. M. Shahidehpour, M. Alomoush, "Restructured electrical power systems: operation, trading and volatility," New York: Marcel Dekker Publishers, 2001.
3. D. K. Chaturvedi, M. Mohan, R. K. Singh, and P. K. Kalra. "Improved Generalized Neuron Model for Short-Term Load Forecasting" Soft Computing - A Fusion of Foundations, Methodologies and Applications 8, no. 5 April 1, 2004.
4. D. K. Chaturvedi, P. S. Satsangi, and Prem K. Kalra. "New neuron models for simulating rotating electrical machines and load forecasting problems." Electric Power Systems Research 52, vol no. 2 pp. 123–131, 1999.
5. D. K. Chaturvedi, O. P. Malik, and P. K. Kalra. "Experimental studies with a generalized neuron-based power system stabilizer." Power Systems, IEEE Transactions on 19, vol no. 3, pp. 1445–1453, 2004.
6. D. K. Chaturvedi, and O. P. Malik. "A Generalized Neuron Based Adaptive Power System Stabilizer for Multi-machine Environment." Int. J. Soft Computing-A Fusion of Foundations, Methodologies and Applications vol 11 pp. 149–155 (2006).
7. A. Faruqui, B. K. Eakin, "Pricing in competitive electricity markets," Kluwer Academic Publishers, 2000.
8. M. D. Ilic, F. D. Galiana, L. H. Fink, "Power system restructuring: engineering and economics," Kluwer Academic Publishers, 1998.
9. Mallat, Stephane G. "A Theory for Multi-resolution Signal Decomposition: The Wavelet Representation." Pattern Analysis and Machine Intelligence, IEEE Transactions on 11, no. 7 pp. 674–693, 1989.
10. A. R. Reis, & A. A. da Silva, "Feature extraction via multiresolution analysis for short-term load forecasting". *IEEE Transactions on Power Systems*, *20*(1), 189–198, 2005.
11. Sanjeev Kumar Aggarwal, Lalit Mohan Saini, and Ashwani Kumar. "Electricity Price Forecasting in Ontario Electricity Market Using Wavelet Transform in Artificial Neural Network Based Model." International Journal of Control, Automation, and Systems 6, no. 5 pp. 639–650, 2008.
12. [Online] Available: http://www.aemo.com.au/.
13. C. Hamzacebi, "Improving artificial neural networks' Performance in seasonal time series forecasting", Information Sciences 178(2008), pp. 4550–4559.

# Part IV
# Evolutionary and Soft Computing

# Multiple Instance Learning Based on Twin Support Vector Machine

**Divya Tomar and Sonali Agarwal**

**Abstract** Each input object in multiple instance learning (MIL) is represented by a set of instances, referred to as 'bag.' Therefore, in MIL, class labels are associated with each bag instead of individual instance. This study proposes a classifier for multiple instance learning based on Twin Support Vector Machine, termed as MIL-TWSVM. The proposed approach is trained at bag level, where each bag is represented by a vector of its dissimilarities to other bags in the training set. A comparative analysis of MIL-TWSVM approach is performed with the instance-level and noisy-or (NOR) learning approaches based on TWSVM. The performance of the proposed MIL-TWSVM approach has also been compared with several existing approaches of multiple instance learning. The experiments on eight multiple instance benchmark datasets have shown the superiority of the proposed approach. The significance of experimental results has been tested via statistical analysis conducted by using Friedman's statistic and Nemenyi post hoc tests.

**Keywords** Single instance learning · Multiple instance learning · Bag dissimilarity · Twin support vector machine

## 1 Introduction

In single instance (SI) learning, the object to be classified is denoted by an instance $x$, which represents m-dimensional features vector. The training dataset $T = \{x^i, l^i\}_{i=1}^{N}$ consists of N such instances and their corresponding class labels. Here $l^i$ denotes the class label of instance $x^i$. For binary classification problem, the class label takes only two values, i.e., $l^i = +1$ or $-1$. The objective of such type of

D. Tomar (✉) · S. Agarwal
Indian Institute of Information Technology, Allahabad, India
e-mail: divyatomar26@gmail.com

S. Agarwal
e-mail: sonali@iiita.ac.in



S.K. Bhatia et al. (eds.), *Advances in Computer and Computational Sciences*,
Advances in Intelligent Systems and Computing 553,
DOI 10.1007/978-981-10-3770-2_46

problem is to learn a classifier $f(x^i)$ using this training dataset and predict the class label for new instances. In pattern recognition problems, some objects are complex due to which it is difficult to represent them using a single instance. For example, a protein molecule has several amino acid subsequences, an image can be represented by various objects and a document may contain several paragraphs [1]. The traditional supervised learning approaches handle such kind of problems by representing complex objects as a single feature vector or instance. This reduction may lose significant information which further degrades the performance of supervised learning approaches [1–4]. Complex object can be represented by multiple feature vectors or multiple instances instead of single feature vector for their better understanding. This representation can preserve more information of a complex object than a single instance representation. MIL is a variation of supervised learning in which a classifier is learnt by multiple instances known as bag, instead of the individual instance. For binary classification problem, a bag is labeled as positive or negative. A bag is labeled with positive class if there is at least one positive instance while with the negative class when all of its instances are negative [1, 5]. Figure 1 shows the traditional supervised learning framework and multiple instance learning framework.

In MIL framework, each object is represented by a bag $b^i$ instead of individual instance. A bag contains a set of instances and different bags may have different number of instances. Let bag $b^i$ contains $n^i$ number of instances then a bag can be denoted as $b^i = \{x_1^i, x_2^i, \ldots, x_{n^i}^i\}$. The training dataset for this type of framework is denoted as $X = \{b^i, l^i\}_{i=1}^N$, where $l^i = \{-1, +1\}$ is the class label of bag $b^i$. Thus, in MIL case, the classifier works at bag level rather than instance-level and takes a bag as an input [6]. The objective of MIL is to develop a classifier which generates a decision function for the bag. MIL has many applications such as drug activity recognition, image indexing, text classification, etc. [1, 7, 8]. Axis parallel rectangles algorithm, Diverse Density (DD), and Expectation Maximization Diverse Density (EMDD) are some popular approaches of MIL [1, 3, 9]. Researchers also extended several regular supervised classifiers to multiple instance learning scenarios, for example, Bayesian kNN, Citation kNN, multiple instance learning

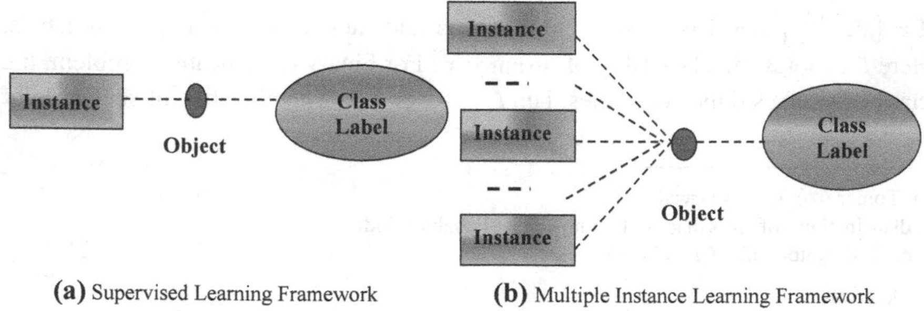


**Fig. 1** Machine learning frameworks

Support Vector Machine (MI-SVM), MIL Boost, Neural Network, and Logistic Regression [10–14].

This study has adopted TWSVM for multiple instance learning scenarios. TWSVM is an advanced machine learning approach which has better generalization ability and faster computational speed [15]. TWSVM has shown its superiority over other existing machine learning approaches [15, 16]. Shao et al. extends TWSVM to MIL scenario which generates a positive and a negative hyperplane in such a way that the previous one is proximal to at least one instance in every positive bag and is far from all instances belonging to negative bags [17]. The negative hyperplane is close to all instances belonging to negative bags and is far from the instances in positive bags. They have not considered the bag dissimilarity into consideration. In this study, we propose a multiple instance learning TWSVM approach using bag dissimilarities. The proposed approach is trained with the summarized information of instances in each bag, where each bag is represented by a feature vector. Feature vector contains the dissimilarity scores of a bag from other bags in the training set. In this research work, the experiment is conducted on 8 MIL benchmark datasets. Initially, the performance and usefulness of proposed MIL-TWSVM is compared with the Single Instance Learning TWSVM and Noisy-or TWSVM. Then the experimental result of proposed approach has been compared with several existing MIL approaches, such as EMDD, DD, MI-SVM, and MILR. The effectiveness of proposed MIL-TWSVM approach has also been analyzed by using nonparametric Friedman test statistic and Nemenyi post hoc test. The statistical inferences are made from the observed difference in the experimental results.

The content of the paper is organized into five sections. The second section gives the brief outline of TWSVM. The third section describes the formulation of proposed MIL approach which is based on bag dissimilarities. The experimental results are discussed and the conclusion is drawn in sections four and five respectively.

## 2 Twin Support Vector Machine

TWSVM is a binary classification approach which predicts the class of data instances by determining nonparallel hyperplane for each class [15]. The planes are generated in such a manner that the data instances of each class maintain closeness to their respective hyperplane while farthest from the other plane. The effectiveness of TWSVM over other existing machine learning approaches has been validated on several benchmark datasets [15]. Consider two matrices $A \in R^{n_1 \times m}$ and $B \in R^{n_2 \times m}$ consist of the data instances of positive and negative class correspondingly. TWSVM determines following two hyperplanes (one plane for each class):

$$f_{+1}(x) = x^T w_{+1} + b_{+1} = 0 \quad \text{and} \quad f_{-1}(x) = x^T w_{-1} + b_{-1} = 0 \tag{1}$$

by solving two Quadratic Programming Problems given as follows:

$$\begin{aligned} &\min(w_{+1}, b_{+1}, \xi)\frac{1}{2}\|Aw_{+1} + e_{+1}b_{+1}\|^2 + c_1 e_{-1}^T \xi \\ &\text{s.t.} \quad -(Bw_{+1} + e_{-1}b_{+1}) + \xi \geq e_{-1}, \xi \geq 0 \end{aligned} \tag{2}$$

$$\begin{aligned} &\min(w_{-1}, b_{-1}, \eta)\frac{1}{2}\|Bw_{-1} + e_{-1}b_{-1}\|^2 + c_2 e_{+1}^T \eta \\ &\text{s.t.} \quad (Aw_{-1} + e_{+1}b_{-1}) + \eta \geq e_{+1}, \eta \geq 0 \end{aligned} \tag{3}$$

where $w_{+1}$ and $w_{-1}$ are the normal vectors to the hyperplane $f_{+1}(x)$ and $f_{-1}(x)$ respectively. $b_{+1}$ and $b_{-1}$ are bias terms. $e_{+1} \in R^{n_1}$ and $e_{-1} \in R^{n_2}$ represent vector of one's of suitable dimension. $c_1, c_2 > 0$ are penalty parameters and $\xi \in R^{n_1}$ and $\eta \in R^{n_2}$ are slack variables. The hyperplane parameters are determined using Eqs. (2) and (3) as follows:

$$u_{+1} = -(M^T M)^{-1} N^T \alpha_{+1}, u_{-1} = (N^T N)^{-1} M^T \alpha_{-1}, \tag{4}$$

where $M = [A\ e_{+1}]$, $N = [B\ e_{-1}]$ and $u_{+1} = \begin{bmatrix} w_{+1} \\ b_{+1} \end{bmatrix}$, $u_{-1} = \begin{bmatrix} w_{-1} \\ b_{-1} \end{bmatrix}$. In this way, we can get hyperplane parameters which are used to construct the hyperplane for each class according to Eq. 1. TWSVM predicts the class label of new data instance on the basis of its distance from each hyperplane. The new instance is labeled with a class which lies closest to it. TWSVM determines the class label of an instance by using following decision function:

$$f(x) = \underset{i=+1,-1}{\operatorname{argmin}} \frac{|w_i x + b_i|}{\|w_i\|} \tag{5}$$

TWSVM is also extended to nonlinear cases where it uses kernel trick for transforming the data instances into higher dimension. Nonlinear TWSVM generates following kernel surfaces instead of planes:

$$K(x^T, C^T)\mu_{+1} + b_{+1} = 0 \quad \text{and} \quad K(x^T, C^T)\mu_{-1} + b_{-1} = 0 \tag{6}$$

where K represents the kernel function and $C^T = [A\ B]^T$. The optimization problems for nonlinear TWSVM can be constructed as:

$$\begin{aligned} &\min(\mu_{+1}, b_{+1}, \xi)\frac{1}{2}\left\|K(A, C^T)\mu_{+1} + e_{+1}b_{+1}\right\|^2 + c_1 e_{-1}^T \xi \\ &\text{s.t.} \quad -\left(K(B, C^T)\mu_{+1} + e_{-1}b_{+1}\right) + \xi \geq e_{-1}, \xi \geq 0 \end{aligned} \tag{7}$$

$$\begin{aligned} &\min(\mu_{-1}, b_{-1}, \eta) \quad \frac{1}{2}\left\|K(B, C^T)\mu_{-1} + e_{-1}b_{-1}\right\|^2 + c_2 e_{+1}^T \eta \\ &\text{s.t.} \quad (K(A, C^T)\mu_{-1} + e_{+1}b_{-1}) + \eta \geq e_{+1}, \eta \geq 0 \end{aligned} \tag{8}$$

Similar to (5), kernel surface parameters are obtained as:

$$\begin{bmatrix} \mu_{+1} \\ b_{+1} \end{bmatrix} = -(P^T P)^{-1} Q^T \alpha \quad \text{and} \quad \begin{bmatrix} \mu_{-1} \\ b_{-1} \end{bmatrix} = (Q^T Q)^{-1} P^T \gamma \tag{9}$$

Here, $P = [K(A, C^T)e_{+1}]$ and $Q = [K(B, C^T)e_{-1}]$. New data instance is assigned to a class in a manner similar to the linear case.

## 3 Multiple Instance Learning TWSVM

This research work constructs the three adaptation of TWSVM to multiple instance learning scenarios MIL-TWSVM based on bag dissimilarity, SIL-TWSVM and NOR-TWSVM. MIL-TWSVM is obtained by extending the recently proposed TWSVM classifier to MIL scenarios. In the proposed approach, each bag is denoted by a feature vector containing the dissimilarity scores of a bag derived from the other bags in the training set. The dissimilarity of a bag from all other bags represents a feature vector. If there are N bags in the training set, then the *i*th bag is represented as:

$$v^i = [d(b^i, b^1), d(b^i, b^2), \ldots, d(b^i, b^N)]^T \tag{10}$$

Consider *i*th bag contains $n^i$ number of instances. The dissimilarity between two bags $b^1$ and $b^2$ is defined as:

$$d(b^1, b^2) = \max\{d_{dir}(b^1, b^2), d_{dir}(b^2, b^1)\} \tag{11}$$

Here, $d_{dir}(b^1, b^2)$ represents the directed distance between two bags $b^1$ and $b^2$. In detail, given two bags $b^1 = \{x_1^1, x_2^1, \ldots, x_{n^1}^1\}$ and $b^2 = \{x_1^2, x_2^2, \ldots, x_{n^2}^2\}$, the directed distance between $b^1$ and $b^2$ is calculated as:

$$d_{dir}(b^1, b^2) = \max_{x^1 \in b^1} \min_{x^2 \in b^2} \left\|x^1 - x^2\right\|_2 \tag{12}$$

Simply, $\left\|x^1 - x^2\right\|_2$ measures the Euclidean distance between the instances $x^1$ and $x^2$. The N-dimensional representation of *i*th bag, $v^i$, is formed as a vector containing dissimilarity scores between *i*th bag and all the other bags in the training set. In this way, we can calculate the dissimilarity score of a bag from rest of the other bags. The new vector representation of each bag $v^i$ is act as an input to the

TWSVM classifier which now works at the bag level $f(v^i)$. Now the problem has been converted to the binary classification problem in which a bag has either 1 or −1 class label.

Single instance learning TWSVM (SIL-TWSVM) solves the MIL problems by considering them as single instance-level problems. In this case, an instance 'i' is assigned with its bag label, i.e., all the instances of a bag take the label of their corresponding bag and then instance-level TWSVM classifier is constructed. SI learning TWSVM classifier is trained with these instances. Same procedure is repeated during testing phase in which each test instance get the label of its bag. In this way, the bag information is totally discarded during training and testing phase. If SIL-TWSVM works fine that means the bag information does not contribute too much and therefore, we can solve MI learning problem as SI learning.

Noisy-or (NOR) TWSVM is another approach which has the same training procedure in SIL-TWSVM, i.e., the instances carry the label of their corresponding bag and the classifier is trained with these instances. But during testing phase, the decision is made at the bag level. The decision of SIL-TWSVM $f(x_i^t)$ is combined to get a bag-level decision:

$$g(b^t) = \varnothing\left(f(x_1^t), f(x_2^t), \ldots, f(x_{n^t}^t)\right) \tag{13}$$

A bag gets positive label if at least one of the decisions at instance level is positive (i.e., positive bags consists of at least one positive instance otherwise negative label is assigned to the bag).

## 4 Results and Discussion

This section presents the experimental results of proposed MIL-TWSVM on eight benchmark datasets taken from KEEL dataset repository [18]. The detailed description of each dataset is given in the Table 1.

**Table 1** Dataset details

| Datasets | # bags | | | # Attributes | # Instances | Average bag size |
|---|---|---|---|---|---|---|
| | +bags | −bags | Total | | | |
| eastWest | 10 | 10 | 20 | 25 | 213 | 10.65 |
| westEast | 10 | 10 | 20 | 25 | 213 | 10.65 |
| Fox | 100 | 100 | 200 | 230 | 1320 | 6.60 |
| Elephant | 100 | 100 | 200 | 230 | 1391 | 6.96 |
| Tiger | 100 | 100 | 200 | 230 | 1220 | 6.20 |
| Mutagenesis-atoms (Muta) | 125 | 63 | 188 | 11 | 1618 | 8.61 |
| Musk1 | 47 | 45 | 92 | 167 | 476 | 5.17 |
| Musk2 | 39 | 63 | 102 | 167 | 6598 | 64.69 |

This study has used Gaussian Kernel function $\mathrm{K}(x_i, x_j) = \exp\left(-\frac{\|\mathrm{x}_i - \mathrm{x}_j\|^2}{2\sigma^2}\right)$ for nonlinear case and Grid Search method for optimal parameters selection. The penalty parameters and sigma are chosen from the following range: $c_1$, $c_2 \in \{10^{-4}, \ldots, 10^3\}$ and $\sigma \in \{2^{-2}, \ldots, 2^4\}$. The experiment is conducted by using 10-fold cross-validation method. The performance of MIL-TWSVM has been compared with SIL-TWSVM and NOR-TWSVM for both linear and nonlinear cases on eight benchmark datasets as shown in Table 2. The result includes the mean and standard deviation of classification accuracies of the 10-folds. In the table, we have mentioned the best performance of each approach. Bold figures indicate better performance of an approach for each dataset. It is observed that the SIL-TWSVM gives better performance with eastWest and Fox datasets for linear cases. However, the proposed MIL-TWSVM approach performs better for rest of other datasets. For nonlinear case, MIL-TWSVM shows better predictive accuracy on seven datasets out of eight datasets. These results confirm the suitability of multi-instance learning approach.

This study has also compared the performance of proposed MIL-TWSVM approach with the existing MIL approaches—EMDD, DD, Multi-instance Logistic Regression (MILR), and multi-instance Support Vector Machine (MISVM) as shown in Table 3. In this table, we have mentioned the best predictive accuracy of the proposed MIL-TWSVM approach for each dataset which is observed in linear and nonlinear cases. The results of other MIL approaches have been taken from the [19]. From Table 3, it is analyzed that the proposed MIL-TWSVM approach has achieved highest predictive accuracy on all eight benchmark datasets and thus performs better with respect to the other existing MI learning approaches.

This study has also analyzed the significance of experimental results using Friedman's statistic and Nemenyi post hoc tests. Friedman test statistic assigns rank to each classification approach on the basis of their predictive performance on each dataset independently [20, 21]. Let $\mathrm{r}_\mathrm{i}^\mathrm{j}$ be the rank of $j$th classifier on $i$th dataset. Friedman test statistic is calculated as:

$$\chi_\mathrm{F}^2 = \frac{\mathbf{12D}}{\mathbf{K(K+1)}}\left[\sum_{\mathrm{j}=1}^{\mathrm{K}} \mathbf{AR}_\mathrm{j}^2 - \frac{\mathbf{K(K+1)}^2}{\mathbf{4}}\right], \text{ where } \mathbf{AR}_\mathrm{j} = \frac{\mathbf{1}}{\mathbf{D}}\sum_{\mathrm{i}=1}^{\mathrm{D}} \mathbf{r}_\mathrm{i}^\mathrm{j} \tag{14}$$

Here, D represents the number of datasets used in the study for comparison purpose, K denotes the number of classification approaches and $\mathrm{AR}_\mathrm{j}$ is the average rank of $j$th classifier. If the value of Friedman test statistic is large as compared to the critical value corresponds to K-1 degrees of freedom then we can accept or reject the null hypothesis which states that there is no difference in classification approaches. The Nemenyi post hoc test reports that the two classification approaches are significantly different if their average rank differs by at least the critical difference which is obtained as $\mathrm{CD} = \mathrm{q}_\alpha\sqrt{\frac{\mathrm{K(K+1)}}{6\mathrm{D}}}$, where $\mathrm{q}_\alpha$ is calculated on the basis of Studentized range statistic. The Friedman test statistic results are plotted by

**Table 2** Performance comparison of classifiers

| Datasets | SIL-TWSVM Acc ± std (%) | NOR-TWSVM Acc ± std (%) | MIL-TWSVM Acc ± std (%) | SIL-TWSVM Acc ± std (%) | NOR-TWSVM Acc ± std (%) | MIL-TWSVM Acc ± std (%) |
|---|---|---|---|---|---|---|
| | Linear | | | Non-linear | | |
| Eastwest | **73.24 ± 2.48** | 65.0 ± 7.24 | 60.00 ± 6.32 | 76.79 ± 2.61 | 65.00 ± 7.24 | **80.00 ± 7.74** |
| Westeast | 70.83 ± 2.06 | 50.0 ± 0.00 | **75.00 ± 7.90** | 74.63 ± 4.01 | 70.00 ± 7.74 | **85.00 ± 7.24** |
| Fox | **62.24 ± 1.26** | 58.0 ± 1.62 | 58.50 ± 1.74 | 66.59 ± 1.31 | **68.00 ± 2.26** | 61.50 ± 1.01 |
| Elephant | 81.33 ± 2.12 | 72.5 ± 3.01 | **82.00 ± 3.01** | 83.06 ± 1.49 | 73.00 ± 2.37 | **84.00 ± 3.37** |
| Tiger | 79.56 ± 1.51 | 83.00 ± 1.61 | **84.50 ± 2.45** | 80.29 ± 2.07 | 84.50 ± 1.49 | **86.00 ± 1.22** |
| Musk1 | 78.67 ± 2.69 | 67.78 ± 3.31 | **93.33 ± 2.34** | 43.79 ± 4.92 | 44.44 ± 0.00 | **55.56 ± 0.00** |
| Musk2 | 76.36 ± 3.27 | 68.22 ± 4.62 | **90.02 ± 4.56** | 54.03 ± 5.12 | 52.84 ± 2.36 | **58.25 ± 3.45** |
| Muta | 68.49 ± 0.853 | 68.42 ± 0.0 | **89.47 ± 0.00** | 69.16 ± 0.59 | 68.42 ± 0.00 | **85.52 ± 1.28** |

**Table 3** Performance comparison of multiple instance learning approaches

| Datasets | EM-DD Acc ± std (%) rank | DD Acc ± std (%) rank | MILR Acc ± std (%) rank | MI-SVM Acc ± std (%) rank | MIL-TWSVM Acc ± std (%) rank |
|---|---|---|---|---|---|
| Eastwest | 64.00 ± 28.50 (4) | 64.50 ± 32.79 (3) | 67.00 ± 36.39 (2) | 60.50 ± 21.67 (5) | **80.0** ± **7.74**(1) |
| Westeast | 38.00 ± 27.63 (2.5) | 36.00 ± 29.37 (4) | 35.50 ± 34.30 (5) | 38.00 ± 25.74 (2.5) | **85.0** ± **7.24**(1) |
| Fox | 59.65 ± 9.16 (2) | 59.40 ± 9.88 (3) | 57.35 ± 10.81 (4) | 49.35 ± 5.71 (5) | **61.5** ± **1.01**(1) |
| Elephant | 75.25 ± 10.57 (5) | 81.45 ± 10.23 (2) | 78.90 ± 9.28 (4) | 79.45 ± 9.32 (3) | **84.0** ± **3.37**(1) |
| Tiger | 71.35 ± 9.45 (5) | 72.20 ± 8.80 (4) | 75.30 ± 8.98 (3) | 80.70 ± 7.42 (2) | **86.0** ± **1.22**(1) |
| Musk1 | 83.61 ± 13.16 (4) | 84.65 ± 11.83 (3) | 73.51 ± 13.43 (5) | 89.21 ± 9.84 (2) | **93.33** ± **2.34**(1) |
| Musk2 | 85.57 ± 10.28 (2) | 80.39 ± 13.44 (4) | 78.51 ± 13.06 (5) | 83.92 ± 10.36 (3) | **90.02** ± **4.56**(1) |
| Muta | 68.89 ± 12.29 (4) | 72.89 ± 9.34 (3) | 74.00 ± 10.37 (2) | 66.54 ± 1.85 (5) | **85.52** ± **1.28**(1) |
| Average Rank | 3.56 | 3.25 | 3.75 | 3.44 | 1 |
| Friedman test statistic = 16.42 | | | | | |

using modified Demsar significance diagram. The average rank of each MI approach is obtained on the basis of its performance on each dataset and then Friedman test statistic is calculated as shown in Table 3. Table 3 indicates that MIL-TWSVM gets highest average rank among all MI learning approaches. The critical value for 4-degree of freedom is 9.488 and the value of $\chi_F^2$ is 16.42. The high value of $\chi_F^2$ leads to the rejection of null hypothesis. The critical difference $CD = 2.728\sqrt{\frac{5\times6}{6\times8}} = 2.157$ for $\alpha = 0.05$. ($q_\alpha$ for five classifiers is 2.728). Figure 2 indicates the significance diagram where MIL classification approaches are

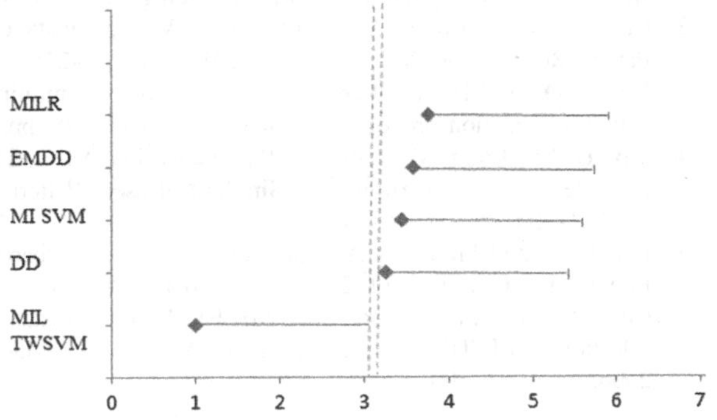


**Fig. 2** Average rank comparison of MI learning approaches

arranged in increasing order of their average rank on the y-axis and their average ranks are mentioned on the x-axis.

Then critical value is added to the average rank of each classifier. From the Fig. 2, it is clear that the proposed approach is significantly better than the other MIL approaches. Therefore, it can be concluded that MIL-TWSVM classification approach is an effective approach for multiple instance learning scenario, which is ensured by the fact that its performance is better than those of other four approaches.

## 5 Conclusion

This study has focused on MIL framework in which a classifier learns from a set of instances (bag) instead of single instance and proposed a MI learning approach based on TWSVM, termed MIL-TWSVM. In the proposed approach, each bag is represented by a vector containing the dissimilarity scores of that bag to the other bags. Initially, the performance of proposed MIL-TWSVM classifier has been compared with SIL-TWSVM and NOR-TWSVM to show its effectiveness in multiple learning scenarios. We have also compared the performance of MIL-TWSVM with four existing MI learning approaches. Experimental results shows that MIL-TWSVM approach has achieved highest predictive accuracy as compared to the other existing MIL approaches on all eight datasets which further support the suitability of MIL-TWSVM in multiple instance learning scenarios. The findings of experimental results are also supported by the statistical analysis performed by using Friedman test. The test shows that the MIL-TWSVM is significantly better than the EMDD, DD, MI-SVM, and MILR. In the future, we are interested to extend MIL-TWSVM to multi-instance multi label scenario.

## References

1. Dietterich, T. G., Lathrop, R. H., and Lozano-Pérez, T. Solving the multiple instance problem with axis-parallel rectangles. Artificial intelligence, Vol. *89*, no. 1, 1997, pp. 31–71.
2. Cheplygina, V., Tax, D. M., and Loog, M. Multiple instance learning with bag dissimilarities. Pattern Recognition, Vol. 48, no. 1, 2015, pp. 264–275.
3. Maron, O., and Lozano-Pérez, T. A framework for multiple-instance learning. Advances in neural information processing systems, Vol. 10, 1998, pp. 570–576.
4. Tax, D. M., Loog, M., Duin, R. P., Cheplygina, V., and Lee, W. J. Bag dissimilarities for multiple instance learning. In Similarity-Based Pattern Recognition, Vol. 7005, 2011, pp. 222–234.
5. Foulds, J., and Frank, E. A review of multi-instance learning assumptions. The Knowledge Engineering Review, Vol. 25, no. 01, 2010, pp. 1–25.
6. Chen, Y., Bi, J., and Wang, J. Z. MILES: Multiple-instance learning via embedded instance selection. IEEE Transactions on Pattern Analysis and Machine Intelligence, Vol. 28, no. 12, 2006, pp. 1931–1947.

7. Ramon J., De Raedt L., Multi instance neural networks, In Proceedings of the ICML-2000 Workshop on Attribute-Value and Relational Learning, 2000, pp. 53–60.
8. Wang, H., Nie, F., and Huang, H. Learning Instance Specific Distance for Multi-Instance Classification. In 25th AAAI Conference on Artificial Intelligence, 2011, pp. 507–512.
9. Zhang, Q., and Goldman, S. A. EM-DD: An improved multiple-instance learning technique. In Advances in neural information processing systems, Vol. 14, 2001, pp. 1073–1080.
10. Wang, J., and Zucker, J. D. Solving multiple-instance problem: A lazy learning approach. In Proceedings of the 17th International Conference on Machine Learning, San Francisco, 2000, pp. 1119–1125.
11. Andrews, S., Tsochantaridis, I., and Hofmann, T. Support vector machines for multiple-instance learning. In Advances in neural information processing systems, Vol. 15, 2003, pp. 561–568.
12. P. A. Viola, J. C. Platt, and C. Zhang, Multiple Instance boosting for object detection. In Advances in neural information processing systems (NIPS), Vol. 18, 2006, pp. 1419–1426.
13. Zhou, Z. H., and Zhang, M. L. Neural networks for multi-instance learning. In Proceedings of the International Conference on Intelligent Information Technology, Beijing, China, 2002, pp. 455–459.
14. Xu, X., and Frank, E., Logistic regression and boosting for labeled bags of instances. In Advances in knowledge discovery and data mining, Springer Berlin Heidelberg, 2004, pp. 272–281.
15. Jayadeva, Khemchandani, R., and Chandra, S. Twin support vector machines for pattern classification. IEEE Transactions on Pattern Analysis and Machine Intelligence, Vol. 29, no. 5, 2007, pp. 905–910.
16. Tomar, D., and Agarwal, S. Twin support vector machine: a review from 2007 to 2014. Egyptian Informatics Journal, Vol. 16, no. 1, 2015, pp. 55–69.
17. Shao, Y. H., Yang Z. X., Wang X. B. and Deng N. Y. Deng. Multiple instance twin support vector machines. Lect Note Oper Res, Vol. 12, 2010, pp. 433–442.
18. multi-instance dataset, http://sci2s.ugr.es/keel/. Accessed on May 2015.
19. Dong, L. A comparison of multi-instance learning algorithms (Doctoral dissertation, The University of Waikato), 2006.
20. Demšar, J. Statistical comparisons of classifiers over multiple data sets. The Journal of Machine Learning Research, Vol. 7, 2006, pp. 1–30.
21. Nemenyi, P. Distribution-free multiple comparisons. Ph.D. Thesis. Princeton University, 1963.

# Optimized Task Scheduling Using Differential Evolutionary Algorithm

**Somesh Singh Thakur, Siddharth Singh, Pratibha Singh and Abhishek Goyal**

**Abstract** Task scheduling plays a key role for efficiently assigning resources to tasks and performing multitasking. In heterogeneous environments, hard computing task scheduling does not give optimal solution. There are many soft computing techniques used for task scheduling such as evolutionary algorithm which includes genetic algorithm, Differential Evolution (DE), metaheuristic, and swarm intelligence like particle swarm intelligence and ant colony optimization. Genetic Algorithms give locally optimum solution but get stuck in nonoptimal conditions and suffers from quick convergence. DE does not get stuck in local minima and gives a globally optimum solution. Rate of convergence of DE is also slower than GAs and increases with problem size. We have implemented DE for solving task scheduling problem and results demonstrated significant improvement in the fitness of solution with varying parameters as mutation factor, crossover probability, number of iterations, and population. The main aim of this paper is to visualize the effect of variation in various parameters of DE algorithm on the solution of task allocation problem.

**Keywords** Differential evolution · Evolutionary computation · Task scheduling

## 1 Introduction

Task scheduling is the important step to effectively exploit the capabilities of distributed heterogeneous computing systems. In the optimization process of a difficult task, the method of first choice will usually be a problem-specific heuristics. These techniques using expert knowledge achieve a superior performance. Evolutionary Algorithms (EAs) are capable to overcome the limitations of multiple local minima

S.S. Thakur · S. Singh · P. Singh (✉) · A. Goyal
ABES Engineering College, Ghaziabad, India
e-mail: pratibha.singh@abes.ac.in

A. Goyal
e-mail: abhishek.goyal@abes.ac.in



S.K. Bhatia et al. (eds.), *Advances in Computer and Computational Sciences*,
Advances in Intelligent Systems and Computing 553,
DOI 10.1007/978-981-10-3770-2_47

and unknown system parameters. EAs are conventional direct search algorithm generating the variations of design parameter vectors and once a variant is generated, then the new parameter vector is accepted or not depending on the fact whether or not the new parameters increases the objective value. This method is a typical greedy search method which converges fast but has a disadvantage of getting trapped into local minima which can be reduced by Differential Evolution (DE) algorithms. It generates several vectors simultaneously. One of the most popular EAs is genetic algorithm. Although many versions of genetic algorithm have been developed, they are still time consuming. It was observed from the simulation results that convergence of DE is much better than various versions of GA [1]. Therefore, in this paper we are focusing on visualizing the affects of variation of various parameters of differential evolution algorithm for getting an optimized result of task scheduling problem.

### *1.1 Problem Introduction*

A task is defined to be a program segment that can be individually scheduled. The problem of obtaining an optimal assignment of tasks to processors in any distributed environment is well recognized as a NP-hard problem even when the tasks are independent. The problem is much more difficult when these tasks have dependencies because the order of task execution as well as task to processor pairing affects overall completion time. Dynamic tasks assignment assumes a continuous stochastic stream of incoming tasks. There are very little parameters, known in advance for dynamic tasks assignment. Thus dynamic dependent tasks assignment is much more complex than static independent tasks assignment to solve. Also, it is the most desired because of the application demand.

## 2 Literature Survey

For proper utilization of heterogeneous computing system, task scheduling is very important. The aim is to provide the method of task assignment such that there should be minimum response time and turnaround time and increased throughput. Differential Evolution is a metaheuristic algorithm like Genetic Algorithms, meaning it aims at finding an optimized solution to a given instance of a problem instead of a general approach which gives optimized solutions in all instances. Differential Evolution is used to address multiprocessor scheduling problem for parallel program represented by Directed Acyclic Graph [2]. The proposed method solves the problem and outperformed various greedy approaches by numerically optimizing priorities of tasks which is a NP-hard problem. The differential scheduler was compared with HEFT (Heterogeneous Earliest Finish Time) and a slightly modified Genetic Algorithm and it has shown improvement on average results.

The approach of novel discrete differential evolution algorithm attained high performance for tasks in scheduling problem with different architecture and computational speed interconnected by high-speed network. DE algorithm, a population-based metaheuristic algorithm for global optimization outperformed over GA in terms of both solution quality and computational time. Improvement of DE over GA is an average of 19.96% [3]. DE has been used for task scheduling in heterogeneous setup of environment minimizing the makespan [4]. An extensive review of five metaheuristic techniques for scheduling namely BAT algorithm, Particle Swarm Optimization (PSO), Ant Colony Optimization (ACO), GA, and League Championship Algorithm (LCA) and its results suggested that a lot of more contribution is required towards the improvement of convergence of metaheuristic algorithms and most importantly in quality of solution [5].

## 3 Proposed Methodology

The main aim of this paper is the optimization of heterogeneous task scheduling in a multiprocessor environment. The approach of focus is called Differential Evolution.

The steps followed during the implementation can be broadly summarized as below:

1. Problem Specification: The input specifications in which the number of tasks, processors, population size, and execution time matrix are specified by the user.
2. Factor Selection: The DE approach is mainly affected by three factors, namely: Crossover factor (F), Crossover probability (CR), and number of iterations (I). These factors must be carefully selected to maximize efficiency.
3. Random Population Generation: A random population of individuals is generated where each individual specifies a unique combination of processor-task mapping and has a corresponding fitness value.
4. Selection: For each individual of the population, a certain number of random individuals (three in our case) are selected which are termed as "parents."
5. Crossover: These "parents" are combined in a predefined manner to generate a "child" individual known as the mutated individual.
6. Replacement: If the child proves to be "fitter" than the parent, then the parent is replaced by the child in the population else the child is discarded (this is Sparta!!!).
7. Iteration: The process is carried out over and over again until a specified fitness value is reached or a certain number of iterations have been performed.
8. Evaluation: The maximum fitness value in the latest population obtained at the end of the process is matched up against the maximum value in the original population, and the resulting gain is calculated.

## *3.1 Algorithm*

1. Input of the number of tasks (t) versus the number of processors (p).
2. Input of the execution time matrix E = tXp (times of each task corresponding to each processor).
3. Input of the population size (s).
4. Input of the number of DE iterations (I).
5. Generation of a random population matrix P = sXt (each element $P_{ij}$ is a processor number allocated to task '$T_j$' under population vector '$V_i$').
6. Calculate the fitness value (simply the reciprocal of the sum of the execution times corresponding to processor $P_{ij}$ with task $T_j$) of each vector and pick the maximum value.
7. Pick three random vectors (a, b, c) from the population corresponding to the first vector ($P_i$ = 1).
8. Also select a random probability rj and random crossover index R and apply crossover formula: If ($r_j$ < CR OR i = R) M = (floor [a + F * abs(b − c)] modulo p) + 1 Where M is the mutated vector and F is the mutation factor ($0 < F \leq 2$).
9. Calculate the fitness values for P1 and M.
10. If fitness value of M exceeds that of P1, replace P1 by M in the population matrix P
11. Apply steps 5–7 for every vector of the population matrix.
12. Calculate the maximum fitness value among all vectors of the latest population matrix obtained at the end of step 8 and calculate the gain % on the basis of formulae:

$$\text{Gain \%} = (\text{final max fitness} - \text{initial max fitness})/\text{initial max fitness} * 100$$

13. Take the final population as the initial population for the next iteration and repeat steps 6–12, I-1 times more.
14. The gain percentage obtained at the end of step 12 is the final gain of the differential evolution algorithm.

## *3.2 Flowchart*

Figure 1a shows the flowchart of initialization of parameters as Input: the number of tasks (t), number of processors (p), execution time matrix (E), population size (s) and number of Iterations (I), F (mutation factor), CR (crossover). Figure 1b shows the part of flowchart of selecting: three random vectors from given population, random crossover probability $r_j$, random crossover, mutated vector and calculating fitness (M), comparing the value to previous population fitness, calculating final max fitness = Max(fitness($P_j$)), j = 1 to s and Gain percentage.

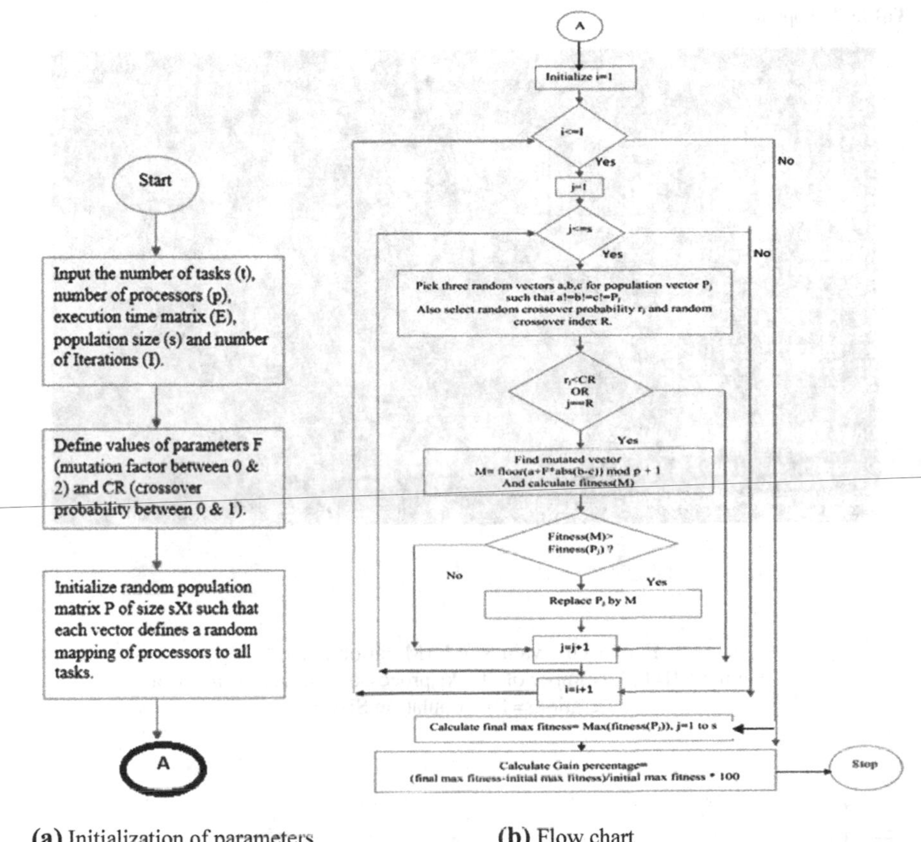


(a) Initialization of parameters (b) Flow chart

**Fig. 1** **a** Initialization of parameters, **b** Flowchart

**Table 1** Execution time matrix

| Processor/Task | $T_1$ | $T_2$ | $T_3$ | $T_4$ | $T_5$ | $T_6$ | $T_7$ | $T_8$ |
|---|---|---|---|---|---|---|---|---|
| P1 | 12 | 45 | 4 | 48 | 75 | 12 | 96 | 23 |
| P2 | 96 | 48 | 52 | 48 | 36 | 96 | 19 | 76 |
| P3 | 1 | 9 | 6 | 7 | 53 | 8 | 86 | 45 |
| P4 | 8 | 5 | 11 | 30 | 81 | 25 | 33 | 17 |
| P5 | 32 | 34 | 35 | 36 | 37 | 62 | 61 | 60 |

# 4 Implementation Results

Software use for the implementation of the above mentioned algorithm is Code Blocks open-source cross platform IDE 13.12. Table 1 is showing execution time matrix which is one of the various possible configurations evaluated and Table 2 is demonstrating initial population generated along with cost calculated and

**Table 2** Population matrix

```
Values for parameters F(mutation factor) & CR(crossover probability)
are taken to be 1.8 & 0.8.

Initial population matrix:
P6 P6 P5 P5 P6 P5 P1 P1 cost=203
P5 P3 P6 P6 P2 P4 P2 P6 cost=202
P2 P3 P4 P1 P4 P1 P3 P4 cost=190
P5 P5 P4 P3 P3 P6 P6 P1 cost=201
P6 P1 P4 P5 P6 P2 P2 P1 cost=195

Initial optimum cost and fitness value:
190 0.00526316

Population matrix at the end of iteration 1:
P5 P6 P2 P1 P1 P6 P2 P5
P5 P3 P5 P5 P4 P3 P2 P5
P2 P3 P4 P1 P4 P1 P3 P4
P1 P1 P4 P1 P3 P2 P3 P2
P6 P1 P4 P5 P6 P2 P2 P1

Optimum allocation matrix:
P1 P1 P4 P1 P3 P2 P3 P2
Optimum (minimum) cost and optimum (maximum) fitness value are:
185 0.00540541
Fitness Gain percentage:
2.7027
```

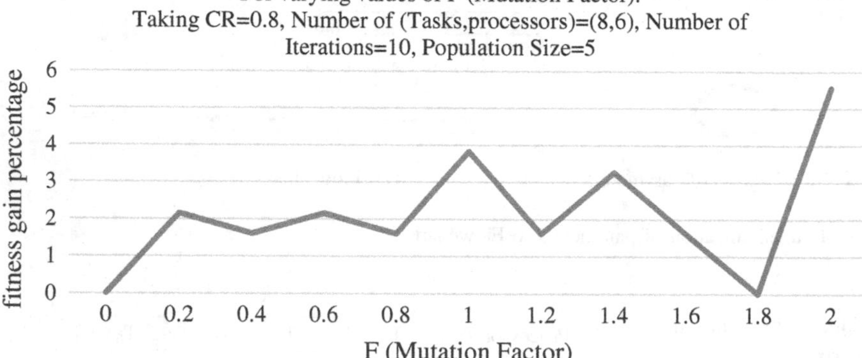


**Fig. 2** Fitness gain percentage for varying values of F (Mutation Factor)

population landed after first iteration. Figure 2 is showing that no defined pattern can be seen in the graph as it zigzags between highs and lows many times in the mutation factor (F) range of 0–2. Thus, the exact relation between F and gain is difficult to estimate and we have to estimate the optimum value of F through trial and error. Figure 3 shows that the fitness gain percentage increases between CR value 0 and 0.3, and then drops to 0 at CR = 0.5. It again increases, attains a uniform value between CR = 0.6 and CR = 0.8, then drops down to 0 at 0.9. Thereafter, it increases. Thus, increasing the CR value increases the gain.

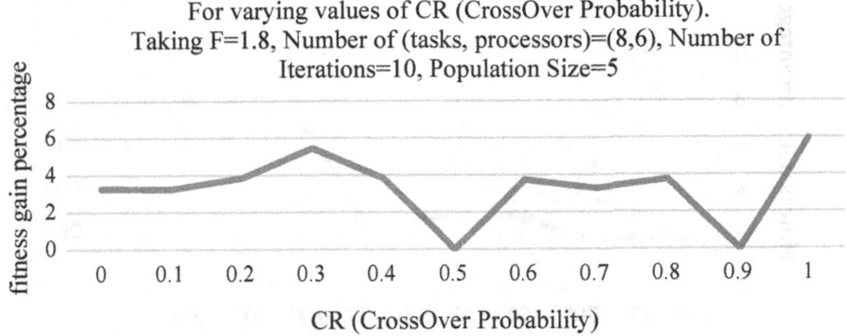


**Fig. 3** Fitness gain percentage for varying values of CR (Crossover Probability)

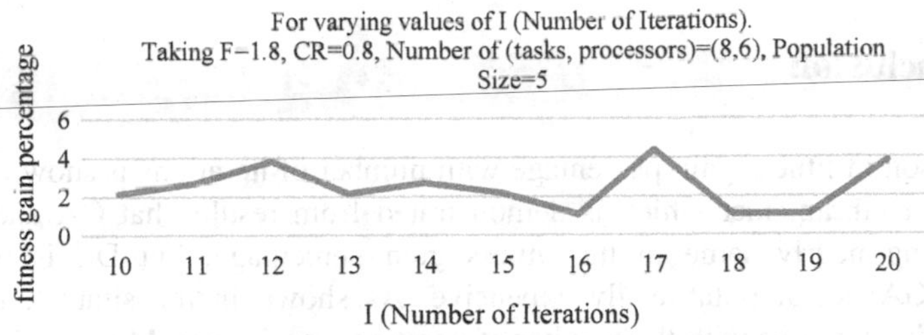


**Fig. 4** Fitness gain percentage for varying values of I (Number of Iterations)

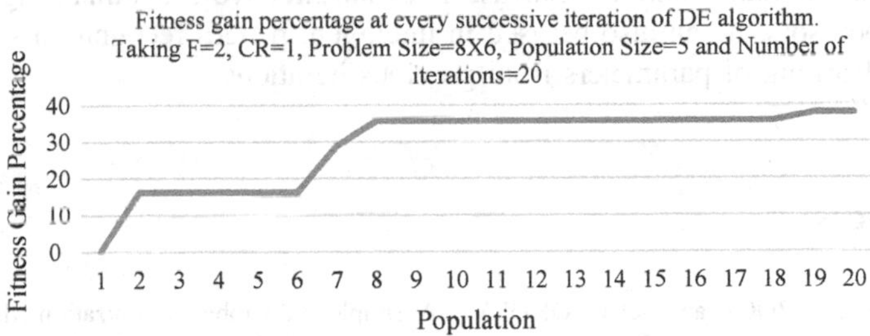


**Fig. 5** Fitness gain percentage at every successive iteration of DE algorithm

Figure 4 is showing the fitness gain percentage increases up to 12 iterations and then gradually decreases up to 1% at 16th iteration. Then it rises up to a maximum of 4.5% at 17th iteration and again drops to 1% at 18th iteration, maintains that value up to 19, and then increases beyond it before saturation state. Thus, increasing the number of iterations increases the gain and finally reaches to the saturation state with maximum fitness. Figure 5 is showing that the fitness gain percentage increases with population and gradually attains a consistent value after a certain population size. Thus, the gain percentage increases and after exhaustive search for the solution, it stops changing.

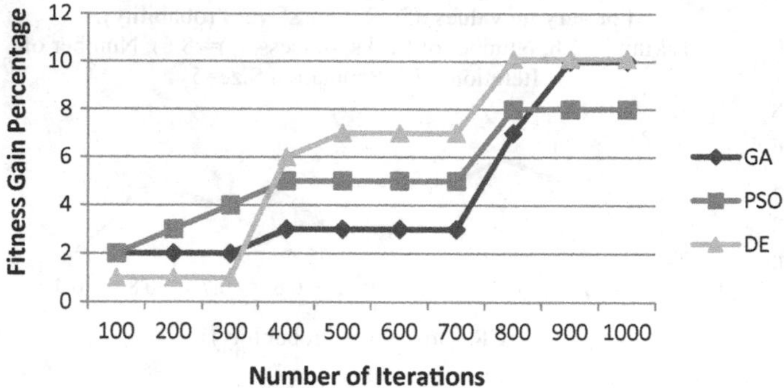


**Fig. 6** Comparison analysis of GA, PSO, DE

# 5 Conclusion

Comparison in fitness gain percentage with number of iterations is shown in Fig. 6 and it is a vibrant fact which is demonstrated from results that GA and DE are contributing nearly same in the fitness gain percentage, but DE is preferable because GA is computationally expensive. As shown in the simulation results, fitness gain increases with the number of iterations, population Mutation Factor (F). Fitness gain is more or less independent of the value of the crossover probability (CR). We can further suggest the selection criterion for the discussed parameters to optimize the solution and we can use it as an effective scheduling algorithm in future processors. It can also be used in intelligent pattern recognition systems by adapting learning of parameters from previous iterations.

# References

1. Dervis KARABOGA and Selcuk ÖKDEM; "A Simple and Global Optimization Algorithm for Engineering Problems: Differential Evolution Algorithm.", Turk J Elec Engin, VOL. 12, NO. 1 2004, TUBITAK
2. Krzysztof Rzadca and Franciszek Seredynski; "Heterogeneous multiprocessor scheduling with differential evolution.", 2005 IEEE Congress on Evolutionary Computation (Volume: 3)., Date of Conference: 2–5 Sept. 2005, Page(s):2840–2847 Vol. 3 ISSN: 1089-778X, Print ISBN: 0-7803-9363-5
3. Qinma Kang and Hong He; "A Novel Discrete Differential Evolution Algorithm for Task Scheduling in Heterogeneous Computing Systems", Proceedings of the 2009 IEEE International Conference on Systems, Man, and Cybernetics San Antonio, TX, USA - October 2009
4. Krömer, Pavel, et al. "Scheduling Independent Tasks on Heterogeneous Distributed Environments by Differential Evolution." *INCoS*. 2009
5. Kalra, Mala, and Sarbjeet Singh. "A review of metaheuristic scheduling techniques in cloud computing." Egyptian Informatics Journal 16.3 (2015): 275–295

# Adaptive Krill Herd Algorithm for Global Numerical Optimization

**Indrajit N. Trivedi, Amir H. Gandomi, Pradeep Jangir, Arvind Kumar, Narottam Jangir and Rahul Totlani**

**Abstract** A recent bio-inspired optimization algorithm, that is, based on the Lagrangian and evolutionary behavior of krill individuals in nature is called the Krill Herd (KH) Algorithm. Randomization has a key role in both exploration and exploitation of a problem using KH algorithm. A new randomization technique termed adaptive technique is integrated with Krill Herd algorithm and tested on several global numerical functions. The KH uses Lagrangian movement which includes induced movement, random diffusion, and foraging motion, and therefore, it covers a vast area in the exploration phase. And then adding the powerful adaptive randomization technique potent the adaptive KH (AKH) algorithm to attain global optimal solution with faster convergence as well as less parameter dependency. The proposed AKH outperforms the standard KH in terms of both statistical results and best solution.

I.N. Trivedi (✉)
Electrical Engineering Department, G.E. College, Gandhinagar, Gujarat, India
e-mail: forumtrivedi@gmail.com

A.H. Gandomi
Department of Civil Engineering, BEACON Center for the Study of Evolution in Action, Michigan State University, East Lansing, MI 48824, USA
e-mail: a.h.gandomi@gmail.com

P. Jangir · N. Jangir
Electrical Engineering Department, LEC, Morbi, Gujarat, India
e-mail: pkjmtech@gmail.com

N. Jangir
e-mail: nkjmtech@gmail.com

A. Kumar
Electrical Engineering Department, S.S.E.C, Bhavnagar, Gujarat, India
e-mail: akbharia8@gmail.com

R. Totlani
Electrical Engineering Department, JECRC, Jaipur, Rajasthan, India
e-mail: rhl.totlani@gmail.com



S.K. Bhatia et al. (eds.), *Advances in Computer and Computational Sciences*, Advances in Intelligent Systems and Computing 553,
DOI 10.1007/978-981-10-3770-2_48

**Keywords** Meta-heuristic · Krill Herd algorithm · Adaptive Krill Herd · Numerical optimization · Benchmark function

## 1 Introduction

In the meta-heuristic algorithms, randomization plays a very important role in both exploration and exploitation. Based on this fact, several randomization techniques such as Markov chains, Levy flights, Gaussian (or normal distribution) random number, and several new techniques have been used in meta-heuristics. In general, meta-heuristic algorithms that integrated with adaptive technique results in less computational time to reach an optimum solution, local minima avoidance, and faster convergence.

Population-based Krill Herd Algorithm [1] is a meta-heuristic optimization algorithm which has an ability to avoid local optima and get a globally optimal solution which makes it appropriate for practical applications without structural modifications in the algorithm for solving different constrained or unconstraint optimization problems. This novel algorithm is based on the Lagrangian and evolutionary behavior of krill individuals in nature. KH is able to do both exploration and exploitation in optimization problem simultaneously. In this algorithm random value plays very important role and, therefore, coupling it with an adaptive technique which changes the positions of current solutions towards global optimum, according to its fitness function, could be beneficial.

Recent trend of optimization is to improve the performance of meta-heuristic algorithms by integrating with chaos theory, Levy flights strategy, adaptive randomization technique, evolutionary boundary handling scheme, and genetic operators such as crossover and mutation. Popular genetic operators have been already used in KH [1] which can accelerate its global convergence speed and improve the best solutions.

There are several adaptive algorithms that have been proposed in the literature [2–5]. Adaptive Cuckoo Search Algorithm (ACSA) [6–8] is a new algorithm proposed recently. In the ACSA-related paper, the authors have compared Cuckoo Search Algorithm (CSA) that is purely based on Levy Flights Distribution with ACSA. Adaptive Cuckoo Search Algorithm outperformed CSA and it is faster in terms of convergence rate as well as it requires less parameters compared to Levy flight-based Cuckoo search algorithm.

In this paper, the performance of KH is improved by incorporating an adaptive technique and a new method is proposed which is called Adaptive KH (AKH). The results show that implementation of adaptive technique into KH reduces the computational times for benchmark problems.

The remainder of this paper is organized as follows: The next Section describes the Krill Herd algorithm and its algebraic equations are given in Sect. 2. Section 3 includes a description of the Adaptive technique. Section 4 includes Numerical optimization and Sect. 5 consists of simulation results of unconstrained benchmark test function, convergence curve and tables of results compared with Krill Herd algorithm.

## 2 Krill Herd Algorithm

The Krill Herd (KH) algorithm [1] was first proposed by Gandomi and Alavi in 2012. KH algorithm is based on nature inspiration of the mimics the behavior of krill individuals in krill herds. KH algorithm is inspired from activities of krill such as:

i. movement induced by other krill individuals;
ii. foraging activity;
iii. random diffusion.

This optimization algorithm has a capability of searching of unknown search space.

- Lagrangian model is generalized to an $n$-dimensional decision space:

$$\frac{dX_i}{dt} = N_i + F_i + D_i \tag{1}$$

where $N_i$ the motion is induced by other krill individuals; $F_i$ is the foraging motion, and $D_i$ is the physical diffusion of the $i$th krill individuals.

1. The induced movement expresses the density preservation of herd by every individual. The algebraic equation expresses this behavior formulated as:

$$N_i^{next} = N^{max}\alpha_i + \omega_n N_i^{present} \tag{2}$$

$$\alpha_i = \alpha_i^{local} + \alpha_i^{target} \tag{3}$$

where $N^{max}$ is highest induced speed, $\omega_n$ expresses inertia weight, $\alpha_i^{local}$ and $\alpha_i^{target}$ express the local effect of neighbors, best solution directs respectively of $i$th individual.

$\alpha_i^{target}$ is formulated by equation:

$$\alpha_i^{target} = C^{best}\hat{K}_{i,best}\hat{X}_{i,best} \tag{4}$$

$$C^{best} = 2\left(r_1 + \frac{I}{I_{max}}\right) \tag{5}$$

where $C^{best}$ is the effective coefficient of the krill individual with the best fitness to the $i$th krill individual, $r_1$ is a random number having values in between 0 and

1 and it is for enhancing exploration, $I$ is the current iteration number, and $I_{max}$ is maximum iterations.

2. Foraging activity/motion is mathematically calculated as:
   The foraging motion is formulated with two main parameters: first is the food location and second is the previous experience about the food location.

$$F_i^{next} = V_f \beta_i + \omega_f F_i^{previous} \tag{6}$$

$$\beta_i = \beta_i^{food} + \beta_i^{best} \tag{7}$$

   and $V_f$ is the foraging speed, $\omega_f$ is the inertia weight of the foraging motion in the range [0, 1], $F_i^{previous}$ is the last foraging motion, $\beta_i^{food}$ is the food attractive, and $\beta_i^{best}$ is the effect of the best fitness of the $i$th krill so far. According to the measured values of the foraging speed, it is taken as 0.02 ($ms^{-1}$).

3. Random/Physical diffusion
   Physical diffusion is calculated in terms highest diffusion speed and a random directional vector given as:

$$D_i = D^{max} \delta \tag{8}$$

$$D_i = D^{max} \left(1 - \frac{I}{I_{max}}\right) \delta \tag{9}$$

   where $D^{max}$ is highest induced speed and $\delta$ is random directional vector [0, 1]. Finally, position of every krill is updated as

$$X_i^{next} = X_i^{current} + \Delta x_i(t) \tag{10}$$

$$\Delta x_i(t) = N_i(t) + F_i(t) + D_i(t) \tag{11}$$

## 3 Adaptive Krill Herd Algorithm

In the meta-heuristic algorithms, randomization plays a very important role in both exploration and exploitation where more randomization techniques are Markov chains, Levy flights and Gaussian or normal distribution and a new technique is adaptive technique. One of the adaptive techniques recently proposed by Ong in Cuckoo Search Algorithm [6] and the results show a clear improvement in performance of CSAs. The Adaptive technique [7] includes best features like it consists of less parameter dependency, not required to define the initial parameter and step size or position towards an optimum solution is adaptively changes according to its functional fitness value over the course of the iteration. Therefore, KH algorithm integrated with an adaptive technique called Adaptive Krill Herd Algorithm (AKH) to improve the KH. This adaptive technique can reduce the

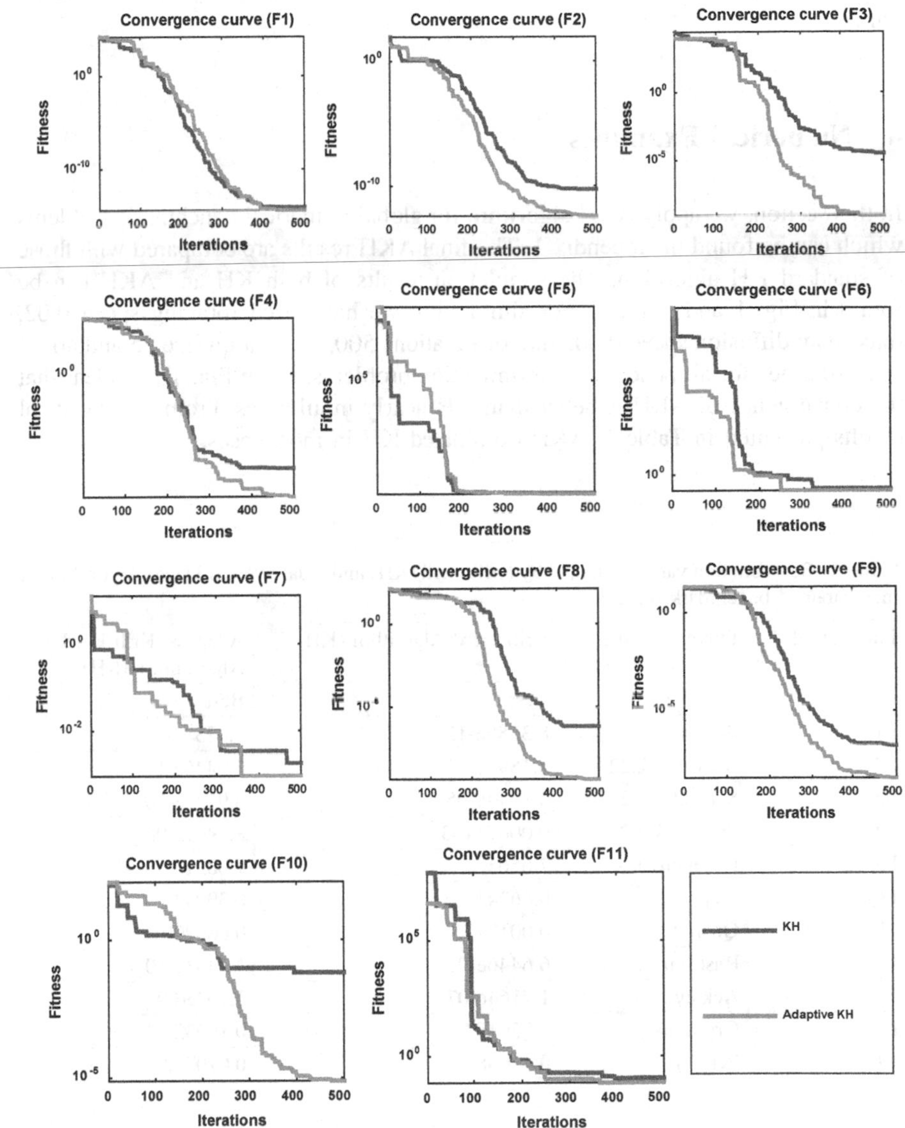


**Fig. 1** Function values versus iterations for the Adaptive Krill Herd (AKH) Algorithm and Krill Herd (KH) Algorithm

computational time to reach an optimum solution, avoid local minima, and have faster convergence. The adaptive strategy for KH is formulated as:

$$X_i^{t+1} = X_i^t + randn * \left(\frac{1}{t}\right)^{|((bestf(t) - fi(t))/(bestf(t) - worstf(t)))|} \quad (12)$$

where $X_i^{t+1}$ a new solution of $i$th dimension in the $t$-th iteration $f(t)$ is the fitness value.

## 4 Numerical Examples

In this section, we apply AKH algorithm for global numerical benchmark problems which can be found in Appendix 1. The final AKH results are compared with those of standard KH algorithm. The simulation results of both KH and AKH can be found in Fig. 1 and Table 1. For simulation, we have used foraging speed 0.02, maximum diffusion speed 0.005, no. of iterations 500, population size 25 and no. of runs 10 times for all benchmark optimization problems. From Fig. 1 it is clear that the convergence of AKH is better than KH nearly in all cases. From the statistical results presented in Table 1, AKH dominated KH in most cases.

**Table 1** Best function values obtained by Krill Herd (KH) and Adaptive Krill Herd Algorithm for unconstrained benchmark function

| Function id | Function name | Krill Herd Algorithm (KH) | Adaptive Krill Herd Algorithm (AKH) |
|---|---|---|---|
| | | Best | Best |
| F1 | Sphere | 8.3555e-15 | 4.0258e-15 |
| F2 | Schwefel 2.22 | 5.1897e-11 | 3.9446e-13 |
| F3 | Schwefel 1.2 | 3.3344e-05 | 8.9735e-10 |
| F4 | Schwefel 2.21 | 0.00071183 | 8.1901e-05 |
| F5 | Rosenbrock's | 7.6561 | 7.0821 |
| F6 | Step | 0.46241 | 0.39943 |
| F7 | Quartic | 0.0017773 | 0.0009001 |
| F8 | Rastrigin | 6.6446e-07 | 4.2483e-10 |
| F9 | Ackley's | 1.7164e-07 | 3.702e-09 |
| F10 | Griewank | 0.25018 | 0.16032 |
| F11 | Penalty 1 | 0.12648 | 0.070762 |

## 5 Conclusion

Randomization plays an important role in both exploration and exploitation phase in every efficient algorithm. The results show that adaptive Krill Herd algorithm (AKH) has the ability to find out an optimum solution for global numerical optimization problems. Also, Adaptive technique in Krill Herd causes faster convergence, randomness, and stochastic behavior for improving solutions in comparison with basic KH algorithm. The adaptive technique was also used for random walk in search space when no neighboring solution exits to converse towards optimal solution, therefore, the AKH results prove that it is also an effective method for various unconstrained problems in unknown search space.

For future scope it is recommended to integrate AKH algorithm with other optimization algorithms, build up binary version with different S-shaped and V-shaped transfer function, multi-objective version, chaotic version as well as testing it with real challenging engineering optimization problems.

## Appendix 1

**F1: Sphere Function**

n (dimension): 10, lb: −100, ub: 100, fmin: 0

$$f(x) = \sum_{i=1}^{n} x_i^2 * R(x)$$

**F2: Schwefel 2.22 Function**

n (dimension): 10, lb: −10, ub: 10, fmin: 0

$$f(x) = \sum_{i=1}^{n} |x_i| + \prod_{i=1}^{n} |x_i| * R(x)$$

**F3: Schwefel 1.2 Function**

n (dimension): 10, lb: −100, ub: 100, fmin: 0

$$f(x) = \sum_{i=1}^{n} \left( \sum_{j-1}^{i} x_j \right)^2 * R(x)$$

**F4: Schwefel 2.21 Function**

n (dimension): 10, lb: −100, ub: 100, fmin: 0

$$f(x) = \max_i \{|x_i|, 1 \le i \le n\}$$

**F5: Rosenbrock's Function**

n (dimension): 10, lb: −30, ub: 30, fmin: 0

$$f(x) = \sum_{i=1}^{n-1}\left[100\left(x_{i+1} - x_i^2\right)^2 + (x_i - 1)^2\right] * R(x)$$

**F6: Step Function**

n (dimension): 10, lb: −100, ub: 100, fmin: 0

$$f(x) = \sum_{i=1}^{n} ([x_i + 0.5])^2 * R(x)$$

**F7: Quartic Function**

n (dimension): 10, lb: −1.28, ub: 1.28, fmin: 0

$$f(x) = \sum_{i=1}^{n} ix_i^4 + random[0, 1) * R(x)$$

**F8: Rastrigin Function**

n (dimension): 10, lb: −5.12, ub: 5.12, fmin: 0

$$F(x) = \sum_{i=1}^{n} \left[x_i^2 - 10\cos(2\pi x_i) + 10\right] * R(x)$$

**F9: Ackley's Function**

n (dimension): 10, lb: −32, ub: 32, fmin: 0

$$F(x) = -20\,exp\left(-0.2\sqrt{\frac{1}{n}\sum_{i=1}^{n} x_i^2}\right) - exp\left(\frac{1}{n}\sum_{i=1}^{n}\cos(2\pi x_i)\right) + 20 + e * R(x)$$

**F10: Griewank Function**

n (dimension): 10, lb: −600, ub: 600, fmin: 0

$$F(x) = \frac{1}{4000}\sum_{i=1}^{n} x_i^2 - \prod_{i=1}^{n}\cos\left(\frac{x_i}{\sqrt{i}}\right) + 1 * R(x)$$

**F11: Penalty 1 Function**

n (dimension): 10, lb: −50, ub: 50, fmin: 0

$$F(x) = \frac{\pi}{n}\left\{\begin{array}{l} 10\sin(\pi y_1) + \sum_{i=1}^{n-1}(y_i - 1)^2 \\ \left[1 + 10\sin^2(\pi y_{i+1})\right] + (y_n - 1)^2 \end{array}\right\}, y_i = 1 + \frac{x_i + 1}{4}$$

$$u(x_i, a, k, m) = \begin{cases} k(x_i - a)^m & x_i > a \\ 0 & -a < x_i < a \\ k(-x_i - a)^m & x_i < -a \end{cases}$$

## References


1. A.H. Gandomi, A.H. Alavi, Krill Herd: a new bio-inspired optimization algorithm, Common Nonlinear Sci. Numer. Simul. 17 (12) (2012) 4831–4845.
2. Das, S. Mandal, A. Mukherjee, R. An Adaptive Differential Evolution Algorithm for Global Optimization in Dynamic Environments. IEEE Transactions on Cybernetics, 44, 6, 966–978, 2014.
3. Costa L, Oliveira P, An Adaptive Sharing Elitist Evolution Strategy for Multiobjective Optimization. Evolutionary Computation, 2003, 11, 4, 417–438.
4. Dai Y, Li Y; Wei L; Wang J; Zheng D. Adaptive immune-genetic algorithm for global optimization to multivariable function. Journal of Systems Engineering and Electronics, 18, 3, 655–660, 2007.
5. Lim WH, Isa NAM, An adaptive two-layer particle swarm optimization with elitist learning strategy. Information Sciences, 273, 49–72, 2014.
6. P. Ong, "Adaptive Cuckoo search algorithm for unconstrained optimization," The Scientific World Journal, Hindawi Publication, vol. 2014, pp. 1–8, 2014.
7. Naik MK, Panda R, A novel adaptive cuckoo search algorithm for intrinsic discriminant analysis based face recognition, Applied Soft Computing, 38, 661–675, 2016.
8. A.H. Gandomi, X.S. Yang, S. Talatahari, A.H. Alavi, Metaheuristic Applications in Structures and Infrastructures, Elsevier, 2013.

# Sliding Mode Control of Uncertain Nonlinear Discrete Delayed Time System Using Chebyshev Neural Network

**Parmendra Singh, Vishal Goyal, Vinay Kumar Deolia and Tripti Nath Sharma**

**Abstract** This paper investigates a Chebyshev Neural Network (CNN) sliding mode controller for stabilization of time-delayed version of system with uncertainty and nonlinearity. The nonlinearity in the system is unknown but bounded and has been approximated with the help of CNN. The input delay has been balanced and further converted into regular form and the original system is converted into a delayed free version with the help of Smith Predictor. Now, the predicted states of the system and "Gao's reaching law" are used to derive the robust control law. Further, to prove the stability analysis Lyapunov–Krasovskii candidates has been chosen according to the proposed system. A numerical example is provided to illustrate the stability of the system in the presence of uncertainty, time delay and nonlinearity.

**Keywords** Chebyshev neural network · Sliding mode control · Smith predictor

## 1 Introduction

The real-time dynamical systems are not easy to deal within the presence of uncertainties, nonlinearity, disturbances and time delay. So, the dynamics of the system has been affected by the uncertainties, nonlinearity and disturbances that should be taken into account in the design of the controller as it degrade the system performance and sometimes may tend the system towards instability. The uncertainties and nonlinearity have been included in the real-world control systems due to inaccuracy in modelling, errors in measurement and some unavoidable external conditions. Another real-time problem that should also be taken into account is time delay which can be frequently found in real physical systems such as biological systems, aircrafts, rolling mills and economic systems and the main cause of time

P. Singh · V. Goyal (✉) · V.K. Deolia · T.N. Sharma
Department of Electronics and Communication Engineering,
GLA University, Mathura, India
e-mail: vishal.glaitm@gmail.com



S.K. Bhatia et al. (eds.), *Advances in Computer and Computational Sciences*,
Advances in Intelligent Systems and Computing 553,
DOI 10.1007/978-981-10-3770-2_49

delay is the "dead-time" or "transportation delay". Controlling the delayed systems is a challenging task as an achievable control bandwidth is limited [1] and closed-loop stability of the system is also affected. Time delay can appear in the system in the form of state delay, system having input delay and system having both input and state delays. The time delay existence can lead to instability in systems. It is difficult to obtain significant performance and attain stability under the existence of time delay compared to delayed free systems. One of the practical applications of having input delay is Jacket-Type Offshore Structure [29]. Over the past years, the stability analysis and control of time delay in the systems have received significant attention from both mathematical and engineering communities. Many efforts have been done by researchers to address the sliding mode control (SMC) problem for systems having time delay [2–4]. The attractive feature of SMC is its insensitivity towards uncertainties and disturbances in the system but chattering is the major problem [5]. If the boundary layer is considered in the vicinity of the sliding surface, the chattering can be removed and this can be done by incorporating a saturation function in the control law. Due to widespread digitalization, concentration of many researchers has shifted from continuous time SMC to discrete time SMC based on state-space models [6, 7]. Any practical system can be represented mathematically by taking the help of state-space model as it incorporates initial conditions of the system.

In the literature, various approaches such as Riccati equation approach [8], Lyapunov min-max approach [9] and Linear Matrix Inequality approach (LMI) [10–13] have been used in stabilizing uncertain systems having time delay but in these techniques time delay is not compensated. Some of the SMC methods are proposed for state-delayed uncertain systems [25, 26] but they do not guarantee robustness of the system as predictor is not used to compensate the state delay. A predictor to compensate the state delay is proposed in [27]. Nowadays many researchers' deal with predictor-based controller [1, 3, 13]. In [1] the input delay effect is minimized by using a predictor and control law has also been derived to provide sliding mode existence. Smith Predictor has capability to stabilize the system effectively under the presence of input delay, $H_\infty$ control is widely used in stability analysis with multiple delayed inputs [28, 29]. Multilayer neural network schemes show effective results for unknown nonlinear functions over a past few years. Neural network is a powerful tool to approximate the unknown nonlinear systems. In [20], the control results show the effectiveness of Chebyshev Neural Network (CNN). In [15], the transformation technique is proposed to obtain the Chebyshev polynomial neural networks for recurrent/feedforward neural networks by approximating the Chebyshev polynomials. The recursive least square method with forgetting factor is used as a learning algorithm for CNN. The computational time is less for CNN compared to MLP due to its single layer structure.

The motivation of this work is to design a predictor which compensates the input time delay to make the analysis easy. This paper put forwards a sliding mode controller for discrete-time systems consisting of nonlinearity, uncertainty and time delay. The time delay in the input is taken care by Smith Predictor. Then, the analysis is carried forward by taking delayed free system. CNN approximated the unknown but bounded nonlinear function. CNN becomes adaptive by obtaining a weight update law. The stability of the system is proved by selecting proper Lyapunov–Krasovskii functional candidate.

The paper is continued as follows. Chebyshev Neural Network (CNN) is depicted in Sect. 2. Problem is formulated in Sect. 3. The predicted state of the system is described in Sect. 4. A controller is designed in Sect. 5. Section 6, provides the analysis to prove the stability of the system. The validation of the proposed system is shown by an example through simulation results in Sect. 7. At the end, a conclusion is remarked in Sect. 8.

## 2 CNN Structure

Artificial neural network (ANN) comes into view as a robust method to do complicated work in an extremely nonlinear environment [16]. The major advantage of ANN model is to approximate nonlinear functions. Multilayer feedforward network (MLP) architecture is usually used in ANNs, and for network training backpropagation technique is used. As MLP uses multilayer structure, the speed of training is typically inferior as compared to single-layer network. The disadvantage of MLP is the problems like weight interference and local minima trapping occurs so to avoid these problems functional link neural network (FLNN) is used [17]. FLNN technique is modest when compared to MLP as it is a single-layer neural network. Chebyshev neural network (CNN) used in this paper is a FLNN based on Chebyshev polynomials. CNN's architecture contains two parts: learning and numerical transformation part [18] as shown in Fig. 1. The FE (Functional Expansion) is numerical transformation to input patterns containing finite Chebyshev polynomials. It is obtained by a recursive formula [18]:

$$P_{i+1}(x) = 2xP_i(x) - P_{i-1}(x), \quad P_i(x) = 1 \tag{1}$$

where Chebyshev polynomials is shown by $P_i(x)$.

**Fig. 1** CNN model [19]

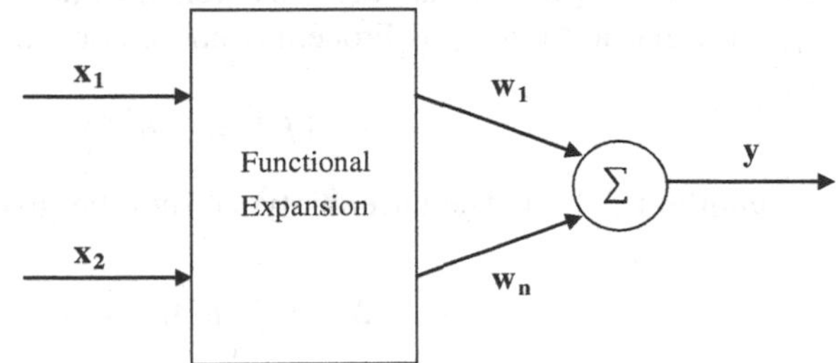

The output of single layer CNN is given by

$$\hat{g}(x) = \hat{w}^T \phi \tag{2}$$

where the NN weights are $w$ and $\phi$ as the appropriate basis function of NN. The unknown nonlinear function $g(x)$ can be approximated by using ideal weights $w$ based on approximation property of CNN as

$$g(x) = w^T \phi + \varepsilon \tag{3}$$

where $\varepsilon$ is a CNN reconstruction error vector and is bounded.

## 3 Problem Formulation

Generally, most of the practical systems consist of time delay, uncertainties caused due to improper modelling and unknown nonlinearity. So, an input delay system with uncertainties and nonlinearity is considered in discrete-time state model as

$$x(k+1) = (A + \Delta A)x(k) + Bg(x(k))u(k-h) \tag{4}$$

where $x(k) \in R^m$, $u(k) \in R^n$ and $g(x(k))$ are the state, control input and unknown nonlinear function respectively. $A$ and $\Delta A$ are real constant matrix (known) having proper dimensions and known real matrix representing parametric uncertainties respectively. $h$ is a constant positive delay.

**Assumption 1** [20]: The input time delay $h$ used in the system (4) is constant for which the upper and lower bound are same.

**Assumption 2** [4]: $A$ and $B$ are controllable in nominal form (4) and of full rank. Controllable means it is possible to transform the existing state of the system to any desired state by applying input within a finite time of interval.

**Assumption 3** [20]: ΔA and $g(x(k))$ are the uncertainty and nonlinear function that is unknown but bounded.

**Assumption 4** [20]: For an ideal NN weight $w$, there is a positive constant bound $w_M$ such that $w \leq w_M.\|.\|$ is Frobenius norm. For a matrix $B$, Frobenius norm is

$$\|B\|^2 = tr(B^T B) \tag{5}$$

**Assumption 5** [21]: The uncertainty satisfies the given assumptions:

$$\Delta A = \sum_{i-1}^{p} \alpha_i A_i \,, \; |\alpha_i| \leq 1 \tag{6}$$

where $\alpha_i$ is unknown scaling parameter but its range is known.

From assumption Eq. (5) suppose $A_i = K_i H_i$, $K_i \in R^{n\times m}$ and $H_i \in R^{n\times m}$, then

$$\Delta A = \sum_{i-1}^{p} \alpha_i A_i = KDH \tag{7}$$

where $K = [K_1 \ldots K_n]$, $H = \left[H_1^T \ldots H_p^T\right]$ and $D = blockdiag\left[\alpha_1 I_{n_1\times n_1} \cdots \alpha_p I_{n_p\times n_p}\right]$.

From Assumption 2, a non-singular matrix $N \in R^{n\times n}$ exists such that

$$NB = \begin{bmatrix} 0_{(n-m)\times m} \\ B_2 \end{bmatrix} \tag{8}$$

where $B_2 \in R^{m\times m}$ is a non-singular matrix. Suppose, a state transformation matrix can be taken as

$$N = \begin{bmatrix} U_2^T \\ U_1^T \end{bmatrix} \tag{9}$$

where $U_1 \in R^{n\times m}$ and $U_2 \in R^{n\times(n-m)}$ are two subordinates of unitary matrix and results by using singular value decomposition of $B$ is given by

$$B = [U_1 \quad U_2] \begin{bmatrix} \Sigma \\ 0_{(n-m)\times m} \end{bmatrix} V^T \tag{10}$$

where $\Sigma \in R^{n\times m}$ and $V \in R^{m\times m}$ are a diagonal positive definite and unitary matrices. Due to improper modelling which is unavoidable, the system parameters get deviated from its original values. So, the transformation of a system is needed to get the regular form of the system. Now the original system is transformed into an appropriate regular form with the help of state transformation [24]. On applying state transformation $z = N\hat{x}$, where $N$ is a regular coordinate transformation $N \in R^{n\times m}$ the system (4) can be written as

$$z(k+1) = (\overline{A} + \Delta\overline{A})z(k) + \begin{bmatrix} 0_{(n-m)\times m} \\ B_2 \end{bmatrix} (g(z(k))u(k)) \tag{11}$$

where $\overline{A} = NAN^{-1}, \Delta\overline{A} = N\Delta AN^{-1}, NB = [0 \quad B_2]^T$ and $z = [z_1 \quad z_2]^T$.

The system (11) can be rewritten as

$$z_1(k+1) = (\overline{A_{11}} + \Delta\overline{A_{11}})z_1(k) + (\overline{A_{12}} + \Delta\overline{A_{12}})z_2(k) \tag{12}$$

$$z_2(k+1) = (\overline{A_{21}} + \Delta\overline{A_{21}})z_1(k) + (\overline{A_{22}} + \Delta\overline{A_{22}})z_2(k) + B(g(z(k))u(k) \tag{13}$$

where

$z_1, z_2 \in R^m, B_2 = \Sigma V^T, \overline{A_{11}} = U_2^T A U_2, \Delta\overline{A_{11}} = U_2^T KDH_1 U_2, \overline{A_{12}} = U_2^T A U_1, \Delta\overline{A_{12}} = U_2^T KDH_1 U_1, \overline{A_{21}} = U_1^T A U_2, \Delta\overline{A_{21}} = U_1^T KDH_1 U_2, \overline{A_{22}} = U_1^T A U_1, \Delta\overline{A_{22}} = U_1^T KDH_1 U_1$

## 4 Predicted SMC

SMC design methodology is divided into two steps: (i) to construct linear sliding manifold (ii) to design control law. So, a system containing uncertainties, unknown nonlinear function and time delay, few special measures has to be taken. If time delay has been introduced in an optimally tuned system then to maintain stability, the gain should be reduced. The remedy is in the Smith predictor algorithm [14] which allows larger gain and avoids poor performance. The Smith Predictor may provide inferior performance in the case of imperfect modelling of the system. Time delay in closed loop causes instability to the system so by using the Smith Predictor, time delay present in closed loop is converted into an open-loop time delay, which is not a matter of concern and does not affect the stability of the system. Smith predictor is helpful in removing the time delay present in closed loop of the system. The design steps of Smith predictor in discrete time is given in [22].

For the sake of simplicity, assume matrix A to be a non-singular matrix. The predicted states $\hat{x}(k) \in R^n$ of the system (4) is defined as

$$\hat{x}(k) = (A + \Delta A)^h + \sum_{i=-k+1}^{0} ((A + \Delta A)B)^{-1} g(x(k + h + i - 1)) \qquad (14)$$

**Proposition 1** *The predictor's dynamics* (14) *for system* (4) *can be easily defined by the system matrices A and B as*

$$\hat{x}(k+1) = (A + \Delta A)\hat{x}(k) + Bg(\hat{x}(k))u(k) \qquad (15)$$

*Proof* The system (4) is applied to the predictor (14) and can be written as

$$\begin{aligned}
\hat{x}(k+1) &= (A + \Delta A)^h x(k+1) + \sum_{i=-k+1}^{0} ((A + \Delta A)B)^{-1} g(x(k+h+i))u(k+i) \\
&= (A + \Delta A)^h ((A + \Delta A)x(k) + Bg(x(k))u(k-h)) \\
&\quad + \sum_{i=-k+1}^{0} ((A + \Delta A)B)^{-1} g(x(k+h+i))u(k+i) \\
&= (A + \Delta A)^h ((A + \Delta A)^h x(k) + \sum_{i=-k+1}^{0} ((A + \Delta A)B)^{-1}) g(x(k+h+i-1))u(k+i-1) \\
&\quad + Bg(x(k+h))u(k)
\end{aligned}$$

$$\hat{x}(k+1) = (A + \Delta A)\hat{x}(k) + Bg(\hat{x}(k))u(k) \tag{16}$$

*Remark 1* After introducing the predictor (14), the system (4) containing delay is converted into delayed free system. Now, with the help of new predicted states the controller is designed. To apply Smith predictor algorithm there is a need of known time delay [3].

## 5 Controller Design

The first step in SMC design methodology is to design sliding manifold and is given as

$$s(k) = \overline{C}z(k) = Cz_1(k) + z_2(k) = 0 \tag{17}$$

where $\overline{C} \in R^{m\times m}$ and $C \in R^{m\times(n-m)}$ are matrices with proper dimensions.

First linear sliding surface is selected such that it is asymptotically stable and then second control law (21) is designed so that it compels the trajectory of the system towards chosen linear sliding surface within a finite time and be there for all time.

$s(k+1)$ can be represented as

$$s(k+1) = \overline{CA}z(k) + \overline{C}\Delta\overline{A}z(k) + \overline{C}B_2 g(z(k))u(k) \tag{18}$$

Subtracting Eqs. (17) from (18) to get

$$s(k+1) - s(k) = \overline{C}(\overline{A} - I)z(k) + \overline{C}B_2 g(z(k))u(k) + f_{di}(k) \tag{19}$$

where $f_{di}(k) = diag(\text{sgn}(s_i))f_d(k)$, $f_d(k) = \sum_{i=1}^{p} \left|\overline{C}TA_iT^{-1}z(k)\right|$ and $f_{di}(k)$ includes all the uncertain terms.

Now, consider Gao's reaching law [23], which is helpful in obtaining control law as it satisfies all the characteristics of the SMC and in discrete time it is given as

$$s(k+1) - s(k) = -qTs(k) - \varepsilon T\text{sgn}(s(k)) \tag{20}$$

where $q > 0$, $\varepsilon > 0$, $(1 - qT) > 0$. $T > 0$ is a sampling time, $\varepsilon$ is the reaching rate and $q$ is an approximation rate.

On comparing Eqs. (19) and (20) and computing the control law $u(k)$

$$u(k) = (1/B_2C\hat{g}(z(k))) \times (qTs(k) + \varepsilon T\text{sgn}(s(k)) + \overline{C}(\overline{A} - I)z(k) - diag(\text{sgn}(s_i))f_d(k)) \tag{21}$$

## 6 Stability Analysis

**Theorem 1** *The system (4) with Assumptions* 1–5 *and designed control law (21), an estimated NN weights are given by:*

$$\|\hat{w}(k+1)\| = \|\hat{w}(k)\| + (z^T(k)\,(\overline{A}-I)^T G^T PG\,(\overline{A}-I)\,z(k))^{1/2}$$
$$+ (z^T(k)(\overline{A}-I)^T P\,(\overline{A}-I)z\,(k))^{1/2} + (z^T(k)\tilde{A}^T P\tilde{A}z(k))^{1/2}$$
$$+ (z^T(k)\,(\overline{A}-I)^T PG\,(\overline{A}-I)z(k))^{1/2} + (z^T(k)(\overline{A}-I)^T G^T P(\overline{A}-I)z(k))^{1/2} \tag{22}$$

*With*

$$H(k) < 0 \tag{23}$$

*where*

$$H(k) = z^T(k)(-\tilde{A}^T PC^{-1} - \tilde{A}^T PGC^{-1} + (\overline{A}-I)^T PC^{-1} + (\overline{A}-I)^T PGC^{-1} + (\overline{A}-I)^T G^T PC^{-1}$$
$$+ (\overline{A}-I)^T G^T PGC^{-1})\mu + \mu^T(-(C^{-1})^T P\tilde{A} + (C^{-1})^T P(\overline{A}-I) + (C^{-1})^T PG\,(\overline{A}-I)$$
$$-(C^{-1})^T G^T P\tilde{A} + (C^{-1})^T G^T P\,(\overline{A}-I) + (C^{-1})^T G^T PG\,(\overline{A}-I))\,z(k)$$
$$+\mu^T(-(C^{-1})^T PC^{-1} + (C^{-1})^T PGC^{-1} + (C^{-1})^T G^T PC^{-1} + (C^{-1})^T G^T PGC^{-1})\mu$$

*where* $P$ *and* $G$ *are* $m \times m$ *positive definite matrix* $G = \tilde{g}(z(k))/\hat{g}(z(k))$, $\mu = qTs\,(k)\;\varepsilon T + \mathrm{sgn}(s(k)) + f_{di}(k)$ and $\tilde{A} = \overline{A} + \Delta\overline{A}$. *Thus on correctly selecting the design parameters, the trajectory would be on defined linear sliding manifold for all time.*

*Proof* The Lyapunov function chosen is

$$V(k) = V_1(k) + V_2(k) \tag{24}$$

where

$$V_1(k) = z^T(k)P\,z(k),\; V_2(k) = tr(\tilde{w}^T(k)\tilde{w}(k))$$
$$V(k) = z^T(k)P\,z(k) + tr(\tilde{w}^T(k)\tilde{w}(k)) \tag{25}$$

The incremental change in Lyapunov function is known as the energy function and for the system to be stable the incremental change in the energy function should be negative and is given by

$$\Delta V(k) = z^T(k+1)Pz(k+1) + tr(\tilde{w}^T(k+1)\tilde{w}(k+1)) - z^T(k)Pz(k) - tr(\tilde{w}^T(k)\tilde{w}(k)) \tag{26}$$

$$\Delta V(k) = (\tilde{A}z(k) + B_2\hat{g}(z(k))u(k) + B_2\tilde{g}(z(k))u(k)^T P(\tilde{A}z(k) + B_2\hat{g}(z(k))u(k) + B_2\tilde{g}(z(k))u(k) + tr(\tilde{w}^T(k+1)\tilde{w}(k+1)) - z^T(k)Pz(k) - tr(\tilde{w}^T(k)\tilde{w}(k)) \tag{27}$$

Putting the control law in (21) and Assumption 4, then after some manipulations

$$\begin{aligned}\Delta V(k) = {} & z^T(k)\tilde{A}^T P\tilde{A}z(k) - z^T(k)Pz(k) - z^T(k)\tilde{A}P(\overline{A}-I)z(k) - z^T(k)(\overline{A}-I)^T P\tilde{A}z(k) \\ & - z^T(k)\tilde{A}^T PG(\overline{A}-I)z(k) - z^T(k)(\overline{A}-I)^T G^T P\tilde{A}z(k) + z^T(k)(\overline{A}-I)^T P(\overline{A}-I)z(k) \\ & + z^T(k)(\overline{A}-I)^T G^T PG(\overline{A}-I)z(k) + z^T(k)(\overline{A}-I)^T PG(\overline{A}-I)z(k) \\ & + z^T(k)(\overline{A}-I)^T G^T P(\overline{A}-I)z(k) - \|\hat{w}(k+1)\| + \|\hat{w}(k)\| + H(k)\end{aligned} \tag{28}$$

Further calculating using inequality $((a+b)/2) \leq \sqrt{ab}$, weight tuning law is obtained (22) and rest of the terms are

$$\begin{aligned}\Delta V(k) = {} & -2\|\hat{w}(k)\|((z^T(k)(\overline{A}-I)^T G^T PG(\overline{A}-I))^{1/2} - (z^T(k)(\overline{A}-I)^T P(\overline{A}-I)z(k))^{1/2} \\ & - (z^T(k)\tilde{A}^T P\tilde{A}z(k))^{1/2} - 2\sqrt{ab}) - 2(z^T(k)(\overline{A}-I)^T G^T PG(\overline{A}-I)z(k))^{1/2}(z^T(k)(\overline{A}-I)^T \\ & \times P(\overline{A}-I)z(k))^{1/2} - 2(z^T(k)(\overline{A}-I)^T P(\overline{A}-I)z(k))^{1/2}(z^T(k)\tilde{A}^T P\tilde{A}z(k))^{1/2} \\ & - z^T(k)Pz(k) - z^T(k)\tilde{A}^T PG(\overline{A}-I)z(k) - z^T(k)(\overline{A}-I)^T G^T P\tilde{A}z(k) \\ & - 4(z^T(k)(\overline{A}-I)^T G^T PG(\overline{A}-I)z(k))^{1/2}\sqrt{ab} - 4(z^T(k)(\overline{A}-I)^T P(\overline{A}-I)z(k))^{1/2}\sqrt{ab} \\ & - 4(z^T(k)\tilde{A}^T P\tilde{A}z(k))^{1/2}\sqrt{ab} - 4\sqrt{ab}\end{aligned} \tag{29}$$

where $a = (z^T(k)(\overline{A}-I)^T PG(\overline{A}-I)z(k))^{1/2}$, $b = (z^T(k)(\overline{A}-I)^T G^T P(\overline{A}-I)z(k))^{1/2}$

Therefore,

$$\Delta V(k) \leq 0 \tag{30}$$

Hence, it can be concluded that the system is stable in certain conditions. The weight tuning law obtained in (22) is helpful in properly approximating the unknown nonlinear function and the control law derived in (21) force the system's trajectory to come on a linear sliding manifold which is at an equilibrium and when the system comes on that surface it will also be stable, that has been proved by taking an example and results has been obtained after simulation. So the system (4) is stable with condition (22) to be satisfied and with Assumptions 1–5.

## 7 Simulation Results

The section presents an example to check the nonlinear uncertain discrete delayed system performance. The parameter sets for the proposed system considered to be are [4]

$$A=\begin{bmatrix}1.0 & -0.4\\ 0.6 & 0.8\end{bmatrix},\ B=\begin{bmatrix}1\\ 0\end{bmatrix},\ \Delta A=\begin{bmatrix}0.02\sin(0.01k\pi) & 0.01\sin(0.01k\pi)\\ 0.01\sin(0.01k\pi) & 0.005\sin(0.01k\pi)\end{bmatrix}$$

$$g(x(k))=\begin{bmatrix}\frac{x_1(k)}{1+x_1^2(k)+x_2^2(k)}\\ \frac{x_2(k)}{1+x_1^2(k)+x_2^2(k)}\\ \frac{x_3(k)}{1+x_3^2(k)+x_4^2(k)}\\ \frac{x_4(k)}{1+x_3^2(k)+x_4^2(k)}\end{bmatrix}$$

The delay in time is constant and let it be $h=2$. The initial conditions of system (4) states $x_1$ and $x_2$ are assumed to be $[2 \quad 1]$. The positive definite matrices are chosen to be

$$P=\begin{bmatrix}3.7420\times10^8 & 0\\ 0 & 3.7420\times10^8\end{bmatrix} G=\begin{bmatrix}3.7420\times10^8 & 0\\ 0 & 3.7420\times10^8\end{bmatrix}$$

The linear sliding surface (17) should be chosen as $s(k)=[1 \quad 0.6454],x(k)=0$. The designed control law that compels the trajectory to come on sliding manifold is

$$u(k)=(1/B_2C\hat{g}(z(k)))\times(qTs(k)+\varepsilon T\mathrm{sgn}(s(k))+\overline{C}(\overline{A}-I)z(k)-diag(\mathrm{sgn}(s_i))f_d(k))$$

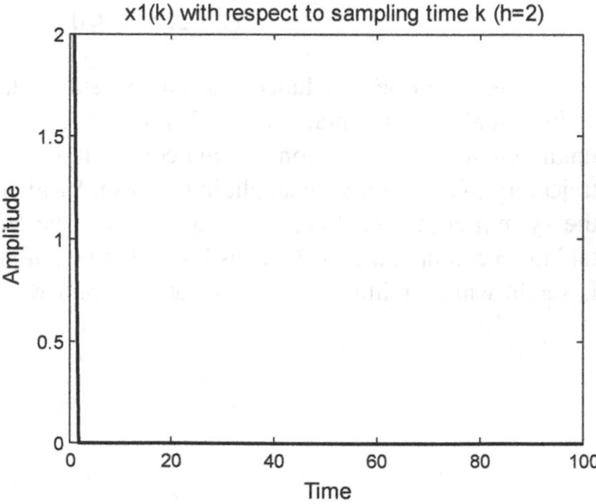


**Fig. 2** State $x_1(k)$

**Fig. 3** State $x_2(k)$

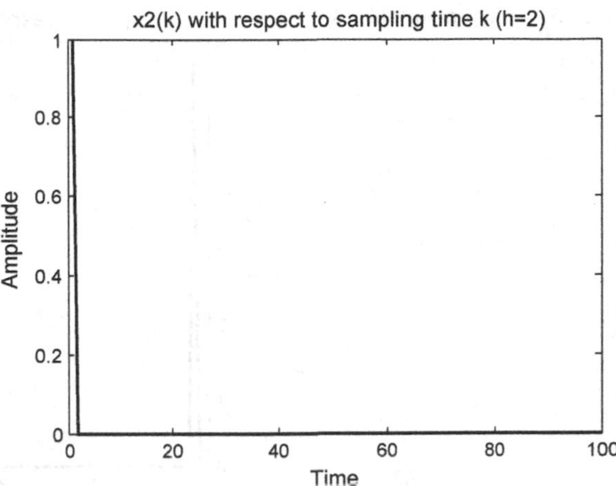


**Fig. 4** Control law $u(k)$

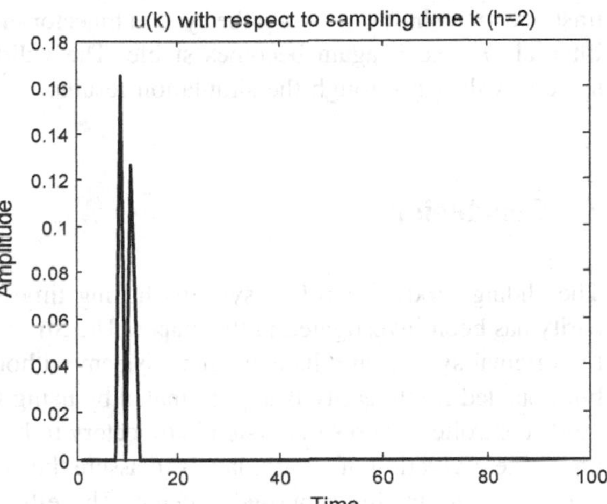


where

$$f_d(k) = 0.49|x_1(k)| + 0.36|x_2(k)| + 0.40|x_1(k-h)| + 0.23|x_2(k-h)|0.001\sin(0.75kh)$$

The system states $x_1$ and $x_2$ are depicted in Figs. 2 and 3. It has been shown that states of the system reaches towards zero rapidly and becomes stable.

The controller is depicted in Fig. 4 that shows robustness of the system in the presence of delay. The control law shows that after some iterations it also becomes stable and now the trajectory of the system is in the vicinity of the stable sliding surface. The sliding surface depicted in Fig. 5. The sliding surface becomes

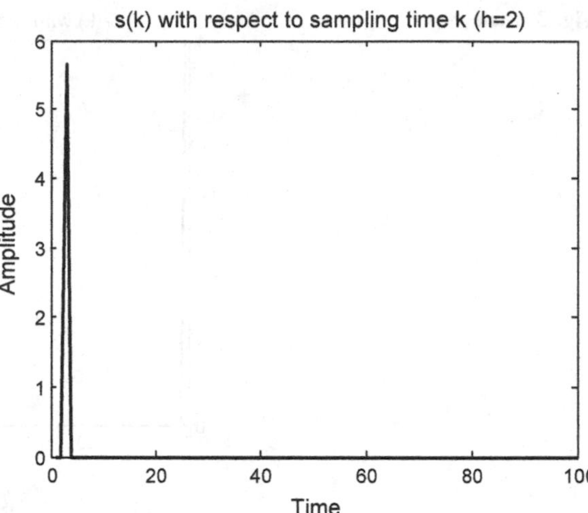


**Fig. 5** Sliding manifold $s(k)$

unstable for a short time when the system trajectory reached on it and within a small interval of time it again becomes stable. The validation and effectiveness of the system is shown through the simulation results.

## 8 Conclusion

The sliding mode control of systems having time delay, uncertainty and nonlinearity has been investigated in this paper. The Smith Predictor is used that converts the original system into its equivalent system without delay in time. The unknown but bounded nonlinearity is approximated by using CNN. This CNN based sliding mode controller assures the system's trajectory to be on sliding surface and remains on it. The derivation of control law is to assure the sliding mode existence. The time delay is fixed in this proposed scheme. The effectiveness and validation of the proposed system is shown in simulation results. In [3], the author also considered the same problem but the nonlinearity was approximated by using multi-layer perceptron and obtained the results. As CNN has faster convergence than MLP so the simulated results that has been obtained in this paper takes less time in computing and on comparing the results, the results of this paper is more robust compared to the results by Xia et al. [3].

## References

1. Young-Hoon and Jun-Ho Oh. "Sliding mode control for robust stabilization of uncertain input-delay systems," *Transaction on Control, Automation and System Engineering*, vol. 2, no. 2, pp. 98–103, June 2000.

2. Y. G. Niu, D. W. C. Ho and J. Lam, "Robust integral sliding mode control for uncertain stochastic systems with time-varying delay," *Automatic*, vol. 41, no. 5, pp. 873–880, May 2005.
3. Y. Xia, G. P. Liu, P. Shi, J. Chen, D. Rees and J. Liang, "Sliding mode control of uncertain linear discrete time systems with input delays," *IET Control Theory Appl.*, vol. 1, no. 4, pp. 1169–1175, Apr. 2007.
4. M. Yan and Y. Shi, "Robust discrete-time sliding mode control for uncertain systems with time-varying state delays," *IET Control Theory Appl.*, vol. 2, no. 8, pp. 662–674, Aug. 2008.
5. K. David Young, Vadim I. Utkin and Umit Ozguner, "A Control Engineer's Guide to Sliding Mode Control," *IEEE Transactions on Control Systems Technology*, vol. 7, no. 3, pp. 328–342, May 1999.
6. C. Y. Chan, "Discrete-time adaptive sliding mode control of a linear system in state-space form," *Internat. J. of Control*, vol. 67, no. 6, pp. 859–868, 1997.
7. A. Bartoszewicz, "Discrete-time quasi-sliding mode control strategies," *IEEE Trans. On Industrial Electronics*, vol. 45, no. 4, pp. 633–637, 1998.
8. E. T. Jeung, D. C. Oh, J. H. Kim and H. B. Park, "Robust controller design for uncertain systems with time-delays: LMI approach," *Automatica*, vol. 35, pp. 1229–1231, 1996.
9. S. O. R. Moheimani, V. A. Savkin, and I. R. Peterson, "Synthesis of min-max optimal controllers for uncertain time delay systems with structured uncertainty," *Int. J. Syst. Sci.* vol. 31, pp. 137–147, 2000.
10. C. E. De Souza, and X. Li, "Delay-dependent robust H infinity control of uncertain linear state-delayed systems," *Automatica*, vol. 35, pp. 1313–1321, 1996.
11. D. Yue, "Robust stabilization of uncertain systems with unknown input delays," *Automatica*, vol. 40, no. 2, pp. 331–336, 2004.
12. M. V. Basin, J. Perez, P. Acosta, and L. Fridman, "Optimal filtering for non-linear polynomial systems over linear observations with delay," *Int. J. Innov. Comput. Inf. Control*, vol. 2, pp. 863–874, 2006.
13. E. K. Boukasand N. F. Al-Muthairi, "Delay-dependent stabilization of singular linear systems with delays," *Int. J. Innov. Comput. Inf. Control*, vol. 2, no. 2, pp. 283–291, 2006.
14. O. J. Smith, "Closer control of loops with dead time," *Chemical Engineering Progress.*, vol. 53, no. 5, pp. 217–219, 1957.
15. Tsu-Tian Lee and Jin-Tsong Jeng, "The Chebyshev Polynomials based Unified Model Neural Networks for Function Approximation," IEEE Transactions on Systems, Man and Cybernetics-Part B: Cybernetics, vol. 28, no. 6, pp. 925–935, 1998.
16. S. Haykin, Ottawa, ON, Canada, "Neural Networks," Maxwell Macmillan, 1994.
17. S. Dehuri and S. B. Cho, "A comprehensive survey on functional link neural networks and an adaptive PSO-BP learning for CFLNN," *Neural Computing and Applications*, vol. 19, pp. 187–205, 2010.
18. T. T. Lee and J. T. Jeng, "The Chebyshev polynomial based unified model neural networks for function approximations," *IEEE Trans. Systems, Man and Cybernetics*, vol. 28, pp. 925–935, 1998.
19. S. Purwar, I. N. Kar, and A. N. Jha, "On-Line System Identification of Complex Systems Using Chebyshev Neural Networks," *Applied Soft Computing*, vol. 7, pp. 364–372, 2005.
20. Vishal Goyal, Vinay Kumar Deolia and Tripti Nath Sharma, "Robust sliding mode control for non-linear discrete time delayed systems based on neural networks," *Intelligent Control and Automation*, vol. 6, pp. 75–83, 2015.
21. Zhiwei Lin, Yuanqing Xia, Peng Shi and Harris Wu, "Robust sliding mode control for uncertain linear discrete systems independent of time-delay," *Int. J. Innov. Comput. Inf. Control*, vol. 7, no. 2, pp. 869–880, 2011.
22. K. Kawaguchi, H. Shibasaki, R. Tanaka, H. Ogawa, T. Murakami and Y. Ishida, "Sliding mode control for a plant with a time delay," *SERSC ASTL*, vol. 25, pp. 278–283, 2013.
23. W. Gao, Y. Wang, and A. Homaifa, "Discrete time variable structure control systems," *IEEE Trans. Electron.*, vol. 42, no. 2, pp. 117–122, 1995.

24. J.P. Richard, F. Gouaisbaut and W. Perruquetti, "Sliding Mode Control in the presence of delay," *Kybernetika*, vol. 37, no. 3, pp. 277–294, 2001.
25. A. J. Koshkouei and A.S.I. Zinober, "Sliding mode time-delay systems," *Proc. Of Int. Workshop on VSS*, Tokyo, pp. 99–101, 1996.
26. K.K. Shyu and J.J. Yah, "Robust stability of uncertain time delay systems and its stabilization by variable structure control," *Internat. J. Control*, vol. 57, pp. 237–246, 1993.
27. Y.H. Roh and J.H. Oh, "Robust stabilization of uncertain input delay systems by sliding mode control with delay compensation," *Automatica*, vol. 35, pp. 1861–1865, 1996.
28. H. Zhang, L. Xie and G. Duan, "$H_\infty$ control of discrete-time systems with multiple input delays," *IEEE Transactions on Automatic Control*, vol. 52, no. 2, pp. 271–283, 2007.
29. B. L. Zhang, L. Ma and Q. L. Han, "Sliding mode $H_\infty$ control of offshore steel jacket platforms subject to non-linear self-excited wave force and external disturbance," *Non-linear Analysis: Real world Applications*, vol. 14, no. 1, pp. 163–178, 2013.

# Artificial Bee Colony as a Frontier in Evolutionary Optimization: A Survey

Divya Kumar and Krishn K. Mishra

**Abstract** Artificial Bee Colony (ABC) algorithm is now a long-familiar example of Swarm Intelligence. It has been consistently drawing the attention of research scholars since last decade. The adept performance of ABC algorithm has already been proved in various researches. Hence this algorithm has been used in wide variety of applications, spanning almost all aspects of engineering optimization. This manuscript details out some of the application areas of ABC algorithm in a concise way and it aims to provide a bird eye view of various application areas for the beginner researchers.

**Keywords** Swarm intelligence · Evolutionary algorithms · Honey bee swarms · Artificial Bee Colony · Numerical optimization

## 1 Introduction

The term “swarm” refers to a collection of species moving together in large number. Swarms can be identified as bird flocks, fishes, ants or bees, performing collectively without any central control or central communication channel. Each member of the swarm group acts independently and scholastically based on his perceptual environmental sensing. This interesting behavior of swarms is noticed and well simulated in the field of computer science chiefly for optimization [1, 2]. Some of the well established Swarm Intelligence based algorithms are Ant Colony Optimization (ACO) [3], Particle Swarm Optimization (PSO) [4], Artificial Bee Colony (ABC) [5], Bat algorithm [6], and Bacterial Foraging algorithms [7]. These algorithms are being extensively deployed for solving wide variety of real life engineering problems

D. Kumar (✉) · K.K. Mishra
Computer Science and Engineering Department, Motilal Nehru National Institute of Technology Allahabad, Allahabad, India
e-mail: divyak@mnnit.ac.in

K.K. Mishra
e-mail: kkm@mnnit.ac.in



S.K. Bhatia et al. (eds.), *Advances in Computer and Computational Sciences*, Advances in Intelligent Systems and Computing 553,
DOI 10.1007/978-981-10-3770-2_50

[8, 9]. The operators of these algorithms exhibit a proper combination of five SI principles, i.e., self-organization, division of labor, natural communication, goal oriented, and satisfaction [10]. All the desirable properties of the operators are clearly observable in Artificial Bee Colony (ABC) algorithm, proposed by Dr. Dervis Karaboga [5, 11] in 2005. ABC is an stochastic optimization technique motivated by the reasoning power of natural honey bees swarm, who search their food sources in a smart and coordinated manner. This manuscript is based on the ABC algorithm, exploring its theory and applications. To meet its objective, this manuscript goes in the following manner: first we have detailed the basics of ABC algorithm, then its application areas are discussed in detail with performance evaluations.

## 2 The Artificial Bee Colony

The logical model of foregoing honey bee swarms was proposed by Tereshko in [12] which consisted of two main components, i.e., food sources and foragers. Foragers were further categorized into two parts, i.e., employed foragers and un-employed foragers. Employed foragers denotes evolutionary exploration of the search space and unemployed foragers represent evolutionary exploitation [13]. In a system there happens to be two types of unemployed bees namely, onlooker bees and scouts. The Artificial Bee Colony algorithm simulates the behavior of natural bees to form the solution of optimization problems.

The natural honey bee swarms arrange themselves in a colony of three types of groups of bees namely: employed bees, onlookers bees, and scouts bees. Each type of bees have different roles and responsibilities for the hive. The employed are the young ones who search the entire space, gather the food information, and come back. They dance on front of the hive to share their information with the onlookers. The better the nectar sources a bee had discovered, the more it dances in happiness. Consecutively more onlookers are attracted toward an employed bee which dances more. Which leads to more exploitation of the regions that previously contained high nectar sources. In this way, gradually all the bees are directed to a single promising area which is expected to be the optimum solution. The employed bees whose food source/solution fitness has been desolated, becomes scout bee. They then begin to look for a new food source randomly, from the fresh initialization. Thus, the ABC algorithm works iteratively over the phases of these three bees to improve the candidate solution. This meta-heuristic is briefly described in Algorithm 1.

## 3 Applications of ABC Algorithm

The Artificial Bee Colony meta-heuristic technique has been successfully utilized for many applications pertaining to many branches of engineering. In the following subsections, we have discussed various applications of ABC.

**Algorithm 1**: ABC General Scheme

Step 1: Initialization phase.
*//random generation of candidate solutions*

Step 2: Repeat step 2.1 through 2.3 until (termination condition)

Step 2.1: Employed bee phase.
*//search new solutions in whole search space*

Step 2.2: Onlooker bee phase.
*//further exploitation of better solutions*

Step 2.3: Scout bee phase.
*//reinitialize a solutions which can not be further evolved*

- **Applications in Electrical Engineering**
ABC is being used by research scholars to solve the optimization problems dwelling in the domains of electrical engineering. Linh and Anh in [14] presented a novel technique based on ABC algorithm for calculating the optimal sectionalizing of power switch to be controlled in order to minimize the distribution system power losses. To establish the validity of the purposed algorithm, the authors carried out the simulations on 14–33 bus systems. Ayan and Kilic in [15] provided a utilitarian strategy for optimizing the reactive power flow through ABC algorithm. The reactive power flow is considered to be a complex and nonlinear optimization problem of energy transmission lines. To demonstrate the validity of the proposed algorithm they used IEEE-11 bus test system and concluded that ABC algorithm provides significantly promising results.
- **Applications in Civil and Mechanical Engineering**
Yao et al. in [16] demonstrated a practical procedure grounded on ABC algorithm to find the optimal subway routes with an aim of maximizing the population density handled by subway routes. Hadidi et al. in [17] brought out an ABC algorithm for space structural optimization of planar and space trusses with the underlying constraints of stress, buckling, and displacement.
- **Applications in Bio-technology**
Motif Discovery Problem (MDP) has been well tackled by David et al. in [18] for discovering TFBS (Transcription Factor Binding Sites) in DNA sequences. The effectiveness of the proposed algorithm has been verified over twelve well-known data sets of MDP problem. Zhang and Wu in [19] applied ABC algorithm on protein folding problems in hydrophobic-hydrophilic lattice model. From their four experiments they concluded that ABC algorithm has higher success rates than Genetic Algorithm.
- **Training Neural Network**
At the first Karaboga and Basturk in [20] used ABC algorithm for the training of multi-layer feedforward and back-propogated neural networks. They apply the ABC algorithm for searching the optimal weights for the neurons connecting hidden layers of neural network. The trained feedforward artificial neural networks (ANNs) are being deployed for classification and pattern recognition purposes.

When the performance of the ABC trained ANNs were compared with the conventional error back propagation algorithm and Genetic Algorithm trained ANNs, it was observed that the ANNs which were trained through ABC were more precise in classifying the inputs.

- **ABC for Theorem Proving**
  Divya and Mishra in [21] proposed a novel ABC-based algorithm for automated theorem proving for the problems described in first order logic. They have experimentally shown the procedure, with clause forms, automate first order reasoning principles of forward chaining and backward chaining using Artificial Bee Colony algorithm. From the results, the potential of ABC algorithm has been observed and it was concluded that ABC meta-heuristic speedily solves the sample logic problems.
- **ABC for Data Mining**
  Data Clustering is currently being deployed in various multidisciplinary applications. Clustering is a crucial tool in all data mining systems. It can be described as a descriptive task in which we strive to group together the objects which are homogeneous. This is done on the basis of the values of object's attributes. Karaboga and Oztruk in [22] utilized the ABC algorithm for data clustering on thirteen benchmark problems chosen from the open access UCI Machine Learning Repository [23].
- **ABC in Image Processing Area**
  ABC algorithm has been thoroughly used for image processing purposes. Benala et al. in [24] reported a new technique for the enhancement of image edges using ABC hybridized smoothening filters. Chidambaram and Heitor Silverio Lopes [25] have used ABC algorithm for the pattern and object recognition in the digital images.

Besides the previously described applications, ABC has many applications in various other research areas also. Some of these as shown in Table 1.

## 4 Performance Evaluation of ABC

The Artificial Bee Colony (ABC) algorithm is a stochastic optimization algorithm. It is a collective arrangement of three modules namely: employed bee, onlooker bee, and scout bee which is purely based on intelligent foraging behavior of natural honeybee swarms. The qualitative as well as quantitative assessment of ABC algorithm was done by Dervis Karaboga et al. in [55, 56]. The performance was compared against the latest state of the art evolutionary algorithms which are Genetic Programming (GP), Evolution Strategy (ES), Evolutionary Programming (EP), Differential Evolution (DE), and Particle Swarm Optimization (PSO). From the simulation results it was ascertained that for benchmark multi-dimensional as well as for benchmark multi-modal numeric problems the performance of ABC algorithm is fairly better or atleast comparable to those of the mentioned algorithms with an added

**Table 1** Some other ABC application areas

| Application areas | References |
|---|---|
| Solving NP hard problems | [26, 27] |
| Sensor networks | [28, 29] |
| Forensics | [30, 31] |
| Electronics circuit design | [32, 33] |
| Resonant frequency calculation | [34, 35] |
| Real parameter optimization | [36, 37] |
| Time tabling problem | [38] |
| Data mining and clustering | [22, 39] |
| Vehicle routing | [40] |
| Load dispatch | [41, 42] |
| Robotics | [43] |
| Test case optimization | [44, 45] |
| Designing distribution network | [46] |
| Signal processing | [47, 48] |
| Scheduling problem | [49, 50] |
| Cloud computing | [51] |
| Stock market forecasting | [52] |
| Transportation problem | [53] |
| Integer programming | [54] |

advantage of employing fewer control parameters. It was also ascertained that ABC can be efficiently employed to solve engineering problems with high dimensionality.

## 5 Conclusion

Artificial Bee Colony (ABC) is one of the trending Swarm Intelligence based algorithm. Also, it is one of the most deployed algorithm in evolutionary literature, used for solving optimization problems related to nearly all realms of science and commerce. It has been well observed that the performance results of ABC are better than or at least comparable to all the other established algorithms of its class. In this manuscript, we have briefly described ABC algorithm with its major application areas and performance evaluation, so as to provide a concise overview of its efficiency and applicability in optimization domain.

## References


1. Christian Blum and Xiaodong Li. *Swarm intelligence in optimization*. Springer, 2008.
2. James Kennedy, James F Kennedy, Russell C Eberhart, and Yuhui Shi. *Swarm intelligence*. Morgan Kaufmann, 2001.
3. Marco Dorigo, Vittorio Maniezzo, Alberto Colorni, and Vittorio Maniezzo. Positive feedback as a search strategy. 1991.
4. Russ C Eberhart and James Kennedy. A new optimizer using particle swarm theory. In *Proceedings of the sixth international symposium on micro machine and human science*, volume 1, pages 39–43. New York, NY, 1995.
5. Dervis Karaboga. An idea based on honey bee swarm for numerical optimization. Technical report, Technical report-tr06, Erciyes university, engineering faculty, computer engineering department, 2005.
6. Xin-She Yang. A new metaheuristic bat-inspired algorithm. In *Nature inspired cooperative strategies for optimization (NICSO 2010)*, pages 65–74. Springer, 2010.
7. Kevin M Passino. Bacterial foraging optimization. *Innovations and Developments of Swarm Intelligence Applications*, page 219, 2012.
8. Divya Kumar, Divya Kashyap, KK Mishra, and AK Mishra. Routing path determination using qos metrics and priority based evolutionary optimization. In *High Performance Computing and Communications (HPCC), 2011 IEEE 13th International Conference on*, pages 615–621. IEEE, 2011.
9. Divya Kumar and Krishn Kumar Mishra. Incorporating logic in artificial bee colony (abc) algorithm to solve first order logic problems: The logical abc. In *Knowledge and Smart Technology (KST), 2015 7th International Conference on*, pages 65–70. IEEE, 2015.
10. Eric Bonabeau, Marco Dorigo, and Guy Theraulaz. *Swarm intelligence: from natural to artificial systems*. Number 1. Oxford university press, 1999.
11. Dervis Karaboga, Bahriye Akay, and Celal Ozturk. Artificial bee colony (abc) optimization algorithm for training feed-forward neural networks. In *Modeling decisions for artificial intelligence*, pages 318–329. Springer, 2007.
12. Valery Tereshko. Reaction-diffusion model of a honeybee colony's foraging behaviour. In *Parallel Problem Solving from Nature PPSN VI*, pages 807–816. Springer, 2000.
13. Agoston E Eiben and Cornelis A Schippers. On evolutionary exploration and exploitation. *Fundamenta Informaticae*, 35(1–4):35–50, 1998.
14. Nguyen Tung Linh and Nguyen Quynh Anh. Application artificial bee colony algorithm (abc) for reconfiguring distribution network. In *Computer Modeling and Simulation, 2010. ICCMS'10. Second International Conference on*, volume 1, pages 102–106. IEEE, 2010.
15. Kursat Ayan and Ulas Kilic. Artificial bee colony algorithm solution for optimal reactive power flow. *Applied Soft Computing*, 12(5):1477–1482, 2012.
16. B Yao, C Yang, J Hu, and B Yu. The optimization of urban subway routes based on artificial bee colony algorithm. *Key technologies of railway engineering high speed railway, heavy haul railway and urban rail transit. Beijing Jiaotong University, Beijing*, pages 747–751, 2010.
17. Ali Hadidi, Sina Kazemzadeh Azad, and Saeid Kazemzadeh Azad. Structural optimization using artificial bee colony algorithm. In *2nd international conference on engineering optimization*, 2010.
18. David L Gonzalez-Alvarez, Miguel A Vega-Rodriguez, Juan A Gomez-Pulido, and Juan M Sanchez-Perez. Finding motifs in dna sequences applying a multiobjective artificial bee colony (moabc) algorithm. In *Evolutionary Computation, Machine Learning and Data Mining in Bioinformatics*, pages 89–100. Springer, 2011.
19. Yudong Zhang and Lenan Wu. Artificial bee colony for two dimensional protein folding. *Advances in Electrical Engineering Systems*, 1(1):19–23, 2012.
20. Dervis Karaboga and Bahriye Basturk. Artificial bee colony (abc) optimization algorithm for solving constrained optimization problems. In *Foundations of Fuzzy Logic and Soft Computing*, pages 789–798. Springer, 2007.

21. D. Kumar and K.K. Mishra. Incorporating logic in artificial bee colony (abc) algorithm to solve first order logic problems: The logical abc. In *Knowledge and Smart Technology (KST), 2015 7th International Conference on*, pages 65–70, Jan 2015.
22. Dervis Karaboga and Celal Ozturk. A novel clustering approach: Artificial bee colony (abc) algorithm. *Applied Soft Computing*, 11(1):652–657, 2011.
23. M. Lichman. UCI machine learning repository, 2013.
24. Tirimula Rao Benala, Sathya Durga Jampala, SH Villa, and Bhargavi Konathala. A novel approach to image edge enhancement using artificial bee colony optimization algorithm for hybridized smoothening filters. In *Nature & Biologically Inspired Computing, 2009. NaBIC 2009. World Congress on*, pages 1071–1076. IEEE, 2009.
25. Chidambaram Chidambaram and Heitor Silverio Lopes. An improved artificial bee colony algorithm for the object recognition problem in complex digital images using template matching. *International Journal of Natural Computing Research (IJNCR)*, 1(2):54–70, 2010.
26. Dervis Karaboga and Beyza Gorkemli. A combinatorial artificial bee colony algorithm for traveling salesman problem. In *Innovations in Intelligent Systems and Applications (INISTA), 2011 International Symposium on*, pages 50–53. IEEE, 2011.
27. Alok Singh. An artificial bee colony algorithm for the leaf-constrained minimum spanning tree problem. *Applied Soft Computing*, 9(2):625–631, 2009.
28. Dervis Karaboga, Selcuk Okdem, and Celal Ozturk. Cluster based wireless sensor network routing using artificial bee colony algorithm. *Wireless Networks*, 18(7):847–860, 2012.
29. Celal Ozturk, Dervis Karaboga, and Beyza Gorkemli. Probabilistic dynamic deployment of wireless sensor networks by artificial bee colony algorithm. *Sensors*, 11(6):6056–6065, 2011.
30. F Ghareh Mohammadi and M Saniee Abadeh. Image steganalysis using a bee colony based feature selection algorithm. *Engineering Applications of Artificial Intelligence*, 31:35–43, 2014.
31. Pei-Wei Tsai, Muhammad Khurram Khan, Jeng-Shyang Pan, and Bin-Yih Liao. Interactive artificial bee colony supported passive continuous authentication system. *Systems Journal, IEEE*, 8(2):395–405, 2014.
32. Y Delican, RA Vural, and T Yildirim. Artificial bee colony optimization based cmos inverter design considering propagation delays. In *Symbolic and Numerical Methods, Modeling and Applications to Circuit Design (SM2ACD), 2010 XIth International Workshop on*, pages 1–5. IEEE, 2010.
33. VJ Manoj and Elizabeth Elias. Artificial bee colony algorithm for the design of multiplier-less nonuniform filter bank transmultiplexer. *Information Sciences*, 192:193–203, 2012.
34. Ali Akdagli, Mustafa Berkan Bicer, and Seda Ermis. A novel expression for resonant length obtained by using artificial bee colony algorithm in calculating resonant frequency of c-shaped compact microstrip antennas. *Turkish Journal of Electrical Engineering & Computer Sciences*, 19(4):597–606, 2011.
35. A Toktas, MB Bicer, A Akdagli, and A Kayabasi. Simple formulas for calculating resonant frequencies of c and h shaped compact microstrip antennas obtained by using artificial bee colony algorithm. *Journal of Electromagnetic Waves and Applications*, 25(11–12):1718–1729, 2011.
36. Bahriye Akay and Dervis Karaboga. A modified artificial bee colony algorithm for real-parameter optimization. *Information Sciences*, 192:120–142, 2012.
37. Dervis Karaboga and Bahriye Basturk. A powerful and efficient algorithm for numerical function optimization: artificial bee colony (abc) algorithm. *Journal of global optimization*, 39(3):459–471, 2007.
38. Malek Alzaqebah and Salwani Abdullah. Hybrid artificial bee colony search algorithm based on disruptive selection for examination timetabling problems. In *COCOA*, pages 31–45. Springer, 2011.
39. Changsheng Zhang, Dantong Ouyang, and Jiaxu Ning. An artificial bee colony approach for clustering. *Expert Systems with Applications*, 37(7):4761–4767, 2010.
40. WY Szeto, Yongzhong Wu, and Sin C Ho. An artificial bee colony algorithm for the capacitated vehicle routing problem. *European Journal of Operational Research*, 215(1):126–135, 2011.

41. NK Garg, Shimpi Singh Jadon, Harish Sharma, and DK Palwalia. Gbest-artificial bee colony algorithm to solve load flow problem. In *Proceedings of the Third International Conference on Soft Computing for Problem Solving*, pages 529–538. Springer, 2014.
42. S Hemamalini and Sishaj P Simon. Artificial bee colony algorithm for economic load dispatch problem with non-smooth cost functions. *Electric Power Components and Systems*, 38(7):786–803, 2010.
43. Preetha Bhattacharjee, Pratyusha Rakshit, Indrani Goswami, Amit Konar, and Atulya K Nagar. Multi-robot path-planning using artificial bee colony optimization algorithm. In *Nature and Biologically Inspired Computing (NaBIC), 2011 Third World Congress on*, pages 219–224. IEEE, 2011.
44. Surender Singh Dahiya, Jitender Kumar Chhabra, and Shakti Kumar. Application of artificial bee colony algorithm to software testing. In *Software Engineering Conference (ASWEC), 2010 21st Australian*, pages 149–154. IEEE, 2010.
45. D Jeya Mala, V Mohan, and M Kamalapriya. Automated software test optimisation framework-an artificial bee colony optimisation-based approach. *Software, IET*, 4(5):334–348, 2010.
46. R Srinivasa Rao, SVL Narasimham, and M Ramalingaraju. Optimization of distribution network configuration for loss reduction using artificial bee colony algorithm. *International Journal of Electrical Power and Energy Systems Engineering*, 1(2):116–122, 2008.
47. Nurhan Karaboga. A new design method based on artificial bee colony algorithm for digital iir filters. *Journal of the Franklin Institute*, 346(4):328–348, 2009.
48. Yudong Zhang, Lenan Wu, and Shuihua Wang. Magnetic resonance brain image classification by an improved artificial bee colony algorithm. *Progress in Electromagnetics Research*, 116:65–79, 2011.
49. Jun-Qing Li, Quan-Ke Pan, and Kai-Zhou Gao. Pareto-based discrete artificial bee colony algorithm for multi-objective flexible job shop scheduling problems. *The International Journal of Advanced Manufacturing Technology*, 55(9–12):1159–1169, 2011.
50. Quan-Ke Pan, M Fatih Tasgetiren, Ponnuthurai N Suganthan, and Tay Jin Chua. A discrete artificial bee colony algorithm for the lot-streaming flow shop scheduling problem. *Information sciences*, 181(12):2455–2468, 2011.
51. Jing Yao and Ju-hou He. Load balancing strategy of cloud computing based on artificial bee algorithm. In *Computing Technology and Information Management (ICCM), 2012 8th International Conference on*, volume 1, pages 185–189. IEEE, 2012.
52. Tsung-Jung Hsieh, Hsiao-Fen Hsiao, and Wei-Chang Yeh. Forecasting stock markets using wavelet transforms and recurrent neural networks: An integrated system based on artificial bee colony algorithm. *Applied soft computing*, 11(2):2510–2525, 2011.
53. Dusan Teodorovic and Mauro Dellorco. Bee colony optimization–a cooperative learning approach to complex transportation problems. In *Advanced OR and AI Methods in Transportation: Proceedings of 16th Mini–EURO Conference and 10th Meeting of EWGT (13–16 September 2005).–Poznan: Publishing House of the Polish Operational and System Research*, pages 51–60, 2005.
54. Bahriye Akay and Dervis Karaboga. Solving integer programming problems by using artificial bee colony algorithm. In *AI* IA 2009: Emergent Perspectives in Artificial Intelligence*, pages 355–364. Springer, 2009.
55. Dervis Karaboga and Bahriye Basturk. On the performance of artificial bee colony (abc) algorithm. *Applied soft computing*, 8(1):687–697, 2008.
56. Dervis Karaboga and Bahriye Akay. A comparative study of artificial bee colony algorithm. *Applied Mathematics and Computation*, 214(1):108–132, 2009.

# Selection of Best State for Tourism in India by Fuzzy Approach

**Shalini Singh, Varsha Mundepi, Deeksha Hatwal, Vidhi Raturi, Mukesh Chand, Rashmi, Sanjay Sharma and Shwetank Avikal**

**Abstract** India has always been an attraction seeker to tourist from all over the world. India indeed stands through its tittle "INCREDIBLE INDIA" because of its diversity in culture and religion. Tourism in India is economically important and is growing rapidly. About 22.57 million tourist arrived in India in 2014, compared to 19.80 million in 2013. In terms of foreign tourist arrivals, India ranked as the 38th country in the world. With the help of Fuzzy-AHP technique, i.e. Fuzzy Analytical Hierarchy Process is the best method decided for finding the most influential tourist place from the tourist point of view. The purpose of this work is to present a multi criteria decision making (MCDM) model for management of tourists across various tourist places in India. In this work five (5) criteria from various literature reviews and practical investigations has been taken. Fuzzy-AHP techniques is used to ample decision makers assesments about criteria weightings. Finally, a factual study is done for identifying the best tourist place across India. In this work, about 30 states are taken and various survey are conducted among different groups of people and then final decision is made by the computational process and effectiveness of Fuzzy-AHP.

**Keywords** Tourism management · MCDM · AHP · Fuzzy set

## 1 Introduction

Tourism industry is one of the most influential industries around the world. Tourism not only enchants tourists to a country but also brings foreign exchange, domestic exchange, and job opportunities. The vast it spreads, wider picture of country is exposed to the world. In recent years, India has shown its great concern in tourism,

S. Singh · V. Mundepi · D. Hatwal · V. Raturi · M. Chand · Rashmi · S. Sharma · S. Avikal (✉)
Department of Mechanical Engineering, Graphic Era Hill University, Dehradun, India
e-mail: Shwetank.avikal@gmail.com



S.K. Bhatia et al. (eds.), *Advances in Computer and Computational Sciences*, Advances in Intelligent Systems and Computing 553,
DOI 10.1007/978-981-10-3770-2_51

"Incredible India", is a slogan adopted by Indian government to attract tourists. India as a whole country not only supports tourism, but each state of the country has sensed the importance of tourism in their own perspectives. So every state of India is creating different sources to tug domestic as well as foreign tourist toward it. When a tourist decides to travel a place, budget, ease of travelling, and security problems are the multi criteria structures. In the context of selecting a place for travel, it is important to include certain elements that provide attributes and make tourist's decision easier, safe, and comfortable leading to a holiday or travel plan. People check different sources like internet, tourist offices, and question the locals to attain a decision of travel. For most people, visual attraction and safety are their top priority but it differs through each individual. Travelling a place is a decision-making problem and reflects the preferences of the traveler. In this study, "The best tourist state in India" case has been handled using MCDM-based approach, i.e., Fuzzy-AHP. The study contains thirty different states of India, with five different criteria. The criteria used here are like visual attraction, ease of access, safety, etc. Among these states the best one will be selected at the end of the study. The report has different sections like literature review, methodology, case study including problem definitions, calculations, and result. Finally the conclusion is discussed in the end.

## 2 Literature Review

This study has been used in selecting the best tourist place in India. As we all know that the tourism is one of the fastest growing industry today, thus within the tourism industry events are getting more and more important. Although tourism in India growing rapidly but from tourist point of view the selection of best tourist state is still difficult. A number of research has been conducted on smart tourism involves artificial intelligence, cloud computing, and internet of things while dealing with major decision-making problem, i.e., tourist place and destination selection which requires some important process.

The only path for tourism industry is customer needs. Expectations and attitude of mind, likes and dislikes such changes in tourist needs have brought the competition to the tourism sector.

The various researches have been discussed as follows:

Morrison et al. [1] reviewed the applied balance score card (Bsc.) method in tourism study and develop the modified Bsc method for future development.

Higham [2] has discussed the interest center on the management of tourism in natural areas. In his research he examined international visitor's perception wildness in New Zealand and published article addressing ecotourism, wildness management, and the impact of tourism on wildlife.

Carr [3] has focused on the level of discrimination felt by dog owners in terms of their ability to access the tourist spaces with their pets and now this process undertaken by dog owners to enable them to access leisure and tourism space with their pets.

Hall [4] has examined that how tourism affects and is affected by various dimensions of environmental change as climate change, urbanization, diseases, globalization, etc. The smart systems have been introduced to many areas including public safety, health services, infrastructure constructions, water saving, sustainable development, and environmental protection according to tourist point of view the smart tourism is fairly limited.

Chou [5] have discussed a fuzzy multi criteria decision model and created 21 criteria for selecting the international tourist hotels location.

Buhalis and Law [6] have analyzed e-tourism related studies in past 20 years and predicts the future of e-tourism for next 10 years. He explained that it is difficult to understand the mind of new tourists and their needs. He also noticed the development of (Information and Communication Technology) ICT and where the internet has produced the new group of tourists who are more experienced and demanding. He also discuss the main tourist demands such as booking online tickets and room reservations, making online purchase, providing mobile facilities and application such as wifi, SMS, etc.

Sevrani and Elmazi [7] have given the several new trends in tourist behavior by (Information and Communication Technology) ICT development, i.e., king for better service, wanting more special offer, and becoming more knowledgeable. The new tourists have become more dependent on information technology and self service.

Hsu et al. [8] identifies the factors that influence the tourists' choice of destination and evaluates the preferences of tourists for destinations. A 4-level AHP model, consisting of 22 attributes on the 4th level, was proposed and tested using data collected from tourists visiting Taiwan to establish the relative importance of preselected factors (criteria).

Zhang et al. [9] have described the smart tourism destination, which analyzes the definition critical technologies and of smart tourism destination and gives the essential concepts of smart tourism attraction according to the characteristics of tourism resource conservation, tourism development, and public services.

Wang et al. [10] have discussed the strength and weakness of the smart tourism attraction. The first factor was used to find out tourists key evaluation items of smart tourism attraction. Next factor was using the analytic hierarchy process for finding the smart tourism evaluation. In their research they included the, "intelligent tourism management," "smart sightseeing," "e-commerce system," and "smart safety".

Dickinson et al. [11] have discussed smart phone application on tourism which focuses on the functionalities of smart phone apps.

Zhu et al. [12] have argued that the development of Smart Tourism Destinations benefits tourism industry by providing convenient access to information for both tourism organizations and tourists through integrated and centralized data platform. Smart Tourism Destinations also harnessing the true essence of technology by building framework to facilitate multiple visualizations in a common direction.

Lamsfus et al. [13] have discussed the smart tourisms and smart city, which explains the application of the "smart city" concept and technologies into a tourism context.

Chang and Chang [14] have developed a model to investigate the tourists' preference. Ten attributes of tourist destinations were used in this study. Fuzzy set theory was adopted as the main analysis method to find the tourists' preference. In this study, 248 pieces of data were used. Besides the evaluations for the factors, the overall evaluations (namely, satisfied, neutral, and dissatisfied) for every tourism destination were also inquired.

It has been seen that number of research paper are published in this field of tourism industry and these studies are the overview of the historical development in the tourism sector.

## 3 Methodology

In this paper, the problem has been taken as a Multi Criteria Decision Making (MCDM) problem. Fuzzy-AHP has been used to calculate the weight of each criterion and then ranking has been done. Five criteria (described in the Table 1) have been taken for this study which are C1-Visual value, C2-No. of attraction, C3-Ease of access, C4-Security, and C5-Environmental impact.

### *3.1 AHP and Fuzzy-AHP Approach*

In AHP, the pair wise comparisons are done for each level with respect to chosen criteria and then ranged over a scale of 1–9. Each comparison gives the value of importance among the compared criteria Avikal et al. [15]. The compared values are always in real numbers. Even though AHP has easiness and simplicity in its

**Table 1** Various criteria and their definition

| No. | Criteria | Definition |
|---|---|---|
| C1 | Visual value | There are certain attractions that have the ability to attract tourist and appeal to them, such as natural, cultural, historical, or man made attractions |
| C2 | No. of attractions | Quantity of tourist attractions. For, e.g., no.of natural and cultural attractions |
| C3 | Ease of access | Access to tourist destinations. How can you reach the desired destination either by plane, car, taxi, or train |
| C4 | Security | Crime rates in tourism destinations, night security, especially for women, and also the presence of police forces to provide security |
| C5 | Environmental impact | Environmental impact like waste disposal system, noise pollution, and environmental pollution |

decision-making, and it gives out a decent result but when the complexity of the problem increases its ability to give more accuracy diminishes, and sometimes precision exclusively become an important characteristic. Therefore conventional AHP seems to be inadequate for this research work as such exact pairwise comparison is almost impossible to determine in this case where the information and data are uncertain. Hence for such case new MCDM techniques are more accurate just like Fuzzy-AHP.

In Fuzzy set theory the members work between the range of real numbers [0, 1]. Each member function and information describes fuzzy set. The fuzzy set elements are ranged between an interval which is usually [0, 1]. In fuzzy set, every individual is grouped but no sharp boundaries are mentioned. The fuzzy set theory described by Avikal et al. [16, 17] has been taken as the reference for the computation.

## 4 Case Study

### 4.1 Problem Definition

In this study the focus is on the mental clarity of the tourist that where to go and where not to. While planning for a vacations everybody has a dilemma in his/her mind to choose a suitable place for trip. But in our study, Fuzzy-AHP has been used which is a problem representation in the form of hierarchy/ladder.

Firstly five (5) different criteria for the study was choosen (Table 1). On the basis of these criterion, the survey has been conducted among the tourist experts and travelling lover people. In the survey procedure, the problem was explained to everyone and then ratings was taken from them following the standard table of Satty (1980). This ratings were represented in the matrix form. Then after collecting the data/ratings from approx 10–12 experts related to the discussed area, one main matrix of each criteria has been constructed. While considering the main matrix, weights of all criteria have been calculated. Using the calculated weights, further calculation will be carried and finally the states will be ranked.

### 4.2 Calculation

Table 2 shows the ratings of the 30 states on the basis of criteria taken from different sites (https://en.wikipedia.org/wiki/Indian_states_and_territories_ranked_by_safety_of_women,://www.tripadvisor.com/TravelersChoice). Table 3 contains the pairwise comparison matrix of Fuzzy values using triangular membership function. The fuzzy membership values have been converted or defuzzified using Fuzzy set theory. The weights of each criterian has been calculated using pairwise

**Table 2** Common data

| | C1 | C2 | C3 | C4 | C5 |
|---|---|---|---|---|---|
| Uttarkhnad | 4.5 | 12 | 4 | 4 | 4.5 |
| Madhya Pradesh | 4 | 8 | 4.5 | 3 | 3.5 |
| Maharashtra | 4.5 | 18 | 5 | 4 | 3.5 |
| Kerala | 4 | 15 | 4 | 3.5 | 4 |
| Jammu Kashmir | 4.5 | 6 | 3.5 | 3 | 4.5 |
| Delhi | 4 | 11 | 4.5 | 3 | 4 |
| Andhra Pradesh | 3.5 | 7 | 4 | 3.2 | 3.4 |
| Arunachal Pradesh | 4 | 9 | 4.5 | 4 | 4 |
| Assam | 3 | 8 | 4.5 | 3.6 | 4 |
| Bihar | 2 | 6 | 3.5 | 4.2 | 2 |
| Chhattisgarh | 2 | 8 | 3 | 4.3 | 5 |
| Goa | 4.5 | 10 | 4.5 | 3.5 | 4 |
| Haryana | 2.5 | 6 | 4 | 3.8 | 3.6 |
| Himachal Pradesh | 4.5 | 12 | 4 | 4 | 3.4 |
| Jharkhand | 3 | 11 | 4 | 3.5 | 3 |
| Karnataka | 4 | 10 | 4.5 | 4 | 3 |
| Gujarat | 4.5 | 16 | 4.5 | 3 | 3 |
| Manipur | 4 | 14 | 3.5 | 4 | 4.5 |
| Meghalaya | 4 | 13 | 3 | 4 | 4 |
| Mizoram | 3.5 | 15 | 3.5 | 3.8 | 2.5 |
| Nagaland | 4.5 | 15 | 3 | 4.2 | 4.2 |
| Odisha | 3.3 | 10 | 4 | 3 | 3 |
| Punjab | 4 | 10 | 4 | 2.8 | 3 |
| Rajasthan | 3.5 | 13 | 4 | 3 | 3.5 |
| Sikkim | 4 | 9 | 3.5 | 3.2 | 4 |
| Tamil Nadu | 3 | 15 | 4.5 | 2.9 | 4 |
| Tealangana | 2.5 | 6 | 3 | 2 | 3 |
| Tripura | 3 | 8 | 3 | 3.1 | 3 |
| Uttarpradesh | 3.5 | 13 | 4.5 | 4.6 | 4.5 |
| West Bengal | 3 | 7 | 4.5 | 3.8 | 4 |

**Table 3** Fuzzy data matrix

| | C1 | C2 | C3 | C4 | C5 |
|---|---|---|---|---|---|
| C1 | 1 | 3 | 1/3 | 1/3 | 3 |
| C2 | 1/3 | 1 | 1/7 | 1/5 | 1/3 |
| C3 | 3 | 7 | 1 | 3 | 5 |
| C4 | 3 | 5 | 1/3 | 1 | 3 |
| C5 | 1/3 | 3 | 1/5 | 1/3 | 1 |

comparison matrix given in Table 4. The data of Table 2 has been normalized for further calculation. The normalized data has been multiplied by the weight of each criterion. In presented problem, all the criteria are the beneficiary criteria. So that all

**Table 4** Results obtained with Fuzzy-AHP

| Criteria | Weights | $\lambda_{max}$, CI, RI | CR |
|---|---|---|---|
| C1 | 0.1566 | | |
| C2 | 0.0485 | $\lambda_{max} = 5.4176$ | |
| C3 | 0.4479 | CI = 0.1044 | CR = 0.0932 |
| C4 | 0.2545 | RI = 1.12 | |
| C5 | 0.0923 | | |

**Table 5** Final rank of states in India

| | C1 | C2 | C3 | C4 | C5 | ∑(C1-C5) | Rank |
|---|---|---|---|---|---|---|---|
| Uttarakhand | 0.1566 | 0.0320 | 0.3583 | 0.2188 | 0.0830 | 0.8487 | 5th |
| Madhyapradesh | 0.1378 | 0.0213 | 0.4031 | 0.1654 | 0.0646 | 0.7922 | 13th |
| Maharashtra | 0.1566 | 0.0485 | 0.4479 | 0.2188 | 0.0646 | 0.9364 | 1st |
| Kerala | 0.1378 | 0.0402 | 0.3583 | 0.1934 | 0.0738 | 0.8035 | 11th |
| Jammu Kashmir | 0.1566 | 0.0160 | 0.3135 | 0.1654 | 0.0830 | 0.7345 | 21th |
| Delhi | 0.1378 | 0.0295 | 0.4031 | 0.1654 | 0.0738 | 0.8096 | 9th |
| Andhra Pradesh | 0.1205 | 0.0184 | 0.3583 | 0.1756 | 0.0627 | 0.7355 | 19th |
| Arunachal Pradesh | 0.1378 | 0.0242 | 0.4031 | 0.2188 | 0.0738 | 0.8577 | 3rd |
| Assam | 0.1033 | 0.0213 | 0.4031 | 0.1985 | 0.0738 | 0.8000 | 12th |
| Bihar | 0.0689 | 0.0160 | 0.3135 | 0.2315 | 0.0369 | 0.6668 | 28th |
| Chhattisgarh | 0.0689 | 0.0213 | 0.2687 | 0.2366 | 0.0923 | 0.6878 | 27th |
| Goa | 0.1566 | 0.0266 | 0.4031 | 0.1934 | 0.0738 | 0.8535 | 4th |
| Haryana | 0.0861 | 0.0160 | 0.3583 | 0.2086 | 0.0664 | 0.7354 | 20th |
| Himachal Pradesh | 0.1566 | 0.0320 | 0.3583 | 0.2188 | 0.0627 | 0.8284 | 7th |
| Jharkhand | 0.1033 | 0.0295 | 0.3583 | 0.1934 | 0.0553 | 0.7398 | 18th |
| Karnataka | 0.1378 | 0.0266 | 0.4031 | 0.2188 | 0.0553 | 0.8416 | 6th |
| Gujarat | 0.1566 | 0.0426 | 0.4031 | 0.1654 | 0.0553 | 0.8230 | 8th |
| Manipur | 0.1378 | 0.0373 | 0.3135 | 0.2188 | 0.0830 | 0.7904 | 14th |
| Meghalaya | 0.1378 | 0.0349 | 0.2687 | 0.2188 | 0.0738 | 0.7340 | 22nd |
| Mizoram | 0.1205 | 0.0402 | 0.3135 | 0.2086 | 0.0184 | 0.7012 | 26th |
| Nagaland | 0.1566 | 0.0402 | 0.2687 | 0.2315 | 0.0775 | 0.7745 | 16th |
| Odisha | 0.1143 | 0.0266 | 0.3583 | 0.1654 | 0.0553 | 0.7199 | 25th |
| Punjab | 0.1378 | 0.0266 | 0.3583 | 0.1527 | 0.0553 | 0.7307 | 23th |
| Rajasthan | 0.1205 | 0.0349 | 0.3583 | 0.1654 | 0.0646 | 0.7437 | 17th |
| Sikkim | 0.1378 | 0.0242 | 0.3135 | 0.1756 | 0.0738 | 0.7249 | 24th |
| Tamil Nadu | 0.1033 | 0.0402 | 0.4031 | 0.1603 | 0.0738 | 0.7807 | 15th |
| Telangana | 0.0861 | 0.0160 | 0.2687 | 0.1094 | 0.0553 | 0.5355 | 30th |
| Tripura | 0.1033 | 0.0213 | 0.2687 | 0.1705 | 0.0553 | 0.6191 | 29th |
| Uttarpradesh | 0.1205 | 0.0349 | 0.4031 | 0.2545 | 0.0830 | 0.8960 | 2nd |
| West Bengal | 0.1033 | 0.0184 | 0.4031 | 0.2086 | 0.738 | 0.8072 | 10th |

the criteria have the positive impact on the calculation of rank of all states. The Final ranks of all the states have been calculated by taking row wise summation of all the alternatives and presented in Table 5.

## 5 Conclusion

This study analyzes the application of the MCDM technique which is Fuzzy-AHP to find out the most influential tourist state in India. Application of this MCDM technique enables a comparative analysis of alternative ranking and weights of criteria based on the reviews and survey. The selection of Fuzzy-AHP was done due to its capability to give the most precise result.

In this study, first of all we have collected the common data of all states from various websites and searches which gave out the different values of criteria. After common data table we have taken the reviews of various people about the criteria ratings out of 5 and then the values are input on the basis of Fuzzy-AHP. The real values of fuzzy matrix are input in next table, with the help of this weights for each criteria are found out along with consistency index and random index values. Now the data is normalized by dividing the maximum value with each of the values in particular column. After normalization the values of each column are multiplied by its criteria weight and in next step the summation of criteria $\sum$C1-C5 is taken along each row. On the basis of that the maximum value after summation is ranked as 1st and so on. This is how we are able to calculate the rank of each state on the basis of the review and ratings.

Finally our research concluded the finding rank of all chosen states; among which Maharashtra ranked at 1st (as having the highest rating in every criteria like 4.5 for visual value, 18 for no. of attraction, 5 for ease of access, 4 for security, and finally environmental impact with 3.5 rating) and Telangana at last with lowest criteria ratings (visual value-2.5, no. of attraction-6, ease of access-3, security-2, environmental impact-3).

## References

1. Morrison A.M., Taylor S., Morrison A.J., & Morrisob A.D.: Marketing small hotels on the world wide web. Information Technology and Tourism, 2(2) 97–113 (1999).
2. Higham J.E.S.: Perceptions of International Visitors to New Zealand Wilderness. In The State of Wilderness in New Zealand. Special Issue edited by G. Cessford. Science Publications, Science and Research Unit. Department of Conservation. Wellington, New Zealand (2001).
3. Carr N.: The Myth and Reality of Freedom and the Consequences for Leisure. Australian and New Zealand Association of Leisure Studies (ANZALS) 7th biennial conference. Tauranga City, Bay of Plenty, New Zealand (2005).

4. Hall C.M.: Biodiversity and global environmental change, in Tourism and Global Environmental Change: Ecological, Economic, Social and Political Interrelationships, eds Gössling S., & Hall, C.M., Routledge, London, (2006).
5. Chou T.Y.: A fuzzy multi-criteria decision model for international tourist hotels location selection, International Journal of Hospitality Management., 27, 293–301 (2008).
6. Buhalis D., & Law R.: Progress in information technology and tourism management: 20 years on and 10 years after the internet-the state of e-tourism research, Tourism Management. 29 (4), 609–623 (2008).
7. Sevrani K., & Elmazi l.: ICT and the changing landscape of tourism distribution: a new dimension of tourism in the global condition, Revista de turismstudii si cercetari in turism, 6, 22–29 (2008).
8. Hsu T.K., Tsai Y.F., & Wu H.H.: The preference analysis for tourist choice of destination: A case study of Taiwan, Tourism Management, 30, 288–297 (2009).
9. Zhang l., Li N., & Li Mu.: On the basic concept of smarter tourism and its theoretical system. Tourism Tribune, 27(5), 66–73 (2012).
10. Wang D., Li X., & Li Y.: China smart tourism destination initiative: a taste of service dominant logic, Journal of Destination Marketing and Management, 2(2), 59–61 (2013).
11. Dickinson J.E., Ghali K., Cherrett T., Speed C., Davies N., & Norgate S.,: Tourism and the smartphone app: capabilities emerging practice and scope in the travel domain", Current Issues in Tourism, 17(1) 84–101 (2014).
12. Zhu, W., Zhang, L. & Li, N. Challenges: Function Changing of Government and Enterprises in Chinese Smart Tourism. Dublin, IFITT (2014).
13. Lamsfus C., Martin D., Alzua-Sorzabal, A, & Torres Manzanera E.: Smart tourism destination; an extended conception of smart cities focusing on human mobility. In L Tussyadiah & A Inversini (Eds.), Information and communication technologies in tourism Cham New York: Springer, 363–375 (2015).
14. Chang J.R. & Chang B.: The development of a tourism attraction model by using fuzzy theory, Mathematical Problems in Engineering, 1–10 (2015).
15. Avikal S., Mishra P.K., & Jain R.: A Fuzzy AHP and PROMETHEE method-based environment friendly heuristic for disassembly line balancing problems. Interdisciplinary Environmental Review, 14(1), 69–85 (2013).
16. Avikal S., Mishra P.K., & Jain R.: A Fuzzy AHP and PROMETHEE method-based heuristic for disassembly line balancing problems. International Journal of Production Research, 52 (5), 1306–1317 (2013).
17. Avikal S., Jain R., & Mishra P.K.: A Kano model, AHP and TOPSIS method-based technique for disassembly line balancing problems under fuzzy environment. Applied Soft Computing (25), 519–529 (2014).

# Impact of Memory Space Optimization Technique on Fast Network Motif Search Algorithm

**Himanshu and Sarika Jain**

**Abstract** In this paper we propose PATCOMP—a PARTICIA-based novel approach for Network motif search. The algorithm takes advantage of compression and speed of PATRICIA data structure to store the collection of subgraphs in memory and search for classification and census of network. Paper also describes the structure of PATRICIA nodes and how data structure is developed for using it for counting of subgraphs. The main benefit of this approach is significant reduction in memory space requirement particularly for larger network motifs with acceptable time performance. To assess the effectiveness of PATRICIA-based approach we compared the performance (memory and time) of this proposed approach with QuateXelero. The experiments with different networks like ecoli and yeast validate the advantage of PATRICIA-based approach in terms of reduction in memory usage by 4.4–20% for *E. coli* and 5.8–23.2% for yeast networks.

**Keywords** Bioinformatics · Network motifs · Biological networks · Complex networks · Algorithms · Data structures · PATRICIA · Optimization

## 1 Introduction

Network motif searching has many applications including its use in bioinformatics in identification of genes responsible for diseases in humans and animals, and for biotic and abiotic stress in plants, etc. Network motifs were defined for the first time as pattern of interconnections occurring in complex networks in numbers that are significantly higher than those in the randomized networks. Some network motifs

Himanshu (✉) · S. Jain
Amity Institute of Information Technology, Amity University, Noida, UP, India
e-mail: himanshu@icar.org.in

S. Jain
e-mail: sjain@amity.edu



S.K. Bhatia et al. (eds.), *Advances in Computer and Computational Sciences*,
Advances in Intelligent Systems and Computing 553,
DOI 10.1007/978-981-10-3770-2_52

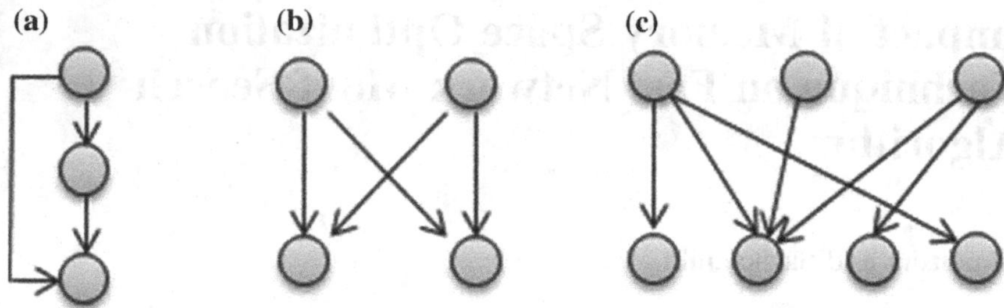


**Fig. 1** Examples of network motif. **a** Feedforward loop, **b** bi-fan, **c** multi input module

are depicted in Fig. 1. In graph theoretic terms, networks are described as graphs with nodes representing vertices and interconnections represented as edges. Therefore, network motif is a subgraph in a larger graph which is statistically over represented than in a set of random networks having similar properties. Network motif is considered as structural building block of several natural networks specifically in biological networks like protein-protein interaction, transcription regulation network, disease gene networks, and gene regulation networks. They are also believed to have important functional roles besides being structural components of complex networks. Study of network motifs in [1] focused not only on biological networks but also on other networks like biochemical, neurobiological, electronic, and ecological networks. They observed that network motifs present in different type of networks were unique to them and therefore network motif analysis could be important approach to uncover basic building blocks of most of the networks.

Network motif searching is np-hard problem. It requires subgraphs to be matched with desired motifs, which involves graph isomorphism checking. Graph isomorphism checking in turn is a problem in np and is not known to be complete, whereas subgraph isomorphism checking is known to be np-complete. Network motif search solutions can be broadly categorized into 'exact motif search' and 'approximate motif search' depending upon the classification and enumeration of subgraphs carried out with or without sampling performed on main graph. Exact motif search is more computation intensive but results are accurate, whereas approximate approach-based solutions are fast with accuracy compromised. Further the solutions have been classified as 'network centric' or 'motif centric' depending upon whether the k-size subgraphs to be searched are enumerated from original network and then used for census on original and random networks as in former class of solution. The alternative approach for later class of solutions is to first generate all different non-isomorphic classes for the specified motif size and then calculate the frequency of each in the network (i.e., count the number of matches of each class in the network). The drawback of network-centric approach is that it requires checking the isomorphism of each enumerated subgraph and the number of non-isomorphic classes grows exponentially with the increasing size of the subgraph. Drawback of motif-centric solutions is that it may spend unnecessary time for checking subgraphs that may not be present in the target network [2]. FANMOD [3], Kavosh [4], G-Tries [5], and QuateXelero [6] are the tools developed based on exact network motif search approach. The algorithm proposed in this paper is also based upon the exact network motif search. FANMOD can detect motifs of size up to k = 8 very fast. It is based upon Rand-ESU algorithm. FANMOD also implements the canonical graph labelling algorithm called NAUTY [7] for grouping subgraphs into isomorphic subgraph classes. Kavosh is an algorithm developed for fast k-motif search with less memory usage. Its four steps include enumeration, classification, random graph generation, and motif identification. It could search network motifs up to size 10 for *E. coli*, *S. cerevisiae* (Yeast) and social networks and upto size 12 for electronic network in satisfactory time. GTries enumerates using ESU and GTrie method and builds graph trie data structure for the subgraphs. It is used for classification of the original network and random networks. It can search network motifs of size k = 3 and more. Its memory usage also increases noticeably as the motif size increases.

QuateXelero makes use of Quaternary tree data structure during enumeration. The quaternary tree is built as the enumeration progresses and it is also used for classification and census on original and random networks. Although QuateXelero is always faster in census on original networks as compared to ESU of G-Tries, it is also generally faster in census on random networks for smaller motifs [6]. Gtries is fast for census on random network. Overall, QuateXelero has best time performance, however, it trades speed for high memory usage of algorithm [8]. Due to high memory usage, it is possible to use QuateXelero only on high end server for large motif (k = 11, 12) discovery. This paper explores impact of memory space optimization using PATRICIA (Practical Algorithm to Retrieve Information Coded in Alphanumeric) trie for network motif search so that the tool could be used for large motif detection on mid-size computers with less memory within satisfactory time. The PATCOMP algorithm has been implemented using C++. PATRICIA has been implemented as described in [9]. The tool would benefit researchers who do not have access to high-end computing resources for motif search requirement. The tool could also be useful for motif search for size unachievable before with the same computing resources.

## 2 Material and Method

PATRICIA is a radix trie which has exactly ‘n’ number of nodes for ‘n’ elements, so it stores the key exactly once. PATRICIA search is a binary algorithm and while searching for the key, branching happens based upon the bit of the key being searched. Search path is followed based on the bits present at the position of the bit index of each node along the path. If the bit is ‘1’ right branch is followed, in case

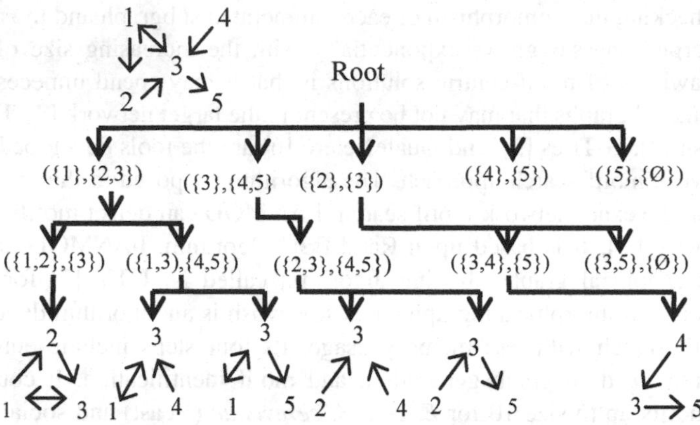


**Fig. 2** Recursive search tree as generated by implementation of ESU algorithm for the given 5-node graph (Adapted from [10])

of '0' left branch is followed. Bit index of the key is the first position in the binary key where the key differs from its parent key in the trie. In the proposed algorithm each node of the trie has one binary key representing the subgraph, a bit index and a pointer which points to the leaf of corresponding subgraph in the binary tree containing the counter of that subgraph. As in QuateXelero, the algorithm uses a binary tree to store the classified non-isomorphic subgraphs in the original and random graphs (Figs 2 and 3).

## 2.1 Enumeration

For enumerating all subgraphs of size 'k' in a given network, PATCOMP algorithm uses procedure similar to the one in FANMOD algorithm. In FANMOD the subgraph is extended by one vertex in each step. As the subgraphs are enumerated one by one using ESU, a keystring is built using the character code for each type of four possible connectivity among the vertices. Character code '1' indicates one way connection from the existing vertex to added vertex, '3' stands for a one way connection in the reverse direction, '4' indicates no connection between them, and '2' represents a two way connection. Table 1 lists the motif size and number of characters in the keystring formed during enumeration.

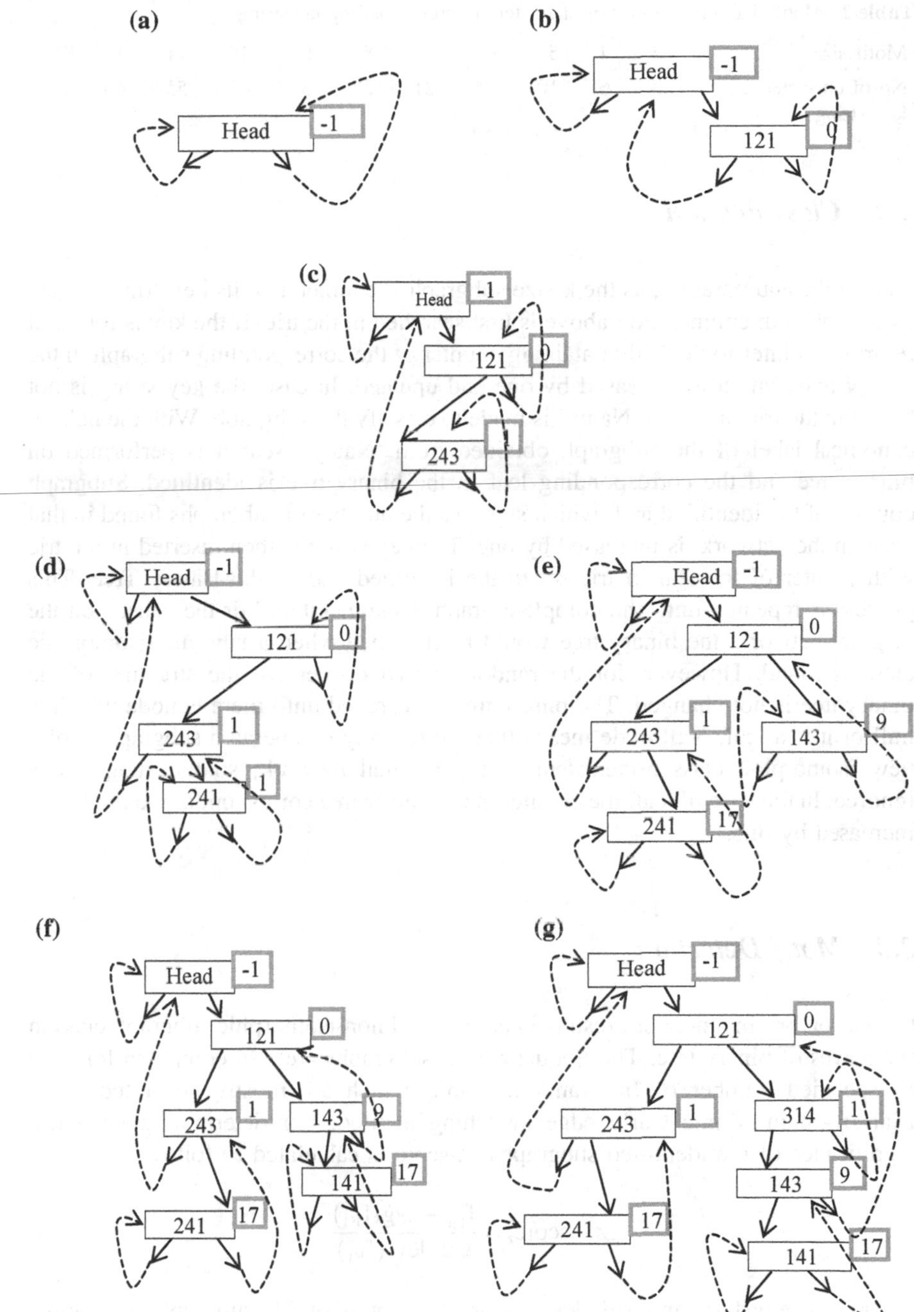


**Fig. 3** **a–g** Step-by-step demonstration of process of building PATRICIA trie during enumeration and classification of a sample graph for 3-node motifs. Keystring is composed of character codes representing each type of connectivity present among the nodes in the subgraph being enumerated. *Number* in the *square* at *the top corner* of keystring indicate bit index

**Table 1** Motif size and number of characters in corresponding keystring

| Motif size | 3 | 4 | 5 | 6 | 7 | 8 | 9 | 10 | 11 | 12 | 13 |
|---|---|---|---|---|---|---|---|---|---|---|---|
| No of characters in key string | 3 | 6 | 10 | 15 | 21 | 28 | 36 | 45 | 55 | 66 | 78 |

## 2.2 Classification

During the enumeration, as the k-size subgraph is enumerated, its keystring formed as described in enumeration above is first searched in the trie. If the key is found, it returns a pointer to the leaf containing counter of the corresponding subgraph in the binary tree, which is increased by one and updated. In case, the key string is not found in the trie, a call to 'Nauty' is made to classify the subgraph. With the help of canonical label of the subgraph obtained from 'Nauty,' search is performed on binary tree and the corresponding leaf in the binary tree is identified. Subgraph counter of the identified leaf, which signifies the number of subgraphs found in that class in the network, is increased by one. The key string is then inserted in the trie with pointer of the leaf in trie set to the identified leaf of the Binary Tree. This process is repeated until the complete graph is enumerated. For the census on the original network, the binary tree would be modified when a new non-isomorphic class is found. However, for the random networks' census, the structure of the binary tree is not changed. The binary tree is searched until there is node which is null or it is a leaf. Null node means that the recently enumerated subgraph is of a new isomorphic class nonexistent in the original network; so the subgraph is ignored. In the case of leaf, the counter in the binary tree corresponding to the leaf is increased by one.

## 2.3 Motif Detection

Census on original network results in count of all non-isomorphic subgraph class in the leaves of binary tree. The frequencies of subgraphs are also computed for a set of specified number of 'm' random graphs which are mostly generated using Markov-chain Monte Carlo edge switching method. For determining statistical significance of the identified subgraphs, Z-score is calculated as follows:

$$z - \text{score}_i = \frac{f_{Gi} - \text{avg}(f_{Ri})}{\text{std dev }(f_{Ri})}$$

where $f_{Gi}$, $\text{avg}(f_{Ri})$, and std $\text{dev}(f_{Ri})$ are the count of ith subgraph in original network, average number of occurrences of ith subgraph in random graphs and standard deviation of occurrences of ith subgraph in 'n' random networks, respectively. Higher z-score indicates higher possibility of isomorphism class being the motif.

### 2.4 Datasets

We have used two standard directed networks viz. the metabolic pathway of bacteria *E. coli* [11] and transcription network of Yeast (*S. cerevisiae*) [12], for validating our algorithm. Currently the comparison of PATCOMP, for small and large motifs with QuateXelero has been carried out. Work is underway to compare the PATCOMP with GTries.

## 3 Results and Discussion

For comparing our algorithm PATCOMP with QuateXelero both the tools were executed on the same computer with Intel Core i5 CPU 2.6 GHz and 4 GB RAM, Windows 8.1 (64-bit) installed with cygwyn (64-bit). For larger experiments, a server with Quad-Core Intel Xeon Processor ES2650, 2.6 GHz, 64 GB main memory, and MS Windows 2008 R2 installed with Cygwyn (64-bit) was used. The work is in progress to evaluate both the tools for other standard networks and large motif sizes for yeast and ecoli as well. The results are illustrated in Table 2. It is seen that, while QuateXelero is still faster as compared with PATCOMP in all cases, the superiority of PATCOMP algorithms is visible in memory space utilization for both the networks. Memory space requirement for PATCOMP is lesser than QuateXelero in general and the savings in memory increase with increased motif size. For example, whereas QuateXelero for searching network motif of size 10 for ecoli requires 18.83 GB memory; PATCOMP is able to complete the search with 17.16 GB memory only. Similar trend is expected for other standard networks also.

It may be noted that the memory requirement of PATRICIA trie depends upon the number of non-isomorphic classes present in the given network for specified network motif size.

There is further scope for reducing the memory requirement of the algorithm using string compression on the PATRICIA trie node key string. Work is in progress to implement the same in the present algorithm.

**Table 2** Comparison of PATCOMP with QuateXelero vis-e-vis directed networks

| Network | Motif size | No. of random networks | Total time (s) | | Memory | | Memory saving (vs QuateXelero) |
|---|---|---|---|---|---|---|---|
| | | | QuateXelero | PATCOMP | QuateXelero | PATCOMP | PATCOMP |
| Ecoli | 5 | 100 | 0.765 | 1.54 | 11.6 MB | 11.6 MB | 0 MB |
| | 6 | 100 | 4.79 | 13.01 | 22.7 MB | 21.7 MB | 1 MB |
| | 7 | 100 | 39.84 | 122.3 | 137.3 MB | 118.3 MB | 19 MB |
| | 8 | 100 | 486.32 | 1624.03 | 1.11 GB | 1.03 GB | 0.08 GB |
| | 9 | 5 | 150.13 | 333.95 | 2.34 GB | 1.87 GB | 0.47 GB |
| | 10 | 5 | 1391.2 | 3267 | 18.83 GB | 17.16 GB | 1.67 GB |
| Yeast | 5 | 10 | 1.34 | 3.51 | 1.9 MB | 1.7 MB | 0.2 MB |
| | 6 | 10 | 28.15 | 69.96 | 3.4 MB | 3.2 MB | 0.2 MB |
| | 7 | 10 | 372.76 | 1125.34 | 17.1 MB | 13.1 MB | 4 MB |

## 4 Conclusion and Future Work

In general the PATCOMP utilizes less memory by 4.4–20% over QuateXelero for directed networks of *E. coli*, and memory utilized by PATCOMP is less by 5.8–23.3% for directed network of yeast. Time taken by PATCOMP is more by $2\times$–$3.33\times$ that of QuateXelero. Processing time for census on original network is in general less than GTries (as per published data of GTries) the second best solution for Network motif finding for directed networks. PATCOMP is more suitable for use on computers with less memory than QuateXelero for the networks analyzed. Work is in progress to analyze the performance of the PATCOMP algorithm for other standard networks like social, yeast PPI and electronic, and for larger motif sizes for *E. coli* and yeast.

**Acknowledgements** Authors express their deep sense of gratitude to the Founder President of Amity University, Dr. Ashok K. Chauhan, for his keen interest in promoting research in the Amity University and has always been an inspiration for achieving greater heights.

## References

1. R, Milo, S Shen-Orr, S Itzkovitz, N Kashtan, D Chklovskii, and U Alon. "Network motifs: Simple building blocks of complex networks." *Science* 298 (2002): 824–827.
2. Wong, E, B Baur, S Quader, et al. "Biological network motif detection: principles and practice." *BriefBioinform* 13, no. 2 (2001): 202–15.
3. Wernicke, S, and F Rasche. "FANMOD: a tool for fast network motif detection." *Bioinformatics* 22 (2006): 1152–1153.
4. Kashani, Z R, H Ahrabian, E Elahi, A Nowzari-Dalini, and et al. "Kavosh: a new algorithm for finding network motifs." *BMC bioinformatics* 10 (2009): 318.
5. Ribeiro, P, and F Silva. "Efficient subgraph frequency estimation with g-tries." *International workshop on algorithms in bioinformatics (WABI), LNCS*. Springer, 2010. 238–249.
6. Khakabimamaghani, S, I Sharafuddin, N Dichter, et al. "QuateXelero: an accelerated exact network motif detection algorithm." *PLoSOne* 8, no. 7 (2013).
7. McKay, B D. "Practical graph isomorphism." *10th Manitoba conference on numerical mathematics and computing*. Congressus Numerantium, 1981. 45–87.
8. Tran, NgocTam L, Sominder Mohan, Zhuoqing Xu, and Chun-Hsi Huang. "Current innovations and future challenges of network motif detection." *Briefings in Bioinformatics*, 2014: 1–29.
9. Robert, Sedgewick. *Algorithms*. Addison Wesley, 1984.
10. Warnicke, S. "Efficient detection of network motifs." *IEEE/ACM Transactions on Computational Biology and Bioinformatics* 3, no. 4 (2006): 347–359.
11. The E.coli Database. Available: http://www.kegg.com/.
12. The S. cerevisiae Database. Available: http://www.weizmann.ac.il/mcb/UriAlon/.

# A Novel Hybrid Approach Particle Swarm Optimizer with Moth-Flame Optimizer Algorithm

**R.H. Bhesdadiya, Indrajit N. Trivedi, Pradeep Jangir, Arvind Kumar, Narottam Jangir and Rahul Totlani**

**Abstract** Recent trend of research is to hybridize two and more algorithms to obtain superior solution in the field of optimization problems. In this context, a new method hybrid PSO (Particle Swarm Optimization)—MFO (Moth-Flame Optimizer) is exercised on some unconstraint benchmark test functions and overcurrent relay coordination optimization problems in contrast to test results on constrained/complex design problem. Hybrid PSO-MFO is combination of PSO used for exploitation phase and MFO for exploration phase in uncertain environment. Position and Velocity of particle is updated according to Moth and flame position in each iteration. Analysis of competitive results obtained from PSO-MFO validates its effectiveness compare to standard PSO and MFO algorithm.

**Keywords** Heuristic · Moth-flame optimizer · Particle swarm optimization · HPSO-MFO · Overcurrent relay

R.H. Bhesdadiya (✉)
Electrical Engineering Department, School of Engineering, RK University, Rajkot, Gujarat, India
e-mail: rhbhesdadiya@gmail.com

I.N. Trivedi
Electrical Engineering Department, G.E. College, Gandhinagar, Gujarat, India
e-mail: forumtrivedi@gmail.com

P. Jangir · N. Jangir
Electrical Engineering Department, LEC, Morbi, Gujarat, India
e-mail: pkjmtech@gmail.com

N. Jangir
e-mail: nkjmtech@gmail.com

A. Kumar
Electrical Engineering Department, S.S.E.C., Bhavnagar, Gujarat, India
e-mail: akbharia8@gmail.com

R. Totlani
Electrical Engineering Department, JECRC, Jaipur, Rajasthan, India
e-mail: rhl.totlani@gmail.com



S.K. Bhatia et al. (eds.), *Advances in Computer and Computational Sciences*, Advances in Intelligent Systems and Computing 553,
DOI 10.1007/978-981-10-3770-2_53

# 1 Introduction

HPSO-MFO comprises of best characteristic of both Particle Swarm Optimization [1] and Moth-Flame Optimizer [2] algorithm. HPSO-MFO result expresses that it has ability to converse faster with comparatively optimum solution for both unconstrained and constrained function.

Population-based algorithms, based on randomization consists of two main phases for obtaining better results that are exploration (unknown search space) and exploitation (best solution). In this HPSO-MFO, MFO is applied for exploration as it uses logarithmic spiral path so covers large uncertain search space with less computational time to explore possible solution or to converse particle toward optimum value. Most popular PSO algorithms have ability to attain near optimal solution avoiding local solution.

Contemporary works in hybridization are PBIL-KH [3] the population-based incremental learning (PBIL) with KH, a type of elitism is applied to memorize the krill with the best fitness when finding the best solution, KH-QPSO [4] is intended for enhancing the ability of the local search and increasing the individual diversity in the population, HS/FA [5] the exploration of HS and the exploitation of FA are fully exerted, CKH [6] the chaos theory into the KH optimization process with the aim of accelerating its global convergence speed, HS/BA [7], CSKH [8], DEKH [9], HS/CS [10], HSBBO [11] are used for the speeding up convergence, thus making the approach more feasible for a wider range of real-world applications.

Recently, trend of optimization is to improve performance of meta-heuristic algorithms [12] by integrating with chaos theory, levy flights strategy, adaptive randomization technique, evolutionary boundary handling scheme, and genetic operators like as crossover and mutation. Popular genetic operators used in KH [13] that can accelerate its global convergence speed. Evolutionary constraint handling scheme is used in Interior Search Algorithm (ISA) [14] that avoid upper and lower limits of variables.

The structure of the paper can be given as follows: Introduction; description of participated algorithms; competitive results analysis of unconstraint test benchmark problem and constrained relay coordination problem finally acknowledgement, and conclusion based on results is drawn.

# 2 Standard PSO and Standard MFO

## 2.1 PSO (Particle Swarm Optimization)

The PSO (particle swarm optimization) algorithm was discovered by James Kennedy and Russell C. Eberhart in 1995 [1]. This algorithm is inspired by simulation of sociological expression of birds and fishes. PSO includes two terms P best and G best. Position and velocity are updated over the course of iteration from these mathematical equations:

$$v_{ij}^{t+1} = wv_{ij}^{t} + c_1 * R_1(P_{best}^{t} - (X)^{t}) + c_2 * R_2(G_{best}^{t} - (X)^{t}) \tag{1}$$

$$\begin{aligned} &(X)^{t+1} = X^{t} + v^{t+1}, (\mathrm{i} = 1, 2, \ldots, \text{No. of Particles}) \\ &\mathit{and}\ (\mathrm{j} = 1, 2, \ldots, \text{No. of Generators.}) \end{aligned} \tag{2}$$

where,

$$w = w^{\max} - \frac{(w^{\max imum} - w^{\min imum}) * iteration}{\max imum\ iteration} \tag{3}$$

$w^{max} = 0.4$ and $w^{min} = 0.9.v_{ij}^{t}$, $v_{ij}^{t+1}$ is the velocity of *jth* member of *ith* particle at iteration number (t) and (t + 1). (Usually C1 = C2 = 2), r1 and r2 Random number (0, 1).

Flow Chart for PSO Algorithm is shown in Fig. 1.

## 2.2 Moth-Flame Optimizer

Moth-Flame optimizer was first introduced by Seyedali Mirjalili in 2015 [2]. MFO is a population-based meta-heuristic algorithm. The MFO algorithm is three-rows that approximate the global solution of the problems defined as follows:

$$\text{Moth Flame Optimizer} = [\mathrm{I},\ \mathrm{P},\ \mathrm{T}], \tag{4}$$

*I* is the function that yield an uncertain population of moths and corresponding fitness values. Considering these points, we define a log (logarithmic scale) spiral for the MFO algorithm as follows:

$$S(M_i, F_j) = D_i * e^{bt} \cos(2\pi t) + F_j \tag{5}$$

where $D_i$ expresses the distance of the moth for the *jth* flame, *b* is a constant for expressing the shape of the log (logarithmic) spiral, and *t* is a random value in [− 1, 1].

$$Di = |Fj - Mi|Z \tag{6}$$

where $M_i$ indicate the *ith* moth, $F_j$ indicates the *jth* flame, and where expresses the path length of the *ith* moth for the *jth* flame.

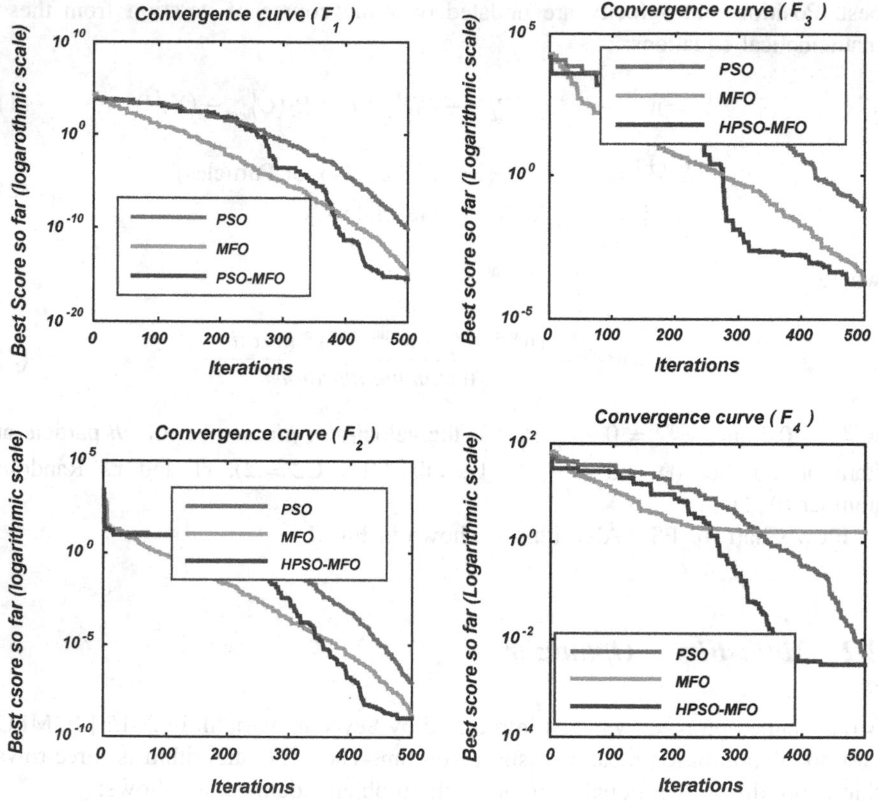


**Fig. 1** Convergence characteristics of benchmark test functions

The number of flames is adaptively decreased over the course of iterations. We use the following formula:

$$no.\,of\,flame = round\left(N - l * \frac{N-1}{T}\right) \tag{7}$$

where $l$ indicates the current number of iteration, $N$ indicates the maximum number of flames, and $T$ is the maximum number of iterations.

## 2.3 The Hybrid PSO-MFO Algorithm

A set of Hybrid PSO-MFO is combination of separate PSO and MFO. The drawback of PSO is the limitation to cover small search space while solving higher order or complex design problem due to constant inertia weight. This problem can be tackled with Hybrid PSO-MFO as it extracts the quality characteristics of both PSO and MFO. Moth-Flame Optimizer is used for exploration phase as it uses logarithmic spiral function so it covers broader area in uncertain search space. Because both of the algorithms are randomization techniques so we use term uncertain search space during the computation over the course of iteration from starting to maximum iteration limit. Exploration phase means capability of algorithm to try out large number of possible solutions. Position of particle that is responsible for finding the optimum solution of the complex nonlinear problem is replaced with the position of Moths that is equivalent to position of particle but highly efficient to move solution toward optimal one. MFO directs the particles faster toward optimal value, reduces computational time. As we know that PSO is a well-known algorithm that exploits the best possible solution from its unknown search space. So combination of best characteristic (exploration with MFO and exploitation with PSO) guarantees to obtain best possible optimal solution of the problem that also avoids local stagnation or local optima of problem. Hybrid PSO-MFO merges the best strength of both PSO in exploitation and MFO in exploration phase toward the targeted optimum solution.

$$v_{ij}^{t+1} = w\, v_{ij}^{t} + c_1 R_1 (Moth\_Pos^t - X^t) + c_2 R_2 (Gbest^t - X^t) \tag{8}$$

## 3 Simulation Results for Unconstraint Test Benchmark Function

Unconstraint benchmark test functions are solved using HPSO-MFO algorithm. Four benchmark test functions (F1-F4) are performed to verify the HPSO-MFO algorithm in terms of exploration and exploitation. These test functions are shown in Table 1. Results are shown in Table 2, HPSO-MFO algorithm able to given more competitive results compared to standard PSO and MFO algorithm. The convergence characteristics of HPSO-MFO is shown in Fig. 1. Search agent no. is 30 and maximum iteration no. is 500 used for all Unconstraint benchmark test functions.

**Table 1** Unconstraint benchmark test functions

| Function | Dim | Range | $F_{min}$ |
|---|---|---|---|
| $f_1(x) = \sum_{k=1}^{n} (x_k)^2 * R(x)$ | 10 | [− 100, 100] | 0 |
| $f_2(x) = \sum_{k=1}^{n} \left[ \lvert x_k \rvert + \prod_{k=1}^{n} (\lvert x_k \rvert) \right] * R(x)$ | 10 | [− 10, 10] | 0 |
| $f_3(x) = \sum_{k=1}^{n} \left[ \sum_{m-1}^{k} (x_m) \right]^2 * R(x)$ | 10 | [− 100, 100] | 0 |
| $f_4(x) = \max_k \{ \lvert x_k \rvert, 1 \le k \le n \}$ | 10 | [− 100, 100] | 0 |

**Table 2** Result for unconstraint benchmark test functions

| Fun. | Std. PSO | | | Std. MFO | | | PSO-MFO | | |
|---|---|---|---|---|---|---|---|---|---|
| | Ave | Best | S.D. | Ave | Best | S.D. | Ave | Best | S.D. |
| F1 | 7.687E-11 | 7.553E-11 | 1.889E-12 | 4.6425E-14 | 1.2818E-15 | 6.3842E-14 | 2.3464E-14 | **3.9336E-16** | 3.2627E-14 |
| F2 | 2.416E-07 | 9.005E-08 | 2.144E-07 | 1.5500E-09 | 1.5491E-09 | 1.2605E-12 | 1.7256E-09 | **1.0854E-09** | 9.0540E-10 |
| F3 | 0.1510412 | 0.691860 | 0.115760 | 0.3145 | 2.1568E-04 | 0.4444 | 0.0010 | **1.7681E-04** | 0.0012 |
| F4 | 0.025 | 0.004 | 0.029 | 1.8100 | 1.6334 | 0.2498 | 1.0127 | **0.0031** | 1.427 |

The significance of bold represents best value of newly proposed Hybrid PSO-MFO algorithm with respect other algorithm

## 4 Overcurrent Relay Coordination with Common Configuration in Power System

Overcurrent relay is used for primary and backup protection in distribution power systems. To minimize the total operating time relays should be coordinated and set at the optimum values [15, 16].

All relays used in this paper are identical and they show the normal IDMT (Inverse Definite Minimum Time) characteristics represented in terms of equations are as follows:

$$t = \frac{0.14 * (TMS)}{PSM^{(0.02)} - 1} \tag{9}$$

where $t$ is the operating time of relay, $PSM$ is plug setting multiplier, and $TMS$ represents time multiplier setting.

$$PSM = \frac{I_{relay}}{PS} \tag{10}$$

For linear problem $PSM$ is constant, so $t$ decreases to

$$t = \alpha_p * (TMS) \tag{11}$$

$$\alpha_p = \frac{0.14}{PSM^{(0.02)} - 1} \tag{12}$$

The target is to minimize the objective function given by:

$$F_{\min} = \sum_{p=1}^{n} \alpha_p * (TMS)_p \tag{13}$$

The optimal results are given in Table 3. Figure 2 show the Convergence Characteristics of Overcurrent Relay Coordination for Parallel Feeder, fed from a single end. The constraints are taken from [15]. Search agent no. is 30 and maximum iteration no. is 500 used for solve the Over current relay coordination problem.

**Table 3** Values of TMS for parallel feeder system, fed from a single end

| Relay | TMS | MFO | PSO | HPSO-MFO |
|---|---|---|---|---|
| $R_1$ | $TMS_1$ | 0.09383 | 0.099141 | 0.069729 |
| $R_2$ | $TMS_2$ | 0.026826 | 0.025 | 0.025219 |
| $R_3$ | $TMS_3$ | 0.034959 | 0.050317 | 0.03895 |
| $R_4$ | $TMS_4$ | 0.08049 | 0.076716 | 0.072356 |
| $R_5$ | $TMS_5$ | 0.057816 | 0.060471 | 0.051094 |
| Total operating time | | 0.293921 | 0.311645 | **0.257348** |

The significance of bold represent best value of newly proposed Hybrid PSO-MFO algorithm with respect other algorithm

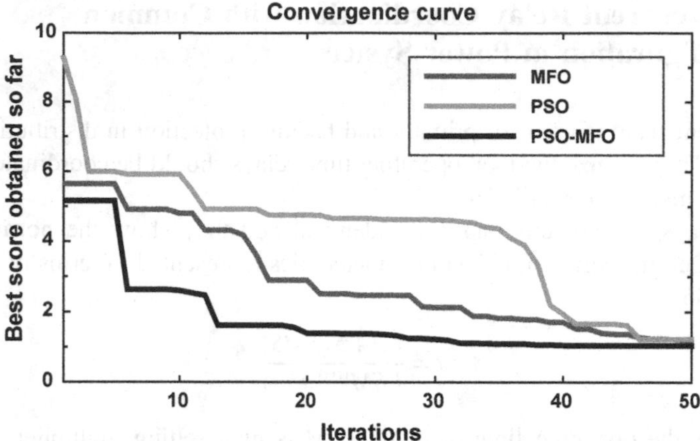


**Fig. 2** Convergence characteristics of overcurrent relay coordination for parallel feeder, fed from a single end

Minimize

$$Z = \left\{ \begin{array}{l} 3.106X_1 + 6.265X_2 + 3.106X_3 \\ + 6.265X_4 + 2.004X_5 \end{array} \right\} \tag{14}$$

$$6.265X_4 - 3.106X_2 \geq 0.2, \tag{15}$$

$$6.265X_1 - 3.106X_3 \geq 0.2, \tag{16}$$

$$4.341X_1 - 2.004X_5 \geq 0.2, \tag{17}$$

$$4.341X_4 - 2.004X_5 \geq 0.2, \tag{18}$$

# 5 Conclusions

The drawback of PSO is the limitation to cover small search space while solving higher order or complex design problem due to constant inertia weight. This problem can be tackled with Hybrid PSO-MFO as it extracts the quality characteristics of both PSO and MFO. MFO is used for exploration phase as it uses logarithmic spiral function so it covers broader area in uncertain search space. So MFO directs the particles faster toward optimal value, reduces computational time. HPSO-MFO is tested on four unconstrained and one overcurrent relay as constrained problems. HPSO-MFO gives optimal results in most of the cases and in some cases results are inferior that demonstrate the enhanced performance with respect to original PSO and MFO.

## References


1. J. Kennedy, R. Eberhart, Particle swarm optimization, in: Proceedings of the IEEE International Conference on Neural Networks, Perth, Australia, 1995, pp. 1942–1948.
2. Seyedali Mirjalili, "Moth-flame optimization algorithm: A novel nature-inspired heuristic paradigm," Knowledge-Based System, vol. 89, pages 228–249, 2015.
3. Gai-Ge Wang, Amir H. Gandomi, Amir H. Alavi, Suash Deb, A hybrid PBIL-based Krill Herd Algorithm, December 2015.
4. Gai-Ge Wang, Amir H. Gandomi, Amir H. Alavi, Suash Deb, A hybrid method based on krill herd and quantum-behaved particle swarm optimization, Neural Computing and Applications, 2015, doi:10.1007/s00521-015-1914-z.
5. Lihong Guo, Gai-Ge Wang, Heqi Wang, and Dinan Wang, An Effective Hybrid Firefly Algorithm with Harmony Search for Global Numerical Optimization, Hindawi Publishing Corporation The Scientific World Journal Volume 2013, Article ID 125625, 9 pages 10.1155/2013/125625.
6. Gai-Ge Wang, Lihong Guo, Amir Hossein Gandomi, Guo-Sheng Hao, Heqi Wang. Chaotic krill herd algorithm. Information Sciences, Vol. 274, pp. 17–34, 2014.
7. GaigeWang and Lihong Guo, A Novel Hybrid Bat Algorithm with Harmony Search for Global Numerical Optimization, Hindawi Publishing Corporation Journal of Applied Mathematics Volume 2013, Article ID 696491, 21 pages http://dx.doi.org/10.1155/2013/696491.
8. Gai-Ge Wang, Amir H. Gandomi, Xin-She Yang, Amir H. Alavi, A new hybrid method based on krill herd and cuckoo search for global optimization tasks. Int J of Bio-Inspired Computation, 2012, in press.
9. Gai-Ge Wang, Amir Hossein Gandomi, Amir Hossein Alavi, Guo-Sheng Hao. Hybrid krill herd algorithm with differential evolution for global numerical optimization. Neural Computing & Applications, Vol. 25, No. 2, pp. 297–308, 2014.
10. Gai-Ge Wang, Amir Hossein Gandomi, Xiangjun Zhao, HaiCheng Eric Chu. Hybridizing harmony search algorithm with cuckoo search for global numerical optimization. Soft Computing, 2014. doi:10.1007/s00500-014-1502-7.
11. Gaige Wang, Lihong Guo, Hong Duan, Heqi Wang, Luo Liu, and Mingzhen Shao, Hybridizing Harmony Search with Biogeography Based Optimization for Global Numerical Optimization, Journal of Computational and Theoretical Nanoscience Vol. 10, 2312–2322, 2013.
12. A.H. Gandomi, X.S. Yang, S. Talatahari, A.H. Alavi, Metaheuristic Applications in Structures and Infrastructures, Elsevier, 2013.
13. A.H. Gandomi, A.H. Alavi, Krill Herd: a new bio-inspired optimization algorithm, Common Nonlinear Sci. Numer. Simul. 17 (12) (2012) 4831–4845.
14. Gandomi A.H. "Interior Search Algorithm (ISA): A Novel Approach for Global Optimization." ISA Transactions, Elsevier, 53(4), 1168–1183, 2014.
15. S. S. Gokhle, Dr. V. S. Kale, "Application of the Firefly Algorithm to Optimal Overcurrent Relay Coordination", IEEE Conference on Optimization of Electrical and Electronic equipment, Bran, 2014.
16. I.N. Trivedi, S.V. Purani, Pradeep Jangir, "Optimized over-current relay coordination using Flower Pollination Algorithm", "Advance Computing Conference (IACC), 2015 IEEE International", pages 72–77.

# Differential Evolution Algorithm Using Population-Based Homeostasis Difference Vector

**Shailendra Pratap Singh and Anoj Kumar**

**Abstract** For the last two decades, the differential evolution is considered as one of the powerful nature inspired algorithm which is used to solve real-world problems. DE takes minimum number of function evaluations to reach close to global optimum solution. The performance is very good, but it suffers from the problem of stagnation when tested on multi-modal functions. In this paper, the population-based homeostasis difference vector strategy has been used to improve the performance of differential evolution algorithms. Here we propose two independent difference random vectors named as best difference vector and random difference vector which helps in avoiding stagnation problem of multi-modal functions. The performance of proposed algorithm is compared with other state-of-the-art algorithms on COCO (Comparing Continuous Optimizers) framework. The result verifies that our proposed population-based homeostasis difference vector strategy outperform most of the state-of-the-art DE variants.

**Keywords** Differential evolution algorithm · Homeostasis · COCO platform

## 1 Introduction

Practical optimization problem solving techniques should be able to achieve the following [1]:

1. True global minimum.
2. Fast convergence.
3. Minimum number of control parameters, which are easy to use.

S.P. Singh (✉) · A. Kumar
Department of Computer Science and Engineering, Motilal Nehru National Institute of Technology Allahabad, Allahabad, UP, India
e-mail: shail2007singh@gmail.com

A. Kumar
e-mail: anojk@mnnit.ac.in

 
S.K. Bhatia et al. (eds.), *Advances in Computer and Computational Sciences*, Advances in Intelligent Systems and Computing 553,
DOI 10.1007/978-981-10-3770-2_54

With the above requirements in mind, Storn and Price introduced the Differential Evolution algorithm in 1995, which is a population-based stochastic search algorithm for unconstrained continuous optimization problems. It is similar to Genetic algorithm (GA), but differs in the following:

- It uses floating point representations.
- It uses a unique arithmetic reproduction operator for altering the internal representation of individuals in a population.

Stron and Price [1] introduced DE as a "Simple and efficient heuristics for global optimization over continuous spaces". It searches for global optimum solution in D-dimension real parameter space in four steps, i.e., initialization, mutation, crossover, and selection. The proposed approach of Stron and Price is explained in Algorithm 1.

**Algorithm 1** Basic Differential Evolution

```
procedure DEA
Initialization: Number of population size(NP), G is the generation and D is the dimension.
Control parameter: Mutation Factor(δ) and crossover rate Cr.
For = 1 to NP do
Mutation Strategy: Generate a donor vector
```

$$\gamma_{i,G} = \alpha_{\text{best},G} + \delta 1 \cdot (\alpha_{r_1^i,G} - \alpha_{r_2^i,G})$$

```
Crossover Strategy: Generate a trial vector
Selection Strategy
    A better selection scheme is used
    Choose best fitness as solution
End for
end procedure
```

DE is an evolutionary algorithm using iterative progress. To let DE adapt to changing environment, an attempt has been made in the present paper to enhance its performance using population-based homeostasis difference vector. The stability of every species and also human body in case of changes in external or internal environment conditions is maintained by a process known as homeostasis. Following the same homeostasis phenomena we have devised a new approach.

The rest of the paper has been organized as follows: Related work has been discussed and described in Sect. 2. The proposed homeostasis-based strategy for DE has been described in Sect. 3. Experimental setup and result analysis have been given in Sect. 4 to prove the credibility of proposed approach. Section 5 concludes the paper.

## 2 Related Work

Differential Evolution (DE) given by Storn and Price in 1995 performs well in standard test functions and real-world optimization problems [2] such as multi-objective, non-convex, non-differential, nonlinear constraints, and dynamic components [3, 4].

For enhancing the performance and application of DE, many variants of DE for continuous, Single-objective, and Multi-objective optimization problems and with different choice rules have been developed which are briefly described below.

In 2006, Janez Brest et al. [5], proposed a new version of the DE algorithm for obtaining self-adaptive control parameter settings. The idea "Evolution of the evolution" has been used to find the values of control parameters F and CR. The proposed algorithm has been shown to be better than or at least comparable to the standard DE algorithms. In 2006, Efren Mezura-Montes et al. [6] have presented an empirical comparison of eight DE variants to solve global optimization problems and shown that regardless of the characteristics of the problem to be solved, the best/1/bin was the most competitive approach. In 2011, Nasimul Noman et al. [7] have proposed a crossover-based adaptive local search operation for enhancing the performance of DE. In 2011, Swagatam Das et al. [3] have done a state-of-the-art survey and review of the literature concerning "Differential Evolution" and its variants. It includes the different schemes for controlling and adaption of parameters, modifications of DE for "Tackling constraints, multi-objective, uncertain, and large scale optimization problems" of various domains. They have also mentioned many future directions for further research. The parameter selection and assignment of values itself is a tedious task. Rahul A. Sarker et al. [8] in 2014, proposed a new mechanism to dynamically select the best performing combination of parameters for a problem during the course of a single run. Zhou et al. [9] in 2014 proposed a DE variant with novel role assignment (RA) scheme, which utilizes both fitness and personal information of each individual in the population for dynamically dividing the population in three groups. In 2015, Brown et al. [10], proposed a new mutation operator "current-by-rand-to-best" which is applied on less than ten population size for unconstrained continuous optimization problems. In 2016, Hu et al. [11] suggested that DE cannot guarantee global convergence of multi-modal functions because, the optimum solution of the multi-modal functions lies near the boundary of the search space as well as they have larger deceptive optima in the search space.

## 3 Proposed Approach

Performance of differential evolution (DE) algorithms depends on its parameter and the operators, i.e., crossover, mutation, and selection. Out of these operators, mutation ($\delta$) is used for exploring and exploiting the search area around the possible solutions. To improve the convergence rate, to reduce the number of fitness evaluations and to prevent stagnation caused by premature convergence, it has been suggested [12, 13] to explore and exploit around the best solutions, rather than all the solutions. Further, solutions might get entrapped around the local and global minima and start exploiting, as we see in the cases of both unimodal and multimodal problems. To avoid this situation, one worst solution has also been considered in the present proposed approach for the mutation operator to maintain the diversity. The convergence rate is improved as exploring and exploitation is happening around the best solutions in such a way that the solutions do not stagnate in mutation. If yet it stagnates on a multimodal functions, then to avoid this problem we have used the concept of population-based homeostasis difference vector (PBHDV).

**Population-based homeostasis difference vector (PBHDV)**: For PBHDV, the whole current population has been divided into two parts. The first part consists of all the better individuals of the populations, based on their fitness values and the rest of the individuals are kept in the other part. The division point is chosen randomly. The first part is called as best population and the second part is called as random population. The division of population has been done to improve the diversity and convergence, because the best population consists of the better solutions and it is expected that all the good vectors will be available from starting. From the above two population segments two difference vectors named as best difference vector and random difference vector are generated using Eqs. 1 and 2, respectively.

$$\text{best difference vector} = (\alpha_{BRV_1^i,G} - \alpha_{BRV_2^i,G}). \quad (1)$$

Where BRV1 is the Best random vector1 and BRV2 is the Best random vector2.

$$\text{random difference vector} = (\alpha_{r_2^i,G} - \alpha_{r_3^i,G}). \quad (2)$$

Where r2 is the random vector2 and r3 is the random vector3.

The above two difference vectors have been designed to provide better diversity among solutions so that global search area can be exploited and better solutions from all areas can be identified during selection. During initial generations, the above two difference vectors provide enhanced diversity, but as the generation increases, the difference vectors which always try to select better solutions become ineffective. As such, after few generations these solutions start converging to a particular region and loose their diversity. At this point of time, the random population is used, which may also consists of some good solutions.

The donor vector is generated as the sum of two vectors as donor vector = best difference vector + random difference vector, and this strategy has been applied to DE and its variant, i.e., HADE-Best (HADE-B) have been proposed as given in Algorithm 2.

## 4 Experimental and Analysis

The proposed variants of DE algorithm using Population-based homeostasis difference vector (HADE) have been tested using COCO platform for the Black-Box Optimization Benchmarking (BBOB) on following 24 noiseless test functions [14–16, 21]. These function are grouped into five groups which include Separable Functions (f1–f5), Functions with low or moderate conditioning (f6–f9), Functions with high conditioning and unimodal (f10–f14), Multimodal functions with adequate global structure (f15–f19) and Multimodal functions with weak global structure (f20–f24) [14].

**Algorithm 2** First Strategy- $(HADE - B)$

**procedure** START
Control Parameter:
    Cr = 0.7
    $\delta = 0.8$
For each target vector generate from new population.
For I = 1 to NP do
Choose three best random vectors BRV1, BRV2 and BRV3 and individual random vectors r1, r2 and r3 from new population.
Mutation Strategy: Generate a donor vector: $\gamma_{i,G} =$

$$\alpha_{\text{best},G} + \delta 1 \cdot (\alpha_{BRV^i_1,G} - \alpha_{BRV^i_2,G}) + \delta 2 \cdot (\alpha_{r^i_2,G} - \alpha_{r^i_3,G})$$

Apply the crossover
Apply the selection
    Choose best Vectors
    Termination conditions met
    End for
**end procedure**

## *4.1 Testing Framework*

Search space is $[-5, 5]^D$ and fifteen instances of each test function have been taken. The terminal conditions are either function evaluations or a precision of greater than $10^{-8}$. Majority of benchmark functions have optima in [−4, 4] domain. The population size is taken as 40 is fixed and maximum number of functional evaluations(***FE***) is 10000 * D, where D is the dimension. Further, expected running time loss has been calculated in all the variants of proposed HADE algorithm.

The experiments have been conducted on a PC with Intel Core i5 CPU with a speed of 3.40 GHz, installed memory (RAM) 4 GB, operating system Windows 7 Pro 64 bit, and x64-based processor.

## *4.2 Analysis of the Results*

Newly created Differential Evolution Algorithm Using Population-Based Homeostasis Difference Vector (HADE), i.e., HADE-B (best) are compared with existing DE algorithms like JADEb [17], DE [1], DEb [22], DE-PAL [18], DEAE [15], and MVDE [19]. Figure 1 shows the comparison for 2-D, 3-D, 5-D, 10-D, 20-D, and 40-D. It also shows the numbers preceding the various algorithm is the position of that optimization with respect to other optimization algorithms. These observations

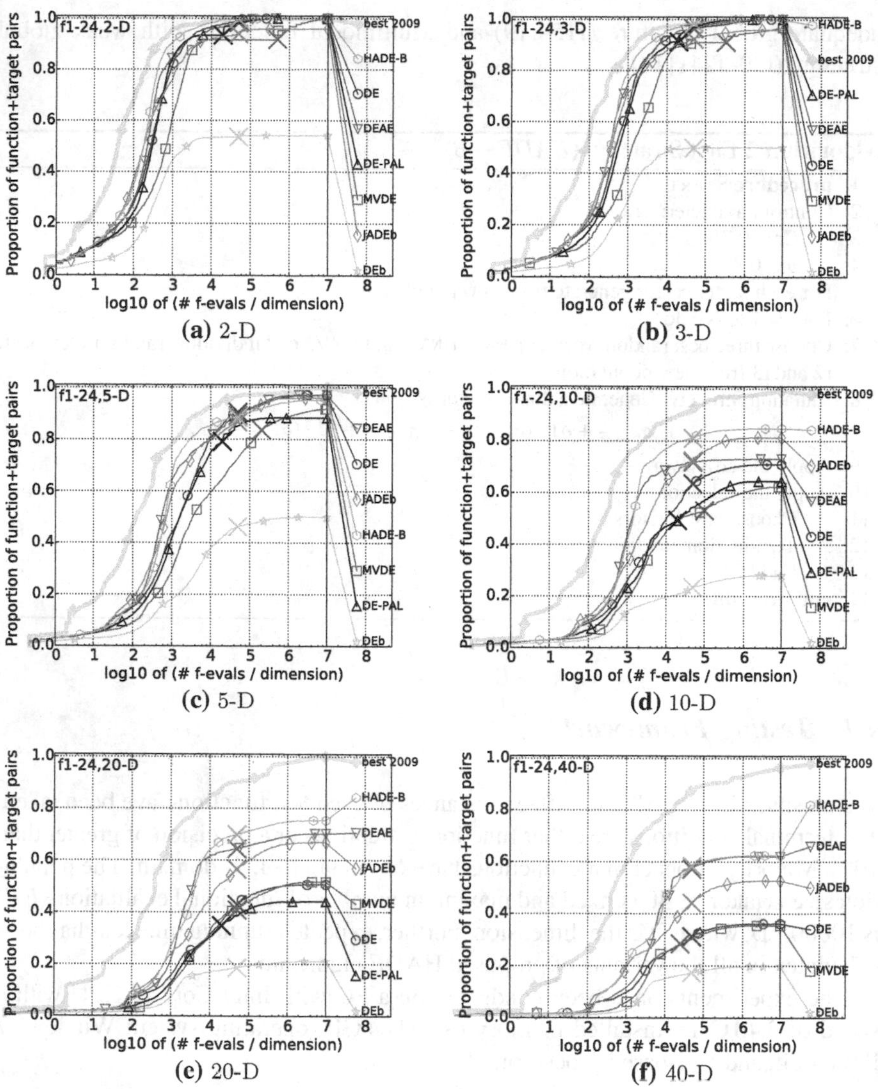


**(a)** 2-D **(b)** 3-D

**(c)** 5-D **(d)** 10-D

**(c)** 20-D **(f)** 40-D

**Fig. 1** Comparison of HADE variants with other state-of-the-art algorithms in [2-D, 3-D, 5-D, 10-D, 20-D and 40-D] on all function

have been made: better convergence rate have been obtained for HADE-B in 2-D, 3-D, 10-D, 20-D, and 40-D, as evident from Fig. 1 and Table 1.

**Table 1** Position of HADE-B on 7 state-of-the-art algorithm over BBOB benchmark functions

| Group | Dimension | Proposed approach | DE variants |
|---|---|---|---|
| All function (f1–f24) | 2 | 1-HADE-B | 2-DE |
| | 3 | 1-HADE-B | 2-DE-PAL |
| | 10 | 1-HADE-B | 2-JADEb |
| | 20 | 1-HADE-B | 2-DEAE |
| | 40 | 1-HADE-B | 2-DEAE |

### *4.3 Effect of Various Parameters on the Performance of Algorithm*

**Population Size:** Population size is very important parameter for this algorithm. For small dimension like 2, 3, and 5 small size 40 is sufficient for capturing targeted solutions in all functions. However, for higher dimensions 10, 20, and 40 large population size 100 is required. This is due to the design of adaptation operator. If we take large size we may do exploitation and exploration around many areas.
**Number of Iterations**: Number of iteration is not that important for this algorithm. A small number of iteration like 500 may work with any number of dimensions.
**Tuning parameter**: Generally the values of mutation factor($\delta 1$ and $\delta 2$) and crossover $Cr$ are fixed during problem optimization and due to which it is not possible to get good results on certain dimensions [20]. Thus to obtain better solutions and to improve the convergence rate, these two parameters are changed and tuned accordingly.

## 5 Conclusion

DE using proposed population-based homeostasis difference vector strategy have been seen to have improved performance in all dimensions (lower to higher). The two proposed independent difference random vectors help in avoiding stagnation problem of multi-modal functions. A comparison with existing DE algorithms like JADEb, DE, DEb, DE-PAL, DEAE, and MVDE in 2, 3, 5, 10, 20, and 40 dimension have been made and it shows an enhanced performance. We have also reduced the number of fitness evaluations and improve the convergence rate by the use of population-based homeostasis difference vector. The performance of this variants were almost very good in all kinds of functions, and a little downfall have been seen in very high dimensions.

## References


1. Storn, Rainer, and Kenneth Price, Differential EvolutionA Simple and Efficient Adaptive Scheme for Global Optimization Over Continuous Spaces, International Computer Science Institute, Berkeley. Berkeley, CA (1995), 1995.
2. Qinqin Fan and Xuefeng Yan, Self-adaptive differential evolution algorithm with discrete mutation control parameters, pp. 1551–1572, Expert Systems with Applications: An International Journal, Volume 42 Issue 3, February 2015.
3. Das, Swagatam, and Ponnuthurai Nagaratnam Suganthan, Differential evolution: a survey of the state-of-the-art Evolutionary Computation, pp. 4–31, IEEE Trans. on Evolutionary Computation, Vol. 15, No. 1, Feb. 2011.
4. J. Ronkkonen, S. Kukkonen, and K. V. Price, Real parameter optimization with differential evolution, pp. 506–513 Proc. IEEE CEC, vol. 1, 2005.
5. J. Brest, S. Greiner, B. Bo skovi c, M. Mernik, and V. Zumer, Selfadapting control parameters in differential evolution: A comparative study on numerical benchmark problems, pp. 646–657 IEEE Trans. Evol. Comput., vol. 10, no. 6, Dec. 2006.
6. E. Mezura-Montes, J. Velazquez-Reyes, and C. A. Coello, A comparative study of differential evolution variants for global optimization, pp. 485–492, in Proc. Genet. Evol. Comput. Conf., 2006.
7. Nasimul Noman, Danushka Bollegala and Hitoshi Iba, An Adaptive Differential Evolution Algorithm, pp. 2229–2236, IEEE Congress on Evolutionary Computation, 2011.
8. Ruhul A. Sarker, Saber M. Elsayed, and Tapabrata Ray, Differential Evolution With Dynamic Parameters Selection for Optimization Problems, pp. 689–707, IEEE Transaction on Evolutionary Computation, VOL. 18, NO. 5, OCTOBER 2014.
9. Xinyu Zhou ,Zhijian Wu ,Hui Wang and Shahryar Rahnamayan, Enhancing differential evolution with role assignment scheme, pp. 2209–2225, Soft Comput(2014) 18, 2014.
10. Craig Brown, Yaochu Jin, Matthew Leach and Martin Hodgson, "$\mu$JADE: Adaptive Differential Evolution with a Small Population", Soft computing, 27 jun 2015.
11. Zhongbo Hu, Qinghua Su, Xianshan Yang and Zenggang Xiong, "Not guaranteeing convergence of differential evolution on a class of multimodal functions", Applied Soft Computing (41), pp. 479–487, 2016.
12. J. Andre, P. Siarry, and T. Dognon, An improvement of the standard genetic algorithm fighting premature convergence in continuous optimization, pp. 49–60, Advance in Engineering Software 32, pp. 2001.
13. O. Hrstka and A. Kucerov, Improvement of real coded genetic algorithm based on differential operators preventing premature convergence, pp. 237–246, Advance in Engineering Software 35, 2004.
14. D. Brockho, Comparison of the MATSuMoTo Library for Expensive Optimization on the Noiseless Black-Box Optimization Benchmarking Testbed, in Congress on Evolutionary Computation (CEC 2015), IEEE Press, 2015 (accepted).
15. http://coco.gforge.inria.fr/
16. N. Hansen, S. Finck, R. Ros, and A. Auger, Real-parameter black-box optimization benchmarking 2009: Noiseless functions definitions, INRIA, Tech. Rep. RR-6829, 2009, updated February 2010.
17. J. Zhang, A. C. Sanderson, JADE: Adaptive differential evolution with optional external archive, pp. 945–958, IEEE Trans. Evol. Comput., vol. 13, no. 5, Oct. 2009.
18. Lszl Pl, Benchmarking a Hybrid Multi Level Single Linkage Algorithm on the BBOB Noiseless Testbed, GECCO13, Amsterdam, Netherlands, July 2013.
19. Vincius Veloso de Melo, Benchmarking the Multi-View Differential Evolution on the Noiseless BBOB-2012 Function Testbed, GECCO12, Philadelphia, PA, USA, July 2012.
20. D. Zaharie, Critical values for the control parameters of differential evolution algorithms, pp. 62–67, Proc. 8th Int. Mendel Conf. Soft Comput., 2002.
21. Neal Holtschulte, Melanie Moses, Benchmarking Cellular Genetic Algorithms on the BBOB Noiseless Testbed, GECCO13 Companion, Amsterdam, Netherlands, July 610, 2013.

22. Ryoji Tanabe and Alex Fukunaga, Tuning Differential Evolution for Cheap, Medium, and Expensive Computational Budgets, IEEE CEC-2015, Press, July 2015.

# Cost Effective Parameter Analysis of Real-Time Multi Core Algorithms

**Avantika Agarwal and Nitin Rakesh**

**Abstract** Cost effective and reliable scheduling of tasks on real-time systems has always been a challenge and with multiprocessor platforms penetrating into the real-time systems environment, the job of efficient scheduling has become even more problematic. Over the years several techniques on the lines of Rate Monotonic Scheduling algorithm have been suggested in the literature to optimize the scheduling process on multicore systems. This paper presents the comparison of three such optimal techniques called Compatibility Aware Task Partition (CATP), Group-wise Compatibility Aware Task Partition (GCATP) and Rate Monotonic Least Splitting (RMLS) algorithms in an attempt to identify the most efficient approach of scheduling. The efficiency of the algorithms has been evaluated in terms of CPU utilization, Throughput, Average Turnaround time, Acceptance Ratio, Average Waiting time and Deadline Miss Ratio. The results of the comparative analysis have shown RMLS to be more effective then CATP and GCATP. Therefore, on this basis, RMLS algorithm has been chosen to be worked upon in future for further optimization to guarantee even higher task scheduling reliability.

**Keywords** CPU time · Utilization · Deadline · Acceptance ratio

## 1 Introduction

A real-time system is a system where not only the logical correctness of the output is required but also the time at which the output is delivered is equally important. Real-time systems can be categorized as Hard Real-time systems and Soft Real-time systems depending upon the urgency with which the output should be

A. Agarwal (✉) · N. Rakesh
Department of Computer Science and Engineering, Amity University, Noida, UP, India
e-mail: avantika.aggarwal@gmail.com

N. Rakesh
e-mail: nitin.rakesh@gmail.com



S.K. Bhatia et al. (eds.), *Advances in Computer and Computational Sciences*,
Advances in Intelligent Systems and Computing 553,
DOI 10.1007/978-981-10-3770-2_55

delivered. In many safety critical systems like avionics and flight control systems where the result should be evaluated and delivered within a set deadline come under hard real-time systems as the failure of delivery of output at the right time can lead to catastrophic results. Such systems should therefore, be built effectively so that they ensure schedulability of the entire task set within a set deadline. The soft real-time system is, however, a bit relaxed with timing constraints as the delay in delivery of output does not lead to tragic results [1, 2].

Several traditional algorithms used to schedule tasks on these systems are Rate Monotonic Scheduling, Earliest Deadline First, Deadline Monotonic Scheduling, Least Laxity First, etc. [1, 3]. Of the above, the Rate Monotonic Scheduling algorithm is a very popular and widely used static-priority scheduling technique. The static priorities are assigned on the basis of the cycle duration of the job: the shorter the cycle duration is the higher is the job's priority. Introduced by Liu and Layland et al. [4], the algorithm was originally developed for effectively scheduling tasks in uniprocessor real-time environment. However, with the real-time tasks becoming more memory intensive and requiring higher processing capabilities, a shift from single core to multicore processors has been made to ensure better system utilization, reliability, and scalability. Modern day architecture of multicore and multiprocessor systems can be divided on the basis of their homogenous or heterogeneous nature. The homogenous behavior of processors suggests same processor speed as all processors involved in the system are identical. On the other hand, heterogeneous systems deal with processors of different configurations so scheduling and processing speed depend on both processors and tasks [1, 2]. This paper, however, restricts the research on homogenous multicore systems.

Multicore systems address two decision-making problems. First is to decide the order of tasks in which they will be scheduled on a processor. Second is to decide the processor to which a particular set of tasks will be assigned. Furthermore, the use of multicores and multiprocessors has posed some issues. The decreased size and mass integration of the transistors on a single chip has led to increase in the power density increasing the temperature of the chip resulting in high failure rates [5]. This affects the reliability of the system. In the past, attempts have been made to map traditional real-time scheduling algorithms in multiprocessor environment, but they lack in efficiency. Therefore, to combat the increasing complex nature of the real-time tasks and multiprocessor environment, several new heuristics and techniques have been suggested over the past years. The efforts have been made to improve processor utilization and ensure reliability of hard real-time systems as reliability continues to be a major concern [1, 5]. In this paper, various real-time multicore system scheduling algorithms were studied. The three best techniques in terms of processor utilization and system reliability offered so far, belonging to partitioned and semi-partitioned approaches has been identified. The three algorithms are Compatibility aware task partition, Group-wise compatibility aware task partition and Rate monotonic least splitting algorithm. The comparison of these algorithms on parameters like CPU utilization, Turnaround time, average waiting time, throughput, etc. has been done in an attempt to identify the most optimal technique to assign and schedule tasks on real-time multi-processor systems. Many other related algorithms exist in literature which worked on this optimization [6–8].

The rest of the paper is organized as follows. Section 2 gives the literature review and describes all the algorithms that have been used in the paper to perform comparison analyses. Section 3 presents the details of the comparative analysis and includes all the preliminaries, notations and parameters used in evaluation process. Section 4 presents the result of the conducted analyses and Sect. 5 concludes the paper and discusses the future work to be done.

## 2 Related Work

All real-time multiprocessor system scheduling techniques can be categorized on the basis of two metrics. The first is prioritization metrics which classifies the scheduling algorithms on the basis of their priority assignment level. This can happen in two ways—static (job level and task level) and dynamic priority assignment. The second is migration metrics which classifies the algorithms on extent of which they permit the tasks to migrate between different processors [3]. When no migration is permitted, then partitioned approach is used, where a set of tasks are assigned to and executed by a single processor only. Since only one processor executes a set of tasks, this orientation hence has minimum overheads and migration costs [1, 2]. It also benefits from the simplicity of uniprocessor scheduling but is not very optimal in processor utilization.

The migration at job level means jobs of a task can be moved among different processors but cannot be executed in parallel on them. The migration at task level means jobs of a task can execute on different processors but cannot move among them. When the migration happens at job and task level, then unrestricted and restricted global approach is used. The study on global approach reveals that this technique uses a single queue for all the tasks to reside and hence they easily migrate among different processors according to their priority level. Though this approach has highly optimal schedules, the cost of its task overheads and migrations is extremely high. Therefore, the limitations of partitioned and global approach gave rise to the third kind of approach called semi-partitioned or hybrid. In this, advantages of both techniques are employed to overcome their limitations. Here a group of tasks is assigned to a cluster of processors instead of a single processor. So tasks can migrate only over the cluster and not over all the processors. This decrease the overheads cost substantially and improve utilization of the processors [1, 2]. The algorithms belonging to different real-time multicore system approaches used in the paper for comparative analysis have been described further.

Rate Monotonic Least Splitting (RMLS) algorithm works in two phases, partitioning phase and scheduling phase. In the partitioning phase, the tasks are assigned to the appropriate processors. If a task is entirely allocated on one processor, it is called a non-split task and it is executed by that processor completely. If a task is split into two subtasks then it is called a split task and is assigned to two consecutive processors. The interesting feature of this algorithm is that the processors running a split task cannot execute them in parallel and the processor with lower index say 'm' is allowed to run more than its share of task than the processor with higher index say 'm + 1'. Also for processor m, the priority of subtask is lowest while for processor m + 1 the priority is highest. Therefore, the chances of them executing the tasks together are minimal [9]. The scheduling phase decides when to run the allocated tasks and for how long they should be run. Here the scheduling is done according to the RMS procedure. The algorithm can have at most m-1 split tasks if there are m processors in the system. The algorithm has proven to be most optimal in terms of the number of processors required to schedule tasks and can attain a utilization bound of 80% [9].

Compatibility Aware Task Partition (CATP) algorithm follows a greedy approach and works to allocate a task to the most compatible core. It is based on the principle that harmonic task sets need not always be the best combination to allocate on a processor in order to attain maximum processor utilization. The compatibility of task sets is a much better parameter in selecting the best combination of task sets to allocate on a processor to ensure maximum processor utilization. Therefore, this algorithm uses a compatibility index and iteratively looks for most compatible core for the allocation of current task so that the CPU utilization of the core becomes maximum and value of compatibility index is minimum [5]. The partitioning algorithm then schedules the tasks according to Rate Monotonic Scheduling scheme as out of all traditional scheduling algorithms, it is the most optimal one. Group-wise Compatibility Aware Task Partition (G-CATP) algorithm is based on partitioned approach and iteratively works to form the most compatible task set which can then be allocated to the first available empty core. The algorithm takes a base task and then the compatibility of the remaining tasks in the task set is checked with it. The most compatible task is then added to the set. Then again, the compatibility of remaining tasks is checked with the task set and the compatible tasks are added to the set till the processor utilization bound is not exceeded [5]. Following this scheme, the allocated tasks are then scheduled according by RMS algorithm.

## 3 Comparative Analysis

This section presents all the preliminaries and notations used in the paper. The real-time system being analyzed in the paper consists of 'n' sporadic tasks denoted by $\tau_i$, where $i = \{1, 2,\ldots, n\}$ to be scheduled on 'm' homogenous processors denoted by $P_i$, where $i = \{1, 2,\ldots, m\}$. Each task in the task set is represented by $\{C_i, T_i\}$, representing worst case execution time and minimum inter arrival time called the time period of the task. In the paper we have taken deadline of each task $\tau_i$, to be same as its time period. The effective utilization of each task represented by $U_i$, has been calculated by the formula: $i = \frac{Ci}{Ti}$. The comparison has been performed

**Table 1** Contains the task set used in the first scenario

| $\tau_i$ | $C_i$ | $T_i$ | $U_i$ |
|---|---|---|---|
| 1 | 3.5 | 10 | 0.35 |
| 2 | 3.1 | 10 | 0.31 |
| 3 | 6 | 19 | 0.32 |
| 4 | 3 | 19 | 0.16 |
| 5 | 4 | 19 | 0.21 |
| 6 | 2.6 | 19 | 0.13 |

**Table 2** Contains the task set used in second scenario

| $\tau_i$ | *Ci* | *Ti* | *Ui* |
|---|---|---|---|
| 1 | 1 | 5 | 0.2 |
| 2 | 1 | 5 | 0.2 |
| 3 | 2 | 5 | 0.4 |
| 4 | 4 | 10 | 0.4 |
| 5 | 3 | 10 | 0.3 |
| 6 | 3.5 | 10 | 0.35 |
| 7 | 3.1 | 10 | 0.31 |
| 8 | 5 | 13 | 0.38 |
| 9 | 5 | 15 | 0.33 |
| 10 | 6 | 19 | 0.32 |
| 11 | 3 | 19 | 0.16 |
| 12 | 2.6 | 19 | 0.13 |

using two different scenarios. In the first scenario, the task set contains six fixed priority sporadic tasks to be scheduled on two homogenous processors.

In the second scenario, the tasks and number of processors used in the previous scenario were doubled. The task set contains twelve fixed priority sporadic tasks to be scheduled on four homogenous processors. The RMLS, CATP and GCATP algorithms mentioned in the previous section are used to allocate task set in Table 1 on dual core system and task set in Table 2 on quad core system. First, the worst case execution time and time period of the tasks are used to calculate the utilization of each task in task set. Then after the allocation procedure is done, tasks are scheduled using rate monotonic scheduling algorithm (Figs. 1, 2, 3, 4, 5, 6, 7, 8, 9, 10, 11, 12, 13, 14, 15, 16, 17 and 18).

Scenario 1: CATP Scheduling

See Figs. 1 and 2.

GCATP

See Figs. 3 and 4.

RMLS Scheduling

See Figs. 5 and 6.

Scenario 2: CATP Scheduling

See Figs. 7, 8, 9 and 10.

GCATP Scheduling

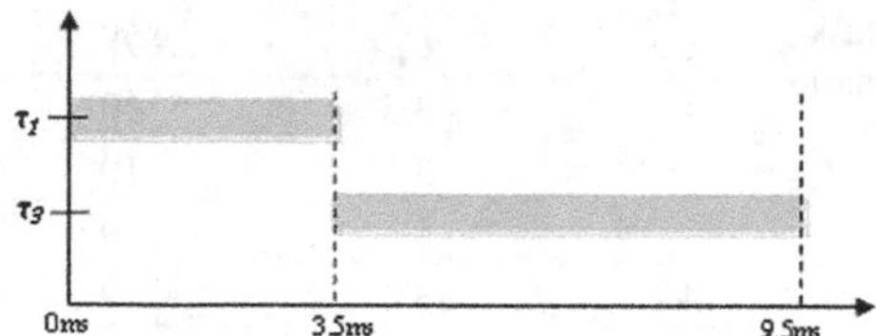


**Fig. 1** Tasks being scheduled on processor 1 using RMS (CORE 1)

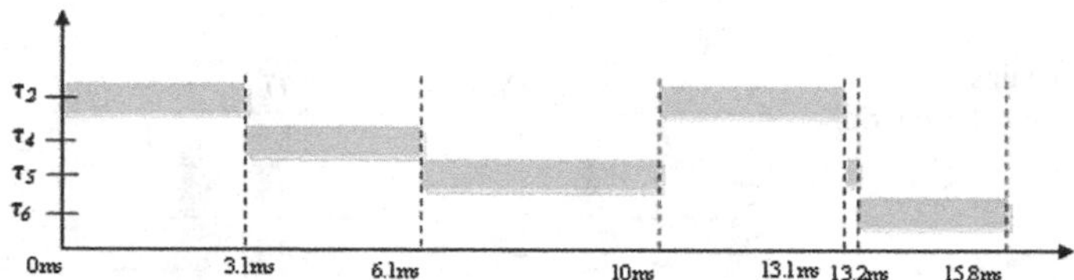


**Fig. 2** Tasks being scheduled on processor 2 using RMS (CORE 2)

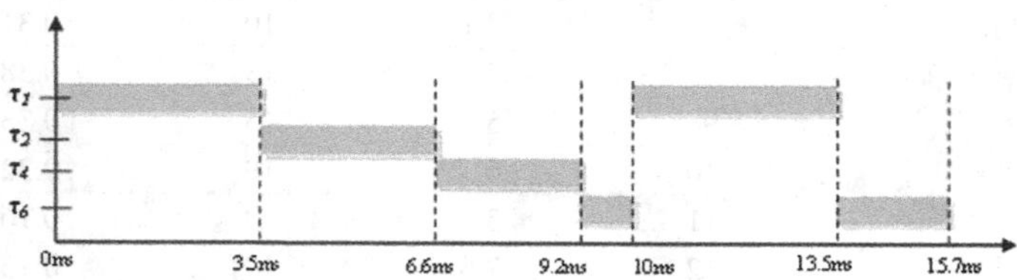


**Fig. 3** Tasks being scheduled on processor 1 using RMS (CORE 1)

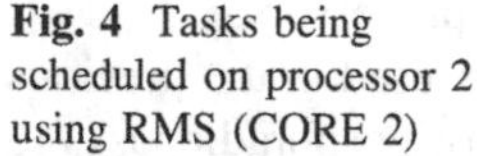

**Fig. 4** Tasks being scheduled on processor 2 using RMS (CORE 2)

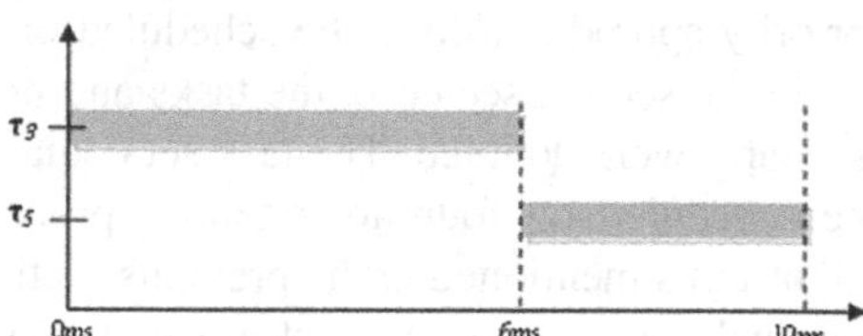


**Fig. 5** Tasks being scheduled on processor 1 using RMS (CORE 1)

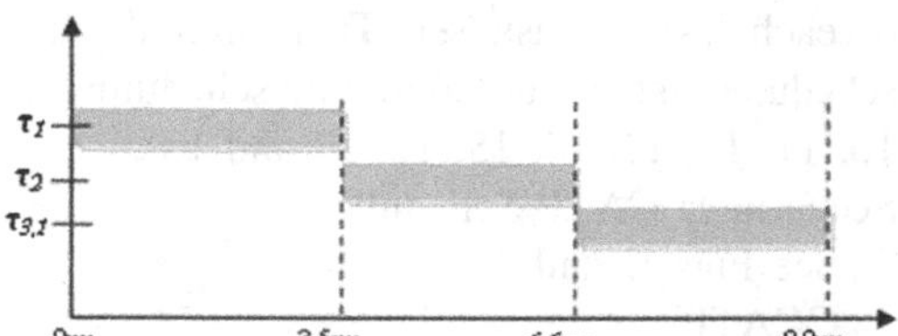


**Fig. 6** Tasks being scheduled on processor 2 using RMS (CORE 2)

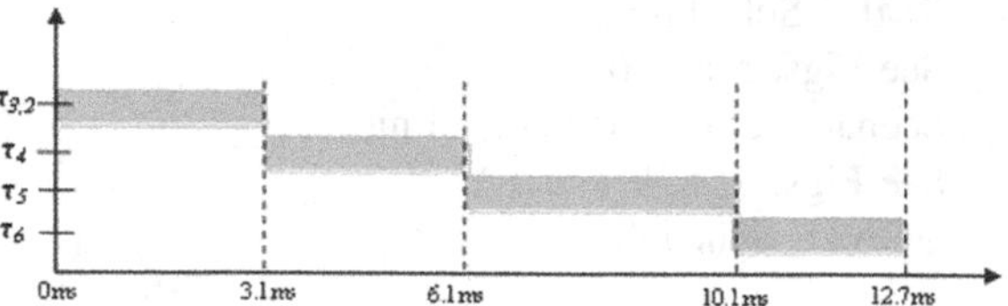

See Figs. 11, 12, 13 and 14.
RMLS Scheduling
See Figs. 15, 16, 17 and 18.

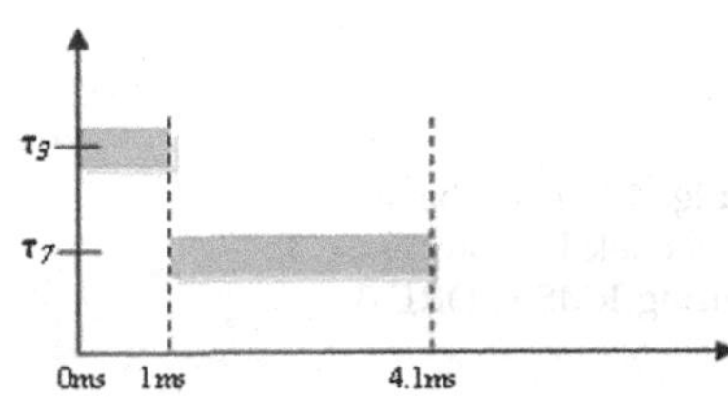


**Fig. 7** Tasks being scheduled on processor 1 using RMS (CORE 1)

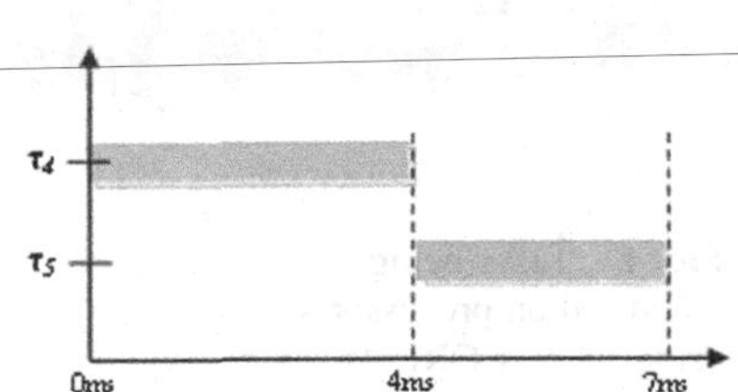


**Fig. 8** Tasks being scheduled on processor 2 using RMS (CORE 2)

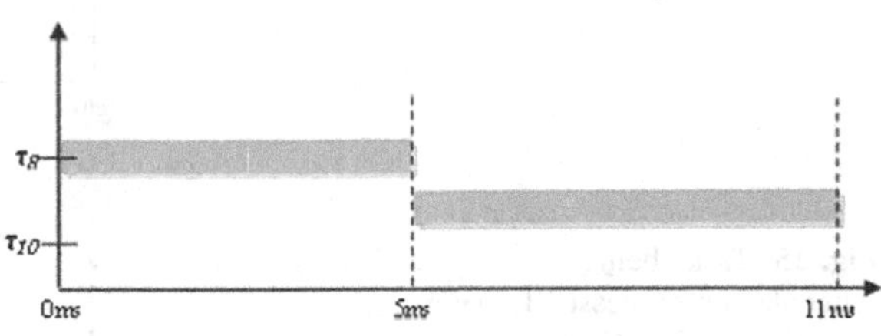


**Fig. 9** Tasks being scheduled on processor 3 using RMS (CORE 3)

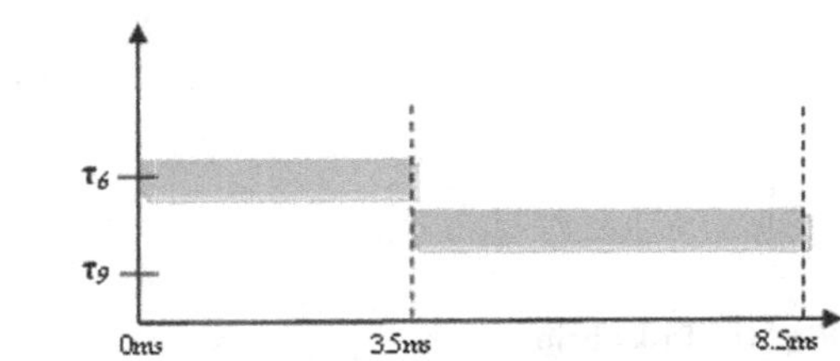


**Fig. 10** Tasks being scheduled on processor 4 using RMS (CORE 4)

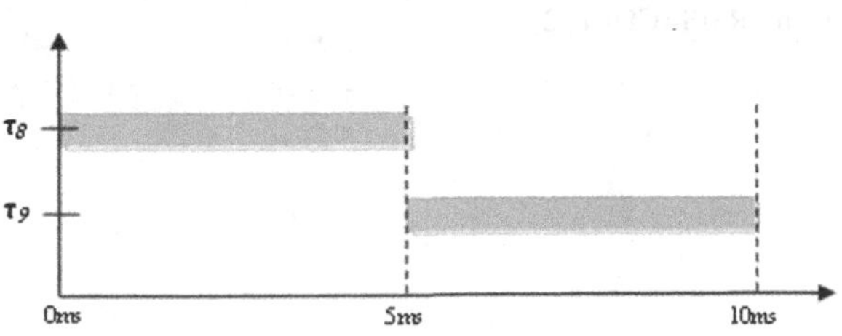


**Fig. 11** Tasks being scheduled on processor 1 using RMS (CORE 1)

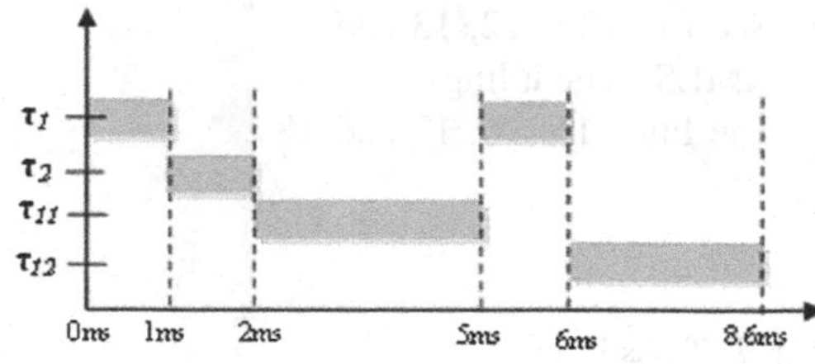


**Fig. 12** Tasks being scheduled on processor 2 using RMS (CORE 2)

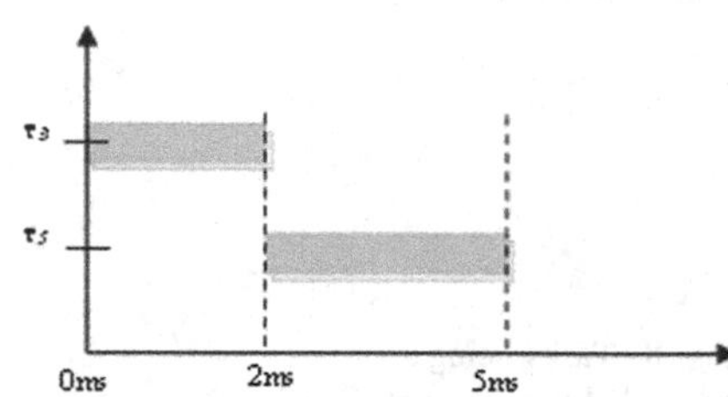


**Fig. 13** Tasks being scheduled on processor 3 using RMS (CORE 3)

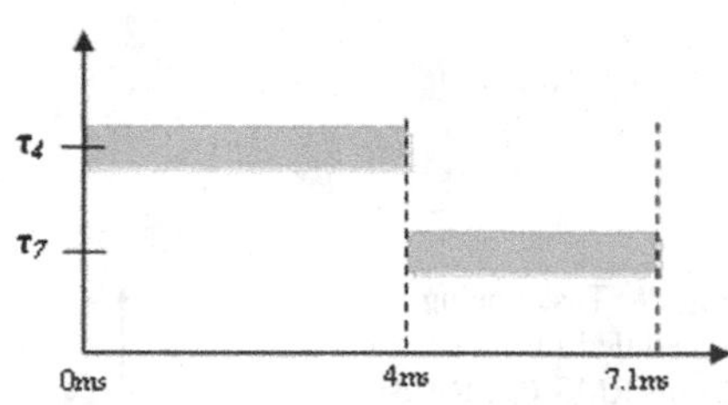


**Fig. 14** Tasks being scheduled on processor 4 using RMS (CORE 4)

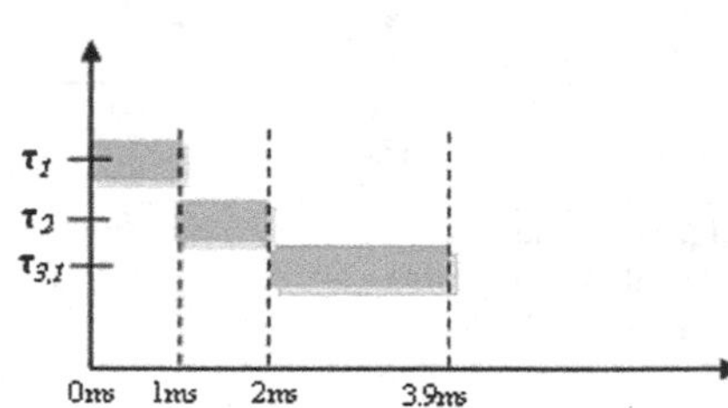


**Fig. 15** Tasks being scheduled on processor 1 using RMS (CORE 1)

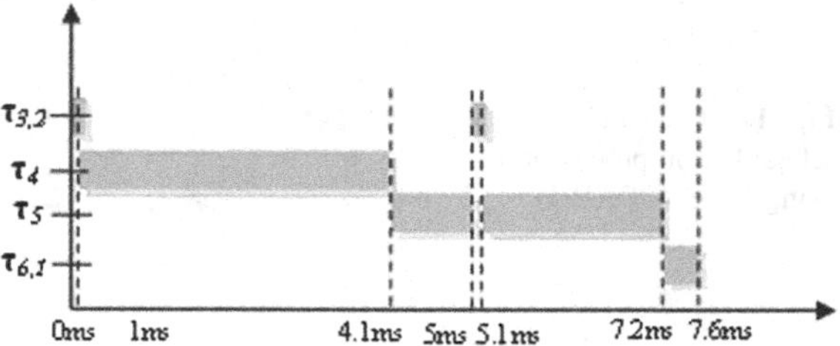


**Fig. 16** Tasks being scheduled on processor 2 using RMS (CORE 2)

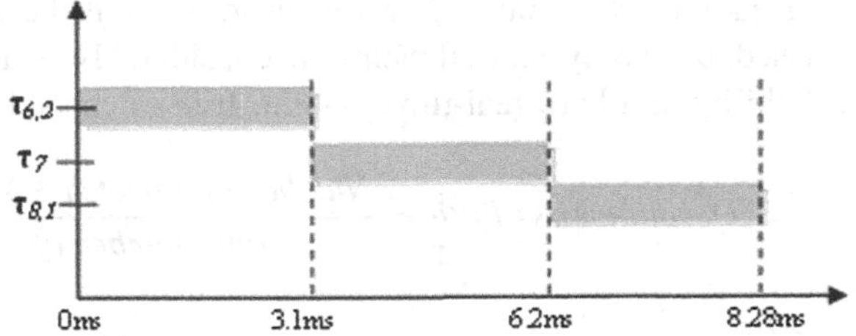


**Fig. 17** Tasks being scheduled on processor 3 using RMS (CORE 3)

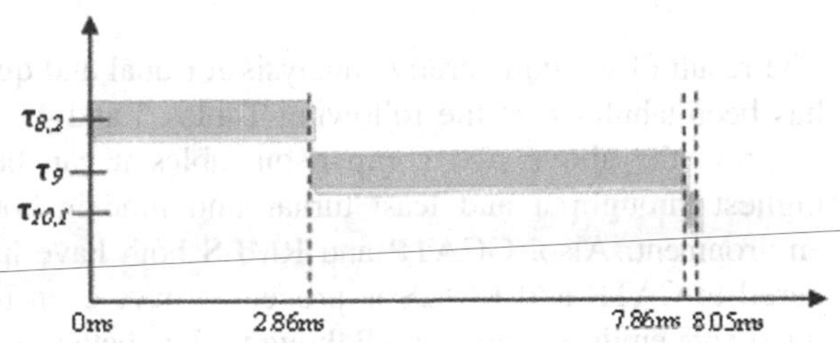


**Fig. 18** Tasks being scheduled on processor 4 using RMS (CORE 4)

After scheduling, the three different techniques used were compared on following parameters and final results were tabulated. Several other web-based approached exists in literature which are working on optimization [6–8, 10, 11].

Processor Utilization: Ideally, a CPU must be utilized to its maximum that is 100% so that the desired output is evaluated fast and not many CPU cycles are wasted. Also, if a single processor is utilized to its maximum, the requirement of more and more processors would be reduced lowering down the cost associated with system. Hence, processor utilization gives a fair idea about the efficiency of the system.

Throughput: Number of processes completed per unit time. In our analysis, 10 ms have been taken as the unit.

Turnaround Time: Time required for a particular process to complete, from submission time to the time it gets completed. The average turnaround time has been calculated for the system.

Waiting Time: The amount of time a task waits in ready queue to get processed by CPU. The average waiting time has been calculated for the system.

More throughput, less turnaround time and less waiting time means higher efficiency as more work is completed in less time. Thus, in operation a more efficient system is more cost effective.

Acceptance Ratio: It gives a fair idea about the reliability of the system as it suggests the ratio/percentage of the tasks that are executed by the system. It is calculated using the following formula [1]:

$$Acceptance\ Ratio = \frac{Number\ of\ tasks\ executed\ by\ the\ system}{Total\ number\ of\ tasks\ in\ the\ system}$$

Deadline Miss Ratio: It gives an idea about the number of tasks rejected and missed by the system. It helps in considerably evaluating the effectiveness and reliability of a hard real-time system. It is calculated using the following formula:

$$\textit{Deadline Miss Ratio} = \frac{\textit{Number of rejected} + \textit{Number of missed tasks}}{\textit{Total number of tasks in the system}}$$

## 4 Result

The result of the comparative analysis for dual and quad core environment models has been tabulated in the following Tables 3 and 4.

From the above two comparison tables it can be realized that RMLS gives highest throughput and least turnaround time in both dual core and quad core environment. Also, GCATP and RMLS both have high CPU utilization as compared to CATP and RMLS is proved to have even better processor utilization in quad core environment. CATP though offers better average waiting time than other two algorithms in both scenarios but lags behind considerably in all other parameters. Also, in the dual core environment all three algorithms are able to schedule all tasks guaranteeing full reliability but as the task load and number of processors increase, the reliability of the three algorithms decreases [12].

The GCATP algorithm is found to be most reliable out of the three having an acceptance ratio of 83.3%, while RMLS follows at 75%. Therefore, from over all analyses, it has been found that RMLS is most efficient algorithm out of the three. It is a semi-partitioned approach and hence it can be clearly observed that a semi-partitioned approach proves to be more effective then partitioned approach.

**Table 3** Comparative analyses for scenario 1

| Parameters | CATP | GCATP | RMLS |
|---|---|---|---|
| Processor utilization | 0.77 | 0.74 | 0.74 |
| Throughput | 2.32 per 10 ms | 2.27 per 10 ms | 3.05 per 10 ms |
| Average turnaround time | 8.025 ms | 8.37 ms | 7.34 ms |
| Average waiting time | 4.06 ms | 4.35 ms | 4.095 ms |

**Table 4** Comparative analyses for scenario 1

| Parameters | CATP | GCATP | RMLS |
|---|---|---|---|
| Processor utilization | 0.69 | 0.703 | 0.78 |
| Throughput | 2.9 per 10 ms | 3.37 per 10 ms | 5.1 per 10 ms |
| Average turn around time | 5.51 ms | 5.18 ms | 4.79 ms |
| Average waiting time | 1.69 ms | 1.94 ms | 2.4 ms |
| Acceptance ratio (%) | 66.7 | 83.3 | 75 |
| Deadline miss ratio (%) | 33.3 | 16.7 | 25 |

## 5 Conclusion and Future Work

In this paper we were able to effectively compare three techniques used for scheduling tasks in real-time multicore systems. Through comparative analysis and calculations it was found that Rate Monotonic Least Splitting (RMLS), a semi-partitioned approach algorithm outperforms the two other algorithms in parameters like throughput, processor utilization and average turnaround time thus making it more cost effective. However, as the load increases it gives slightly low reliability as compared to Group-wise Compatibility Aware Task Partition (G-CATP) algorithm. Therefore, work may be carried out to optimize RMLS further to increase its reliability.

## References


1. MehrinRouhifar and Reza Ravanmehr, "A Survey on Scheduling Approaches for Hard Real-Time Systems", International Journal of Computer Applications, Vol. 131, No.17, pp. 0975–8887, December 2015.
2. AjitRamachandran, JegadishManoharan and SomanathanChandrakumar, "Real-Time Scheduling methods for High Performance Signal Processing Applications on Multicore platform", Master's Thesis in Embedded and Intelligent Systems, IDE 1260, August 2012.
3. Dirk Müller and Matthias Werner, "Genealogy of Hard Real-Time Preemptive Scheduling Algorithms for Identical Multiprocessors", Central European Journal of Computer Science, 7 September 2011.
4. LuiSha,Mark H. Klein and John B. Goodenough, "Rate Monotonic Analysis for Real-Time Systems", Technical Report CMU/SEI-91-TR-006 ESD-91-TR-006, March 1991.
5. Qiushi Han, Tianyi Wang and Gang Quan, "Enhanced Fault-Tolerant Fixed-Priority Scheduling of Hard Real-Time Tasks on Multi-Core Platforms", 2015 IEEE 21st International Conference on Embedded and Real-Time Computing Systems and Applications.
6. Praveen K Gupta, Nitin Rakesh, "Different job scheduling methodologies for web application and web server in a cloud computing environment", 2010 3rd International Conference on Emerging Trends in Engineering and Technology (ICETET), pp. 569–572, 2010.
7. Nitin Rakesh, VipinTyagi,"Failure recovery in XOR'ed networks", 2012 IEEE International Conference on Signal Processing, Computing and Control (ISPCC), pp. 1–6, 2012.
8. Kinjal Shah, GaganDua, Dharmendar Sharma, Priyanka Mishra, Nitin Rakesh,"Transmission of Successful Route Error Message (RERR) in Routing Aware Multiple Description Video Coding over Mobile Ad-Hoc Network", International Journal of Multimedia & Its Applications (IJMA), Vol. 3, No. 3, 51–59, August 2011.
9. Mahmoud Naghibzadeh, PeymanNeamatollahi, Reza Ramezani, Amin Rezaeian and ToktamDehghani, "Efficient Semi-Partitioning and Rate-Monotonic Scheduling Hard Real-Time Tasks on Multi-Core Systems", 8th IEEE International Symposium on Industrial Embedded Systems (SIES 2013).
10. AnkitMundra, Nitin Rakesh, (2013) "Online Hybrid Model for Online Fraud Prevention and Detection," International Conference on Advance Computing, Networking, and Informatics – ICACNI-2013, Springer, pp. 805–815.

11. SandeepPratap Singh, Shiv Shankar P. Shukla, Nitin Rakesh and VipinTyagi, “Problem reduction in online payment system using hybrid model,” International Journal of Managing Information Technology, 3 (3) 62–71, August 2011.
12. Guillaume Vigeant, Alain Beaulieu and Sidney N. Givigi, “Hard Real-Time Scheduling on a Multicore Platform”, 2015 9th Annual IEEE International Systems Conference (SysCon).

# Variant of Differential Evolution Algorithm

**Richa Shukla, Bramah Hazela, Shashwat Shukla, Ravi Prakash and Krishn K. Mishra**

**Abstract** Differential evolution is a nature-inspired optimization technique. It has achieved best solutions on large area of test suits. DE algorithm is efficient in programming and it has broad applicability in engineering. This paper presents modified mutation vector generation strategy of basic DE for solving stagnation problem. A new variant of differential evolution that is DE_New has been proposed and the performance of DE_New is tested on Comparing Continuous Optimisers (COCO) framework composed of 24 benchmark functions and found DE_New has better exploration capability inside the given search space in comparison to GA, DE-PSO, DE-AUTO on Black-Box Optimization Benchmarking (BBOB) 2015 devised by COCO.

**Keywords** DE_New · Exploitation · Differential evolution · Mean cost

R. Shukla · B. Hazela · S. Shukla
Computer Science and Engineering Department, Amity University,
Lucknow 226028, India
e-mail: richamity22@gmail.com

B. Hazela
e-mail: bhazela@lko.amity.edu

S. Shukla
e-mail: shashwatshukla10aug@gmail.com

R. Prakash · K.K. Mishra (✉)
Computer Science and Engineering Department, MNNIT Allahabad,
Allahabad 211004, India
e-mail: kkm@mnnit.ac.in

R. Prakash
e-mail: raviprakashguddu@gmail.com

 
S.K. Bhatia et al. (eds.), *Advances in Computer and Computational Sciences*,
Advances in Intelligent Systems and Computing 553,
DOI 10.1007/978-981-10-3770-2_56

## 1 Introduction

Optimization problem can be solved with the help of different types of algorithms [1, 2]. In this scope reputation of evolutionary algorithm (EA) is unbeatable [3–5]. DE is an EA that is used for solving many modal optimization problems in continuous search space [6, 7]. DE algorithm is efficient and based on natural selection with globally strong optimization technique in continuous search space [8]. DE logic works with the help of DE operators (mutation, crossover and selection) and associated parameters [9–11]. Myths related to GA and DE are created due to similarity of the operators. In GA, crossover is the main operator but in DE, mutation operator plays lead role for solving global optimization problem [12, 13]. DE is better when compared with GA because it does not use binary encoding [5, 14] and probability density function [15]. DE variants are very helpful in enhancing the vector position thereby resolving the stagnation problem and increasing the convergence speed. DE basically works on global search space. So for better exploration of solution, here DE_New navigates to investigate the local optimum problem for solving blocked solutions. Further content of this paper has been ordered as below, Sect. 2 defines differential evolution, Sect. 3 states existing and modified algorithm, Sect. 4 presents proposed work, practical approach of result analysis has been elaborated in Sect. 5 and finally conclusions are drawn in Sect. 5.

## 2 Differential Evolution

Differential Evolution is a randomized, population based, meta-heuristic and nature-inspired [16] global optimization technique [17] given by Storn and Price in 1995 [18]. Mutation, crossover and selection are the DE operators [19] but DE algorithm fully depends on mutation operator. The role of mutation operator is to generate a mutant vector and the role of crossover is to develop trial vector and the aim of selection operator is to select best between trial vector and mutant vector for each generation.

Best example of DE is that it uses in mobile robot systems [20] for determining the position and orientations of the robot. The way of initialization of the robot position and track the position is the concept of Robot kidnapping (particle kidnapping) [21–23]. Robot kidnapping understands the position and movement of robot path without telling the new position [17].

Brief description of DE operators are given below:

1. **Mutation:** In this, operator generates a mutant vector. It selects three vectors randomly from the search space and multiply the difference of two vectors with a constant value 'F'(range of F is in between 0–2) and add with third vector that produces a mutant vector $\boldsymbol{m}_{i,g}$, this process is called mutation. Mutant vector $\boldsymbol{m}_{i,g}$ for each generation is stated as:

$$x_{i,g} = \boldsymbol{x}_p + \mathrm{F}(\boldsymbol{x}_q - \boldsymbol{x}_r) \tag{1}$$

*Notations:*
1. i = 1, 2, ...... population size.
2. p, q, r are the three random vectors of the search space for 'g' generation [24, 25].

2. **Crossover:** After generation of mutant vector, trial vector is created with the help of mutant operator by recombination of two different vectors.

$$\boldsymbol{t}_{i,g+1} = \boldsymbol{t}_{1i,g}, \boldsymbol{t}_{2i,g}, \boldsymbol{t}_{3i,g} \ldots\ldots\ldots\ldots\ldots\ldots\ldots\ldots \boldsymbol{t}_{ni,g} \tag{2}$$

The overall equation of trial vector for crossover operation is divided in two phases in which we compare randomly generated number with crossover constant value. If random number generated is $\leq CR$ then we will use mutant vector $\boldsymbol{m}_{i,g}$ for each generation otherwise choose target vector without altering parent vector [26–28].

3. **Selection:** Selection of new member is totally dependent on selection operator. Comparing between trial vector and mutant vector and selecting the best between these two vectors $\boldsymbol{t}_{i,g}$ and $\boldsymbol{x}_{i,g}$ is the main aim of selection operator.

## 3 Basic and Proposed DE Algorithm

**Algorithm 1** Basic Differential Evolution Algorithm

**for** Each particle *i* **do**
    **for** each vectors $\boldsymbol{x}_{i,g}$ of population N **do**
        MUTATION
        select three random vectors from search space for mutation
        $x_{i,g}=\boldsymbol{x}_p+\mathrm{F}(\boldsymbol{x}_q-\boldsymbol{x}_r)$
        CROSS-OVER
        **if** random number generated is $\leq CR$ **then** select $\boldsymbol{m}_{i,g}$
        **else** select target vector
        **end if**
        SELECTION
        evaluate fitness of $x_{i,g}$ and $t_{i,g+1}$
        choose best fitness as solution and discard previous
        End if termination conditions met
    **end for**
**end for**

**Algorithm 2** Proposed Differential Evolution Algorithm

**for** Each particle *i* **do**<br>
    **for** each vectors $\boldsymbol{x}_{i,g}$ of population N **do**<br>
        MUTATION<br>
        select $\boldsymbol{x}_p, \boldsymbol{x}_q, \boldsymbol{x}_r$<br>
        Fix first vector $\boldsymbol{x}_p$<br>
        where $\boldsymbol{x}_q >$ mean-cost and $\boldsymbol{x}_r <$ mean-cost<br>
        $x_{i,g}=\boldsymbol{x}_p+F(\boldsymbol{x}_q-\boldsymbol{x}_r)$<br>
        CROSS-OVER<br>
        **if** random number generated is $\leq CR$ **then**<br>
            select $\boldsymbol{m}_{i,g}$ for each generation<br>
        **else** select target vector<br>
        **end if**<br>
        SELECTION<br>
        evaluate fitness of $x_{i,g}$ and $t_{i,g+1}$<br>
        choose best fitness as solution and discard previous<br>
        End if termination conditions met<br>
    **end for**<br>
**end for**

## 4 Proposed Work

In DE algorithm, mutation and associated parameters maintain the diversity of search space. There are many solutions in search space that are blocked, stagnated, trapped, and unable to give better solution for multimodal problems in global search space. So for solving this problem DE_NEW algorithm has been proposed. In this algorithm, strategy of basic mutation operator has been changed and a new variant of DE algorithm (DE_NEW) has been proposed in this paper. The logic of proposed DE_NEW algorithm is to select three vectors from the search space $\boldsymbol{x}_p, \boldsymbol{x}_q, \boldsymbol{x}_r$. Calculate the mean cost and on the basis of mean cost, search space has been divided into two parts. Select $\boldsymbol{x}_p$ vector randomly from the total population size and store in index number_c, which will be the fixed vector in each iteration. Then select next two vectors $\boldsymbol{x}_q$ and $\boldsymbol{x}_r$. Choose $\boldsymbol{x}_q$ randomly from the above of mean cost and store in index number_a and select $\boldsymbol{x}_r$ randomly from the below of mean cost and store in index number_b. The new strategy of mutation operator has been used for providing better exploration of solution and greater diversity of solutions in search space in comparison with GA, DE-PSO, DE-AUTO on Black-Box Optimization Benchmarking (BBOB) [29] 2015 devised by COCO (Figs. 1 and 2).

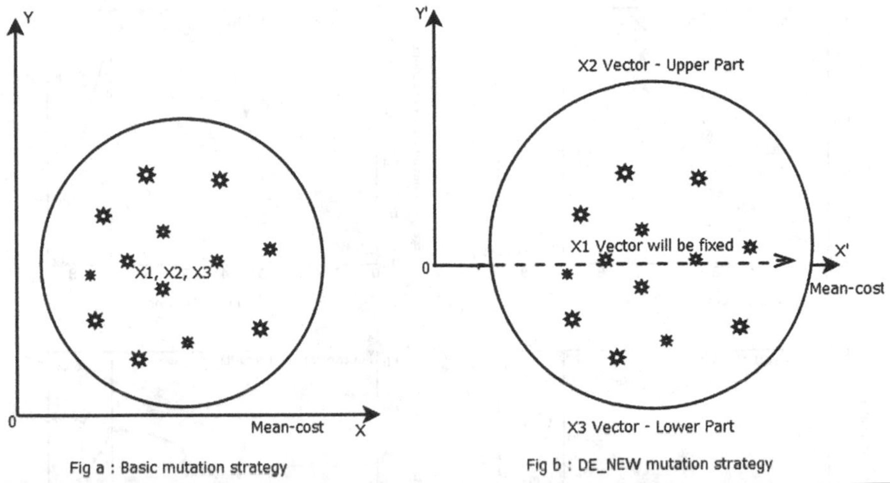


**Fig. 1** DE_NEW

## 5 Result Analysis

The three popular algorithms have been compared with tested performance of DE_New. In the following table, the rank has been allotted for proposed variant and noiseless results have been attached. Its performance decreased which might be due to lower value of randomized F, Cr etc. But in this paper, DE_New has been succeeded to give better results for each modal in 24 benchmark functions in 2D, 3D, 5D, 10D, 20D and 40D.

## 6 Conclusion

In this paper, proposed variant of DE algorithm has been based on modified mutation strategy. It gives best performance for unimodal and multimodal problems and increases the convergence rate and resolves the stagnation problems. In DE_New unlike basic DE and other algorithms, it can get best optimization performance. DE_New is able to give satisfactory performance for DE-AUTO and best performance for DE-PSO, GA in 2D, 3D, 5D, 10D, 20D and 40D on various benchmark functions, however there are a lot of ways for improving it. In future work, DE algorithm can be used in cost estimation of the software and implemented in many applications. Also merge two or more evolutionary algorithm for achieving desired results (Table 1).

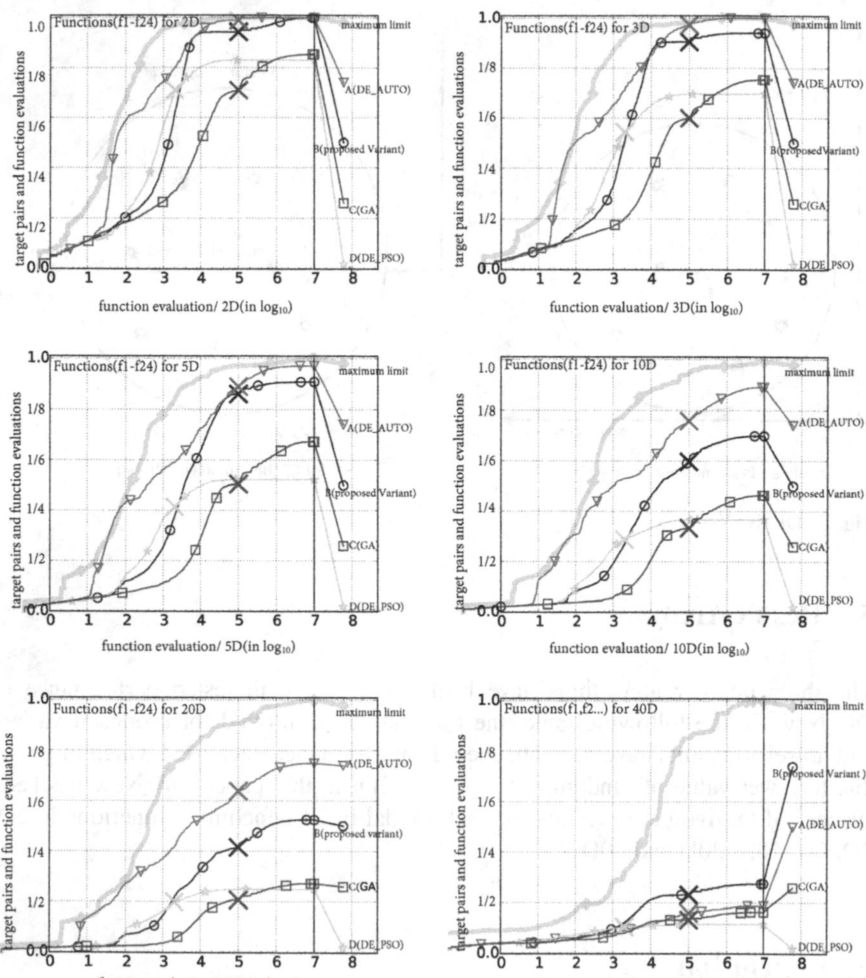


**Fig. 2** Comparison results

**Table 1** Ranking of Proposed Variant

| DIMENSIONS | Hcond | Lcond | Mult2 | Multi | Separable |
|---|---|---|---|---|---|
| 2D | 2nd | 2nd | 2nd | 1st | 1st |
| 3D | 2nd | 2nd | 3rd | 2nd | 1st |
| 5D | 2nd | 2nd | 2nd | 2nd | 2nd |
| 10D | 2nd | 2nd | 2nd | 2nd | 2nd |
| 20D | 2nd | 2nd | 2nd | 1st | 2nd |
| 40D | 2nd | 2nd | 4th | 1st | 2nd |

## References

1. Z. W. Geem, J. H. Kim, and G. Loganathan, "A new heuristic optimization algorithm: harmony search," Simulation, vol. 76, no. 2, pp. 60–68, 2001.
2. S. Das and P. N. Suganthan, "Differential evolution: a survey of the stateof-the-art," Evolutionary Computation, IEEE Transactions on, vol. 15, no. 1, pp. 4–31, 2011.
3. R. Storn and K. Price, "Differential evolution–a simple and efficient heuristic for global optimization over continuous spaces," Journal of global optimization, vol. 11, no. 4, pp. 341–359, 1997.
4. S. Tiwari, K. Mishra, and A. K. Misra, "Test case generation for modified code using a variant of particle swarm optimization (pso) algorithm," in Information Technology: New Generations (ITNG), 2013 Tenth International Conference on. IEEE, 2013, pp. 363–368.
5. K. Mishra, S. Tiwari, and A. Misra, "Combining non revisiting genetic algorithm and neural network to generate test cases for white box testing," in Practical Applications of Intelligent Systems. Springer, 2011, pp. 373–380.
6. E. Mezura-Montes, J. Vel azquez-Reyes, and C. A. Coello Coello, "A comparative study of differential evolution variants for global optimization," in Proceedings of the 8th annual conference on Genetic and evolutionary computation. ACM, 2006, pp. 485–492.
7. D. Zaharie, "Critical values for the control parameters of differential evolution algorithms," in Proceedings of MENDEL, vol. 2, 2002, p. 6267.
8. Y. Ao and H. Chi, "Experimental study on differential evolution strategies," in Intelligent Systems, 2009. GCIS'09. WRI Global Congress on, vol. 2. IEEE, 2009, pp. 19–24.
9. M. Daoudi, S. Hamena, Z. Benmounah, and M. Batouche, "Parallel diffrential evolution clustering algorithm based on mapreduce," in Soft Computing and Pattern Recognition (SoCPaR), 2014 6th International Conference of. IEEE, 2014, pp. 337–341.
10. M. F. Tasgetiren, O. Bulut, Q.-K. Pan, and P. N. Suganthan, "A differential evolution algorithm for the median cycle problem," in Differential Evolution (SDE), 2011 IEEE Symposium on. IEEE, 2011, pp. 1–7.
11. F. Neri and V. Tirronen, "Recent advances in differential evolution: a survey and experimental analysis," Artificial Intelligence Review, vol. 33, no. 1–2, pp. 61–106, 2010.
12. R. Joshi and A. C. Sanderson, "Minimal representation multisensor fusion using differential evolution," Systems, Man and Cybernetics, Part A: Systems and Humans, IEEE Transactions on, vol. 29, no. 1, pp. 63–76, 1999.
13. J. Holland, "Adaption in natural and artificial systems," Ann Arbor, MI: University of Michigan Press, 1975.
14. D. E. Goldberg et al., Genetic algorithms in search optimization and machine learning. Addison-wesley Reading Menlo Park, 1989, vol. 412.
15. H.-P. Schwefel, "Evolution and optimization seeking," John Wiley Sons, 1995.
16. X.-S. Yang, Nature-inspired metaheuristic algorithms. Luniver press, 2010.
17. A. R. Vahdat, N. NourAshrafoddin, and S. S. Ghidary, "Mobile robot global localization using differential evolution and particle swarm optimization," in Evolutionary Computation, 2007. CEC 2007. IEEE Congress on. IEEE, 2007, pp. 1527–1534.
18. R. Storn and K. Price, Differential evolution-a simple and efficient adaptive scheme for global optimization over continuous spaces. ICSI Berkeley, 1995, vol. 3.
19. D. Kumar and K. K. Mishra, "Incorporating logic in artificial bee colony (abc) algorithm to solve first order logic problems: The logical abc," in Knowledge and Smart Technology (KST), 2015 7th International Conference on. IEEE, 2015, pp. 65–70.
20. C. Gonz alez, D. Blanco, and L. Moreno, "Optimum robot manipulator path generation using differential evolution," in Evolutionary Computation, 2009. CEC'09. IEEE Congress on. IEEE, 2009, pp. 3322–3329.
21. J. Gu and G. Gu, "Differential evolution with a local search operator," in Informatics in Control, Automation and Robotics (CAR), 2010 2nd International Asia Conference on, vol. 2. IEEE, 2010, pp. 480–483.

22. J. M. Hereford, M. Siebold, and S. Nichols, "Using the particle swarm optimization algorithm for robotic search applications," in Swarm Intelligence Symposium, 2007. SIS 2007. IEEE. IEEE, 2007, pp. 53–59.
23. V. Pano and P. R. Ouyang, "Comparative study of ga, pso, and de for tuning position domain pid controller," in Robotics and Biomimetics (ROBIO), 2014 IEEE International Conference on. IEEE, 2014, pp. 1254–1259.
24. M. G. Epitropakis, D. K. Tasoulis, N. G. Pavlidis, V. P. Plagianakos, and M. N. Vrahatis, "Enhancing differential evolution utilizing proximitybased mutation operators," Evolutionary Computation, IEEE Transactions on, vol. 15, no. 1, pp. 99–119, 2011.
25. P. Melin, F. Olivas, O. Castillo, F. Valdez, J. Soria, and M. Valdez, "Optimal design of fuzzy classification systems using pso with dynamic parameter adaptation through fuzzy logic," Expert Systems with Applications, vol. 40, no. 8, pp. 3196–3206, 2013.
26. C. Sun, H. Zhou, and L. Chen, "Improved differential evolution algorithms," in Computer Science and Automation Engineering (CSAE), 2012 IEEE International Conference on, vol. 3. IEEE, 2012, pp. 142–145.
27. A. Tripathi, N. Saxena, K. K. Mishra, and A. K. Misra, "An environmental adaption method with real parameter encoding for dynamic environment," Journal of Intelligent Fuzzy Systems, no. Preprint, pp. 1–13.
28. Y.-W. Chen, C.-L. Lin, and A. Mimori, "Multimodal medical image registration using particle swarm optimization," in Intelligent Systems Design and Applications, 2008. ISDA'08. Eighth International Conference on, vol.3. IEEE, 2008, pp. 127–131.
29. Brockho. (n.d.). Cec-bbob-2015 [COmparing Continuous Optimisers: COCO]. Retrieved June 11, 2016, from http://coco.gforge.inria.fr/

# Novel Apparition Attributes to Improve Interactive Visualization

**Khushbu Gulabani and Anil Kumar Dubey**

**Abstract** Scientific visualization is the research era for vision measurement and formation of object as per the higher interaction. Various television shows, movies, etc., are designed to improve the better visualization by using key frames, frame rate, layering and compute their vision perspective in real-time systems. Most of the movies are using visualization for better attraction to the users and increase their usability for the rating as well as cost. I appreciate the previous survey and find out that the eye visualization and photo metrics are a necessity of people to show various pictures. Much of the research is done in this field and much more remains to be explored. Domains of interactive visualization are focused and the problems in derived parameters of visualization are found out. Novel parameters have been introduced to improve the quality of interactive visualization for better results.

**Keywords** Interactive visualization (IV) · Age group · Color context · Resolution quality

## 1 Introduction

Visualization is a process to manipulate the images as per the visual aspects, considering contextual parameters for easily understanding attractive look of objects and so on. Visualization is collaborated with human interaction for multiple application areas such as medical, scientific computing, architecture, and other social applications. Interaction is the term showing connection between two entities

---

K. Gulabani (✉)
Government Engineering College Ajmer, Ajmer, India
e-mail: khushbulakhani@gmail.com

A.K. Dubey
Senior Member IEEE, Government Engineering College Ajmer,
Ajmer 305001, Rajasthan, India
e-mail: anildudenish@gmail.com

 
S.K. Bhatia et al. (eds.), *Advances in Computer and Computational Sciences*,
Advances in Intelligent Systems and Computing 553,
DOI 10.1007/978-981-10-3770-2_57

for their interacting methods, approach, and techniques to enhance the attraction for better interaction.

The primary objective of research is to analyze the visualization and improve the quality of interactive visualization.

## 2 Related Work

Daniel F. Keefe et al. [1] recommend an interactive framework for the purpose of exploring space-time and form–function relationships in high-resolution biomechanical datasets collected experimentally. It demonstrates a new interactive approach for constructing small multiple visualizations in a collection of more than 100 repeated cyclic motions. Work may be extended for challenges like improving scalability, implementing framework for new data sources and refining visual representations. Daniel F. Keefe et al. [2] examine interaction in specific visualization scenarios. It also brings interaction research into the context of visualization research for the purpose of best describing design, model, and implementing interactive visualizations that facilitate complex data analysis. Daniel Acevedo and David H. Laidlaw [3] appraise a parameterized set of 2D icon-based visualization methods. The capability of 120 different methods for visualization was characterized to represent 2D scalar fields effectively. It successfully applies a new methodology for evaluating perpetual interactions involving multiple visual elements and underlines the fact that with change in one of the element, the user's reading of data in visualization is affected.

Lingyun Yu et al. [4] suggest technique which provides the feature of touch interaction with 3D scientific data spaces in 7DOF. The unique feature of this interaction is that it does not require the presence of dedicated objects for the purpose of constrain in the mapping and a design decision which is important for many scientific datasets for particle simulations in astronomy or physics. Hiroaki Ohtani et al. [5] develop the virtual reality (VR) technology for the purpose of analyzing simulation results in 3-D VR space. Using VR Technology, it becomes possible to check the distance between the plasma and the device in the vessel. This success in this work opens up new paths in the contribution to the experiments in related fields. Mario Ohlberger and Martin Rumpf [6] discover multiresolution visualization methods for real-time interactive processing. Daniel F. Keefe et al. [7] illustrate four design cases of process and tools for collaborative design of VR visualizations and proposed Scientific Sketching, a formal methodology for collaboration. David Borland and Alan Huber [8] confer color mapping, the most common visualization techniques and suggest the selection of a color map apt for specific datasets which can greatly enhance the effectiveness of visualization. Daniel Weiskopf et al. [9] converse an interactive technique for dense texture-based visualization of unsteady 3D flow for enhancing computational efficiency and visual perception.

Martin Hafner et al. [10] pioneer an interactive post processing for allowing a dynamic interaction with finite element solutions, a mesh cutting, and data exploring solutions for simplifying complex solutions. Keir Smith [11] presents interfacing between Viuser (any visual information user) and televisual data. It opens up new dimensions for human televisual data interaction. Matthew McGinity et al. [12] confer the design, technical challenges, novel features as well as current and future applications of the system for 360 degree cylindrical stereo virtual reality theater. Tim Barker [13] investigates a process of creating a specific type of reality. Aesthetic is performed for significant new media forms, creation of new processes, and new sites of creativity. Timothy Barker [14] positions the aesthetic with MR environments as a hybrid process mixed reality (MR) environments present a radical shift in aesthetics. Interaction is proposed as the collaboration of two conditions; the condition of the machine and the condition of 'userness'. Volker Kuchelmeister [15] suggests a method for describing a novel way of recording and documenting motion. It enables detailed analyzation after the happening of the event. Neil C.M. Brown et al. [16] sets up a framework and identifies three key functions responsible for interactive narrative in new media. It motivates the future researchers to develop a relational theory that plans the agency of functions into a relational map of descriptive formation. Glorianna Davenport [17] confers interactive multimedia as a user-directed form of storytelling. It also examines the nature of cinematic storytelling. It proposes the representation of content in layers. Pascal Volino [18] integrates framework with innovative tools aiming toward efficiency and quality in the process of garment designing and prototyping, taking advantage of state-of-the-art algorithms from the field of mechanical simulation, animation, and rendering.

Paulo. Rech Wagner [19] contributes the benefits achieved by the availability of visual interactive simulation and modeling facilities simultaneous during the experiment and not the facilities themselves. Mirjam Vosmeer and Ben Schouten [20] illustrate interactive cinematography for new domain of media entertainment for the purpose of enhancing production and user experience. Robert B. Trelease et al. [21] initiate the QuickTime multimedia environment, VR extensions, basic linear and nonlinear digital video technologies, image attainment, and other specialized QuickTime VR production methods. Mark O. Riedl et al. [22] explain machinima, a technique used to produce computer animated movies with the help of manipulation of gaming technology with the help of two intelligent support tools for authoring and producing "Machinima". Eva Oliveira et al. [23] categorize accesses, explore, and visualize movies on emotional features with iFelt, which is an interactive web video application. It explores the design and evaluates various ways to browses, access and visualizes movies and their contents.

## 3 Problem Identification

It is more crucial to measure the interaction level between any user and the system objects because it depends on their attraction as different people have different attraction for the same object. Therefore it is more critical for the researchers to deploy the interactive visualization of object as per the user demands; even though the identical twins are not having the same visualization for objects. Researchers are planning to resolve such problems by following different approaches but yet much needs to be explored.

## 4 Research Methodology

We put up the age parameter, distance, and environmental factors to measure the above task and enhance the attraction to target interactive visualization. Here, we focus on the aspect visualization through derived visual parameters contain is distance color angle group rotations etc. The proposal is demonstrated for human vision and their requirements. The scientific computing process is performed for qualitative research and research methodology. Two different phases are derived in the methodological approach to resolve the appropriate visualization problem and find out the reality in the visual aspects.

## 5 Novel Apparition Attributes for IV

To resolve the above problem, we proposed novel apparition attributes for interactive visualization in mathematical terms as set of distance, color, and resolution in computation techniques.

### *5.1 Age Group*

Visual aspects according to age group: As we know that the human eye vision capability depends on their age limit, different age group people are able to see the different visualization of object visualized in their eyes. It means the same object is visualized in different ways due to the age factor/ability of object reorganization via eye. Therefore we include the age parameter in the visualization aspect. Let us consider three different age people such as below 25, between 25 and 50, and above 50. These groups of people are seeing the same images keeping all other parameters constant. But the visual aspect of images differs because of the capabilities of retina recognition. Therefore the modification is done as integrity of visualization using human age parameters.

Let us suppose interactive visualization is function set of color context of visualization, angle, or rotation of object from recognizer and distance between recognizer and object and other constrains, then

$$\mathbf{IV} \approx \mathbf{f_1}\{\mathbf{D}, \mathbf{C}, \mathbf{Rq}, \mathbf{A}, \mathbf{A^*}\} \tag{1}$$

IV Interactive visualization
D Distance between recognizer and object
C Color context of visualization (RGB)
Rq Resolution quality mode
$A^*$ Other constraints

Methodology 1 focuses on age group, therefore equations are as:

a. Interactive visualization for people whose age is below 25 is

$$\mathbf{IV_{b25}} \approx \mathbf{f_1}\{\mathbf{D_h}, \mathbf{C_{ma}}, \mathbf{Rq_{ma}}, \mathbf{A}, \mathbf{A^*}\} \tag{2}$$

Here the designer required the focus on C, D, and E as C is more than average, D becomes high and R is more than average, so that user receives the interactive visualization quality.

b. Interactive visualization for people whose age is above 25 and below 50 is

$$\mathbf{IV_{bw50}} \approx \mathbf{f_1}\{\mathbf{D_h}, \mathbf{C_{avg}}, \mathbf{Rq_a}, \mathbf{A}, \mathbf{A^*}\} \tag{3}$$

Here the C is average, D becomes high and R is average so that user visual interaction is excellent.

c. Interactive visualization for people whose age is above 50 is

$$\mathbf{IV_{a50}} \approx \mathbf{f_1}\{\mathbf{D_l}, \mathbf{C_{ma}}, \mathbf{Rq_h}, \mathbf{A}, \mathbf{A^*}\} \tag{4}$$

Here the C is maximum, D becomes less and R is higher so that user gained the interactive visualization quality.

## 5.2 *Environmental Effects*

Visual aspects according to environmental effect. The researcher focuses to detect the problems behind the quality of visualization due to environmental effects. May be the surrounding molecule or change the vision context of visual object or temperature effect the angle or shear the one side of image increasing or decreasing order. To resolve such a problem, we use the environmental effect as the context parameter of visualization so that interactive visualization is easily performed. It means other constrains are angle or rotation of object from recognizer and size of object.

$$\mathbf{IV} \approx \mathbf{f_2}\{\mathbf{R},\ \mathbf{S}\} \tag{5}$$

where

R Angle or rotation of object from recognizer
S Size of object

Now, determination of both equation (1 and 5), we find that

$$\mathbf{IV} = \{\mathbf{D}, \mathbf{C}, \mathbf{R}, \mathbf{Rq}, \mathbf{S}, \mathbf{A}\} \tag{6}$$

IV Interactive visualization
D Distance between recognizer and object
C Color context of visualization (RGB)
R Angle or rotation of object from recognizer
Rq Resolution quality mode
S Size of object
A Age factor

These two constraints of proposed methodology improve the visualization toward interactive vision of user. The computation of visual interaction is performed using dimensions and coordinates where object is situated and mode of object (1D, 2D, 3D…) in the dimension is derived. Object is distributed in different frames and each individual frame of object is used to compute and enhance the visualization of center coordinates of key frame objects and measured to point out that the object is in which coordinate (x, y, −x, −y). According to their coordinate the computation will be initialized and assured that the shearing will be evident due to their coordinates or object of coordinate.

Interactive virtualization of images or objects is according to the usability of user in different cases as

(a) The object is generally seen by user in their own embedded items(i.e., computer system, laptops, etc.) in that case the variance/integration of object quality parameters is derived from 9 to 1 m according to user application
(b) During the projection of objects in wide range of users to spread the objects visualization then the parameters of interactive visualization are considered for 5–30 m and for a specific task like projection of movies the average (from 5 to 30 m) averages 20 m and computed for approximate visualization.

To visualize the objects in open environment like screening the pictures or objects along roadside then the integration of interactive virtualization follows the distance parameter of visualization from 3 to 40 m and maximizes all the quality factors that are majorly impacted in these environments.

## 5.3 Simulation Result

The simulation is done under deep learning tools of Ubuntu environments using python syntax and results show that proposed formula for novel apparition attributes to interactive visualization is applicable for society.

Through python script, we find out that the proposal is suiting the interactive effects for both age factor and environmental factor.

## 6 Conclusion

It is more crucial to measure attraction of different people for different objects. In this paper, we deeply analyzed the interactive visualization and follow the scientific visualization as per the scientific computing. To achieve higher attraction, we propose novel apparition attributes that support to enhance the quality of interactive visualization. Interactive visualization is the proportional concern of distance between recognizer and object, color context, resolution quality modes, and age factor with other constraints. Based on these, we derived a formulation to compute the interactive visualization and studied their effect by changing the values of their dependencies in proposed function. Different age groups differ in attraction for the same object and same age group may differ in interaction for randomly changed multiple color context of object. We simulated the proposal by using python syntax and results show that proposed formula for novel apparition attributes to interactive visualization for applications everywhere in the deployment for interactive visualization for TV shows, movies, and other entities. Researchers are invited to compute the proposal for real-time application and measure the usability as maximum possible population through qualitative and quantitative research methodology using samples of user attraction.

**Acknowledgements** The author is thankful to investigator for investigating new era of research and also thankful to the all researchers who contributed in the same.

## References

1. Daniel F. Keefe, Marcus Ewert, William Ribarsky, and Remco Chang, "Interactive Coordinated Multiple-View Visualization of Biomechanical Motion Data", IEEE Transactions On Visualization And Computer Graphics, Vol. 15, No. 6, November/December 2009, Published by the IEEE Computer Society, 1077-2626/09/$25.00 © 2009 IEEE, pp: 1383–1390.
2. Daniel F. Keefe, "Integrating Visualization and Interaction Research to Improve Scientific Workflows", IEEE Computer Graphics and Applications: Visualization Viewpoints, Published by the IEEE Computer Society, March/April 2010, 0272-1716/10/$26.00 © 2010 IEEE, pp: 8–13.
3. Daniel Acevedo and David H. Laidlaw, "Subjective Quantification of Perceptual Interactions among some 2D Scientific Visualization Methods", IEEE Transactions On Visualization And Computer Graphics, Vol. 12, No. 5, September/October 2006, Published by the IEEE Computer Society, 1077-2626/06/$20.00 © 2006 IEEE, pp: 1133–1140.
4. Lingyun Yu, Pjotr Svetachov, Petra Isenberg, Maarten H. Everts, and Tobias Isenberg, "FI3D: Direct-Touch Interaction for the Exploration of 3D Scientific Visualization Spaces", IEEE Transactions On Visualization And Computer Graphics, Vol. 16, No. 6, November/December 2010, Published by the IEEE Computer Society, 1077-2626/10/$26.00 © 2010 IEEE, pp: 1613–1622.
5. Hiroaki Ohtani, Yuichi Tamura, Akira Kageyama, and Seiji Ishiguro, "Scientific Visualization of Plasma Simulation Results and Device Data in Virtual-Reality Space", IEEE

Transactions On Plasma Science, Vol. 39, No. 11, November 2011, 0093-3813/$26.00 © 2011 IEEE, DOI:10.1109/TPS.2011.2157174.
6. Mario Ohlberger and Martin Rumpf, "Adaptive Projection Operators in Multiresolution Scientific Visualization", IEEE Transactions On Visualization And Computer Graphics, Vol. 5, No. 1, January-March 1999, 1077-2626/99/$10.000 1999 IEEE, pp: 74–94.
7. Daniel F. Keefe, Daniel Acevedo, Jadrian Miles, Fritz Drury, Sharon M. Swartz, and David H. Laidlaw, "Scientific Sketching for Collaborative VR Visualization Design", IEEE Transactions On Visualization And Computer Graphics, Vol. 14, No. 4, July/August 2008, Published by the IEEE Computer Society, 1077-2626/08/$25.00_2008 IEEE, DOI:10.1109/TVCG.2008.31, pp: 835–847.
8. David Borland, Alan Huber, "Collaboration-Specific Color-Map Design", IEEE Computer Graphics and Applications: Visualization Viewpoints, Published by the IEEE Computer Society, 0272-1716/11/$26.00 © 2011 IEEE, pp: 7–11.
9. Daniel Weiskopf, Tobias Schafhitzel, and Thomas Ertl, "Texture-Based Visualization of Unsteady 3D Flow by Real-Time Advection and Volumetric Illumination", IEEE Transactions On Visualization And Computer Graphics, Vol. 13, No. 3, May/June 2007, Published by the IEEE Computer Society, 1077-2626/07/$25.00_2007 IEEE, DOI:10.1109/TVCG.2007.1014, pp: 569–582.
10. Martin Hafner, Marc Schöning, Marcin Antczak, Andrzej Demenko, and Kay Hameyer, "Interactive Postprocessing in 3D Electromagnetics", IEEE Transactions On Magnetics, Vol. 46, No. 8, August 2010, 0018-9464/$26.00 © 2010 IEEE, DOI:10.1109/TMAG.2010.2043821, pp: 3437–3440.
11. Keir Smith, "Rewarding the Viuser: A Human-Televisual Data Interface Application", http://www.icinema.unsw.edu.au/assets/163/rewarding_viuser.pdf.
12. Matthew McGinity, Jeffrey Shaw Volker Kuchelmeister Ardrian Hardjono Dennis Del Favero, "AVIE: A Versatile Multi-User Stereo 360° Interactive VR Theatre: Advanced Visualization and Interaction Environment (AVIE)", EDT 2007, San Diego, California, August 04, 2007. © 2007 ACM 978-1-59593-669-1/07/0008 $5.00.
13. Tim Barker, "Toward a Process Philosophy for Digital Aesthetics", Proceedings of the International Symposium on Electronic Arts 09 (ISEA09), Belfast, 23rd August–1st September 2009.
14. Timothy Barker, "Process and (Mixed) Reality: A Process Philosophy for Interaction in Mixed Reality Environments", IEEE International Symposium on Mixed and Augmented Reality 2009 Arts, Media and Humanities Proceedings 19–22 October, Orlando, Florida, USA 978-1-4244-5463-1/09/$25.00 © 2009 IEEE, pp: 17–23.
15. Volker Kuchelmeister, "Universal capture through stereographic multiperspective recording and scene reconstruction", http://www.icinema.unsw.edu.au/assets/167/Universal_Capture_PCM2009.pdf.
16. Neil C.M. Brown, Timothy S. Barker and Dennis Del Favero, "Performing Digital Aesthetics: The Framework for a Theory of the Formation of Interactive Narratives", Leonardo, Vol. 44, No. 3, pp. 212–219, 2011, © 2011 ISAST, pp: 213–219.
17. Glorianna Davenport, Thomas Aguirre Smith, Natalio Pincever, "Cinematic Primitives for Multimedia", IEEE Computer Graphics Books, Published by the IEEE Computer Society, Issue No. 04 - July/August (1991 vol. 11), pp: 67–74, DOI Bookmark: http://doi.ieeecomputersociety.org/10.1109/38.126883.
18. Pascal Volino, Frederic Cordier, Nadia Magnenat-Thalmann, "From early virtual garment simulation to interactive fashion design", Computer-Aided Design, Volume 37, Issue 6, May 2005, Pages 593–608, CAD Methods in Garment Design, Science Direct © Elsevier.
19. Paulo Rech Wagner, "A New Paradigm for Visual Interactive Modeling And Simulation", European Simulation Symposium, 1996.
20. Mirjam Vosmeer, Ben Schouten, "Interactive Cinema: Engagement and Interaction", Proceedings of 7th International Conference on Interactive Digital Storytelling, ICIDS 2014, Singapore, November 3–6, 2014, Chapter: Interactive Storytelling of Book Interactive Storytelling, Volume 8832 of the series Lecture Notes in Computer Science pp 140–147,

ISBN: 978-3-319-12337-0, DOI:10.1007/978-3-319-12337-0_14, Springer International Publishing.

21. Robert B. Trelease, Gary L. Nieder, Jens Dørup and Michael Schacht Hansen, "Going virtual with quicktime VR: New methods and standardized tools for interactive dynamic visualization of anatomical structures", The Anatomical Record, Volume 261, Issue 2, pages 64–77, April 2000, DOI:10.1002/(SICI)1097-0185(20000415)261:2<64::AID-AR6>3.0.CO;2-O, © 2000 Wiley-Liss, Inc.
22. Mark O. Riedl, Jonathan P. Rowe, David K. Elson, "Toward intelligent support of authoring machinima media content: story and visualization", Proceeding INTETAIN '08 Proceedings of the 2nd international conference on INtelligent TEchnologies for interactive enterTAINment, articles 4, ICST (Institute for Computer Sciences, Social-Informatics and Telecommunications Engineering), Brussels, Belgium, Belgium © 2007, ISBN: 978-963-9799-13-4, Copyright © 2007 ACM.
23. Eva Oliveira, Pedro Martins, Teresa Chambel, "Ifelt: accessing movies through our emotions", Proceeding EuroITV '11 Proceedings of the 9th international interactive conference on Interactive television, ACM New York, NY, USA © 2011, ISBN: 978-1-4503-0602-7, Pages 105–114, DOI:10.1145/2000119.2000141.

# Five-Layered Neural Fuzzy Closed-Loop Hybrid Control System with Compound Bayesian Decision-Making Process for Classification Cum Identification of Mixed Connective Conjunct Consonants and Numerals

**Santosh Kumar Henge and B. Rama**

**Abstract** The OCR generation systems are most sophisticated active field and interesting conversional discovery for digitalization of handwritten and typed imprecise data into machine detectable characters. The fuzzy logic system processes the data with help of primary-based bunch set of knowledge. Fuzzy logic closed-loop system having very good successful rate of frame work for decision-based functions, can derive the fuzzy rules to build the decision-making procedure and detect the letters human to system to human. Artificial neural networks are compatible and the best area to solve the pattern cum text recognition tasks. The innovative combinational based characteristic of neural fuzzy-based closed-loop hybrid system proposing a five-layered approach with technical ideas, solutions solve the critical problems in the field of character, face, symbol recognition procedure, and estimating the density ratio. Recognition of single text, numbers is easy than the recognition of mixed connective conjunct consonants. Because of their variations, various handwritten pen-stroke pulses, tuning the initial and end position of each conjunct consonant, some consonants are connected and mixed with their left-cum right-side placed conjuncts, numerals, and symbols. Many languages such as Arabic, Hindi, Urdu, Telugu, and Tamil represent syllabic, symbol scripted form, and most of words formed with the mixed conjuncts, mixed cum touched consonants, mixed conjuncts with numerals in their representation. This research approach has proposed the five-layered neural fuzzy closed-loop hybrid system with compound Bayesian decision-making process holding good outcome for classification cum identification of mixed connective conjunct consonants with their numerals. The recognition process can start with categorizing

S.K. Henge (✉) · B. Rama
Computer Science Department, University College, Kakatiya University, Hanamkonda 506001, Warangal District, Telangana, India
e-mail: hingesanthosh@gmail.com

B. Rama
e-mail: rama.abbidi@gmail.com



S.K. Bhatia et al. (eds.), *Advances in Computer and Computational Sciences*, Advances in Intelligent Systems and Computing 553,
DOI 10.1007/978-981-10-3770-2_58

total text into two forms; normal conjunct consonants and mixed connective conjunct consonants. The permutation futures of five-layered neural fuzzy closed-loop hybrid system represent inputs as neurons, convert it into fuzzy set inputs, and then apply the fuzzification process with desirable fuzzy knowledge based rules to produce the required output through responses. The compound Bayesian decision-making process is used to perform the operations of probability of sum to unity to reduce the recognition problems in the mixed connective conjunct consonants.

**Keywords** Bayesian decision-making process (BDMP) · Mixed connective conjunct consonants (MCCC) · Fuzzy logic controller (FLC) · Mixed conjunct consonants (MCjC) · Fuzzy artificial neural hybrid system (FANHS) · OCR (Optical character recognition)

## 1 Introduction

The optical character recognition is the fast-growing innovational attractive exploration in the development of template, pattern with image-based processing, and text cum character recognition. In present life, the OCR inventions are more actively and accurately using in many industries, organizations, and home-based appliances especially in banking, financial, postal tracking, office automation, security based, and authorizing the ownership documents. The OCR generation systems are most sophisticated active field and interesting conversional discovery for digitalization of handwritten cum typewritten letters with imprecise data into machine detectable characters. The offline and online [1] are the major categories in character recognition procedure. Offline OCR method is innovational investigation in expansion of technical cum algorithmic development in handwritten cum printed character recognition process [2]. The decision-making states, acquisitioned segmentation processing, preprocessing acquisition, preprocessing-classification, postprocessing, extracting the future sets with their responses are the processing stages in OCR approach [3].

The scanned picture cum image or photograph is the input sample of initial image acquisition stage to start the process of OCR approach [4, 3]. The conversion of primary imprecise data into precise data sets with their co-related rules cum calculations is active mode in preprocessing stage to help the sensor for capturing text-based images. In addition to this stage, the other basic steps like as text stroke rate detection, detection cum removal of noise problems, and stages of normalization. Text extra space separation cum removal, text start ending positions, categorizing, and grouping of individuals are processed in segmentation stage [5]. The process of analyzing, finding and exploring the output result sets are composed based on the identification of mixed conjunct consonants and deriving group of future sets which representing the shape of the underlying character as precisely and uniquely as possible in the future extraction stage.

## 2 Handwritten Mixed Connective Conjunct Consonants and Variations

Recognition of single consonant text along with the numbers [6, 7] are easy than the recognition of mixed connective conjunct consonants, because of these type of characters are framed with various handwritten pen-stroke pulses, mixed conjuncts, numerals, mixing range of left cum right most side placed symbols and finding the initial cum end position of each conjunct consonant. Many languages such as Arabic, Hindi, Urdu, Telugu, Tamil represent syllabic, symbol scripted form and most of word formed with the mixed conjuncts, mixed cum touched consonants, mixed conjuncts with numerals in their representation, and position of conjunct consonants. Conjunct consonant symbols are presented either on left, or on to the top, or at the bottom, or to the right or a combination of all these types. The recognition of mixed connective conjunct consonants is complex to recognize, analyze, and identify by the OCR system. Some of the following problems are identified in recognition of handwritten conjunct mixed consonants.

- Most of conjuncts will appear top, bottom, and along with their consonants, difficult to recognize number of conjunct for a particular consonant.
- Finding the initial cum ending point of connective conjunct consonants.
- Some consonants holding both top cum bottom placed conjunct consonants.
- Some conjunct consonants mixed with pre-positioned, post-positioned consonants along with their original consonant.
- Some numerals are mixed with their nearer conjunct consonants.
- Variation and alignment of mixed connective conjunct consonants.
- Mixed connective conjunct consonants input thickness and stroke levels.
- Start cum end point of each consonant with it contains the conjunctives.
- Mixed connective conjunct consonants improper extrusions and ending point corners.
- Some conjuncts are overlapping with their original consonants.
- Different shapes and sizes of single consonants and conjuncts.
- Unconnected and irrelative line pixels, segments and curves.

Because of above-mentioned variations, the recognition and categorizations of mixed connective conjunct consonants are complex to the OCR machine. This approach proposes ideological methodology to overcome the problems.

## 3 Manuscript Objectives

The ideological objective of this paper is proposing five-layered neural fuzzy closed-loop hybrid control system with functionalities of compound Bayesian decision-making method for classification cum identification and to develop the deference prototyping of handwritten mixed connective conjunct consonants, numerals and the compound Bayesian decision-making process used to perform the operations of probability sum to unity to reduce the recognition problems in the mixed connective conjunct consonants.

## 4 Five Layer of Neural Fuzzy Closed-Loop Hybrid Control System (FNFHCS)

The five layer of neural fuzzy closed-loop hybrid control system (FNFHCS) is deriving the combinational futures and functionalities of artificial neural networks and closed-loop variation of fuzzy logic control system (FLCS). Fuzzy rules can be interpreted as imprecise prototypes of training data [8, 9]. The underlying fuzzy system and leaning procedure in build for ensuring the semantic properties. In fuzzy logic processing the hug set of prior basic knowledge is needed for processing the data, the leaning process is not possible, linguistic description rules of are needed for input and output variables with certain desirable inference rules [10–12]. Fuzzy logic controller (FLC) very good at implementations of decision functions but it can good at automatically acquiring the fuzzy rules used for decision-making procedure [5]. Neural networks are very good in pattern recognition but they are not that good to reach their mentioned decisions and explanations. In neural networks, the learning process can be achieved from basic scratch data, several learning–based algorithms have been designed and the back box behavior is used to train the observations (Fig. 1).

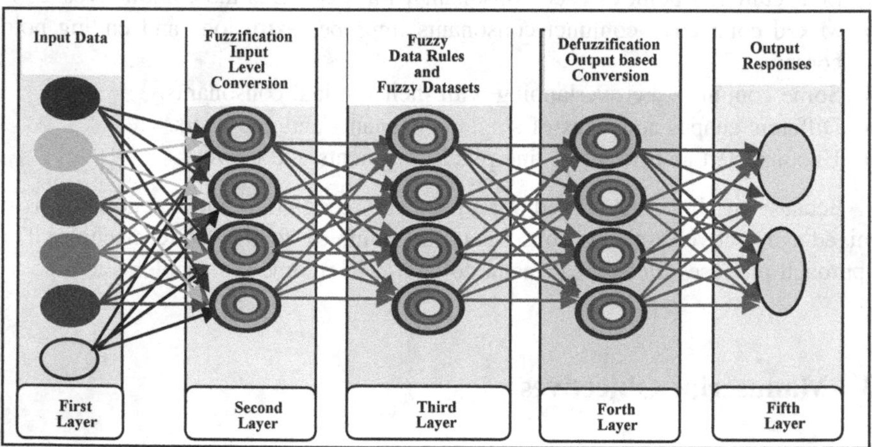


**Fig. 1** Five-layered neural fuzzy closed-loop hybrid control system descriptive methodology

### *4.1 Advantages and Major Functionalities of Neural Networks and Fuzzy Logic Control System in OCR Approach*

FLCS and artificial neural network (ANN) are having the individual advantages. In OCR process, neural networks and fuzzy logic control system perform individual levels of task regions.

Learning and Training process in OCR: In neural networks, the learning–training process can be potential, it can perform through the basic scrape data, but in FLCS the trained samples are essential for data processing.

Formation of prior fundamental knowledge in OCR: Prior knowledge is not required about the problem for extracting desirable rules from the neural networks and It deals with implicit knowledge that can be acquired by learning, but in fuzzy system prior knowledge is required and needed for further processing and the fuzzy system can be tuned in a systematic way when the knowledge is insufficient or holding wrong assumptions, it deals with explicit knowledge.

Decision-making process in OCR: ANN are having expert representation at pattern recognition, it is poor to reach their expected decisions and FLC excellent at classification and their explanations and fuzzy logic not that much good at automatically acquiring the rules for decision-makings.

Data processing mode in OCR: ANN is simplified advance of human nervous system with memorizing ability to adjust certain desirable situations, to be trained from the past-precedent familiarity, but the FLCS deal with imprecise or uncertainty data in the existing system, to form the fuzzy logic rules and to calculate the equations for finding various solutions to the problem.

Rule configuration and generation in OCR: In NN, the trained enough observations with applicable samples are prepared for the problem; it is trained with required neural-based mechanism. FLC required the linguistic rules and linguistic description for the input and output variables.

Precise restrictions in OCR: ANN can activate on perfect limitations with negligible fluctuations but the FLC cannot operate perfect limitations. It provides a transition between the membership and non-membership functions.

Method of tuning task in OCR: In ANN, the tuning functions can be derived as optimization problem, handled by the approach of Hopfield neural networks and in FLCS, tuning is the important process handled by the fuzzy membership functions.

### *4.2 Advantages and Major Functionalities of Five Layers Approach of Neural Fuzzy Closed-Loop Hybrid Control System*

ANNs and closed-loop approach of FLCS having important advantages and common functionalities such as mathematical approach are not needed. The arrangement individuality of five layers based neural fuzzy closed-loop hybrid control system can overcome the difficulties, drawback of individual systems by combining the both. The innovative combinational based characteristic of neural fuzzy based closed-loop hybrid system proposing a five-layered approach with technical ideas, solutions to solve the critical problems in the field of character face symbol recognition procedure. NFCLHS takes the input values in the form neural-represented principles through the first-layer, neural-based input values converted into fuzzy-represented sets with help fuzzification process through the second-layer, fuzzy knowledge represented implication rules applied to fuzzy represented sets through the third-layer, the linguistic values generated from fuzzy-represented input sets through the fourth layer and finally the fifth layer generates co-related output responses. The major effectives of five layers based neural fuzzy closed-loop hybrid control system:

- Fuzzy sets, fuzzy based functions can quickly perform the task-based calculations and derive the equations in simple manner
- It can handle all kinds of data like as logical, numeric, and linguistic improper data, partially mentioned data and half-derived imprecise data
- The NFCLHS can produce unfailing responses
- The neural fuzzy closed-loop hybrid system saves the time implementation cost. NFCLHS is more active in its operations like as self-learning, self-tuning, and self-organizing processes
- The primary approach of data relationship method is not needed for deriving the NFCLHS.

### 4.3 Arrangement of Fuzzy Closed-Loop Control System and Formation of Compound Bayesian Classification Process (CBCP)

The closed-loop approach of fuzzy logic system exists with the top-priority and middle-priority panels. The top priority panel derives the dynamic filters and middle priority panel derives the static mapping method as shown in the Fig. 2. Bayesian decision-making method is operating based on computing the segmentation points between the various required classifications by using possibilities and its outlays to go together with such required decisions. And also it builds the hypothesis, which the decision-based problem is derived in the terms of grouped possibilities with recognized corresponding probability values. These rules are similar to the human-based natural language representations and the future states of the nature are characterized as probability events, conventionally the probability sum to unity. The main benefit is gathering group of rules collective simultaneously that already formed by the individual layer-based.

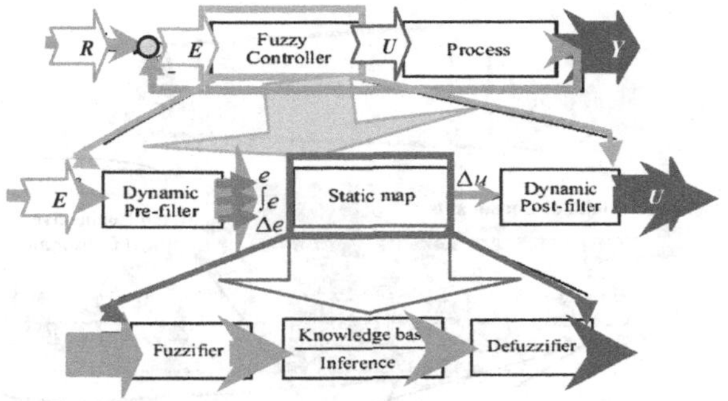


**Fig. 2** Fuzzy closed-loop control system configuration with dynamic filters of *top panel* and static map of *middle panel*

$$A = \{A(k_1),\ A(k_2),\ A(k_3),\ \ldots\ldots\ A(k_n)\} \quad \text{where} \quad 1 = \sum_{i-1}^{y} A(i)$$

In posterior probability, the probability of certain state of nature with given observations such as A(k|y).

$$A(k, y) = A(k|y)P(y) = A(y|k)A(y) \quad \text{where} \quad A\left(\frac{k}{y}\right) = \frac{A\left(\frac{y}{k}\right)A(k)}{A(y)} = \frac{A\left(\frac{y}{k}\right)A(k)}{\sum_i A\left(\frac{y}{ki}\right)A(ki)}$$

## 5 Methodology Execution

This paper proposed the five-layered neural fuzzy closed-loop hybrid control system to implement the classification cum identification process for handwritten cum printed characters taken from the scanned page or captured photograph, processing the image with observations of page intensity test with help of quality ration, scrutinize the direction, style, blow level of handwritten text and foundation, production level of printed characters. Extort page as basic input image with higher quality when recognition ability is lesser than or equal to page superiority allocation.

Recognition process can start with categorizing total text into two forms; normal conjunct consonants and mixed connective conjunct consonants. The permutation futures of FLNFCLHS represent inputs as neurons, convert it into fuzzy set inputs, and then apply the fuzzification process with desirable fuzzy knowledge cum rule based to the input fuzzy sets, and to produce the required output through responses. The compound Bayesian decision-making process used to perform the operations of probability sum to unity to reduce the recognition complexity problems in the mixed connective conjunct consonants. The problems of determining the connection point of conjunct consonants, slot correction, and thinning reduced by using CBCP by categorizing the input image has into original form and black-and-white

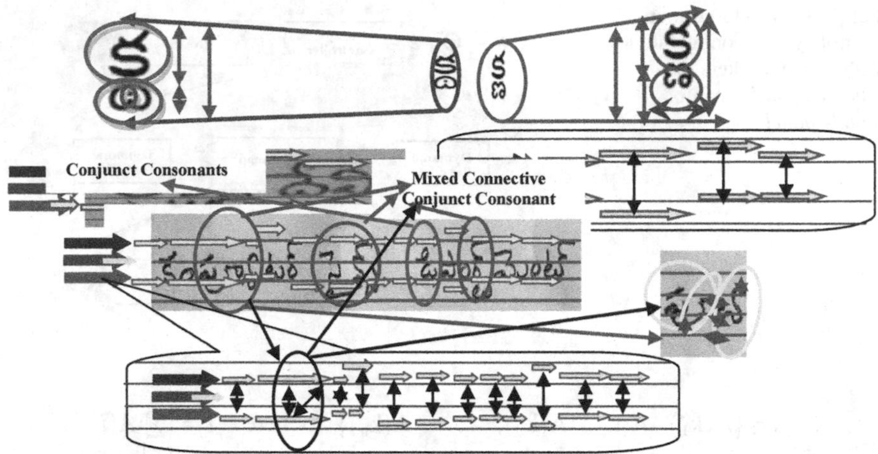


**Fig. 3** Print cum Handwritten mixed connective conjunct consonants with sensor detection flow of *middle-top-bottom*, *top-middle-bottom* for classification of consonants, mixed connective conjunct consonants and connective paths between the conjunct of original or other consonants

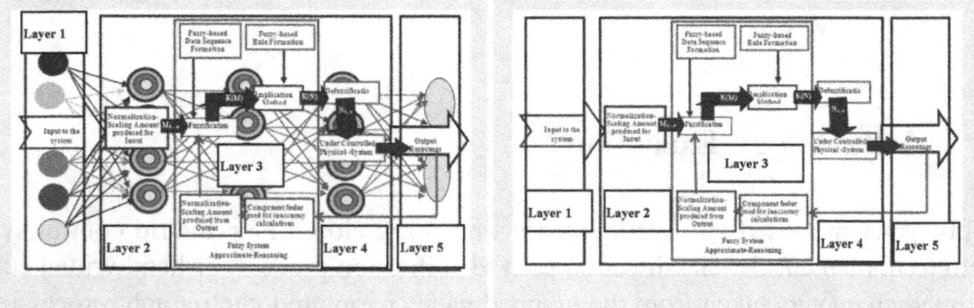


**Fig. 4** The NFCLHS approach with error calculation method passing through the five layers

conversion form as shown in Fig. 3. The black-and-white converted image especially used for classification and identification of connective conjunct consonants (Fig. 4).

The input levels of connective character recognition, preprocessing consonants, segmentation, and postprocessing technical modules were discussed in the analysis improvement process of the algorithm. The NFCLHS mechanism derives all learning possibility parameters and generates presumption calculations based on possible parameters with co-related fuzzy rules generated by fuzzy controller with available knowledge and it integrated conditional If-then-Else based fuzzy rules (Fig. 5).

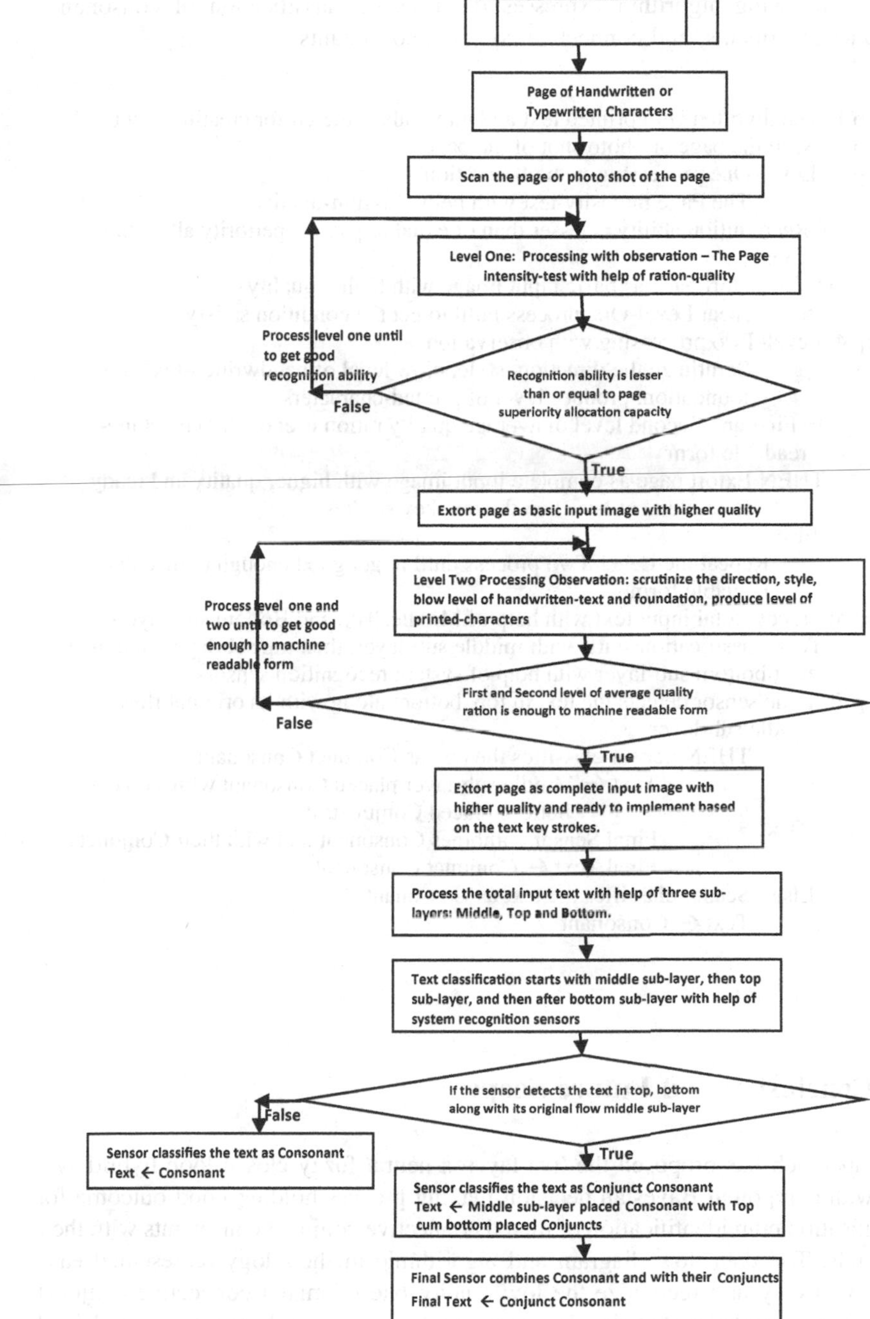


**Fig. 5** Data flow representation of processing, preprocessing and postprocessing stages of OCR approach with classification of consonants and mixed connective conjunct consonants

The following algorithm expresses the flow of classification of consonants, conjunct consonants, and connective conjunct consonants.

Step 1: Handwritten cum printed text and numerals gathered for creating input task
Step 2 : Scan the page or photo shot of the page
Step 3 : Level-One processing with observation
The Page intensity-test with help of ration-quality
IF Recognition ability is lesser than or equal to page-superiority allocation capacity
THEN Extort page as basic input image with higher quality
Else Repeat Level-One process until to get the condition satisfy
Step 4: Level-Two processing with observation
Scrutinize the direction, style, blow level of handwritten-text and foundation, produce level of printed-characters
IF First and Second level of average quality ration is enough to machine-readable form
THEN Extort page as complete input image with higher quality and ready to implement based on the text key strokes.
Else
Repeat the Level-Two process until to get good enough to machine-readable form
Step 5: Process total input text with help of Middle, Top and Bottom sub-layers
Step 6: Text classification starts with middle sub-layer, then top sub-layer, and then after bottom sub-layer with help of system recognition sensors
Step 7: IF the sensor detects the text in top, bottom along with its original flow middle sub-layer
THEN Sensor classifies the text as Conjunct Consonant
Text ← Middle sub-layer placed Consonant with Top cum bottom placed Conjuncts
Final Sensor combines Consonant and with their Conjuncts
Final Text ← Conjunct Consonant
Else Sensor classifies the text as Consonant
Text ← Consonant

## 6 Conclusion and Future Scope

This approach has proposed the five-layered neural fuzzy closed-loop hybrid system with compound Bayesian decision-making process holding good outcome for classification cum identification of mixed connective conjunct consonants with their numerals. The data flow diagram and algorithmic methodology represented easy way to classify and recognize the low quality rate of mixed connective conjunct consonants with help of their alignment and orientation and also it has explained how the sensor can detect and categorize the consonants and mixed connective conjunct consonants based on text variation flow of top-middle-bottom and middle-top-bottom. The OCR generations are most sophisticated active field and interesting conversional discovery for digitalization of hand cum typewritten imprecise data into machine detectable characters. In future, this invention ideology framework of five-layered NFCLHS with compound Bayesian decision-making process for classification cum identification of mixed connective conjunct consonants and numerals is scheduling to analyze, conduct experiments with their test results to get good success rate by using either Python Dato or MATLAB or the LabVIEW GUI VI version or math tool. Future work may include with further technical, algorithmic, result-oriented improvements for better classification with segmentation process of handwritten cum printed connective and touching conjunct consonants recognition.

## References


1. Razali Bin Abu Bakar,: Development of Online Unconstrained Handwritten Character Recognition Using Fuzzy Logic, Universiti Teknologi MARA.
2. P. Phokharatkul, K. Sankhuangaw, S. Somkuarnpanit, S. Phaiboon, and C. Kimpan: Off-Line Hand Written Thai Character Recognition using Ant-Miner Algorithm. World Academy of Science, Engineering and Technology, 8, 2005, Pg 276–281.
3. Gunvantsinh Gohil, Rekha, Mahesh Goyani: Chain code and holistic features based OCR System for printed devanagari script using ANN and SVM, IJAIA, Vol. 3, No. 1, Jan 2012.
4. P.Vanaja Ranjan,: Efficient Zone Based Feature Extraction Algorithm for Hand Written Numeral Recognition of Four Popular South Indian Scripts, Journal of Theoretical and Applied Information Technology. pg 1171–1181.
5. Santosh Kumar Henge, Laxmikanth, Niranjan: Advanced Fuzzy Logic controller based Mirror-Image-Character-Recognition OCR, LAICEEE. 3101/01/32-32. Pg 261–268.
6. J. Bharathi, Dr. P. Chandrasekar Reddy: Segmentation of Telugu Touching Conjunct Consonants using Overlapping Bounding Boxes, IJCSE, Vol. 5 No. 06 Jun 2013.
7. J.Bharathi, Dr.P.Chandrasekar Reddy: Improvement of Telugu OCR by segmentation of Touching Characters, IJRET, Volume: 03 Issue: 10th Oct-2014.
8. Lotfi A. Zadeh.: Is there a need for fuzzy logic? Department of EECS, University of California, Berkeley, CA 94720–1776, United States, 8 February 2008; 25 February 2008.
9. I.Perfilieya: Fuzzy transforms: a challenge to conventional transforms, in: P.W. Hawkes(Ed), AIEP, vol. 147, Elsevier Academic Press, San Diego 2007, pp. 137–196.
10. L.A. Zadeh: A fuzzy-algorithmic approach to the definition of complex or imprecise concepts, International Journal of Man–Machine Studies 8 (1976) 249–291.
11. Fuzzy Logic Toolbox User's Guide, The MathWorks Inc., 2001.
12. L.A. Zadeh,: Toward a perception-based theory of probabilistic reasoning with imprecise Probabilities, Journal of Statistical Planning and Inference 105 (2002) 233–264.
13. Mr.Danish Nadeem, Miss. Saleha Rizvi,: Character Recognition using Template Matching. Department of Computer Science, Jamia Millia Islamia, New Delhi.
14. Panyam Narahari, Sastry, Ramakrishnan, Bhagavatula Venkata Sanker Ram: Classification and identification of Telugu Handwritten Characters Extraction from Plam Leaves using Decision Tree Approach", ARPNJEAS, Vol. 5, No. 3, March 2010.

# UMEED-A Fuzzy Rule-Based Legal Expert System to Address Domestic Violence Against Women

**Chandra Prakash, Gour Sundar Mitra Thakur, Natasha Vashisht and Rajesh Kumar**

**Abstract** Viciousness against women has become an alarming topic and a serious concern for the government of India in recent years. Pending cases related to domestic violence in different courts need to be resolved faster to give proper justice to the victims and their families. In this paper a fuzzy logic-based legal expert system, 'UMEED', is proposed that emulates the decision-making ability of a legal expert, to address different legal issues regarding domestic violence. The system is proposed in such a way that it produces the expert opinion on this subject within the legal bounds. Being a rule-based system it acts as a legal advisor for the lawyers and ensures faster delivery of the legal decisions, and hence reduces overall decision-making time and effort. The model is tested on few historical cases regarding domestic violence and its performance proved to be satisfactory when it is compared with the actual legal decisions.

**Keywords** Legal expert system · Fuzzy rule-based system · Domestic violence · Fuzzy logic

C. Prakash (✉) · R. Kumar
Malaviya National Institute of Technology, Jaipur, India
e-mail: cse.cprakash@gmail.com

R. Kumar
e-mail: rkumar.ee@gmail.com

G.S.M. Thakur
Dr. B.C. Roy Engineering College, Durgapur, India
e-mail: cse.gsmt@gmail.com

N. Vashisht
Lovely Professional University, Phagwara, Punjab, India
e-mail: er.natasha09@gmail.com



S.K. Bhatia et al. (eds.), *Advances in Computer and Computational Sciences*,
Advances in Intelligent Systems and Computing 553,
DOI 10.1007/978-981-10-3770-2_59

## 1 Introduction

People in different parts of the world are anguish from a larger number of crimes like murder, robbery, illegal activities like smuggling, gambling, violence, etc. The greatest sufferer of all these crimes is the women. The women that once held a place of respect and honor in the Indian society is today afraid of that very society [1]. Many women in different parts of the world are subjected to violence that includes domestic violence, exploitation at work, etc.

According to the national survey conducted in 2012, 8% of the married women are subjected to sexual abuse, 31% of them are victims of physical abuse such as hitting, slapping, etc., and 14% of the women have suffered from emotional abuse. According to the 'United Population Funds Report' around two-third of the married women are victims of domestic violence and as many as 70% of the married women in India between the age of 15 and 49 are victims of domestic violence [2]. More than 55% of the women suffer from violence especially in the states of Bihar, Uttar Pradesh, Madhya Pradesh, and other Northern states. Only few of them are protected against this crime and others are not even subjected to know what their rights to protect themselves are. More than 70% of the cases related to domestic violence are pending in different courts [2].

Survey of literature reveals that legal expert systems are developed mainly to ensure quick delivery of the legal decisions, to reduce time spent in the labor-intensive legal tasks and to reduce overhead and labor cost [3]. Expert system emulates the knowledge of expert person in a particular domain [4].

Legal Expert Systems supports administrative processes, decision-making, rule-based analysis and exchange of the information among the users [5]. The legal expert systems vary in different ways. These include architectural variations, theoretical variations, and the functional variations. These variations led to different challenges in making the expert systems for law. In spite of these variations, most of the recent expert systems are not adopted for the practice of law in real life. One of the major reasons behind this is the lack of uncertainty and vagueness handling capabilities in the existing models. Recently growing interest in adopting various soft computing techniques in the legal expert system model is noticed. In a work [6] deductive rules within the knowledge base are used to represent the information. In another work [7] case-based reasoning model is used to store and manipulate the cases. This model draws conclusions through the known experiences for similar problems. Artificial neural network (ANN), which emulated the learning process of human brain up to a certain extent, is another popular soft computing tool used for classification and pattern recognition in various problem domains. ANN is used for recognizing and classifying patterns within the realm of legal knowledge and dealing with imprecise inputs [8]. The use of fuzzy logic and fuzzy system has become most popular decision-making tool in recent years specially where the inherent uncertainty and vagueness of the problem domain the major concern. The scope of fuzzy logic in legal expert system is discussed in [9].

A Legal Expert System is a domain-specific expert system that emulates the decision-making ability of a human expert from the legal domain. Such systems use a rule base called as knowledge base and an inference engine to produce expert knowledge within the legal bounds. [9–11] are two examples of this kind legal expert systems.

In this paper a fuzzy rule-based legal expert system named 'UMEED' is proposed to assist the legal decision-making regarding domestic violence against women in India. The main focus regarding domestic violence is given in different types of abuses like physical abuse, verbal or nonverbal abuse, sexual abuse, economic or financial abuse [12]. In India, violence against women has achieved utmost attention but the biggest hindrance is that most of the cases are under-reported and even most of the cases are pending seeking a proper decision. High fees of the lawyers is another major constraint. Poor people who cannot afford this fees are being deprived of justice. The expert system proposed here may reduce the decision-making time considerably and also will reduce the total consultation fee paid by any client before the le of any case.

The rest of the paper is organized as follows. Section. 2 describes the design and development of proposed expert system. Section 3 covers the result analysis in terms of case studies. Finally, conclusions and future scope are discussed in Sect. 4.

## 2 Design and Implementation of 'UMEED'

As no any legal expert system has been yet built to address domestic violence, here an attempt has been made to develop a fuzzy logic-based system that acts as a legal advisory system for lawyers for quick building of legal decisions. Fuzzy rule base here helps to handle uncertainty and for dealing with the ambiguous data.

Fuzzy logic can work in this scenario with reasoning algorithms to simulate human thinking and decision-making in machines. These algorithms let researcher to build expert system in the areas where data cannot be represented in binary form. Fuzzy logic lets expert systems perform optimally with uncertain or ambiguous data and knowledge This concept of fuzzy logic was proposed by Lot A. Zadeh in 1965 (Zadeh 1965) [13]. In contrast to conventional boolean logic where each element must have either 0 or 1 as the membership degree, fuzzy logic can be thought of as gray logic, whose members may have degrees of membership between 0 and 1. In binary logic if value is 0, the element is completely outside the set; if 1, the element is completely in the set. Fuzzy logic associates a grade or level, with a data range, giving it a value of 1 at its maximum and 0 at its minimum. It is used in modelling imprecise concepts and dependences (set of rules). Fuzzy logic surmounted the problem of classical logic by allowing statements to be interpreted as both true and false.

There are many factors like physical abuse, verbal or nonverbal abuse, sexual abuse, economic abuse that cause domestic violence. For making a prototype system, all of these factors are not taken into consideration [14]. The proposed

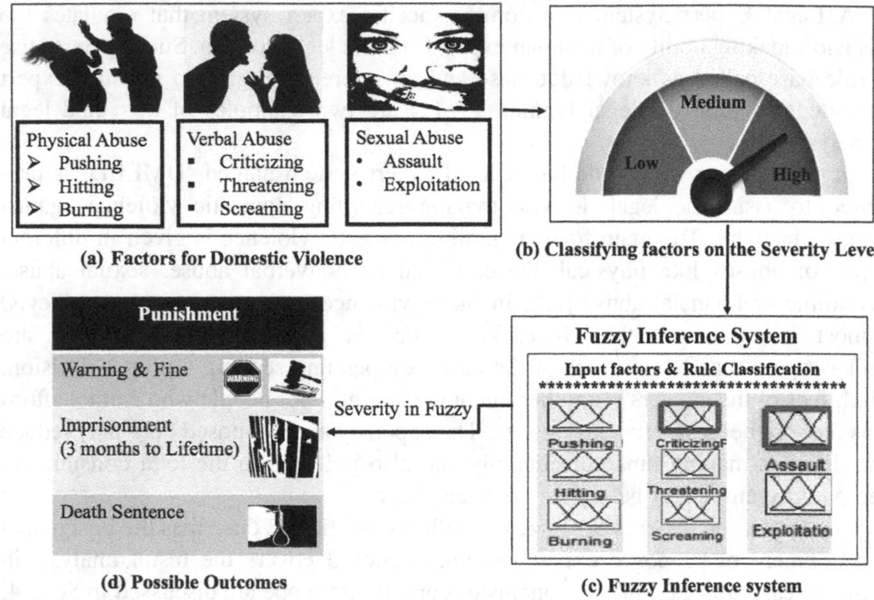


**Fig. 1** Methodology of the proposed system—UMEED

system is built upon using three factors and their corresponding subfactors. Severity of physical abuse, verbal abuse, and sexual abuse are considered here as fuzzy linguistic input for the proposed legal expert system. Structure of the proposed methodology is depicted in Fig. 1.

These factors are further divided into three subfactors. Figure 1a shows that pushing, hitting, and burning are considered as subfactors for physical abuse. Similarly, criticizing, threatening, and screaming are the subfactors of verbal abuse. Sexual abuse is considered of assault and sexual exploitation. These linguistic variables are mapped using three membership functions, namely low, medium, and high as shown in Fig. 1b.

Corresponding to physical and verbal abuse factor, a set of 27 rules are made for each abuse and the sample fuzzy rule base for sexual abuse is shown in Fig. 2. These rule bases are red by the fuzzy inference engine based on the crisp input given by the user. Fuzzy logic used here creates concepts or rules which are indexed and then retrieved by the legal expert system built to get a specific outcome for the case registered and thus it also acts as an advisory system for the lawyers as well. The implementation of fuzzy inferencing method on each factor is shown in Fig. 1c and the possible output is given in Fig. 1d. Table 1 shows the set of fuzzy rules for sexual exploitation. Similarly, rules for physical and verbal abuse can be note down after discussion with the legal expert. The fuzzy system designed for this work is a Mamdani system. The defuzzification scheme used is centroid-based. That is, the output is just the consequence of a specific condition of the factors consider.

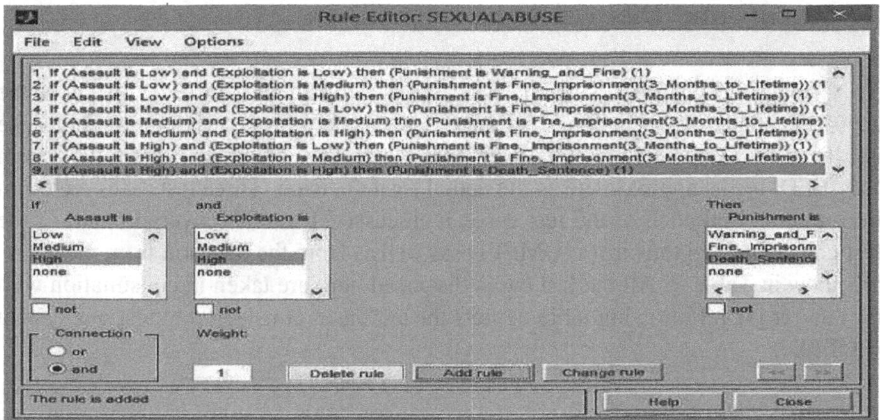


**Fig. 2** Rule base for sexual abuse

**Table 1** Set of fuzzy rules for sexual exploitation

| S.No | Sexual exploitation | Sexual assault | Punishment |
|---|---|---|---|
| 1 | Low | Low | Warning and fine |
| 2 | *Low* | *Medium* | *Imprisonment (3 Months to Lifetime)* |
| *3* | *Low* | *High* | *Imprisonment (3 Months to Lifetime)* |
| *4* | *Medium* | *Low* | *Imprisonment (3 Months to Lifetime)* |
| *5* | *Medium* | *Medium* | *Imprisonment (3 Months to Lifetime)* |
| 6 | *Medium* | *High* | *Imprisonment (3 Months to Lifetime)* |
| *7r* | *High* | *Low* | *Imprisonment (3 Months to Lifetime)* |
| *8* | *High* | *Medium* | *Imprisonment (3 Months to Lifetime)* |
| *9* | *High* | *High* | *Death sentence* |

The linguistic values considered for each fuzzy input variables are *Low*, *Medium,* and *High*. The output *punishment* is considered on the basis of the defuzzified output value. The punishment for any abuse depending upon their severity is *Warning* and *Fine*, *Lifetime Imprisonment,* and *Death Sentence*. These punishments act as membership variables and thus every factor is mapped on these also. Following is an example how fuzzy rules are implemented into the system.

> ***IF*** *Slapping is* ***Medium*** *and Criticizing is* ***Medium*** *and Hitting is* ***Low THEN*** *Punishment is Warning and Fine.*

In this way the output of each factor is further fed into the fuzzy inference engine to get the overall result.

## 3 Results and Discussion

To validate the proposed system test cases are considered. In order to know the accuracy, some historical data of some cases are taken into consideration under the guidance of a lawyer and it is seen that the results obtained using our system 'UMEED' are as approximate as the actual legal decision. Three test cases are taken into account and each of the test cases is discussed here. The overall result of the test cases obtained thus from 'UMEED' as well as from the decision from the courts are show in Table 2. All the test cases discussed here are taken in consultation with the lawyer [15]. Following table depicts the test cases considered to test our system UMEED.

### *3.1 Test Case 1*

In the High Court of Karnataka, dated 25 October, 2013, an appeal was led under Section 347(2) against the judgment dated 27 June, 2006 passed by the session Judge, fast track court at Mandya convicting appellant accused of offenses punishable under Section 498-A and 306-IPC sentencing him to undergo Lifetime Imprisonment and Fine.

The accused charged for conduction severe violence against women, where his wife was physically abused and criticized and due to this, she along with her two children committed suicide. The husband was charged for murder as he tortured his wife. When this was tested by our legal system UMEED, the judgments were almost same as the Judge gave. So the punishment given to that person was Fine and a Lifetime Imprisonment.

### *3.2 Test Case 2*

This test case deals with the abuse suffered by a working lady who, due to narrow thinking of her husband used to be often beaten up and threatened up. The lady suffered from abuse like Slapping, Hitting, and Criticizing. All these abuse come under the category of Physical as well as Verbal abuse. So when the case was taken to the court, the punishment to the husband given was Warning and a fine of

**Table 2** Comparison of UMEED results and survey results

| Test cases | UMEED results (severity) (%) | Survey result (%) |
|---|---|---|
| Case 1 | 50.1 | 51 |
| Case 2 | 34.33 | 35 |
| Case 3 | 76.3 | 77 |

20,000/- rupees. On testing through 'UMEED', the result was also the same that is Warning and Fine where the rule that was justified by our System 'UMEED' is:

*IF Slapping is Medium and Criticizing is Medium and Hitting is Low THEN Punishment is Warning and Fine.*

### *3.3 Test Case 3*

This case deals with the acid attack on a girl. The case came into existence on fast track basis after the involvement of media. The punishment given to the accuse person is Lifetime Imprisonment and approximate results were obtained through 'UMEED'. In this case the severity of each of the abuse factor was high [2]. The result of the test case is shown in Table 2.

Through all the results and test cases, it is found that approximate results are achieved using the legal expert system 'UMEED'. As, the judgment varies from person to person (Judge to Judge), which is the biggest challenge, so we are getting approximate results. But this system ensures faster delivery of legal decisions.

## 4 Conclusions and Future Scope

The work discussed in this paper is an initial study of implementing fuzzy-based legal decision support system to address domestic violence against woman. The main advantage of the proposed system is that such a system ensures faster delivery of the legal decisions by assisting lawyers or other decision makers and therefore the pending cases can be resolved out efficiently and effectively. When analyzing and comparing the results obtained from the lawyers for the cases registered for domestic violence in the courts with our proposed system 'UMEED', it is found that the results obtained are very close approximation to the actual decisions.

'UMEED' not only can be used by the lawyers but also can act like an advisory system for the laypeople and also as an awareness system to make women and girls aware of the consequences of the abuses they are facing. The uncertainty and vagueness handling capability of the fuzzy systems has made the proposed model robust and e ective.

Though the system here is proposed to deal with domestic violence against women, similar prototype model can be prepared to address the consequences of any kind of crime but the number and types of the input factors can vary.

To further enhance the robustness and the adaptability of the proposed model reasoning with neuro-fuzzy system can be incorporated in the proposed model. Dempster Shafer evidence theory can also be used in calculating the degree of belief for each legal decision.

**Acknowledgements** The authors would like to acknowledge Ms. Pooja Shastry, lawyer, for the support and guidance in the work.

## References

1. Baldwin K, Canada best G20 country to be a woman, India worst—TrustLaw poll, Thomson Reuters, 13 June, 2012.
2. http://www.indiankanoon.com/ Accessed on 11th May 2015.
3. Schafer, Burkhard, ZombAIs: Legal Expert Systems as Representatives "Beyond the Grave". SCRIPTed 7 Section 498 of the Indian Penal Code (Dowry Death), Government of India, 2010.
4. Patterson, D.W. Introduction to Arti cial Intelligence and Expert Systems. Prentice-Hall: New Delhi, 2004.
5. Aikenhead, M. "Legal Knowledge-Based Systems: some observations on the future". Web JCLI2, 1995.
6. Pal, Kamalendu; John A. Campbell, "An Application of Rule-Based and Case-Based Reasoning within a Single Legal Knowledge-Based System". The DATA BASE for Advances in Information System 28(4), 1997.
7. Main, Julie; Sankar K Pal, et al.; Tharam Dillon and SimonShiu. "A Tutorial on Case-Based Reasoning". in Soft Computing in Case Based Reasoning (4th ed.). London: (Ltd), 2001.
8. Jackson, Peter, Introduction to Expert systems (3 ed.), Addison Wesley, p. 2, 1998.
9. Susskind, Richard, "Expert Systems in Law: A Jurisprudential Approach to Artificial Intelligence and Legal Reasoning", Modern Law Review 1986.
10. Schweighofer Erich; Werner Winiwarter."Legal Expert System KONTERM- Automatic Representation of Document Structure and Contents". DDEXA '93 Proceedings of the 4th International Conference on Database and Expert Systems Applications: 486, 1993.
11. Groothuis, Marga M.; Jorgen S. Svenson, "Expert system support and juridical quality?" in Legal Knowledge and Information Systems, Amsterdam, The Thirteenth Annual Conference, p. 9, 2000.
12. The Protection Act for Women against Domestic Violence, Ministry of Law, India, 2005.
13. Zadeh, L.A. Fuzzy logic= computing with words. IEEE transactions on fuzzy systems, 4(2), pp. 103–111. Vancouver, 1996.
14. Dowry Prohibition Act, Government of India, 1961.
15. http://www.legalindia.com. Accessed on 11th Dec 2015.

# Medical Image Defects Investigation Through Reliability Computing

**Reshma Parveen, Satyanarayan Tazi and Anil Kumar Dubey**

**Abstract** Defect is an important issue in medical image; we have investigated the defects in medical images. The missing data in an image is computed through error detection mechanism. Defect is creating a major problem in the medical science. When a patient with disease has any internal infections like cancer, kidney stone and external infections like fungal infection, skin disease, etc., patient follows the doctor's advice. Doctor checks his patient and his disease and he suggests many types of tests like X-ray. Theses tests are done by patients, but some defects are caused in the machine internal or external, which makes the result improper, hence the problem cannot be identified by the doctorclearly. In this paper calculate the percentage of defected image with healthy image and comparison of healthy image.

**Keywords** Defect · Reliability · Medical image · Defect image · Healthy image

## 1 Introduction

Image processing is a method to change an image into digital form and perform some operations on it. This is a type of signal processing and in this process the input is like image, video, frame and photograph and output may be image or feature adjuvant with that an image. The aim of image processing is divided into five types, first is visualizing and observing the object that are not visible. Second is image sharpening and restoration to create a better image. Third one is image

R. Parveen (✉) · S. Tazi · A.K. Dubey
Department of Computer Science and Engineering,
Government Engineering College Ajmer, Ajmer, Rajasthan, India
e-mail: reshmaparveen034@gmail.com

S. Tazi
e-mail: satya.tazi@gmail.com

A.K. Dubey
e-mail: anildudenish@gmail.com



S.K. Bhatia et al. (eds.), *Advances in Computer and Computational Sciences*,
Advances in Intelligent Systems and Computing 553,
DOI 10.1007/978-981-10-3770-2_60

retrieval seeks for the image of interest and fourth type is measuring pattern of various objects present in an image. Fifth type is recognizing image and distinguishing the object in an image.

Two types of methods used in image processing are: (a) Analog image processing (b) Digital image processing. Analog image processing is used for the hard copies like printout, photograph. Digital image processing is dealt with manipulation of digital images by a digital computer image processing and it usesa pixel, "Little bit element of an image is known as pixel," in an 8-bit grayscale image the value of pixel is between 0 and 255, it is a PEL.

### *1.1 Edge Detection*

"A significant transition in an image is called as edges", many types of edge detection like

(i) Horizontal edges
(ii) Vertical edges
(iii) Diagonal edges

**Definition of defect:** When X-ray, etc., test is performed then machine gives output. These outputs create some fault by machine, these faults are not the patient disease, while these faults in any internal and external problem caused by other fault machine show it is called defect.

## 2 Literature Survey

Papers that introduced defect detection problems to identify the image defect have been studied.

In this paper Detection of defect in road surface by vision system. Authors tried to qualify the road scanner of MACR. These papers show comparative result of methods and three types of road surface [1].

In this paper Automatic defect detection and classification technique from image a special cause using ceramic tiles. In this paper authors have worked on an effective defect detection and classifying technique. Method proposed is used to find out defect in an image with high rate of accuracy [2].

In this paper Detection of defect in digital texture image using segmentation. Authors of paper introduced an algorithm which can easily compare one defect with another. The algorithm was used for feature extraction and segmentation and recognizes the defects in image [3].

In this paper, paper recognizes fabric image mistake and methods are applied using MATLAB. Fault identification is performed by authors in their work such as hole scratch finding, etc. [4].

In this paper, Color image-based defect detection method and steel bridge paper authors have proposed defect recognition method which was developed by processing coating images with colors directly without converting to grayscale image. This paper presents the recognition of rust defect on highway steel bridge rust defect is commonly observed defects on coating surface [5].

In the paper Comparison analysis for efficient defect detection algorithm for gray-level digital images using median filter Gabor filter and ICA. In this paper authors introduced a method on defect detection in digital image utilization of various different Feature extraction parameter and segmentation [6].

In this paper, a paper on automatic fabric fault processing using image processing technique in MATLAB. Authors of paper have used method of liver segmentation and using this method enhancement is done using CT SCAN image. Main area of interest is the method segment of the liver using global threshold [7].

In this paper liver segmentation and enhancement is done using CT image the proposed method segmenting the liver using global threshold and them by identifying the largest area. Proposed a detection method by using weight texture dissimilarities to measure perceptual texture distortion of a pixel is defined the dissimilarity between the original feature value and the represented one of pixel [8].

The objectives of our research are to calculate percentage of defective image with healthy images. Right prescription and suggestion given doctor to patient in this paper give a new algorithm to find the defect of image.

## 3 Image Defect Removal Technique

### 3.1 *Noise*

The meaning of "noise" is unwanted electrical fluctuations in signal received by AM radios produce audible sonic noise.

#### 3.1.1 Gaussian Noise

Gaussian noise law states that probability density function is equal to normal distribution. Normal distribution is also referred as Gaussian distribution. The probability density function of P is a Gaussian random variable Z is given by formula,

$$PG(Z) = \frac{1}{\sigma\sqrt{2\Pi}} e^{-\frac{Z-\mu}{2\sigma^2}} \tag{1}$$

where Z represents the gray level $\mu$ the mean value and $\sigma$ the standard deviation

#### 3.1.2 Salt-and-Pepper Noise

When any image contains darker pixels in light regions and lighter pixel in dark regions it is called impulsive noise or salt-and-pepper noise (spike noise).

#### 3.1.3 Shot Noise

Electronic noise have a part poisson noise can be modeled from a Poisson process in electronics short noise emerge from the separate nature of electric change

$$\mathrm{SNR} = \frac{N}{\sqrt{N}} = \sqrt{N}. \tag{2}$$

#### 3.1.4 Speckle Noise

Speckel noise is modeled as multiplying random value with pixel values of the image and it can be expressed as follows:

$$\mathrm{J} = \mathrm{I} + \mathrm{n} * \mathrm{I} \tag{3}$$

Here, J is the spackle noise distribution image I is the input image and n is the uniform noise image by mean o and variance v.

### *3.2 Filter*

It is a process that removes from a signal some unwanted component or features it is called filter.

#### 3.2.1 Average Filter (Mean Filter)

These filter computed the average value of the useless image it is called an averaging linear filter.

#### 3.2.2 Order Static Filter

Effect of input noise value is removed by median filter with excessively huge magnitude.

### 3.2.3 Wiener Filter

The wiener filter is also known as MSE optimal stationary linear filter images by additive noise and blurring.

## 4 Proposed Algorithm

```
STEP1: Select an image/capture from original image database
STEP2: Divided the image into abstract level
STEP3: Get any one part of image
STEP4: Search the probable area of image
STEP5: Collect the printed data our information of image clearance in a table
If (image clearance is more efficient)
    {
     P (ID) = 0;
          (Print probable area of image defect is,"Low");
    }
       Else if
           (Image clearance is some crucial)
          {
            P (ID) = ½;
                 (Print probable area of image defect is "common/
average");
            }
            Else
            {
             P (ID) =1
   (Print probable area of image defect is "maximum");
                   }
 STEP6: go to step 3 till all part complete image process
 STEP7: Exit
```

## 5 Working Process

**MATLAB**: MATLAB is a tool (Matrix Laboratory) which is known for its high performance and technical computing. MATLAB is a fourth-generation programming language and it is popularly known for its multi-pattern arithmetic computing environment. MATLAB permits matrix data plotting and manipulating function and data. MATLAB is simple access matrix software which is developed by LINPACK and EISPACK project. MATLAB works as external library that allow integrating C and FORTRAN languages with it. It has facilities like calling routines from dynamic matlab link, calling MATLAB as a mathematical engine, reading, and writing MAT file.

We generally find four types of defects in the existing defect detection method (Table 1).

**Table 1** Types of Defect image

| Name of defects | Description |
|---|---|
| Most dangerous | Very large spot show in an image |
| Major | Large spot show in an image |
| Little | Very little spot show in an image |
| Negligible | Not consider spot in an image |

**Table 2** Percentage of X-ray

| S. No. | **Image name and type of defect** | **X-ray** |
|---|---|---|
| 1 | Most dangerous | 91.8090 |
| 2 | Major | 81.6547 |
| 3 | Little | 25.9374 |
| 4 | Negligible | 3.8656 |

# 6 Simulation and Result

In our paper analyzed and measurement of image defected and calculates percentage of image defect take a healthy image with defected image is my subject for our work download data and images of medical X-ray finger image and calculate area of defected image with healthy image and tool is used MATLAB, this tool help to generate a histogram of all images and result shows in this paper calculate the percentage and give a formula.

$$percentage = \text{mod}\left|\frac{original - defected}{original}\right| \times 100 \tag{4}$$

These formulas are used to calculate percentage of defected image, mod function is also used, these signs show all the negative value change to positive value so it is the sign used to convert values negative to positive because of medical image defect percentage never calculate negative. The X-ray healthy and defected image shows and histogram shows. Procedure to calculate the percentage take a value of healthy image and defected image, both values create difference value, these formula used and calculate percentage after that create a 3D graph these graph shows the value of percentage. In this table shows percentage healthy and defected image percentage many types of defect is divided into four categories, these categories divided according to percentage, how much defect in an image first calculate healthy and defected image values and these values put up formula and manually calculate the percentage. Take a medical image and find a defect this paper purpose of doctor give a better suggestion to patient and no confused doctor any situation and create any defected of image unwanted machine defect perfectly handle doctor to patient (Table 2).

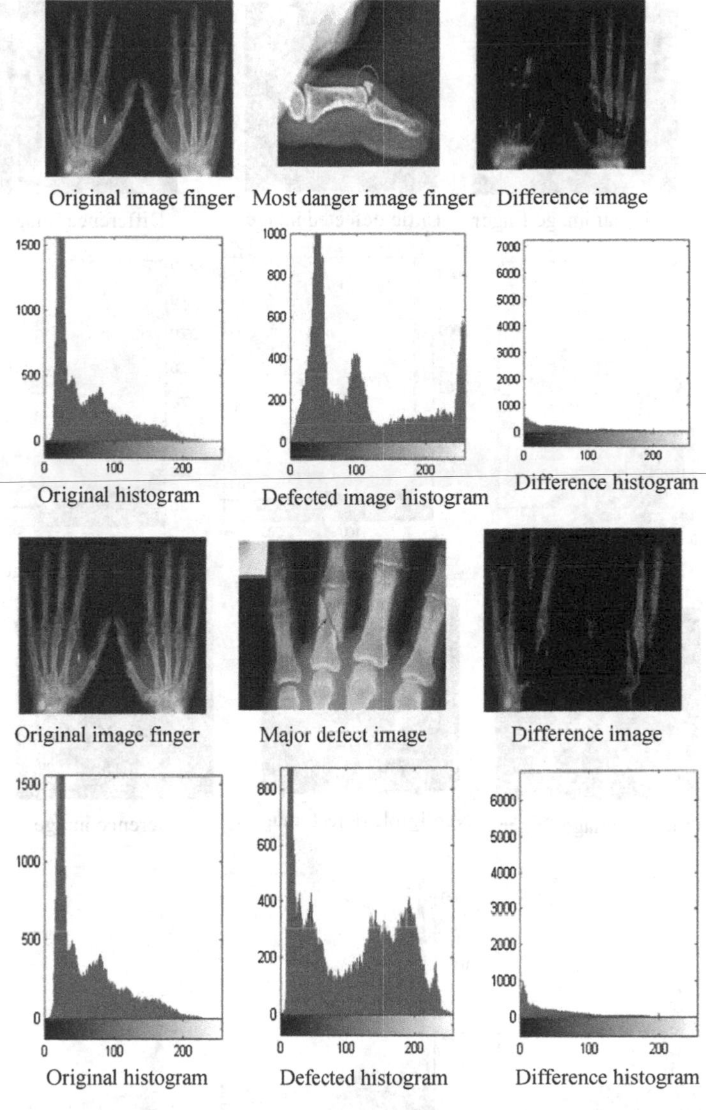
Original image finger
Most danger image finger
Difference image
1500
1000
500
0
0 100 200
Original histogram
1000
800
600
400
200
0
0 100 200
Defected image histogram
7000
6000
5000
4000
3000
2000
1000
0
0 100 200
Difference histogram
Original image finger
Major defect image
Difference image
1500
1000
500
0
0 100 200
Original histogram
800
600
400
200
0
0 100 200
Defected histogram
6000
5000
4000
3000
2000
1000
0
0 100 200
Difference histogram

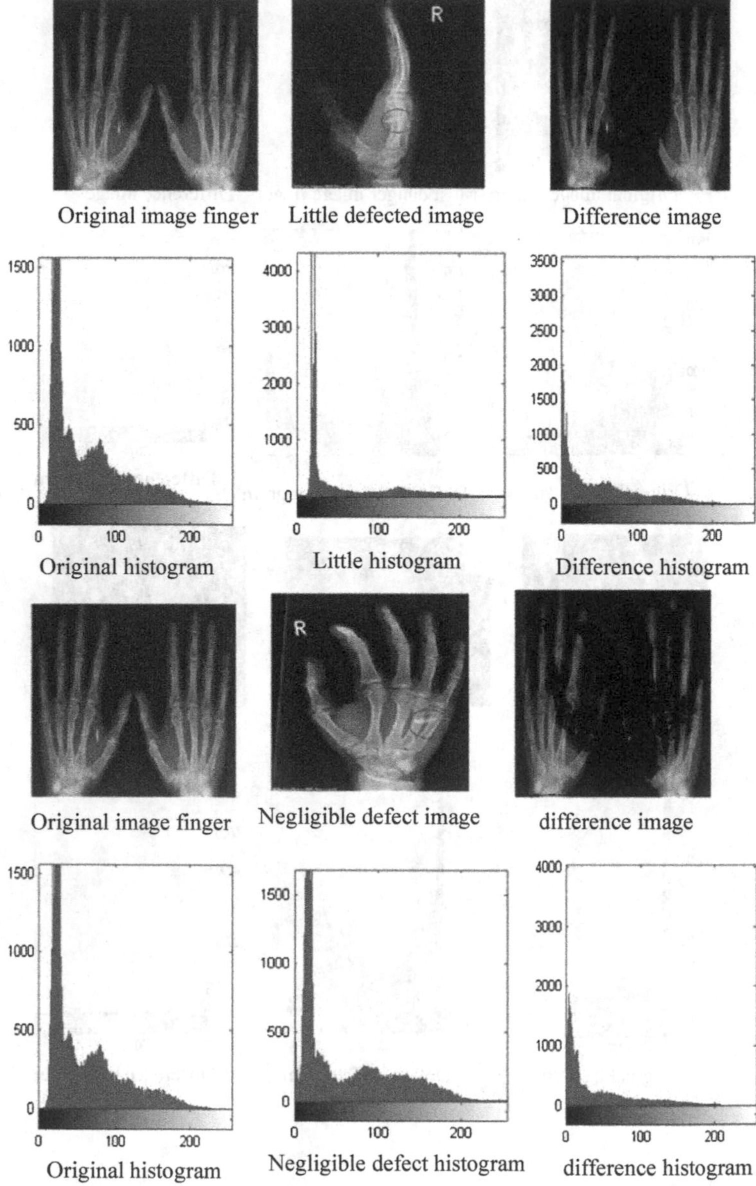

Original image finger Little defected image Difference image

Original histogram Little histogram Difference histogram

Original image finger Negligible defect image difference image

Original histogram Negligible defect histogram difference histogram

**Fig. 1** Percentage of X-ray Finger plot

## 7 Conclusion and Future Work

In this paper we investigate and discuss different types of image defects in an image. Image processing is a method to convert an image into digital form and important issues on medical images. In this paper we find a problem of image defect in medical images take an original and defected images and calculate difference between original and defected image and compute the percentage of missing data, this paper outcomes show take X-ray hand finger images and find types of defect Most dangerous, Major, Little, Negligible percentage of defect take a X-ray images Percentage 91.8090%, 81.6547%, 25.9374%, 3.8656% special purpose of calculate medical image defect percentage main purpose of paper doctor given to patient right advised of any kind of disease and right medicine stipulate. Future work should be done to take a satellite image and find the percentage of defect Fig. 1.

## References

1. N. T. Sy M. Avila S. Begot and J. C. Bardet Detection of Defects in Road Surface by a Vision System Author manuscript, published in "Electrotechnical Conference, The 14th IEEE Mediterranean, AJACCIO: France (2008).
2. G. M. Atiqur Rahaman and Md. Mobarak Hossain AUTOMATIC DEFECT DETECTION AND CLASSIFICATION TECHNIQUE FROM IMAGE: A SPECIAL CASE USING CERAMIC TILES (IJCSIS) International Journal of Computer Science and Information Security, Vol. 1, No. 1, May 2009.
3. K. N. SIVABALAN and Dr. D. GHANADURAI Detection of defects in digital texture images using segmentation International Journal of Engineering Science and Technology Vol. 2(10), 2010, 5187–5191.

4. R. Thilepa and M. Thanikachalam A PAPER ON AUTOMATIC FABRICS FAULT PROCESSING USING IMAGE PROCESSING TECHNIQUE IN MATLAB Signal& Image Processing: An International Journal(SIPIJ) Vol.1, No. 2, December 2010.
5. Sangwook Lee Color Image-based Defect Detection Method and Steel Bridge Coating 47th ASC Annual International Conference Proceedings Copyright 2011 by the Associated Schools of Construction.
6. Rashmi S Deshmukh and Dr P R Deshmukh Comparison Analysis For Efficient Defect Detection Algorithm For Gray Level Digital Images Using Median Filters Gabor Filter and ICA Volume 2, Issue 1, January 2012 ISSN: 2277 128X International Journal of Advanced Research in Computer Science and Software Engineering.
7. Vinita Dixit, Jyotika Pruthi Review of Image Processing Techniques for Automatic Detection of Tumor in Human Liver International Journal of Computer Science and Mobile Computing, Vol. 3 Issue. 3, March- 2014, pg. 371–378.
8. QiangGuo, Caiming Zhang, Hui Liu, and Xiaofeng Zhang Defect Detection in Tire XRay Images Using Weighted Texture Dissimilarity Volume 2016, Article ID 4140175.

# An Improved Apriori Algorithm with Prejudging and Screening

**Xuejian Zhao, Dongjun Li, Yuan Yuan, Zhixin Sun and Yong Chen**

**Abstract** Association rule analysis, as one of the significant means of data mining, plays an important role in discovering the implicit knowledge in massive transaction data. Aiming at the inherent defects of the classic Apriori algorithm, this paper proposes IAPS (Improved Apriori with Prejudging and Screening) algorithm. IAPS algorithm adds a prejudging and screening procedure on the basis of the self-join and pruning progress in Apriori algorithm which can reduce and optimize the k-frequent item sets using prior probability. IAPS algorithm simplifies the operation process of mining frequent item sets. Experimental results show that the improved algorithm can effectively reduce the number of scanning databases and reduce the running time of the algorithm.

**Keywords** Association rules · Transaction database · Prejudging and screening · Apriori

X. Zhao (✉) · D. Li · Z. Sun
Key Lab of Broadband Wireless Communication and Sensor Network Technology of Ministry of Education, Nanjing University of Posts and Telecommunications, Nanjing, China
e-mail: zhaoxj@njupt.edu.cn

D. Li
e-mail: lidongjun@njupt.edu.cn

X. Zhao · Y. Yuan
Jiangsu Posts & Telecommunications Planning and Designing Institute Co. LTD, Nanjing, China
e-mail: yuanyuan@jsptpd.com

Y. Chen
Nanjing Longyuan Microelectronic Co. LTD, Nanjing, China
e-mail: sunzx@njupt.edu.cn



S.K. Bhatia et al. (eds.), *Advances in Computer and Computational Sciences*, Advances in Intelligent Systems and Computing 553,
DOI 10.1007/978-981-10-3770-2_61

## 1 Introduction

Nowadays, with the rapid development of big data technologies, an increasing number of people realize that data is wealth, especially business data. As one of the most effective methods of data mining, the association rule mining algorithm is an integral part of data mining technology [1, 2]. The association rule mining algorithm is mainly used to discover the hidden association rules in massive transactions. Therefore, the research on association rule mining algorithms is of important practical significance.

As early as 1993, Agrawal, a computer scientist of IBM, found the purchase law in transaction database, and proposed correlation modes, i.e., association rules between transactions. Association rules are usually not complicated but of great practical value. Through the analysis of association rules, we can dig out the relationship between transaction item sets. The most typical application of association rule analysis is market basket analysis on transactional retail data, such as the classic {beer} -> {diaper} rule. And additionally, association rule analysis also has a wide range of applications in other fields, such as personalized recommendation, financial services, advertising planning, etc. [3–5].

Apriori algorithm was the first association rule mining algorithm proposed by Agrawal and Swami in 1994 [6]. In recent years,a lot of association rule mining algorithms are proposed based on Apriori algorithm [7–12]. Apriori-TFP algorithm was proposed in [7]. In order to obtain the association rules, this algorithm stores the original data in a partial support tree. Through the effective preprocessing process, the algorithm reduces the running time of mining association rules. In [8], the GP-Apriori algorithm was proposed, which used GPU (Graphical Processing Unit) to do parallel support counting, and stored vertical transaction column as linear ordered array. Compared with the traditional CPU running Apriori algorithm, GP-Apriori algorithm improved the running efficiency due to the adoption of advanced GPU. However, the complexity of GP-Apriori had increased. In [9], another improved algorithm (Mend Algorithm Apriori) based on Apriori is proposed. Initially it generates frequent itemset using perfect Hash function in the database. The user has to specify the minimum support to prune the database item sets and deletes the unwanted item sets. This algorithm is found to be more admirable than the traditional method Apriori algorithm in terms of efficiency. Execution time is increased. Ning et al. propose a parallel Apriori algorithm based on MapReduce in [10]. This algorithm is implemented based on MapReduce, which creates it applicable to mine association rules from large databases of transactions. It can scale fine and efficiently processed large datasets on commodity hardware. It requires more computation power and memory to find association rules. Sulianta et al. improves Apriori algorithm for multidimensional data in [11]. In this improved Apriori algorithm, the problem focuses on the process of preparation of data. Cleaning and integration are done after collecting the data from the system. Selection and transformation is used which includes procedures to handle the data which is transformed to standardize data for mining process. In [12], the M-Apriori algorithm is proposed. It avoids the wasting time for scanning the whole database searching on the frequent item sets, and presents an improvement on Apriori by reducing that wasted time depending on scanning only some transactions. The improvement makes the Apriori algorithm more efficient and less time consuming.

Through the above analysis, Apriori algorithm suffers from some weakness in spite of being clear and simple. The main limitation is costly wasting of time to hold a vast number of candidate sets with much frequent item sets, low minimum support or large item sets [13]. It will scan database many times repeatedly for finding candidate item sets. The above-mentioned algorithms improved the efficiency to some extent, but with the cost of more memory capacity or better hardware equipment.

In order to avoid the inherent defects of the classical Apriori algorithm, the IAPS (Improved Apriori with Prejudging and Screening) algorithm is proposed in this paper. The IAPS algorithm adds a prejudging and screening procedure following the self-join and pruning progress in Apriori algorithm. It can reduce and optimize the k-frequent item sets to a large extent.

## 2 Theoretical Basis

Before going into details of IAPS algorithm, we will see the definitions of some common terminologies which are used in IAPS algorithm first.

**Definition 1** (*Support Degree*) The number of transactions containing the item set X in database D is recorded as the supporting number of item set X, denoted as $\sigma_X$. The support degree of the item set X is denoted as S(X):

$$S(X) = \sigma_X / |D| \quad (1)$$

|D| is the number of transactions in database D. If S(x) is greater than a certain threshold (Minimum Support Degree), then the item set X is called a frequent item set, otherwise it is called a non-frequent item set.

**Definition 2** (*False Alarm Rate*) There is a database denoted as D, and the specified Minimum Support Degree is min_s. The actual set of frequent item sets is L, which contains N members. However, the set of frequent item sets obtained by running a specified algorithm is $L_a$, which contains $N_a$ members. If $L_a$ contains $N_f$ members whose support degree is smaller than min_s, the false alarm rate of the algorithm, denoted as FAR, can be calculated by formula (2)

$$FAR = N_f / N \quad (2)$$

**Definition 3** (*Omission Factor*) There is a database denoted as D, and the specified Minimum Support Degree is min_s. The actual set of frequent item sets is L, which contains N members. However, the set of frequent item sets obtained by running a specified algorithm is $L_a$, which contains $N_a$ members. If there are $N_o$ item sets which belong to L but not belong to $L_a$, the omission factor of the algorithm, denoted as OF, can be calculated by formula (3)

$$OF = N_o/N \tag{3}$$

**Theorem 1** *It is assumed that* X *and* Y *are two nonempty item sets, and* $X \cap Y = \emptyset$, *the occurrence probability of item set* X *and* Y *are* P(X) *and* P(Y) *respectively. The occurrence probability of the item set* $X \cup Y$ *denoted as* $P(X \cup Y)$ *will be in the range of* [P(X) P(Y) − 0.25, P(X) P(Y) + 0.25].

*Proof*

$$\begin{aligned} \text{Since } P(X) &= P(X \cup Y) + P(X \cup Y-) \\ &= P(Y)P(X/Y) + P(Y-)P(X/Y-), \\ P(X) - P(X/Y) &= P(Y)\,P(X/Y) + P(Y-)P(X/Y-) - P(X/Y) \\ &= P(Y-)\,P(X/Y-) - P(Y-)\,P(X/Y) \\ &= P(Y-)(P(X/Y-) - P(X/Y)). \end{aligned}$$

Thus

$$\begin{aligned} P(X \cup Y) &= P(Y)P(X/Y) \\ &= P(Y)\,P(X) - P(Y)P(Y-)[P(X/Y-) - P(X/Y)]. \end{aligned}$$

In the above formulas, Y-is the complement set of Y. As the value of P (X/Y −) − P(X/Y) varies from −1 to 1, and P(Y) P(Y −) = P(Y)(1 − P(Y))gets the maximum value 0.25 when P(Y) = 0.5 according to the derivation. Theorem 1 is proved.

# 3 IAPS Algorithm

## 3.1 The Idea

Base on the analysis of Apriori and its improved algorithms, it is found that there are too many members in the set of candidate frequent k item sets which are generated by self-join procedure of frequent k − 1 item sets. In order to reduce the number of members in the set of candidate frequent k item sets, the prior probability of candidate item sets will be calculated for a prejudging and screening procedure following the self-join and pruning progress. In the prejudging and screening procedure, according to theorem 1, if the prior probability of a candidate item set is too much smaller than the given minimum support degree, the candidate item set can be discarded directly without scanning the database. In another hand, if the prior probability of a candidate item set is too much bigger than the given minimum support degree, the candidate item set can be added to frequent item sets directly without scanning the database too. Consequently, the prejudging and screening procedure can reduce the times of scanning database, reduce the running time and improve the algorithm efficiency.

## 3.2 *The IAPS Algorithm*

In order to mining association rules, all the frequent item sets must be found beforehand. IAPS algorithm contains the following steps:

Step 1 Scan the transaction database D and record the occurrence number $N_i$ of every item $I_i$. If and only if $S(X) = S(\{I_i\}) = N_i/|D|$ is greater than a given minimum support degree, denoted as min_s, $\{I_i\}$ belongs to the set of frequent 1 item sets $L_1$, In the above description, $i \in \{1, 2,\ldots,n_1\}$, $n_1$ is the number of the items included in the database D, and |D| is the number of transactions included in database D.

Step 2 Generate the set of candidate frequent k item sets $C_k$ by self-join of the set of frequent k − 1 item sets $L_{k-1}$, where $K \in \{2, 3,\ldots\}$. The self-join process is as follows. Assuming that $m_1$ and $m_2$ are two members of the set of frequent k − 1 item sets $L_{K-1}$, the items in the frequent K − 1 item sets are sorted according to the dictionary order. In other words, for the member $m_i$, $m_i[1] < m_i[2] < \cdots < m_i[K-1]$, where $m_i[j]$ represents the jth item in member $m_i$, $i \in \{0,1\}$, $j \in \{1, 2,\ldots, K-1\}$. If $m_1[1] = m_2[1])$ && $(m_1[2] = m_2[2])$&&…&& $(m_1[k-2] = m_2[k-2])$&&$(m_1[k-1] < m_2[k-1]$, then $m_1$ and $m_2$ are joinable. The self-join result of $m_1$ and $m_2$ is $\{m_1[1], m_1[2],\ldots, m_1[k-1], m_2[k-1]\}$.

Step 3 Implement the pruning process on the set of candidate frequent k item sets $C_k$ according to Apriori property [6]. The pruning process is as follows: for each member $m_i$ in the set of candidate frequent k item sets $C_k$, $i \in \{1, 2, 3\ldots\}$, if it contains a nonempty subset whose support degree is smaller than the specified minimum support degree min_s, it can be judged that $m_i$ is not a member of the set of frequent k item sets $L_k$. Therefore, delete it from $C_k$. Otherwise, keep it.

Step 4 Carry out the prejudging and screening procedure. The prejudging and screening procedure is as follows. Calculate prejudging support degree of the members in the set of candidate frequent k item sets $C_k$ using the formula $P(c_i) = \sum P(c)P(c_i - c)/n$, $i \in \{1, 2, 3,\ldots\}$. In this formula, $c_i$ is a member of $C_k$, c is the single element subset of member $c_i$, and n is the number of items in member $c_i$. P(c) can be obtained by step. As for P $(c_i - c)$, if $c_i - c$ contains just one item, it also can be obtained by step 1.

Otherwise, it should be obtained by step 5 in last cycle. If $P(c_i) > min\_s + 0.25\alpha$, $c_i$ will be directly added to the frequent k item sets $L_K$ without scanning the database. If $P(c_i) < min\_s - 0.25\beta$, $c_i$ will be discarded directly without scanning the database.

Step 5 Determine the support degree of the members in the set of candidate frequent k item sets $C_k$ by scanning the database D. As for the member $c_i$, if $P(c_i) < min\_s$, discard it; Or keep it.

Step 6 Repeat the above steps 2–5 until you cannot generate a larger frequent item set.

## 4 Experimental Results and Analysis

Due to the prejudging and careening procedure of IAPS algorithm, the occurrence of false alarm and omission is inevitable. In this paper, we intend to study the impact of prejudging and careening procedure on false alarm rate and omission factor. All the experiments are carried out on 10 datasets from supermarket transaction data which contain $2 \times 10^3$, $4 \times 10^3$, $6 \times 10^3$, $8 \times 10^3$, $1 \times 10^4$, $2 \times 10^4$, $4 \times 10^4$, $6 \times 10^4$, $8 \times 10^4$, $1 \times 10^5$ transactions, respectively.

The first group of experiments is carried out to analysis the false alarm rate with the variation of parameter α when the minimum support min_s is set to 0.05. As shown in Fig. 1, when the value of parameter α varies from 0.05 to 1, the false

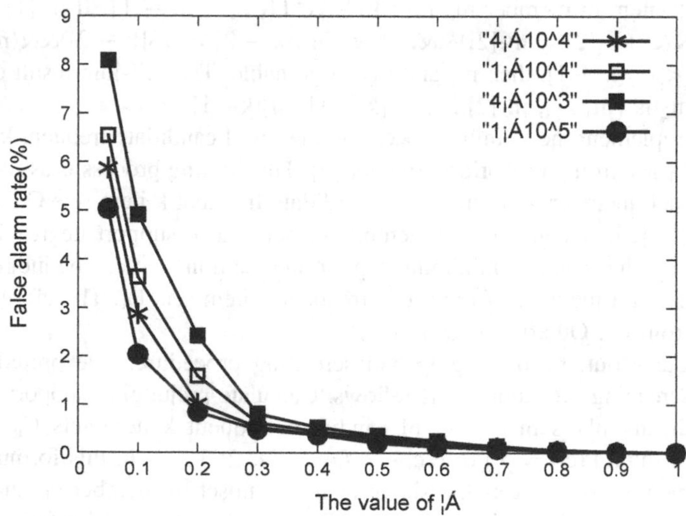


**Fig. 1** The false alarm rate varies with the value of α

alarm rate decreases gradually. Particularly, when the value of parameter α increases from 0.05 to 0.1, the false alarm rate declines sharply from about 15% to 4%. Consequently, in order to ensure the false alarm rate is lower than 5%, the parameter α can be set to 0.1. In another hand, we can see that with the increase of the number of transactions in the datasets, the false alarm rate have a slight decrease, which is not significant.

The second group of experiments is carried out to analysis the omission factor with the variation of parameter β when the minimum support min_s is set to 0.05. As shown in Fig. 2, when the value of parameter β varies from 0.05 to 1, the omission factor decreases gradually too. In particular, when the value of parameter β increases from 0.02 to 0.1, the omission factor declines sharply from about 25% to 10%. Consequently, in order to ensure the omission factor is lower than 10%, the parameter β can be set to 0.1 too. Similarly in Fig. 1, we can see from Fig. 2 that with the increase of the number of transactions in the datasets, the omission factor has a slight decrease too.

The next experiment compares the time consumed of the original Apriori and the proposed IAPS algorithm. The result is shown in Fig. 3. As we can observe in Fig. 3, that the time consuming in improved Apriori in each group of transactions is less than it in the original Apriori, and the difference increases more and more as the number of transactions increases.

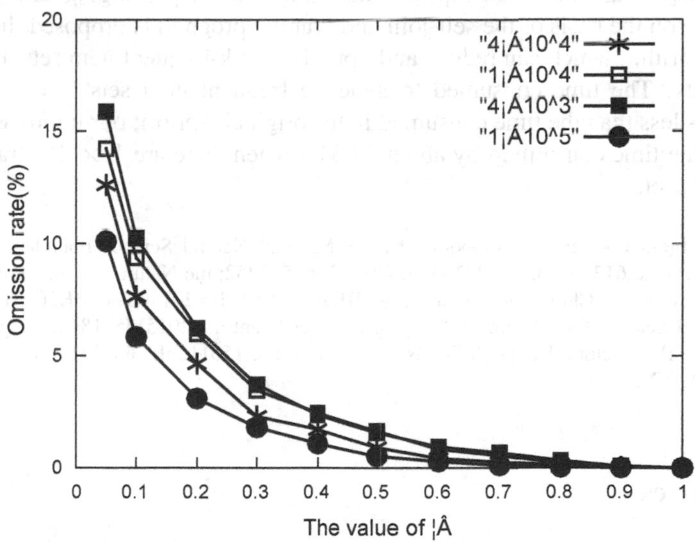


**Fig. 2** The omission rate varies with the value of β

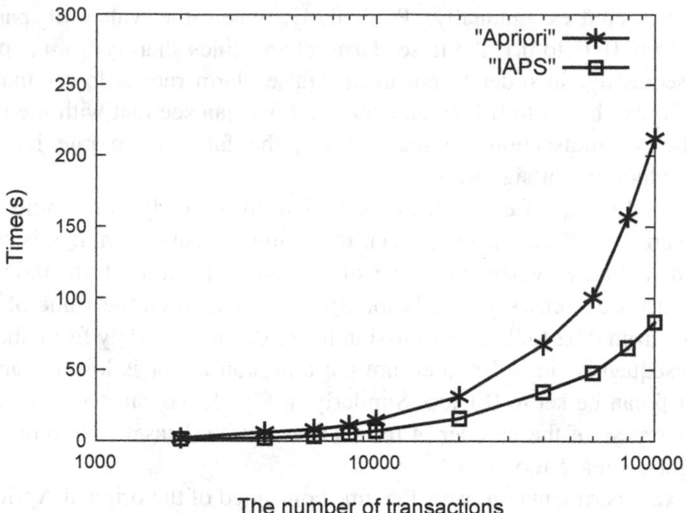


**Fig. 3** The running time varies with the number of transactions

# 5 Conclusions

In this paper, an improved Apriori algorithm with a prejudging and screening procedure on the basis of the self-join and pruning progress is proposed. It is named IAPS algorithm which can reduce and optimize the k-frequent item sets using prior probability. The time consumed to generate frequent item sets in our improved Apriori is less than the time consumed in the original Apriori; our improved Apriori reduces the time consuming by about 57.34% when there are $1 \times 10^5$ transactions in the dataset.

**Acknowledgements** Project supported by the National Natural Science Foundation of China under Grant nos. 61373135, 61300240, 61401225, 61502252; the Natural Science Foundation of Jiangsu Province of China under Grant nos. BK20140883, BK20140894, BK20131377; China Postdoctoral Science Foundation funded project under Grant no. 2015M581844; Jiangsu Planned Projects for Postdoctoral Research Funds under Grant no. 1501125B; NUPTSF under Grant no. NY214101, NY215147.

# References

1. Singla, S., Malik, A.: Survey on various improved Apriori algorithms. International Journal of Advanced Research in Computer and Communication Engineering. 3(11), 8528–851 (2014).
2. Minal, G.I., Suryavanshi, N.Y.: Association rule mining using improved Apriori algorithm. International Journal of Computer Applications. 112(4), 37–42 (2015).

3. Rajeswari, K.: Improved Apriori algorithm – a comparative study using different objective measures. International Journal of Computer Science and Information Technologies. 6(3), 3185–3191 (2015).
4. Achar, A., Laxman, S., Sastry, P.S.: A unified view of the Apriori-based algorithms for frequent episode discovery. Knowledge & Information Systems. 31(2), 223–250 (2012).
5. Peng, L., Xiaoyang, Y., Boyu, S.: Video recommendation method based on group user behavior analysis. Journal of Electronics & Information Technology. 36(6), 1484–1491 (2014).
6. Agrawal, R., Srikant, R.: Fast algorithms for mining association rules. In: VLDB '94 Proceedings of the 20th International Conference on Very Large Data Bases, pp. 487–499. Santiago (1994).
7. Yang, Z., Tang, W., Shintemirov, A., et al.: Association rule mining-based dissolved gas analysis for fault diagnosis of power transformers. IEEE Transactions on Systems, Man, and Cybernetics, Part C: Applications and Reviews. 39(6):597–610 (2009).
8. Zhang, F., Zhang, Y., Bakos, J.D.: Gpapriori: Gpu-accelerated frequent itemset mining. In: 2011 IEEE International Conference on Cluster Computing, pp. 590–594, Austin, TX, USA (2011).
9. Angeline, M.D., James S.P.: Association rule generation using Apriori mend algorithm for student's placement. International Journal of Emerging Sciences. 2(1),78–86 (2012).
10. Li, N., Zeng, L., He, Q.: Parallel implementation of apriori algorithm based on MapReduce. In: 13th ACIS International Conference on Software Engineering, Artificial Intelligence, Networking and Parallel Distributed Computing (SNPD), pp. 236–241(2012).
11. Sulianta, F., Liong, T.H., Atastina, I.: Mining food industry's multidimensional data to produce association rules using Apriori algorithm as a basis of business strategy. In: 2013 International Conference of Information and Communication Technology (ICoICT), pp. 176–181, Bandung, 2013.
12. Lin, G., Xinsheng, J., Tao J.: Discovery of network information content security incidents based on association rules and its implementation in Map-Reduce. Journal of Electronics & Information Technology. 36(8), 1831–1837(2014).
13. Rao, S., Gupta, R.: Implementing improved algorithm over Apriori data mining association rule algorithm. International Journal of Computer Science and Technology. 34(3), 489–493 (2012).

# Design and Implementation for Massively Parallel Automated Localization of Neurons for Brain Circuits

**Dan Zou, Hong Ye, Min Zhu, Xiaoqian Zhu, Liangyuan Zhou, Fei Xia and Lina Lu**

**Abstract** Automatic localization of neurons is the foundation of tracing and reconstructing the neuronal connections from the brain image stacks. With rapid development of fluorescence labeling and imaging at submicron resolution, a huge amount of data were generated, making it challenging to efficiently locate neurons from massive multidimensional images. In this manuscript, we present the implementation of an efficient parallel neuronal localization algorithm that is based on NeuroGPS. We split the image stack with a space overlapping scheme to eliminate the communication overhead among computing nodes. On this basis, we develop a hybrid parallel automated neuronal localization algorithm. We evaluate this implementation on the TianHe-2 supercomputer. The preliminary results on a terabyte-sized image stack indicate that it is capable of handle large data sets and obtains good scalability and computing performance.

D. Zou (✉) · M. Zhu · X. Zhu · L. Zhou
Academy of Ocean Science and Engineering, National University of Defense Technology, Changsha, China
e-mail: zoudan@nudt.edu.cn

M. Zhu
e-mail: zm@nudt.edu.cn

X. Zhu
e-mail: zhu_xiaoqian@nudt.edu.cn

L. Zhou
e-mail: llyzhou@nudt.edu.cn

H. Ye
Network Center, Beijing Technology and Business University, Beijing, China
e-mail: yehong@btbu.edu.cn

F. Xia
Electronic Engineering College, Naval University of Engineering, Wuhan, China
e-mail: xcyphoenix@nudt.edu.cn

L. Lu
College of Mechatronic Engineering and Automation,
National University of Defense Technology, Changsha, China
e-mail: lulina.nudt@gmail.com

 
S.K. Bhatia et al. (eds.), *Advances in Computer and Computational Sciences*,
Advances in Intelligent Systems and Computing 553,
DOI 10.1007/978-981-10-3770-2_62

**Keywords** Brain imaging · Neuronal localization · Parallel algorithm

## 1 Introduction

Characterizing the structure of the neural circuit at high resolution is crucial for understanding the physical basis of the brain functions [1–3]. Automatic neuronal localization is the foundation of tracing and reconstructing the neuronal connections from the brain image stacks [4]. There are increasing efforts to develop algorithms in automatically locating and segmenting cells from the image stacks, such as watershed algorithm [5, 6], multiscale filters [7, 8], minimum-model [9] and gradient flow tracking [10, 11], which lay particular emphasis on probing cells with regular shape. In a recent study, a novel neuronal localization method named NeuroGPS was presented based on L1 minimization model, which is robust to the broad diversity of shape, size, and density of the neurons in a mouse brain and allows locating the neurons across different brain areas without human intervention [12].

Complicated neural circuits are composed of tens of billions of neurons in higher mammals, thus challenging state-of-art image analysis methods [13]. How to efficiently locate the neurons from terabyte-sized image stacks has become a major challenge. In this manuscript we present the implementation of Massively Automated Localization Of NEurons (MALONE), an efficient parallel neuronal localization algorithm that is based on NeuroGPS. Being based on the block scheme, a hybrid parallelization strategy and the three-stage parallel pipeline, MALONE overcomes the constraint of local memory, and enables concurrent I/O accesses and parallel computing of multiple processes. We evaluate this implementation on the TianHe-2 supercomputer. The preliminary results show that MALONE is capable of handling a 5.5 TB image set in about 408 min using 128 computing nodes. It indicates that MALONE is capable of handling large data sets and obtains good scalability and high performance.

## 2 Related Work

The research of La Torre et al. puts forward an algorithm for segmenting neuronal cells [14]. The algorithm reuses the information obtained in the 2D segmentation and attempts to correct some typical mistakes made by the previous 2D segmentation step. This method is tested on three small stacks ($512 \times 512 \times 45$, ~12 MVoxels) from different layers of the rat cerebral cortex. Quan et al. proposes NeuroGPS, a neuron soma localization method which is based on a minimization problem [12]. NeuroGPS is evaluated on an image stack of brain coronal profile of Thy-1-eGFP-H and Thy1-YFP-M Transgenic fluorescence mice ($1300 \times 1850 \times 150$ voxels, ~361 MVoxels). Vousden et al. presents an automated algorithm for obtaining cellular-level, whole-brain maps of behaviourally induced neural activity in a particular region of interest [15]. Their test set is obtained by scanning 24 mouse brains at the mesoscale using the two-photon tomography (832 × 832 pixels, 16-bit tiff file, ~13 GVoxels).

The analysis of these references show that they are all tested on small datasets. Recent advances in fluorescence labeling and imaging have made it possible to analyze the whole brain of a rodent at submicron-resolution, which have produced terabyte-sized multidimensional images [16, 17]. Nevertheless, most of the existing neuronal localization methods are based on a stand-alone computing platform and can only be adapted to megabyte-sized or gigabyte-sized datasets.

## 3 Methodology

MALONE is designed to operate in HPC clusters with up to tens of millions of processors. The typical architecture of HPC cluster is shown in Fig. 1, which consists of one or more management nodes, multiple computing nodes, and disk arrays. MALONE is built on the basis of NeuroGPS and we have rewritten the control module and I/O module to make use of the massive parallelism of HPC cluster. It also employs the L1 minimization model to adapt to the diverse shape, size, and density of neurons. The image stack is stored in disk array. The management node is responsible for task scheduling and visualization, while the computation is executed on computing nodes. MALONE utilizes block algorithm to ensure high parallelism and good scalability. First, the program scans the complete

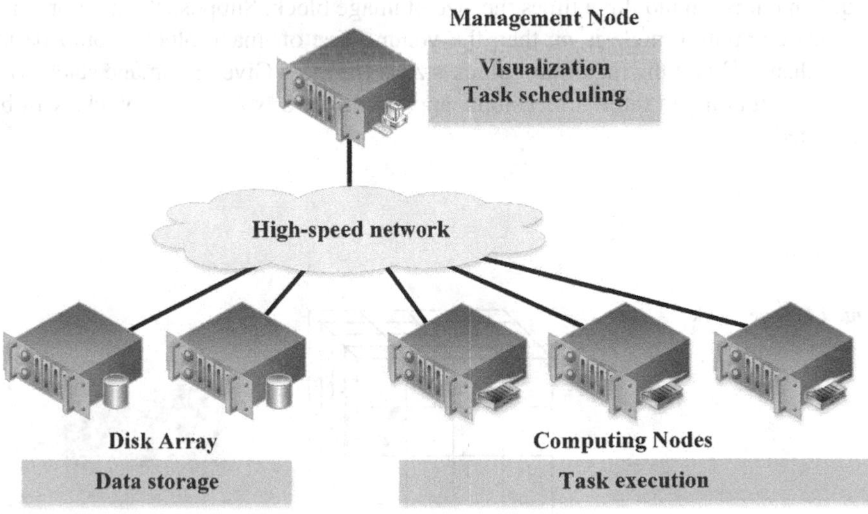


**Fig. 1** MALONE in HPC cluster

image stack from disk array to generate parameters for block partition. This is followed by launching multiple computing kernels, where the neuronal localization tasks of image blocks are mapped to computing nodes and computations are performed in parallel. Once the kernel finishes, the results are transferred from computing nodes to disk array. The following sections describe the details of the steps performed in MALONE.

## 3.1 Image Stack Partition

As the size of the image stack exceeds the memory capacity of the single computing node, it is inevitable to split the large image stack into small blocks that can fit into the memory of each computing node. Meanwhile, the neuronal localization process in the massive image stack is reduced to a series of subtasks on image blocks. In this case, each computing node just needs to load one image block into local memory at a time.

However, when it comes to calculate the position of neurons in marginal area, the computing node needs data from adjacent nodes. This introduces communication overhead, thereby reducing the performance. In order to reduce the data transferring load, we extend the image blocks to partially overlap with adjacent ones. The overlap size is determined by the maximum radius $r$ of neuronal soma. In our scheme, the image block within range $[x_1{:}x_2, y_1{:}y_2, z_1{:}z_2]$ is extended to range $[x_1 - r{:}x_2 + r, y_1 - r{:}y_2 + r, z_1 - r{:}z_2 + r]$, as shown in Fig. 2. This eliminates the communication overhead during computing at the cost of introducing additional computing and storage overhead.

Taking into account the intermediate data and temporary variables, the memory requirement is around three times the size of image block. Suppose the memory size of each computing node is $m$, then the volume size of image block should be no more than $m/3$ and the maximum block size is $(m/3)^{1/3}$. Given an image stack with equal resolution $s$ in three dimensions, approximately $3*s^3/m$ image blocks will be generated.

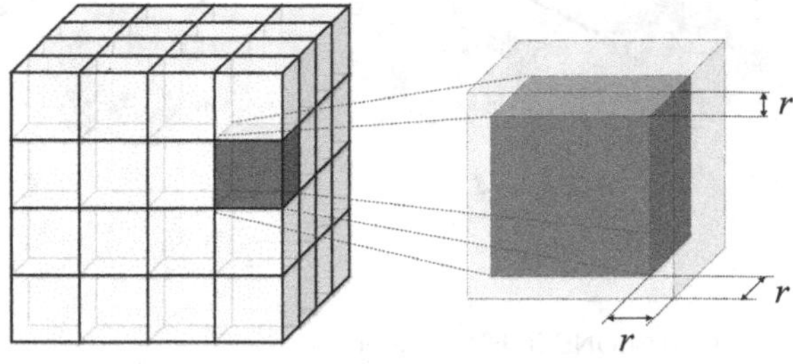


**Fig. 2** Image block extension

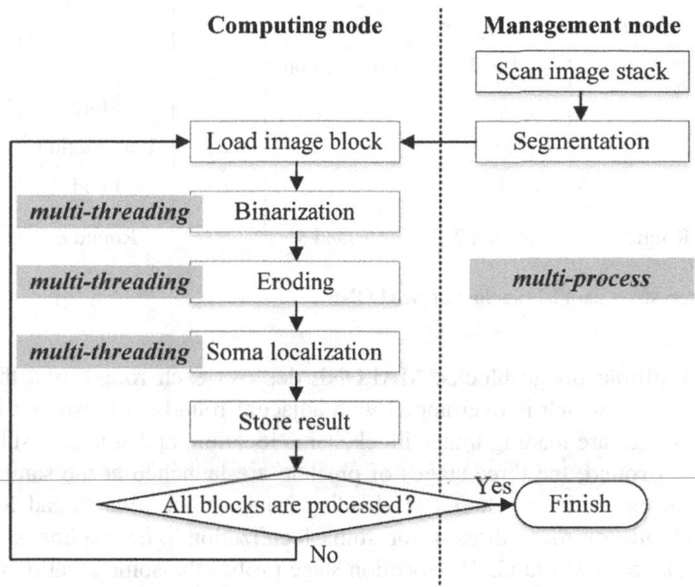


**Fig. 3** Hybrid parallelization scheme of MALONE

## 3.2 Hybrid Parallelization Algorithm

MALONE employs a hybrid parallelization model that assigns each process to each computing node, in which multiple threads are parallelized, as shown in Fig. 3. The analysis procedure for image block on computing nodes includes binarization, eroding, and localization of neurons using L1 minimization. The binarization and eroding process are typical embarrassingly parallel algorithms, which can be easily implemented on multi-core computing nodes. Due to a sparse distribution of neurons in the image blocks, the soma localization algorithm extracts subregions and analyzing the extracted signals. Accordingly, we assign each subregion to one computing thread, thus perform independent optimizations on different extracted regions.

## 3.3 Three-Stage Parallel Pipeline

The I/O operations of large image block are very time consuming compared to computing stage. In order to improve the utilization of computing nodes, we propose a three-stage parallel pipeline. As described above, the image stack is split into multiple image blocks and MALONE generated a series of processes whose number is equal to that of the available computing nodes. In general, the image blocks are more than available computing nodes, thus each task is in charge of

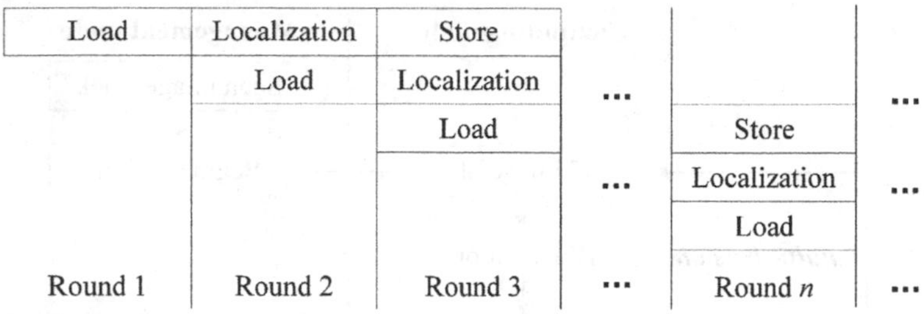

| Load | Localization | Store | | | |
|---|---|---|---|---|---|
| | Load | Localization | ... | | ... |
| | | Load | | Store | |
| | | | ... | Localization | ... |
| | | | | Load | |
| Round 1 | Round 2 | Round 3 | ... | Round *n* | ... |

**Fig. 4** Three-stage parallel pipeline of MALONE

handling multiple image blocks. MALONE deploys each round as a three-stage parallel pipeline which is overlapped with adjacent round, as illustrated in Fig. 4. The three stages are loading image block, soma location, and storing results. At the start of each round, the three stages of pipeline are launched at the same time via multi-threading, with one thread for loading image block, one thread for storing results and one or more threads for soma localization. The loading stage loads image block for next round. The location stage probes the soma location of current round. The storing stage writes results of last round. Thus, the processes of image block I/O and soma localization can be overlapped except for the first and last round in each process, resulting in an improved utilization of computing nodes.

## 4 Experimental Results

We implemented MALONE on Tianhe-2 supercomputer equipped with two 12-core Intel Xeon E5-2692 2.2 GHz processors and 64 GB memory per computing node. The system runs on 64-bit Kylin Linux (kernel version 2.6.32). We applied MALONE to a terabyte-sized image stack which contains 23400 × 32200 × 7320 volume pixels and the overall size is approximately 5.5 TB. The pixel resolution is 0.3 μm/pixel in *x*/*y* dimension and 1 μm/pixel in *z* dimension. The shape of neuronal cells has a variety of patterns and the size of the neuronal soma varies from 4 μm to 12 μm. Therefore, the overlap size is 40 and 12 pixels in *x*/*y* dimension and *z* dimension respectively. The image block size is set to 3980*4105*1244, and 288 blocks are generated.

The process time and parallel efficiency obtained for different computing node number is shown in Fig. 5. The results show that MALONE is capable of handling a 5.5 TB image set in about 408 min using 128 computing nodes. The overall wall time is shorter than the sum of load, computation, and store time, which is due to overlap of computation and I/O. As the number of computing node increases, the computation time decreases significantly while the load/store time drops at a much slower pace, which is due to the I/O bandwidth saturation of disk arrays. As a result, the parallel efficiency declines with the increasing scale.

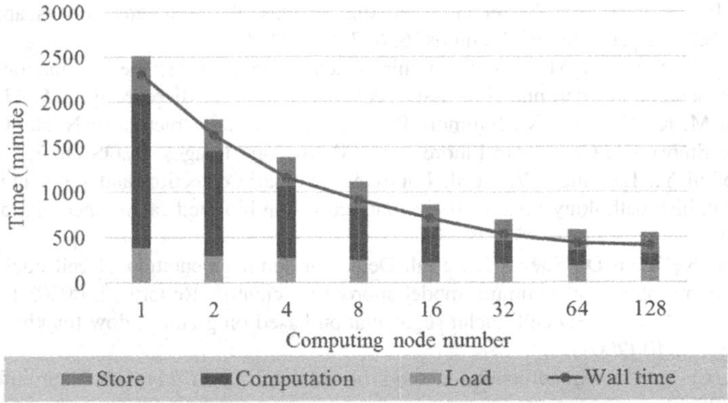


**Fig. 5** Performance evaluation of MALONE

## 5 Conclusions

In this manuscript, we present the implementation of MALONE, an efficient distributed parallel neuronal localization algorithm on the basis of NeuroGPS. MALONE utilizes the massive memory and parallelism of HPC cluster to process terabyte-sized image stack. Being evaluated on a, MALONE is capable of handling a 5.5 TB image stack in about 408 min using 128 computing nodes, which delivers an effective capacity of processing large data sets and obtains good scalability as well as computing performance.

Numerous avenues can be undertaken for future work. We aim to deploy our algorithm on more computing nodes for solving larger problems. Moreover, we plan to deploy MALONE on many-core accelerators such as GPU and Intel Xeon Phi.

**Acknowledgements** This research is partially supported by the Chinese 973 Program (2015CB755604) and the National Natural Science Foundation of China (61502516, 61572515). We greatly thank Hang Zhou for his constructive comments and useful discussions on details of NeuroGPS.

## References


1. Lu J. Neuronal Tracing for Connectomic Studies. Neuroinformatics, 9, 159–166 (2011).
2. Lichtman J.W., Denk W. The Big and the Small: Challenges of Imaging the Brain's Circuits. Science, 334, 618–623 (2011).
3. Alivisatos, A.P., Chun, M., Church, G.M., et al. The brain activity map project and the challenge of functional connectomics. Neuron, 74(6), 970–974 (2012).
4. Briggman K.L., Bock D.D. Volume electron microscopy for neuronal circuit reconstruction. Current opinion in neurobiology, 22, 154–161 (2011).

5. Long F., Peng H., Liu X., et al. A 3d digital atlas of C. elegans and its application to single-cell analyses. Nature Methods, 6, 667–672 (2009).
6. Wahlby C., Sintorn I. M., et al. Combining intensity, edge and shape information for 2D and 3D segmentation of cell nuclei in tissue sections. Journal of Microscopy, 215, 67–76 (2004).
7. Bashar M. K., Komatsu K., Fujimori T., et al. Automatic Extraction of Nuclei Centroids of Mouse Embryonic Cells from Fluorescence Microscopy Images. PLOS ONE, 7 (2012).
8. Al-Kofahi Y., Lassoued W., et al. Improved automatic detection and segmentation of cell nuclei in histopathology images. IEEE transactions on bio-medical engineering, 57, 841–852 (2010).
9. Wienert S., Heim D., Saeger K., et al. Detection and segmentation of cell nuclei in virtual microscopy images: a minimum-model approach. Scientific Reports, 2, 00503 (2012).
10. Li G., Liu T., et al. 3D cell nuclei segmentation based on gradient flow tracking. BMC Cell Biology, 8, 40 (2007).
11. Liu T., Li G., et al. An automated method for cell detection in zebrafish. Neuroinformatics, 6, 5–21 (2008).
12. Quan T., Zheng T., Yang Z., et al. NeuroGPS: automated localization of neurons for brain circuits using L1 minimization model. Scientific Reports, 3, 1414 (2013).
13. Yuan J., Gong H., Li A., et al. Visible rodent brain-wide networks at single-neuron resolution. Frontiers in neuroanatomy, 9, 70 (2015).
14. A. LaTorre, L. Alonso-Nanclares, et al. 3D segmentations of neuronal nuclei from confocal microscope image stacks. Frontiers in neuroanatomy, 7, 1–10 (2013).
15. D. A. Vousden, J. Epp, et al., Whole-brainmapping of behaviourally induced neural activation in mice. Brain Structure and Function, 220(4), 2043–2057 (2015).
16. Gong H., Zeng S., Yan C., et al. Continuously tracing brain-wide long-distance axonal projections in mice at a one-micron voxel resolution. NeuroImage, 87–98 (2013).
17. Ragan T., Kadiri L. R., Venkataraju K. U., et al. Serial two-photon tomography for automated ex vivo mouse brain imaging. Nature Methods, 9, 255–258 (2012).

# Research on Cross-Connect Technology for Large-Capacity and High-Speed SDH Signal

**Zhen Zuo, Qi Lu, Yi-Meng Zhang and Yang-Yi Chen**

**Abstract** With the development of optical fiber communication based on Synchronous Digital Hierarchy (SDH), network communication and network data processing become increasingly prominent, which brings serious challenges to network management. Cross-connect technology plays an important role in network management. This paper proposes the scheme of cross-connect system which is used for decreasing the number of routes and solving the problem about network management. Both high-order (VC-4) and low-order (VC-12) cross-connect system are designed, from the aspect of hardware and software. Test results show that the proposed scheme can meet the demand of cross-connect for large-capacity and high-speed SDH signal.

**Keywords** Cross-connect · Synchronous digital hierarchy (SDH) · Network management · Time-space-time

## 1 Introduction

Modern communication technology is developing rapidly, which requires better performance of modern transmission system, especially the transmission system for high-speed, large-capacity, and multi-type complex signal. According to the 36th report of China Internet Network Information Center (CNNIC) [1], with growth of the number of the Internet users and development of Internet and communication technology, Chinese international export bandwidth is increasing year by year. Figure 1 shows the development of international export bandwidth of China in recent years.

In the era of Big Data, the density of valuable data is very low. Because of this, how to discover useful information from great capacity of facts becomes an urgent

Z. Zuo (✉) · Q. Lu · Y.-M. Zhang · Y.-Y. Chen
College of Mechatronic Engineering and Automation, National University of Defense Technology, Hunan Province, Changsha 410073, People's Republic of China
e-mail: qq466729206@icloud.com



S.K. Bhatia et al. (eds.), *Advances in Computer and Computational Sciences*, Advances in Intelligent Systems and Computing 553,
DOI 10.1007/978-981-10-3770-2_63

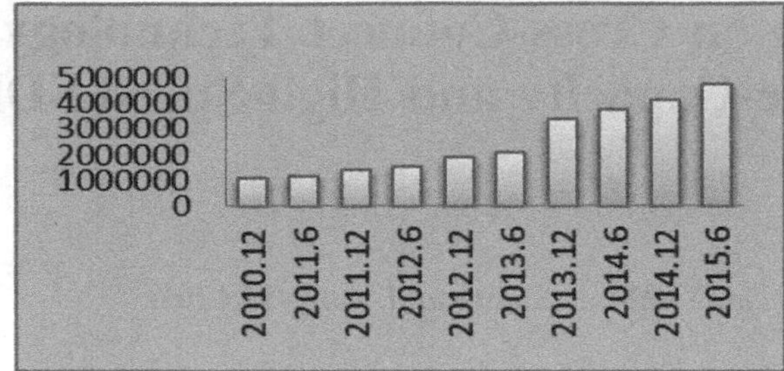


**Fig. 1** International export bandwidth of China

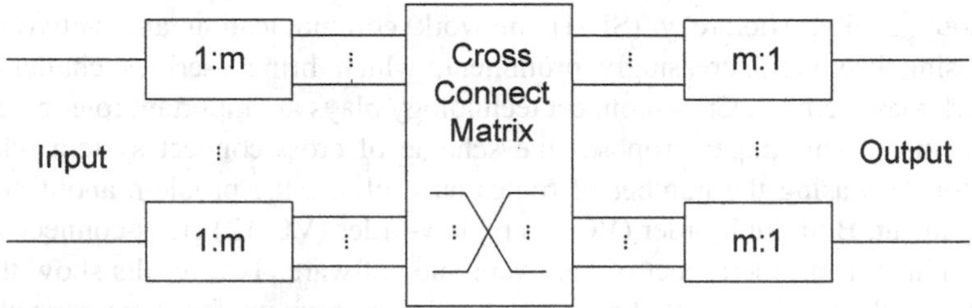


**Fig. 2** Simplified structure of SDXC

problem. Meanwhile, SDH business cross-connect technology can decrease the number of communication lines and cost of processing, which is very important in network management. Cross-connect technology is becoming one of the hottest issues in network management.

## 2 Principle of SDH Cross-Connect

A traditional simplified structure of SDH digital cross-connect (SDXC) is shown in Fig. 2. Cross-connect matrix is the core function of SDXC [2].

SDXC can be realized by time division switching and space division switching [3].

Time division switching is a kind of time-slot process. It consists of data storage and control storage. Data memory, also called buffer memory, is used for storing a frame of exchanged information. Control memory, also called address memory, is used for storing the address of storage cell in the data memory. Their capacities mainly depend on the number of time slots. The control methods of control memory mainly include read-out control and write-in control [4–6].

Space division switching uses space division matrixes to cross-connect a group of signal of the same kind. It consists of electronic cross point matrix and control memory. There are two control methods, namely input control and output control.

## 3 Design Scheme of SDH Cross-Connect System

### 3.1 Overall Design

The design principle of cross-connect matrix is to design the structure and order of matrix in accordance with the principle of cross-connect. Due to the limitation of the capacity of data memory and control memory, one-stage time division switching is usually used for the switching of single signal with different time slot, rather than the large-capacity signal. On the other hand, one-stage space division switching is used when multichannel signals exchange in the same time slot, which is lack of flexibility. For the two-stage cross network, neither T-T cross nor S-T cross can meet the requirements of large-capacity and complex data exchange [7–10].

The three-stage (time-space-time) cross network is qualified for the above requirements. For the consideration of flexibility, time division exchange is employed at the input and output terminals and space division switching is used in the middle stage of the cross network in order to realize large-capacity exchange.

Figure 3 shows the overall system design scheme. The whole system consists of three parts: blade server, switch board, and business boards. Blade server controls the whole system by sending commands to each business board. Switch board mainly exchanges the data from each business board.

The design of each business board is the key of the whole system. In the current research, we design a high-order business board for VC-4 switching and low-order business board for VC-12 switching. Every high-order board can input 16 routes of SDH signal, including eight routes of 10 Gbps signal and eight routes of 2.5 Gbps

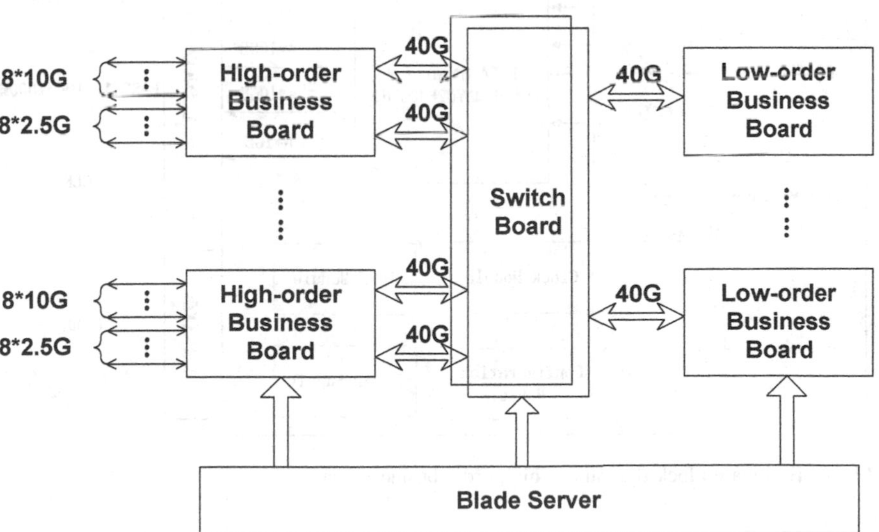


**Fig. 3** SDH cross-connect system

signal, and output two routes of 40 Gbps signal. Every low-order board can exchange total amount of 40 Gbps signal with any other high-order board [11, 12].

## 3.2 Design of High-Order Business Board

High-order business board must have enough SDH photoelectric conversion modules, powerful processing chips, rich clock resource, and secure stable power supply system [13, 14]. The hardware block diagram of high-order business board is shown in Fig. 4.

The signal processing procedure of high-order signal is shown in Fig. 5, which should possess photoelectric conversion, homologous treatment, cross-connect module, and back board transport.

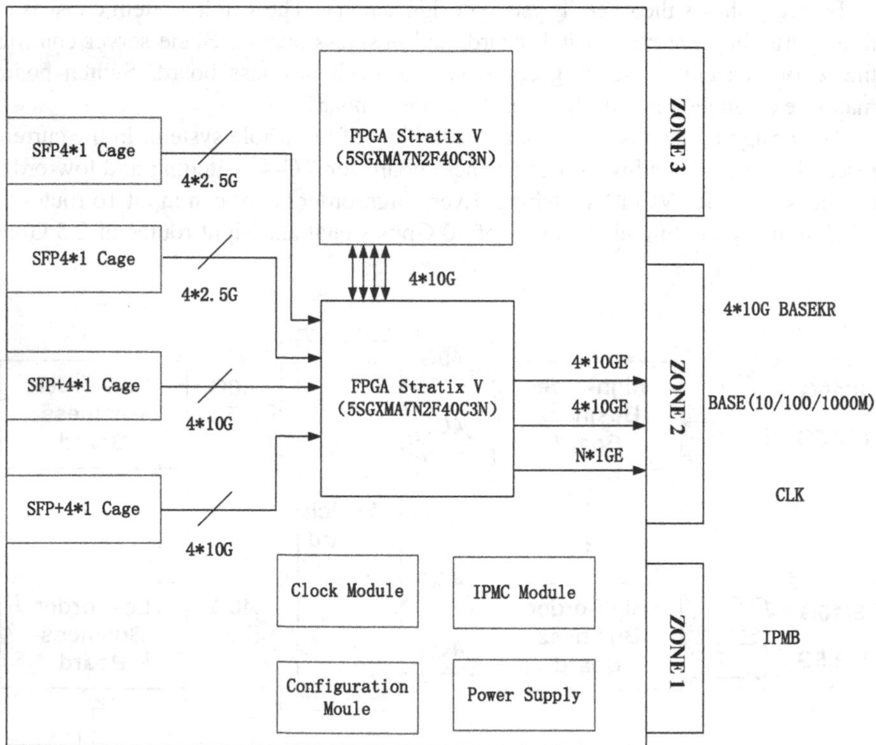


**Fig. 4** Hardware block diagram of high-order business board

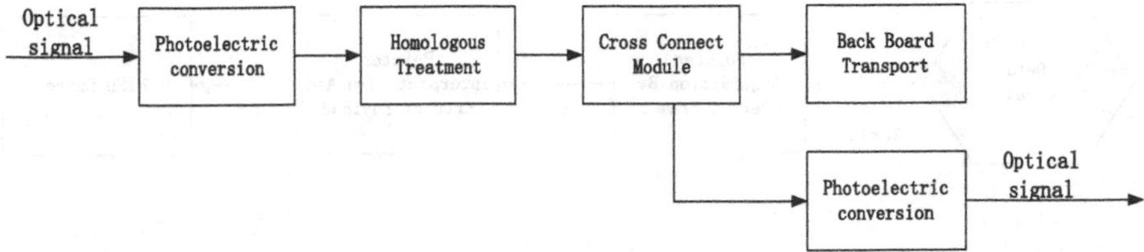


**Fig. 5** Software flow chart of high-order signal processing

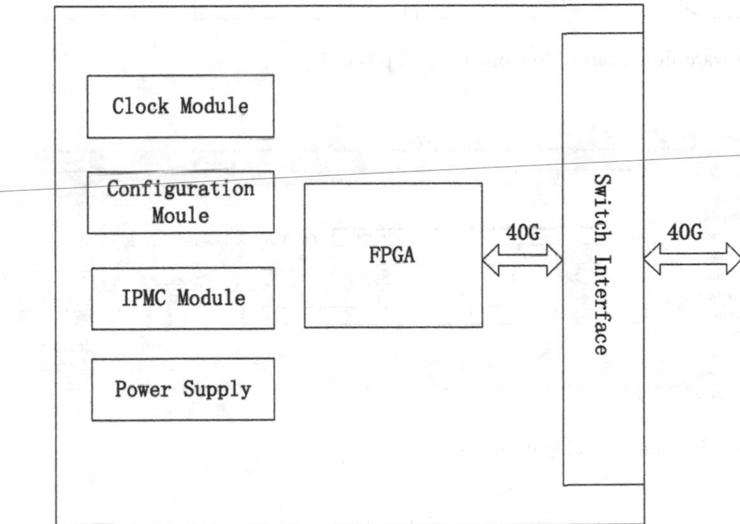


**Fig. 6** Hardware block diagram of low-order business board

## *3.3 Design of Low-Order Business Board*

The hardware design of low-order business board is shown in Fig. 6, which mainly consists of FPGA, clock module, configuration module, IPMC module, power supply module, and switching interface [15].

The signal processing procedure of low-order signal is shown in Fig. 7. The procedure mainly consists of data access, business classification, pointer acquisition, pointer interpretation, extract payload, FIFO cache, pointer generation, cross–connect, and MAC framing.

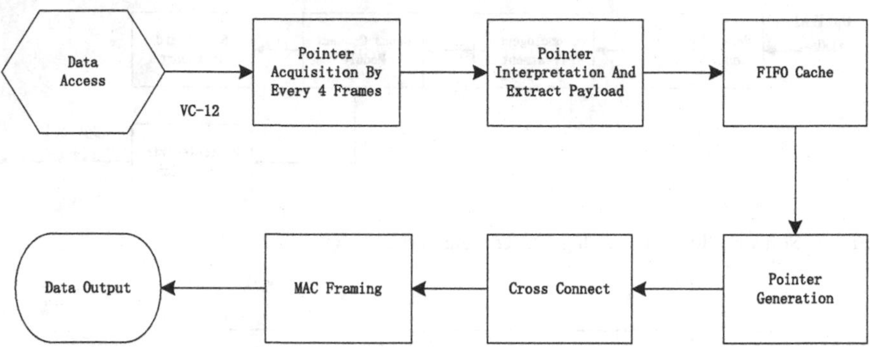


**Fig. 7** Software flow chart of low-order signal processing

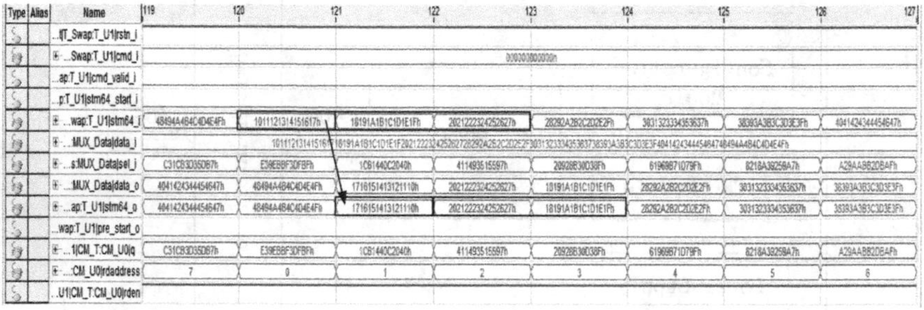

**Fig. 8** Time division switching of VC-4

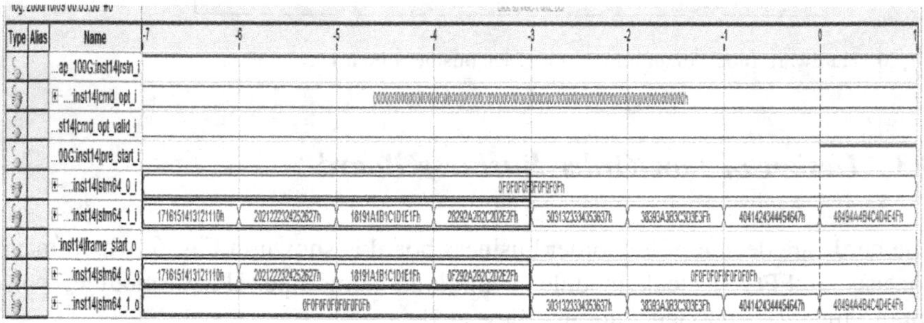

**Fig. 9** Space division switching of VC-4

# 4 Test Results

It's effective to verify if cross-connect function is accomplished through the waveform of Signal Tap. Figures 8 and 9 show the test results of high-order business board, including time division switching and space division switching. Figure 10 shows the test result of low-order business board.

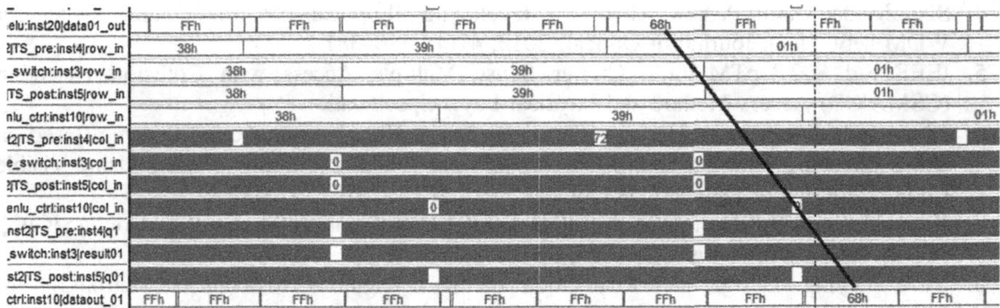


**Fig. 10** Waveform of every signal line (VC-12)

Figures 8, 9 and 10 confirm the correctness of the proposed cross-connect scheme for large-capacity and high-speed SDH Signal.

# 5 Conclusion

This paper introduces the basic knowledge of SDH, including the main characteristic of SDH, the multiplexing principle and equipment constitute. Then introduce SDXC briefly. Based on this, it discussed the basic function and modules of SDH digital cross-connect system, and shows the technology route, design scheme of the paper and the simulation environment of the design. Its emphases is the fundamental and design method of main modules in SDXC system. The design is based on the Verilog HDL, and Top-Down design methodology is adopted. The FPGA verification of the key modules in the platform of Altera FPGA device.

At present, cross-connect matrix is undergoing high-speed development and its function has greatly improved. The future cross-connect equipment will develop towards the direction of diversification, high-speed and large capacity, which makes it possible to meet the development requirements of network management. The research on SDH cross-connect technology is carried on in this paper, which will, to a great degree, promote the development, maturity, and universality of network management.

# References


1. Bowen Sun, Research on Merge Technology of Large-Capacity SDH Business Signal[D], National University of Defense Technology, 2014.
2. Yangyi Chen, Research on Converging Technology of Large-Capacity SDH Business Signal [D], National University of Defense Technology, 2015.
3. Yangyi Chen, Shaojing Su, Zhen Zuo, Zhiping Huan, Yimeng Zhang. Design and achievement of an efficient low-order pointer multiplexing and mapping scheme[C]. SPIE Applied Optics and Photonic China, 2015.

4. Sisir Kumar Garai. A new scheme of developing all-optical 4x4 cross-connect switch for WDM network[J]. Journal of Optical, 2013, 42(4): 376–381.
5. IEEE Std 802.3apTM-2007, Carrier Sense Multiple Access with Collision Detection (CSMA/CD) Access Method and Physical Layer Specifications[S].
6. Yinan Hou, Shilin Xiao, Haifeng Zheng, Weisheng Hu. Multiple Access Scheme Based on Block Encoding Time Division Multiplexing in an Indoor Positioning System Using Visible Light[J]. Journal of Optical Communication Network, 2015,7(5): 489–495.
7. Yuan Yuying. The Design of the SDH Digital Cross-connect Matrix Based on ASIC[C]. World Automation Congress(WAC), 2012:24–28P.
8. Yuan Yuying. Design and Realization of SDH Digital Cross-connect Matrix[C]. 3rd International Conference on Wireless Communication, Network and Mobile Computation (WiCOM2007), 2007:692–695P.
9. Liu Huiyong, Meng Luoming. Multi-weight path routing algorithm of SDH transport network based on principle of the least jumper wire operations[J]. Tongxin Xuebao/Journal on Communications, 2006, 27:37–43.
10. B. Melian, M. Laguna, J.A. Moreno-Perez. Capacity expansion of fiber optic networks with WDN systems: Problem formulation and comparative analysis[J]. Computers&Operations Research, 2004, 3(31):461–472.
11. Jin Cao, William S, Cleveland, Dong Lin, Don X. Sun, Internet tend to Poisson and Independence as the Load Increases[R]. Murray Hill, NJ:Bell Labs, 2001.
12. Karagiannis T, Molle M, Faloutsos M, Broido A. A nonstationary Poisson view of Internet traffic[C]. In: Proc. of the 23rd Annual Joint Conf. of the IEEE Computer and Communications Societies, Hong Kong,2004:1558–1569.
13. Barbara Gonzalez-Arevalo, Julie Roy. Simulating a Poisson cluster process for Internet traffic packet arrivals[J]. Computer Communications, 2010(33):612–618.
14. J.Mogul, et al. IP MTU Discovery Options[EB/OL]. http://www.ietf.org/rfc/rfc1063.txt,1988-7.
15. K.Papagiannaki, N.Taft, et al. A Pragnetic Definition of Elephants in Internet Backbone Traffic. In: Proc. of the 2nd Internet Measurement Workshop, Marseille, France, 2002: 175–176.

# Design and Optimization of an Intelligent Evacuation Light System

WenXia Liu and Chenfei Qu

**Abstract** CAN bus and Modbus protocol is commonly used in intelligent building designed. This paper analyzes the different characteristics between them. Based on the analysis, this article proposes an overall concept design of intelligent evacuation and light system and power management solution, and further optimizes the communication topology and software design. The experience in intelligent building designed shows that the system is reliable and meets the design requirements.

**Keywords** CAN bus · Modbus protocol · Evacuation and light system · Power management

## 1 Introduction

At present, with the rapid development of urban construction, massive public building is increasing. These building have a complex internal structure and a high concentration of people. In case of fire, it is prone to injury or death. In addition, most of the evacuation flag that is widely used in current projects is fixed directional lighting [1]. When a fire occurs near the emergency exits, personnel cannot be guided correctly. Therefore, the correct evacuation and light system is an important aspect for safe evacuation when unexpected incidents [2].

With the development of computer and communication technology, and evacuation lighting system is also moving in the direction of development of artificial intelligence. According to the site conditions, preset programmed program. In case

W. Liu (✉) · C. Qu
Tianjin Architecture Design Institute, Meteorological Station Road No. 95, Tianjin 300074, China
e-mail: Liuwenxia1986@163.com

C. Qu
e-mail: chinaqchf@163.com



S.K. Bhatia et al. (eds.), *Advances in Computer and Computational Sciences*, Advances in Intelligent Systems and Computing 553,
DOI 10.1007/978-981-10-3770-2_64

of a fire, automatic switch [3]. After switching, when the fire is changed, the programme cannot be randomly changed.

With the development of CAN bus, bus technology is used by people from the automotive sector industry. Based on bus technology evolved the E-bus technology. And further analysis, integration, and the gateway monitor [4]. Line data congestion in this way is a problem.

Adaptive Ant Colony algorithm proposed in recent years, it has been suggested this algorithm used in the industrial field intelligent evacuating system, dynamic characteristics of evacuation system, tracking environmental changes in a timely manner, through the analysis of environmental parameters for finding the optimal evacuation route [5]. Directly in the actual project using this algorithm is vulnerable to local optimization, and data stagnation phenomenon.

This paper is based on the CAN bus protocol, combined with MODBUS Protocol, analysis, integration, optimization design of intelligent evacuating system. Through the use of a data recovery function of gateway, reasonable packet interval set, the data congestion problem in communication is solved. Using improved Adaptive Ant Colony algorithm, tracking and analyzing the change of environmental parameters, find the current overall architecture the optimal evacuation route. Compared with the traditional evacuation system, this system has tremendous technological advantages and practical value. To change the status quo existing fire safety evacuation has a very important meaning.

## 2 Comparative Study Between Modbus Protocol and CAN Bus

Modbus is an application layer packet transmission protocol on layer 7 of the OSI model, which uses twisted-pair communication between multiple devices [6]. It provides master–slave question-and-answer communication protocol between the different types connection of network devices or bus, and often using RS232/RS485 [7–9]. Master–slave station's working method is the query and response.

RS232's transmission baud rate is up to 115200 bps, but Modbus protocol need to define the address, CRC parity bit, etc. So the actual transmission baud rate is approximately 8 KB/s. At the same time, unbalanced circuits that using RS232 are easily affected by base point voltage between different equipment, and also RS232 has a poor ability to control the signal when rising and falling. Therefore, RS232 is recommended as a short distance (generally less than 15 m) communication [10].

The definition layer of CAN and OSI is consistent, it can be divided into application layer, data link layer and physical layer. Each layer is transparent; one layer can only communicate with the equipment that in the same layer, but the actual communication occurs between adjacent layers on each device; each device can only be linked together through the physical layer via the communication medium [11–13]. CAN achieve a complete distributed control, simplify the system's structure and improve control system's reliability [14].

With its high reliability, high performance, and real-time ability, CAN bus is widely used in many fields. As a multi-Master bus system, it can transfer a long distance (up to 10 km), and have excellent anti-jamming capability. Despite the harsh environment, it can also ensure a safely and fast signal transmission [15].

# 3 Intelligent System Overall Design Scheme

## *3.1 The Topology Structure of the Intelligent System*

Intelligent Evacuation and Light System as an important part of intelligent building can dynamically indicate a secure evacuation passageway in case of fire. Thus, using evacuation signs can guide people in the dark, smoke and other unfavorable environment to identify safe evacuation direction.

The system including host computer, gateways, Ethernet switch, combined intelligent control host, smart-point controller extension, intelligent battery master station, exit light, bidirectional light, emergency light, and other security hardware. The topology structure of the intelligent evacuation and light system is shown in Fig. 1.

Host computer including monitoring computer and configuration software, can manage and monitor the whole evacuation light system. It is the main unit of human–computer interaction. In the PC, application of improved Ant Colony algorithm to programmed a corresponding program. Analysis of smoke movement, environmental parameters, node parameters of dispersion, human behavior, and making the appropriate parameters, calculate the survival path. System will select a larger safety factors and the shortest path, to ensure as soon as possible in the limited available safe egress time reaches a safe place.

Ethernet switch can achieve transparent transmission between TCP network packet and RS232/RS485 interface data. Integrated within the TCP/IP protocol stack and 10/100 M adaptive Ethernet interface, you can use it to complete the embedded device's networking capability. Using the Ethernet Converter in this paper has data recovery capabilities. By configuring the appropriate packet interval, and achieve the purpose of achieving stable transmission of data. The typical schematic diagram is shown in Fig. 2.

Combined intelligent control host has system equipment and lighting setting, display, control, storage, and exchange of information on FAS system center. Intelligent control host's communication bus mode is CAN-Second bus when communicating with internal devices. Each communication loop can accommodate 32 addresses such as DC battery master or controller extension. Intelligent control host using CAN-Second bus as a communication network connects controller

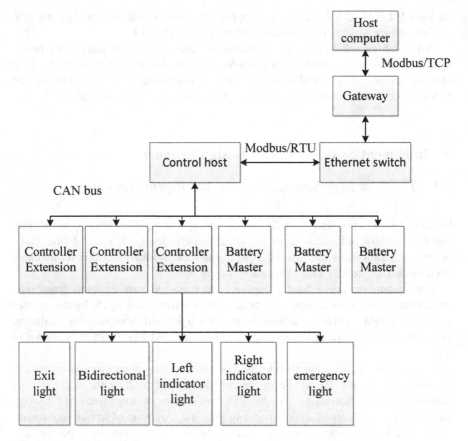

**Fig. 1** System structure diagram

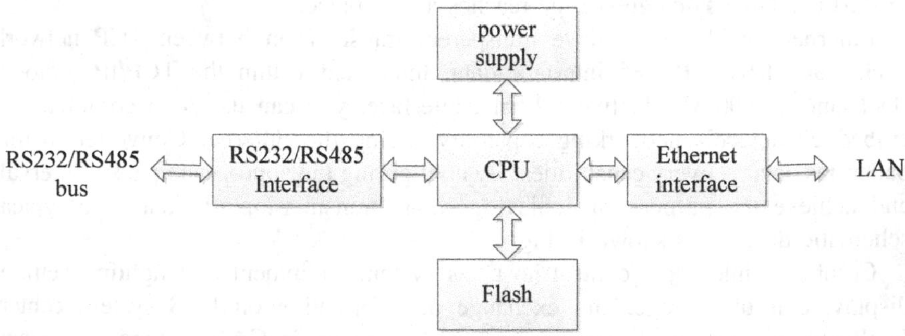


**Fig. 2** Ethernet switch's typical schematic diagram

extension, intelligent battery master station, and other devices into a distributed intelligent monitoring system.

Intelligent control host's communication interface mode is RS232 bus when communicating with external devices. It can only communicate in short distance. Therefore, the intelligent system using Ethernet switch converts the RS232 to network interface when intelligent control host and host computer communicate with each other. It converts the communication protocol from Modbus RTU to Modbus TCP in order to achieve remote connection between intelligent control host and Host computer.

Smart-point controller extension is type of safety voltage controller extension, it can connect with the intelligent control host through CAN bus. It can accommodate 64 device addresses or 6400 lamp addresses.

### 3.2 Power Management Scheme Design

Power management is an important part of intelligent evacuation and light system and its structure is shown in Fig. 3.

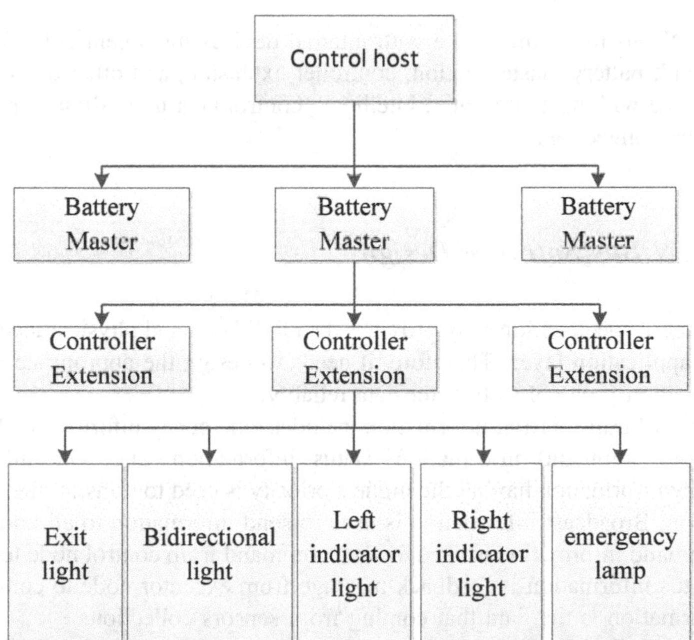


**Fig. 3** Power management structure

Each part's function is as follows:

- Controller host communicates with intelligent battery master station via CAN bus, and control the battery master station to complete state switching and data acquisition.
- Intelligent battery master station provides power for controller extension. It has a normal/emergency output states, normal output AC220 V, emergency output DC216 V. When emergency, intelligent battery master station uses 18 series of lead–acid battery pack to provide emergency output. The master station can automatically detect the battery's status, and automatically charge or discharge the battery in order to extend the emergency battery pocket's life and reduce maintenance costs.
- Controller extension has a dual-input function that normal power is AC220 V, emergency power is DC216 V, and its output power is DC24 V, Having loop control function, it can directly control the various emergency lights on and off.
- While using a dedicated high brightness LED lamps, all emergency lights are not equipped with a battery, so that, it can reduce post-maintenance effort.

## 4 System Software Design

Using CAN bus to communicate with internal devices intelligent control host can connect with battery master station, controller extension, and other devices. When communicate with host computer, intelligent control host uses Modbus protocol to achieve the connection.

### *4.1 CAN Bus Software Design*

CAN protocol specification only provides data link layer and physical layer, but not provides application layer. Therefore, it needs to design the appropriate CAN bus communication protocol to transfer data reliably.

The type of transmission information includes emergency information, broadcast information, command information, status information, and data information. Emergency information having the highest priority is used to transmit the important information; Broadcast information is used to send information to all nodes on the bus; Command information is used to send command from control node to executor node; Status information is feedback message from executor node to control node; Data information is the data that coming from sensors collection.

Software design for CAN bus communication protocol includes data transmission program and receiving program. Send data program flow diagram is shown as Fig. 4. First, check whether the transmit buffer is meeting the requirement through

Fig. 4 Send data flow diagram

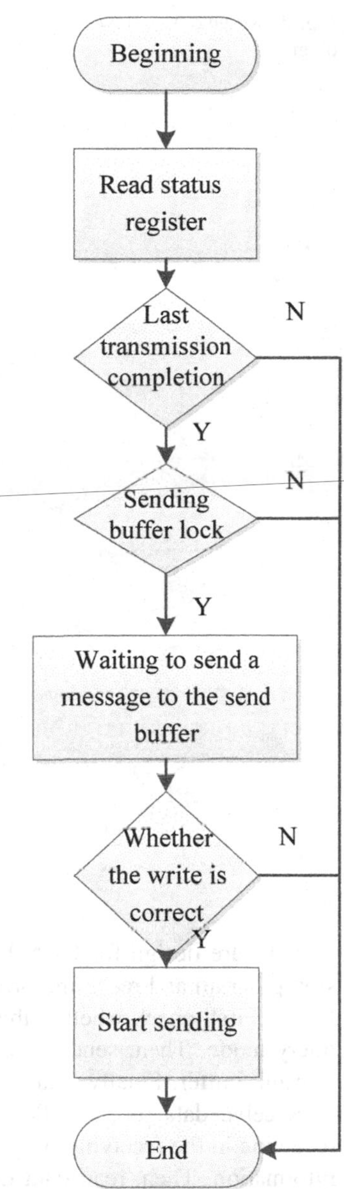


query mode. Then, send a message that combined from transmission data to the transmit buffer. Finally, start sending command.

Receive data program flow diagram is shown as Fig. 5. First, open interrupt reception in the receiving program. Second, use interrupt reception mode to receive information. Then, read data from receive buffer after the CPU interrupt and deal with the data. Finally, clear the receive buffer.

**Fig. 5** Receive data flow diagram

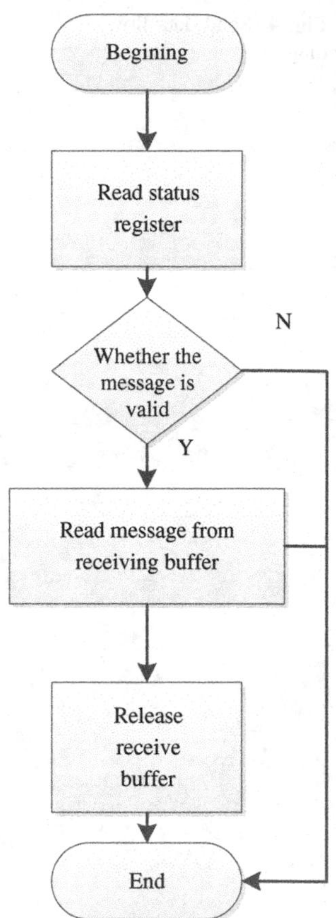


Software design for CAN bus communication protocol includes data transmission program and receiving program. Send data program flow diagram is shown as Fig. 4. First, check whether the transmit buffer is meeting the requirement through query mode. Then, send a message that combined from transmission data to the transmit buffer. Finally, start sending command.

Receive data program flow diagram is shown as Fig. 5. First, open interrupt reception in the receiving program. Second, use interrupt reception mode to receive information. Then, read data from receive buffer after the CPU interrupt and deal with the data. Finally, clear the receive buffer.

## 4.2 Modbus Communication Protocol Software Design

Ethernet switch converts the message between Modbus/TCP protocol and Modbus/RTU protocol. Based on such inter-conversion, it is able to complete the

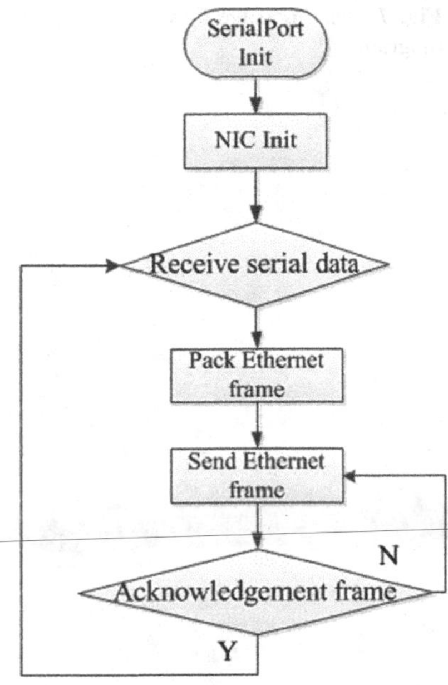


**Fig. 6** Ethernet switch network transmission flow diagram

transmission between the Ethernet and serial port using the Ethernet switch. The main Ethernet switch's internal program includes network transmission program and the serial communication program. Network transmission program diagram is shown in Fig. 6. First, the relevant registers of serial and NIC are initialized. Then, according to requires for protocol format, the message is packed into special format. Finally, Ethernet switch transmits the message to Ethernet through the NIC. The serial communication program is similar to the network transmission program.

Intelligent evacuation and light system's host computer connect with Ethernet switch through communication software, then, send Modbus TCP protocol message to Ethernet switch by intranet. The protocol message can translate into Modbus RTU protocol through Ethernet switch to transmit to the controller host.

Host computer send message flow diagram is shown in Fig. 7. According to protocol address, host computer sends protocol message to slave computer, then judges whether the slave computer replies. When slave has replied, host computer send command message immediately. Slave computer receives message flow diagram is shown in Fig. 8. First, slave computer judges whether the protocol address that received from host computer is the same as local address. Second, Slave computer verifies the CRC parity bit and analyzes the function number from protocol message. Then, according to the different function numbers slave computer can read register, write register, and write switch status. Finally, the slave computer returns to the waiting state.

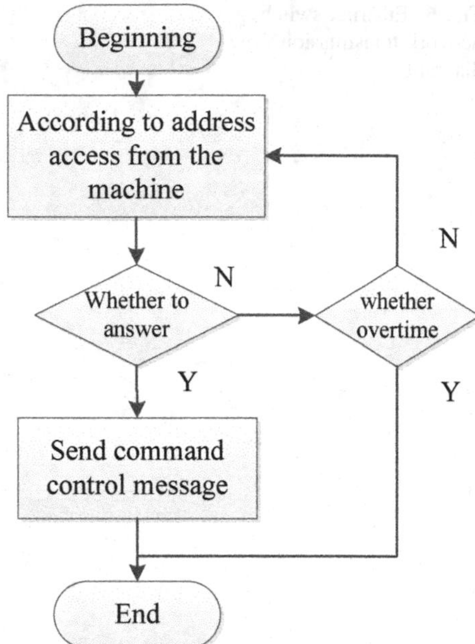


**Fig. 7** Send message flow diagram

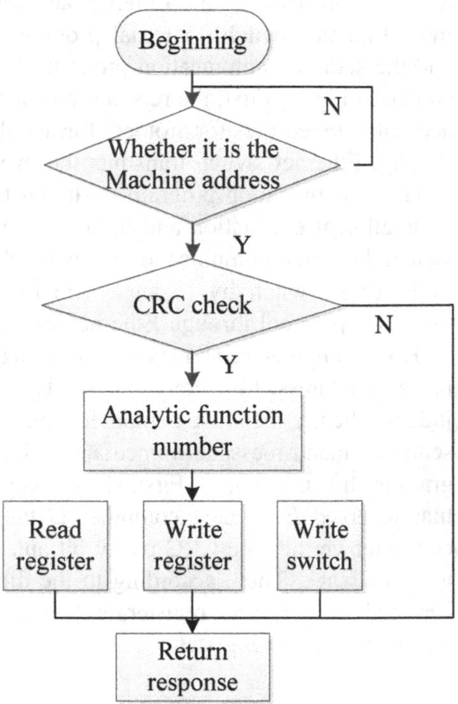


**Fig. 8** Receive message software flow diagram

## 5 Conclusion

Intelligent evacuation and light system is an important part of intelligent building. This paper establishes an overall concept design of intelligent evacuation and light system based on CAN bus and Modbus protocol. Intelligent controller host connects with internal devices through CAN communication protocol, and forms the bottom intelligent control system. The bottom intelligent control system can achieve real-time monitoring of multiple devices and distributed control. Intelligent evacuation and light system communicates with host computer via Modbus protocol. Combined with Ethernet switch, it can achieve long-distance communication.

The intelligent evacuation and light system has been already applied in intelligent building of a design institute of Tianjin. According to the practical application, this system has the characteristics of real-time, reliable, low false alarm rate and easy to maintain, and can detect various devices effectively. Thus, this system can meet the requirements for intelligent building of evacuation and light.

## References

1. GB 50116-2013. Code for design of fire automatic alarm system. S.
2. GB 17945-2010. Fire emergency lighting and evacuation instructions system. S.
3. Wu Zhouxiong, He Weijie, Liu Qingrui. Intelligent system of fire emergency lighting and evacuation instructions. J. Study on fire-fighting equipment, 2014, pp. 549–551.
4. Wang Dan. Research on the intelligent emergency evacuation system. D. Shenyang Institute of aeronautical engineering master's thesis, 2009.
5. Wang Xia, Cheng Naiwei. Ant Colony algorithm in the application of dynamic evacuation route optimization. J. Research and Exploration, 2009, pp. 7–9.
6. Yang Fen, Xu Zhao, Cao Maohong. Increased analysis and test of CAN bus transport distance. J. Industry and Mine Automation, 2007, pp. 30–32.
7. MODBUS over Serial Line Specification & Implementation guide V1.0[DB/OL]. http://www.Modbus-IDA.org, 2004.
8. Qian Zhang, Chen Tao, Lv Xianzhi. Building intelligent evacuating system structure. J. Fire science and technology, 2011, pp. 205–207.
9. Ye Xin, Chen Wenyi, Zhao Jian Based Modbus protocol Matlab Things gateway. J. Measurement & Control Technology, 2013, pp. 77–80.
10. Kong XiangTong. Design of embedded equipment based on CAN bus. J. 2013, pp. 85–87.
11. Zhao Qiang. Design and Implementation of CAN Bus Controller IP. J. Computer Technology and Development, 2013, pp. 77–80.
12. Jiang pupil, Liu Yuming, Yangchu Ping, CAN bus physics research and modeling. J. Applied Science and Technology, 2007, pp. 35–38.
13. Wang Qiong Based on Ecological CAN bus intelligent building control system. J. Design of public works, 2011, pp. 111–117.
14. Wang Zhong, Sun Haoqin. Realizes communication protocol based on CAN Bus Intelligent Building Monitoring System. J. Electronic Science and Technology, 2010, pp. 63–65.
15. Pan Hongyue. Design and implementation of communication based MODBUS protocol. J. Measurement & Equipment, 2002, pp. 35–36.

# Forty Years of BioFETOLOGY: A Research Review

**Jiten Chandra Dutta, Purnima Kumari Sharma and Hiranya Ranjan Thakur**

**Abstract** The past 40 years since the introduction of enzyme field-effect transistor (ENFET) in 1976 has been invaluable towards the development of biological sensors. Many devices came up with its own merits and demerits which made this area of research very popular worldwide. When the biological materials such as living organisms, cells, enzymes, DNA, etc., were combined with ISFET, BioFETs came up. By detailed study of the BioFETs one finds that most of the devices were formed of Si-based ISFETs. Though these devices have many advantages but when it came to sub 22 nm range, scaling problems arose which led to power dissipation, leakage, short channel effects, etc. To overcome these problems researchers opted for the use of carbon nanotubes (CNT) as channel material which gave better scalability, reduced power dissipation, better control over channel formation, etc. The complex fabrication process of the traditional Si-based devices was also simplified by introduction of junctionless CNTFETs. This paper puts forward a study of advances and developments of various BioFETs starting with ENFET and continuing with junctionless CNTFET till date.

**Keywords** ENFET · BIOFET · ISFET · CNTFET

J.C. Dutta (✉) · P.K. Sharma · H.R. Thakur
Department of Electronics and Communication Engineering, Tezpur University, Napaam, Tezpur 784028, Assam, India
e-mail: jitend@tezu.ernet.in

P.K. Sharma
e-mail: purnimasoni4018@gmail.com

H.R. Thakur
e-mail: hrtnerist@gmail.com



S.K. Bhatia et al. (eds.), *Advances in Computer and Computational Sciences*,
Advances in Intelligent Systems and Computing 553,
DOI 10.1007/978-981-10-3770-2_65

## 1 Introduction

Last few decades have been very important towards the research and development of biosensors. The first biosensor was developed by Clark in 1962 [1]. He developed an amperometric oxygen electrode biosensor by immobilizing it with glucose oxidase enzyme. Since then, many varieties of ideas were put forward towards the advancement of biosensors with the combined knowledge of various branches of applied science and engineering. Among those ideas, the one that caught the eyes of researchers the most was the combination of biologically active materials with an ISFET (ion-sensitive FET). ISFET was first introduced by Bergveld in 1970 [2] and called as the first miniaturized silicon-based chemical sensor. This invention gave a very vast area of research to the researchers. In the midst of many problems faced while applications, ISFET-based biosensors which are also termed as biologically modified field-effect transistors (BioFETs) gained the most interest. It has given many publications which are still on progress. BioFETs became so popular because of the use of FET as the basic structure. This helped in miniaturization of devices. A lot more advantages came into light like small size, low weight, fast response, better fabrication and packaging, high reliability, low cost, etc. [3].

## 2 Theory of BIOFETS

Biomolecules and living biological systems are very specific and sensitive. This is the reason why they are very often opted for sensor function. A biosensor is a combination of two blocks in series connection: a biochemical recognition system and a physicochemical transducer. Here, the recognition is done by a chemoreceptive cell and so it is termed as bioreceptor. The materials which are biologically sensitive can be used for recognition. These materials can be either biological molecular species like enzymes, proteins, antigens, antibodies, etc., or living biological systems like cells, tissues, plants, etc. The main task of the biological recognition system is to convert the information that is received from the biochemical part which is an analyte into a physical or chemical or biochemical response signal. Specific molecular interaction occurs which thereby changes one or more physicochemical parameters. As a result of these changes electrons, ions, gases, heat, light, etc., are produced. These quantities are then quantified (generally converted to an electrical signal) with the use of transducer, whose output is fed to the signal processor to give the desired result.

The basic structure of BioFET is ISFET. If the gate of an ISFET is modified or coupled with bioreceptor, we get the BioFET as shown in Fig. 1. The charge variations result in transduction of the recognizing phenomena. The main information transfer occurs at the interface of the biological recognition section and transducer section. To understand the basic principle of a BioFET one should know the ISFET.

For instance, an n-channel structure of ISFET is depicted in Fig. 2, which comprises of a p-doped silicon substrate and n-doped source and drain regions. These two regions are separated by a short channel which has a gate insulator on its top. Generally, the gate insulator is made up of $SiO_2$. Also, this $SiO_2$ layer has on top another insulator layer formed of materials like $Si_3N_4$, $Al_2O_3$, $Ta_2O_5$ which are typically pH-sensitive materials. This forms a double layer gate insulator and is used for pH-sensitive ISFETs. An ISFET starts operating when the gate voltage ($V_G$) is applied. This voltage is provided by a reference electrode, generally Ag/AgCl electrode is used. This electrode gives a closed circuit and bias to the analyte. A positive bias voltage with respect to the bulk, results in formation of an

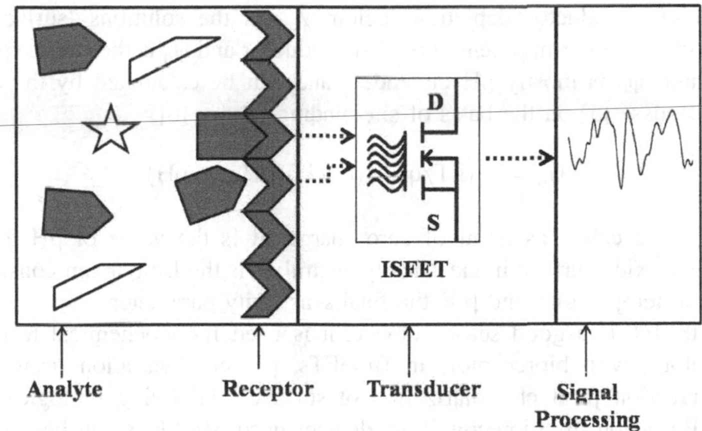


**Fig. 1** Schematic showing the principle of a biochemical sensor

**Fig. 2** The n-channel structure of an ISFET

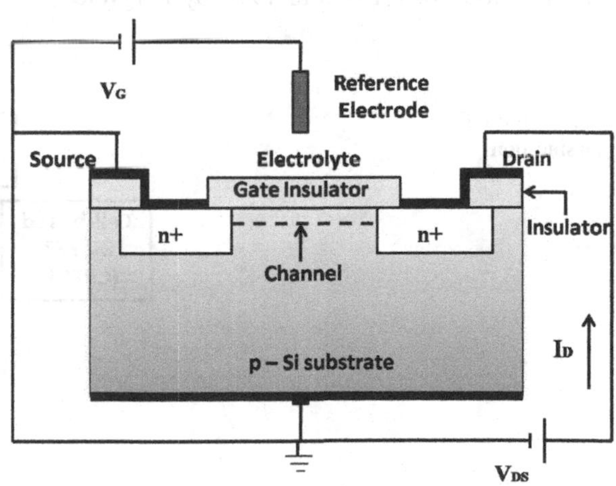

n-type inversion layer in the channel between source and drain. The drain current depends on the potential variations at the various interfaces and is given by [4]

$$I_D = \mu C_i (W/L) V_{DS} [(V_{GS} - V_{TH}) - 0.5 V_{DS}] \quad (1)$$

$$V_{TH} = E_{ref} - \psi_0 + \chi_{sol} - (\Phi_{Si}/q) - (Q_i + Q_{ss})/C_i - (Q_B/C_i) + 2\Phi_f \quad (2)$$

In the above equations μ is the mobility of electrons in the channel; W is the width of the channel; L is the length of the channel; $V_{TH}$ is the threshold voltage; $E_{ref}$ is the potential of reference electrode; $\Phi_{Si}$ is the bulk semiconductor work function; Ci is the gate insulator capacitance; q is the basic charge; $Q_{ss}$ is the insulator semiconductor interface charge; $Q_i$ is the insulator charge; $Q_B$ is the charge in semiconductor depletion region; $\chi_{sol}$ is the solutions' surface dipole potential; $\Phi_f$ is the Fermi potential of semiconductor and $\psi_0$ is the surface potential. The potential $\psi_0$ is mostly pH dependent and can be calculated by the equation given by Bousse [5] on the basis of site-binding theory [6]:

$$\psi_0 = 2.3(kT/q)[\beta/(\beta + 1)](pH_{pzc} - pH) \quad (3)$$

where $pH_{pzc}$ is called as point of zero charge. It is the value of pH for which $\psi_0 = 0$, i.e., oxide surface in electrically neutral; k is the Boltzmann constant; T is the absolute temperature and β is the final sensitivity parameter.

Since ISFET is a good sensor device, it is used for biochemical to electrical changes along with bioreceptor. In BioFETs, potential variation occurs due to catalytic reaction product, polarization of surface, and living biological system changes. Based on the biorecognition element used, BioFETs can be categorized into various types as shown in Fig. 3. [7]. Table 1 puts forward major various inventions in the development of BioFET devices in chronological order starting with invention of ISFETs in 1970 by Bergveld.

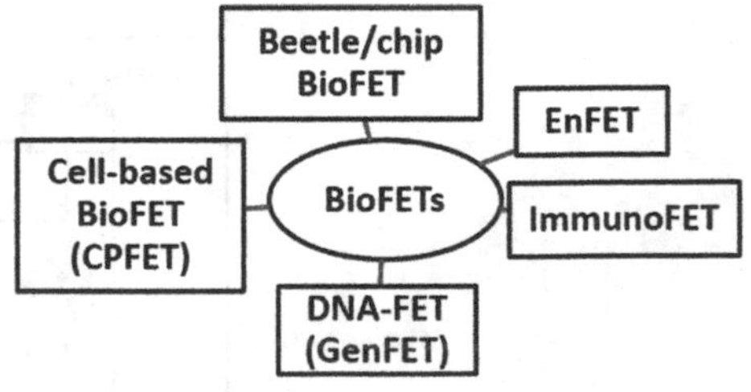


**Fig. 3** BioFETs classification

## 3 BioFETs Classification and Advancements

### 3.1 ENFET

ENFET is a bioelectronic device which is formed by incorporating an enzyme layer with an ISFET. The enzyme stimulates the biocatalytic processes thereby altering the pH value at the ISFET's gate surface. This variation in pH occurs because of either by consumption or generation of protons in accordance with the site-binding theory. The change in pH alters the potential at the surface of the ISFET and results in variation in current flowing through the channel. Hence, from signal transduction point of view, an ENFET can be termed as a bioelectronic device. Its main purpose is to convert a biochemical or biological signal into an electrical signal.

The idea of ENFET was first given by Janata and Moss in 1976 [8]. Taking this idea, Caras and Janata realized the first ENFET in 1980 [9] which was a penicillin-sensitive biosensor as shown in Fig. 4. In this biosensor, the hydrolysis of penicillin was used as the basic enzymatic reaction catalyzed by penicillinase

**Table 1** Various developments of BioFET devices in chronological order

| Development | Year |
|---|---|
| First ISFET introduced [2] | 1970 |
| Idea of the first BioFET which was an ENFET [8] | 1976 |
| Realization of first ENFET [9] | 1980 |
| Idea of ImmunoFET [10] | 1980 |
| MOSFET coupling cells [14] | 1981 |
| First neuron transistor or CPFET [12] | 1991 |
| First experiment on detection using direct DNA-hybridisation [15] | 1997 |
| Beetle chip BioFET [11] | 1997 |
| First CNTFET demonstrated [16] | 1998 |

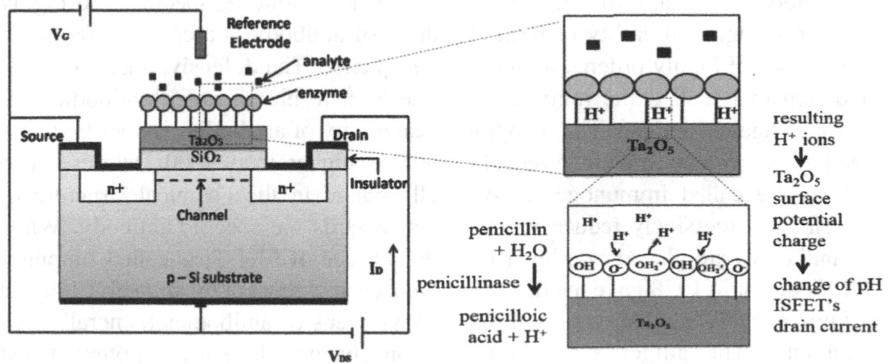


**Fig. 4** PENFET (penicillin sensitive ENFET) structure and functional principle

**Table 2** The enzymes used as bioreceptors and analyte used for detection [7]

| Sl. No. | Analyte | Enzymes used as Bioreceptors |
|---|---|---|
| 1 | Urea | Urease, Penicillin G acylase, Penicillin penicillinase |
| 2 | Glucose | Glucose dehydrogenase, Glucose oxidase/$MnO_2$ powder, Glucose oxidase |
| 3 | Sucrose | Invertase/glucose dehydrogenase, Invertase/glucose oxidase/mutarotase |
| 4 | Ethanol | Aldehyde dehydrogenase, Alcohol dehydrogenase |
| 5 | Maltose | Maltase, Glucose dehydrogenase |
| 6 | Lactose | β-Galactosidase, glucose dehydrogenase |
| 7 | Ascorbic acid | Peroxidase |
| 8 | Acetylcholine | Acetylcholinesterase |
| 9 | Creatinine | Creatinine deiminase |
| 10 | Formaldehyde | Alcohol oxidase |
| 11 | Fluorine containing organophosphates | Organophophorus acid anhydrolase |
| 12 | Organophosphate compound (paraoxon) | Organophosphate hydrolase |

enzyme. This resulted in production of penicilloic acid as product which thereby became the cause of pH variation of the ISFET. $Ta_2O_5$ was used as pH-sensitive membrane on top of gate insulator. The enzyme penicillinase was immobilized on top of $Ta_2O_5$. The output directly varied with the concentration penicillin in the given solution. In the similar way, various enzymes can be utilized for ENFET development as shown in Table 2.

## 3.2 ImmunoFET

The antibody–antigen (Ab–Ag) interaction shows the unique, specific, and highly sensitive recognition ability of biomolecules. An antibody is a complex biomolecule formed of highly ordered amino acids sequence. The defensive mechanism of an organism is called the immune response as a result of which antibodies are generated against antigen. The recognition capability of antibodies is very high and specific for proteins having molecular weights higher than 5000 Daltons. Such proteins are called immunogenic. A small change in the chemical structure of antigen can excessively reduce its attraction towards the specific antibody. When the antibodies are placed in a layer over the surface of FET, it is called ImmunoFET. Schenck in 1978 gave the direct immunosensing concept by an ISFET [10]. In an ImmunoFET, the gate is immobilized by antigens or antibodies (generally in a membrane). The antigen–antibody interaction changes the surface potential and hence, the drain current.

### 3.3 GENFET

GENFET uses genetic blocks, i.e., DNA and RNA as biological recognition elements. Such devices enable simple, speedy, and cheaper nucleic acid sample analysis in real-life medical applications like the genetic diseases diagnosis, infectious agent's detection, screening of drugs, etc. Mostly the DNA detection is done using DNA-hybridisation process. Here, the target, i.e., unknown single-stranded DNA called ssDNA is detected by another molecule with which it forms a double-stranded helix structure called dsDNA. Many complementary and non-complementary nucleic acids are present but the probe molecule identifies the complementary one with high efficiency and specificity. This unique complementary nature called base pairing helps in the biorecognition process. Few examples of base pairs are: adenine–thymine and cytosine–guanine pairs. A DNA-FET (also called GENFET) for hybridization detection is shown in Fig. 5. A GENFET is be obtained by immobilizing a membrane consisting of sequences of ssDNA onto a transducer (say ISFET) [11]. After hybridisation dsDNA is formed which is detected by the ISFET.

### 3.4 Cell-Based BioFET

Recognition in cell-based BioFETs is done using the biological cells which are known probably as the smallest unit. If a single cell or a system of cells is placed on the FET sensor, a 'cell–transistor' hybrid is formed [12]. Such kind of sensors is formed mostly because they utilize living cells and using such device, one can get direct response from any external phenomena which affects the cells. The outputs of such sensors are very informative and efficient from clinical diagnostics point of view. Lots of study on toxic elements and compounds, pollutants, etc., which effect cell metabolism activities can be made using cell-based BioFETs.

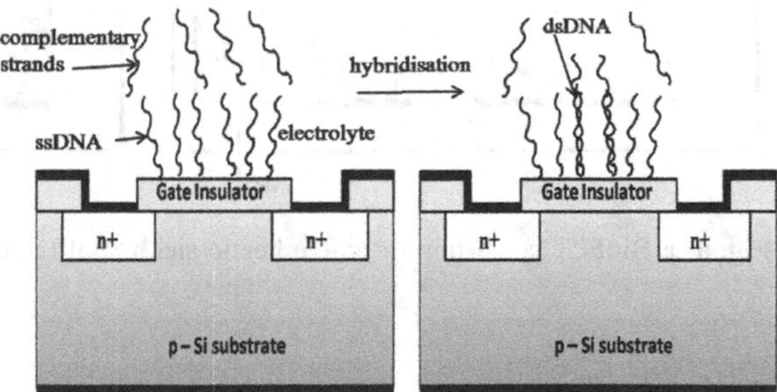


**Fig. 5** The structure and DNA-hybridisation principle of a DNA-FET

### 3.5 Beetle Chip FET

The idea of using whole organism or a part of the organism, i.e., the sensory organs as a biorecognition element was first given by Rechnitz in 1986. Using his idea, a new type of bioelectronic sensor is developed. For instance, the specialty of insects is their sense of smell. Insects are very selective and sensitive towards sensing different odors. If the insect antenna is coupled with a FET device, a new biosensor is formed called as beetle chip FET sensor. Such a sensor was first realized in 1997 [11]. Here, the voltage generated in the antenna was used for driving a FET device. The voltage in the antenna changes when it detects some odor and this voltage varies the drain current of the FET sensor. For this purpose, the FET device must be connected with the insect antenna forming an interface. This is done in two possible ways, either using the whole organism (say the whole beetle) or dip the tip of insect antenna in the electrolyte as shown in Fig. 6a with the reference electrode (say Pt wire) placed at an appropriate point on its body (generally between the neck and the head) or only using the isolated antenna placed on both sides of the plate with the electrolyte solution as shown in Fig. 6b.

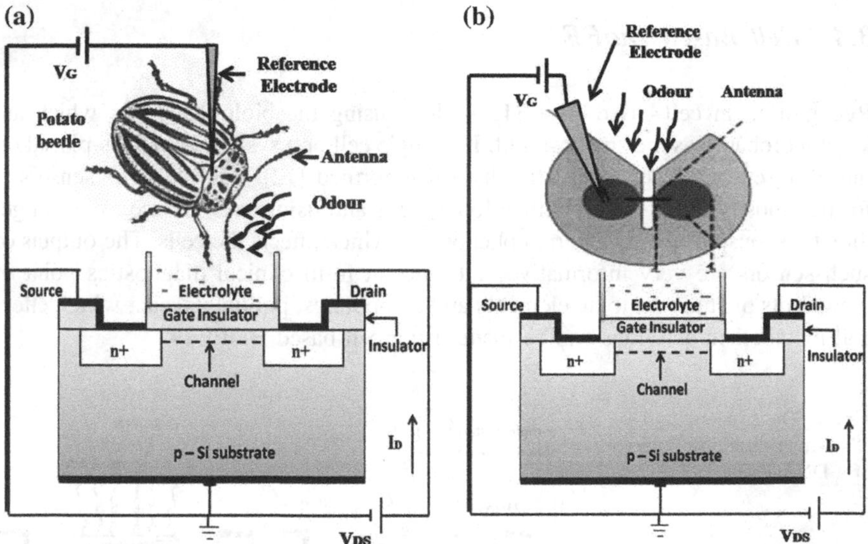


**Fig. 6** **a** BioFET consisting of whole beetle and **b** BioFET formed of isolated antenna

## 4 BioFETs Based on CNT Technology

Despite of many advantages, Si-based devices have lot of draw backs such as: technology used for fabrication is IC technology which requires huge number of sophisticated instruments and therefore cannot be easily set up in ordinary laboratory; have several junctions resulting into large internal contact resistance; high threshold voltage; small on–off current ratio; low sensitivity and scaling limitations for which cannot be easily pushed to nanotechnology domain.

Use of junctionless FET can overcome many of the limitations of conventional BioFET. Among the variety of concepts and different materials introduced to fabricate junctionless FET, use of CNT as transporting channel is one of the most attractive approach. It is because CNT has excellent properties including high carrier mobility (due to ballistic transport), high chemical stability and robustness, high ON/OFF current ratios, steep switching and compatibility with high dielectric materials [17, 18] facilitating device miniaturization. The biocompatibility and size compatibility make CNT suitable for biosensing applications. Moreover, instead of IC technology, in case of CNT, it is possible to use chemical solution technique (electrochemical and electrophoretic deposition, spin coating, dip coating techniques) for fabrication of BioFET devices. These techniques require minimal instrumentation, low cost and simple to use, low power, inherent miniaturization, portability and high degree of compatibility with advanced micromachining technologies. It is expected that in near future, the IC technology can be replaced by wet lab.

Figure 7 shows the complete schematic of one of the fabricated JLCNTFET device structure. It is a new concept with no PN or $N^{+}N$ or $P^{+}P$ junction in the

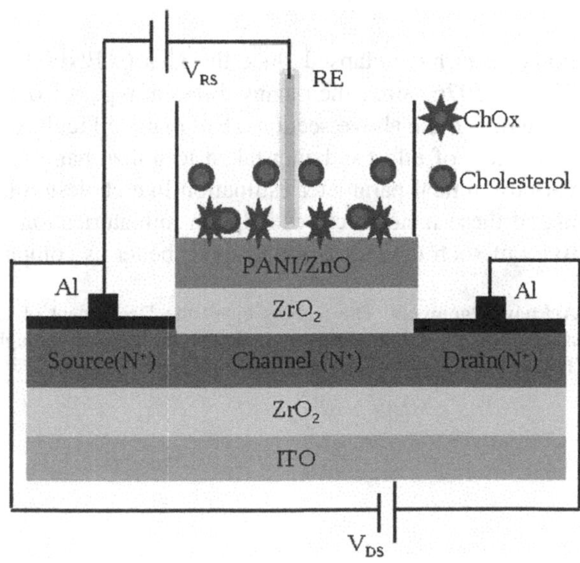


**Fig. 7** Schematic of junctionless CNTFET for detection of Cholesterol

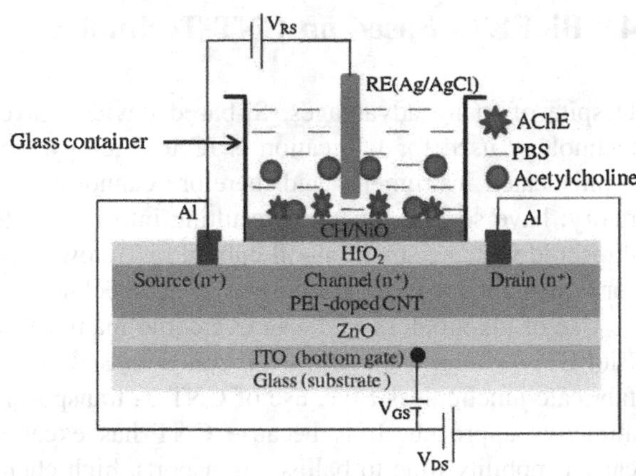


**Fig. 8** Schematic of junctionless dual-gated CNTFET for detection of Acetylcholine

source-channel-drain regions. It has either N-type or P-type material through the body just like a uniform resistor. The difference of work function between semiconductor channel and gate depletes the channel and the device behaves as OFF state. In the ON state, there is large drain current due to heavily doped charge carriers.

Dual-gated junctionless CNTFET provide better sensitivity than single gated [13]. At its off state, large work function difference between metals and semiconductor lowers off state drain current and thus, on–off current ratio becomes high. A dual-gated junctionless FET for detection of acetylcholine is shown in Fig. 8.

## 5 Conclusion

Forty years have elapsed since the idea of ENFET was first given by Janata and Moss in 1976. Since then many different types of BioFETs have been developed as discussed in the above sections. But many difficulties are still to be overcomed. For which lots of effort is being taken to utilize nano material based FET devices for various clinical parameter estimation like cholesterol, acetylcholine, urea, etc. The use of these nanostructures helps in miniaturization of devices and also the sensitivity of such devices is found to be better as compared to traditional FETs.

**Acknowledgements** The authors thank the Department of Electronics and Information Technology, MCIT, Government of India for the support through Visvesvaraya PhD scheme and Tezpur University for providing the laboratory facilities.

## References


1. Clark, L. C., Lyons, C.: Electrode systems for continuous monitoring in cardiovascular surgery. Ann N Y Acad Sci., 102, 29–45 (1962).
2. Bergveld, P.: Development of an ion-sensitive solid-state device for neurophysiological measurements. IEEE Trans. Biomed. Eng., BME–17(1), 70–71 (1970).
3. Dutta, J. C.: Ion Sensitive Field Effect Transistor for Applications in Bioelectronic Sensors: A Research Review. In: National Conference on Computational Intelligence and Signal Processing (CISP), IEEE, pp. 185–191, India (2012).
4. Bergveld, P.: Thirty years of ISFETOLOGY what happened in the past thirty years and what may happen in the next thirty years. Sensors and Actuators B, 88, 1–20 (2003).
5. Bousse, L., Rooij, N.F., Bergveld, P.: Operation of Chemically sensitive field effect sensors as a function of the properties of the insulator/electrolyte interface. IEEE Trans. Electron Devices, 30, 1263–1270 (1983).
6. Yates, D. E., Levine, S., Healy, T. W.: Site binding model of the electrical double layer at the oxide/water interface. J. Chem. Soc. Faraday Trans., 70, 1807 – 1819 (1974).
7. Schoning, M. J., Poghossian, A.: Recent advances in biologically sensitive field-effect transistors (BioFETs). Analyst, 127, 1137–1151 (2002).
8. Janata, J., Moss, S.: Chemically sensitive field effect transistors: Biomed. Eng., 6, 241–245 (1976).
9. Caras, S., Janata, J.: Field effect transistor sensitive to penicillin. Anal. Chem., 52, 1935–1937 (1980).
10. Schenck, J. F.: Technical difficulties remaining to the application of ISFET devices. in Theory, Design and Biomedical Applications of Solid State Chemical Sensors, ed. P. W. Cheung, CRC Press, Boca Raton, 165–173 (1978).
11. Schutz, S., et al.: Microscaled Living Bioelectronic Systems – Coupling Beetles to Silicon Transducers, Naturwissenschaften, 84, 86–88 (1997).
12. Romherz, P., et al.: A neuronsilicon junction: A retzius cell of the leech on insulated gate field effect transistor. Science, 252, 1290–1293(1991).
13. Barik, M. A., et al.: Carbon Nanotube-Based Dual-Gated Junctionless Field-Effect Transistor for Acetylcholine Detection. IEEE Sensors Journal,16 (2), 280–286, (2016).
14. Jobling, D. T., et al.: Active Microelectrode Array to record from The Mammalian Central Nervous System in Vitro. Med. Biol. Eng. Comput., 19, 553–560 (1981).
15. Souteyrand, E., et al.: Direct Detection of the Hybridization of Synthetic Homo-Oligomer DNA Sequences by Field Effect. J. Phys. Chem. B, 101, 2980–2985 (1997).
16. Tans, S. J., et al.: Room-temperature transistor based on a single carbon nanotube. Nature, 393, 49–52 (1998).
17. Javey, A., et al.: High-κ dielectrics for advanced carbon nanotube transistors and logic gates. Nat. Mater., 1, 241–246 (2002).
18. Barik M. A., Dutta, J. C.: Fabrication and characterization of junctionless carbon nanotube field effect transistor for cholesterol detection, App. Phys. Lett.,105, 053509(1–5) (2014).

# Domain Based Assessment of Users' Dependency on Search Engines

**Nidhi Bajpai and Deepak Arora**

**Abstract** A search engine plays a pivotal role in finding information for the purpose of education, business, and entertainment. A users' quest for any topic or browsing any website starts from a search engine, revolves around the search engine and in most of the cases ends with their preferred choice of search engine. The dependency of users upon search engine for their professional, personal, academic tasks appears to be ever increasing. This research work aims to determine users' dependency on search engine. The authors present a method to analyze and evaluate users' dependency on search engines based on an experiment which was conducted on working professionals employed in various domains like software companies, law firms, banks, educational institutes, government, etc.

**Keywords** Search engine · Dependency · Internet · Time dependency

## 1 Introduction

The influence of search engines on internet users is vast. A search engine is a primary tool in the hands of internet users to sift through vast amounts of information available on the internet. In this new age of information technology, user accesses the internet for diverse purposes ranging from professional to personal to academic queries.

Internet is, today, the fastest growing medium for dissemination of information. Information by itself is a very wide and varied term. In order to appreciate the true impact of the internet, it is necessary to study the dependency of users in relation to the productive aspects of the internet. Most internet users access its bountiful

N. Bajpai (✉) · D. Arora
Department of Computer Science & Engineering, Amity School of Engineering,
Amity University, Lucknow, India
e-mail: nidhibajpai07@gmail.com

D. Arora
e-mail: deepakarorainbox@gmail.com

 
S.K. Bhatia et al. (eds.), *Advances in Computer and Computational Sciences*,
Advances in Intelligent Systems and Computing 553,
DOI 10.1007/978-981-10-3770-2_66

potential relying on the power of search engines. Thus, a fair and critical assessment of the dependency of users upon search engines is a must for understanding the usage of internet by common users.

This study aims to determine the dependency upon search engines among working professionals in various domains, where access to the internet is fundamentally engrained in the job profile. When it comes to professionals working in organizations which provide full time access to the internet and search engines, the influence of search engine on the professionals is tremendous. Greater and more convenient access to the internet available in offices, homes, educational institutions, and via mobile devices provides a dependent, easy, and fast mode of information retrieval to users. Search engines have become a vital tool for the users to retrieve information for anything at anytime from anywhere. This ubiquitous nature of search engine is creating tremendous influence on users in variety of ways.

The authors have presented different parameters for determining users' dependency on search engines. These parameters are studied and analyzed and their domain wise analysis is performed. Different domains include software companies, law firms, banks, education, government offices, and 'others'.

## 2 Background Study

In 1998, when Google search engine was introduced, it used to serve 10000 queries per day [1] while today this number has increased to 3.5 billion searches per day and 1.2 trillion searches per year worldwide [2]. Lewandowski in his paper [3] has emphasized on the importance of search engines for acquiring knowledge. The dependency of users on search engines can be seen in various ways, for example, if in a conversation an internet user feels the urge to check on facts his first recourse is to a search engine, similarly, most users instead of remembering the full web address of website, they prefer to simply search the website via keyword using search engine and then visit the website with a click on search results. This paper also outlines the challenges that library and information science and information retrieval systems intent to face with the emergence of search engines. In the study [4], "Most Americans Suffer From 'Digital Amnesia'", it is stated that the dependency on search engines is so critical that it sometimes creates a situation of digital amnesia in which the users tend to forget information that they think they can easily found online by using web search engines [4]. Krieger and Lisa M. in their work [5] also explain the point of digital amnesia and growing dependency on search engines. Behrends, Shawn has done study about library versus search engines in their paper [6] and stated that library and search engines both are great source of information. Libraries have been the traditional gatekeeper of information while search engines are the vast ocean of information which is available and accessible to user all the time and everywhere. Search engines are fast, vast, and available as per users' need round the clock but the credibility of information available on internet. On the other hand, libraries are considered to be trustworthy but for the users of today's era of information technology, they have their own different ways to be sure about the credibility of the information.

## 3 Identification of Dependency Parameters and Experimental Setup

In order to determine dependency parameters, the authors conducted an online experiment on working professionals employed in different domains as indicated above. In this experiment, few questions related to search engine were presented to the participants. Based on the input received from the participants, the authors have determined different parameters based on which users' dependency on search engines can be established. These different parameters of users' dependency on search engine are discussed in the headings below:

### *3.1 Time Dependency*

Time dependency parameter will directly translate to users' dependency on search engines. Time parameter for dependency of users on search engine covers three aspects outlined as below:

- Number of hours spent by a user using search engines in a day
- Frequency of search engine usage (daily, weekly, monthly, or rarely)
- Number of years for which a user has been using search engine.

**Number of Hours Spent By a User Using Search Engine in a Day**. The authors analyzed the users for number of hours spent using search engine in a day. Table 1. shows the results for this aspect of dependency parameter. In order to calculate other descriptive statistic for this parameter, variable encoding as shown in Table 1 is done.

The average number of hours spent per day using search engine by a working professional is 2.5042 h as shown in Table 2. This calculation is based on variable encoding as shown in Table 1. This shows the dependency of users on search engines.

**Table 1** Hours spent by a user using search engines in a day

| Hours (h) spent in a day | Frequency | Percent | Variable encoding |
|---|---|---|---|
| Less than 1 h | 29 | 24.2 | 0.50 |
| 1–3 h | 49 | 40.8 | 2.00 |
| 3–5 h | 22 | 18.3 | 4.00 |
| More than 5 h | 20 | 16.7 | 5.00 |

**Table 2** Average number of hours spent by a user using search engine

| N | Range | Minimum | Maximum | Mean | Std. deviation | Variance |
|---|---|---|---|---|---|---|
| 120 | 4.50 | 0.50 | 5.00 | 2.5042 | 1.59502 | 2.544 |

**Table 3** Frequency of search engine usage

| | Frequency | Percentage | Variable encoding |
|---|---|---|---|
| Daily | 110 | 91.7 | 365 |
| Weekly | 10 | 8.3 | 52 |
| Monthly | 0 | 0 | 12 |
| Rarely | 0 | 0 | 0 |

**Table 4** Average number of days spent by a user using search engine in a year

| N | Range | Minimum | Maximum | Mean | Std. deviation | Variance |
|---|---|---|---|---|---|---|
| 120 | 313.00 | 52.00 | 365.00 | 338.9167 | 86.87135 | 7546.632 |

**Table 5** Number of years for which a user has been using search engine

| | Frequency | Percent | Variable encoding |
|---|---|---|---|
| Less than 1 year | 2 | 1.7 | 1 |
| Between 1 and 3 years | 7 | 5.8 | 2 |
| Between 3 and 5 years | 10 | 8.3 | 4 |
| Greater than 5 years | 101 | 84.2 | 5 |

**Frequency of Search Engine Usage**. It determines how often the users use the search engine, i.e., on daily basis, weekly basis, monthly basis, or rarely. The results for this aspect of dependency parameter are shown in Table 3. This statistic shows that 91.70% professionals use the search engines on daily basis and rest of the professionals use them on weekly basis. In order to calculate other descriptive statistic for this parameter, variable encoding as shown in Table 3 is done.

Based on variable encoding, the results for mean, variance, and standard deviation are shown in Table 4. It shows that average number of days spent by a user using search engine in a year is 338.91 days.

**Number of Years for which a User has been using Search Engine**. Search engines history dates back to the year 1982 when Whois search engine was introduced which was termed as debut of internet search engine [7]. In all these years, search engines have completely revolutionized the way users search. The authors analyzed the working professionals about number of years for which a user has been using search engine. The findings for this aspect are shown in Table 5. This statistic shows that 84.20% of working professionals have been using search engines for more than 5 years. In order to calculate other descriptive statistic for this parameter, variable encoding as shown in Table 5 is done.

Based on variable encoding, the results for mean, variance, and standard deviation are shown in Table 6. It shows that average number of years for which users have been using search engines is 4.675 years.

### 3.2 *Most Frequently Used Search Engine*

The results for this aspect are shown in Table 7. According to this research, 99.2% of users prefer Google as the most frequently used search engine as shown in Table 1. For many users the term Google is synonymous to search engine [8]. This shows great dependency of users on a single search engine.

### 3.3 *Second Choice of Search Engine*

During literature review phase, it was well stated that Google is the most preferred choice of users [9]. Hence, the authors asked the professionals that other than Google which is the second most frequently used search engine. The results for this aspect are shown in Table 8. With 41.7% of professionals saying that they use only Google as their search engine which shows great dependency of users on Google search engine and establishes that Google is an incredible force on the internet.

**Table 6** Number of years for which users have been using search engines

| N | Range | Minimum | Maximum | Mean | Std. deviation | Variance |
|---|---|---|---|---|---|---|
| 120 | 4.00 | 1.00 | 5.00 | 4.675 | 0.88082 | 0.776 |

**Table 7** Which search engine do you use most frequently?

| | Frequency | Percent |
|---|---|---|
| Google | 119 | 99.2 |
| Bing | 0 | 0 |
| Yahoo! search | 0 | 0 |
| Ask.com | 0 | 0 |
| AOL | 0 | 0 |
| Other | 1 | 0.8 |

**Table 8** Other than Google, which search engine do you use?

| | Frequency | Percent |
|---|---|---|
| Ask.com | 3 | 2.5 |
| Bing | 24 | 20.0 |
| Only Google | 50 | 41.7 |
| Other | 18 | 15.0 |
| Yahoo! | 25 | 20.8 |

**Table 9** Do you think that search engines have replaced use of library?

| | Frequency | Percent | Cumulative percent |
|---|---|---|---|
| No | 21 | 17.5 | 17.5 |
| Yes | 99 | 82.5 | 82.5 |
| Total | 120 | 100.0 | 100.0 |

**Table 10** What is your first choice to acquire knowledge for a particular domain?

| | Frequency | Percent | Cumulative percent |
|---|---|---|---|
| Ask your friends/relatives | 16 | 13.3 | 13.3 |
| Library access | 11 | 9.2 | 22.5 |
| Other media | 2 | 1.7 | 24.2 |
| Search engine | 91 | 75.8 | 100.0 |
| Total | 120 | 100.0 | |

### 3.4 Search Engine Versus Library

Library and search engines both are great sources of information. Libraries have been the traditional gatekeeper of information while search engines are the vast ocean of information which is available and accessible to user all the time and everywhere. In order to determine the increasing dependency of users on search engines for seeking information and the relevance of library for the user, the authors asked the users if they think that search engines have replaced library use. As shown in Table 9, 82.5% of working professionals in different domains feels that search engine has replaced libraries as a method to retrieve information.

### 3.5 Preferred Option to Get Information

The different modes through which a user can acquire information include accessing libraries, consulting friends, relatives, and colleagues, search engines and other media such as radio, magazines, television, newspapers, etc. The users were analyzed for their preferred option to get information. The results are shown in Table 10 which shows that search engines are the first choice to acquire knowledge for a particular domain and 75.80% of users prefer to use search engines as compared to other options for acquiring knowledge.

### 3.6 Complete Dependency on Search Engine

This statistic reveals the growing influence of users on search engines. The professionals were asked if their work is completely dependent on search engines for

**Table 11** Is your work completely dependent on search engines?

| | Frequency | Percent | Cumulative percent |
|---|---|---|---|
| No | 81 | 67.5 | 67.5 |
| Yes | 39 | 32.5 | 100.0 |
| Total | 120 | 100.0 | |

their professional expertise. The results are shown in Table 11. According to which 32.5% of users agree for complete dependency of their work on search engine.

# 4 Methodology Used

## *4.1 Stage I—Qualitative Research*

In this study, first the authors have used qualitative research methodology by doing detailed study about search engines. The exploratory research was done aimed at achieving better understanding of the topic. Literature review technique was used to gain better understanding of the topic which helped to frame research questions. This exploratory research step will aid in more in-depth analysis of the subject at later stages. This step is important to have better understanding of the topic, users' perspectives and future implications of the topic and provides better insight into the topic [10].

## *4.2 Stage II—Quantitative Research*

A quantitative research methodology was thereafter followed which used the online survey method. A questionnaire was designed using Google forms and circulated online to working professionals in different domains like Software Company, Government, Banking, Legal firms, Education, etc. The responses collected from the working professionals were analyzed using the software GNU PSPP software [10].

# 5 Results and Discussions

This research work is based on an experiment conducted online among 120 working professionals from different domains which include 55 software professionals, 13 law professionals, 13 government officials, 12 professionals working in education field, 11 professionals from different banks, and 16 professionals in 'other' domains. The sample includes 76 male professionals and 44 female professionals [10].

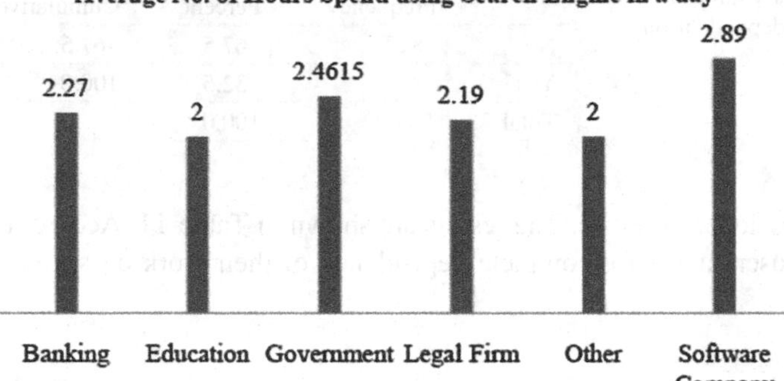


**Fig. 1** Average number of hours spent by a user using search engine in different domains

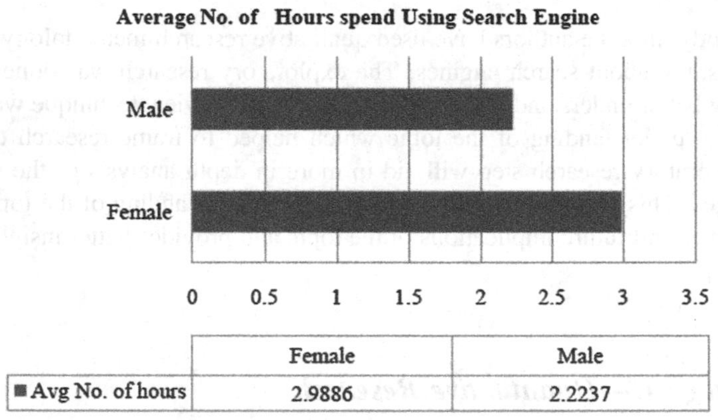


**Fig. 2** Figure showing difference between average number of hours spent using search engine by male users and female users

The result for time dependency parameter states that average number of hours spent by a user using search engine in a day is 2.5 h (Table 2). The analysis of this parameter shows that the users in software companies have maximum average number of hours spent by a user using search engine than any other domain, i.e., 2.89 h as shown in Fig. 1.

There is considerable difference in the time dependency parameter between male users and female users as shown in Fig. 2. The average number of hours spent using search engines by a female user is 2.98 h while for a male user is 2.22 h.

The analysis for time dependency parameter among different age groups shows that average number of hours spent using search engine in a day is maximum, i.e., 3.1 h for users belonging to the age group of 18–25 years. The value of parameter decreases with increasing age till 55 years as shown in Fig. 3. The value of parameter again increases for users above 55 years.

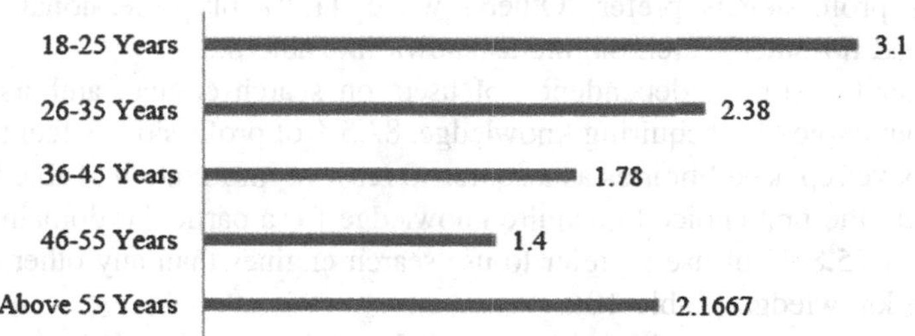


**Fig. 3** Figure showing difference between average number of hours spent using search engine by users in different age groups

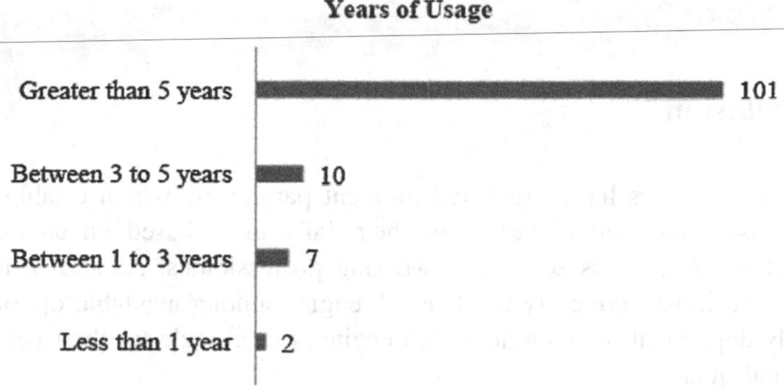


**Fig. 4** Figure showing number of years since when users have been using search engine

Analysis of time dependency parameter described above shows that search engine is used on a daily basis by 91.7% of internet users and remaining 8.3% of internet users use it on a weekly basis (Table 3). This analysis states that average number of days on which a user uses search engines in a year is 338.9 days (Table 4).

The authors state the need for knowing since when individual users have been using search engines. Average number of years since when individual users have been using search engines comes to 4.6 years (Table 6). The domain wise analysis of this parameter shows that average number of years since when users are using search engines in all domain remains approximately same, i.e., 4.6 years. There is no difference between male users and female users for this parameter. The results for this aspect are shown in Fig. 4 which states that 84.20% of users are using search engines for more than 5 years.

Among different search engines available in market today, 99.2% of users prefer Google as the most frequently used search engine as shown in Table 7. As the second choice for search engine after Google, 20.8% of professionals prefer Yahoo!

search, 20% of professionals prefer Bing, 2.5% of professionals prefer Ask.com, and 15% professionals prefer 'Others' while 41.7% of professionals use only Google and no other search engine as shown in Table 8.

In order to estimate dependency of users on search engine, authors analyzed users about choice for acquiring knowledge. 82.5% of professionals feel that search engines have replaced libraries as a source to retrieve information (Table 9). Search engines are the first choice to acquire knowledge for a particular domain. It is also shown that 75.80% of users prefer to use search engines than any other option for acquiring knowledge (Table 10).

In order to go deeper into this aspect, the authors analyzed the users further which established that 32.50% of working professionals said that their work is completely dependent on search engine (Table 11) and 85% of professionals want access to search engines in their organization and they are not fine if their organization blocks access to search engines.

## 6 Conclusion

In this study, authors have presented different parameters which establish dependency of users on search engine for their daily tasks based on an experiment performed across a cross section of working professionals. As shown in results, Google is the most frequently used search engine among available options. Users are greatly dependent on a single search engine, i.e., Google for their personal and professional queries.

Users spend a lot of time using search engine which comes to an average value of 2.5 h per day. The users in software companies tend to spend maximum amount of time, i.e., 2.89 h using search engine as compared to users in other domains. Female users spend average value of 2.98 h using search engine as compared to male users who use search engine for an average value of 2.22 h. The users in the age group of 18–25 years tend to spend maximum amount of time, i.e., 3.1 h using search engine. This great dependency of users on search engine has been since last 4.6 years.

This study also shows that search engines have replaced the use of library to a large extent and search engine is the first choice to acquire knowledge. 32.5% of users even agree about the complete dependence of their work on search engines.

All these results establish the increasing dependency of users on search engines. The results of this experiment will be very helpful for the upcoming researchers toward increasing the efficiency and opening new dimensions of different search algorithm/techniques, by adding more intelligence to upcoming browsers in the market. The future scope of this study includes conducting the same experiment on larger sample size by including different domains for further analysis and domain specific understanding.

## References

1. Battelle, J.: The search. Portfolio, New York (2005).
2. google-search-statistics, http://www.internetlivestats.com/google-search-statistics/.
3. Lewandowski., Dirk. New Perspectives On Web Search Engine Research. UK: Emerald Publishing, 2012.
4. "Study: Most Americans Suffer From 'Digital Amnesia'" http://wtop.com/health/2015/07/study-most-americans-suffer-from-digital-amnesia/. N.p.,2015. Web. 5 Mar. 2016.
5. Krieger, Lisa M. "Google Is Changing Your Brain, Study Says, And Don't You Forget It". N. p., 2011. Web. 5 Mar. 2016.
6. Behrends, Shawn. "Libraries Vs. Google In The 21St Century". Idaho Librarian 62.2 (2012): 13.
7. "RFC 812 - NICNAME/WHOIS". Tools.ietf.org. N.p., 1982. Web.
8. Levene, Mark. An Introduction To Search Engines And Web Navigation. 2nd ed. Hoboken, New Jersey: John Wiley & Sons, Inc, 2010.
9. "Google Search Statistics - Internet Live Stats". Internetlivestats.com. N.p., 2016. Web. 14 Apr. 2016.
10. Bajpai, N. and Arora, D.: An estimation of user preferences for search engine results and usage. Advances in Intelligent Systems and computing series Springer, n.d. (in-press).

# Literature Review on Knowledge Harvesting and Management System

**Shachi Pathak and Shalini Nigam**

**Abstract** The present study is focusing on the characteristics of knowledge harvesting continuity management, and knowledge management. This begins from reviewing the previous research papers on knowledge management and information systems following with some recommendations to an ideal knowledge management system. This paper worked on an organized analysis of 32 empirical refereed articles on knowledge harvesting, continuity management, and knowledge management. Findings of the study state that an organization must integrate its knowledge management system, knowledge harvesting, and management functioning to decrease the loss of knowledge at the time of knowledge worker leaving the organization.

**Keywords** Knowledge harvesting · Knowledge management system · Knowledge harvesting process

## 1 Introduction

As the baby boomers approaching toward retirement, domestic, and international business organizations are facing challenge of future knowledge loss. Organization's most experienced workers with their critical expertise will soon say goodbye and their replacement will be tough if not impossible [1]. Firms accept that their productivity and success may be badly affected by this loss of key knowledge. Yet most organizations are not properly extracting and preserving this knowledge before their key employees head for the door. The Bureau of Labor Statistics of U. S. proposes that the consequence of current workforce retirements will have 60% of job openings in 2010-2020, resulting in a major loss of experienced employees [2].

S. Pathak (✉) · S. Nigam
Department of Management, Dayalbagh Educational Institute,
Dayalbagh, Agra, India
e-mail: mspathak31@gmail.com



S.K. Bhatia et al. (eds.), *Advances in Computer and Computational Sciences*,
Advances in Intelligent Systems and Computing 553,
DOI 10.1007/978-981-10-3770-2_67

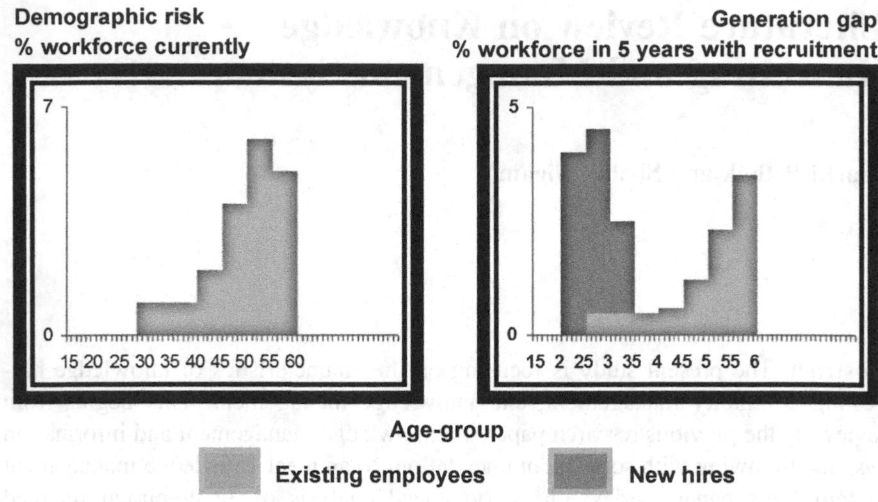


Fig. 1 Yawning generation gap. *Sources* BCG analysis

In India, only 12% of a swelling workforce will be above 60 which include technical, scientific, and professional people by 2030 [3].

According to a survey and report presented by The Boston Consulting Group (BCG), Indian Banks' Association (IBA) and Federation of Indian Chambers of Commerce and Industry (FICCI), in next coming 10 years 50% of junior and 80% of intermediate managers will no more be with organizations because of retirement. Critical capabilities and loss of knowledge will be due to the leave-taking officers. This is attached with necessity for big level skill regaining, inviting and maintaining new workforce, monitoring increasing costs of workers, and building performance control are the major problems. Other side of the problem is shown in Fig. 1 [4].

As experts and long-standing executives retire, they take with them a breadth of experience, a wealth of organizational knowledge, and the kind of linkage that helps work getting done most efficiently. This loss of knowledge and wide experience may cause weak organization with undefined future and performance decreases because valued information has to be reattained. Valuable time is again invested to restoring products or processes that already exist. Learning from past experience disappears as failure to hold 'knowledge learned' from critical projects, most of whose consequences have never been documented. The knowledge concerning key inspirations' about 'how things get done in this organization' evaporates. Therefore, many organizations have started focusing to preserve this critical knowledge before they go.

Based on a survey of 171 central managers, 78% of managers reported being worried about their organization's existing knowledge management (KM) practices. According to the McKinsey Quarterly survey, people between 52 to 70 years of age (baby boomers) are most—knowledgeable, capable experienced personnel in the United State's past [5].

Today in most organizations and industries, experienced employees are retiring at an exceptional rate. Younger employees job-hop to somewhere new. In both cases, the current employees to the organization's success are knowledge important, resulting 'brain-drain' in an organization. Consequently, organizational knowledge is washed out and newly entered employees are not prepared or capable to perform at the levels developed by their more knowledgeable and experienced precursors. Though companies cannot limit qualified officials from retiring. Neither can bound younger employees' aspiration to search for career opportunities or alter their job situations. However what they can do is to recognize, reproduce and make practical and effective use of the essential information which can guide to improved business outcomes [2].

Knowledge harvesting (KH) is a solution to the threat of loss of critical knowledge when senior employees leave the organization. KH helps to hold up the expertise of experts and experienced employee and use it when needed. This knowledge can be reused in many circumstances such as: To seizure expertise of people those are quitting their jobs, when a company wants to 'know about its knowledge' and expertise required for a particular project or work, to support on a process of modification or development, and to encourage a KM program by speedily creating an expert and accepted ontology with reference to a matter and providing access within the organization [6].

Before struggling to create a blueprint and build up the solution personalized for a particular company, it is necessary to inquire about the characteristics and system's required features. The paper represents the characteristics of KH, CM, & KM which begins with reviewing the previous research papers on knowledge management and information systems following with some recommendations to an ideal knowledge management system.

## 2 Literature Review

### 2.1 Knowledge

Knowledge can be defined as information that alters something or someone [7]. Knowledge is the result of information. After its evaluation, processing and situated in a situation, turns into knowledge. Knowledge can be defined as information possessed in the minds of individual [8, 9].

### 2.2 Knowledge Management (KM)

It is concerning to the collective knowledge of the total staff to reach definite organizational goals. The purpose of KM is not to manage all knowledge, just to manage the knowledge that is crucial to the organization [10]. Distributing "the right knowledge to the right person at the right time" is KM [11]. KM is a logical process of obtaining, organizing, and transferring knowledge to members of organization to be more effective and productive by using this knowledge [8].

KM usually emphasizes on tacit and explicit knowledge, explicit knowledge is seized in the form of documents whereas tacit knowledge is implied to individuals, based on experience and so goes on in their heads. Hence, in nature, it is personal, particular, and experimental. Thus, it is fast changing as compare to knowledge which is documented [12].

## 2.3 *Continuity Management (CM)*

At the time of interruption, the skill to confirm continuity is the outcome of CM. Basically, ensuring non- interrupted operations in a unit comprises of evaluating present level of ongoing processes and resources required to uphold operations, finding the events that may cause a threat, its impact as well as the extent of risk, willingness of the organization to recognize any hazardous acts. The outcome of all of above-written evaluation is a planning of CM which deals with necessary resources and processes required for restoring non-interruption if noncontinuity occurs. According to a researcher, CM can be implemented nevertheless of a threatening act. Any disruption can be piloted [13], operations stay steady [14] and risks can be managed [15]. Which mean, proper CM knowledge is capable enough to resist the noncontinuity [16] with diminishing the bad affect of noncontinuity on the particular unit or department.

CM as holistic course of action should involve all the parts of the organizations such as all resources, operations of the organization, interactions with other organizations [17, 18], the part investors play in attaining organizational objectives [19].

As a preservative process continuity management should be capable to defend organization's ability to meet its purpose, goals, customer support, and manage profits [13]. This can be attained by preserving knowledge related to current risks and the probable disruption as well as past interruptions, resources, and operations [19]. As a continuous process it should be joined with organization administration [17].

## 2.4 *Knowledge Harvesting (KH)*

Organizations capture knowledge and tossed it into some file and forget. One way to be sure that this crucial and expensive know-how should be beneficial for the staff, which is required for them, is to engage them with KH that is a logical, made easy collecting and transmission of expertise. This approach conducted by Intel's IT consulting arm, Intel Solution Services (ISS) helped to accelerate gather and communication, and has enhanced its possibility that this know-how creatively and innovatively used again. Before the harvest instigates it is important to recognize knowledge seekers, who can get benefit of this information, and engage these employees in collecting expensive learning [20].

KM has become an essential tool for several organizations and a key element of KM is KH. The company LearnerFirst in 1992 originated the methodology of KH and marketed money-making new product with the expertise of knowledgeable workers and packed it to application of practical learning. They sold KH applications to over 4000 organizations. In 1995, they started production of customaries educational systems and its applications for big companies and followed with producing comprehensive manuscript based on converting implicit knowledge to explicit. In 1998, they started mentoring KM proficient workers for pioneering organizations. Finally in 2000, they internationally revealed KH methods through trained advisors than in 2004 LearnersFirst incorporated KH with project based, expertise preservation programs. In general, customer assignments were based on strategic problems of structural design of know-how, knowledge collection and use it again, combined ideas or product R&D after it in 2009 they again introduced it as knowledge harvesting lite. Now KH is a registered Trademark of the company "Knowledge Harvesting, Inc" [21].

KH is the complete knowledge retention methodology that supports implicit knowledge to be uttered and turned into knowledge assets that help an organization to develop and improve [21, 22]. Whereas Keyes argued, KH should include a range of procedure which emphasis on repossess information resources from several types of tacit knowledge present in a company [23]. On the contrary, others argued that it is an integration of different practices wherever implicit know-how of a subject can be transformed to explicit, actable knowledge converted to learners with the help of technology and personal communication that can be used to intensely gain competitiveness, improve corporate performance and valuation [24]. In addition, the organization is sheltered from costly workforce risks and defects, from the inaccessibility of knowledge wherever it is required [21].

Mostly, as one hears of losing key employees and try to gather their knowledge, they start using techniques like exit interviews. Suddenly trying to capture this knowledge just before a staff member leaves is possibly a flawed process. The moment a staff member recognizes he is being tapped for information, he will probably repel and try to sabotage the process. In reality, knowledge preservation should be combined with how the organization drives and starts before crucial employee leaves as it is important for long-term organizational success [22].

Larry Todd Wilson is the founder and CTO of Knowledge Harvesting, Inc. (formerly LearnerFirst, Inc.), in Brimingham, Ala, explains his knowledge gathering approach, followed by many organizations and researchers, so that information can be used for getting competitive advantage [21, 25]. Methodology involves seven modules (Fig. 2) [26] which can be repeated several times in a course of a project and each module is a complete learning program to put into practice KH with any company:

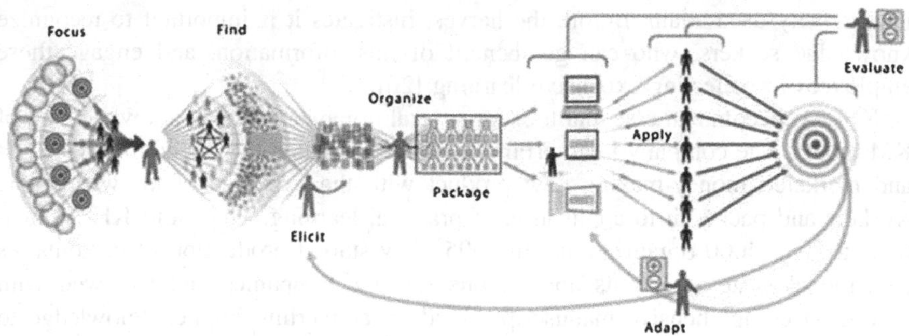


**Fig. 2** Process of knowledge harvesting. *Source* Knowledge Harvesting Inc. (2014)

i. ***Focus*** is used to recognize essential knowledge which may be very critical and significant to gather.
ii. ***Find*** gives direction to locate proficient and accessible, valuable credentials, and knowledge.
iii. ***Elicit*** explains to perform useful discussion meetings.
iv. ***Organize*** educates to develop a good judgment about the knowledge gathered during discursions and credentials. At the point, categorize blueprint and arrange information to structural sets of maintain information and directions.
v. ***Package*** is to find out most relevant medium to make a package of usable and transferable information to other people. Establish the finest approach to relate this knowledge.
vi. ***Evaluate*** offers techniques and assistance to measure the usefulness of the knowledge resources.
vii. ***Adapt*** offers techniques and assistance to become familiarized with the outputs for the benefit of the talented requirement of a large number of information seekers.

The emphasis should be on the process that enhances knowledge: the capacity for effective performance. Technology should play a critical role in the achievement of knowledge management [27–31].

## 2.5 KH and Management System

In the course of the above analysis, lots of characteristics of CM and KH are identified that should be assimilated into a system [32].

First, it must focus on workforce and plan to help client interaction. The system must be able to sustain client interface for navigation so that he can access and retrive the available knowledge and maintaine it. A persone who is new for the system and not familier with the exixting one can also easily access it.

Second, it must be capable of extracting the information. For making knowledge available in future to use stimulating improve its expression. This must contain query which can fortify this recall of information, make the most of easy language and vocabulary, and allow knowledge to appear. The sequence of questions should be general-to-specific.

Third, the system should have broad scope about the knowledge harvested. Knowledge harvesting should begin as the employee joins the organization because separation of an employee occurs very often in the organization. Employee may leave the organization without warning that is why the system should allow preserving as much as possible work-related knowledge from employee. As employee makes his mind to get separated from organization, the interest of participation reduces and so as the sharing of knowledge. However it is advised to include people with knowledge in the process or experts.

Fourth, it needs to be holistic in nature. The system should include all subunits, units, resources, or inputs used, the operations executed and the outputs created as well as duties of affected persons, current threats, and potential interruptions. As continuity means continuation over time, this should involve information related to previously held, currently ongoing, and upcoming formation and purpose. A company must manuscript or text all previously held disturbances to any part of organization, effect of all disturbances and actions taken to regain non-interruptions.

Additionally, this system must gather and accumulate information in support of upcoming events of the company. The integration of a system with administrative functions would completely maintain the non-interruption within the organization. This strategic placement of a system in the organization shows the importance of continuity to whole organization, consolidates the process of knowledge harvesting, and supports the continuance of a system with the intention of current and relevant information.

Next, the system should be responsive. The response should be accurate and appropriate because it maintains collected information that will support the organizational continuity and knowledge harvesting process. The system and its content should be accessible to the employees at any time and location. The employee can access the knowledge according to his convenience from the system. The system should create relationships among parts of information to manage the encoded and stored knowledge that supports to filter the knowledge related to requirements presented through the client. Finally, this system must be able to share out this knowledge and with the use of all above characteristics information in the system would be updated.

## 3 Conclusion and Future Scope

Performance of organizations and today's economy are mostly related to information, the leaving staff can going to impact upon company's capability of being information producer. The threat of losing the expertise due to top performers' disconnection is never-ending and will continue till the absence of its true solutions. The major consequences of present demographic improvement give birth to some challenges to any institutions and top management: losing the expert's speciality due to no proper use of know-how of experienced workforce. In support of literature review in this present study, the below written challenges must be taking into account for future work:

1. To determine and compare the difference between the contribution of aged or young workforce for the processing of producing the information. To specify competitive advantage of expert knowledgeable employee.
2. To recognize the implementation of a particular competency of a aged employee for developing Human Resources policies and professional skill enhancement programs.
3. To get benefited by the linkage between two generations to invent new knowledge methods and to encourage learning of new generation from old one.
4. To motivate sharing of information from aged to new workforce

A response needs to be facilitated by KH and KM, which collects and accumulates information, when employee leaves the organization and also affects non-interrupted production within a unit.

These characteristics recognized here in the present study are related to the literature review and may give much information an organization should be considered when creating an in-house solution. In a system, CM, KH, and KM must have integration with middle management of the organization and work together to combine better working of all subunits. This type of information system must have the capability to address all necessary requirement of all the subunits to accumulate this cropped information and on the same time guide to construct a past record of the organization, article the present principles and resources to produce, and assemble all the know-how which is useful in enhancing operating functions or productivity and efficiency of presently employed workforce for better future and growth.

## References

1. Seidman, W. & McCauley, M., 2005. Saving Retiring Knowledge Workers' "Secret Sauce". *Performance Improvement,* 44(8), pp. 34–38.
2. Minnick, D. & Stixrud, E., 2013. *Capturing Organizational Insight: Knowledge Harvesting & Management.* [Online] Available at: http://www.ysc.com/our-thinking/article/the-foresight-to-capture-organizational-insight [Accessed 6 1 2016].
3. International Center for Peace and Development, 2010. *Employment Trends in the 21st Century.* [Online] Available at: http://www.icpd.org/employment/Empltrends21century.htm [Accessed 26 12 2015].
4. Shah, A. et al., 2010. *Indian Banking 2020 Making the Decade's Promise Come True,* Mumbai: The Boston Consulting Group, Inc.

5. Schings, S., 2016. *Capturing the Knowledge: Federal Agencies Work to Retain Baby Boomers' Wisdom.* [Online] Available at: http://www.siop.org/Media/News/capturing.aspx [Accessed 12 2 2016].
6. Giniwala, N. et al., 2010. *knowledge harvesting.* [Online] Available at: http://knowledgeharvesting1.wikidot.com/knowledge-harvesting [Accessed 20 12 2016].
7. Hasanzadeh, S., Sarkari, M. & Hasiri, A., 2014. Evaluation of Organizational Factors Affecting Knowledge Sharing in Work Teams. *Indian Journal of Science and Research,* 4(6), pp. 259–263.
8. Alavi, M. & Leidner, D. E., 2001. Review: Knowledge Management and Knowledge Management Systems: Conceptual Foundations and Research Issues. *MIS Quarterly,* 25(1), pp. 107–136.
9. Islam, M. S. & Khan, R. H., 2014. Exploring the Factors Affecting Knowledge Sharing Practices in Dhaka University Library. *Library Philosophy and Practice (e-journal),* pp. 1–11.
10. Servin, G., 2005. *ABC of Knowledge Management.* [Online] Available at: Servin, G. (2005). ABC of Kn http://www.fao.org/fileadmin/user_upload/knowledge/docs/ABC_of_KM.pdf [Accessed 5 1 2016].
11. Beazley, H., Boenisch, J. & Harden, D., 2003. Knowledge continuity: The new management function. *Journal of Organizational Excellence,* 22(3), pp. 65–81.
12. DEAN, G., 2009. *Organisational Knowledge: Capturing its diversity and sharing its power,* Australia: Australian Emergency Management Knowledge Hub.
13. Hiles, A., 2010. *The definitive handbook of business continuity management.* 3rd ed. s.l.: A John Wiley & Sons, Ltd., Publication.
14. Cline, M., 2007. *Institutions of higher education continuity of operations (COOP) planning manua,* Virginia: Richmond: Commonwealth of Virginia.
15. Gibb, F. & Buchanan, S., 2006. A framework for business continuity management. *International Journal of Information Management,* 26(2), pp. 128–141.
16. Bertalanffy, L., 1972. The history and status of general systems theory. *The Academy of Management Journal,* 15(4), pp. 407–426.
17. McCrackan, A., 2005. *Practical guide to business continuity assurance.* Boston: Artech House, Inc.
18. Vancoppenolle, G., 1999. What are we planning for?. In: *In The definitive handbook of business continuity management.* New York: John Wiley & Sons, Ltd., pp. 3–23.
19. Elliott, D., Swartz, E. & Herbane, B., 2010. *Business Continuity Management 2e: A Crisis Management Approach.* 2e ed. New York: Routledge.
20. Pugh, K. & Dixon, N. M., 2008. Don't Just Capture Knowledge– Put It to Work. *Harvard Business Review*, 28 May, pp. 90–92.
21. Knowledge Harvesting In., 2011. *History.* [Online] Available at: http://knowledgeharvesting.com/Methodology.html [Accessed 31 Oct 2015].
22. Liebowitz, J., 2010. Strategic intelligence, social networking and knowledge retention. *IEEE Computer Society,* 43(2), pp. 87–89.
23. Keyes, J., 2006. *Knowledge Management, Business Intelligence, and Content Management: The IT Practitioner's Guide.* MA, USA: Auerbach Publications Boston.
24. Liu, C., Rama, D. & Becerra-Fernandez, I., 2007. *The Proposal of Conditions of Personal Engagement in Knowledge Harvesting.* Hong Kong, IEEE Computer Society.
25. Eisenhart, M., 2001. Gathering Knowledge While it's Ripe. *Knowledge Management*, April, p. 48.
26. Knowledge Harvesting In., 2011. *Methodology.* [Online] Available at: http://knowledgeharvesting.com/Methodology.html [Accessed 31 Oct 2015].
27. Frappaolo, C. & Wilson, L. T., 2000. *Implicit Knowledge Management The New Frontier of Corporate Capability.* [Online].
28. Hanson, K. T. & Kararach, G., 2011. *The Challenges of Knowledge Harvesting and the Promotion of Sustainable Development for the Achievement of the MDGs in Africa,* Africa: African Capacity Building Foundation.

29. North, K. & Kumta, G., 2014. Knowledge in Organisations. In: *Knowledge Management.* Switzerland: Springer International Publishing, pp. 31–61.
30. Serrat, O., 2010. *Harvesting Knowledge.* [Online] Available at: http://adb.org/sites/default/files/pub/2010/harvesting-knowledge.pdf [Accessed 5 April 2016].
31. *Case Study: Knowledge Harvesting During the Big Crew Change* (2016) Jeffrey E. Stemke; Larry Todd Wilson.
32. Pierson, M. E. & Tech, V., 2013. Characteristics of a Knowledge Harvesting and Management System. *Capturing What Employees Know: TechTrends,* 57(2), pp. 26–32.

# Adaptive Network Based Fuzzy Inference System for Early Diagnosis of Dengue Disease

**Darshana Saikia and Jiten Chandra Dutta**

**Abstract** There is always an increasing demand for the development of new soft computing technologies for medical diagnosis in regular clinical use. With the advent of soft computing technologies, the use of intelligent methods and algorithms provides a viable alternative for vague, uncertain and complex real life problems such as diagnosis of diseases, for which mathematical model is not available. In this work, a hybrid artificial intelligence system namely Adaptive Neuro-Fuzzy Inference System (ANFIS) based model is developed for early diagnosis of dengue disease. Dengue fever, caused by the dengue virus is an infectious tropical disease. Dengue disease has been considered as a fatal disease and delay in diagnosis may increase its severity as well as life risk of the patients. The signs and symptoms of early dengue disease are nonspecific and overlap with the other infectious diseases. So, the principal aim of this study was to develop an acceptable diagnostic system for early diagnosis of dengue disease.

**Keywords** Dengue disease • Artificial intelligence • Fuzzy logic • Neural network • ANFIS

## 1 Introduction

Dengue fever is a viral infection, known as break bone fever caused by Aedes aegypti and Aedes albopictus, the female mosquitoes. In recent years, dengue fever has been considered as a vital public health problem and widespread in many tropical and subtropical regions around the world [1–3].

D. Saikia (✉) · J.C. Dutta
Department of Electronics and Communication Engineering, Tezpur University, Napaam, Tezpur 784028, Assam, India
e-mail: darshanasaikia316@gmail.com

J.C. Dutta
e-mail: jitend@tezu.ernet.in

 
S.K. Bhatia et al. (eds.), *Advances in Computer and Computational Sciences*, Advances in Intelligent Systems and Computing 553,
DOI 10.1007/978-981-10-3770-2_68

The world health organization (WHO) set up a goal to reduce the cases of global dengue mortality by 50% by 2020. This aim can be achieved by improving the diagnosis and dengue patient management systems in early stage. It is because, early diagnosis of dengue disease helps the healthcare professional to take better care for dengue patients and reduced the risk level of dengue disease. A range of laboratory diagnostic methods including detection of virus, viral nucleic acid, antigens or antibodies, or a combination of these techniques have been developed to support patient management and disease control [1]. However, till date there are numerous challenges for busy primary care clinicians in making a diagnosis of dengue within the first few days of illness and also in rural areas (or) remote areas, where medical experts are not easily available. The diagnostic performance of early dengue depends on nonspecific signs and symptoms and it is difficult to differentiate these from other infectious diseases and also confirmatory diagnosis tests are not available in all clinical care settings due to greater cost. Because of these, early diagnosis of dengue disease is difficult and delay of diagnosis may increase the severity of disease at high risk level.

To address the above-mentioned challenges, soft computing techniques with artificial intelligent methods and algorithms have been used to diagnose the dengue disease in early stage. Research in the field of artificial intelligence (AI) gives promising research results in the area of medical decision-making, especially in last few years. With the advent of soft computer technology, fuzzy set theory and neural networks become important and prime techniques for solving complex and uncertain problems such as diagnosis of diseases. Both have certain advantages over conventional approaches, in medical diagnosis, especially when the cases involving vague data or prior knowledge are concerned. However as individual models, their applicability suffered from several weaknesses. Certain limitations of the individual models introduced the concept of hybrid intelligence in medical diagnosis. A hybrid intelligent system is obtained from neuro-fuzzy hybridization that synergizes these two techniques by combining the human-like reasoning style of fuzzy systems with the learning and connectionist structure of neural networks [4]. One type of neuro-fuzzy classifier is the Adaptive Neuro-Fuzzy Inference System (ANFIS), which is the recent advance in the field of artificial intelligence systems, have facilitated a wider usage of soft computing techniques, in medical diagnosis. Takagi-Sugeno type FIS is used in ANFIS and this fuzzy inference system was developed in 1993 by Shing and Jang [5].

From the literature investigation, diverse types of studies based on artificial intelligence (AI) techniques have been viewed. Guillaume investigated an interpretability-oriented review on developing Fuzzy Inference Systems using data [6]. Lim et al. showed an approach using neuro-fuzzy method for diagnosis of antibody deficiency syndrome [7]. Furthermore, Singh et al. worked on diagnosis of arthritis through Fuzzy Inference System [8]. Ziasabounchi and Askerzade employed a model using ANFIS for heart disease prediction [9]. Some of the studies related to dengue disease are listed as follow: Faisal et al. developed ANFIS model for classification of dengue patients based on their risk level [10]. On the other hands Afan et al. worked on computational intelligence method that is fuzzy on android-based mobile device for diagnosis of dengue haemorrhagic fever in early stage of illness [11].

# 2 Methodology

This study used patients' dataset recorded by Ref. [12] from the clinical study in Southern Vietnam. At first, the patients' baseline characteristics of clinical and simple laboratory features were stratified based on their range for laboratory confirmed dengue and non-dengue cases of participants. Then, the dataset was prepared for training and testing the ANFIS model. Finally, the ANFIS model for early dengue diagnosis was developed employing MATLAB software package (Math-Works, 2013) [13].

## 2.1 Data Set

The information related to dengue disease was acquired from expert's knowledge and various study about dengue disease [1–3]. The dataset was collected from the data record center of Guwahati Medical College and Hospital (GMCH) and from the clinical based study from 2010 to 2012 in Southern Vietnam [12]. Basically the data is the demographic, clinical signs, and symptoms of patients with some medical tests report. The dataset obtained from Ref. [12], from which 405 patients were analyzed. Out of 405 cases, 102 cases are of dengue positive and 303 cases are of dengue negative. By consulting doctor and analyzing the data of patients from several hospitals based studies, eleven sensitive symptoms were finalized as the inputs for diagnosis of the dengue disease in early stage. The eleven symptoms which are relevant and most significant for early diagnosis of dengue disease, used as inputs for this developed ANFIS model are shown in Table 1.

**Table 1** Information about input variables

| History and physical symptoms | Laboratory findings |
|---|---|
| • Age of the patient<br>• Day of illness (Day1/Day2/Day3)<br>• Vomiting ('1' for yes and '2' for no)<br>• Abdominal pain ('1' for yes and '2' for no)<br>• Mucosal bleeding ('1' for yes and '2' for no)<br>• Skin rash ('1' for yes and '2' for no)<br>• Temperature (°C) | • White blood cell count (WBC)<br>• Hematocrit (HCT)<br>• Platelet count (PLT)<br>• Aspartate aminotransferase (AST) |

## 2.2 ANFIS Model for Dengue Disease Classification

The ANFIS architecture is shown in Fig. 1. There are five individual layers of the ANFIS system which are described in [14]. Two steps are employed for developing the model: defining the initial ANFIS model architecture is the first step, which requires detailing of model's parameters such as the number of inputs and outputs membership' function and the number of rules. The subtractive clustering technique of ANFIS has been carried out for developing the model. The optimal number of membership functions and fuzzy rules for the initial ANFIS model was determined by employing the algorithm of subtractive clustering. In this technique, the number of the clusters was optimized based on the variation of clustering radius. Training the initial model uses the patients' data and modifies the membership function parameters of initial model so that the error due to the differences between the actual output and the output obtained from the model are minimized is the second step. For this task, hybrid learning algorithm was utilized [14]. To determine the optimal performance of the model, the parameter, i.e., number of iterations was investigated.

Finally, ROC curves have been computed for evaluating the optimal ANFIS model's performance. From the ROC curve, it is possible to determine the optimal threshold value for the model.

### 2.2.1 Optimization and Evaluation of Parameters

Optimization of parameters was carried out experimentally. This optimization technique was processed by randomly splitting the datasets into two sets: training set and testing set. A fivefold cross-validation technique was employed to minimize

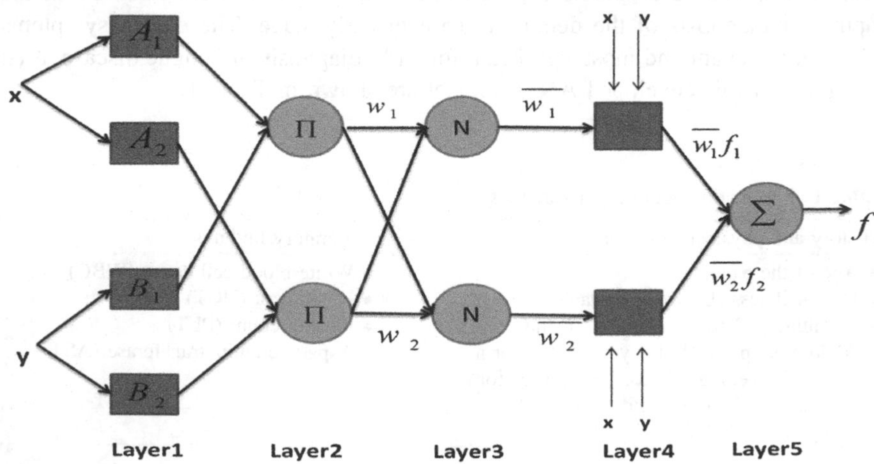


**Fig. 1** ANFIS architecture

the bias associated with the random sampling of data (training and testing) [15]. The fivefold cross validation (CV) was implemented by dividing the whole dataset into five sets. Out of five sets, four sets (80%) were used as training sets and remaining set (20%) was used as testing set. The training process was repeated for five times in this case of fivefold cross validation technique, at each time one set from the five sets was used as training set.

For optimal parameters, the average root mean square error (RMS) of training data and average diagnosis accuracy of the testing data were employed as selection criteria.

## 3 Results and Discussions

### 3.1 *Optimization of ANFIS Model*

The results for optimizing the ANFIS model based on subtractive clustering algorithm are shown in Figs. 2 and 3. Optimization of the clustering radius is

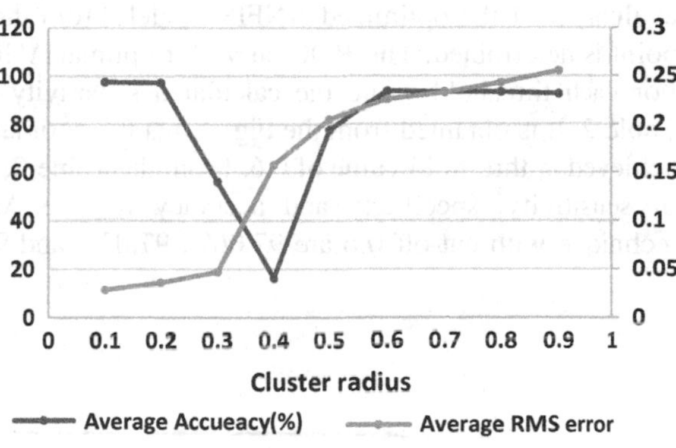


**Fig. 2** Average diagnostic accuracy and RMS with the clustering radius variation

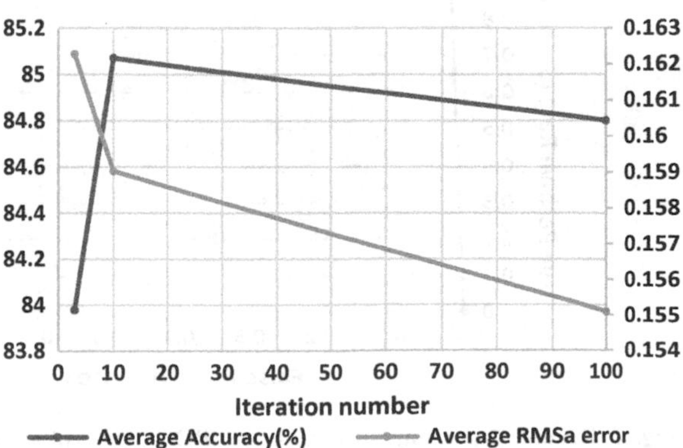


**Fig. 3** Average diagnostic accuracy and RMS with varying the iterations number

shown in Fig. 2 and it shows that highest diagnostic accuracy of 97.52% and lowest RMS value of 0.028 is produced by the cluster radius value of 0.1. Figure 3 shows the ANFIS model performances as a function of iteration number. Due to the highest average accuracy of 85.07% with lower RMS of 0.15902, ten iteration numbers was considered to be the optimal number.

### 3.2 Testing of ANFIS Model

The effectiveness of ANFIS model was determined by providing the dataset of 405 patients to the optimized ANFIS model and the result of each patient obtained from the proposed model is compared with the results of laboratory confirmation tests for dengue disease of the same patient.

### 3.3 Evaluation of the ANFIS Model Performance

Finally, computations of ROC curves were accomplished for the performance evaluation of the optimized ANFIS model. From the ROC curve, the best cut-off point is determined. The ROC curve for optimal ANFIS model is shown in Fig. 4. For each threshold value the calculated sensitivity and specificity are shown in Table 2. It is obtained from the Fig. 4 that the optimal performance for the model is achieved at threshold value of 0.6. From the Table 2, it is determined that the value of sensitivity, specificity and accuracy for the ANFIS-based fuzzy clustering technique with cut-off 0.6 are 97.06%, 97.03% and 97.04% respectively.

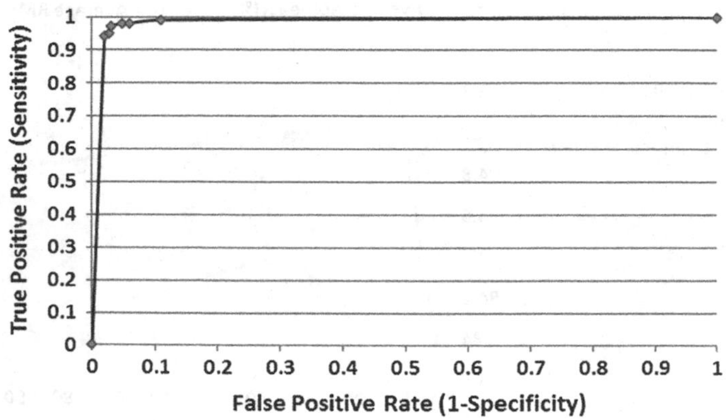


**Fig. 4** Receiver operating characteristics (ROC) curve of optimal ANFIS model

**Table 2** Performance evaluation of optimal ANFIS model

| Threshold values | TP cases | FP cases | TN cases | FN cases | Sensitivity (%) | Specificity (%) |
|---|---|---|---|---|---|---|
| 0.1 | 96 | 6 | 297 | 6 | 94.12 | 98.02 |
| 0.5 | 97 | 8 | 295 | 5 | 95.09 | 97.35 |
| 0.6 | 99 | 9 | 294 | 3 | 97.06 | 97.03 |
| 0.7 | 99 | 12 | 291 | 3 | 97.06 | 96.04 |
| 0.8 | 100 | 18 | 285 | 2 | 98.04 | 94.04 |
| 0.9 | 101 | 33 | 270 | 1 | 99.02 | 89.10 |

## 4 Conclusion

Early diagnosis of dengue disease using adaptive neuro-fuzzy inference system was designed with patients' data, defining initial ANFIS model architecture, train the initial FIS model and testing the system. This proposed system may be used for dengue diagnosis of patients, having physical signs and symptoms with some laboratory findings in early stage. This will be helpful to guide them to take proper curative measure before severity of dengue disease increases. Experimental results showed that this developed system can able to early diagnosis of dengue disease with overall accuracy of 97.04%. Since this study focused on patients with maximum three days of illness, hence the results obtained from the proposed system will not be applicable for the patients beyond that time period. Reliably identifying dengue patient early could help patient management and also reduce the transmission of dengue virus in a community. This dengue diagnostic system can provide a contribution to achieving the goal of World Health Organization of reducing the dengue case mortality by 50% by 2020.

**Acknowledgements** The authors thank Dr. B.K. Bezbaruah, the Superintendent of Guwahati Medical College and Hospital, Guwahati, Assam, who helped a lot to collect patients' data, which helped us in the development of the dengue diagnostic system.

## References

1. World Health Organization.: Dengue: Guidelines for Diagnosis, Treatment, prevention, & Control. A joint publication of the World Health Organization (WHO) and the Special Programme for Research and Training in Tropical Diseases (TDR), New edition (2009)
2. James, A.P., Alan, L.R.: Clinical and laboratory features that distinguish dengue from other febrile illnesses in endemic populations. Trop Med Int Health, 13(11), 1328–1340 November (2008)
3. Lo, C.H., Ben, R.J., Chen, C.D., Hsueh, C.W., Feng, N.H.: Clinically Experience of Dengue Fever in A Regional Teaching Hospital in Southern Taiwan. Centre for Disease Control, Kaohsiung City, ch. 20, pp. 248–254 (2009)

4. Neshat, M., Yaghobi, M.: Designing a Fuzzy Expert System of Diagnosing the Hepatitis B Intensity Rate and Comparing in with Adaptive Neural Network Fuzzy System. Proceeding of the World Congress on Engineering and Computer Science (2009)
5. Shing, J., Jang, R.: ANFIS: Adaptive Neuro Fuzzy Inference System, computer methods and programs in biomedicine. IEEE Transaction on Systems (1993)
6. Guillaume, S.: Designing Fuzzy Inference Systems from Data: An Interpretability-Oriented Review. IEEE Transactions On Fuzzy Systems, vol. 9, no. 3, June (2001)
7. Lim, J.S., Wang, D., Kim, Y.S., Gupta, S.: A neuro-fuzzy approach for diagnosis of antibody deficiency syndrome. Neurocomputing, 69, 969–974 (2006)
8. Singh, S., Kumar, A., Panneerselvam, K., Venilla, J.J.: Diagnosis of Arthritis Through Fuzzy Inference System. J Med Syst, 36(3), 1459–1468 (2012)
9. Ziasabounchi, N., Askerzade, I.: ANFIS Based Classification Model for Heart Disease Prediction. International Journal of Electrical & Computer Sciences IJECS-IJENS, vol. 14 no. 02 (2014)
10. Faisal, T., Taib, M.N., Ibrahim, F.: Adaptive Neuro-Fuzzy Inference System for diagnosis risk in dengue patients. Expert Systems with Applications, 39, 4483–4495 (2012)
11. Afan, S., Yen, L., Christian, S.: Computational Intelligence Method for Early Diagnosis Dengue Haemorrhagic Fever Using Fuzzy on Mobile Device. EPJ Web of Conferences, 68, 00003 (2014)
12. Taun et al.: Sensitivity and Specificity of a Novel Classifier for the Early Diagnosis of Dengue. PLOS Neglected Tropical Diseases, doi:10.1371/journal.pntd.0003638, April 2 (2015)
13. Bystrov, D., Westin, J.: Practice. neuro-fuzzy logic systems. Matlab' toolbox GUI. ch. 2, pp. 8–39
14. Jang J.S.R., Sun, C.T., Mizutani. E.: NeuroFuzzy and Soft Computing: A computational approach to learning and machine intelligence, Prentice Hall Inc (1997)
15. Delen, D., Sharda, R., Bessonov, M.: Identifying significant predictors of injury severity in traffic accidents using a series of artificial neural networks. Accident Analysis and Prevention, 38(3), 434–444 (2006)

# Empowering Agile Method Feature-Driven Development by Extending It in RUP Shell

**Rinky Dwivedi and Vinita Rohilla**

**Abstract** The System Engineering Methods (SEM) can consider as standardization for the development of information systems often demands a methodical approach to knowing which steps have to be taken in which order and at which time in the development process. These methods usually designed for delimited parts of software development life cycle, leaving other phases of SDLC in an ad hoc environment. This results in a need to extend SEMethods and empowered them with the potential to support other phases of software development as well. The paper addresses this issue and shows how popular agile method Feature-Driven Development (FDD) can be extended to satisfy case organizational requirements.

**Keywords** Software engineering · Agile methods · Feature-driven development

## 1 Introduction

Standard systems engineering method can define as "A system engineering method that has its origin outside the target organization and is used by more than one organization, and the development of which the target organization does not control" [1].

The major problem with these methods is that they do not originate in the target organization. The target organization needs to adapt these methods in a way that is similar to implement a standard system [1]. Another major problem is that these

R. Dwivedi (✉) · V. Rohilla
Computer Science Engineering Department, Maharaja Surajmal Institute of Technology, New Delhi, India
e-mail: rinkydwivedi@gmail.com

V. Rohilla
e-mail: vinita.rohilla@gmail.com

 
S.K. Bhatia et al. (eds.), *Advances in Computer and Computational Sciences*, Advances in Intelligent Systems and Computing 553,
DOI 10.1007/978-981-10-3770-2_69

methods are very rigid and inflexible resulting in their incapability to manage complexity and anticipate in a new type of applications [2].

For all these reasons, the standard system engineering methods do not allow users to be guided effectively in their work or to use their expertise in a systematic way.

Method engineering (ME) is a field emerged to form situation/organization specific methods and provides flexibility to the users to share their view of a method-to-be.

Method Engineering aims to bring appropriate solutions to the construction, improvement, and modification of methods used to develop information and software systems.

The basis of method engineering is **meta-model** instantiation. A meta-model is an abstract representation of the method model. It is a system of meta-concepts and draws the interrelationships and constraints between them.

The area of **situational method engineering** is another variant of method engineering that developed for constructing information system development methods that address specific project needs. Harmsen [3] has defined SME as "the discipline to build project-specific methods, called situational methods, from parts of existing methods, called methods fragments".

One of the goals of ME is to improve the use of methods [4]. This objective does not always reach, especially because the methods were not always well adapted to projects specificities.

In the last few years, many agile methods have been introduced to create people focussed and communications-oriented methods [5]. To get maximum benefit from these methods, present research proposes a framework for extending these light-weight methods. During the exploration, it was observed that there may be some requirements that may not be covered by the original method alone [6, 7]. To satisfy the complete set of project requirements, the candidate method needs to extend with another concept. The method extension is a detailed and tedious task; it starts from the process framework of the method, followed by the actors responsible for the process and management and finally to the practices need to add that are required to implement the extended phase.

Hence, the goal of the current research is to "Develop a Method Extension process to extend the agile method Feature-Driven Development that adapts the required practices of other agile methodologies to support entire set of project requirements".

In the next section, history and evolution of agile methodology is discussed followed by the motivation behind the present research. In Sect. 3 Feature-Driven development (FDD) is described, since FDD is the method under consideration so before moving further it prefer to have a close look on the method. Sections 4 and 5 explains the extended FDD method.

## 2 Background and Motivation

Agile software development is now a familiar name in software development industry, and some agile proposals were introduced [8–10] and are used currently. To avert a repetition of arguments in the research and to present the effort contextually, in a nutshell, I avoid exhibiting a review of all the above methods. Interested readers are referred to [11] to get a detailed overview of the agile methods.

Although agile methods may differ in appearance and target modules, they are grossly established on gentle guidance for agile software development, defined by the agile manifesto [12] that offers four essential beliefs.

- Individuals and interactions over processes and tools
- Working software over comprehensive documentation
- Customer collaboration
- Responding to change

In addition to these four beliefs, the agile manifesto further exemplifies few conventions that design guidelines for the development. These repeatedly used design a set of practices in various agile methods.

In the last few years, many researchers paid their attention in collecting the quantitative and qualitative data about the successfulness of agile methods that further helps to evolutes these methods [13].

In 2005, Karlstrom and Runeson [14] threw light on the environment compatibility of the agile method with traditional plan-driven methods, their findings reveal that initially there may arise some resistance, but agile methods can adapt themselves to the traditional plan-based environment.

In 2008, Salo and Abrahamsson [15] explored that agile method Scrum is very successful in developing embedded software projects. They concluded the results by surveying much organization established in various European countries.

In 2009, [16] identified factors that will influence the success of projects that adopt Agile Software development practices. Authors identified 14 influence factors and after a detailed survey found that nine out of 14 factors have a statistically blooming relationship with practicing Agile Software Development as compared to plan-driven software.

On the other side, Jokela and Abrahamson [17] found that agile method Extreme Programming (popularly known as XP) is poor in explicitly attending to the usability of the software. Jordan et al. [18] identified the weakness and limitation of XP in medium and large projects and extended the method capabilities for these projects. Another limitation of XP is that it proposes for simple and small-scale projects, [19] address this issue and modifies the XP model for large and medium projects.

From the last few years, Agile software development methods are gaining importance. Software practitioners are switching to adopt agile practices and require solutions that address their need for an extended method capable of delivering situation precise method. Agile methods individually may not fulfill the complete set of requirements, in the present research a method extension approach is presented that integrates the practices of many agile methods and provide an a la carte package for the developer. This improves the software development process as it empowers the original method by extending it capabilities and also delimits the ad hoc approach for fulfilling the imaginary part by the original method.

Integrating agile practices from various methods is not a new issue previous proposals shows its applicability in the industrial domain. In [20] presents a successful integration of practices from a software product line engineering and agile software development. In [21, 22], authors presents that how XP and Scrum are integrated to improve development process that integrated method formed will successfully deliver the quality software with right cost and in given time frame.

## 3 Feature-Driven Development Method

The problem with FDD is that it mainly concentrates on design and building phases of software development life cycle while ignoring or giving less attention to other phases. Another problem is agile teams are working on product development using very high levels of abstractions that are very far away from the real projects [23].

Invariably teams need to look to other methods to fill in the process gaps that FDD deliberately ignores. When considering alternatives, there is considerable overlap and conflicting terminology that can be confusing to practitioners as well as outside stakeholders. To address these challenges, we extend FDD and present it as Extended FDD (ExFDD) framework that is cohesive and flexible method. Before going further, the paper presents an overview of the FDD.

FDD consists of five chronologically ordered processes during which the entire process is passed out. The design and build phase are the iterative parts of the process and solves the purpose of swift alterations. The period required for each release depends upon the complexity, but it lies between 1 and 3 week period. In the following section, all the five phases of the FDD process is described according to the [24].

The first process, develop an overall model has a pre-requisite that domain experts should be already aware of extent, situation, and constraints of the system to be build. However, FDD does not explicitly address the issue of gathering and managing the requirements. The domain experts present a so-called "walkthrough" in which the team member and the chief architect inform of the high-level description of the system.

After these walkthroughs, object models are produced. These object models give as an input to the second phase that is **Build a Feature list**. In this phase, a comprehensive feature list is a build for the system develops. The feature sets in the feature list are planned and sequenced according to their priorities and dependencies in the third phase. The fourth and fifth process of design and building can consider as a unit. The prime purpose is to design and build the feature sets in an iterative manner.

**Comparison of FDD with Other Approaches**

As mentioned above, FDD is a five-phase process, supports the object-oriented view, simple to implement but focuses on design and build phase hence, need other supporting approaches. Since, for the complete method delivery it needs to be extending with other methods. To understand the major pitfalls we compared FDD with other preferred approaches the great outcomes are:

- XP says that the overall architecture should emerge from a process of the incremental story (feature) implementation and continuous refactoring. FDD significantly eschews refactoring and advocates front-end skeletal modeling (an object model in this case).
- Another area in which FDD differs from XP, Scrum, and Adaptive Software Development (ASD) [25], in particular, is the scheduling of features to deliver. Within the boundaries of technical reasonableness, XP, Scrum, and ASD insist on the customer's setting development priorities at the beginning of each iteration. The patron determines which features to develop next based on their current business value assessment. FDD assumes that the overall value of the features actuate early in the project and that scheduling those features, should be primarily a technical decision.
- For scheduling, some methods support customer scheduling and other support technical scheduling. There are advantages and disadvantages of both. Letting customers schedule features on a short-cycle basis could increase development costs by causing additional rework because of technical dependencies. Conversely, technical scheduling can allow the project to drift off course from the customer's perspective. FDD offers a potential path for a "best-of-both" approach—customer scheduling at the business activity level and technical scheduling at the feature level.

Unlike Scrum sprints, FDD support for short iterations gives another potential reason to opt it for the development process. But as mentioned earlier, it needs to be extended with a complete framework that helps FDD to perform in a better manner and make it suitable for various situations.

## 4 Extending FDD in an RUP Shell

The present research embeds FDD life cycle in an RUP shell. RUP recommend for large, long-term, enterprise-level projects with medium-to-high complexity. RUP subdivides the project lifecycle into four major phases. Even though it encourages concurrent workflows across the entire cycle, the general understanding is that certain activities will peak during certain phases. From starting, RUP supports some guidelines for each phase that is not agile but in 2006, IBM created a subset of RUP, tailored for delivering the agile projects and releases it as an open source method called open UP through the Eclipse website. A recent study shows that the adoption phase of RUP has a great potential to being agile as it is a six phase iteration cycle that is repeated until the new process has been completely implemented. Each increment brings in new practices to implement and adjusts the previously added practices. Based on feedback from the previous cycle, the development case is updated if needed.

**The prime question at this point is why we are extending FDD in an RUP shell**. The reason behind is RUP supports the complete life cycle of the software development. It is well established and is successfully implementing in many organizations. There are some tools available like Rose, Visio, etc. that implements RUP, so the framework helps to form a technical support for automating the ExFDD process that FDD currently lacks.

In spite of the four phases in RUP, we implement the ExFDD in three phases to make it analogous with the Method Development Life Cycle (MDLC). The MDLC introduce in [26] and supports the basic life cycle operations i.e. Initiate, Construct, and Deploy. The Fig. 1 shows the complete life cycle of ExFDD consists of Inception phase, Transition Phase, and Construction Phase.

For each phase, ExFDD provides some Product Factors, for example in the Inception Phase the defined product factors are- perform basic activities, explore initial scope, develop the model, secure funding, and plan release. These product factors are the goals that may need to be achieved by the developer; they are designed to help and facilitate the developers in the decision-making process for the current situation or the unique context of the situation.

In comparison to FDD, ExFDD is more user-friendly and flexible approach, it provides the developer a set of alternative solutions to address the unique conditions. For example, to address *Dynamic or Changing requirements*, domain experts in Basic FDD are communicating in processes 1 (develop an overall design) and then again in process 4 (design by feature). ExFDD however, provides a set of goals for each phase, these goals are further explored to draw alternative paths by

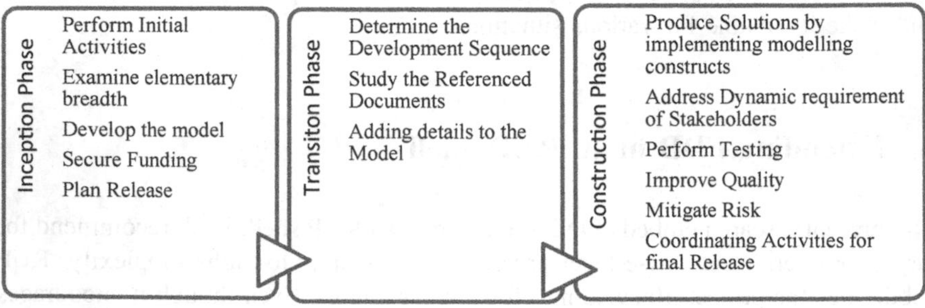


**Fig. 1** Basic version of ExFDD

**Table 1** Exploring FDD

| Product factors (goals) | |
|---|---|
| Perform initial activities | Domain walkthrough |
| | Study documents |
| | Align with enterprise direction |
| | Identify initial technical strategy |
| Examine elementary breadth | Level of details |
| | View types |
| | Modeling strategy |
| Develop the model | Develop the initial release plan |
| | Refine the model |
| | Write model notes |
| | Formal/informal modeling sessions |
| Secure funding | Identify risk |
| | Internal and external assessments |
| | Cost estimation |
| Plan releases | Decompose method into small releases |
| | Ensure the cohesiveness of module |
| | Determine the priority and sequence |
| | Change management |

integrating the concepts and solutions equipped with the popular established methods. The goals of Inception phase are analyzed as follows (Table 1).

ExFDD doesn't believe that one framework best used for all scaling situation. For example for the goal "Examining Elementary Breadth", a goal that should address at the beginning of a project during the Inception phase (remember ExFDD promotes a complete life cycle, not just construction life cycle). Where, FDD will intangibly advise you to model object diagram with some initial requirements the goal diagram Fig. 2 makes it clear that you might want to be a bit more sophisticated in your approach i.e.

- What level of details should you capture.
- What view types should you consider (use-cases are one approach to usage modeling, but there are other approaches to explore the data).
- Non-functional requirements need to capture such as reliability, availability and security requirements.

One of the great advantages of agile is the wealth of practices, techniques and strategies available. ExFDD took the benefit of it and described what to choose and how to fit them together. Instead of simple feature list in FDD we suggest *work item pool* in ExFDD that adopts practices and strategies from existing sources and provides advice for when and how to apply them together.

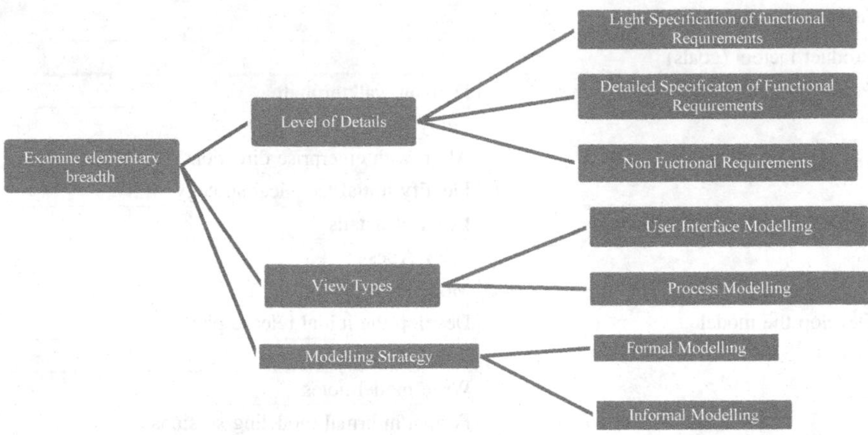


**Fig. 2** Exploring product factors in ExFDD

## 5 Why ExFDD

**ExFDD does not prescribe a use-case driven approach, or insist that OOAD is rigorously applied to build out services/components**—A use-case-driven approach is a potential practice to apply, but there is a danger that this could lead to an exhaustive requirements specification that is not particularly agile. User stories might be an option, but the key point is that you will have to adapt to the situation that you find yourself in. This is why [we prioritize the team's work with a work item pool, rather than Scrum's backlog comprised of user stories. Using a work item pool allows us the flexibility to put any work on our feature list, extending the applicability of ExFDD to many types of projects beyond those for which RUP or FDD would be ideally suited.

**ExFDD is goal-driven, not artefact-driven**. It does not prescribe practices or specific artifacts. Rather, it suggests alternative strategies that can be applied at certain parts of the lifecycle, but which ones you choose is up to you. ExFDD picks several way of doing the same thing by various agile method and provides a-la-carte package for selecting the process like for changing requirements

**ExFDD picks up where FDD leaves off**—the ExFDD describes how different practices from various Agile methods fit together, going far beyond FDD it provides a complete delivery life cycle. Like FDD the ExFDD framework addresses planning, designing and construction unlike FDD ExFDD does not stop here, it also addresses detailed elicitation process of requirements, addresses changing requirements, incorporating different perspectives, release prioritized schedules, etc. in short ExFDD provides a much broader understanding of how agile development works in practice, doing a lot of the "heavy process lifting" that FDD leaves up to you.

**Extending FDD as a Goal-Oriented Approach**

ExFDD goal-driven approach enables ExFDD to avoid being prescriptive and thereby be more flexible and easier to scale than other agile methods. For example, where FDD prescribes an object modeling approach for managing requirements ExFDD instead says that during construction you have the goal of addressing changing stakeholders need. ExFDD then indicates that there are several issues surrounding that goal that you need to consider, and there are several techniques/practices that you should consider adopting to do so. ExFDD can extend further to describe the advantages and disadvantages of each technique and in what situation it is best for.

Presenting ExFDD as goal-oriented approach ends up in following advantages for configuring a method

- A goal-driven approach supports process tailoring by making process decision explicit.
- It makes your process options very clear and thereby making it easier to identify the appropriate strategy for the situation in hand.
- It enables the developer to focus on the actual job that is to provide value to the stakeholders.

## 6 Conclusion

The paper presents the Method Extension process- the process supports the blend of different methods based on the rich knowledge of the past usage of these methods under different requirement sets. The applicability of the method thus formed will be significantly improved than the existing methods because the extended method contains the required constituent of more than one method.

In this research, a different point of the beginning has been introduced, for example, earlier method engineering proposals based on modular method construction they proceed by integrating several small methods. Whereas, the present research starts with a rigorous method, the prime intention is to tailor this method to the current situation in hand.

"One-size-fits-all," suggests that every problem fits into a single methodological framework. The correct phrase should be "one-size-fits-one". However, while one size doesn't fit all projects, does not lead us to the conclusion that every project team needs to start from scratch and design its process- previously developed method elements can be reused [10]. This motivates to create a blend of different agile methods based on the rich knowledge of the past usage of these methods under different requirement sets.

Extending an agile method means balancing the consistency needs of business enterprises with the flexibility required by project teams. The applicability of the method thus formed will be significantly improved than the existing methods because the extended method thus formed contains the required constituent of a different method.

## References


1. Karlsson F. "Meta Method for method configuration- A Rational Unified Process Case", ISBN 91-7373-474-8, Faculty of art and science, Sweden, 2002.
2. Dwivedi, R. and Gupta, D. 'A framework to support project-in-hand and selection of software development method', *Journal of Theoretical and Applied Information Technology*, Vol. 73, No. 1, pp. 137–148, 2015.
3. Harmsen A.F., Situational Method Engineering. Moret Ernst & Young, 1997.
4. Bajec M. et.al., "Practice driven approach for creating project-specific software development methods", Information and Software technology, Vol. 49, 345–365, 2007.
5. Karlsson, F. and Ågerfalk, P.J. 'Exploring agile values in method configuration', *European Journal of Information Systems*, Vol. 18, No. 4, pp. 300–316, 2009.
6. Dwivedi, R. and Gupta, D. 'Applying machine learning for configuring agile methods', *International Journal of Software Engineering and its applications*, Vol. 9(3), pp. 29–40, (2015).
7. Dwivedi, R. and Gupta, D. 'the agile method engineering: Applying fuzzy logic for evaluating and configuring agile methods in practice', *Int. J. Computer Aided Engineering and Technology*, (In Production).
8. Beck, K. 'Embracing change with extreme programming', *IEEE Computer Society Press*, Vol. 32, No. 10, pp. 70–77, (1999).
9. Dwivedi, R. 'Configuring issues and effort for configuring agile approaches-situational based method engineering', *International Journal of Computer Application*, January, Vol. 16, No. 17, pp 23–27, 2013.
10. Schwaber, K. and Beedle, M. *Agile Software Development with Scrum*, Nouvelle editions, (2002).
11. Abrahamsson, P. et al. 'Agile Software Development Methods Review and Analysis', *VIT Publications*, Juhani Warsta, University of Oulu, (2002).
12. Agile Manifesto *Manifesto for Agile Software Development*, [online] http://www.agilealliance.org/the-alliance/the-agile-manifesto/. (2001).
13. Jim Highsmith, "Agile software development ecosystems", publisher Addisson-wesley, Pub-date May 26, 2002, ISBN 0-201-76043-6.
14. Karlstrom, D., Runeson, P.,. Combining agile methods with StarGate project management. IEEE Software 22 (3), 43–49, 2005.
15. Salo, O., Abrahamsson, P., 2008. Agile methods in European embedded development organizations: a survey study of extreme programming and scrum. IET Software Vol. 2 (1), 58–64.
16. Misra S. C. et al. "Identifying some important success factors in adopting agile software development practices", Journal of system and software, Vol. 82, pp 1869–1890, 2009.
17. Jokela, T. & Abrahamsson, P. (2004) "Usability assessment of an Extreme Programming Project: Close Co-Operation with the Customer Does Not Equal to Good Usability". In PROFES 2004, Keihanna-Plaza, Kansai Science City, Japan.
18. Jordan et. al. "Overview and guidance on agile development in large organisation", Communications of the association for information systems, Vol. 29, Article 2, 2011.
19. Qureshi M. R. J. "Agile Software development methodology for medium and large projects", IET software, Vol 6 (4), pp. 358–363.

20. Hanssen G. K., Faegi T.E., "Process fusion: An industrial case study on agile software product line engineering", Journal of systems and software, Vol. 81, pp. 843–854, 2008.
21. Dwivedi, R. and Gupta, D. 'Customizing agile methods using genetic algorithm', Proceedings of *International Conference on Advances in Communication, Network and Computing*, Elsevier, 2014.
22. Fitzgerald, B., Hartnett, G. and Conboy, K. 'Customizing agile methods to software practices at Intel Shannon', *European Journal of Information Systems*, Vol. 15, No. 2, pp. 197–210, 2006.
23. Hunt, J. 'Feature driven development', *Agile Software Construction*, pp. 161–182, Springer, ISBN: 978-1-85233-944-9 (Print) 978-1-84628-262-1 (Online), 2006.
24. Palmer, S.R. and Felsing, J.M. (2002) 'A Practical Guide to Feature-Driven-Development', *Prentice-Hall*, Upper Saddle River, NJ, (2008).
25. Highsmith, J. A. "Adaptive software development: A collaborative approach to managing complex systems New York, NY, Dorset house publishing.
26. Prakash N. and Goyal S.B., 'Towards a Life Cycle for Method Engineering', *Proceedings Eleventh International Workshop on Exploring Modelling Methods in Systems Analysis and Design (EMMSAD'07)*, 27–36, (2007).

# Social Media Trends and Prediction of Subjective Well-Being: A Literature Review

**Simarpreet Singh and Pankaj Deep Kaur**

**Abstract** Everybody is now addicted to the online social media. Social media sites have been used by millions of people globally. Each individual expresses his thoughts, daily life events, and opinions on social media. The individual's expressions on social media are mostly in the text form. The text contains sentiments, opinions, attitudes and emotions of the individuals, which are largely related to the happiness in the personal life of individuals. Extensive usage of social media affects the happiness, which can be either on the positive or negative level. Happiness level is normally measured by self-report and often been indirectly characterized by more readily quantifiable economic indicators such as gross development product (GDP) or genuine progress indicator (GPI). However, the growing importance of linguistic text analysis of social media gives a direction to predict the happiness of individuals and is termed as subjective well-being (SWB). SWB is the scientific term used to describe happiness and quality of life of individuals. It includes emotional reactions and cognitive judgments and is of great use to public policy-makers as well as economic, sociological, and psychological research. The richness and availability of social media make it an ideal platform to conduct psychological research in the topic SWB. In this paper, at last, the evidence of the importance of the social media analytics has been provided followed by identification of major factors involved in SWB. Further, the effects of social media usage on the SWB of individuals have been elaborated.

**Keywords** Well-being · Happiness · Social media · Social well-being · Subjective well-being · Social networking · Social happiness · Literature review

---

S. Singh (✉) · P.D. Kaur
Department of Computer Science & Engineering, Guru Nanak Dev University, Regional Campus, Jalandhar 144007, Punjab, India
e-mail: simarpreetsinghbangar@gmail.com

 
S.K. Bhatia et al. (eds.), *Advances in Computer and Computational Sciences*, Advances in Intelligent Systems and Computing 553,
DOI 10.1007/978-981-10-3770-2_70

## 1 Introduction

The advancement of big data tools and techniques in a short time has reduced the many problems of large data stored on cloud servers. Huge data stored on servers mostly come from web services, such as blogs, social media sites, forums, etc. Social media is a combination of two words, the first part social refers to the interaction between people by sharing information to and fro, and the later part media refers to a medium of communication, like the Internet. Social media can be defined as forms of electronic communication (as Web sites for social networking and microblogging) through which users create online communities to share information, ideas, personal messages, and other content (as videos).[1] The Internet, mobile technologies, and Internet of Things (IoT) drastically diffuse the social media. It is the array of web-based applications which define the way social media operates. Examples include social networking, microblogs, weblogs, online forums, podcasts, and 3-D virtual reality.

Social media word is used vaguely, every website has been considered as social media. Many popular websites dedicated to social media are Twitter (social networking and microblogging), Facebook and Google+ (social networking), YouTube and Netflix (media content sharing), Blog and Blogger (web blogging), MozVR and Second Life (virtual reality). Social media websites generally contain heterogeneous data, like text, pictures, videos, etc. Facebook is the first social media website to cross the 1 billion mark of registered user (January 2016), Facebook had 1.55 billion active users.[2] The Ice Bucket Challenge[3] on Facebook in 2014 was one of the biggest trends, ALS.net raised $4 million and every penny was spent on research. The attraction of people to these types of trends, quickly adopt and spread it to others on social media. Social media attracted every age group, mainly teens. Teens spend 9 h[4] daily on social media, which have both positive and negative impact on the personal life. Also, the study [18] provides guidelines for understanding the social media websites usage motives, which is very beneficial for scientists to choose the specific social media platform for research.

Social media gain widespread adoption and become part of the ecosystem, attracting users, consumers, businesses, governments, and non-profit organizations. People have been able to communicate with local and people outside the country at the same time. People have leveraged the social media in smart ways, like the formation of online communities that allowed them to get moral support, education, latest information, and even promote and sell products. Similar way the impact of social media changed the whole view and promises to accelerate innovation, cost-saving mechanisms, and strengthens the brand value of industries. Every company is using the social media to hype new products and services and also

[1]Social Media Definition, “Merriam-Webster”.

[2]http://www.statista.com/statistics/272014/global-social-networks-ranked-by-number-of-users.

[3]https://www.als.net/icebucketchallenge/.

[4]http://edition.cnn.com/2015/11/03/health/teens-tweens-media-screen-use-report/.

monitor what people are saying about the product. Firm's performance with respect to business networks can be analyzed to predict the price movement in the stock market [28]. Individuals and organizations need to transform information into decision-making intelligence and also to measure the value of their business with social media analytics.

## 2 Social Media Analytics

Social media analytics can be described as the process of collecting data from the social media websites and analyzing that data to make business decisions.[5] Social media analytics is mostly used to mine customer sentiment in order to support marketing and customer service activities. Data analytics can be real-time or offline analysis, including factors such as influence, reach, and relevance of suitable measurements. Time considerations are important to understanding the context of data being analyzed. The importance of social media analytics can be seen as the researchers at AT&T developed an analytic software to eavesdrop customers network problem complaints on Twitter. The crews will be sent to fix the problem by extracting time, location, and type, from the tweet [11]. Organization's dedication to serving the mass with this level of priority makes it more interesting and creates competition among the organization. Organizations have been focusing on research and innovation in analytics based on the resources they already possess.

### 2.1 *Offline Social Media Analytics*

Offline data analytics refers to the passive analysis of data, which is generally used for digital marketing channels. The offline data is the specific data captured, which is generated by the user or from offline sources such as CRM data files. The captured offline data of a particular user from social media have been very useful and the outcomes of the data analysis shed light to the uncovered variables. The importance of offline analytics can be seen as the biggest presidential elections occurs in the USA, where candidates mostly campaign through social media. Researchers present a reliable forecasting system for US presidential elections and US house race, named Competitive Vector Auto Regression (CVAR) [26]. CVAR compares the popularity of multiple competing candidates by combining visual information with textual information from the Flicker social media. This type of system can provide campaign insights to the candidate so that candidate can work on their weaknesses and can further improve their self.

[5]http://searchbusinessanalytics.techtarget.com/denition/social-media-analytics.

Apart from the elections, analytics have been used to predict stock market prices. The stock market determines the economic value of the country, many people daily share ups and downs of stock market prices on social media. Researchers suggested that stock market price movements can be predicted through social media analytics by proposing an Energy Cascade Model (ECM) [27]. ECM can effectively predict middle-term directional stock market price movements, achieving an average accuracy of 67.7% for upward stock price movements. A similar approach [20] was used to explore the two major events of stock market price change and trade volume. Trust information extracted from the Twitter group was compared with Dow Jones Industrial Average (DJIA). By keeping trust information into account, the results show that price change and trade volume are more related than just counting the number of tweets and trade volume is stronger correlated than price change.

Yet another analysis has been carried out through the public microblogging social networking site Twitter. The performance and psychological well-being of runners [10] had been tracked, by monitoring Twitter tweets of runners group. The 925,825 messages of runners who used Nike + fitness tracking device were analyzed. Researchers found that (1) fitness devices were most popular in North America (2) less than 2% runners consistently ran for at least 150 min a week, which is recommended by Centers for Disease Control and Prevention (3) physical activity lowered on Friday as the users may need to be relaxed. The runners have been recorded for 3 months long, this somehow indicates that the old records can be used for the analytics purposes. But it may be a big challenge to convert the records into some useful form before analysis.

From the perspective of language and history, digitization of more than 15 million historical books and an analysis of the past 200 years had been done by the tech giant Google,[6] showing a change of language usage, dynamics of fame, censorship, and time compression of collective memory. The mysteries of social media before the Internet age can be solved through more efforts in this field. The large offline data analytics are sometimes easier to perform because data and noise present in the data are consistent. The large volume and high velocity of data can be a real challenge, and many researchers have done marvelous work in the real-time social media analytics.

## 2.2 *Real-Time Social Media Analytics*

Real-time analytics denote the capacity to use all available data and resources when needed. The analysis of data is carried out dynamically and reports are generated with no delay. Mostly real-time analytics is used for geographic location and tracking purposes. Nowadays, people instantly share on social media about

---

[6]http://books.google.com/googlebooks/library/index.html.

situations like natural disasters, hence the real-time analytics of social media may provide life-saving information. Real-time social media analytics of streams and graphs called as Milano Design Week (MSW13) [1], it recommends venues to visitors of geo- and temporally bounded city-scale events in Milan featured 681 venues for hosting 1,127 events attended by 500,000 visitors in one week. By combining deductive and inductive stream reasoning techniques, this system analyzes Twitter's tweets and Foursquare checked in's to produce high-quality link predictions. As mentioned earlier, people spend more time on social media and share whatever is happening in the surrounding, whether it is an earthquake, car accident, tsunami, or landslide.

A multilevel content analysis of Twitter tweets collected about landfall suggests that actionable information was easier to find when searching along hashtags. The use of Twitter in precrisis stages of a weather event can be beneficial for the emergency management agencies [14]. Another example of real time social media analytics is the monitoring of outbreaks through the proxy of users search. Googles Flu Trends and Dengue Trends provide estimates of flu and dengue based on search patterns.[7] Also, Google Trends can accurately predict the box offline success based on the rating of online mentions of individual movies and the count of the search made on YouTube. From the above studies, few things are concerned, such as the reactions of the people to the situation. The behavior of peoples varies according to the situation, which can decide that on whom they will trust blindly or may take the risk to trust others.

A rigorous and quantitative meta-analysis was conducted to investigate the empirical evidence of the most influential factors, trust, and risk which affect the individual behavior toward social media platforms [23]. The findings suggested that both risk and trust had significant effects, but trust had a stronger effect. The effects of risk and trust have been clearly visible on the social media. Trust is closely related to the happiness of the human behavior and mostly happier persons are more trustworthy.

At a growth rate of ~8%, Internet users are now more than 40% of entire world population. Social media continuously playing a great role to reach that mark and has touched the many aspects of human life. With this, the social media is responsible for a radical new trend that is of interest to various organizations for finding emerging and unique trends in human behavior.

## 3 Human Behavior

Human living style, languages, and behavior change after a few hundred kilometers. It becomes very difficult to predict the human behavior as there are many factors like genetic, socioeconomic, etc. Since humans are very comfortable with

[7]https://www.google.co.in/trends/.

each other and try to live a social life. Being social is the common characteristics and it can be useful to predict the human behavior with respect to certain conditions. The impact of the Internet on the social life of humans can be seen well. Social media is forcing the people to use it and has become part of everyday life.

From the past 10 years, the tremendous rise has been seen in exploring social networking sites for extracting information related to human psychology. Some surprising findings [13] were obtained such as popular ideas have some opponent of a certain percent and within a large group, only a few have leadership qualities. The dataset was collected from YouTube in real-time anonymous random undergraduate students. The visualization reflects that the intrinsic complexity and obscure characteristics of web data, sometimes, make difficult to establish relationships among certain attributes.

## 3.1 *Factors and Theories of Human Behavior*

There are many factors which influence the human behavior such as genetic, socioeconomic, physical environment and psychological factors. The socioeconomic and psychological factors play a vital role in social behavior. The factors like education, family, culture, self-concept, fear, anxiety, etc., directly associated with the social life. There are certain theories related to human behavior named as Bandura, Bibb Latane.

The Bandura theory[8] is based on the concept of learning from others, which involves attention, retention, reproduction, and motivation. Bibb Latane theory[9] has three basic rules, the first rule considers how individuals can be "sources or targets" of social influences, the second is the social impact is the result of social forces including the strength of sources of impact, the immediacy of the event and the number of sources exerting the impact. Final is, the more targets of the impact that exists, the less impact each individual target has. The impact of social influences on individuals has many forms and are visible in the comments and opinions given by the individuals to the others.

Many theories have been used to predict sentiments from the personality of the individuals. Big-Five [19] personality traits have been the most popular theory used by researchers. To measure the individual's behavior the theory defined five personality traits: extraversion, agreeableness, conscientiousness, neuroticism, experience. Also, the emotional rating of English language words has been helpful from the aspect of measuring emotions. The Affective Norms for English Words (ANEW) [3] defines a set of words that have been rated in terms of pleasure, arousal, and dominance.

---

[8]http://web.stanford.edu/dept/psychology/bandura/pajares/Bandura2004Me-dia.pdf.

[9]10.1037/0003-066X.36.4.343.

## 3.2 *Opinion Mining*

Opinions are common to all human activities and are the key factors of our behaviors. Our beliefs and perceptions of the real world, and the choices individuals make, to a larger extent, depends on how others see and evaluate the world [6]. Opinions related concepts such as sentiments, evaluations, attitudes, and emotions come under the study of sentiment analysis and opinion mining. The rapid growth of social media on the web makes sentiment analysis as the most active research area in natural language processing. It is also the major part of data mining, Web mining, and text mining, and is widely spread from computer sciences to social sciences. Sentiment analysis has applications in almost every business and social domain due to its importance to business and society. Sentiment analysis can be broadly classified into three levels [16] based on the existing research problems: document, sentence, and entity-aspect level.

# 4 Subjective Well-Being

Opinions with emotions play such an important role in human decision-making. Variations in human mood states have become a matter of considerable interest. The increasing importance of social media has made the researchers think about that how the mixing of psychological states affects in situations in the absence of physical contact. A system for sensing social systems has been introduced 10 years back, with data collected from 100 Bluetooth-enabled mobile phones. Using user modeling techniques to recognize social patterns in daily user activity, various features were given such as infer relationships, identify socially significant locations [8]. In the present world of social media, another attempt had been done [2] to measure the happiness level of Twitter users. The results indicate that the online social media may be equally important subject to social mechanisms that cause assortative mixing in real social networks. The increasing prevalence of online social media may be an important factor in how positive and negative sentiments are sustained and spread through human society.

The sentiment analysis projects the perception of individuals and is an indirect measure of the personality. The personality of the individual, which determines the well-being of that individual. Well-being based on the personality of individuals is known as SWB. SWB is a measurement of individuals' evaluations of one's own life, which are largely based on the emotional well-being and positive functioning. Emotional well-being refers to a long-term assessment towards life and positive functioning encompasses of psychological, social well-being, and psychological well-being. Researchers defined multidimensional parameters by developing positive and negative affect scale (PANAS) [24] and psychological well-being scale (PWBS) [21]. PWBS based on positive functioning defines various parameters such as self-acceptance, the purpose of life, environmental mastery, positive relation with others, personal growth and autonomy items.

A state-of-the-art prediction model [9] was established to automatically sense individual SWB from the active users of Sina Weibo social network. The method was based on emotional well-being and positive functioning. The large data set from microblogs was used to sense individual SWB by training machine learning models. The model had very high prediction accuracy and can be applied to identify a large amount of social media users SWB in real time with low cost. The outcomes of the research invoke more researchers to study the happiness of individuals through the SWB.

Individual happiness is a fundamental metric of social well-being. An attempt had been made to measure happiness from the social media. Expressions made on the online social networking service Twitter were examined, which revealed the temporal variations in happiness and information levels over time. The text-based Hedonometer [7] measures happiness which is highly robust, tunable, remote-sensing and noninvasive. Hedonometer has been made available publically on the Internet and is very useful to the researchers for future references. In the above research, it is largely based on the happiness of individuals from the personal perspective. If the happiness needs to be measured based on the geographical location of individuals, effortless research, and novel architectures have been required.

Prediction of well-being was attempted by learning linear regression models on word counts in lexicons of emotionally charged words with a large Twitter dataset. The choropleth maps produced, either at a state or at a country level, that show how well-being varies across the U.S continental [17]. SWB of individuals highly depends on the type of language the individuals use on social media. So, it is the words and phrases of the language used on social media, which characterize the SWB.

Recently, researchers used differential language analysis with particular open-vocabulary analysis to nd language features that distinguish demographic and psychological attributes. The dataset contains over 15.4 million Facebook messages, further extracted into 700 million instances of words, phrases, and automatically generated topics and correlated with gender, age, and personality. The analysis shed new light on psychological processes that suggest the relation of personality to the language used on social media [22]. The above study analyzes every user on social media, therefore, lacks in the value of data and contains a high amount of noise in the dataset. To reduce these factors, the dataset should be precise, and to raise the value of data, active users of social media should be targeted.

The prediction [15] of the personality of active users on microblogging platform Sina Weibo was carried out by Big Five personality traits. Total 845 microblogging behavioral features were extracted and classification models utilizing Support Vector Machine (SVM) was trained. Participants scores were predicted on each dimension of Big-Five Factor Inventory with classification accuracy ranging from 84 to 92%. The results indicated that it was possible to predict the active users personality with high accuracy and the microblogging services of non-US based social media sites. The above examples of research indicate the positive effect of the usage of social media on SWB of individuals. However, as discussed earlier the social media also have negative consequences on SWB. High usage of social media may affect the personal life and behavior of individuals toward to others.

The social and psychological well-being (social success, normalcy, and self-control) are key factors in human behavior. A recent study examines the impact of media multitasking behavior on university students' well-being [25]. The study characterized media multitasking behavior by motivations, characteristics, and contexts. The findings suggested that synchronous social interactions are significant and positively associated with social success, normalcy, and self-control. However, high usage of social media in synchronous social interactions reduced the individuals' social success. Social networking sites o er a lot of features, which may be the strong reason behind the usage addiction of social media.

The usage addiction of social media can lower the performance of an individual! To study this, a classroom task environment was created to measure the usage of social media and task performance [4]. It was found that higher amounts of social media usage led to a lower performance on the task, as well as higher level of technostress and lower happiness. The results suggested that the usage of personal social media during professional times can lead to negative consequences. The negative effects of high social media usage have a huge impact on adults which are highly active users of social media.

Social media theory suggests that adults evaluate good and bad consequences of social relationships they experience, so a study was carried out to report good and bad perceptions of social media, with perceptions varying according to demographic and psychological characteristics [12]. The demographic variables revealed that younger individuals had good perceptions and bad perceptions by those who had health problems. Analysis of psychological variables suggests that good perceptions were reported by angry individuals with strong friend supports and bad perceptions by angry individuals with low self-esteem.

Also, a study [5] investigated (1) patterns of media use for social sharing and (2) effects of social sharing on sharers well-being. The results revealed the positive events were more shared on online social media and negative events via face-to-face. Regardless of the medium, users shared positive event experienced positive affect and who shared negative event had increased the negative effect.

## 5 Conclusion and Future Work

In this paper, we reviewed the research papers of social media focused on individuals' subjective well-being. In the first section, we discussed that the big data is the backbone of social media. In the next section, we have discussed the well-known theories based on human well-being. These theories helped the researchers to set the parameters to predict the human well-being. The trend of social media is very popular and people share their feelings on social media, provides large data related to subjective well-being. The researchers used that data to study the well-being of individuals. The higher accuracy of results and predicted outcomes was very impressive, thus gained a lot of attraction. More researchers are now attempting to get into new insights of social media analytics focusing on subjective well-being. As the data is changing rapidly, so many scientists faced different challenges such as high noise, real-time analysis of data, etc. The real-time prediction of subjective well-being from social media may be the hot topic of future research. The real-time predicted happiness of people will be shared with friends and family. Hence, it may provide a better environment as people will adjust their mood levels according to the other individuals' behavior.

## References


1. Balduini, M., Bozzon, A., Della Valle, E., Huang, Y., Houben, G.J.: Recommending venues using continuous predictive social media analytics. Internet Computing, IEEE 18(5), 28–35 (Sept 2014)
2. Bollen, J., Goncalves, B., Ruan, G., Mao, H.: Happiness is assortative in online social networks. CoRR abs/1103.0784 (2011), http://arxiv.org/abs/1103.0784
3. Bradley, M., Lang, P.: Affective norms for English words (anew): Instruction manual and affective ratings (1999)
4. Brooks, S.: Does personal social media usage affect efficiency and well-being? Computers in Human Behavior 46, 26–37 (2015), http://www.sciencedirect.com/science/article/pii/S0747563215000096
5. Choi, M., Toma, C.L.: Social sharing through interpersonal media: Patterns and effects on emotional well-being. Computers in Human Behavior 36, 530–541 (2014), http://www.sciencedirect.com/science/article/pii/S0747563214002350
6. Dave, K., Lawrence, S., Pennock, D.M.: Mining the peanut gallery: Opinion extraction and semantic classification of product reviews. In: Proceedings of the 12th International Conference on World Wide Web. pp. 519–528. WWW'03, ACM, New York, NY, USA (2003), doi:10.1145/775152.775226
7. Dodds, P.S., Harris, K.D., Kloumann, I.M., Bliss, C.A., Danforth, C.M.: Temporal patterns of happiness and information in a global social network: Hedonometrics and twitter. PLoS ONE 6(12), 1–1 (12 2011), doi:10.1371/journal.pone.0026752
8. Eagle, N., (Sandy) Pentland, A.: Reality mining: sensing complex social systems. Personal and Ubiquitous Computing 10(4), 255–268 (2005), doi:10.1007/s00779-005-0046-3
9. Hao, B., Li, L., Gao, R., Li, A., Zhu, T.: Active Media Technology: 10th International Conference, AMT 2014, Warsaw, Poland, August 11–14, 2014. Proceedings, chap. Sensing Subjective Well-Being from Social Media, pp. 324–335. Springer International Publishing, Cham (2014), doi:10.1007/978-3-319-09912-5_27
10. He, Q., Agu, E., Strong, D., Tulu, B., Pedersen, P.: Characterizing the performance and behaviors of runners using twitter. In: Healthcare Informatics (ICHI), 2013 IEEE International Conference on. pp. 406–414 (Sept 2013)
11. Ikeda, K., Hattori, G., Ono, C., Asoh, H., Higashino, T.: Early detection method of service quality reduction based on linguistic and time series analysis of twitter. In: Advanced Information Networking and Applications Workshops (WAINA), 2013 27th International Conference on. pp. 825–830 (March 2013)
12. Keating, R.T., Hendy, H.M., Can, S.H.: Demographic and psychosocial variables associated with good and bad perceptions of social media use. Computers in Human Behavior 57, 93–98 (2016), http://www.sciencedirect.com/science/article/pii/S0747563215302740

13. Kohli, S., Gupta, A.: Modeling anonymous human behavior using social media. In: Internet Technology and Secured Transactions (ICITST), 2014 9th International Conference for. pp. 409–412 (Dec 2014)
14. Lachlan, K.A., Spence, P.R., Lin, X., Najarian, K., Greco, M.D.: Social media and crisis management: Cerc, search strategies, and twitter con-tent. Computers in Human Behavior 54, 647–652 (2016), http://www.sciencedirect.com/science/article/pii/S0747563215003982
15. Li, L., Li, A., Hao, B., Guan, Z., Zhu, T.: Predicting active users' personality based on micro-blogging behaviors. PLoS ONE 9(1), 1–11 (01 2014), doi:10.1371/journal.pone.0084997
16. Liu, B.: Sentiment Analysis and Opinion Mining. Morgan and Clay-pool (2012), http://ieeexplore.ieee.org/xpl/articleDetails.jsp?arnumber=6812968
17. Lo, J.a., Reis, M., Martins, B.: Predicting well-being with geo-referenced data collected from social media platforms. In: Proceedings of the 30th Annual ACM Symposium on Applied Computing. pp. 1167–1173. SAC'15, ACM, New York, NY, USA (2015), doi:10.1145/2695664.2695939
18. Luchman, J.N., Bergstrom, J., Krulikowski, C.: A motives framework of social media website use: A survey of young Americans. Computers in Human Behavior 38, 136–141 (2014), http://www.sciencedirect.com/science/article/pii/S0747563214002945
19. Norman, W.T.: Toward an adequate taxonomy of personality attributes: Replicated factor structure in peer nomination personality ratings. The Journal of Abnormal and Social Psychology 66(6), 574–583 (June 1963)
20. Ruan, Y., Alfantoukh, L., Durresi, A.: Exploring stock market using twitter trust network. In: Advanced Information Networking and Applications (AINA), 2015 IEEE 29th International Conference on. pp. 428–433 (March 2015)
21. Ry, C.D., Keyes, Corey, L.M.: The structure of psychological well-being revisited. Journal of Personality and Social Psychology 69(6), 719–727 (1995), doi:10.1037/0022-3514.69.4.719
22. Schwartz, H.A., Eichstaedt, J.C., Kern, M.L., Dziurzynski, L., Ramones, S.M., Agrawal, M., Shah, A., Kosinski, M., Stillwell, D., Seligman, M.E.P., Ungar, L.H.: Personality, gender, and age in the language of social media: The open-vocabulary approach. PLoS ONE 8(9), 1–16 (09 2013), doi:10.1371/journal.pone.0073791
23. Wang, Y., Min, Q., Han, S.: Understanding the effects of trust and risk on individual behavior toward social media platforms: A meta-analysis of the empirical evidence. Computers in Human Behavior 56, 34–44 (2016), http://www.sciencedirect.com/science/article/pii/S0747563215302260
24. Watson, D., Clark, L., Tellegen, A.: Development and validation of brief measures of positive and negative affect: The panas scales. Journal of Personality and Social Psychology 54(6), 1063–1070 (1988), doi:10.1037/0022-3514.54.6.1063
25. Xu, S., Wang, Z.J., David, P.: Media multitasking and well-being of university students. Computers in Human Behavior 55, Part A, 242–250 (2016), http://www.sciencedirect.com/science/article/pii/S0747563215301163
26. You, Q., Cao, L., Cong, Y., Zhang, X., Luo, J.: A multifaceted approach to social multimedia-based prediction of elections. IEEE Transactions on Multimedia 17(12), 2271–2280 (Dec 2015)
27. Zhang, W., Li, C., Ye, Y., Li, W., Ngai, E.W.T.: Dynamic business network analysis for correlated stock price movement prediction. IEEE Intelligent Systems 30(2), 26–33 (Mar 2015)
28. Zhang, W., Li, C., Ye, Y., Li, W., Ngai, E.W.T.: Dynamic business network analysis for correlated stock price movement prediction. IEEE Intelligent Systems 30(2), 26–33 (Mar 2015)

# Chattering Free Trajectory Tracking Control of a Robotic Manipulator Using High Order Sliding Mode

**Ankur Goel and Akhilesh Swarup**

**Abstract** This paper proposes a novel chattering free, finite-time convergent, robust high order super-twisting sliding mode controller for trajectory tracking of a robotic manipulator in presence of unknown structured uncertainties, parametric uncertainties and time varying external disturbances. The control method is designed using homogeneous sliding manifold and super-twisting sliding mode control (STC). Next, unmeasured states are estimated by a robust exact differentiator. The stability of the proposed controller is analyzed by Lyapunov stability theory and its efficacy is examined by performing simulations on 2-DoF planar robot manipulator system in presence of inertial uncertainty and external disturbances. The proposed controller judiciously eliminates the chattering and successfully overcomes the effect of external disturbances and inertia uncertainty.

**Keywords** Finite-time stability · High order sliding mode control · Robotic manipulator · Super-twisting algorithm · Uncertainty

## 1 Introduction

Robotic manipulators (RM) are widely involved in various sophisticated tasks such as drilling, welding, painting, etc. where highly precise pre-specified trajectory tracking [1] is required. However, strong nonlinear dynamics, inherent uncertainties and disturbances such as payload variation, coulomb friction, etc. [2] have to be compensated accurately otherwise the overall performance will be degraded and further could leads to system instability [3]. To address these issues, among existing elegant nonlinear control algorithms, sliding mode control (SMC) [4] algorithm is more popular because of its special attributes such as strong robustness against uncer-

A. Goel (✉) · A. Swarup
National Institute of Technology, Kurukshetra, India
e-mail: anckurgoel@gmail.com

A. Swarup
e-mail: akhilesh.swarup@gmail.com



S.K. Bhatia et al. (eds.), *Advances in Computer and Computational Sciences*,
Advances in Intelligent Systems and Computing 553,
DOI 10.1007/978-981-10-3770-2_71

tainties, fast transient response, and simpler design [4]. But it has some significant limitations also such as asymptotic convergence, chattering phenomenon and need of prior information of uncertainty bounds. The undesirable chattering phenomenon appears due to high control gains for disturbance compensation and infinite switching action on the sliding surface. This problem has been tried to solved using boundary layer method [5] but presence of steady state error degrades the system performance. Furthermore, the asymptotic stability has been found unsuitable for precise robotic performance [6] and recently, has been tackled by terminal sliding mode (TSM) [7]. However, TSM in conventional sliding mode is unable to provide chattering free robust control. Hence, to overcome aforementioned limitations, SMC has been upgraded by a novel approach known as high order sliding mode control (HOSMC). Twisting algorithm, super-twisting control (STC), Integral sliding mode control (I-SMC) and second order TSM are few examples of HOSM which have been proposed recently.

In HOSMC, the sliding variable $s$ is designed to have relative degree greater than one with respect to discontinuous control, applied on higher derivative of $s$ to start sliding motion on $s = 0$ and hence actual control becomes continuous. The HOSMC algorithms provide effective chattering reduction, show better robustness and ensures finite-time state convergence. However, only few HOSMC applications have been reported for RM systems [8, 9] where controllers are effective in disturbance rejection but problem of chattering is not fully resolved. Henceforth improvement in HOSMC design is required to obtain intelligent and effective RM control system.

In this paper, extending our work in [10], we propose a method for robust, high order STC with robust exact differentiator for RM control when all states are not available for measurement which is required for HOSMC application. Firstly, homogeneous sliding surface is selected to make STC viable for relative degree more than one system. Then, a robust exact differentiator is used for deriving higher order derivative of the states. In control design, discontinuous input is made available at higher order and hence actual control becomes continuous. Finally, a finite-time robust, chattering free trajectory tracking control is obtained. The rest of the article is organized as follows. RM system dynamics and control objective is presented in Sect. 2. The control problem is formulated in Sect. 3. The proposed controller design with stability proof is presented in Sect. 4. In Sect. 5, comparative simulations results are presented and discussed. Finally, concluding remarks given in Sect. 6 end the paper.

## 2 System Description

The dynamic model of uncertain n-DoF robotic manipulator is represented as [11]

$$M(q)\ddot{q} + C(q, \dot{q})\,\dot{q} + G(q) + F(q, \dot{q}) = \tau + \tau_d \tag{1}$$

where, $q,\ \dot{q},\ \ddot{q} \in R^n$ denote the vectors of joint positions, velocities and accelerations, respectively, $M(q) \in R^{n \times n}$ is symmetric, positive-definite inertia matrix, $C(q, \dot{q}) \in R^{n \times n}$ denotes centripetal Coriolis matrix, $G(q) \in R^n$ denotes gravitational vector, $F(q, \dot{q}) \in R^n$ includes friction and other disturbances, $\tau$ and $\tau_d \in R^n$ denotes joint input torque and external disturbances, respectively. Considering the practical imperfection in measurements, the uncertain dynamic model (1) is represented as

$$M_0(q)\ddot{q} + C_0(q, \dot{q})\dot{q} + G_0(q) = \tau + \tau_d + F_d(q, \dot{q}, \ddot{q}) \tag{2}$$

where, $F_d(q, \dot{q}, \ddot{q}) = -\Delta M(q)\,\ddot{q} - \Delta C(q, \dot{q})\,\dot{q} - \Delta G(q) - F(q, \dot{q}) \in R^n$ is total system uncertainty, $M_0(q)$, $C_0(q, \dot{q})$, $G_0(q)$ are nominal matrices, and $\Delta M(q)$, $\Delta C(q, \dot{q})$, $\Delta G(q)$ are system perturbations. For convenience, general assumptions are considered about the robot dynamics [12].

### 2.1 Control Objective

The control objective is to design a robust finite-time control law to ensure tracking error $e = (q - q_d)$ reaches zero in finite time in spite of external disturbances and uncertainties, where, $q_d \in R^n$ is double differentiable desired trajectory.

## 3 Problem Formulation

The problem of robust finite-time trajectory tracking of RM is initiated with the design of sliding surface $\sigma \in R^n$ such that

$$\sigma = e = (q - q_d) \tag{3}$$

Clearly, the second time derivative of (3) is represented as-

$$\begin{aligned}\ddot{\sigma} = \ddot{e} = (\ddot{q} - \ddot{q}_d) &= M_0^{-1}(q)[\tau + \tau_d + F_d(q, \dot{q}, \ddot{q}) - C_0(q, \dot{q})\dot{q} - G_0(q)] - \ddot{q}_d \\ &= -M_0^{-1}(q)[\ C_0(q, \dot{q})\dot{q} + G_0(q)] - \ddot{q}_d + M_0^{-1}(q)\tau + \tilde{d} \\ &= H_{nom}(t, q) + G_{nom}(t, q)\tau + \tilde{d}\end{aligned} \tag{4}$$

where, $H_{nom}(t, q) = -M_0^{-1}(q)[C_0(q, \dot{q})\dot{q} + G_0(q)] - \ddot{q}_d$, $G_{nom}(t, q) = M_0^{-1}$, and $\tilde{d} = M_0^{-1}(q)[\tau_d + F_d(q, \dot{q}, \ddot{q})] \in R^n$ is unknown, lumped uncertainties and external disturbances, such that $|\dot{\tilde{d}}| \leq \rho,\ \rho > 0$.

*Remark 1* Eq. (4) could be expressed as $\ddot{\sigma} \in [-C, C] + [K_m, K_M]u$, where bounds of constants $C, K_m, K_M$ are state dependent. (see [13] for more details)

*Remark 2* The relative degree $r$ of the system (2) with respect to (4) is two and remains constant even in presence of disturbance term $\tilde{d}$.

Now, taking time derivative of (4), we get

$$\ddot{\sigma} = \dot{H}_{nom}(t,q) + \dot{G}_{nom}(t,q)\tau + G_{nom}(t,q)\dot{\tau} + \dot{\tilde{d}} = \bar{H}_{nom} + G_{nom}(t,q)v + \dot{\tilde{d}} \quad (5)$$

where, $\bar{H}_{nom} = \dot{H}_{nom}(t,q) + \dot{G}_{nom}(t,q)\tau$ and $v = \dot{\tau}$ is a control input. Using new system variables $\xi_1, \xi_2, \xi_3 \in R^3$, the sliding vector dynamics could be written as:

$$\dot{\xi}_1 = \dot{\sigma} = \xi_2; \quad \dot{\xi}_2 = \ddot{\sigma} = \xi_3; \quad \dot{\xi}_3 = \ddot{\sigma} = \bar{H}_{nom} + G_{nom}(t,q)v + \dot{\tilde{d}} \quad (6)$$

Under nominal condition, i.e. ($\tilde{d} = 0$), the system (6) will have following feedback control-

$$v = G_{nom}^{-1}(t,q)(-\bar{H}_{nom} + \alpha(t)) \quad (7)$$

where $\alpha(t) \in R^n$ is an auxiliary input. Hence, under lumped uncertainty condition, using (6) and (7), we get

$$\dot{\xi}_1 = \xi_2; \quad \dot{\xi}_2 = \xi_3; \quad \dot{\xi}_3 = \alpha(t) + \dot{\tilde{d}} \quad (8)$$

Finally, the control objective is modified to design an effective control input $\alpha(t)$, which will act on the higher derivative of sliding variable and provides continuous control in actual due to integral action and force the system (8) states $(\xi_1, \xi_2, \xi_3)$ to converge at their respective equilibrium in finite time under the presence of uncertainty and external disturbances.

## 4 Proposed Controller Design

Here, a novel high order robust controller is proposed by employing STC with a new homogeneous sliding surface $s_1$

$$s_1 = \xi_3 + K_4\lfloor\xi_2\rceil^{\gamma_1} + K_3\lfloor\xi_1\rceil^{\gamma_2} \quad (9)$$

The proposed control strategy based on STC is given by:

$$\alpha(t) = -K_1|s_1|^{1/2}sign(s_1) + u_1; \dot{u}_1 = \begin{cases} -v, & |v| > U_M \\ -K_2 sign(s_1) & |v| \le U_M \end{cases} \quad (10)$$

where, $K_1 > 0$, $K_2 > 0$ are proposed controller gains and $K_3, k_4 > 0$ are some constant, and $0 < \gamma_1, \gamma_2 < 1$.

*Remark 3* Equation (10) is a state feedback control with state $\xi_1$ only available.

*Remark 4* Equation (10) is a homogeneous of degree $\delta_f = -1$ with weights $\alpha = [4, 3, 2, 1]$ and its solution is considered understood in the Filippov sense.

*Remark 5* Integral of a discontinuous term $u_1$ in (10) provides continuous control in actual and hence gives an extra disturbance rejection property.

*Remark 6* The disturbance $\tilde{d}$ is not known a priori but assumed that its first derivative is bounded and exist everywhere.

## 4.1 Design of Robust Exact Differentiator

The measurement of high order terms are not easy task and if the sensors are employed to measure them then it will add sensor noises in the closed loop. To tackle this issue, real-time robust finite-time convergent differentiator has been applied [14].

$$\begin{aligned} \dot{z_0} &= -L_1|z_0 - \xi_1|^{2/3} sign(z_0 - \xi_1) + z_1 \\ \dot{z_1} &= -L_2|z_1 - z_0|^{1/2} sign(z_1 - z_0) + z_2 \\ \dot{z_2} &= -L_3 sign(z_2 - z_1) \end{aligned} \tag{11}$$

where $L_2 > L_1, L_3$ and are sufficiently large, and $z_0$, $z_1$ and $z_3$ are outputs of the estimator for $\xi_1$, $\xi_2$ and $\xi_3$, respectively.

## 4.2 Convergence and Stability Analysis

To prove the finite-time convergence of uncertain system states onto sliding surface and tracking error reachability to zero, following theorem is given.

**Theorem 1** *For the system* (8), *in presence of uncertainties and external disturbances, the control law* (10) *along with* (11) *will ensure the state convergence in finite time.*

*Proof* Since the closed loop system obtained from system (8) and controller (10) is homogeneous of degree 1, we consider a continuous and homogeneous Lyapunov candidate function [15] as:

$$\begin{aligned} V(\xi) = {} & p_1|\xi_1| - p_{12}\lfloor\xi_1\rceil^{1/2}\lfloor\xi_2\rceil^{2/3} + p_2\lfloor\xi_1\rceil^{4/3} + p_3\lfloor s_1\rceil^2 + p_4\lfloor\xi_3\rceil^4 - p_{13}\lfloor\xi_1\rceil^{1/2}s_1 \\ & - p_{23}\lfloor\xi_2\rceil^{2/3}s_1 - p_{24}\lfloor\xi_2\rceil^{2/3}\lfloor\xi_3\rceil^2 - p_{34}s_1\lfloor\xi_3\rceil^2 - p_{21}\lfloor\xi_1\rceil^{1/2}\lfloor\xi_3\rceil^2 \dots \\ & + p_2\lfloor\xi_2\rceil^{4/3} \end{aligned} \tag{12}$$

Here, Eq. (12) is homogeneous of degree $\delta_v = 4$ with weights $\alpha = [4, 3, 2, 1]$, differentiable almost everywhere except at $\xi_2 = 0$ due to the term $\lfloor \xi_2 \rceil^{2/3}$ and hence not locally Lipschitz at $\xi_2 = 0$. From [10], by properly choosing set of constants, the function $V > 0$ and its time derivative $\dot{V} < 0$ is negative definite and $\dot{V}(\xi)$ satisfies the differential inequality $\dot{V} \leq -\varpi V^{3/4}$ for some positive $\varpi$ and it is a Lyapunov function for the system (8), whose trajectories, starting from any initial condition $(\xi_0)$ converge to the origin in presence of disturbances in finite time which is calculated as [15]:

$$T(\xi_0) \leq \frac{4}{\varpi} V^{\frac{1}{4}}(\xi_0) \tag{13}$$

Hence it is proved that state trajectories of the system (8) will converge to the origin in finite time after originating from any initial condition, with the help of the proposed controller (10) under the presence of disturbances. □

## 5 Simulation Results and Discussion

Here, proposed control is implemented on a classical example of two-link planar RM system which emulates like a human arm. It is widely researched example because of its importance in industries and humanoids robotic. The efficacy is tested using MATLAB-Simulink environment with ODE-4 solver and a fixed step size of $10^{-2}s$. The dynamic equation of the 2-link manipulator system with uncertainty is given in (2). The different matrices are represented as

$$M(q) = \begin{bmatrix} M_{11}(q) \; M_{12}(q) \\ M_{21}(q) \; M_{22}(q) \end{bmatrix}; C(q, \dot{q})\dot{q} = \begin{bmatrix} C_{11} \\ C_{21} \end{bmatrix}; G(q) = \begin{bmatrix} G_{11} \\ G_{21} \end{bmatrix} \tag{14}$$

where,

$$\begin{cases} M_{11}(q) = m_2\, a_1^2 + 2m_2\, a_1\, a_2\, cos(q_2) + (m_1 + m_2)\, a_1{}^2 + I_1 \\ M_{12}(q) = M_{21}(q) = m_2\, a_2{}^2 + m_2\, a_1\, a_2\, cos(q_2) \\ M_{22}(q) = m_2\, a_2{}^2 + I_2 \\ C_{11}(q) = -m_2\, a_1\, a_2\, sin(q_2)\, \dot{q_2}{}^2 - 2\, m_2\, a_1\, a_2\, sin(q_2)\, \dot{q}_1\, \dot{q}_2 \\ C_{21}(q) = m_2\, a_1\, a_2\, sin(q_2)\, \dot{q_2}{}^2 \\ G_{11}(q) = m_2\, a_2\, g\, cos(q_1 + q_2) + (m_1 + m_2)\, a_1\, g\, cos(q_1) \\ G_{21}(q) = m_2\, a_2\, g\, cos(q_1 + q_2) \end{cases}$$

Here, $q(t) = [q_1(t), q_2(t)]^T$ is the position vector, $\tau = [\tau_1, \tau_2]^T$ is applied torque vector, the states of manipulator are $[q_1(t), \dot{q}_1(t), q_2(t), \dot{q}_2(t)]^T$, two output states are $[q_1(t), q_2(t)]^T$ and two inputs are $[\tau_1, \tau_2]^T$. The viscous and Coulomb friction torque is $F(q,\dot{q}) = [8\ sin(\dot{q}_1) + 1.3\ sign(\dot{q}_1), 8\ sin(\dot{q}_2) + 1.3\ sign(\dot{q}_2)].^T$ The external disturbance in each joint is selected as $\tau_d = [0.5\ sin(2t), 0.8\ cos(2t)].^T$. The physical parameters of the manipulator are $m_1$ = nominal mass of link−1 = 1.0 kg, $m_2$ = nominal mass of link−2 = 1.5 kg, $l_1$ = length of link−1 = 1.0 m, $l_2$ = length of link−2 = 0.8 m, $I_1$ = moment of inertia(y − axis) = 12.0 kg.m$^2$, $I_2$ = moment of inertia(z − axis) = 26.0 kg.m$^2$, $g$ = gravitationalconstant = 9.8 m/s$^2$. The uncertainty in mass of both the links are considered at 80%. The reference signals are chosen as $q_{1d} = 0.3\ sin(t); q_{d2} = 0.5\ sin(t)$. For simulation, the initial states of robotic manipulator are selected as $q_1(0) = 0.1,\ q_2(0) = 0.1,\ \dot{q}_1(0) = 0,\ \dot{q}_2(0) = 0$. The design parameters of the proposed controller are $K_1 = diag(250, 250)$, $K_2 = diag(255, 255)$, $K_3 = diag(10, 10)$, $K_4 = diag(1, 1)$, $\gamma_1 = 2/3$, $\gamma_2 = 1/2$, $U_M = 0.5$, $L_1 = 35$, $L_2 = 97$, $L_3 = 30$. The STC [14] tuning parameters are selected as $K_1 = 10$, $K_2 = 1.2$, $K_3 = 0.12$.

Figures 1, 2 and 3 exhibits the simulation results obtained from proposed controller (10) and compared with super-twisting controller (STC). Figure 1a, b show that proposed controller works efficiently and better than STC for precise trajectory tracking. Moreover, the proposed controller is capable to handle uncertainties and disturbances, resulting fast position error convergence to zero in finite time of 0.6242 sec and steady state precision of 0.0001908. Furthermore, the proposed controller provides better control input (see Fig. 2) than STC without chattering which is an important claim of this paper. Lastly, Fig. 3 shows the phase portrait of the tracking error and trajectory tracking of the end-effector indicating the better results in case of proposed controller. Hence, above results reveal that the proposed controller supports our important claim of having high performance characteristics and chattering-less control under the presence of uncertainties and external disturbances.

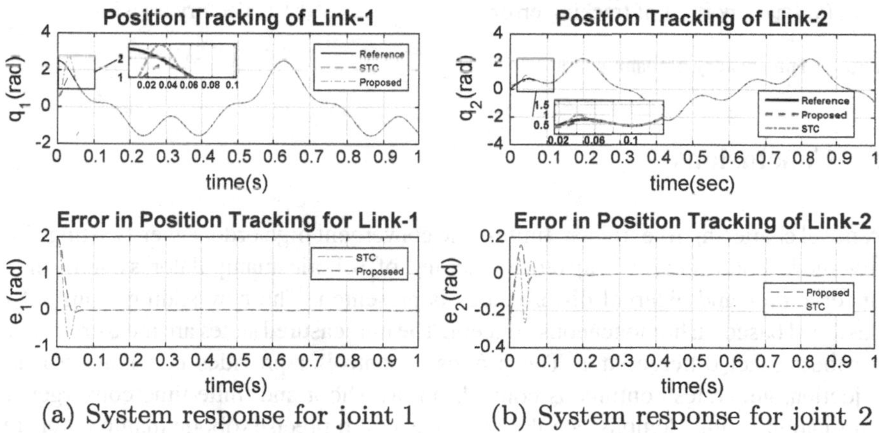


(a) System response for joint 1

(b) System response for joint 2

**Fig. 1** Position tracking responses

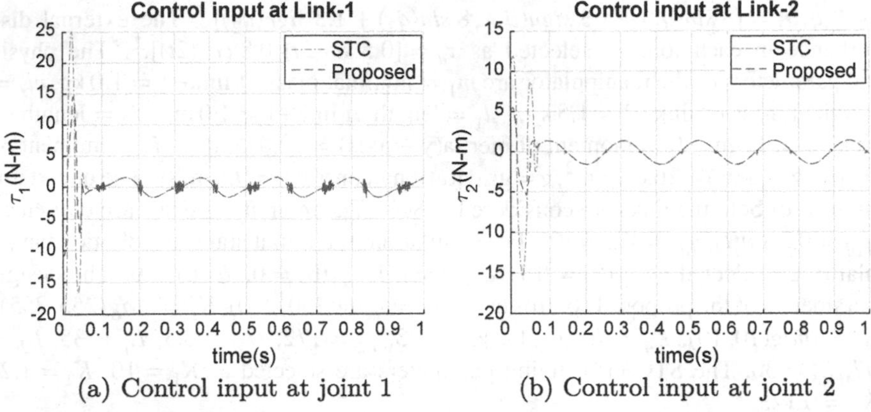


(a) Control input at joint 1 (b) Control input at joint 2

**Fig. 2** Control input responses

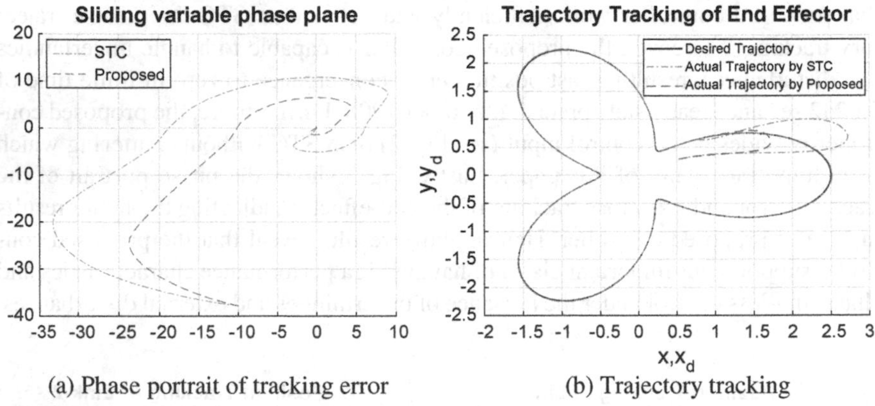


(a) Phase portrait of tracking error (b) Trajectory tracking

**Fig. 3** Time history of manipulator

# 6 Conclusions

A novel chattering free, robust, finite-time convergent high order super-twisting sliding mode controller for trajectory tracking of robotic manipulator system under uncertainties and external disturbances is presented. The new sliding manifold is designed based on homogeneous concept. The unmeasured states are measured using a robust exact differentiator. The proposed controller provides better disturbance rejection, generates continuous control, no overshoot and finite-time convergence. Simulation results demonstrate the effectiveness in precise robotic manipulator trajectory tracking and cancelling the effects of external disturbances and inertia uncertainty. The novel control scheme can be applied to the other higher complex uncertain nonlinear systems.

## References


1. Craig, J.J.: Introduction to robotics: mechanics and control. Volume 3. Pearson Prentice Hall Upper Saddle River (2005)
2. Corless, M.: Control of uncertain nonlinear systems. Journal of Dynamic Systems, Measurement, and Control **115** (1993) 362–372
3. Zhao, D., Li, S., Zhu, Q., Gao, F.: Robust finite-time control approach for robotic manipulators. IET control theory & applications **4** (2010) 1–15
4. Utkin, V.I.: Sliding modes in control and optimization. Springer Science & Business Media (2013)
5. Utkin, V., Guldner, J., Shi, J.: Sliding mode control in electro-mechanical systems. Volume 34. CRC press (2009)
6. Doulgeri, Z.: Sliding regime of a nonlinear robust controller for robot manipulators. IEE Proceedings-Control Theory and Applications **146** (1999) 493–498
7. Venkataraman, S., Gulati, S.: Terminal sliding modes: a new approach to nonlinear control synthesis. In: Advanced Robotics, 1991. 'Robots in Unstructured Environments', 91 ICAR., Fifth International Conference on, IEEE (1991) 443–448
8. Capisani, L.M., Ferrara, A., Magnani, L.: Second order sliding mode motion control of rigid robot manipulators. In: 2007 46th IEEE Conference on Decision and Control, IEEE (2007) 3691–3696
9. González-Jiménez, L.E., Loukianov, A., Bayro-Corrochano, E.: Fully nested super-twisting algorithm for uncertain robotic manipulators. In: Robotics and Automation (ICRA), 2011 IEEE International Conference on, IEEE (2011) 5807–5812
10. Goel, A., Swarup, A.: High order super twisting sliding mode control of robotic manipulator. ICIC express letters. Part B, Applications **6** (2015) 3095–3101
11. Spong, M.W., Hutchinson, S., Vidyasagar, M.: Robot modeling and control. Volume 3. Wiley New York (2006)
12. Zhihong, M., Paplinski, A., Wu, H.: A robust mimo terminal sliding mode control scheme for rigid robotic manipulators. Automatic Control, IEEE Transactions on **39** (1994) 2464–2469
13. Levant, A.: Sliding order and sliding accuracy in sliding mode control. International Journal of Control **58** (1993) 1247–1263
14. Levant, A.: Principles of 2-sliding mode design. Automatica **43** (2007) 576–586
15. Moreno, J.A.: Lyapunov function for levant's second order differentiator. In: 2012 IEEE 51st IEEE Conference on Decision and Control (CDC). (2012) 6448–6453

# Extraction and Enhancement of Moving Objects in a Video

**Sumati Manchanda and Shanu Sharma**

**Abstract** Detection of objects for relocation in a video is a vital as well as initial step for many computer vision-based applications like moving object extraction, video surveillance, pattern classification, etc. The traditional methods used for detection of foreground objects include background subtraction, optical flow and frame differencing techniques. These methods are found to be advantageous only if the extraction of the moving object is precise and clearly visible that it is, the object must be of good quality. This paper emphasizes on the detection as well as the enhancement of the foreground objects. The proposed method uses the amalgam of two traditional techniques background subtraction and motion vector-based optical flow method along with morphological operators to extricate the nonstationary objects from the videos followed by the enhancement of the extracted object to be of better quality in terms of visibility. The proposed algorithm is executed over the videos having frame dimension of 640 × 360 along with the frame rate of 30 frames/second using MATLAB R2013.

**Keywords** Object extraction · Enhancement · Optical flow · Background subtraction · Median filter · Morphological operators

## 1 Introduction

Detection of nonstationary objects in a video sequence is a significant step in numerous applications where the main aim is to identify the physical change in the position of the object present in the video [1–3]. Motion detection can be characterized as the procedure of recognizing and analyzing the position of the foreground

S. Manchanda (✉) · S. Sharma
Department of Computer Science & Engineering, ASET Amity University, Noida, UP, India
e-mail: matimanchanda@gmail.com

S. Sharma
e-mail: shanu.sharma16@gmail.com



S.K. Bhatia et al. (eds.), *Advances in Computer and Computational Sciences*, Advances in Intelligent Systems and Computing 553,
DOI 10.1007/978-981-10-3770-2_72

objects for better understanding as it is an initial step in various computer vision applications [4] and the detection of nonstationary object in a video is basically an initial step in various applications like video surveillance and security [5], traffic monitoring, event detection, pattern recognition [4], object recognition and tracking [6], and many more. But most of the times, the extraction of the moving objects from the videos becomes very arduous due to presence of noise, dynamic changes in the scene and illumination effects; therefore enhancing the features of the detected object is an important concern for the efficient extraction of the moving object [7] and in order to obtain results with better visibility.

The traditional motion detection algorithms can be broadly classified into three categories: background subtraction, frame differencing and the motion vector-based optical flow method. Amongst these approaches the background subtraction is the most widely used approach due to its simplicity and the ability to handle the dynamic changes. The other method being optical flow technique has also gained popularity due to its ability to produce better and efficient results in case of real time application. The motion detection algorithms hold beneficial only if they are able to detect the nonstationary objects precisely and efficiently. Therefore, in this paper an algorithm has been proposed that combines the background subtraction model and the optical flow method to draw benefits from both the approaches and to generate better results as well as frame enhancement technique is used to obtain the outputs with higher quality along with the detailed information about the boundary of the extracted object.

The rest of the paper is organized in following way: Sect. 2 describes the traditional approaches for motion detection and the related work done in the related field. The step wise explanation of methodology of the proposed algorithm is given in Sects. 3, 4 discussed about the image enhancement process and the results of the proposed algorithm are presented in Sects. 5, 6 concludes the paper.

## 2 Related Work

Till now various algorithms for identification of moving objects are very well implemented but still the challenges like low implementation speed, complete contour detection, noise sensitivity, illumination effects and many more. The background subtraction has gained popularity and it is the technique that is widely used for extraction of foreground object from the static background because of its simple calculations and the use of Gaussian mixture model which makes the detection more reliable in case of illumination effects and camera noise [1]. In order to make the extraction of foreground objects more reliable and accurate the optical flow method has the ability to perceive the moving objects by estimating the motion vector [7, 8].

In case of background subtraction method, the basic objective is to isolate the moving foreground object from the static background. It is done by evaluating the difference between the reference frame that is the background frame and the current frame [9–11]. The reference frame should be updated on regular interval of time [12]. The Gaussian mixture model has gained popularity, due to its ability to handle illumination changes, slow, camera noise and its ability to handle the multi modal distribution [13, 14]. Optical flow method is a motion vector-based technique that is used for analyzing the motion of an image. It is not only a method that estimates the motion of the nonstationary objects in a video but also provide the information about the direction in which the object is moving [10]. In this method the motion is estimated by evaluating the motion vector and the optical field vector using optical flow distribution based clustering [15, 16]. It enables the efficient and reliable detection of the nonstationary objects by giving complete information about the object [15]. Frame differencing is a technique for detection of foreground objects by calculating the pixel wise difference between the two succeeding frames. The detailed information about the object in motion can also be obtained by combining frame differencing method with various threshold based methods [14]. The foreground object identified must be free from noise as noise may cause illumination variation in the pixels value which may hinder the efficient detection as well. Therefore with the help of image enhancement techniques, quality of the image can be increased like its sharpness, contrast, etc. Image enhancement is classified into two groups that are spatial domain method and frequency domain method [9]. In Spatial domain method, the pixels of the image are directly dealt with whereas in case of frequency domain method FT of image is modified.

## 3 Proposed Algorithm

The proposed algorithm combines the background subtraction model and the optical flow method to draw benefits from both the approaches and to generate better results. The background subtraction model is used for tracking the pixels of the object in motion and the optical flow being a vector-based method is used to estimate motion vector of the objects in the videos over numerous frames. The various steps of proposed algorithm for detection and enhancement of non- stationary objects are briefly discussed below (Fig. 1).

### *3.1 Preprocessing of Frames*

The camera captured video is generally in RGB format, which needs to be converted into gray scale format to enhance the speed of the detection process as the processing of gray pixels is faster as compared to RGB pixels. When a video is seized there is probability that noise get mixed with the captured video due to various factors like illumination changes, dynamic changes in the environment, etc., due to which the quality of the video gets degraded therefore median filter has been used to de-noise the frames of the video.

## 3.2 Detection of Objects in Motion

The detection of the nonstationary object in the video is performed with the help of background subtraction method. There are various methods to obtain the reference frame or initial background frame like the initial frames can be treated as reference frames or with the help of average of pixels brightness can be evaluated for the background frame [5].

The second step in this procedure is to identify the nonstationary objects in the video. It is done by subtracting the reference frame from the current frame in order to capture the distinctness between the frames. The areas in the image where the pixels can be located are the pixels of the moving object [5]. If the differences between the pixels is found to greater than the set threshold value then the pixel belongs to the group of pixels constituting for the foreground object [17]. The expression for background modeling is given in Eq. 1

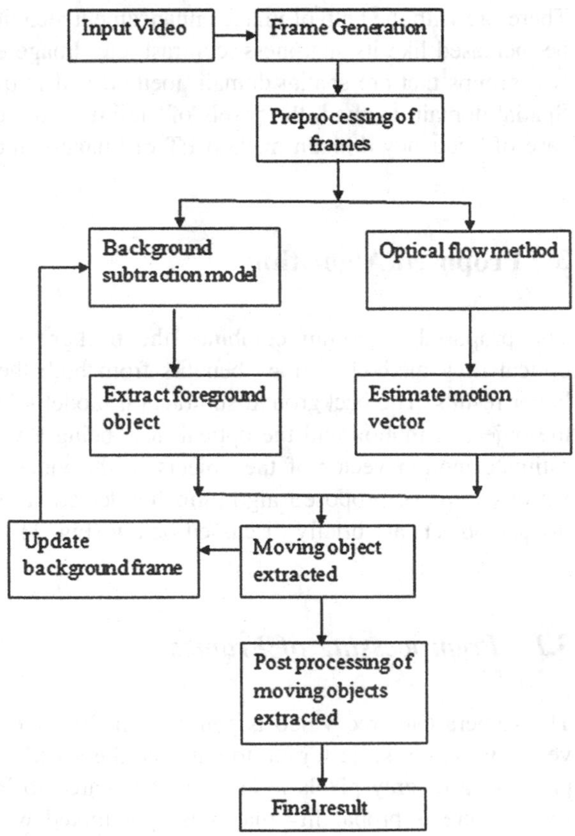


**Fig. 1** Flowchart of proposed algorithm

$$\begin{aligned} dk &= 1 \quad if \quad |Fk\,(1,m) - Bk\,(1,m)| \geq T \\ \mathrm{dk} &= 0 \quad \text{otherwise} \end{aligned} \tag{1}$$

The other methodology used for the identification of a moving object is optical flow estimation where vector field of the image sequence is calculated. In this technique optical flow is estimated with the help of two frames differential method to identify the moving object. The final step is to combine the results to obtain the accurate foreground object from the video. Only those objects are considered to be moving whose amplitude values and the direction of optical field lie within the specified range.

### *3.3 Post Processing of Extracted Objects*

The nonstationary objects are detected but at times the region of motion is also eroded due to the presence of noise. The noise from the frames was removed by running median filter across the frames. There are also situations when the morphological operators also needed to be used in order to attain the desired results, therefore dilation and closing operations are performed.

## 4 Image Enhancement

Image enhancement is process that increases the quality of the image in many aspects like brightness, contrast, sharpness, noise removal and many more. The requirement to improve the quality of the images may vary according to the application in which they are deployed [10] (Fig. 2).

## 5 Experiments and Results

The effectiveness of the proposed work is demonstrated with the help of the videos captured in different environmental conditions. The algorithm is tested and executed in MATLABR2013 with the help of a video titled as 'atrium.avi' that is seized with the help of a static camera. The results below are demonstrating the effectiveness of the proposed algorithm.

Figure 3a shows the edge detection using the Sobel edge detection method and Fig. 3b show segmentation of the frame using threshold in order to detect the moving object in the video. Figure 4a, b show the cleaning process of the

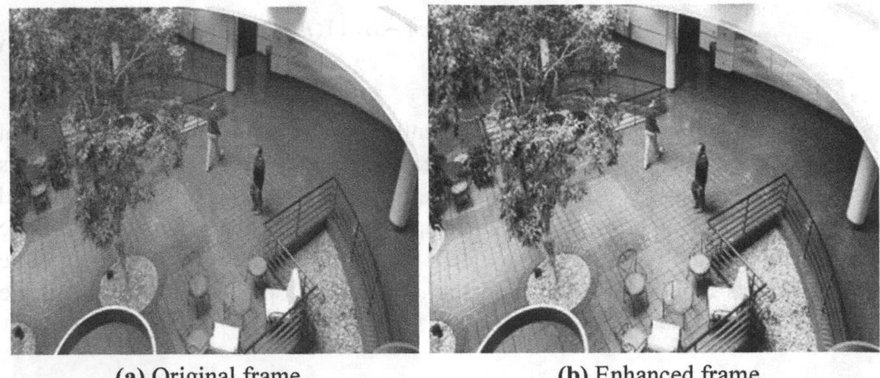


**Fig. 2** **a** Original frame, **b** Enhanced frame

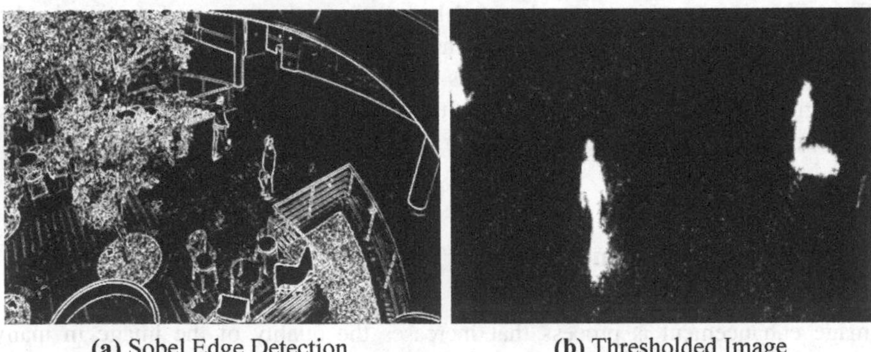


**Fig. 3** **a** Sobel edge detection, **b** Thresholded image

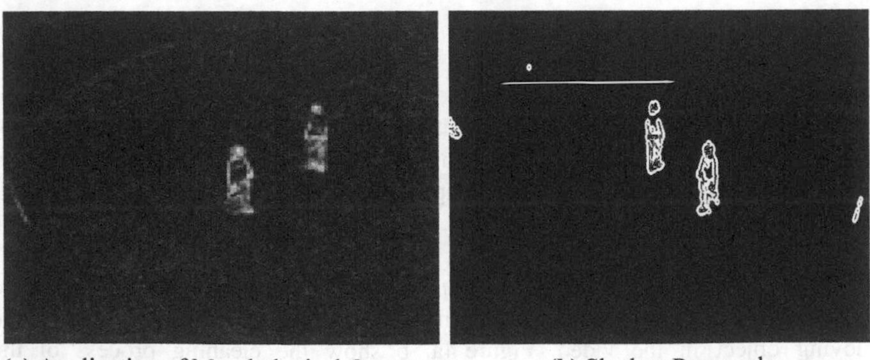


**Fig. 4** **a** Application of morphological operators, **b** Shadow removal

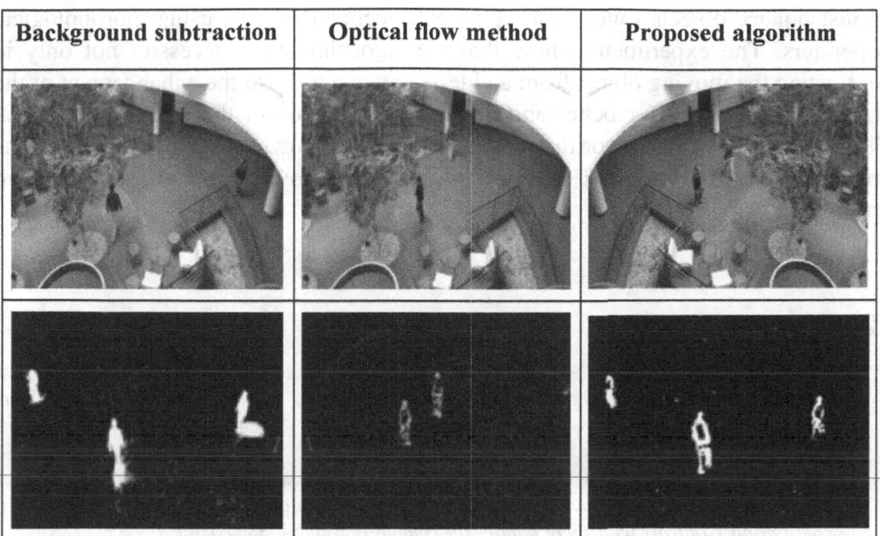


**Fig. 5** Comparative analysis of the obtained results

**Table 1** Quantitative results

| Frames | Std deviation | PSNR(db) | MSE |
|---|---|---|---|
| 11.png | 48.1838 | 58.97 | 31.72 |
| 81.png | 53.7006 | 56.68 | 48.78 |
| 171.png | 56.1420 | 55.44 | 27.8 |
| 204.png | 27.6639 | 30.41 | 42.17 |

foreground and removal of shaodow as well in oder to make the detection process more efficient.

Figure 5 is used to represent the comparison of the proposed algorithm with the techniques named Background subtraction and the optical flow method.

The evaluation of the proposed work is done with the help of Peak signal noise ratio (PSNR) and the mean square error (MSE) and standard deviation (Table 1). The PSNR and MSE are the quantitative measure that is used to measure the quality of the image or frame. The higher the value of PSNR better is quality of the frame and the value of MSE should be lower to reduce the chances of error in the resultant frame.

## 6 Conclusion

In this paper an approach for the extraction and enhancement of the nonstationary objects in a video is proposed. The amalgam of two approaches namely background modeling techniques along with the optical flow methods are used to identify the nonstationary objects followed by post enhancement process using morphological operators. The experiments show that the algorithm was successful not only in extracting the moving object from a video sequence but also the enhancement of the object is performed for better and precise identification of the foreground object. The advantage of this algorithm was that the prior knowledge is not required. The results obtained were good enough and can be improvised in future in extended problems.

## References

1. Soharab Hossain Shaikh, Khalid Saeed and Nabendu Chaki, "Moving Object Detection Using Background Subtraction" Springer Briefs in Computer Science, Springer International Publishing, 2014.
2. Kaur, R. and Singh,S. "Background modelling, detection and tracking of human in video surveillance system", *Innovative Applications of Computational Intelligence on Power, Energy and Controls with their impact on Humanity*, pp. 54–58, 2014.
3. Karamiani, Aziz, and Nacer Farajzadeh."Optimal feature points for tracking multiple moving objects in active camera model", Multimedia Tools and Applications, 2015.
4. Sighla Nishu, "Motion Detection Based on Frame Difference Method", International Journal of Information & Computation Technology, Vol 4 No. 15, pp. 1559–1565, 2014.
5. Rakibe Rupali S. and Patil Bharti D, "Background Subtraction Algorithm Based Human Motion Detection", International Journal of Scientific and Research Publication, Vol 3, No. 5, May 2013.
6. Wu-Chih, Chao-Ho Chen, Tsong-Yi Chen, Deng-Yuan Huang and ZongChe Wu, "Moving object detection and tracking from video captured by moving camera", Journal of Visual Communication and Image Representation, Vol 30 Pages 164–180, July 2015.
7. K. Kalirajan and M. Sudha, "Moving Object Detection for Video Surveillance" Hindawi Publishing Corporation Scientific World Journal Volume 2015, Article ID 907469. 2015.
8. P.N. Pathirana, A.E.K. Lim, J. Carminati, M. Premaratne, "Simultaneous estimation of optical flow and object state, A modified approach to optical flow calculation". In Proceeding of IEEE International Conference on Networking, Sensing and Control, pp. 634–638, 2007.
9. Manvi, R.S. Chauhan and M. Singh, "Image contrast enhancement using histogram equalization", International Journal of Computing & Business Research, I-Society12, no. 33, 2012.
10. Sharma Urvashi, Sharma Tripti And Jain Trisha, "Efficient Object Detection With Its Enhancement", International Conference On Computing, Communication And Automation (ICCCA2015).
11. Khare, Manish, Rajneesh Kumar Srivastava, and Ashish Khare. "Object tracking using combination of daubechies complex wavelet transform and Zernike moment", Multimedia Tools and Applications, 2015.
12. Arwa Darwish Alzughaibi, Hanadi Ahmed Hakami and Zenon Chacxzko, "Review of Human Motion Detection Based On Background Subtraction Techniques", International Journal of Computer Allocations Volume 122, No 13, July 2015.
13. Asim R. Aldhaheri And Eran A. Edirisinghe, "Detection And Classification Of A Moving Object In A Video Stream", Proc Of The International Conference On Advances In Computing And Information Technology-ACIT 2014.
14. Hu, W., Tan, T., Wang, L., Maybank, S., 2004. A survey on visual surveillance of object motion and behaviors. IEEE Trans. On System Man Cybernetics August 2012.

15. Abhishek Kumar Chauhan, Prashant Krishan, "Moving Object Tracking Using Gaussian Mixture Model And Optical Flow", proc Of International Journal Of Advanced Research And Software Engineering, April 2013.
16. Dongxiang Zhou; Hong Zhang, "Modified GMM background modeling and optical flow for detection of moving objects", International Conference on in Systems, Man and Cybernetics, vol. 3, pp. 2224–2229, Oct. 2005.
17. Dongxiang Zhou; Hong Zhang, "Modified GMM background modeling and optical flow for detection of moving objects", International Conference on in Systems, Man and Cybernetics, vol.3, pp. 2224–2229, Oct. 2005.

# Author Index





S.K. Bhatia et al. (eds.), *Advances in Computer and Computational Sciences*,
Advances in Intelligent Systems and Computing 553,
DOI 10.1007/978-981-10-3770-2

Printed in the United States
By Bookmasters